When large or small units are involved, it is convenient to use the follo̶w̶i̶n̶g̶ and oral presentations:

Prefix	Symbol	Indicated Multiplication Factor
giga-	G	10^9 (one billion times greater)
mega-	M	10^6 (one million times greater)
kilo-	k	10^3 (one thousand times greater)
centi-	c	10^{-2} (a hundredth part)
milli-	m	10^{-3} (a thousandth part)
micro-	μ	10^{-6} (a millionth part)
nano-	n	10^{-9} (a billionth part)

Thus, the term *kilonewton* indicates a force of one thousand newtons, the term *millimeter* indicates a length of one one-thousandth of a meter, and so on. The use of the well-established prefix *centi-* is discouraged in the SI system.

Symbols—Greek Alphabet

Greek Letter		Greek Name	Greek Letter			Greek Name
A	α	Alpha	N	ν		Nu
B	β	Beta	Ξ	ξ		Xi
Γ	γ	Gamma	O	o		Omicron
Δ	δ	Delta	Π	π		Pi
E	ε	Epsilon	P	ρ		Rho
Z	ζ	Zeta	Σ	σ	ς	Sigma
H	η	Eta	T	τ		Tau
Θ	θ ϑ	Theta	Y	υ		Upsilon
I	ι	Iota	Φ	φ	φ	Phi
K	κ	Kappa	X	χ		Chi
Λ	λ	Lambda	Ψ	ψ		Psi
M	μ	Mu	Ω	ω		Omega

(continued inside back cover)

Essentials of Soil Mechanics and Foundations

Essentials of Soil Mechanics and Foundations
Basic Geotechnics

FIFTH EDITION

David F. McCarthy, P.E.

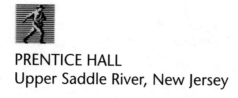

PRENTICE HALL
Upper Saddle River, New Jersey   Columbus, Ohio

Library of Congress Cataloging-in-Publication Data
McCarthy, David F.
   Essentials of soil mechanics and foundations : basic geotechnics /
David F. McCarthy. — 5th ed.
     p. cm.
   Includes bibliographical references and index.
   ISBN 0-13-506932-7
   1. Soil mechanics. 2. Foundations. I. Title.
TA710.M29 1998
624.1'5—dc21   97-14326
                                         CIP

Cover photo: © Westlight
Editor: Ed Francis
Production Editor: Linda Hillis Bayma
Production Coordination: WordCrafters Editorial Services, Inc.
Design Coordinator: Julia Zonneveld Van Hook
Cover Designer: Rod Harris
Production Manager: Deidra M. Schwartz
Marketing Manager: Danny Hoyt
Illustrations: Academy Artworks, Inc.

This book was set in Times Roman by BookMasters, Inc. and was printed by Quebecor Printing/Book Press. The cover was printed by Phoenix Color Corp.

 © 1998, 1993, 1988, 1982, 1977 by Prentice-Hall, Inc.
Simon & Schuster/A Viacom Company
Upper Saddle River, New Jersey 07458

ISBN 0-13-506932-7

Printed in the United States of America

10 9 8 7 6 5 4 3 2

Prentice-Hall International (UK) Limited, *London*
Prentice-Hall of Australia Pty. Limited, *Sydney*
Prentice-Hall Canada Inc., *Toronto*
Prentice-Hall Hispanoamericana, S.A., *Mexico*
Prentice-Hall of India Private Limited, *New Delhi*
Prentice-Hall of Japan, Inc., *Tokyo*
Simon & Schuster Asia Pte. Ltd., *Singapore*
Editora Prentice-Hall Do Brasil, Ltda., *Rio de Janeiro*

Contents

Chapter 8 Combined Stresses in Soil Masses: Stress at a Point and Mohr's Circle 281

Chapter 9 Subsurface Stresses 295

Chapter 10 Settlement: Soil Compression, Volume Distortion, Consolidation 316

Chapter 14 Earth Moving, Compaction, and Stabilization   511

Chapter 15   Stability of Unsupported Slopes   553

Chapter 16   Lateral Pressures and Retaining Structures   613

Preface

The broad range of topics included within the scope of geotechnics (the field of soil mechanics and foundations) surprises many. The comprehensive nature of this text serves to introduce those topics to students in engineering, architectural, and construction-related academic programs. An intention of this fifth edition is to provide present-day, state-of-the-art information as accepted by, and conventionally used by, the engineering profession; theories basic to understanding soil behavior are covered and then associated with practical applications. The text remains an introductory-level learning instrument for a first course in geotechnics in a planned academic program and does not progress into topics considered to be on the advanced level or of professional specialty; however, the extent of presentations and emphasis on application should also make this edition a worthy reference for practitioners.

Topics carried forward from previous editions to the present edition have generally been updated throughout. Topics at the core of geotechnics such as settlement, soil compression and consolidation, foundation design criteria, earthquake effects, site investigation, problem soil categories, subsurface water flow, subsurface stress conditions, slope stability, and retaining walls are new, changed, or expanded. Compared to previous editions, a reorganization in the order of topic presentation has resulted. The SI system is now the precedent system for units, but with data and illustrations also presented in U.S. Customary units. The illustrated problems within chapters, which show applications of theory, are realistic although often simplified. The number of end-of-chapter problems has been increased significantly for benefit of classroom instruction and student practice.

This text is a comprehensive presentation of topics in the field of soil mechanics and foundations. It includes more material than can be covered within the time allocated to the typical 3- or 4-credit-hour college course. Accordingly, it is expected that chapter topics selected for use by classroom instructors at different campuses will vary, guided by the requirements or intent of the institution program. However, it is also the author's experience that after students have learned the basic materials, other text topics, particularly those relating to applications, can be self-taught to obtain working knowledge as may be required for subsequent courses.

The library staff at the author's home institution, Mohawk Valley Community College, SUNY, proved exceptionally helpful in obtaining many of the new references. In perspective, the past efforts of many different individuals are incorporated into the pages of this edition, and the dedication of those advancing the level of our knowledge deserves recognition. For this edition, the task of manuscript preparation was accomplished by Office Technologies of Whitesboro, New York. Illustrations, photographs, and text material provided by others are acknowledged where presented but are due a repeated thank-you. As with previous editions, the tolerance and support of the author's family was important, as well as appreciated.

ACKNOWLEDGMENTS

I wish to thank the colleagues who served as reviewers for this edition: Charles A. Matrosic, Ferris State University; Mohammad Najafi, Missouri Western State College; and Jerry Steffen, Northeast Wisconsin Technical College.

PART 1

Background and Basics

CHAPTER 1

The Soil and Rock of Planet Earth

Geologic Overview

It has long been known that the earth is a dynamic planet, constantly changing, but facts uncovered by scientists only throughout the last half-century indicate that these ongoing changes involve factors of nature that are more profound than ever imagined previously. The major constituents of the earth's outer zone—the soil, rock, and water—are continually subjected to forces that incite change. The changes in rock and soil deposits are often slow and subtle, and may not be recognized over short geologic periods such as the human lifetime. However, rapid (sometimes instantaneous) changes also occur because of natural phenomena or human activities (mass movements during earthquakes, landslides, and tidal waves, or effects of war and large construction projects). Related to the struggle to survive and the drive to improve, humans have learned about many of the factors responsible for changes that occur in the earth's rock and soil deposits and the factors that affect the (geologically) short-term behavior.

Humans have always realized that soil influences their survival. Soil is the ground on which we stand. We build with soil. We have relied on soil to support structures and the paths of transportation. We depend on soil to grow our food and provide the products we use for living, protection, and comfort. We have trusted the soil to be stable and permanent. In some of these presumptions, we have been incorrect.

This text studies soil as a material that is used to build with or on, but also as a material of the environment that may act in combination with other forces of nature or of civilization to affect landforms, our structures, and the state of our environment. This study is the field of *geotechnics*. This chapter presents a concise overview of important geologic-based information, to provide an understanding of the factors responsible for the formation and behavior of today's rock and soil deposits and the factors that will cause changes in the future. Subsequent chapters provide detailed information about the engineering properties of soils and the principles and methods relevant to practical applications in construction projects, land stabilization projects, and environmental protection projects. In this introductory study of geotechnics, the agricultural aspects of soil are not considered, with the exception of discussions relating to vegetation used as a means of erosion control and as a factor influencing the creation of present-day soil types.

The earth's crust is composed of soil and rock. *Rock* can be defined as a natural aggregate of minerals that are connected by strong bonding or attractive forces; for this reason, rock is often considered a consolidated material. *Soil* may be defined as the unconsolidated sediments and deposits of solid particles that have resulted from the disintegration of rock. To the construction industry and the engineering profession, however, soil is also assumed to include the residue of vegetable and animal life, including civilization's buried trash, garbage, and industrial wastes.

Soil is a *particulate* material, which means that a soil mass consists of an accumulation of individual particles that are bonded together by mechanical or attractive means, though not as strongly as with rock. In soil (and in most rock), voids exist between particles, and the voids may be filled with a liquid, usually water, or a gas, usually air. As a result, soil deposits are often referred to as a three-phase material or system (solids plus liquid plus gas).

1.1 ROCK: THE SOURCE OF SOILS

Most of the nonorganic materials that are identified as soil originated from rock as the parent material. Rock types are grouped into three major classes—igneous, sedimentary, and metamorphic—determined by their origin or method of formation. The type of soil that subsequently develops relates to the rock type, its mineral components, and the climatic regime of the area.

Igneous rock resulted from the cooling and hardening of molten rock called *magma,* which originated deep within the earth. Molten magma that escaped to and near to the surface of the earth through volcanoes and fissures in the earth's crust (termed *lava*) cooled quickly. As a result of rapid cooling, the mineral components solidified into small crystals and possessed a fine, interlocking texture. In some situations, the cooling was so rapid that a crystal-free, glassy texture resulted. The molten materials (lava) that cooled rapidly at or near the earth's surface are called extrusive or volcanic rock types and include the *basalts, rhyolites,* and *andesites.*

Molten rock trapped deep below the surface of the earth (magma) cooled slowly. The mineral components formed in large interlocking crystals, and coarse-textured rocks resulted. These rocks are classified as intrusive or plutonic types and include the *granites,* the most common, as well as the *syenites, diorites,* and *gabbros.*

Many of the mineral combinations in igneous rocks are unstable in the environment existing at the earth's surface. Upon exposure to air, water, chemicals in solution in water, freezing temperatures, varying temperatures, and erosive factors, the rock minerals break down to the soil types existing today. Rock whose chief mineral is quartz or orthoclase (potassium feldspar), minerals with high silica content, decomposes to predominantly sandy or gravelly soil with little clay.[1] Granites, syenites, and rhyolites are in this category. Because of the high silica content, these rocks are classified as acidic.

---

[1]According to the engineering definition, coarser soils include sands and gravels and are particles larger than 0.074 mm. The fine-textured silt and clay soils are smaller than 0.074 mm. The 0.074 mm size is close to the smallest particle size observable with the unaided eye under normal conditions. Most clay particles are smaller than 0.002 mm. Chapters 3 and 4 further discuss soil type and classification.

Rocks whose minerals contain iron, magnesium, calcium, or sodium, but little silica, such as the gabbros, diabases, and basalts, are classified as basic rocks. These rocks decompose to the fine-textured silt and clay soils.

Generally, the acidic rocks are light-colored, while the basic rocks are very dark. Intermediate colors reflect an intermediate chemical composition. Rock types intermediate between acidic and basic include the *trachytes, diorites,* and *andesites.* Because of their mineral components, diorite and andesite easily break down into the fine-textured soils.

The clay portion of fine-textured soil is the result of primary rock minerals decomposing to form secondary minerals. The clays are *not* small fragments of the original minerals that existed in the parent rock. Because of this change, the properties and behavior of clay soils are different from those of the gravel, sand, and silt soils, which *are* still composed of the primary rock minerals.

Rocks that are acidic (not basic) such as the granites are considered to be good construction materials.

Sedimentary rocks are formed from accumulated deposits of soil particles or remains of certain organisms that have become hardened by pressure or cemented by minerals. Pressure sufficient to harden or solidify a soil deposit results from the weight of great thicknesses of overlying material. Under this pressure, the deposit is compacted and consolidated, and strong attractive bonds are established. Cementing minerals such as silica, calcium carbonate, and the iron oxides are abundant in soil as a result of rock weathering, and when dissolved in the water circulating through a soil deposit, they precipitate out onto the soil particles. Other cementing may be obtained from within the mass by solution or chemical change of materials. Sedimentary-type rocks include the limestones and dolostone (dolomites),[2] shale (claystone, siltstone, mudstone), sandstone, conglomerate, and breccia.

Geologic conditions in past historic times have had a very significant effect on the location and type of sedimentary rocks that exist across North America today. In early prehistoric times, most of what is now the United States was under water (Fig. 1-1). Gradually, much of the land rose. Accumulated sediments in these shallow seas eventually became the limestones, shales, and sandstones of today. As a consequence of the layered manner of soil deposition, many sedimentary rocks are easily recognized today because of their stratified appearance (Fig. 1-2).

Shales are formed predominantly from deposited clay and silt particles. The degree of hardness varies, depending on the type of minerals, the bonding that developed, and the presence of foreign materials. The hardness is generally due to external pressures and the particle bonds that resulted, not to cementing minerals. Many shales are relatively stable when exposed to the environment, but some expand or delaminate (the layers separate) after contact with water or air. Weathering breaks down shale to fragments of varying sizes. These fragments, in turn, may be quickly reduced back to clay particle sizes. The properties of shale are quite important to the construction industry, for it is estimated that shale represents approximately 50 percent of the rock that is exposed at the earth's surface or closest to the surface under the soil cover. Sound shale can provide a good foundation material. Its use as a construction material is questionable, however, because of its tendency to break down under handling, abuse, and weathering.

---

[2]Historically, the term *dolomite* has referred to both the rock-forming mineral $CaMg(CO_3)_2$ and sedimentary rock. *Dolostone* is a recent proposal for designating the rock material, to avoid confusion.

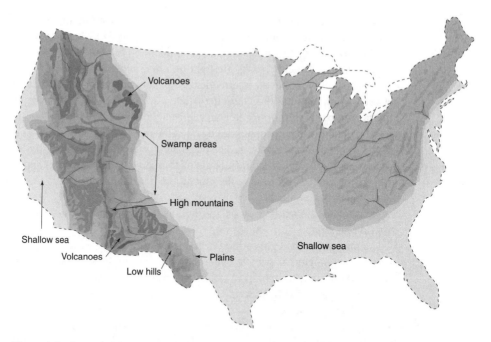

Figure 1-1 Generalized geographic map of the United States for period approximately 100 million years ago. Sediments deposited in shallow seas became sedimentary rock that exists in such areas today. (*Source:* U.S. Department of the Interior, Geological Survey)

Figure 1-2 Sedimentary rock in central New York State showing stratified formation.

Sandstone is predominantly quartz cemented together with mostly silica, but also calcium carbonate, or iron compounds. Sandstones are generally considered good construction materials. *Conglomerate* (cemented sand and gravel) and *breccia* (cemented angular rock fragments) are similar to sandstone.

Limestone is predominantly crystalline calcium carbonate (calcite) formed under water. This rock material forms as a result of chemicals precipitating from solution, and from the remains of marine organisms and action of plant life. Because of their sedimentary nature, limestones frequently include impurities such as clays and organic material. The degree of hardness and durability of limestones varies. Some limestone formations are very sound, but some limestones are very soluble, with the result that the formation contains many cavities.

Dolostone is a variety of limestone, but harder and more durable. *Marl* and *chalk* are softer forms of limestone.

Weathering of limestones can produce a soil that includes a large range of particle sizes, but the fine-grained soils predominate. Limestone is a good foundation material, provided that the formation is sound and free of cavities. Sound limestone is considered a good construction material.

Frequently, because of the manner of deposition of the original rock-forming sediments, sandstone and shale, or limestone and shale, and sometimes sandstone, limestone, and shale are interbedded.

Metamorphic rock results when any type of existing rock is subject to metamorphism, the changes brought about by combinations of heat, pressure, and plastic flow so that the original rock structure and mineral composition are changed. *Plastic flow* for rock refers to slow viscous movement and rearrangement within the rock mass as it changes and adjusts to the pressures created by external forces. Under these conditions, limestone is changed to *marble* and sandstone to *quartzite*. When subject to metamorphism, shale is transformed to *slate* or *phyllite*. Higher levels of these factors change the shale, or slate, to *schist*. Because of the processes affecting their formation, slates and schists become foliated rocks (that is, layered as in a folio). *Gneiss* is a foliated rock with distinctive banding that results from the metamorphosis of sedimentary rock or basalt or granite. Despite their possibly different origin, a distinction between gneisses and schists is not always clear. These two rock types on occasion appear to be gradational.

Metamorphic rocks formed from sound igneous or sedimentary rocks can be good materials for construction. But schist, gneiss, and slate are questionable construction materials because the foliated or banded structure can act to originate planes of weakness that affect strength and durability.

Upon weathering, some metamorphic rocks break down to soil types comparable to that which would be derived from the original igneous or sedimentary-type rock. Others reflect the changes brought about by metamorphism. Gneiss and schist decompose to silt–sand mixtures with mica. Soils from slates and phyllites are more clayey. Soils derived from marble are similar to those resulting from limestone. Decomposition of quartzite generally produces sands and gravels.

The processes of rock changing to soil, soil changing to rock, and alteration of rock are continuous and occur simultaneously. The process of change or alteration takes place over long periods of time, and there is no set sequence in which changes occur.

There are many rock types. To establish a proper perspective, the construction industry's concern is generally not with rocks' names but with their properties. In-place properties such as hardness and possible presence of fractures or fissures affect drilling, blasting, and excavation operations. The suitability for use as a foundation for structures is related to strength, durability, and possible presence of cavities or of fractures and fissures. The commercial value of excavated and crushed rock for fill and as an ingredient of concrete is influenced by soundness and durability. In a general way, desirable and undesirable properties have been associated with the different rock types.

1.2 SOIL CATEGORIES: RESIDUAL, TRANSPORTED

Soils can be grouped into two broad categories—residual or transported—depending on the method of deposition.

Residual Soils

Residual soils (or sedentary soils) have formed from the weathering of rock or accumulation of organic material and remain at the location of their origin. The weathering process may be attributed to mechanical weathering or chemical and solution weathering. Mechanical weathering refers to physical disintegration resulting from the effects of wind, rain, running water, ice and frost wedging, and tectonic forces (earthquakes). Chemical and solution weathering is rock decomposition due to chemical reactions in the rock minerals that occur from exposure to the atmosphere, temperature changes, water, or other materials. Climate topography, drainage, and vegetative cover have great influence on the chemical–solution weathering process. Residual soils are primarily the result of chemical–solution weathering.

In the United States, three geographic areas are known for extensive deposits of residual soils—the Piedmont region of the Southeast, the Pacific Northwest, and the central states south of the Ohio and Missouri rivers.

Residual soils can include particles having a wide range of sizes, shapes, and composition, depending on the amount and type of weathering and the minerals in the parent rock. However, aged formations typically include a preponderance of fine-grained particles (silt or clay) whose mineral composition and behavioral properties relate to the mineral composition of the parent rock. For example, the clays of the Piedmont have derived from deeply weathered metamorphic rock, while the residual clays of the Pacific Northwest and the area south of the Ohio and Missouri rivers have evolved from deeply weathered sedimentary and volcanic rock. The thickness of residual soil existing at any particular location is affected by the rate of rock weathering and the presence or lack of erosive forces to carry the soil away after it is formed. A profile of residual material lying above the unweathered rock often shows the transitional stages of soil formation: Downward from the surficial zone of soil (a material that may bear little mineralogical resemblance to the parent rock), progressively lesser degrees of rock weathering and soil formation exist, until finally the unaltered rock is encountered. See Fig. 1-3.

The rate of weathering is generally greater in warm, humid regions than in cool, dry regions. In the tropics, well-drained regions produce lateritic soils (soils with an absence of silica and alkalines, but enriched with iron and aluminum compounds; see Chapter 3); poorly drained areas show a prevalence of dark expansive montmorillonite clay; and andosols (al-

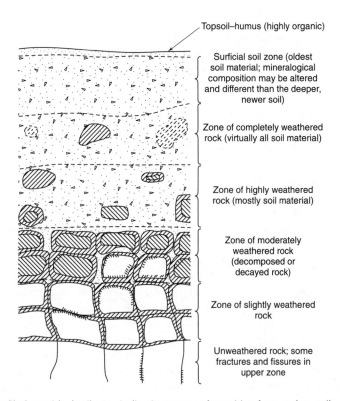

Figure 1-3 Profile for residual soil area, indicating stages of transition from rock to soil.

lophanic soil, a soil with an indefinite mineral composition but rich in silica) develop over volcanic ash and rock. Thick residual soil formations have proved to be prone to landslides in periods of heavy rainfall.

Most residual soils can be used for construction projects, but often some care and special treatment is required. The behavioral properties of disturbed or reworked, wetted then dried soil may be considerably different from those of undisturbed soil.

Transported Soils

Transported soils are those materials that have been moved from their place of origin. Transportation may have resulted from the effects of gravity, wind, water, glaciers, or human activity—either singularly or in combination. Soil particles are often segregated according to size by, or during, the transportation process. The method of transportation and deposition has significant effect on the properties of the resulting soil mass, as discussed in subsequent sections of this chapter.

Gravity- and Wind-Transported Soils.   Gravity is generally capable of transporting aggregate particles only limited distances, such as down a hill or mountain slope, with the result that little change in the soil material is brought about by the transportation procedure.

Wind can move small particles by rolling or carrying them. Soils carried by wind and subsequently deposited are designated *aeolian* deposits. Particles of small sand sizes can be rolled and carried short distances. Accumulations of such wind-deposited sands often form dunes. Dunes are typically characterized by low hill and ridge formation. They generally occur in sandy desert areas and on the down-wind side of bodies of water having sandy beaches. Dune material is a good source of sand for some construction purposes, but if the particles are of uniform size and very weathered and rounded, the sand may not be highly suitable for all construction purposes.

Fine-textured soils, the silts and clays, can be carried great distances by wind. Silt soils in arid regions have no moisture to bond the particles together and are very susceptible to the effects of wind. Clay, however, has sufficient bonding or cohesion to withstand the eroding effects of wind. Deposits of wind-blown silts laid down in a loose condition that has been retained because of particle-bonding or cementing minerals is classified as *loess*. Significant loess deposits are found in North America, Europe, and Asia (Fig. 1-4a.)

In the United States, great thicknesses of loess exist in the vicinity of the Mississippi and Missouri rivers. With these materials, accumulations have built up slowly, and grasses growing at the surface could keep pace with the rate of deposit. The resulting rootholes and grass channels that remain have created a soil that has a high porosity and cleavage in the vertical direction. Natural and man-made cuts in this material will stand with nearly vertical slopes, as illustrated in Figure 1-4b. However, if the soil is exposed to excessive water (becomes saturated or inundated), or is subject to severe ground vibrations, the soil's stable structure can be broken down. Subsidence or settlement results. Consequently, loess formations should be considered as poor foundation soils unless they can be protected from the effects of water and vibrations.

Volcanic eruptions have also produced "wind-transported" soils. The volcanic ash carried into the air with the escaping gases is small fragments of igneous rock. The soil type expected to result will be related to the mineral characteristics of the igneous rock, as discussed earlier. Generally, remains of volcanic ash deposits are limited. Because of surface deposition, they are quickly affected by weathering agents.

Glacial Deposits. Much of Canada and the northern United States, as well as northern Europe and Asia, have been subjected to the past effects of massive moving sheets of ice, the continental glaciers (see Fig. 1-5). The most recent geological period of glaciation is referred to as the Great Ice Age, and scientists estimate that, generally, it covered the span of time extending from about 2 million years ago to about 10,000 years ago. Recent theories supported by evidence consider that extended glaciation occurred periodically even earlier throughout prehistoric times, in cycles apparently related to the significant variation that occurs in the earth's orbit (which gradually varies from nearly circular to elliptical, with the related change in distance between the sun and the earth being approximately 11 million miles), to variations in the tilt of the earth's axis with respect to the plane of the orbit (from about 21.5° to about 24.5°), and to a wobble about the axis of rotation. Conditions conducive to glaciation apparently develop on approximately 100,000-year cycles.

The past periods of glaciation resulted in some present-day areas having been covered once (Fig. 1-5), some areas several times. Glaciers expanded and advanced over the land when climatic conditions permitted or contributed to the formation of ice. Glacial advance

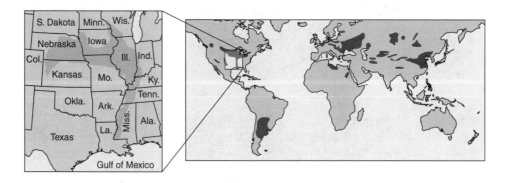

(a)

(b)

Figure 1-4 (a) Major loess deposits of the world (shaded areas); enlargement shows area of loess deposits in central United States. (b) Highway cut through loess deposit in Iowa shows natural vertical cleavage. (*Source:* Courtesy of Dr. H. S. Muskatt, Utica College of Syracuse University)

ceased when melting at its limits equaled the rate of expansion. When, because of climatic or other changes, the rate of melting exceeded the rate of growth, the glaciers receded or shrank. Generally, glacier expansion or shrinkage and movement were slow. At present, only the polar regions of the planet remain covered by glacial ice, and for the present the condition is considered stable.

Considerable quantities of soil have been moved and deposited by or because of glacial action. Such deposits are referred to as *glacial drift*. But although glaciers moved vast quantities of soil and created a surface topography, the major topographical features such as mountain ranges or plains areas are not the direct result of glaciers. Indirectly, however, major topographical features have very likely been affected by the continental glaciers. The glaciers were several thousand feet thick in many areas, and this had two very significant

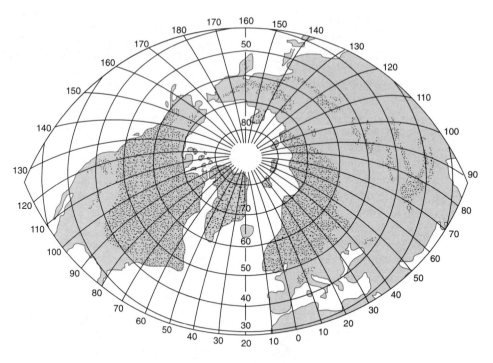

Figure 1-5 Areas of the Northern Hemisphere that have been covered by glacial ice (shaded areas). (*Source:* After Ernst Antevs, "Maps of the Pleistocene Glaciations," Geological Society of America, Bulletin 40, 1929)

effects: The tremendous amount of water taken to form the glaciers lowered the level of the sea by some 125–150 m (400–500 ft), and the tremendous weight of the glaciers caused the land beneath them to depress. As a result of the lowering of the sea level, the continental limits of North America extended beyond their current limits. It is probable that much glacially eroded soil was dropped or washed to areas that are now under the sea. As a result of the land's being depressed, areas of low elevation were subsequently flooded. After the glaciers retreated and the areas were relieved of their weight, the land rose in elevation. Rock and soil deposits that had been carried to and formed in the low flooded areas rose to create new land surfaces. Many lake areas became diminished in size or disappeared entirely. Great Salt Lake in Utah, for instance, is the shrunken remains of a 50,000-km² (20,000-square-mile) glacial lake that once had a depth in excess of 300 m (1000 ft).

As a glacier grew or advanced, it gathered and pushed soil ahead of it or enveloped and gathered the soil into itself. All sizes of particles were picked up and mixed together, with no sorting according to size. Some of the material picked up was subsequently dropped during the advance, either under the glacier or in front of it, and then overrun by the continued movement. When a glacier advance stopped, soil being pushed by the glacier and soil being freed by the melting process accumulated in front of the glacier. When the glacier receded, all soil trapped in the melting ice was dropped. Such direct glacial deposits are a heterogeneous mixture of all soil sizes and are termed *glacial till*.

The land form or topographic surface resulting after a glacier receded is called a *ground moraine* or *till plain*. The hills and ridges of till that formed at the front of the glacier and marked its farthest advance are *terminal moraines* (Fig. 1-6). *Recessional moraines* are hills or ridges that represent deposits along the front of a glacier where it made temporary stops during the recession process. (Long Island, located in the northeastern United States, illustrates such a glacial landform of geographical importance[3]; the western segment provides approximately half the land area for New York City, and the island houses one of the world's highest population densities.) Debris dropped along the side of a glacier as it moved through a valley is termed a *lateral moraine.* Long low hills of till that extend in the direction of the glacial movement are called *drumlins.*

Where the till material was dropped under the glacier and overrun, it became very dense and compact, and can provide excellent foundation support. The suitability of a till material for construction purposes, such as for a compacted earth fill, depends on the quantity and range of sizes of the soil particles; till deposits that have a preponderance of coarse particles are good construction materials, whereas deposits containing large percentages of silt and clay materials are, generally, relatively difficult to handle and compact.

Where an area was subjected to repeated glacial action, early deposits could be overlain by the more recent glacial till, or the original deposits may have been moved and redeposited by the more recent glacier as a new landform. As a result, the original source of a material may be difficult to determine. In some situations, however, the color may provide information about its source. With reference to glacial deposits in the northern United States, material gathered by glaciers from the Hudson Bay area in Canada is gray, whereas soils picked up from glaciers originating in the area northwest of the Great Lakes are red in color as a result of the high iron content in the original soils of that area.

Even while the glaciers covered a land area, there were streams and rivers of water flowing on the surface of the glaciers and in subterranean tunnels eroded within the glaciers. These flowing bodies of water carried material picked up from the land surface or eroded from the glaciers (Fig. 1-7). Some such soil was ultimately carried to the front of the glacier, but much was also dropped along the routes of flow and where the water became trapped within the glacier or between a glacier and a valley wall.

Soils deposited by the surface and subsurface glacial rivers, to remain in the form of long winding ridges (Fig. 1-8), are called *eskers.* These deposits, sorted and stratified, are usually mostly coarse grained (sands and gravels). The range of particle size can vary considerably in cross section or over short distances, however. Nevertheless, eskers can provide a good source of coarse-grained soils for construction purposes, but the uniformity of gradation should not be expected to be consistent at different locations within the esker. Eskers frequently followed along locations of low ground elevation, such as river valleys. As a result, they often represent paths of good foundation soils across areas where poor soil deposits, such as fine-grained soil sediments or marshes, have subsequently accumulated. Such paths of coarse-grained soils ideally suit the requirements for highway subgrades because of their good drainage characteristics, low susceptibility to frost heave, and ease of handling during the construction process.

Kames provide soil deposits somewhat similar to eskers; they are the remains of material dropped along the boundaries of a glacier and a valley wall (kame terrace) or in holes in a glacier. The resulting landform has the shape of knobs or small hills. As with eskers, the gradation of soils from kames should be expected to be variable.

---

[3]Present-day Long Island is actually the result of two glacial actions, one forming the eastern or outer section (Ronkonkoma Ridge) and the other forming the western or inner section (Harbor Hill Ridge).

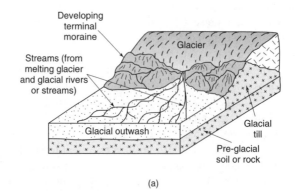

(a)

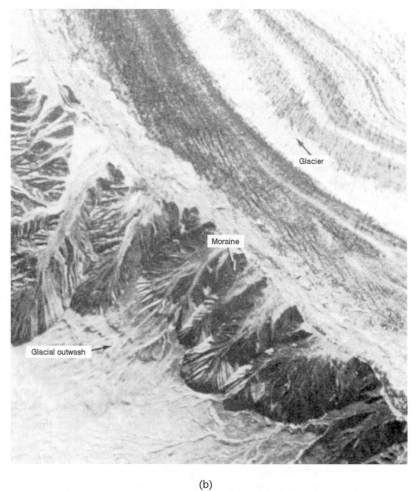

(b)

Figure 1-6 (a) The development of a terminal moraine and outwash plain in front of a glacier; (b) aerial photo of glacier showing developing moraine and outwash area.

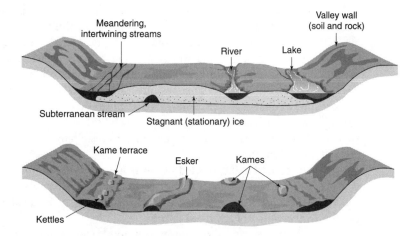

Figure 1-7   Flowing water on a glacier and resulting effect on landform.

Figure 1-8   Esker, a ridgelike deposit of granular soil at the location of a former glacial river. (*Source: Courtesy of Thomas D. Maneen*)

In some areas where kame terraces and outwash plains exist, the surface topography is further affected by depressions called *kettle holes*. Kettle holes apparently formed when great blocks of ice remained after a glacier receded and became buried by the glacial soil. When the ice melted, a hole or depression in the soil surface resulted. When water-filled, the depressions are *kettle lakes*.

River Deposits.   Flowing bodies of water are capable of moving considerable volumes of soil by carrying the particles in suspension or by rolling, sliding, and skipping them along the river bottom. The largest-diameter particle that can be carried in suspension is related to the square of the velocity of the flowing water. Coarser particles being carried in suspension are dropped when a decrease in the water velocity occurs, as when the river deepens, widens, or changes direction. Finer particles remain in suspension to be deposited in quieter waters. Thus, river deposits are segregated according to size. For most rivers, the volume of water that flowed and the volume of soil that was transported were variables that changed as land drainage forms changed or as seasonal variations in precipitation occurred.

All soils carried and deposited by rivers are classified as *alluvial deposits.* However, glacial soils carried by rivers created from melting glacial waters and subsequently sorted and dropped according to size to create deposits of stratified glacial drift are referred to as *glaciofluvial deposits* or *stratified drift.*

Glacially created rivers were developed from water escaping at the edge of a glacier or from trapped water breaking through a recessional moraine left by a receding glacier. Upon leaving the glacier or moraine area, the flowing water rapidly fanned out over a broad area of land, temporarily flooding it. With the resulting decrease in the flow velocity, the larger soil particles dropped out, forming fan- or delta-shaped flat beds of predominantly sand and gravel soils. Overlapping deltas of coarse soils spread over broad areas created land forms classified as *outwash plains* (see Fig. 1-6). The finer soil particles remained in suspension in the escaping water. Subsequently these too settled out where the velocity of flow slowed or the water became ponded.

At locations where a heavily loaded natural or glacial river broadened or encountered flatter terrain so that its velocity decreased, coarse soil particles dropped out to form submerged spreading triangular-shaped deposits termed *alluvial fans.* The alluvial fans are good sources of sand and gravel for construction purposes.

Rivers flowing through broad flat valleys have often overflowed their banks during periods of flooding. When this occurred, the overflow velocity quickly diminished, and the heavier gravel and sand particles dropped out in the vicinity of the bank, forming low ridges termed *natural levees.* The broad lowland areas on either side of the river were also flooded over, but the materials dropped in these areas were the finer-grained soils. These are *floodplain deposits* (Fig. 1-9).

Where rivers bend or curve to change direction, the velocity of the flowing water can vary considerably between the inside and outside edges of the curve. These natural curves are called *meander bends.* Erosion may take place along the outside, while deposition takes place on the inside. The deposits are the coarser soils, generally sorted according to size. Constant erosion along the outside of a bend while the inside is being built up with sediments causes the river to migrate laterally. The old river locations provide good sources of coarse soils for construction and are good foundation sites.

When a river shortcuts a large bend when eroding a new route, the old channel left behind is cut off from new flow, and the trapped water forms an *oxbow lake* (Fig. 1-10). Such lakes eventually fill with predominantly fine-grained soils carried by low-velocity flood waters or surface runoff. These areas become poor foundation sites.

Lake areas were created in natural basins in the topography or, in glacial areas, in depressed reservoirs created between a terminal (or recessional) moraine and a retreating glac-

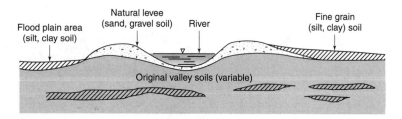

Figure 1-9 Cross section of floodplain deposit.

ier. Natural lakes and glacial lakes often covered vast areas of land. The coarsest soil particles (sands) carried by rivers feeding into the lake would fall out of suspension quickly after entering the lake area because of the sharp decrease in velocity. Such coarse soil deposits, termed *lake deltas* because of the resulting shape of the deposit, are good sources of sand and gravel for construction purposes. Fine-grained particles remained in suspension and were carried to the body of the lake, where they eventually settled out. The larger particles (fine sand and silt) settled out first, while the smaller particles (clay) continued to remain in suspension. After the waters in the lake quieted, as in periods of little or no flow into the lake, the clay settled out. Alternating layers of these fine-grained deposits built up as the variation of water flowing into the lake area continued (Fig. 1-11). If the lake area were extremely large, the coarser silt particles might settle out in areas close to the shore while the clays settled out in the quiet central areas of the lake. Frequently, the basin eventually filled with soil, or drained and left the lake deposit if the trapped water found or eroded an outlet. Soil formations remaining at the locations of former lake areas are termed *lacustrine deposits.* When layers are less than one cm in thickness, the sediments are described as *laminated.* Deposits of such fine sand–silt and clay layering are termed *laminated clay.* Laminations that have been deposited over a one-year period are *varves.* Unless such lake sediments were subject to the weight of a new glacier during a new glacial advance or to other overburden pressures, they are weak and compressible and make poor foundations. However, the deltas of coarse soils, dropped when the flowing waters entered the lake area, provide good foundation support and soils for construction use.

At some former glacial lake locations, the silt–clay soil near the surface has become firm because of drying (desiccation). This can give the illusion of a strong deposit with good foundation capabilities. Such an area may be suitable for carrying roadways and light structures, but foundations for larger and heavier structures generally cannot be satisfactorily supported.

In some areas where flowing waters carried fine-grained soils to ocean or seawater areas, some types of clay particles flocculated in the presence of the saltwater.[4] The silt and clay then settled out of suspension at about the same rate, creating deposits of *marine clays.* Despite their classification, marine clays may actually consist of more silt than clay. The marine clays are typically gray or blue-gray in color. Frequently, these deposits contain the shells or remains of shells from marine life. Much of the seaboard area of the northeastern United States and southeastern Canada has such deposits. Because of the method of

---

[4]*Flocculation* may be defined as the development of an attraction between and bonding of individual particles to form larger particles.

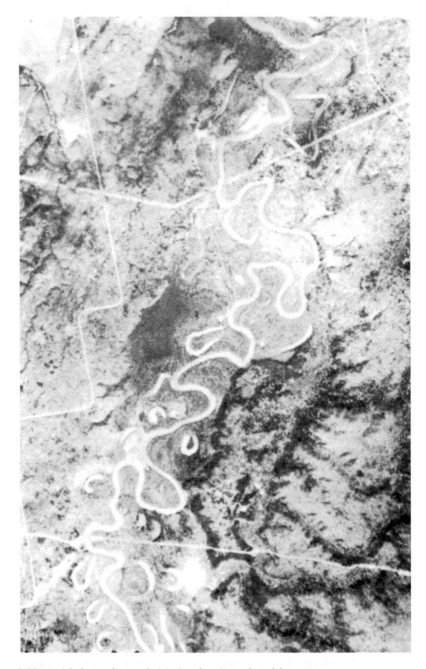

Figure 1-10 Aerial photo of meandering river forming oxbow lakes.

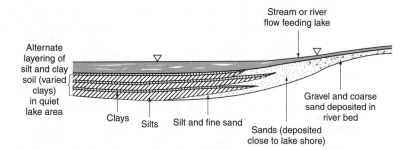

Figure 1-11   Cross section of soil deposits in a lake area, indicating typical alternate layering of silt and clay soil.

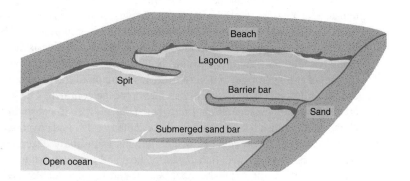

Figure 1-12   Types of sandbar deposits along a shoreline.

deposition, marine clay deposits are generally weak and compressible, and are therefore poor foundation materials. Where the land covered by the marine clays has subsequently elevated to be above sea level, there is an additional danger that the sodium (from the sea-water) that reacted with the clays to cause them to flocculate may have been leached from the soil by percolating rainwater. The result is that the clays suffer a loss in strength and become sensitive to disturbance, and land areas that had been stable become unstable.

Beach Deposits.   Ocean beach deposits are predominantly sand materials and are constantly being changed by the erosive and redistributing effects of currents and wave action. These same currents and wave actions keep silt and clay particles in suspension and carry them to the deeper, quieter, offshore areas, where they eventually settle out. Long ridges of sand that form slightly offshore are termed *bars* (Fig. 1-12). When the formation that develops includes two or more submerged ridges, they are *longshore bars*. Such deposits have been built up by the breaking waves. When a small exposed ridge forms offshore from a gently sloping beach, the formation is termed an *offshore bar* or *island bar* (also called *barrier beach* or *barrier island*). A *barrier bar* is a deposit that almost completely blocks the entrance to a bay. A sand or sand–gravel accumulation that is connected to the shore and extends into open water like a finger is a *spit*.

Marine sands are somewhat rounded and smooth, particularly the larger particles, and the deposit at a particular location may consist of particles of uniform size. Such factors, plus the corrosion potential due to salinity, may affect their usefulness for certain construction purposes, such as making quality concrete. Generally, however, such sands are potentially good sources of materials for construction purposes, particularly for waterfront and marine structures, because excavation is generally uncomplicated and transportation will be economical if barges or hydraulic pumping techniques (through pipelines) can be used.

Swamp and Marsh Deposits.   Swamps and marshes develop in stagnated areas where limited depths of water accumulate, or where periodic inundation and drying occur because of fluctuations in the groundwater level and vegetation has the chance to grow. The soils that subsequently form on the surface of swamp areas are generally of high organic content (from the decaying vegetation) and soft and odoriferous. Accumulations of decomposed or partially decomposed aquatic plants in swamp or marsh areas are termed *muck* or *peat*. Muck, geologically older than peat, is almost fully decomposed vegetation and is relatively dense. Peat includes partially decomposed vegetation and is normally spongy and relatively light. These materials are generally weak and highly compressible. Muck or peat deposits may be buried beneath the ground surface if the marsh area is subsequently overlain by glacial or alluvial materials. Peat has been used as a fuel in many areas of the world.

In an early geologic period of time, when much of the present United States was covered over by shallow seas, an eastern section of the continent was generally a low swampy area where heavy vegetation flourished (Fig. 1-13). Because of subsequent geologic changes, these lush forest areas slowly became flooded and covered over. While they were being destroyed, however, vast accumulations of partially decomposed forest vegetation were also being deposited. Subsequently, heat and pressure due to depth of burial changed these deposits to the extensive coal beds found throughout Pennsylvania, West Virginia, Tennessee, Illinois, and Kentucky.

Sanitary (Solid Waste) Landfills.   The term *sanitary,* or *solid waste, landfill* relates to the technique of using burial methods for disposing of solid waste resulting from human activities. In recent times, many municipalities have accepted the sanitary landfill as a relatively economical and manageable method for handling the large volumes of trash being generated. Landfill sites typically have involved relatively large areas and often thicknesses equivalent to the height of multi-story buildings; a result has been a surficial deposit of material equivalent to or greater than many of nature's soil deposits. The reliance on the buried landfill method for disposing of solid waste has diminished in recent years because of the perceived dangers to the surrounding environment that will occur if thorough planning and careful operational procedures are not followed. Nevertheless, new landfills are being opened while many older landfills remain in operation, and numerous filled or completed landfill sites are already in place.

The solid waste in the modern landfill is commonly the throw-away material generated by commercial business and the private or residential sector (such as garbage and trash), but may also include disposal from industry. Usually, but not always, the solid wastes accepted for burial at the present time are the materials rated nonhazardous and, preferably, degradable. Materials in the hazardous category, such as radioactive waste, medical waste, and petroleum or chemical wastes, usually are not permitted.

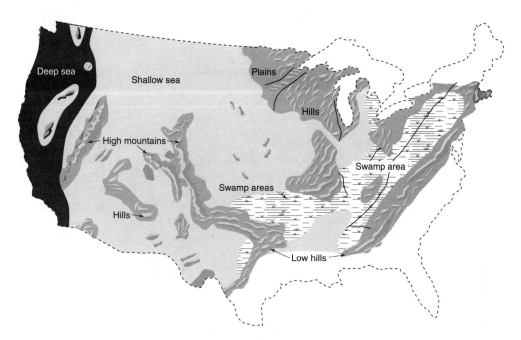

Figure 1-13 Generalized paleogeographic map of the United States approximately 200 million years ago, indicating extensive swamp areas that subsequently formed into coal deposits. (*Source:* U.S. Department of the Interior, Geological Survey)

Typically, modern solid waste landfills have each day's delivery of waste covered by one or multiple layers of soil to help reduce odors, occurrence of flyaway materials, development of fires, and attraction to scavengers, rodents, and insects, and to improve the general appearance of the area being worked. The completed landfill then consists of multiple zones or cells encapsulated by earth (Fig. 1-14). When a landfill site has been filled (totally utilized), the new land area may be recycled to other use; commonly, recreational facilities requiring large surface areas such as playgrounds and golf courses are selected.

Buried solid waste landfills do create some problems which should be recognized. Some can be controlled by proper planning and careful operation. The more significant issues relate to understanding several factors. Surface settlements will vary as the buried waste decomposes (the decomposition and settlement process generally takes many years). In addition, the decomposition of organic materials results in the generation of gases (some of which are combustible and some of which can be hazardous to animal and plant life). Also, surface and subsurface water passing through the buried waste can place into solution undesirable particle matter, bacteria, and chemicals (the resulting polluted liquid is termed *leachate*) which might migrate and contaminate nearby areas. Present-day landfill designs commonly include an installed barrier liner for the base of the landfill area to prevent underground escape of the leachate (which is collected for treatment and proper disposal). At completion the landfill area is sealed or capped to reduce future infiltration of surface water before receiving the topping layer of soil for the final vegetative cover (methods for capping landfills are presented in Chapter 7).

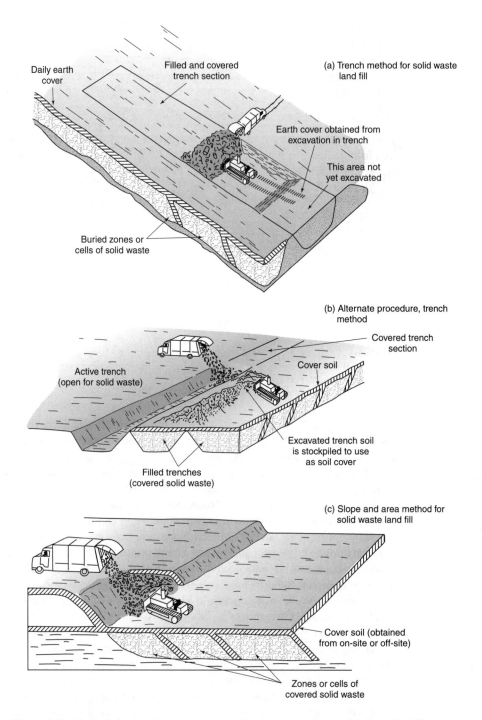

Figure 1-14 Methods for burying and covering sold waste in sanitary (solid waste) landfill.

Soil Gases and Liquids.   Soil deposits, being composed of irregularly shaped particles which have a range of sizes, will include open pore spaces or voids. Typically the void spaces include liquid—commonly water, but not always—and air (gas), whose properties are close to those found in the surface atmosphere. However, gases other than air or liquids other than water may be present as a result of natural conditions or the activities of humankind. Soil overlying deposits of petroleum and natural gas, or radioactive compounds, can include these other liquid or gaseous materials as a result of upward leakage. Radon, a naturally occurring gas, results from the decay of uranium, radium, and polonium and from radioactive waste. As part of the natural decomposition process occurring within masses of buried vegetation, gases such as methane, carbon dioxide, ammonia, hydrogen sulfide, and nitrous oxide are generated, as well as acidic fluids. Foreign liquids and gases find their way into soil deposits as a result of accidents and poorly planned human activities, including the establishment of waste dumps and landfills, the application of agricultural fertilizers and pesticides, the discharge of inadequately treated sewage, and the accidental spills and discharges of industrial and commercial compounds. Frequently, gases or liquids in low concentrations are accepted into the surrounding environment without significant effect and subsequently reduced to a less noxious material or dispersed. However, high concentrations will pollute the soil, the groundwater, and possibly the surface zone of the atmosphere. In sufficient concentrations, methane gas is explosive, hydrogen sulfide is toxic, carbon dioxide causes asphyxiation, and radon is carcinogenic. Additional material on soil gases and liquids is included in Chapters 5 and 7.

For a summary and overview of the interrelationships between rock and soil formations, see Figure 1-15.

1.3 PLATE TECTONICS

The concept of plate tectonics deals with the development, formation, and changes occurring to the earth's land and ocean areas on a long-term, large-scale basis. Plate tectonics relates to the recent recognition that the outer shell of the earth is made up of a small number of large, thick plates that mesh together like pieces of a gigantic puzzle (see Fig. 1-16), but that also move somewhat independently of each other. Continents and oceans are carried (float) on these plates. The phenomenon of plate movement has had a profound effect on what constitutes today's continents, including the type and location of rock and soil materials, mineral deposits, and animal and plant life. Very significantly, areas near plate boundaries coincide with the most active earthquake and volcanic regions of the world.

In macroscopic terms, the earth's interior consists of a core, mantle, and outer crust, concentrically zoned, as depicted in Figure 1-17. The uppermost region of the mantle possesses properties important to the plate tectonics theory. The outer zone, ranging from about 70 to 150 km thick and termed the *lithosphere* (rock sphere), consists of dense and hard brittle rock materials. Generally, the lithosphere is thickest beneath continental mountain masses. Underlying the lithosphere is a 200- to 250-km-thick zone of dense, semisolid, or plastic rock material termed the *asthenosphere* (weak rock). The rock materials making up the lithosphere and asthenosphere actually are similar, with the demarcation between zones being established on the basis of a difference in condition. Material in the lower asthenosphere increases in hardness and rigidity, grading to become the more solid material characteristics of the deeper mantle.

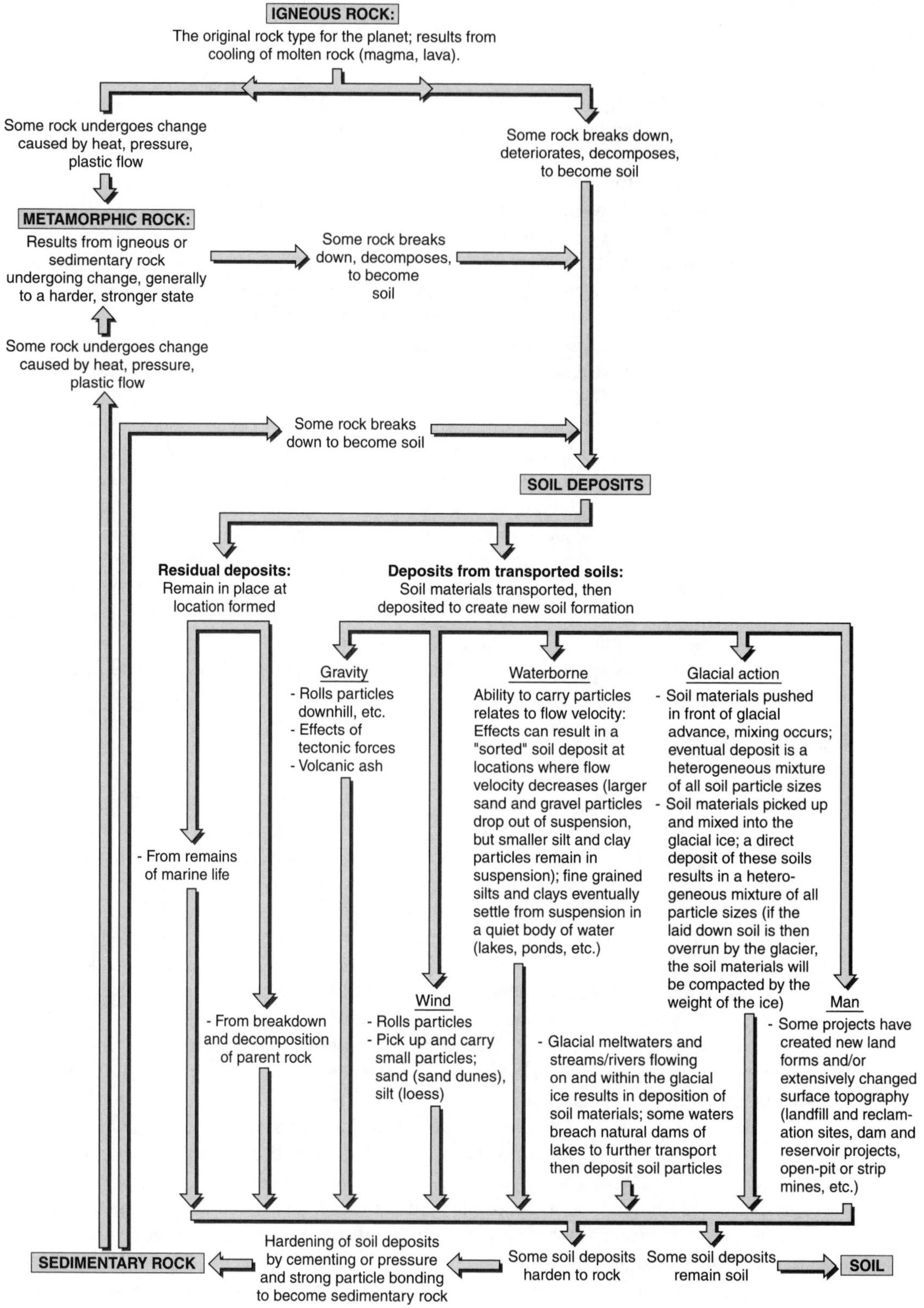

Figure 1-15 Summary and overview: interrelationships between rock and soil formations. (Arrangement: After C. A. Matrosic)

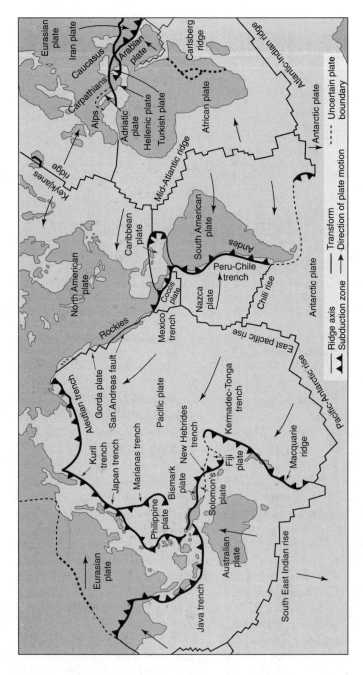

Figure 1-16 Major plates of the lithosphere and general direction of movement. (*Source:* After J. F. Dewey, "Plate Tectonics," copyright May 1972 by Scientific American, Inc. All rights reserved.)

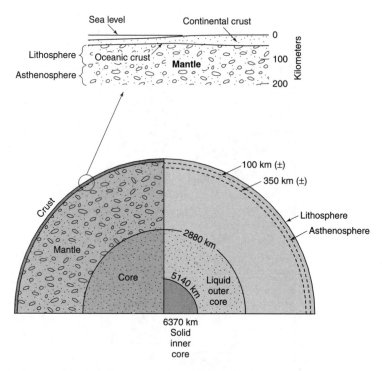

Figure 1-17 Cross section depicting composition of the earth (numerical distances indicate approximate depths from the earth's surface).

Continental and ocean crusts of relatively light-density rock material (in comparison to the rock material of the mantle) top the lithosphere. The crust underlying continental areas is thicker and less dense than oceanic crust; continental crust ranges from about 20 to 60 km in thickness (thickest beneath mountain areas), whereas ocean crust is generally on the order of 10 km thick.

The areal boundaries of the shell-like plates are the result of discontinuities or fractures that have developed in the lithosphere. At some locations, these plates of lithosphere are moving apart (diverging); at other regions the plates are moving toward each other (converging) or sliding past each other. The plates are rigid and tend to not change shape with time. Changes result only at zones where convergence or divergence occurs; the process of plate convergence and divergence will be described in later paragraphs.

The causes of plate movement are not known, but it has been hypothesized that regions of the asthenosphere and underlying mantle are in motion because of a temperature difference between the earth's core zone and the outer mantle (similar to the way heated air moves by convection through a cool room), with the lithosphere being carried along by the asthenosphere. What *is* established is that the plates of lithosphere have been in motion for a long period of the earth's history and that the plates have been moving relative to each other.

Continental and ocean crusts atop the lithosphere move with the lithosphere. The position and orientation of continent-supporting plates have been changing with respect to other continental plates and with respect to location on the earth's surface. Similarly, ocean areas have been changing; some of these changes have been so extensive that oceans have been both created and destroyed. Figure 1-18 illustrates the relative position of continental areas and oceans through the past 200-million-year period, and an expectation for the future.

A *rift* develops where plates of lithosphere are moving apart, creating the condition where molten rock from the asthenosphere rises to create new lithosphere and crust; volcanism is involved. Plate divergence and the creation of new lithosphere in ocean ridge zones may be referred to as *sea-floor spreading*. Plate boundaries identified as *ridges* and *rises* (Fig. 1-16) represent locations where plates are spreading apart. Most of the currently known locations of major plate divergence occur in what is now ocean area. It is likely that early in the earth's history the spreading of some plates began under continental areas, but the widening separations filled with shallow crust, had low surface elevation, and were flooded over.

The surface area of the earth remains approximately constant with time. The creation of new lithosphere where plates diverge must be balanced by an equivalent loss of lithosphere at other plate boundaries. Such losses occur at locations where plates converge, by having one plate subduct (dip) below the other. The subducted section of lithosphere passes into or through the asthenosphere, where it is absorbed. Boundaries where subduction occurs are identified as *trenches* in Figure 1-16. Ocean crust will be carried along when its supporting plate of lithosphere is being subducted, but continental crust does not subduct because of its light density; ocean crust is thus destroyed by subduction and created where plate divergence occurs, but continental crusts tend to be permanent.

Boundaries where plate edges slide past each other are classified as *transform faults*. Such locations are identified as faults on Figure 1-16. Surface area is neither created nor destroyed at such plate edges. Small "filler" plates between the boundaries of main transforming plates apparently rotate in place.

When two plates carrying continental crust converge, a subduction of one lithosphere plate occurs, but its continental crust is too lightweight and thick to be carried along. The result is that the two continental masses stay afloat but undergo a condition of collision over a sustained period, causing the crust material to fold slowly, accordion style, and giving rise to a mountain range (for instance, the Himalayas). Regions near the boundary where two continent-supporting plates converge are earthquake zones but are not volcanically active. Earthquakes relate to the quick release of strains (deformations) that build up within the lithosphere and crust as a result of the plate sections moving past each other.

Where two ocean plates converge, or where an ocean plate dips below a plate carrying continental crust, the boundary region becomes an area of active volcanism and earthquake activity (for instance, the Andes Mountains of South America). Apparently the volcanic activity relates to the surface breakthrough of a rising, relatively low-density magma, which is created when ocean crust, carried into the asthenosphere on the lithosphere plate being subducted, melts (Fig. 1-19). Mountain ranges near the boundary where a continental plate overrides a subducting oceanic plate, such as the Andes, may be the product of volcanism or the result of horizontal compression and folding that continental crust experiences because of the plate convergence.

Transform faults occur at plate boundaries originally created by divergence or convergence. The sliding of one plate past another causes earthquake activity but no volcanism. Transform faults develop where a section of plate boundary is aligned with the general direction of the plate's movement. At some locations the fault is of great length—for example, the San Andreas fault near the west coast of the United States. More prevalently,

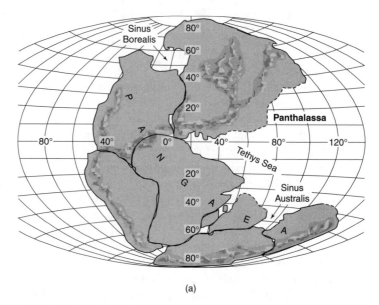

(a)

Figure 1-18 Location and position of continents and oceans over the past 200-million-year period, on the basis of plate movements. (*Source:* After R. S. Dietz and J. C. Holden, "The Breakup of Pangaea," copyright October 1970 by Scientific American, Inc. All rights reserved.) (a) Universal land mass Pangaea may have looked like this 200 million years ago. Panthalassa was the ancestral Pacific Ocean. The Tethys Sea (the ancestral Mediterranean) formed a large bay separating Africa and Eurasia. The Appalachian, Alps, and Ural mountain ranges were already ancient by this time. The Rockies and Andes were in existence but were relatively young. These ranges probably developed as a result of collisions between continental plates predating Pangaea (plates are shown lightly shaded).

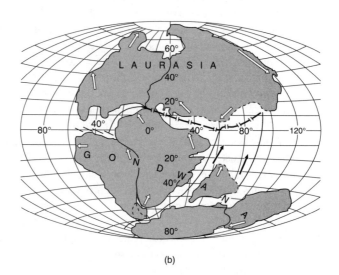

(b)

(b) After 20 million years of drift (180 million years ago), the northern group of continents, collectively known as Laurasia, had split away from the southern group, known as Gondwana. The latter had started to break up: India had been set free by a Y-shaped rift, which had also begun to isolate the Africa–South America land mass from Antarctica–Australia. The Tethyan trench (hatched lines in black), a zone of crustal uptake, runs from Gibraltar to the general area of Borneo. Black lines and black arrows denote megashears, zones of slippage along plate boundaries. The white arrows indicate the vector motions of the continents since drift began.

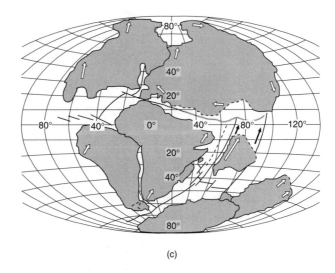

(c)

(c) After 65 million years of drift (135 million years ago), the North Atlantic and Indian oceans had opened considerably. The birth of the South Atlantic had been initiated by a rift. The rotation of the Eurasian land mass had begun to close the eastern end of the Tethys Sea. The Indian plate continued its northerly movement.

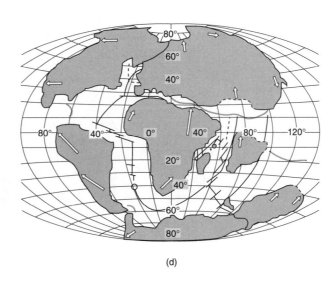

(d)

(d) After 135 million years of drift (65 million years ago), the South Atlantic had widened into a major ocean. A new rift had carved Madagascar away from Africa. The rift in the North Atlantic had switched from the west side to the east side of Greenland. The Mediterranean Sea is clearly recognizable. Australia still remained attached to Antarctica. An extensive north–south trench (not shown) must also have existed in the Pacific to absorb the westward drift of the North American and South American plates.

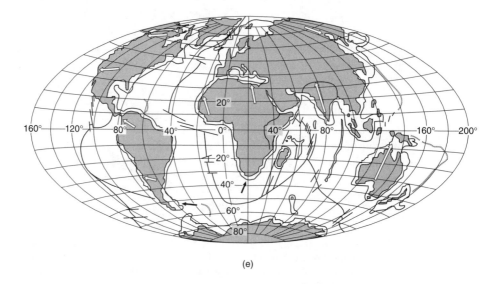

(e)

(**Figure 1-18**, *continued*)   (e) The world as it looks today was produced in the past 65 million years. Nearly half of the Atlantic Ocean floor was created in this geologically brief period. India completed its flight northward by colliding with Asia, and a rift separated Australia from Antarctica. The North Atlantic rift finally entered the Arctic Ocean, fissioning Laurasia. The Antilles and Scotia area now occupy their proper positions with respect to neighboring land masses. Extrapolation indicates that, 50 million years into the future, the Atlantic (particularly the South Atlantic) and Indian oceans will continue to grow at the expense of the Pacific. Australia will drift northward and begin rubbing against the Eurasian plate. The eastern portion of Africa will split off. Baja California and a sliver of California west of the San Andreas fault will be severed from North America, to begin drifting to the northwest.

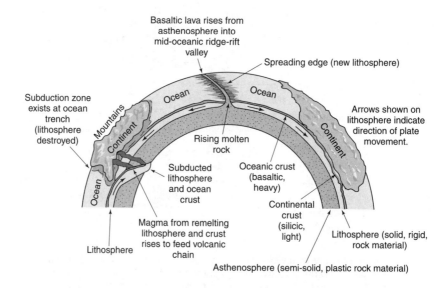

Figure 1-19   Movement of lithospheric plates—plate divergence and plate subduction.

however, numerous limited segments of transform fault are incorporated into the plate boundaries of the ocean ridges and rises identified in Figure 1-16, a condition brought about by the irregular fracturing of lithosphere where plate spreading is taking place.

Recent evidence suggests that plate movement does not occur at a steady rate, as was earlier believed. New crust formed along rift boundaries assumes the north–south magnetism in effect at the time molten magma cools. It has been established that, for reasons as yet unexplained, the magnetism of the earth has reversed numerous times during the planet's existence—possibly 170 times in the last 76 million years. (Recent information hints that the solid core of the earth rotates slightly faster than the surrounding molten outer core and mantle zones, and an association of fluid and electrical currents between the inner core and outer core is related to the generation of the planet's natural magnetic field. If the inner core revolves slightly faster than the outer core, the alignment of the magnetic field will be slowly changed, which, with time, results in periodic reversals of magnetic north–south.) Variations found in the width and directional magnetism of the bands of new crust adjacent to ocean rifts provide a reference to the past rate of plate movement. The causes of plate acceleration and deceleration are not yet understood. The movements of the major plates are related, whereby the plates speed up or slow down together. Of significance, volcanic activity and the mountain building process increases during the periods of plate acceleration. In the past, apparently each period of acceleration and deceleration extended over several millions of years. The earth is probably in a session of accelerated plate movement at the present time.

The plate tectonics theory assumes that the various geologic processes associated with the formation of rock and soil materials as discussed in the early part of this chapter have been ongoing simultaneously with plate movements. If we recognize that the global position of a continent influences the conditions and forces of nature that the land becomes subjected to, such as climate (including temperature and moisture), effects of oceans if nearby (including chances of inundation), and exposure to glacial, volcanic, and earthquake activity, it becomes evident that much of the earth's currently existing rock and soil represents the influence of plate movements.

1.4 EARTHQUAKES

Earthquakes as events of nature have probably been occurring since the early periods of the planet's existence; they will probably continue to occur in the future. Earthquakes unleash awesome amounts of uncontrolled energy, without warning and with typical suddenness. Great destruction is associated with earthquakes; for the human race, fear is the learned response. In recent decades, scientists have made significant advances in understanding earthquakes and have gained some insight into methods of reliable protection; much remains to be discovered, however.

An earthquake is the phenomenon of vibration, shaking, or movement of the ground that occurs when a release of energy into the earth results in the transmission of shock, or seismic, waves. Typically, such release of energy is caused by the sudden fracture of rock in the lithosphere or overlying crust, or by a sudden rupture (shift) at the boundaries where plates of lithosphere meet, or along a fault plane (that is, along a pre-existing fracture in the crust); this type of event is classified as a tectonic earthquake. Energy release capable of causing earthquakes can also occur during volcanic eruptions (volcanic earthquakes) and from man-induced explosions such as the detonation of nuclear devices (explosion earthquakes).

With tectonic earthquakes, the release of energy may be attributed to the elastic rebound property of rock materials located in the zone responsible for the earthquake. Elastic rebound results where rock that is strained and has been deformed (because tectonic movement or thermal or other change occurred) suddenly releases the associated energy of deformation when the rock fractures or slips (analogous to the process of a compressed steel coil springing back toward the original shape when the confining force is released).

Tectonic movements are relative movements between rock zones, usually as adjacent plates of lithosphere or overlying crust shift independently. Deep-rock tectonic movements are also associated with the presence of water, when the micro-cracks which develop in the rock as movement and strain occurs become filled with water. The expanded (dilated) rock cannot close to the original condition as the movement continues (the micro-cracks do not close completely because of the presence of water). Eventually, the loss of strength within the rock mass (because of the loss of internal bonding and inability of the micro-cracks to rebond, and the loss of friction along rough surfaces) results in fracture. (Interestingly, large increases in the volume of deep rock may be reflected by changes in the ground surface level and also by ground tilting, physical features that when monitored may be useful in forecasting the advent of an earthquake.)

The location within the earth where the rock fracture or rupture develops to trigger an earthquake is identified as the focus or hypocenter. The location on the earth's surface directly above the hypocenter is the epicenter (Fig. 1-20).

The majority of earthquakes are in the category of shallow-focus events, where the focus is less than 70 km deep. Most of the planet's shallow earthquakes originate in the trench regions where plates of lithosphere meet (Fig. 1-16).

The energy released into the earth zone surrounding suddenly ruptured or slipped rock, or the zone of a volcanic or human-induced explosion, radiates outward in a wavelike manner, producing seismic waves (i.e., shock waves) and heat. The type of wave movement, and the velocity, depend on the elastic properties of the rock, soil, and water or other liquid material being traversed. Two categories of waves, body waves and surface waves, are used to describe the wave transmission (Fig. 1-20).

Traveling within the earth, two types of waves occur. Body waves, termed *primary* or *P-waves,* consist of compression–dilation (push–pull) types of movements. Shear-type waves, termed *secondary* or *S-waves,* cause up-and-down and side-to-side movements (Fig. 1-21). P-waves travel at higher velocities than S-waves, and will move through solids and liquids. S-waves will not be transmitted through liquid materials. Typically then, when an earthquake occurs, the effects of the P-wave will be the first noted. Building occupants typically first experience vertical jolt, and then horizontal shaking.

Along the earth's surface zone the energy transmission is in the form of *Love* waves and *Rayleigh* waves (Fig. 1-21). Love waves are a shear-type wave similar to S-waves except that only horizontal movement takes place (no vertical movement). Rayleigh waves occur as a rolling or rippling type of wave motion (similar to surface ripples when a stone is dropped into water). Surface waves move at lower velocities than body waves; generally, the Love waves travel at higher velocity than Rayleigh waves.

The transmission of P-waves is similar to the movement of sound waves. P-waves reaching the earth's surface can be passed into the atmosphere as sound waves. However, most of the energy of both P-waves and S-waves is reflected back into the earth. The surface zone thus becomes subjected to seismic waves that tend to go in opposite directions simultaneously, a situation that can result in amplication of the movement (i.e., shaking).

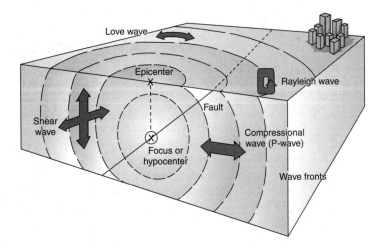

Figure 1-20   Schematic illustration of the directions of vibration caused by body and surface seismic waves generated during an earthquake. When a fault ruptures, seismic waves are propagated in all directions, causing the ground to vibrate at frequencies ranging from about 0.1 to 30 Hertz. Compressional and shear waves mainly cause high-frequency (greater than 1 Hertz) vibrations. Rayleigh and Love waves mainly cause low-frequency vibrations. (*Source: Facing Geologic and Hydrologic Hazards*, W. W. Hays (ed.), Geological Survey Professional Paper 1240-B, 1981)

The vibratory motion experienced by a rock or soil formation relates to the distance from the earthquake focus and the energy released. The *seismograph* is an instrument developed to measure the type and extent of movement at the instrument's point of placement. In principle, the instrument consists of a weight suspended by a wire from a frame (somewhat like a pendulum); when the frame is shaken, the inertia of the suspended weight lags behind the motion of the frame. If a pen or other tracking mechanism attached to the suspended weight continuously plots the relative movement between the vibrating frame and the suspended weight against time, the record of the up-and-down, or back-and-forth, undulations of the earth is a *seismogram*. Typically, surface seismographs will record movements in three directions (e.g., vertical, north-south, and east-west directions). The concept of the seismograph is shown in Figure 1-22, along with an example of a seismogram or accelogram (the accelograph is a strong-motion seismograph that operates only when large, earthquake-size, ground vibrations occur).

Earthquake *magnitude* refers to a numerical scaling system to rate the severity of a seismic event on the basis of the total strain energy released. The magnitude scaling system permits evaluation of earthquakes regardless of the location of the focus or epicenter. The concept of earthquake magnitude was presented by Charles Richter in 1935 and included a numerical scale which ranged upward from zero, with no upper limits (but the severest known earthquake was approximately a magnitude 9). The Richter magnitude remains in widespread usage because of the general population's familiarity with the term but is now identified as local magnitude, M_L.

The local magnitude value is defined as the logarithm (base 10) of the maximum seismic wave amplitude (expressed in thousandths of a mm) recorded on a torsional-type seismograph at a distance of 100 km from an earthquake epicenter. The type of seismic wave

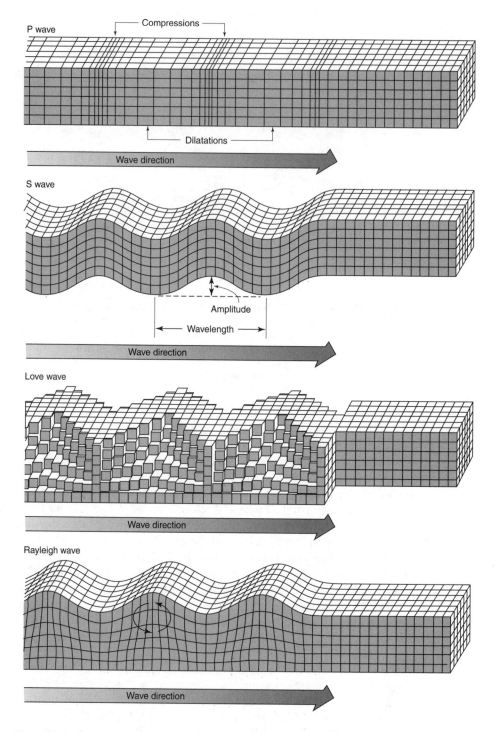

Figure 1-21 Concept of the form of the body and surface waves resulting from earthquake. (*Source: Bruce A. Bolt, Nuclear Explosions and Earthquakes: The Parted Veil,* San Francisco: W. H. Freeman, 1976)

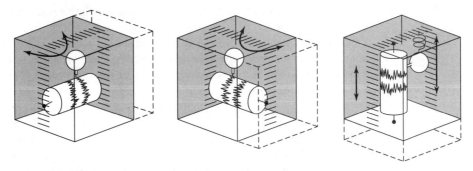

(a) Simple models of pendulum seismographs recording the vertical and horizontal directions of ground motion. (*Source*: Bruce A. Bolt, *Nuclear Explosions and Earthquakes: The Parted Veil,* San Francisco: W.H. Freeman. Copyright © 1976.)

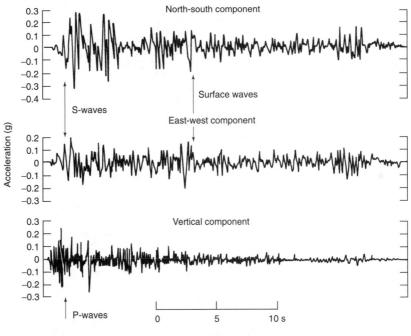

(b) Accelerogram of the 1940 Imperial Valley, California, earthquake recorded at El Centro, California. (*Source*: W.W. Hays, 1980, *Procedures for Estimating Earthquake Ground Motions,* U.S. Geological Survey Professional Paper 1114)

Figure 1-22 (a) Illustration of seismograph equipment. (b) Seismograph or accelogram record produced by the seismograph.

(P-, S-, or surface wave) selected to compute magnitude is the one creating the largest amplitude. The ground motion data recorded by a seismograph will be influenced by its distance from the epicenter, of course; the nomograph shown in Figure 1-23 was developed to include the factor of distance between a seismograph's actual location and the 100-km reference location and permit simple determination of magnitude. Referring to the illustration within Figure 1-23, note that the time difference between the arrival of P-waves and S-waves is used to determine distance from the epicenter.

Since the time of Richter's original work, additional earthquake magnitude scales have been developed in an attempt to obtain better definition of occurrences. These alternative magnitude scales usually are based on measurement of select seismic waves caused by the event.

A listing of various earthquake magnitude scales is presented in Table 1-1.

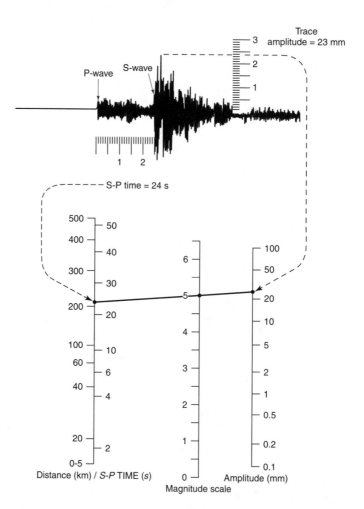

Figure 1-23 Determination of Richter magnitude. Using the maximum amplitude of the seismogram and the difference in arrival times of the P- and S-waves, the value of magnitude can be read from the nomograph. (*Source:* W. W. Hays, *Procedures for Estimating Earthquake Ground Motions,* U.S. Geological Survey Professional Paper 1114, 1980)

Table 1-1 Listing of Earthquake Magnitude Scales

Scale Name or Identitication	Definition	Description of Terms	General Comments
Local Magnitude, M_L (the original Richter scale)	Logarithm (base 10) of the maximum amplitude seismic wave (either P-, S-, or surface wave)	Amplitude expressed in .001 mm	Seismic wave amplitudes as measured by a Wood-Anderson torsional seismograph located 100 km from epicenter of quake. Seismograph must have a natural period of 0.8 seconds, magnification of 2800, with a damping coefficient of 80%.
Surface Magnitude, M_S	$M_S = \log(A/T) + 1.66d + 3.3$ also: $M_S = 1.7 + 0.8M_L - .01M_L^2$ also: $M_S = 2.02 + 1.14 \log L$ also: $M_S \approx 4.15 + \log A_f$	A = horizontal component of the Rayleigh surface wave, for 20-second period, in microns T = period of seismic wave, in seconds d = epicenter distance, in degrees L = length of fault activated by event A_f = fault area (length × depth, in km^2)	Originally developed to determine a magnitude for shallow focus earthquakes at relatively long distances.
Body Wave Magnitude, M_b	$M_b = 0.56M_S + 2.9$ where $M_S = 1.7 + 0.8M_L - 0.1M_L^2$	M_S = surface magnitude M_L = local magnitude	Developed to measure magnitude of deep-focus earthquakes.
Seismic Moment, M_o	$M_o = GA_fD$ (units are ton − m or dyne − cm)	G = modules of rigidity (approximately 3×10^6 ton/m^2 or 3×10^{11} dyne/cm^2 A_f = fault area (length × depth, in m^2) D = longitudinal displacement of fault, in m M_o = seismic moment	Indirect measurement of earthquake energy.
Seismic Moment Magnitude, M_w	$M_w = \tfrac{2}{3} \log M_L - 10.7$		Magnitude based on seismic moment.
Richter Magnitude, M	$M = M_L$ (for $M_L < 5.9$) $M = M_S$ (for $5.9 < M_S < 8.0$) $M = M_w$ (for $8.0 < M_w < 8.3$)		Redefinition, general usage Richter scale.

An expression to roughly approximate the energy E (in ergs) released by an earthquake of local magnitude M_L or Richter magnitude M is

$$\log_{10}E = 11.8 + 1.5M \qquad (1\text{-}1a)$$

$$E = 10^n \text{ ergs}$$

where $n = 11.8 + 1.5M$

(*Note:* 1 erg = 1 dyne-cm = 7.382×10^{-8} ft lb of energy, or 1 ft lb = 13.547×10^6 ergs.)

Another expression proposed for rough evaluation of energy released using the determination of surface magnitude, M_S, is

$$E = 10^{4.8 + 1.5M_s} \text{ joules} \qquad (1\text{-}1b)$$

(Note: 1 joule = 1.36 ft lb; 1 erg = 10^7 joules.)

Illustration 1-1

For a Richter magnitude 5 earthquake, an approximation of the energy released is

$$E \cong 10^n \text{ ergs} \cong 10^{(11.8 + 1.5 \times 5)} \text{ ergs}$$

$$E \cong 10^{19.3} \text{ ergs} \cong 2 \times 10^{19} \text{ ergs} \cong 1.48 \times 10^{12} \text{ ft lb}$$

For a magnitude 6 earthquake, an approximation of the energy released is

$$E \cong 10^{20.9} \text{ ergs} \cong 7.94 \times 10^{20} \text{ ergs} \cong 58.6 \times 10^{12} \text{ ft lb}$$

or approximately 40 times the energy of the magnitude 5 earthquake.

Illustration 1-2

Determine the surface magnitude, M_S, value corresponding to a local (Richter) magnitude M_L 6 earthquake, and, approximate the earthquake energy associated with that M_S value.

Solution For $M_L = 6$, estimate

$$M_S = 1.7 + .8M_L + .01(M_L)^2 \qquad (\text{ref. Table 1 - 2})$$

$$M_S = 1.7 + .8(6) + .01(6)^2 = 6.14$$

and

$$E = 10^{4.8 + 1.5M_s}$$

$$E = 10^{4.8 + 1.5(6.14)} \cong 10^{14} \text{ joules or } 74 \times 10^{12} \text{ ft lb}$$

Body wave transmission (subsurface transmission) becomes complicated at the boundaries where different earth materials meet, such as where a soil material overlies rock or where strata of rock materials join. The direction of a transmitted wave may be altered when it moves from one material into a different material (the phenomenon of refraction), and some of the wave energy can be reflected back into the source material. For either occurrence, the original wave may be converted into a new combination of P-waves and S-waves or S-waves into surface waves. The size of the seismic waves may increase or decrease as transmission occurs from sound dense rock into surficial weathered rock or soil.

A general effort of the various combinations of seismic waves that reach an area is that surface zones tend to shake or move more than the deeper earth zones.

The energy possessed by seismic waves dissipates with distance from the earthquake hypocenter. The faster, high-frequency body waves lose their energy over distance faster than the surface waves. The slow, long-period, long-wavelength, surface waves tend to propagate over long distances. For short-duration earthquakes, the P-waves and S-waves reach an area before the surface waves, but an earthquake of sustained duration can have both body and surface waves moving through a region at the same time.

Ground surface motions resulting from seismic waves are greater in the horizontal direction than in the vertical direction; the combination effects of P-waves and S-waves, or P-waves and S-waves plus surface waves, produce greater horizontal movement than vertical.

On the ground surface, the effect of an earthquake is a vibratory-type motion (i.e., cyclic or oscillating). Movement occurs vertically up and down and also laterally in all directions. Factors affecting the stability of natural and manmade structures located at the earth's surface include the direction of the seismic waves, the duration of the shaking, the distance or displacement of the ground motions, the acceleration of the ground motions, and the time period or frequency of the different seismic waves. Some important basic relationships between seismic wave motion and (ground) movement are shown in Figure 1-24.

The effects of ground acceleration and the duration of seismic vibrations may be more responsible for the damage to a structure during an earthquake than the vertical movements; among other factors, the design and construction of structures typically is based on the everyday gravity (vertical) loads expected, and therefore structures have an inherent capacity for resisting vertical influence. Horizontal ground accelerations transferred into a structure cause an array of potentially damaging horizontal shear forces to cycle vertically up and down through the height of the structure, in accord with the principle of dynamics which identifies the relationship between force, mass, and acceleration (i.e., $F = ma$).

Earthquake *intensity* is the scale of reference based on the effects of an earthquake felt at a particular location. The degree of damage that occurs is related to the distance from the earthquake epicenter and the magnitude, but also relates significantly to the type of facility or structure and the properties of the supporting rock or soil foundation materials. The Modified Mercalli Intensity Scale shown in Table 1-2 includes the experience data obtained from investigating damages and other after-effects of earthquakes. When used with geologic data and seismic probability information, intensity scales such as the Modified Mercalli have become a source of good experience-related information for establishing construction standards appropriate to the geographic area.

Structures that will be located in areas where seismic activity can occur should be designed and constructed to resist the effects of such events. Earthquake-resisting design procedures consider the influence of seismic-induced ground acceleration factors; the most basic approach has been to assign extra horizontal and vertical forces to the static design

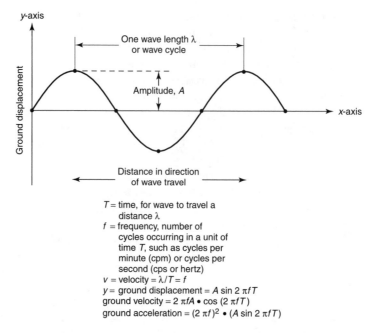

T = time, for wave to travel a distance λ

f = frequency, number of cycles occurring in a unit of time T, such as cycles per minute (cpm) or cycles per second (cps or hertz)

v = velocity = $\lambda/T = f$

y = ground displacement = $A \sin 2\pi f T$

ground velocity = $2\pi f A \cdot \cos(2\pi f T)$

ground acceleration = $(2\pi f)^2 \cdot (A \sin 2\pi f T)$

Figure 1-24 Basic relationships between seismic wave motion and ground movements.

loads so as to simulate the effects resulting from an earthquake (as illustrated in Fig. 1-25).[5] The building codes for many areas of the world have evolved to include procedures for providing a level of earthquake resistance considered appropriate for the region. Building codes, as well as other technical sources, may provide maps indicating zones or levels of seismic risk, or levels of ground motion acceleration or velocity expected from earthquake activity; seismic maps indicating acceleration coefficients and velocity-related acceleration are based on consideration of ground motions, frequency of occurrence, and attenuation with distance. These seismic maps do not indicate the frequency of earthquakes but reflect the extent of ground movement anticipated from significant events which may occur in the area. (Such data are usually based on evaluation of past occurrences.) Building codes may state that the design regulations are primarily intended to achieve a reasonable level of public safety whereby structures should withstand a moderate earthquake without experiencing significant damage and a major earthquake without collapse occurring.

For concept, the following listing represents a summary of seismic-related factors considered by most building codes as important to earthquake-resistant building design:

---

[5]The effect of forces distributed throughout a building structure because of seismic activity can be evaluated using pseudostatic or dynamic analysis. The pseudostatic method involves assigning static-type forces to achieve effects that are considered comparable to those resulting from an earthquake. Dynamic analysis, considered more appropriate for representing a structure's response to ground motion but analytically more rigorous than pseudostatic analysis, is based on applying cyclical loading to represent time-related variations in the magnitude and location of forces resulting from an earthquake. The effect of an earthquake on structures other than buildings (retaining walls, dams, earth structures) is often evaluated using pseudostatic analysis.

Table 1-2 Modified Mercalli Intensity Scale, MMI (Abridged)

Intensity	Effects
I	Not felt except by a very few under especially favorable circumstances.
II	Felt only by a few persons at rest, especially on upper floors of buildings. Delicately suspended objects may swing.
III	Felt quite noticeably indoors, especially on upper floors of buildings, but many people do not recognize it as an earthquake. Standing motor cars may rock slightly. Vibration like passing of truck. Duration estimated.
IV	During the day felt indoors by many, outdoors by few. At night some awakened. Dishes, windows, doors disturbed; walls make cracking sounds. Sensation like heavy truck striking building; standing motor cars rocked noticeably.
V	Felt by nearly everyone; many awakened. Some dishes, windows, etc., broken; a few instances of cracked plaster; unstable objects overturned. Disturbance of trees, poles, and other tall objects sometimes noticed. Pendulum clocks may stop.
VI	Felt by all; many frightened and run outdoors. Some heavy furniture moved; a few instances of fallen plaster or damaged chimneys. Damage slight.
VII	Everybody runs outdoors. Damage negligible in buildings of good design and construction; slight to moderate in well-built ordinary structures; considerable in poorly built or badly designed structures; some chimneys broken. Noticed by persons driving motor cars.
VIII	Damaged slight in specially designed structures; considerable in ordinary substantial buildings, with partial collapse; great in poorly built structures. Panel walls thrown out of frame structures. Fall of chimneys, factory stacks, columns, monuments, walls. Heavy furniture overturned. Sand and mud ejected in small amounts. Changes in well water. Disturbs persons driving motor cars.
IX	Damage considerable in specially designed structures; well-designed frame structures thrown out of plumb; damage great in substantial buildings, with partial collapse. Buildings shifted off foundations. Ground cracked conspicuously. Underground pipes broken.
X	Some well-built wood structures destroyed; most masonry and frame structures with foundations destroyed. Ground badly cracked. Rails bent. Landslides considerable from river banks and steep slopes. Shifted sand and mud. Water splashed (slopped) over banks.
XI	Few, if any (masonry) structures remain standing. Bridges destroyed. Broad fissures in ground. Underground pipelines completely out of service. Earth slumps and land slips in soft ground. Rails bent greatly.
XII	Damage total. Waves seen on ground surfaces. Lines of sight and level distorted. Objects thrown upward into the air.

Source: Wood and Neuman, 1931, by U.S. Geological Survey, 1974, Earthquake Information Bulletin, v. 6, no. 5, p. 28.

1. Level of seismic activity (i.e., ground motion) to assume for the building design procedure, including duration plus peak ground acceleration and peak ground velocity expected.

2. Information relating to the building's structural assembly, including ability of the main structural system to absorb energy and sustain large deformations without major damage; the fundamental time period for the structure in order to help evaluate the tendency to oscillate when subject to ground motions; the dead load and live loadings (weights).

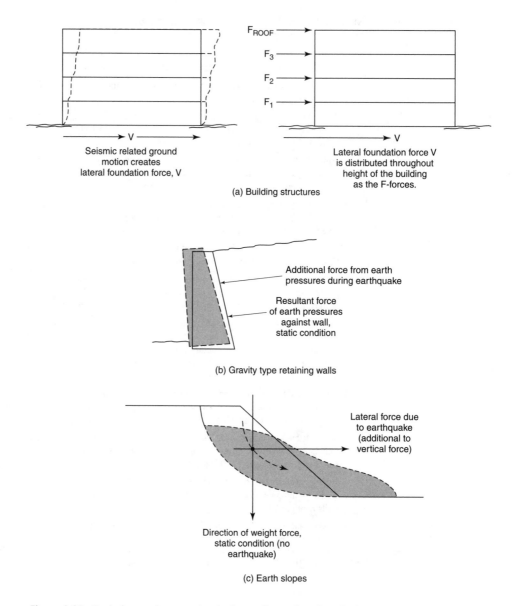

F_{ROOF}

F_3

F_2

F_1

Seismic related ground
motion creates
lateral foundation force, V

Lateral foundation force V
is distributed throughout
height of the building
as the F-forces.

(a) Building structures

Additional force from earth
pressures during earthquake

Resultant force
of earth pressures
against wall,
static condition

(b) Gravity type retaining walls

Lateral force due
to earthquake
(additional to
vertical force)

Direction of weight force,
static condition (no
earthquake)

(c) Earth slopes

Figure 1-25 Typical procedures used to indicate effects of earthquake forces.

3. Subsurface conditions, including indication of soil and rock properties (such as classification, loose or dense condition, natural frequency or vibration period, etc.).

4. Building occupancy importance factors, relating to the need for the building to remain functional after a seismic event (e.g., buildings such as hospitals and those used to house public utility and communications operations have high importance).

In North America, the Uniform Building Code (UBC) and the Building Officials and Code Administrators (BOCA) National Basic Building Code, both U.S., and the National Building Code of Canada (NBCC) are representative of the seismic design criteria incorporated into building codes. These codes provide methods for calculating a lateral shear force V acting at the foundation level of a building when subject to an earthquake, and procedures for distributing this lateral shear force throughout the height of the structure to complete the analysis and design. Expressions for foundation-level lateral shear force V are included herein, along with related maps and tables to illustrate the nature of seismic and subsurface information used for applications.[6]

UBC $$V = \frac{ZICW}{R_w}$$ (1-2)
(ref. Fig. 1-26)

$$\text{where } C = \frac{1.25S}{T^{2/3}} \gamma 2.75$$

BOCA $$V = C_s W$$ (1-3)
(ref. Fig. 1-27)

$$\text{where } C_S = \frac{1.2A_v S}{RT^{2/3}} \gamma \frac{2.5A_a}{R} \text{ (upper limit)}$$

$$\text{and } 1.2A_v = Z, \text{ seismic zone factor}$$

NBCC $$V = \frac{(V_e)U}{R} = \frac{(vSIFW)(0.6)}{R}$$
(ref. Fig 1-28)

(1-4)

In these expressions

Z = seismic zone factor

I = building occupancy importance factor (between 1 and 1.25 for UBC, between 1 and 1.50 for NBCC)

C, C_S = factor relating to the fundamental time period T for the structure

S = soil profile or site coefficient (UBC, BOCA) or seismic response factor relating fundamental period and seismic zoning (NBCC)

F = foundation factor, soil profile coefficient (NBCC)

V_e = equivalent lateral seismic force representing elastic response

U = 0.6, a calibration factor

---

[6]Seismic effects predicted by building codes typically include the influence of factors such as knowledge of local geology and historic experiences. Consequently, building codes from different areas are not necessarily expected to assign similar seismic values to structural designs. For construction projects the designer must apply, as a minimum standard, criteria from the building code having jurisdiction at the site area.

A_v = effective peak velocity-rated acceleration (ground motion) expected from a seismic event

A_a = effective peak acceleration (ground motion) expected from a seismic event

R, R_w = structural factors (between 4 and 12 in UBC, between 1¼ and 8 in BOCA, between 1 and 4 in NBCC), a coefficient reflecting ability of a structure to absorb energy

v = zonal velocity ratio (ratio of peak horizontal ground velocity to 1m/sec)

W = weight of the structure; dead load plus code-specified percent of live load

T = fundamental time period for the structure

For example, in UBC

$$T = C_t(\text{height of building in feet}^{3/4})$$

with C_t = .035 for steel moment-resisting frames

= .030 for reinforced concrete moment-resisting frames

= .020 for all other buildings

and in NBCC, $T = 0.09h/D^{0.5}$

with h = height in m

D = dimension of building measured parallel to direction of seismic force, in m

S, SITE COEFFICIENTS[1] (UBC-1994)

Type	Description	S Factor
S_1	A soil profile with either: (a) A rock-like material characterized by a shear-wave velocity greater than 2,500 feet per second or by other suitable means of classification, or (b) Stiff or dense soil condition where the soil depth is less than 200 feet.	1.0
S_2	A soil profile with dense or stiff soil conditions, where the soil depth exceeds 200 feet.	1.2
S_3	A soil profile 70 feet or more in depth and containing more than 20 feet of soft to medium stiff clay but not more than 40 feet of soft clay.	1.5
S_4	A soil profile containing more than 40 feet of soft clay characterized by a shear-wave velocity less than 500 feet per second	2.0

[1]The site factor shall be established from properly substantiated geotechnical data. In locations where the soil properties are not known in sufficient detail to determine the soil profile type, soil profile S_3 shall be used. Soil profile S_4 need not be assumed unless the building official determines that soil profile S_4 may be present at the site, or in the event that soil profile S_4 is established by geotechnical data.

Figure 1-26(a) UBC code data

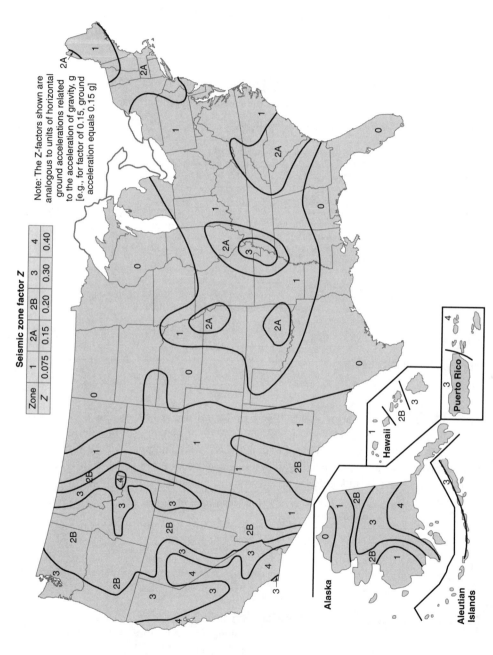

Seismic zone factor Z

Zone	1	2A	2B	3	4
Z	0.075	0.15	0.20	0.30	0.40

Note: The Z-factors shown are analogous to units of horizontal ground accelerations related to the acceleration of gravity, g [e.g., for factor of 0.15, ground acceleration equals 0.15 g]

Figure 1-26(b) UBC code: Seismic zone map of the United States. (Reproduced from the 1994 Uniform Building Code ™, copyright © 1994, with the permission of the publisher, the International Conference of Building Officials.) Note: This map is illustrative, for general information and comparison purposes indicatd in the text. For applications, representatives of International Conference of Building Officials should be consulted for the proper usage of this map including limitations and restrictions.

S, Site Coefficient (BOCA–1996)

Soil-Profile Type	Description[a]	Site Coefficient S
S_1	A soil profile with either: Rock of any characteristic, either shale-like or crystalline in nature, which has a shear wave velocity greater than 2,500 feet per second; or Stiff soil conditions where the soil depth is less than 200 feet and the soil types overlying rock are stable deposits of sands, gravels or stiff clays.	1.0
S_2	A soil profile with deep cohesionless or stiff clay conditions, where the soil depth exceeds 200 feet and soil types overlying rock are stable deposits of sands, gravels or stiff clays.	1.2
S_3	A soil profile containing 20 to 40 feet in thickness of soft to medium-stiff clays with or without intervening layers of cohesionless soils.	1.5
S_4	A soil profile characterized by a shear wave velocity less than 500 feet per second, containing more than 40 feet of soft clays or silts.	2.0

[a]1 foot = 304.8 mm.

Figure 1-27(a) BOCA Code Data

The BOCA criteria have been based on data presented in the 1991 National Earthquake Hazard Reduction Program Provision (NEHRP provision) for the seismic design of new buildings [Building Seismic Safety Council (BSSC), Washington, D.C.]. More recent NEHRP provisions not yet included in all building codes recommend that the method to determine base shear V be modified from Equation 1-3 indicated previously to

$$V = \left(\frac{1.2C_v}{RT^{2/3}}\right)W \leq \left(\frac{2.5C_a}{R}\right)W$$

(1-5)

where $C_a = F_a A_a$

$C_v = F_v A_v$

The values of F_a and F_v are site coefficients for the various generalized soil profiles described in Figure 1-29(a). Values of A_a and A_v, and other terms, are as defined previously for base shear. Values of C_a and C_v are shown in Figure 1-29(b). The second term of the equation represents the upper limit for V.

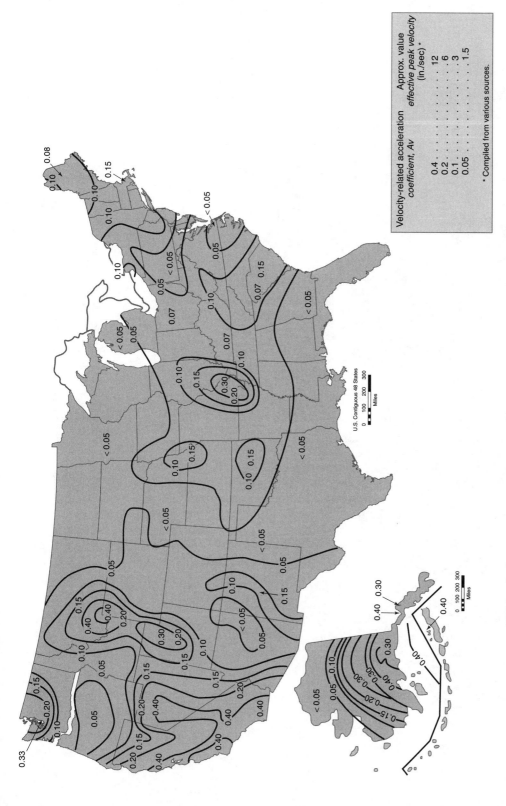

BOCA code: Contour map of effective peak velocity-related acceleration coefficient (A_v) (BOCA–1996)

Figure 1-27(b) Contour map of effective peak velocity-related acceleration coefficient (A_v). (Copyright 1996, Building Officials and Code Administrators International, Inc., Country Club Hills, Illinois. *1996 BOCA National Building Code.* Reprinted with permission of author. All rights reserved.)

Note: This map is illustrative, for general information and comparison purposes indicated in the text. For applications, representatives of BOCA International should be consulted for the proper usage of this map including limitations and restrictions.

47

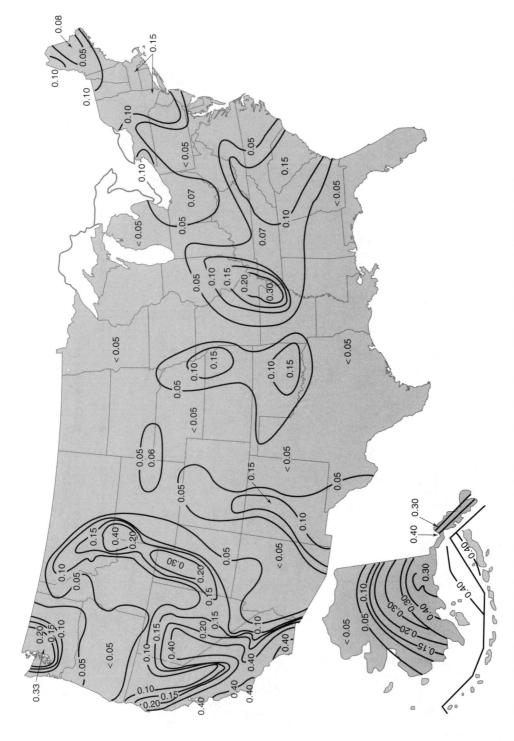

Figure 1-27(c) BOCA Code: Contour map of effective peak acceleration coefficient (A_a). (Copyright 1996, Building Officials and Code Administrators International, Inc., Country Club Hills, Illinois. *1996 BOCA National Building Code*. Reprinted with permission of author. All rights reserved.)

Note: This map is illustrative, for general information and comparison purposes indicated in the text. For applications, representatives of BOCA International should be consulted for the proper usage of this map including limitations and restrictions.

Foundation Factor F (NBCC 1990)

Category	Type of Soil	Depth of Soil (m)	F
1	Rock; dense and very dense coarse-grained soils; very stiff and hard fine-grained soils	Any depth	1.0
	Compact coarse-grained soils; Firm and stiff fine-grained soils	0–15	
2	Compact coarse-grained soils; Firm and stiff fine-grained soils	> 15	1.3
	Very loose and loose coarse-grained soils; Very soft and soft fine-grained soils	0–15	
3	Very loose and loose coarse-grained soils	> 15	1.5
4	Very soft and soft fine-grained soils	> 15	2.0

Figure 1-28 (a)   NBCC Code Data

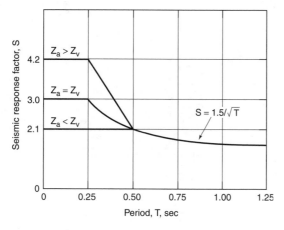

Figure 1-28 (b)   NBCC Code Data

Seismic response factor S (NBCC 1990)

Illustration 1-3

A three-story hospital building is planned for an area indicated by the UBC seismic map to be in a 2A seismic zone. The building will be 200 ft by 150 ft in plan area and 40 ft high. The structural type will be steel, moment-resistant frame construction. The total seismic weight for the building, based on dead loads and percentage of the live loads, is 18,000 kips. Subsurface conditions at the site are not uniform, consisting of variable thicknesses of medium-stiff clay and medium-compact sands (no soft clay or loose sand).

Determine the value for the seismic base shear force, using the UBC criteria

$$V = \frac{ZICW}{R_w}$$

where $C = \dfrac{1.25S}{T^{2/3}}$

UBC code indicates maximum $C = 2.75$, minimum $C/R_w = 0.075$.

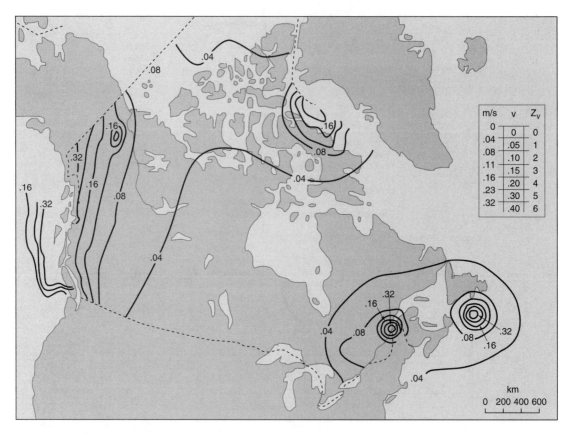

m/s	v	Z_v
0	0	0
.04	.05	1
.08	.10	2
.11	.15	3
.16	.20	4
.23	.30	5
.32	.40	6

Figure 1-28 (c) NBCC Code: Contours of peak horizontal ground velocities, in m/s, having a probabilty of exceedance of 10% in 50 years (NBCC) (Map reproduced from "Supplement to the National Building Code of Canada, 1990", by permission National Research Council of Canada/Institute for Research in Construction.)

Note: This map is illustrative, for general information and comparison purposes indicated in the text. For applications, representatives of National Research Council of Canada should be consulted for the proper usage of this map, including limitations and restrictions.

Solution

W, seismic weight = 18,000 kips

Z, seismic zone 2A = .15

I, hospital (essential facility) = 1.5 (from UBC Code Manual)

R_w, steel structure, moment-resistant frame = 12 (from UBC Code Manual)

S, site factor assuming S_3 category soil conditions = 1.5

$$T = C_t(h)^{3/4} = (.035)(40 \text{ ft})^{3/4} = (.035)(15.91) = .56$$

$$C = \frac{1.25S}{T^{2/3}} = \frac{1.25(1.5)}{(.56)^{2/3}} = \frac{1.875}{.68} = 2.76$$

Check ratio, but use $C = 2.75$ (maximum).

$$\frac{C}{R_w} = 2.75/12 = .229 > .075 \text{ (okay)}$$

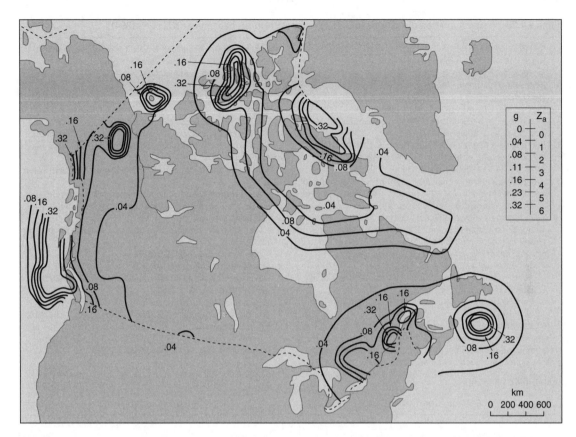

Figure 1-28 (d) NBCC Code: Contours of peak horizontal ground accelerations, in units of g, having a probability of exceedance of 10% in 50 years (NBCC) (Map reproduced from "Supplement to the National Building Code of Canada, 1990", by permission National Research Council of Canada/Institute for Research in Construction.)
Note: This map is illustrative, for general information and comparison purposes indicated in the text. For applications, representatives of National Research Council of Canada should be consulted for the proper usage of this map, including limitations and restrictions.

$$V = \frac{ZICW}{R_W} = \frac{(.15)(1.5)(2.75)(18,000 \text{ kips})}{12} = 928 \text{ kips} = 4170 \text{ kN}$$

Some generalized concepts have evolved from correlating the physical events that occur within the earth during a seismic incident and the effects on structure:

1. Building structures have a natural frequency where harmonic motion results (i.e., a smooth undulating movement occurs along a pattern of flow through the height and length of the structure). Seismic frequencies that are similar to the natural frequency of the foundation soil or rock and that of a structure will cause the greatest movement or vibration in the structure (a seismic frequency similar to the building's natural frequency will amplify motion, while dissimilar frequencies tend to dampen movement). Low-rise buildings tend to have a higher natural frequency than tall buildings.

Figure 1-29 Seismic related conditions and factors for determing base shear (NEHRP, 1994)

(a) Soil Profile Categories

A. Hard rock with measured shear wave velocity, $\bar{v}_s$ > 5000 ft/sec (1520 m/s).

B. Rock with 2500 ft/sec (760 m/s) < $\bar{v}_s$ ≤ 5000 ft/sec (1520 m/s).

C. Very dense soil and soft rock with 1200 ft/sec (365 m/s) < $\bar{v}_s$ ≤ 2500 ft/sec (760 m/s) or with either $\bar{N}$ > 50 or $\bar{s}_u$ > 2000 psf (95.8 kPa).

D. Stiff soil with 600 ft/sec (183 m/s) ≤ $\bar{v}_s$ ≤ 1200 ft/sec (365 m/s) or with either 15 ≤ $\bar{N}$ ≤ 50 or 1000 psf (47.9 kPa) ≤ $\bar{s}_u$ ≤ 2000 psf (95.8 kPa).

E. Any profile with more than 10 ft (3 m) of soft clay defined as soil with PI > 20, w ≥ 40%, and $\bar{s}_u$ < 1000 psf (479 kPa) or a soil profile with $\bar{v}_s$ < 600 ft/sec (183 m/s).

F. Soils requiring site-specific evaluations:

1. Soils vulnerable to potential failure or collapse under seismic loading such as liquefiable soils, quick and highly sensitive clays, and collapsible, weakly cemented soils.
2. Peats and/or highly organic clays (H > 10 ft or 3 m of peat or highly organic clay).
3. Very-high-plasticity clays (H > 25 ft or 7.6 m with PI > 75%).
4. Very thick, soft/medium-stiff clays (H > 120 ft or 37 m).

When soil properties are not known in sufficient detail to determine the soil profile type, Type D is to be used. Soil Profile Type E or F need not be assumed unless the building official determines that Soil Profile Type E or F may be present at the site or in the event that Type E or F is established by geotechnical data.

Notation

A_a = effective peak acceleration coefficient

A_v = effective peak velocity-related acceleration coefficient

C_a = seismic coefficient based on Soil Profile Type and value of A_a

C_v = seismic coefficient based on Soil Profile Type and value of A_v

F_a = acceleration-based site coefficient (at .3 sec period)

F_v = velocity-based site coefficient (at 1.0 sec period)

H = thickness of soil

$\bar{N}$ = average field standard penetration resistance for the top 100 ft (30.5 m)

PI = plasticity index, ASTM D4318

R = response modification factor

$\bar{s}_u$ = average undrained shear strength in top 100 ft (30.5 m), ASTM D2166 or ASTM D2850

S = site coefficient for soil characteristics

T = fundamental period of vibration of structure in direction under consideration

$\bar{v}_s$ = average shear wave velocity in top 100 ft (30.5 m)

w = moisture content (in percent), ASTM D2216

W = total seismic dead load

(b) Seismic Coefficients C_a and C_v

Seismic Coefficient C_a

Soil Type	A_a						
	<.05	=.05	=.10	=.20	=.30	=.40	≥.5
A	A_a	.04	.08	.16	.24	.32	.40
B	A_a	.05	.10	.20	.30	.40	.50
C	A_a	.06	.12	.24	.33	.40	.50
D	A_a	.08	.16	.28	.36	.44	.50
E	A_a	.13	.25	.34	.36	.36	*

Seismic Coefficient C_v

Soil Type	A_a						
	<.05	=.05	=.10	=.20	=.30	=.40	≥.5
A	A_v	.04	.08	.16	.24	.32	.40
B	A_v	.05	.10	.20	.30	.40	.50
C	A_v	.09	.17	.32	.45	.56	.65
D	A_v	.12	.24	.40	.54	.64	.75
E	A_v	.18	.35	.64	.84	.96	*

NOTE: For intermediate values, the higher-value or straight-line interpolation shall be used to determine the value of C_a or C_v. Site-specific geotechnical investigation and dynamic site response analyses should be performed.

2. Low-rise buildings tend to be damaged by the high-frequency body waves, while tall buildings tend to be damaged by the low-frequency surface waves. High-frequency waves (body waves) dissipate energy over distance faster than low-frequency waves (surface waves). As a consequence, low-rise buildings are most prone to damage when located close to the earthquake epicenter, while tall buildings can be damaged at relatively great distances from the epicenter.

Where seismic event velocity and acceleration information is available [e.g., maps such as those in Figures 1-28(c) and 1-28(d)], the ratio of acceleration to velocity for a location can provide an indication of ground motion expected; a low ratio indicates that velocity is the dominant ground motion type (as occurs from a distant, large-magnitude earthquake) while a high ratio indicates that acceleration is the dominant ground motion type (as occurs from an earthquake close to the site).

3. Ground vibrations tend to be amplified when the seismic waves pass from rock into soil (though not always). Commonly, structural damage is greater for buildings constructed on soil than for buildings situated on rock; the potential for greatest damage exists where a building is constructed over deposits of weak soil.

Though the direct effects of earthquake vibrations are a cause of destruction, destructive events greater than those resulting from the primary event may present a more serious danger to an area. *Tsunamis,* tectonically induced tidal waves originating in open ocean areas near the earthquake hypocenter, can swell to tens of meters (hundreds of feet) above normal sea level and travel for hundreds of km (miles); the enormous force from a huge wave of water smashing into the shoreline of a land area can be devastating. Another possible effect is that large areas of land subsidence can occur because of liquefaction (liquefying) of soil deposits and other breakdown of bonds in the soil deposit. Conversely, some land masses may be lifted. Landslides and mud slides in hilly topography are not unusual. Secondary effects, such as collapsed dams releasing reservoirs to cause flooding, fires, destroyed utility conduits, destroyed bridges and transportation routes, breakdown of communication systems, and lack of medical aid, contribute to the havoc resulting from earthquakes.

Large earthquakes are often preceded by a series of smaller foreshocks and a series of aftershocks. Foreshocks can occur over a relatively lengthy period (months), while aftershocks typically occur within hours or days of the main event.

Certainly, some dangers associated with earthquakes, such as loss of life, could be reduced if the ability for prediction existed. Unfortunately, no reliable methods for accurate prediction (place and time) are known at the present. Various methods and items have been proposed for use as indicators of an impending event, some of which have a scientific basis and some that do not. The more promising methods at present involve monitoring various earth conditions for unusual change; predictive methods based on behavioral changes in animals or the weather have not been reliable (although such changes may actually occur). Measurable changes in various earth property parameters have been noted in those earthquake regions of the world where conditions had been monitored in the period leading up to the actual event. The noted occurrences associated with some scientific reasoning include:

1. The velocity of normally ongoing P-waves traveling through the rock in the earthquake region decreases, but shortly prior to the event reverts to normal (a condition attributed to strains taking place in the rock along with the development of small fractures which alter the physical properties of the rock materials).

2. The surfaces of regions underlain by rock that is undergoing strain and movement often experience some uplifting or tilt.

3. The emission of radon gas to the surface increases in the period prior to an earthquake (attributed to a change in the earth's venting process as rock movement and fracture occurs).

4. The electrical resistivity of the rock materials may decrease prior to the event because of the increase in pressure throughout the zone being strained.

5. Large earthquakes show the tendency to be preceded by a series (swarm) of smaller quakes throughout the general region, a phenomenon attributed to the occurrence of rock ruptures on a somewhat limited scale. Unfortunately, some foreshocks have been of relatively large magnitude; at the time of occurrence uncertainty existed about whether that event was the main one or if larger events were yet to come.

Relating to associations between surface changes and earthquakes, *Synthetic Aperture Radar Interferometry* is a developing method which offers opportunities for identifying and monitoring tectonic movements. Satellite radar measurements of the distance to the earth's surface are obtained during the orbital travel and are used to define the relief or topography of sections of the earth's surface. (Radar's long-length radio waves are not affected by factors such as cloud cover which can interfere with camera-like processes based on short-length light waves; the radar procedure becomes a more consistent source of precision data.) Radar readings of distance to the same ground point from different positions along the orbital path can be used to produce a type of stereoscopic or three-dimensional image of the earth's surface. Comparisons of radar measurements obtained from repetitive orbits can be used to indicate topographic changes (movements) occurring across large areas of the earth's surface. It is anticipated that such topographic monitoring will enable identification of changes that just precede or lead to a seismic event, and that can then be used as a prediction tool.

1.5 EFFECT ON DESIGN AND CONSTRUCTION

Many large land areas have been formed with soils deposited primarily by one of the transportation methods described in this chapter. One area may be underlain by glacially deposited soils, another by lake or marine deposits, another by river deposits.

It is not unusual, however, to have stratification of soils that have been deposited by different methods. For instance, a glacial till may be covered with a glacial outwash or other alluvium. Soft lake deposits are known to exist over compact glacial till or river deposits of sands and gravels. Coarse soil alluvium may overlie soft lake deposits or marine clays. In areas that have been subjected to repeated glaciation, stratification of different types of glacially transported soils having different properties can exist. The result is that loose or soft soils may overlie compact or firm soils, and vice versa.

In areas blanketed by residual soil, the character of the material will vary with depth, and both depth and properties can vary across limited horizontal distances. A residual soil formation may be covered by a transported soil, or reduced by erosive and other transportive forces.

The type and condition of soil deposits underlying any proposed construction site must be an important consideration to the engineering and construction personnel concerned with the project, for it is the soil, or rock, that provides the support for the structure. Proper design for a structure includes investigation and evaluation of soil conditions underlying the proposed structure, as well as probability and extent of significant natural occurrences such as seismic activity, site flooding, and so on. However, knowledge of geology and the manner in which land forms and the soils in them have been created, as discussed throughout this chapter, can frequently serve to provide a preliminary evaluation of an area and the potential advantages or problems.

PROBLEMS

1-1. Assemble a listing of all the types of occurrences where serious problems for humans and their developments have resulted when soil deposits (read "solid ground") did not perform as expected.

1-2. Name the three main classes of rock and describe how they originated.

1-3. What factor of formation most influences the texture of igneous rocks?

1-4. Provide generalized comments on the hardness, soundness, and durability associated with the three basic categories of rock.

1-5. What visual properties can frequently be used to distinguish between acidic rocks and basic rocks?

1-6. Describe the two typical processes that occur to transform soil sediments into sedimentary rock.

1-7. The shales represent what percentage (approximately) of rock materials closest to the ground surface? Give probable reasons why this condition exists.

1-8. What processes occur to cause metamorphosis of rocks?

1-9. What type of bedrock formations would be thought to offer good foundation support for structures? What types would be more suspect of being poor bedrock materials?

1-10. If sinkhole topography is related to the underground erosion of rock which results in cavities, what type of rock formations would be considered most susceptible to having the condition develop?

1-11. What are loess soils, and what is the potential danger of loss of stability in loess deposits?

1-12. Indicate the soil types to be expected in a glacial terminal moraine.

1-13. Provide an outline indicating the relationship between soil deposit, glaciation, and sedimentary rock formations.

1-14. Indicate why sand or gravel deposits are frequently found along old river and stream locations.

1-15. What effect does the shape of a channel have on stream flow and the related carried or deposited soil sediments?

1-16. How do natural levees originate? What is the major soil type expected in a natural levee?

1-17. What soil types would be expected in a river or stream delta?

1-18. Describe the process by which a glacial lake is formed.

1-19. How can lakes exist at elevations above the groundwater table?

1-20. What type of transported soil deposit commonly is found at the location of filled-in ancient lakes and basin areas?

1-21. Why are naturally filled-in lake locations often thought of as areas that offer poor support for building foundations?

1-22. How do eskers differ from drumlins with regard to formation and soil types?

1-23. In glacially affected areas, what types of glacial formation represent possible good sources of sands and gravels for the construction industry?

1-24. Does an area covered by glacial till represent a location of advantage or disadvantage to the building and construction industry?

1-25. Do soils typical of beach deposits represent any advantage or disadvantage to the building and construction industry?

1-26. What is the potential danger to stability in areas where the land is formed from marine clays?

1-27. On the basis of known lithosphere plate boundaries:

(a) Identify the major continental areas likely to experience earthquakes in the future.

(b) Identify the major continental areas likely to experience active volcanic activity in the future.

1-28. Indicate the present-day beliefs used to explain the movements of plates of lithosphere.

1-29. Develop a listing that indicates comparative values of specific gravity (or heaviness) for rock materials in the lithosphere, the asthenosphere, the continental crust, and the ocean crust.

1-30. Explain how the boundary regions for the plates of lithosphere relate to the presence of the earth's major mountain ranges.

1-31. Explain how the boundary regions for plates of lithosphere related to the presence of present-day active volcanoes.

1-32. Speculate on possible reasons for the development of the fractures in the lithosphere (the plate boundaries).

1-33. Explain the reason why fossils of cold-region or polar life have been discovered in tropical areas in recent times.

1-34. List the different types of seismic waves produced during an earthquake event, and relate each type to the potential for causing building damage.

1-35. Explain the difference between earthquakes as defined by the Richter scale and by the Modified Mercalli scale.

1-36. Compare the energy released during the earthquake where M_L is equal to 5.5 and the earthquake where M_L is equal to 7.5.

1.37. (a) Determine the Richter magnitude for an earthquake where the following seismograph data have been obtained:
- Maximum registered amplitude of 5 mm
- Time separation between arrival of primary and secondary seismic waves at the seismograph of 30 seconds

(b) Also determine the distance separating the location of the seismograph and the earthquake epicenter.

1-38. A planned building, 45 feet high, will be steel, moment-resisting frame construction. Determine the design value of base shear, V, to apply for the study of effects of an earthquake event if the site is in an area where the seismic zone factor is 0.20 and other conditions are as identified. Assume the UBC applies.
- Building importance factor, $I = 1.25$
- Design building weight for earthquake, $W = 11,000$ kips
- Structural factor, moment-resisting frame, $R_w = 6$
- Soil profile site coefficient, $S = 12$

1-39. A planned building, steel-frame, moment-resistant design, is proposed for an area where seismic and ground conditions are as described below. The design weight for the earthquake condition is 11,000 kips. Determine the value of earthquake-event base shear, V, to apply for the structural design, using the BOCA criteria.
- Structural factor for steel, moment-resisting frame, $R = 4.5$

- Site coefficient $= 1.2$
- Fundamental time period for structure, $T = 0.61$ seconds for building height $= 45$ ft
- Earthquake acceleration factors, A_v and $A_a = 0.15$

1-40. A steel-frame, moment-resisting building design is planned for a site in Canada. For the conditions indicated, calculate the value for base shear, V, to assign for the earthquake condition analysis, applying NBCC criteria.
- Earthquake design weight, $W = 31,000$ kN
- Building dimensions, $D = 45$ m, $h = 15$ m
- Seismic response factor, S, based on $Z_a = Z_v$
- Importance factor for building, $I = 1.3$
- Zonal velocity ratio, $v = 0.10$
- Foundation factor, $F = 1.3$
- Force modification factor relating to steel-frame, moment-resistant design, $R = 4$

1-41. Determine the value for earthquake event base shear, V, for a moment-resistant, steel-frame building, using the NEHRP standards. Building and site condition factors are as follows:
- Design weight for earthquake condition, $W = 11,000$ kips
- Earthquake acceleration factors, $A_v = A_a = 0.15$
- Fundamental time period for structure, $T = 0.61$ seconds
- Structural factor, steel-frame, moment-resistant, $R = 4.5$
- Soil profile conditions, category D

1-42. Determine the earthquake event base shear, V, for a structure where the conditions shown below apply. Calculate V using UBC, BOCA, and NEHRP formulae, then compare results.
- Steel-frame, moment-resistant structural design
- Earthquake design building weight, $W = 5,000$ kips
- Fundamental time period, $T = 0.45$ seconds
- Building importance factor, $I = 1.25$
- Site coefficient or soil profile use S_3 (UBC), S_3 (BOCA), type E (NEHRP)
- Structural factor, $R = 5$ (UBC), $R = 3$ (BOCA)
- Seismic factors, use 2B (UBC); $A_v = A_a = 0.10$ (BOCA, NEHRP)

CHAPTER 2

Soil Composition
Terminology and Definitions

Soil deposits comprise the accumulated solid particles of soil or other materials plus the void spaces that exist between the particles. The void spaces are partially or completely filled with water or other liquid. Void spaces not occupied by fluid are filled with air or other gas. Since the volume occupied by a soil bulk may generally be expected to include material in the three states of matter—solid, liquid, and gas—soil deposits are referred to as three-phase systems.

Significant engineering properties of a soil deposit, such as strength and compressibility, are directly related to or at least affected by basic factors such as how much volume or weight of a bulk soil is solid particles or water or air. Information such as soil density (weight per unit volume),[1] water content, void ratio, degree of saturation—terms defined in the following sections—is used in calculations to determine the bearing capacity for foundations, to estimate foundation settlement, and to determine the stability of earth slopes. In other words, such information helps to define the condition of a soil deposit for its suitability as a foundation or construction material. For this reason, an understanding of the terminology and definitions relating to soil composition is fundamental to the study of soil mechanics.

2.1 SOIL COMPOSITION: ANALYTICAL REPRESENTATION

Bulk soil as it exists in nature is a more or less random accumulations of soil particles, water, and air space, as shown in Figure 2-1(a). For purposes of study and analysis, it is convenient to represent this soil mass by a phase or block diagram, with part of the diagram representing the solid particles, part representing water or other liquid, and another part air or other gas, as shown in Figure 2-1(b).

---

[1]In soil mechanics, the term *density* has frequently been used to indicate unit weight—that is, pounds per cubic foot. For the metric system, unit weights should be expressed as dynes per cubic centimeter, whereas kilonewtons per cubic meter are used with the SI system.

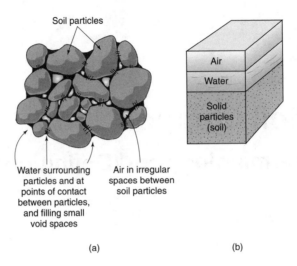

Figure 2-1 (a) Actual soil bulk consisting of soil particles, water, and air; (b) phase diagram representation of soil bulk.

Weight–Volume, Mass–Volume Relationship

On the phase diagram, the interrelationships of weight and mass to the volume that make up the soil system being analyzed can be shown. The relationships are summarized in Figure 2-2.

As the diagram shows, the total *weight* W_T of the soil volume is taken as the sum of the weights of solids W_s plus water W_w. In practical problems, all weighings are made in air, and the weight of air (in the voids) measured in air (the earth's atmosphere) is zero. If the gas is other than air, it may have a measurable weight, but it would normally be very small compared to the total weight of soil plus water and therefore can be neglected without causing serious error.

$$W_T = W_s + W_w \tag{2-1a}$$

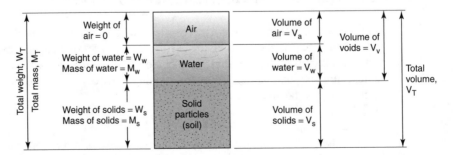

Figure 2-2 Relationship between volume and weight/mass of a soil bulk.

Similarly, for measurement of mass

$$M_T = M_s + M_w \tag{2-1b}$$

The total *volume* of the soil bulk includes the volume occupied by solids plus water (or liquid) plus air (or other gas). The total space occupied by water and air may collectively be indicated as the volume of voids.

$$V_T = V_s + V_w + V_a = V_s + V_v \tag{2-2}$$

where V_T = total volume
$\quad V_s$ = volume of soil solids
$\quad V_w$ = volume of water
$\quad V_a$ = volume of air
$\quad V_v$ = volume of voids

The relationship between weight and volume, for any material (x), is

$$W_x = V_x G_x \gamma_w \tag{2-3a}$$

while that between mass and volume is

$$M_x = V_x G_x \rho_w \tag{2-3b}$$

where W_x = weight of the material (solid, liquid, or gas); lb, N, dyne
$\quad V_x$ = volume occupied by the material; m^3, ft^3
$\quad G_x$ = specific gravity of the material; a dimensionless value
$\quad \gamma_w$ = unit weight of water at the temperature to which the problem refers.
$\qquad$ (In most soils work, γ_w is taken as 62.4 pcf, 980 dynes/cm^3,
$\qquad$ or 9.81 kN/m^3 regarless of temperature. In experimental work, accurate
$\qquad$ values are used.)[2]
$\quad M_x$ = mass of the material; kg, slug
$\quad \rho_w$ = density of water; 1 gm/cm^3, 1 Mg/m^3 (10^{3}kg/m^3), or 1.95 slugs/ft^3 (note 1
$\qquad$ slug = 1 lb-sec^2/ft)

---

[2]The gram is a unit of mass, whereas the pound is a unit of weight or force. A value of grams per cubic centimeter is density (by definition, density is mass per unit volume and is given the symbol ρ). Unit weight is force or weight per unit volume. Unit weight and density are therefore *not* identical, but they are related by the expression $\gamma = \rho g$, where g is the acceleration of gravity. In soil mechanics work, the terms are often taken to have synonymous meaning. Since comparative values of unit weight or density are normally sought to relate to other properties, this practice does not cause complications. To overcome this discrepancy between γ and ρ, unit weights based on a gram measurement can be expressed as gram-force (gf) per unit volume.

For soil mechanics problems, then,

$$W_s = V_s G_s \gamma_w \tag{2-4a}$$

$$M_s = V_s G_s \rho_w \tag{2-4b}$$

$$W_w = V_w G_w \gamma_w = V_w \gamma_w \qquad \text{since } G_w = 1 \tag{2-5a}$$

$$M_w = V_w G_w \rho_w = V_w \rho_w \tag{2-5b}$$

The specific gravity of most commonly occurring rock or soil materials is between 2.30 and 3.10. For many deposits, the specific gravity of soil solids lies within the range of 2.60 to 2.75.

2.2 BASIC TERMS RELATING TO SOIL COMPOSITION AND CONDITION

The unit weight of a soil γ is conventionally expressed as pounds per cubic foot, or kilonewtons per cubic meter (unit weight is weight per one unit of volume). Unit weights are reported as wet unit weight γ_{wet} or dry unit weight γ_{dry}:

$$\text{Wet unit weight,} \qquad \gamma_{wet} = \frac{W_T}{V_T} (\text{kN/m}^3, \text{pcf}) \tag{2-6a}$$

$$\text{Dry unit weight,} \qquad \gamma_{dry} = \frac{W_s}{V_T} (\text{kN/m}^3, \text{pcf}) \tag{2-6b}$$

From the definition, it can be seen that the wet unit weight includes the weight of water as well as soil particles in a soil bulk. The dry unit weight is based on only the weight of soil solids in the accumulation.

Soil density, ρ, expressed in kilograms per cubic meter or grams per cubic centimeter (or other terms of mass per unit volume), also can be in terms of a wet or dry value, whereby

$$\text{Wet density,} \qquad \rho_{wet} = \frac{M_T}{V_T} = (\text{kg/m}^3, \text{gm/cm}^3) \tag{2-7a}$$

$$\text{Dry density,} \qquad \rho_{dry} = \frac{M_s}{V_T} (\text{kg/m}^3, \text{gm/cm}^3) \tag{2-7b}$$

In some applications, it is necessary to relate density and unit weight; basic equivalencies as shown in Table 2-1 can be used to complete conversions.

By definition, water content w is the ratio of the weight of water in a soil volume to the weight of soil solids, or of the mass of water in a soil volume to the mass of solids.

Table 2-1   Equivalencies between Density and Unit Weight

Density[a] (metric)	Density (SI)	Unit Wt.[a] (U.S. customary)	Unit Wt. (SI)
1 gm/cm³	1 Mg/m³	62.4 pcf	9.81 kN/m³

[a]1 gm/cm³ is the equivalent of 62.4 lb/ft³. Since $W = mg$, the weight for one gram of mass is: $W = (1\ gm)(980\ cm/sec^2) = 980$ dynes (where 1 dyne $= 10^{-5}\ N = .2248 \times 10^{-5}\ lb$, and 980 dynes $= .002203\ lb$). Therefore, the weight for 1 cm³ of water is 980 dynes $= .002203\ lb/cm^3$, and the weight for 1 ft³ of water is $(.002203\ lb/cm^3)(28,317\ cm^3/ft^3) = 62.4\ lb/ft^3$.

$$w\% = \frac{W_w}{W_s} \times 100\% \quad \text{or} \quad \frac{M_w}{M_s} \times 100\% \qquad (2\text{-}8)$$

where $w\%$ = water content expressed as a percentage

W_w = weight of water

W_s = weight of dry soil

M_w = mass of water

M_s = mass of soil

(*Cautionary note:* Water content is *not* the weight of water divided by the total weight W_T, nor the mass of water divided by the total mass M_T.)

The relationship of water content and weight of dry soil to total wet weight of a soil volume is as follows:

$$W_T = W_s + W_w \qquad \text{(from Eq. 2-1)}$$

and since

$$W_w = \left(\frac{w\%}{100\%}\right)W_s \qquad \text{(from Eq. 2-8)}$$

$$W_T = W_s + \left(\frac{w\%}{100\%}\right)W_s$$

$$= W_s\left(1 + \frac{w\%}{100\%}\right)$$

then, by rearranging, we obtain

$$W_s = \frac{W_T}{1 + \dfrac{w\%}{100\%}} \qquad (2\text{-}9a)$$

In a similar manner

$$M_s = \frac{M_T}{1 + \dfrac{w\%}{100\%}} \qquad (2\text{-}9b)$$

These equations enable the dry soil weight (or mass) to be easily determined when the wet weight or mass of a large soil sample is known and the water content is determined from a small representative portion taken from the sample. This procedure for determining dry soil weight or mass is frequently used in laboratory and construction work.

Two terms, *void ratio* and *porosity,* express a relationship between the volumes in a soil material occupied by solids and nonsolids. Void ratio *e* is

$$e = \frac{\text{Vol. of voids}}{\text{Vol. of solids}} = \frac{V_v}{V_s} \qquad (2\text{-}10)$$

Void ratio is expressed as a decimal. Porosity *n*, conventionally expressed as a percentage, is

$$n\% = \frac{\text{Vol. of voids}}{\text{Total vol.}} \times 100\% = \frac{V_v}{V_T} \times 100\% \qquad (2\text{-}11)$$

The relationship between void ratio and porosity is

$$e = \frac{\dfrac{n\%}{100\%}}{1 - \dfrac{n\%}{100\%}} \qquad (2\text{-}12a)$$

$$n\% = \frac{e}{1 + e} \times 100\% \qquad (2\text{-}12b)$$

The term *degree of saturation, S,* indicates the portion of the void spaces in a soil material (bulk) that is filled with water. Degree of saturation is expressed as a percentage.

$$S\% = \frac{V_w}{V_v} \times 100\% \qquad (2\text{-}13)$$

Full saturation, or 100 percent saturation, indicates that all voids are filled with water. A soil can remain 100 percent saturated even though its water content is changed if the soil experiences compression or expansion (since compression or expansion indicates a decrease or increase in void spaces).

With reference to a phase diagram, as shown in Figure 2-3, other useful weight–volume relationships can be developed.

Since $V_T = V_s + V_v$, and with $e = \dfrac{V_v}{V_s}$, we obtain

$$V_T = V_s + eV_s = V_s(1 + e) \qquad (2\text{-}14a)$$

or

$$V_s = \frac{V_T}{1 + e}$$ (2-14b)

Also, since $\gamma_{dry} = \frac{W_s}{V_T}$ (from Eq. 2-6), we obtain

$$\gamma_{dry} = \frac{V_s G_s \gamma_w}{V_T} = \frac{V_s G_s \gamma_w}{V_s(1 + e)} = \frac{G_s \gamma_w}{(1 + e)}$$ (2-15)

For a fully saturated soil, the unit weight becomes

$$\gamma_{sat} = \frac{W_T}{V_T} = \frac{W_s + W_w}{V_s(1 + e)}$$ (from Eq. 2-6)

$$= \frac{V_s G_s \gamma_w + V_w \gamma_w}{V_s(1 + e)} = \frac{V_s G_s \gamma_w + e V_s \gamma_w}{V_s(1 + e)}$$

$$= \frac{(G_s + e) \gamma_w}{(1 + e)}$$ (2-16a)

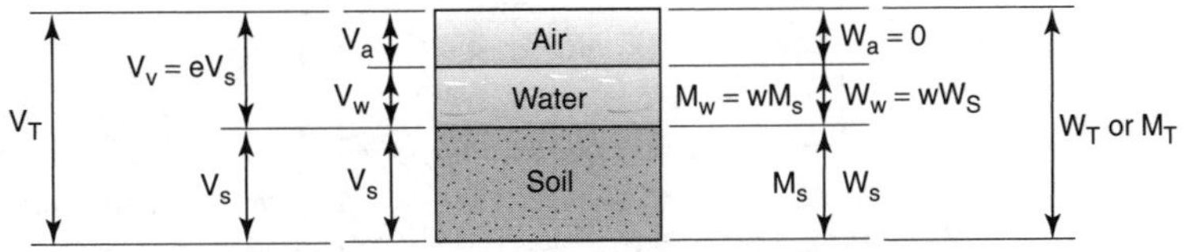

Figure 2-3 Phase diagram used to develop equations applicable to weight–volume relationship.

For a partially saturated soil, the wet unit weight, γ_{wet} (or γ_{total}), becomes

$$\gamma_{wet} \text{ (or } \gamma_{total}) = \frac{W_T}{V_T} = \frac{V_s G_s \gamma_w + V_w \gamma_w}{V_s(1 + e)} = \frac{V_s G_s \gamma_w + (SV_v) \gamma_w}{V_s(1 + e)}$$

$$= \frac{V_s G_s \gamma_w + S(eV_s) \gamma_w}{V_s(1 + e)} = \frac{(G_s + Se) \gamma_w}{1 + e}$$ (2-16b)

A very informative relationship can be obtained by proper substitution of terms into Equation 2-5a.

$$W_w = V_w G_w \gamma_w = V_w \gamma_w$$ (from Eq. 2-5a)

or

$$V_w = \frac{W_w}{\gamma_w}$$

Dividing both sides by V_s gives

$$\frac{V_w}{V_s} = \frac{W_w}{V_s \gamma_w}$$

Multiplying the left term by $\dfrac{V_v}{V_v}$ gives

and

$$\frac{V_w}{V_s} \times \frac{V_v}{V_v} = \frac{W_w}{V_s \gamma_w}$$

$$\frac{V_w}{V_v} \times \frac{V_v}{V_s} = \frac{w W_s}{V_s \gamma_w}$$

$$S \times e = w\left(\frac{W_s}{V_s \gamma_w}\right)$$

$$Se = wG_s \tag{2-17}$$

In this equation, both the S and w terms are expressed as percent. For a given soil and unit weight, this equation shows the relationship between void ratio and water content, and the limiting water content that can be obtained.

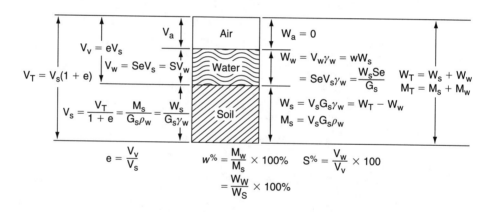

Figure 2-4 Summary of weight–volume and mass–volume equations.

A summary of the weight–volume and mass–volume equations most frequently used in solving problems is presented in Figure 2-4. In performing the analysis to determine weight or volume properties, it is extremely helpful to sketch the phase diagram and indicate on the diagram the information that is known (from given data or as it is developed). This procedure helps to guide the analyst through the proper steps or equations necessary for a complete solution and will eliminate unnecessary computations.

Illustration 2-1

A sample of soil obtained from a test pit 0.0283 m³ (1 ft³) in volume has a mass of 63.56 kg (weight of 140 lb). The entire sample is dried in an oven and found to have a dry mass of 56.75 kg (weight of 125 lb). Calculate the water content, wet density, and dry density.

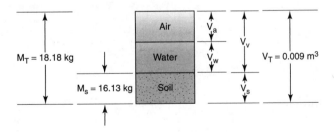

Solution

Mass of water, $\quad M_w = 63.56 - 56.75 = 6.81$ kg
(Weight of water, $\quad W_w = 140 - 125 = 15$ lb)

Mass of dry soil, $\quad M_s = 56.75$ kg
(Weight of dry soil, $\quad W_s = 125$ lb)

Total volume of sample, $\quad V_T = .0283$ m³ $= 1.0$ ft³

Wet density, $\qquad \rho_{wet} = \dfrac{M_T}{V_T} = \dfrac{63.56 \text{ kg}}{.0283 \text{ m}^3} = 2.246$ Mg/m³

(Wet unit weight, $\quad \gamma_{wet} = \dfrac{W_T}{V_T} = \dfrac{140 \text{ lb}}{1.0 \text{ ft}^3} = 140$ pcf)

Dry density $\qquad \rho_{dry} = \dfrac{M_S}{V_T} = \dfrac{56.75 \text{ kg}}{0.283 \text{ m}^3} = 2.005$ Mg/m³

(Dry unit weight, $\quad \gamma_{dry} = \dfrac{W_S}{V_T} = \dfrac{125 \text{ lb}}{1.0 \text{ ft}^3} = 125$ pcf)

Water content, $\quad w\% = \dfrac{M_w}{M_S}(100\%) = \dfrac{W_w}{W_S}(100\%) = 12\%$

Illustration 2-2

Determine the wet density, dry unit weight, void ratio, water content, and degree of saturation for a sample of moist soil which has a mass of 18.18 kg and occupies a total volume of 0.009 m³. When dried in an oven, the dry mass is 16.13 kg. The specific gravity of the soil solids is 2.70. (Acceleration of gravity, $g = 9.81$ m/sec².)

Solution

Wet density, $\rho_{wet} = \dfrac{M_T}{V_T} = \dfrac{18.18 \text{ kg}}{.009 \text{ m}^3} = 2020$ kg/m³

Dry unit weight, $\quad \gamma_{dry} = \dfrac{W_S}{V_T} = \dfrac{M_S g}{V_T} = \dfrac{(16.13 \text{ kg})(9.81 \text{ m/sec}^2)}{.009 \text{ m}^3} = 17.58$ kN/m³

Water content, $\quad w\% = \dfrac{M_S}{M_S} \times 100\% = \dfrac{(18.18 - 16.13)\text{kg}}{16.13\text{ kg}} \times 100\% = 12.7\%$

Void Ratio, $\qquad e = \dfrac{V_V}{V_S} = \dfrac{.0031\text{ m}^3}{.0059\text{ m}^3} = .53$

$\left[\text{where } V_S = \dfrac{M_S}{G_s \rho_w}\right.$

$$= \dfrac{16.13\text{ kg}}{(2.70)\left(1.0\dfrac{\text{gm}}{\text{cm}^3} \times 10^6\dfrac{\text{cm}^3}{\text{m}^3} \times .001\text{ kg/gm}\right)}$$

$$= \dfrac{16.13\text{ kg}}{2.70 \times 10^3\dfrac{\text{kg}}{\text{m}^3}} = .0059\text{ m}^3$$

$\text{and} \quad V_v = V_T - V_S$

$\qquad\qquad = .009\text{ m}^3 - .0059\text{ m}^3$

$\qquad\qquad = .0031\text{ m}^3]$

Degree of saturation, $\quad S\% = \dfrac{w\% G_s}{e} = \dfrac{(12.7\%)(2.70)}{(.53)} = 64.7\%$

Illustration 2-3

A 150-cm³ sample of wet soil has a mass of 250 g when 100 percent saturated. When oven-dried, the mass is 162 g. Calculate the dry density, dry unit weight, water content, void ratio, and G_s.

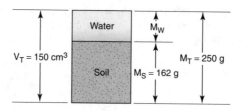

Solution

$$\rho_{dry} = \dfrac{M_s}{V_T} = \dfrac{162\text{ g}}{150\text{ cm}^3} = 1.08\dfrac{\text{g}}{\text{cm}^3}$$

$$\gamma_{dry} = (1.08\text{ g/cm}^3)\left(62.4\dfrac{\text{lb/ft}^3}{\text{g/cm}^3}\right) = 67.5\text{ pcf} \quad(\text{refer to Table 2-1})$$

$$w\% = \left(\dfrac{M_w}{M_s}\right)(100\%) = \left(\dfrac{250\text{ g} - 162\text{ g}}{162\text{ g}}\right)(100\%) = 54.3\%$$

$$V_w = \frac{M_w}{G_w\rho_w} = \frac{250 \text{ g} - 162 \text{ g}}{(1.0)(1.0 \text{ g/cm}^3)} = 88 \text{ cm}^3 = V_v \text{ for this problem, since for}$$

100% saturation all voids are
filled with water

$$V_s = V_T - V_v = 150 \text{ cm}^3 - 88 \text{ cm}^3 = 62 \text{ cm}^3$$

$$e = \frac{V_v}{V_s} = \frac{88 \text{ cm}^3}{62 \text{ cm}^3} = 1.42$$

$$G_s = \frac{M_s}{V_s\rho_w} = \frac{162 \text{ g}}{(62 \text{ cm}^3)(1.0 \text{ g/cm}^3)} = 2.61$$

Illustration 2-4

Laboratory test data on a sample of saturated soil show that the void ratio is 0.45 and the specific gravity of soil solids is 2.65. For these conditions, determine the wet unit weight of the soil and its water
content.

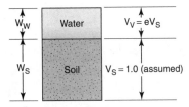

Solution
This sample is saturated; thus all voids are filled with water.

$$e = \frac{V_v}{V_s} = .45$$

but V_v and V_s are not known. On the phase diagram, assume that V_s is unity (i.e., $V_s = 1.0 \text{ ft}^3$).
Therefore

$$V_T = V_s + eV_s = 1.0 \text{ ft}^3 + .45 \text{ ft}^3 = 1.45 \text{ ft}^3$$

As a result, $$W_s = V_sG_s\gamma_w = (1.0 \text{ ft}^3)(2.65)(62.4 \text{ pcf}) = 165 \text{ lb}$$

and $$W_w = V_w\gamma_w = (.45 \text{ ft}^3)(62.4 \text{ pcf}) = 28 \text{ lb}$$

from which $$W_T = W_s + W_w = 165 \text{ lb} + 28 \text{ lb} = 193 \text{ lb}$$

$$\gamma_{wet} = \frac{W_T}{V_T} = \frac{193 \text{ lb}}{1.45 \text{ ft}^3} = 133 \text{ pcf}$$

$$w\% = \frac{W_w}{W_s}(100\%) = \left(\frac{28 \text{ lb}}{165 \text{ lb}}\right)(100\%) = 17\%$$

2.3 SUBMERGED SOIL

In many soil mechanics problems it is necessary to determine the net intergranular weight, or *effective* weight, of a soil when it is below the groundwater table. (In this context, *intergranular* refers to the weight or force that acts at the point, or on the surfaces, where soil particles are in contact. Effective soil weight is used to determine effective stress in a soil deposit, a value that influences factors such as soil shear strength, soil compressibility and settlement, slope stability—topics discussed in later chapters.) For this "underwater" condition, the soil solids are buoyed up by the pressure of the surrounding body of water, and the submerged soil weight becomes less than for the same soil above water. The effective soil weight then becomes the unit weight of the soil material when it is weighed under water. The water in the voids has zero weight (when submerged, all voids can be assumed to be filled with water), and the weight of the soil solids is reduced by the weight of the volume of water they displace. Therefore, a submerged soil weight (W_{sub}) equals the soil weight above water minus the weight of water displaced, or

$$W_{sub} = V_s G_s \gamma_w - V_s G_w \gamma_w = V_s \gamma_w (G_s - G_w)$$

$$W_{sub} = V_s \gamma_w (G_s - 1) \tag{2-18}$$

Since unit weight is total weight divided by total volume,

$$\gamma_{sub} = \frac{W_{sub}}{V_T} = \frac{V_s \gamma_w (G_s - 1)}{V_s (1 + e)}$$

$$\gamma_{sub} = \frac{(G_s - 1)}{(1 + e)} \gamma_w \tag{2-19}$$

Similarly, in terms of density,

$$\rho_{sub} = \frac{G_s - 1}{1 + e} \rho_w$$

Equation 2-19[3] indicates that an accurate determination of the submerged soil weight requires that the specific gravity of the soil solids and the void ratio be known. Unfortunately, in terms of time and expense, some testing or physical analysis is required to determine the specific gravity, which in turn is used to compute the void ratio. Also unfortunately, insofar as analytical studies for practical soil mechanics problems are concerned, void ratios and soil weights vary somewhat even in "uniform deposits." Because of this fact, the effort to make highly accurate determinations of submerged soil weights is rarely undertaken when studies and designs are done. Instead, satisfactory estimates, which can be made from knowing a wet weight, are frequently utilized. For most soils, and fortunately for ease of computation, the submerged weight is on the order of half the wet soil weight above the water table. The most notable exception to this rule is soils containing significant decomposed vegetation or organic material.

$$\rho_{sub} = \frac{1}{2} \rho_{wet} \text{ (approximately)} \tag{2-20a}$$

---

[3]As a word of caution, it is pointed out that γ_{sub} is not equal to γ_{dry}, the dry unit weight of soil. The term γ_{dry} conventionally is restricted to indicate dry soil and does not include the effect of buoyancy as caused by submergence.

$$\rho_{sub} = \frac{1}{2}\rho_{wet} \text{ (approximately)} \qquad (2\text{-}20b)$$

For many practical problems, the effects from applying the simplification are negligible. Where accuracy is required, Equation 2-19 should be used. However, for the situation where the soil above the water table is 100 percent saturated, Equation 2-19 for the submerged unit weight becomes (simple and exact)

$$\gamma_{sub} = \gamma_{sat} - \gamma_w$$

$$\gamma_{sub} = \gamma_{sat} - 62.4 \qquad \text{(in pcf)} \qquad (2\text{-}21a)$$

$$= \gamma_{sat} - 9.81 \quad \text{(in kN/m}^3) \qquad (2\text{-}21b)$$

Illustration 2-5

Undisturbed soil obtained from a test pit 0.0283 m^3 (1 ft^3) in volume is found to have a wet weight of 0.459 kN (103.2 lb). The dry weight of the sample is 0.376 kN (84.5 lb). What would be the effective unit weight of such a soil if it were submerged below the groundwater table? The specific gravity of the soil is determined to be 2.70.

Solution
By Equation 2-20, the effective submerged weight is approximately

$$\gamma_{sub} = \frac{1}{2}\gamma_{wet} \text{ or } \frac{1}{2}(103.2 \text{ pcf}) \cong 52 \text{ pcf} \cong \frac{1}{2}\left(\frac{.459 \text{ kN}}{.0283 \text{m}^3}\right) \cong 8.11 \text{kN/m}^3$$

An accurate determination is as follows:

$$V_s = \frac{W_s}{G_s \gamma_w} = \frac{.376 \text{ kN}}{(2.70)(9.81 \text{ kN/m}^3)} = .0142 \text{ m}^3$$

$$V_v = V_T - V_s = .0283 \text{ m}^3 - .0142 \text{ m}^3 = .0141 \text{ m}^3 \text{ (or .50 ft}^3)$$

$$e = \frac{V_v}{V_s} = \frac{.0141}{.0142} \cong 1.00$$

$$\gamma_{sub} = \left(\frac{G_s - 1}{1 + e}\right)\gamma_w \qquad \text{(from Eq. 2-19)}$$

$$= \left(\frac{2.70 - 1}{1 + 1.0}\right)(9.81 \text{ kN/m}^3) = 8.34 \text{ kN/m}^3 \text{ (or 53.2 pcf)}$$

Illustration 2-6

Assume that a sample of soil similar to the soil from the preceding illustration is excavated from a 1 ft^3 test hole at a location below the water table. The soil is now 100 percent saturated. What saturated weight would be expected?

Solution

$$\gamma_{sub} = \gamma_{sat} - \gamma_w$$

or

$$\gamma_{sat} = \gamma_{sub} + \gamma_w = 53.2 \text{ pcf} + 62.4 \text{ pcf} = 115.6 \text{ pcf} \qquad \text{(from Eq. 2-21a)}$$

PROBLEMS

2-1. A sample of soil taken from a borrow pit has a wet mass of 14.56 kg. Completely dried, the soil dry mass is 11.78 kg. Determine the water content of this soil.

2-2. A sample of soil obtained from a construction site is found to have a wet weight of 29.4 lb. When completely dried, the soil weighs 25.9 lb. What is the water content of the soil sample?

2-3. A large soil sample obtained from a borrow pit has a wet mass of 26.50 kg. The in-place volume occupied by the sample is 0.013 m³. A small portion of the sample is used to determine the water content; the wet mass is 135 g, and after drying in an oven, the mass is 117 g.

(a) Determine the soil's water content.

(b) Determine the soil wet and dry density for conditions at the borrow pit.

2-4. A sample of soil obtained from a borrow pit has a wet weight of 42 lb. The total volume occupied by the sample when in the ground was 0.34 ft³. A small portion of the sample is used to determine the water content. When wet, the sample mass is 150 g; after drying, 125 g.

(a) What is the water content of the sample?

(b) Determine the wet and the dry unit weights of the soil in the borrow pit.

2-5. The mass of a dried soil sample is 250 g, as determined on a laboratory balance (scale). When immersed in water, the soil particles displace 95 cm³, and this is then the volume of soil solids. Using this data, determine the specific gravity of soil solids.

2-6. The following data apply to a soil sample taken from a construction site:

Specific gravity, $G_s = 2.70$

Void ratio, $e = 0.80$

For this soil, determine the dry density (kg/m³) and dry unit weight (in both kN/m³ and pcf).

2-7. An undisturbed soil sample has a dry mass of 59 kg and an in situ (in-ground) volume of 0.035 m³. The specific gravity of the soil particles is 2.65. Determine the void ratio, e.

2-8. A 1-ft³ sample of undisturbed soil is found to have a dry weight of 107 lb. If the specific gravity of soil solids is 2.70, what is the void ratio of the sample? What is the porosity?

2-9. A laboratory test container holds a volume of 1/30 ft³ (0.000943 m³). The container is carefully filled with a dry sand, 1.36 kg, to determine a value for the minimum density. The specific gravity of the soil solids is 2.70. Determine the dry density (kg/m³), the dry unit weight (in both kN/m³ and pcf), and the void ratio.

2-10. A dry sand is placed in a container having a volume of ¼ ft³. The dry soil weight is 27 lb. If the specific gravity of soil solids is 2.75, determine the void ratio of the sand in the container. Also, calculate the dry unit weight in kN/m³ and pcf.

2-11. A dry sand is placed in a container having a volume of 0.30 ft³. The dry weight of the sample is 31 lb. Water is carefully added to the container so as not to disturb the condition of the sand. When the container is filled, the combined weight of soil plus water is 38.2 lb. From these data, compute the void ratio of the soil in the container and the specific gravity of the soil particles. Also, determine the wet and dry unit weights, in kN/m³ and pcf.

2-12. What will be the dry unit weight of a soil (kN/m³ and pcf) whose void ratio is 1.20, where the specific gravity of soil solids is 2.72?

2-13. A volume of undisturbed soil, 0.015 m³, obtained from a construction site has a wet mass of 27.5 kg. The specific gravity of soil solids is 2.71. Determine the void ratio, water content, and degree of saturation for the in-place condition.

2-14. A sand is densified by compaction at a construction site so that the void ratio changes from 0.80 to 0.50. If the specific gravity of solids is 2.70, what is the increase in the dry unit weight of the sand (in kN/m³ and pcf)?

2-15. Demonstrate that $e = n/(1 - n)$ when n is used as a decimal.

2-16. A sample of clay soil taken from a construction site has a wet mass of 45.6 kg and a related in-place volume of 0.021 m³. After being dried in an oven, the dry soil mass is 39.1 kg. The specific gravity of the soil particles is 2.73. Determine the soil water content, void ratio, and degree of saturation.

2-17. An undisturbed sample of clay is found to have a wet weight of 63 lb, a dry weight of 51 lb, and a total volume of 0.50 ft³. If the specific gravity of soil solids is 2.65, determine the water content, void ratio, and degree of saturation.

2-18. A soil sample is taken from a small test hole in a soil borrow pit. The volume of the test hole is determined to be 0.028 m³. The wet mass of the soil is 56 kg and the dry mass is 49 kg. The specific gravity of soil solids is 2.72. Determine wet and dry density, wet and dry unit weight, water content, void ratio, and degree of saturation.

2-19. A clay sample has a wet mass of 417 g and occupies a total volume of 276 cm³. When oven dried, the mass is 225 g. If the specific gravity of soil solids is 2.70, calculate the water content, void ratio, and degree of saturation.

2-20. Given the following data for an undisturbed soil sample,

$$G_s = 2.69, e = 0.65, w = 10\%$$

determine wet unit weight and dry unit weight (in kN/m³ and pcf), wet and dry density (mg/m³), and degree of saturation.

2-21. A saturated sample of undisturbed clay has a wet mass of 700 g. The sample has a total volume of 425 cm³. When dry, the soil mass is 450 g. What is the specific gravity of the soil solids?

2-22. A sample of soil obtained from below the groundwater table is found to have a water content of 20 percent. If it is assumed that the specific gravity of most soil particles is within the range of 2.60 to 2.75, what is the approximate void ratio of the soil sample?

2-23. In an undisturbed soil formation, it is known that the dry unit weight is 18.06 kN/m³. The specific gravity of the soil particles is 2.75.

(a) What is the saturated wet unit weight of the soil, in kN/m³ and pcf?

(b) What is the effective submerged weight of the soil, in kN/m³ and pcf?

(c) What is the effective submerged density (Mg/m³)?

2-24. An undisturbed soil sample has a wet unit weight of 120 pcf when the water content w is 15 percent. The specific gravity of the soil particles is 2.65,

(a) Approximate the effective submerged weight of this soil.

(b) Determine the effective submerged weight, using the formula for the exact value.

2-25. An undisturbed soil sample has a wet density of 2.22 Mg/m³ when the water content is 10 percent. The specific gravity of the soil particles is 2.69.

(a) Approximate the effective submerged density for this soil.

(b) Determine the effective submerged density using the formula for an exact value.

CHAPTER 3

Soil Types and Soil Structure

The term *soil,* as generally used, refers to the accumulation of particles of disintegrated rock and, frequently, also human-made materials. Because of the experienced wide variation in characteristics and behavior, soils have been subdivided into categories based on the materials' physical properties. In nature, soils are made up of particles of varying size and shape. Size and, to some extent, shape are factors that have been found to be related to or to affect the material behavior of soil to some degree. Consequently, soil categories or types have been developed that are basically referenced to size. To distinguish between soils where size cannot be discerned visually (the particles are too small), an additional property, plasticity (or nonplasticity), is used as criterion.

Study has proved that a soil's important behavioral properties are not always controlled by particle size and plasticity. Soil structure and mineralogical composition, and the intereffect with water, may also have significant influence on the properties and behavior deemed important for design and construction. Under certain conditions, simple typing of the soil provides adequate information for design and construction, whereas certain other conditions require that detailed information about the soil's composition and structure be determined.

3.1 MAJOR SOIL TYPES

The major categories of soil are *gravel, sand, silt,* and *clay.* There is not unanimous agreement on the exact division between each of these major soil types, but gravel and sand are universally considered *coarse-grained* soil, for the individual particles are large enough to be distinguished without magnification. Silts and clays are considered *fine-grained* soil because of their small particles—too small, for the most part, to be seen unaided.

The most commonly used divisions for classifying soils for engineering and construction purposes are shown in Table 3-1. On a comparative basis, the division sizes between gravel and sand (4.76 mm or 2.00 mm) and between sand and silt-clay (0.074 mm or 0.05 mm) are actually quite close. As a result, lack of agreement on these division sizes normally does not cause serious problems.

Particles larger than gravel are commonly referred to as *cobbles* or *boulders.* Again, no unanimous agreement exists on range of sizes. When gravel extends up to the 200 mm (8-inch)

Table 3-1 Size Range for Soil Types

Soil Type	Upper Size Limit	Lower Size Limit
Gravel	Varies from 80 mm up to about 200 mm (3 in. to 8 in.)	4.76 mm (about 0.20 in.) (as determined by a #4 U.S. Standard sieve) or 2.00 mm (#10 U.S. Standard sieve)
Sand	4.76 mm or 2.00 mm	0.074 mm (#200 U.S. Standard sieve) or 0.050 mm (#270 U.S. Standard sieve)
Silt and clay	0.074 mm or 0.05 mm	None

size, anything larger would be termed a boulder. Where the 80 mm (3 in.) size, or thereabouts, is taken as the upper size for gravel, the sizes between 80 mm and 200 mm may be designated as cobbles, and anything larger than 200 mm (8 in.) as boulders. However, 150 mm or 300 mm (6 in. or 12 in.) may also be taken as the division between cobbles and boulders. As for sands and gravels, these discrepancies usually do not cause serious problems. Conventionally, when a construction project requires a particular material, it has become standard practice to indicate the soil or aggregate requirements on the basis of size, instead of or in addition to classification.

In conclusion, particle size serves as the basis for classification of sands, gravels, cobbles, and boulders.

The classification of a fine-grained soil as either a silt or a clay is not done on the basis of particle size but, rather, is based on the plasticity or nonplasticity of the material. Clay soil is plastic over a range of water content; that is, the soil can be remolded or deformed without causing cracking, breaking, or change in volume, and will retain the remolded shape. The clays are frequently "sticky." When dried, a clay soil possesses very high strength (resistance to crushing). A silt soil possesses little or no plasticity and, when dried, has little strength. If a small sample of moist silt is shaken easily but rapidly in the palm of the hand, water will appear on the surface of the sample but disappear when shaking stops. This is referred to as *dilatancy*. When a sample of moist clay is similarly shaken, the surface will not become wetted.

The reason for the difference in behavior between clay and silt relates to the difference in mineralogical composition of the soil types and particle shape. Silt soils are very small particles of disintegrated rock, as are sands and gravels, and possess the same general shape and mineralogical composition as sands and gravels (which are nonplastic). The clay minerals, however, represent chemical changes that have resulted from decomposition and alteration of the original rock minerals. The effect is that their size and shape are significantly different from those of other types of soil particles. This is discussed further in a following section.

Naturally occurring soil deposits most generally include more than one soil type. When they are classified, all the soil types actually present should be indicated, but the major constituent soil type should dominate the description, while the soils of lesser percentage are used as modifying terms; for example, a material that is mostly sand but includes silt would be classified as a silty sand, whereas a silt–clay mixture with mostly clay would be termed a silty clay.

Although a soil may be predominantly coarse-grained, the presence of silt or clay can have significant effect on the properties of the mixture. Where the amount of fine-grained material exceeds about one-third of the total soil, the mixture behaves more like a fine-grained soil than a coarse-grained soil.

The condition also exists where small fragments of decomposed vegetation are mixed with the soil, particularly fine-grained soils. Organic material mixed with the nonorganic soil can have striking detrimental effects on the strength and compressibility properties of the material. The presence of organic material should be carefully watched for. A foul odor is characteristic though not always associated with such soils, as is a blackish or dark gray color. Soils in this category are designated as *organic* (e.g., *organic silt* or *organic clay*) in comparison to a *nonorganic* designation for soil free of decomposed vegetation.

3.2 PARTICLE SHAPES AND SIZES

Particles in the sand, gravel, and boulder categories are considered as "bulky grain," indicating that particle dimensions are approximately equal; that is, the dimensions in the length, width, and thickness directions would be of the same order of magnitude (commonly, one dimension is no more than five times larger or smaller than another dimension). Individual particles are frequently very irregular in shape, depending somewhat on the rock they were derived from, their age, and exposure to weathering and transporting processes. Generally, a new particle is "angular" and rough-surfaced, and is then modified with time and exposure to become more smooth-surfaced and rounder. The various stages of transition—angular, subangular, rounded—are illustrated in Figure 3-1. Generally, the angular particles possess better engineering properties, such as higher shear strength, than do weathered and smooth particles.

Particles in the silt category, though classified as "fines" along with the clays, are still angular or bulky in shape and of the same mineralogical composition as the coarse-grained

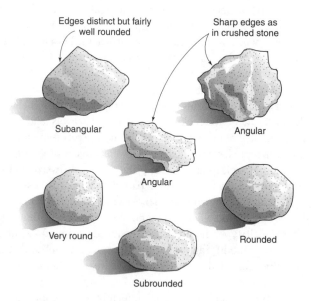

Figure 3-1 Shapes of granular soil particles [1].

soils. Because of the mineralogical composition, such particles rarely break down to less than 2μ or 0.002 mm in size (1 μ, i.e., micron, equals 0.001 mm).

The mineralogical composition of true clay is distinctly different from the mineral components of the other soil types, inviting the distinctions *clay minerals* and *nonclay minerals*. Clay minerals typically result from the alteration of rock minerals (which, unaltered, constitute the coarser soil particles). Almost all clay minerals are crystalline minerals (minerals consisting of an orderly and repetitious arrangement of molecules to produce a sheet-like structure) that are capable of developing cohesion (because of an attraction and bonding between soil sheets and water) and plasticity. Clay particles may be made of many sheets on top of one another. Clay particles are mostly found in sizes less than 2 μ or easily break down to this size. It is emphasized, however, that it is the mineral type and not the small size that is primarily responsible for the high cohesion and plasticity that clays possess. Where particles of nonclay minerals have been broken down to smaller than 2 μ, the clay properties are not developed.

Because of the sedimentary origin of fine-grained soil deposits and the overlap in sizes of the clay and nonclay minerals, it is unusual to find natural deposits of pure clay mineral soils. Very frequently, so-called clay deposits are actually a mixture of clay minerals and nonclay minerals. Because of this, the term *clay mineral* has been used to prevent confusion when one is designating a naturally occurring soil deposit consisting of fine-grained soils that have the general properties of cohesion and plasticity.

The clay minerals themselves can and do vary in their composition and, therefore, in their behavior properties. The building blocks, or constituent sheets, that combine to form most of the different types of clay minerals are the *silica tetrahedral* sheet (Fig. 3-2) and *alumina octahedral* sheet (Fig. 3-3).

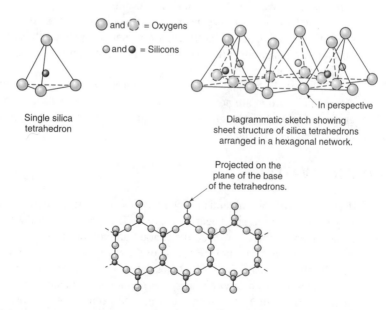

Figure 3-2 Assembly of atoms forming the basic clay mineral sheet silica tetrahedron [106].

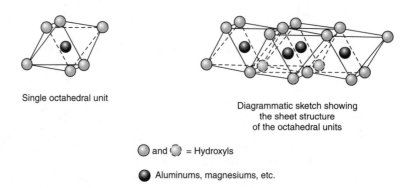

Single octahedral unit

Diagrammatic sketch showing
the sheet structure
of the octahedral units

◯ and ◔ = Hydroxyls

● Aluminums, magnesiums, etc.

Figure 3.3 Assembly of atoms forming the basic clay mineral sheet alumina octahedron [117].

The silica tetrahedron consists of four oxygen ions and one silicon ion. The molecular arrangement is such that the four oxygens are spaced and located at what would be the corners and tip of a three-dimensional, three-sided pyramid, with the silicon located within the pyramid. Oxygen ions at the base are shared by adjacent tetrahedrons, thus combining and forming the sheet. The thickness of a silica sheet is 5×10^{-7} mm, or 5 Angstrom units (one Angstrom unit or $1\ \text{Å} = 1 \times 10^{-7}$ mm).

The alumina octahedron consists of six oxygens and one aluminum. Three of these oxygens are in the top plane of the octahedron, and three are in the bottom plane. The aluminum is within the oxygen grouping. It is possible that the aluminum ion may be replaced with magnesium, iron, or other neutral ions. To obtain a valence balance, some of the oxygens may also carry a hydrogen ion (resulting in a hydroxyl at some oxygen locations). The alumina sheet is also 5×10^{-7} mm or 5 Å units thick.

It also happens that oxygens from the tip of a silica tetrahedron can share in an alumina sheet, thus layering sheets. Different arrangements of sheets then can combine to form the different clay minerals. The composition and typical properties of the more commonly occurring clays are summarized in Table 3-2.

Although the thickness of a clay mineral sheet is limited, the dimensions in the length and width direction are not. As a result, the clay minerals have a flat, platelike shape (like an irregular sheet of paper), where the length and width can be several tens or several hundreds times the thickness.

3.3 CLAY AND WATER

The surfaces of clay mineral particles have a net electrical charge that is negative, while the edges have positive and negative charges. This results from the molecular grouping and arrangement of ions. However, the charges are not uniform, but vary in intensity at different locations on the particle. Because of the extremely small size of clay particles and the very high ratio of particle surface to particle mass, the forces of electrical charge have a profound effect on the behavior of particles coming in association with other particles and water (or other fluids) present in the soil. And because of the manner of development or deposition, clay deposits almost always exist in the presence of some water.

Table 3-2 Basic Properties of Some Typical Clays

Clay Mineral	Composition	Layer Thickness	Shape of Mineral, General Properties, and Comments
Kaolinite	One silica, one alumina sheet. Very strongly bonded together.	7.5 Å	The most prevalent clay mineral. Very stable, with little tendency for volume change when exposed to water. Kaolinite layers stack together to form relatively thick particles. Particles are plate-shaped. Form from crystalline rocks in humid climates.
Halloysite	One silica, one alumina sheet make up the layer. Has sheet of water molecules between layers. (Similar to kaolinite except for sheet of water.)	10 Å	Sheets of halloysite curl into tubes. Strength and plasticity are significantly affected by drying and removal of the water. After drying, the clay mineral will not reinstate a water layer if again exposed to water. Caution is required in identifying this mineral and in using remolded (and rewetted) samples in laboratory testing to determine properties. Dried halloysite has characteristics of kaolinite. Rewetted samples appear stronger and less plastic than naturally wetted halloysite.
Illite	Alumina sheet sandwiched between two silica sheets. Potassium provides the bond between layers.	10 Å	Irregular flake shape. Generally more plastic than kaolinite. Does not expand when exposed to water unless a deficiency in potassium exists. Illite clays seem most prevalent in marine deposits and soil derived from micaceous rock (schists, etc.).
Montmorillonite (also identified as smectite)	Alumina sheet sandwiched between two silica sheets. Iron or magnesium may replace the alumina in the alumina sheet; aluminum may replace some silicons in the silica sheet (isomorphous substitution). Weak bond between layers.	9.5 Å	Irregular plate shapes or fibrous. Because of the weak bond between layers and the negative charge resulting because of isomorphous substitution, the clay readily absorbs water between layers. Has a great tendency for large volume change because of this property. Forms mostly from ferromagnesium rock and develop mostly in semi-arid and temperate climates; also from decomposition of volcanic ash.
Chlorite	Alumina sheet sandwiched between two silica sheets, but layers are bonded together with an alumina sheet.		Irregular plate shapes. Nonexpanding. Formed from well-drained soils and micaceous rocks in humid areas.

Note: 1 Å = 1 x 10^{-7} mm.

In contrast, the engineering behavior of coarse particles is not significantly affected by surface electrical charges because of a comparatively low ratio of particle surface to mass.

The water molecule has a somewhat peculiar arrangement, as illustrated in Figure 3-4. The *electron cloud* configuration shown in Figure 3-4e is responsible for a behavior that is almost unique (and actually responsible for life on earth as we know it). The term *electron cloud* refers to the varying zone of space traveled by the orbiting electrons. Referring to Figure 3-4c and e, note that of the six electrons in the oxygen atom's outer orbit, two are shared or bonded to the hydrogen atoms and two pairs of electrons are unshared in the arms of the

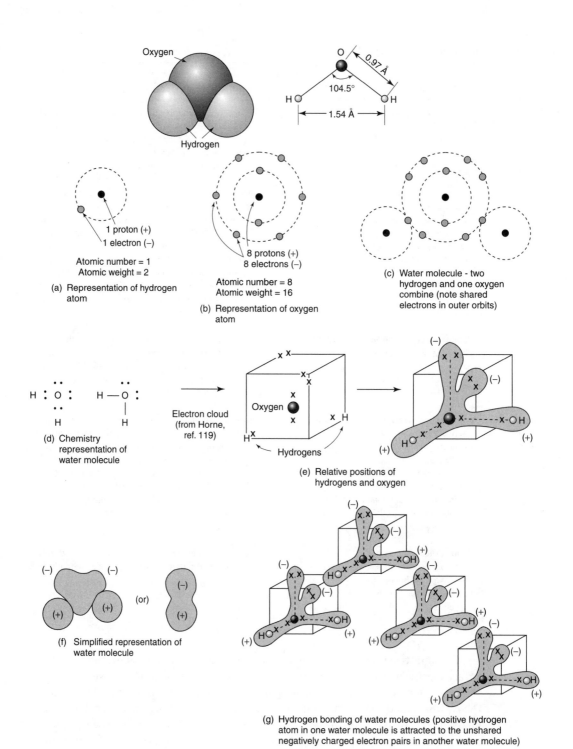

Oxygen

Hydrogen

O

0.97 Å

104.5°

H   H

1.54 Å

1 proton (+)
1 electron (−)

Atomic number = 1
Atomic weight = 2

(a) Representation of hydrogen
atom

8 protons (+)
8 electrons (−)

Atomic number = 8
Atomic weight = 16

(b) Representation of oxygen
atom

(c) Water molecule - two
hydrogen and one oxygen
combine (note shared
electrons in outer orbits)

H : O :   H — O :

H   H

(d) Chemistry
representation of
water molecule

Electron cloud
(from Horne,
ref. 119)

Oxygen

Hydrogens

H

(e) Relative positions of
hydrogens and oxygen

(−)   (−)

(−)

(+)   (+)

(+)

(or)

(f) Simplified representation of
water molecule

(g) Hydrogen bonding of water molecules (positive hydrogen
atom in one water molecule is attracted to the unshared
negatively charged electron pairs in another water molecule)

Figure 3-4 The water molecule.

electron clouds opposite the hydrogens. The center of gravity of the positive and negative electrical charges do not coincide, and the relative positions of the negatively charged electrons result in a molecule that possesses an electrostatic dipole moment.[1] Because the effect is an assembly that has a positive charge at one end and a negative charge at the other, similar to a bar magnet, water molecules are represented as polar molecules (Fig. 3-4f). The two electron cloud arms possess a negative charge which can attract the positive partial charges of the hydrogen atom in a nearby water molecule; water molecules then prefer to be bonded together.

Groundwater is rarely pure, instead containing dissolved gases, minerals, and other compounds in solution or suspension. Groundwater may be acidic (pH less than 7) or alkaline (pH greater than 7). In the presence of groundwater, various rocks and other minerals will disassociate (break down) into the component cations (positively charged ions or groups of atoms) and anions (negatively charged ions).

Clay mineral particles would, by themselves, tend to repel each other because of the net negative charges present on the surface, unless edge-to-surface contact were made (positive to negative would attract). Because of the net negative charge, however, the particles will attract cations (positive ions) such as potassium, sodium, calcium, and aluminum present in the soil moisture (very typically present as a result of solutions from rock weathering), so as to obtain an electrically balanced or equilibrium condition. Further, because of the net positive charge of the cations, they in turn can also attract negative charges. As a result of this phenomenon, water becomes bonded to the cations. The negative tips of water molecules are attracted and held to the cation, which in turn is held by the clay particle. The resulting effect is that significant water (significant with respect to the size and weight of the clay particle) becomes "bonded" to the clay. Water molecules are also held to the particle surface, where they become attracted directly to a location of negative charge. Figure 3-5 depicts the various features of the attraction-bonding.

Additional water molecules then also become attracted to the clay particle because of a chainlike arrangement of negative ends to positive ends of molecules and by hydrogen bonding (the condition where hydrogen atoms in water are shared with hydrogen atoms in the clay).

The state or nature of the water immediately surrounding a clay particle is not clearly understood by soil scientists, but it is generally known to possess properties different from liquid water. It may be in a very dense and viscous state. It *is* certain, however, that this water is very strongly attracted to the clay particle.

The attraction for cations necessary to balance the negative charge of the clay extends beyond the surface layer of molecules surrounding the particle. The further from the particle surface, though, the weaker the attraction becomes, and, therefore, the concentration of cations becomes lessened. The resulting effect is that water molecules are still attracted to the clay particles, indirectly, but the further from the particle, the weaker the attraction. At a distance beyond where cations and, therefore, water molecules are attracted to a clay particle, water in the soil is considered "loose" or "normal" pore water.

---

[1]Without this dipole moment and the negatively charged electron cloud arms, water molecules would not stick together in liquid form but instead would exist in the gaseous state. These features also explain why water remains liquid over a comparatively wide range of temperatures and why the surface tension property important to capillary movement, discussed in Chapter 6, is so great.

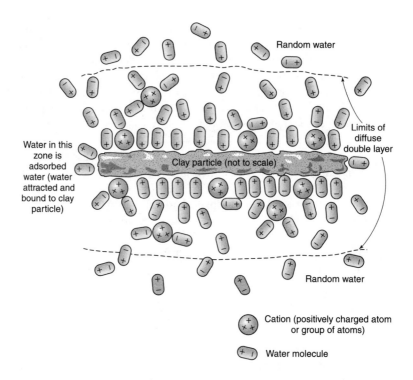

Water in this zone is adsorbed water (water attracted and bound to clay particle)

Random water

Limits of diffuse double layer

Clay particle (not to scale)

Random water

Cation (positively charged atom or group of atoms)

Water molecule

Figure 3-5 Adsorbed water and cations in diffuse double layer surrounding clay particle.

The distance from the clay particle surface to the limit of attraction is termed the *diffuse double layer* (see Fig. 3-5), relating to the negative charges on the particle surface and the distribution of attracted (held) cation charges close to the particle. An effect is that immediately surrounding the particle a thin, very tightly held layer of water, perhaps 1×10^{-6} mm (10 Å) thick, exists, and a second diffused, more mobile zone extends beyond this first layer to the limit of attraction. Molecular movement in the outer layer (or outer zone), and probably also in the tightly held layer, continually occurs, however. The water that is held in the diffuse double layer is frequently termed *adsorbed water* or *oriented water*, to differentiate it from normal pore water, which is not oriented.

The plasticity that clay soils possess is attributed to the attracted and held water. And, as a restatement, water molecules are attracted because of their dipole structure. The unusual properties of plasticity possessed by clays occur because of the unusual molecular structure and the common presence of water in soil deposits. Experiments performed with clay using nonpolar liquid in place of water have resulted in a "no-plasticity" condition similar to that noted for coarse-grained sandy soil.

3.4 SOIL STRUCTURE

The particle arrangement of the equidimensional particles—gravel, sand, and silt—has been likened to arrangements that can be obtained by stacking marbles or oranges. For similar-sized spherical particles, a loose condition (condition with a high void ratio) is ob-

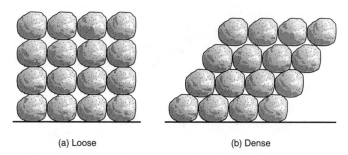

(a) Loose   (b) Dense

Figure 3-6   Schematic diagram of grain arrangement for loose and dense granular soils.

tained from an arrangement as shown in Figure 3-6a. A dense condition (condition with a low void ratio) is obtained from an arrangement as in Figure 3-6b.

Actual soil deposits are made of accumulations of soil particles having at least some variation, but more frequently great variation, in particle size. As a result, the soil structure is not quite like that presented in Figure 3-6. Generally, the greater the range of particle sizes, the smaller the total volume of void spaces there will be. For a given soil deposit, however, a range of conditions between loose and dense is possible. Typical values for different types of soil mixtures are tabulated in Table 3-3.

In relating the volume of void spaces to properties desirable for building construction purposes, it is generally anticipated that the smaller the void ratio (or the denser the material), the higher the strength and the lower the compressibility will be.

Coarse soil in an initially loose condition may be prone to quick volume reductions and loss of strength if subjected to shock or vibrations, unless there is some cementing at points of particle contact or "cohesive strength" provided by moisture menisci (discussed further in Chapter 11).

Experience indicates that it is possible for sands or silts to be deposited in such a manner that an unusually loose or honeycomb structure results. Grains settling slowly in quiet waters, or a loosely dumped moist soil, can develop a particle-to-particle contact that

Table 3-3   Typical Void Ratios and Unit Weights for Cohesionless Soils

| Soil Description | Range of Void Ratio | | Range of Unit Weight | | | | |
	e_{max} (loose)	e_{min} (dense)	Condition (moisture)	$\gamma_{min}\frac{kN}{m^3}$ (loose)	$\gamma_{max}\frac{kN}{m^3}$ (dense)	γ_{min} pcf (loose)	γ_{max} pcf (dense)
Well-graded fine to coarse sand	0.70	0.35	Saturated	19.5	22	125	140
			Dry	15	19	95	120
Uniform fine to medium sand	0.85	0.50	Saturated	19	20.5	120	130
			Dry	14	17.5	85	110
Silty sand and gravel	0.80	0.25	Saturated	18	22.5	115	145
			Dry	14	17	90	130
Micaceous sand with silt	1.25	0.75	Saturated	17	19.5	110	125
			Dry	12	15	75	95

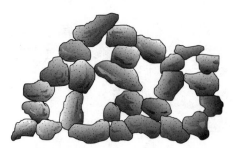

Figure 3-7 Honeycomb structure in a granular soil.

bridges over relatively large void spaces between the aggregates and carries the weight of the overlying material. A possible particle arrangement is shown in Figure 3-7.

The presence of flake-shaped particles, such as mica flakes, in a coarse soil has a significant effect on the void ratio, density, and compressibility of a deposit. The flake-shaped particles are capable of bridging over open spaces so that relatively large void ratios develop. When subject to external loading, however, the flakes are incapable of providing great support, bending or breaking and rearranging under load.

For coarse-grained soils and silts, the mass of an individual particle is relatively great compared to the surface area. Therefore, the effect of gravity has the most influence over the arrangement of deposits of such soils. (The effect of electrical charges on the particle surface is negligible.) Conversely, clay particles, because of their large surface-to-mass ratio, are more affected by the electrical forces acting on their surfaces than by gravity forces.

Clay deposits developed from clay particles that have settled out of suspension in a freshwater or saltwater environment tend toward a *flocculated structure,* whereby the attraction and contact between many of the clay particles is through an edge-to-face arrangement. Clays settling out in a saltwater solution tend to a structure more flocculent than do clays settling out in a freshwater situation. Saltwater (oceans are approximately a 3.5 percent saline solution) acts as an electrolyte in which the repulsion between particles is reduced. With respect to other particles, particle sedimentation then occurs with a random orientation, creating the flocculent structure. Sedimentation in a weaker electrolyte, such as fresh water (fresh surface or subsurface water is not "pure"), produces a structure where some parallel orientation of settled particles occurs, but the overall structure is flocculent; see Figure 3-8a and b. Clay particle sedimentation occurring in ponds and wetlands where organic decay is taking place will result in highly flocculent structures.

Clay deposits with flocculent structures will have high void ratios, low density, and probably high water contents. The structure, however, is quite strong and resistant to external forces because of the attraction between particles. But if the environment surrounding the clays is changed, such as by having the salts leached from the deposit (as has occurred where ocean deposits later rose above sea level and fresh water subsequently percolated through the soil, leaching out the salts with it), the attraction and strength between particles can be markedly decreased.

Clays that have been further transported after being deposited (as from glacial action, or when used in human-made earth fills) are reworked or remolded by the transportation process. The particle structure that develops from remolding is a more parallel arrangement or orientation of particles than existed in the flocculent condition, as shown in Figure 3-9. Such a particle arrangement is considered a *dispersed* or *oriented* structure.

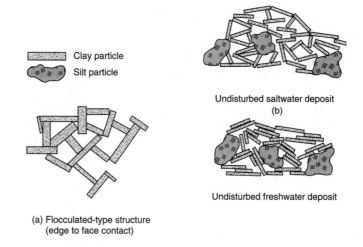

Clay particle
Silt particle

Undisturbed saltwater deposit
(b)

Undisturbed freshwater deposit

(a) Flocculated-type structure
(edge to face contact)

Figure 3-8   Schematic diagram of types of particle orientations [169].

When used for construction, clays that were in a flocculent condition before use generally lose some strength as a result of remolding. Subsequent to remolding, and with the passage of time, however, the strength increases, though not back to the strength of the originally undisturbed clay. Reasons for this increase appear to be related to a time-dependent rebuilding from the remolded, dispersed structure toward a less dispersed, more flocculent structure. In the dispersed condition, the equilibrium of forces between particles is disturbed. With time, the particles become sufficiently reoriented (only very small movements are necessary) to reacquire a structure in which the forces between particles are again in equilibrium (but not as originally structured; see Fig. 3-10). This phenomenon of strength loss–strength gain, with no changes in volume or water content, is termed *thixotropy*. Thixotropy has been defined as a "process of softening caused by remolding, followed by a time-dependent return to the (original) harder state." The degree of difference between the undisturbed strength and remolded strength, and the extent of strength gain after remolding, are affected by the type of clay minerals in the soil. Generally, the clay types that adsorb large quantities of water, such as the montmorillonites, experience greater thixotropic effects than do the more stable clay types, such as kaolinite.

For many construction situations, thixotropy is considered a beneficial phenomenon, since with the passing of time soil structures (dams, highway embankments, etc.) and disturbed foundation soils get stronger and presumably safer. However, the phenomenon also

Figure 3-9   Schematic diagram of particle orientation.

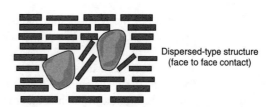

Dispersed-type structure
(face to face contact)

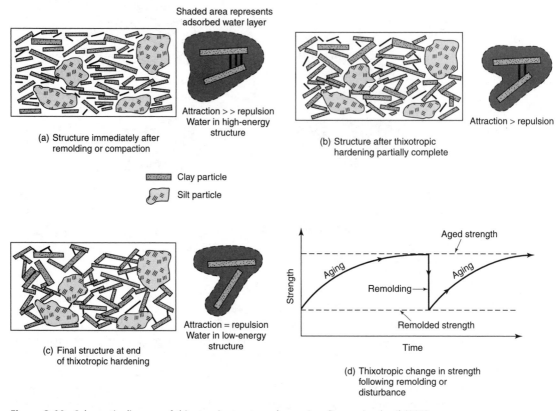

Figure 3-10 Schematic diagram of thixotropic structure change in a fine-grained soil (201).

causes its problems: Construction sites may be quickly transformed into a mire of mud when construction equipment travels across the area, making handling of equipment and materials very difficult. Thixotropic influences have also affected piles driven in clay soils when the driving operation caused remolding and weakening of the clay surrounding the pile. The soil strength is sometimes recovered relatively quickly. This has been experienced where a pile has initially been driven part of its length, and attempts at continued driving after a one- or two-day wait have met with considerable resistance due to the increasing adhesion (directly related to a clay's strength) along the pile surface. This is one reason that piles embedded in cohesive soils should be fully driven whenever possible.

3.5 SOME SPECIAL SOIL CATEGORIES

To those in the construction industry, the term *special soil* implies a soil type with a property or behavior that is considered unusual and capable of causing problems and therefore requires special treatment. *Collapsible soils* are one of the groupings of special soils. The loess soils discussed in Chapter 1 are in this collapsible category, for the in-place structure of these aeolin deposits makes them susceptible to significant volume reduction when in contact with water. Collapsible gravel is also included in this

grouping, due to the significant volume decrease that occurs upon inundation or exposure to large amounts of water. The design and construction of facilities in areas where
collapsible deposits exist should include means for either protecting the foundation soil
from water or otherwise compensating for its adverse effects. Serious ground settlement
can also occur in areas underlain by saturated sand in a loose condition; such deposits
can lose much of their shear strength when significant vibration or seismic shock occurs. The actual occurrence is termed *liquefaction*. (*Note:* Liquefaction more accurately
represents a soil condition rather than a soil type.)

Some clays are prone to large volume changes which are directly related to changes in
water content, shrinking in dry seasons and swelling in wet seasons or when otherwise in
contact with water. Such soils are termed *expansive clays* or *swelling clays*. Clays including the montmorillonite mineral are particularly noted for their high volume change characteristic. The expansive force of a swelling montmorillonite clay can lift pavements and
structures, while volume decrease due to reduction of water content can be responsible for
significant ground settlement. A common construction practice in areas where expansive
clays exist is to keep the soil beneath and surrounding a structure stable by providing protection against changes in the moisture content.

Collapsible soils, liquefaction and expansive clays, are discussed more fully in the sections following. *Dispersive clays* and *lateritic soils* similarly represent materials which require consideration, and also are discussed. *Permafrost,* the permanently frozen ground
which exists in the vicinity of the earth's polar regions, is discussed in Chapter 7.

Collapsible Soils

Collapsible soils refers to the category of soil deposits that experience significant decrease
in volume when exposed to water. Collapsible deposits typically are found in arid regions.
The loess soil deposits (predominantly silt-size particles) discussed in Chapter 1 are in this
category, but granular deposits that include considerable gravel also can be collapsible. One
feature of collapsible soils is the geologic process responsible for the deposition. Wind-
transported silts that formed the present-day loess deposits typically were laid down slowly
in a dry environment. These deposits commonly retain the original loose structure but also
include vertical rootholes and grass channels that function to create a high permeability in
the vertical direction (as well as establishing a natural plane of cleavage for earth slopes and
excavation cuts). Most collapsible gravels are water transported and deposited formations
such as alluvial fans, mud flow or slope wash deposits, or torrential stream deposits, where
rapid deposition was followed by drying. A property of the collapsible gravels is the presence of silt or clay fine-grained materials (typically 25 to 45 percent of the soil weight, but
as low as 10 percent) which act as a binder to produce a fragile metastable structure vulnerable to breakdown in the presence of water.

Granular deposits that are prone to collapse can be difficult to detect with standard subsurface investigation procedures involving soil sampling because high resistance to penetration by the sampling equipment can develop, and also, high density or unit weight can
exist. Collapsible soils will often indicate a low moisture content, however. A procedure for
evaluating collapse potential is to compare the volume and strength for soil samples in the
natural condition and after exposure to water; the review of the geologic history for an area

of proposed construction should include checking for conditions indicative of collapsing soils at neighboring projects (unexpected large settlements, sinkholes, etc.).

Sites underlain by collapsible deposits have been improved by flooding and by using mechanical processes that resulted in collapsing the soil to achieve a stable condition prior to construction. After construction, efforts to prevent structural damage concentrate on preventing water from reaching the collapsible strata. Case histories of construction projects located in areas of collapsible soils, including details of conditions and remedies, are discussed by Rollins, et al. [248] and Mitchell [203].

Liquefaction

Liquefaction is a condition that can occur when saturated cohesionless sand deposits exist in a relatively loose condition. If subject to vibration or shock waves, as can result during earthquake and from explosion or operation of some types of machinery, the soil grains attempt to quickly move (shake) into a denser or more compact arrangement, but the presence of the void space water (i.e., pore water) interferes, and particle-to-particle contact is prevented. Temporarily, much of the soil shear strength is lost, with the result that the deposit assumes the properties of a viscous liquid with little strength. The sand will flow or displace if supporting any type of loading when the event occurs. Structures underlain by sands that undergo liquefaction may experience significant vertical or lateral movements, while unsupported earth slopes tend to slide.

The sand deposits susceptible to liquefaction can be identified from soil boring or penetration resistance testing, which indicates the relatively loose condition, or from relative density testing. Generally, deposits of uniform sands are considered more susceptible than well-graded sands, and fine sands are considered more susceptible than coarse sands. Probability of liquefaction also relates to the severity of shock vibration passing through the deposit; for example, a medium-dense sand deposit may not be affected by minor events but seriously affected by a strong event. Site improvements that can be accomplished to reduce or eliminate the chance of liquefaction occurrence include drainage to remove the saturation condition or a soil densification-compaction procedure.

Expansive Clays

Clays containing the montmorillonite mineral expand in volume if the soil water content is below a stability value[2] when water becomes available. Conversely, these clays experience considerable shrinkage when water content is reduced. The volume change is related to the thickness and mobility of the water film adsorbed onto or surrounding the montmorillonite particle, being increased or decreased relatively easily during natural wetting and drying conditions. [The combination of silica–alumina–silica sheets that comprise the montmorillonite mineral remains intact (refer to Figs. 3-2 and 3-3 and Table 3-2), but the minerals are layered similar to pages assembled in a book to form a clay particle, and it is between layers where the volume of water increases or decreases.]

---

[2]Stability value of water content for clays prone to expanding refers to the condition in which additional water will not be attracted to the particle and (further) expansion does not occur.

The expansive force created by a clay undergoing an increase in water content and volume can be considerable, being capable of lifting heavy structures and imposing lateral pressures that can move retaining walls and basement walls. Swelling pressures in excess of 500 kPa or 10,000 psf have been measured. High swelling pressures can occur even in already high-moisture clays if additional water is adsorbed. Swelled clays can experience shrinkage equal to the volume increase if reduction in water content (drying) occurs. Damage to structures can result when a clay swells or shrinks; the more notorious problems have been associated with the swelling condition, but shrinkage-related ground settlement can result in serious damage to many types of structures. For structures such as roadway and airfield pavements subject to *cycles* of clay expansion and shrinkage, effects can be more damaging than those resulting from conventional loadings.

The existence of clay soils that can develop large swelling pressures or undergo shrinkage damaging to structures apparently is a condition recognized only during the twentieth century. But the presence of expansive clays has now been documented for virtually all continental areas of the planet (North, Central, and South America, Hawaii, Europe, Asia, Africa, and Australia). A generalized distribution of expansive clays in the United States is shown in Figure 3-11, with the most severe conditions reported for Colorado, Texas, and Wyoming. In Canada, expansive clays have been identified in deposits found across the western sector (Canadian prairie regions, Saskatchewan, Alberta, and Manitoba). West-central India and the generally inland regions of south-central China have well-known deposits of expansive soils. The countries of North Africa and Eurasia that border the Mediterranean Sea similarly have identified deposits of expansive soils.

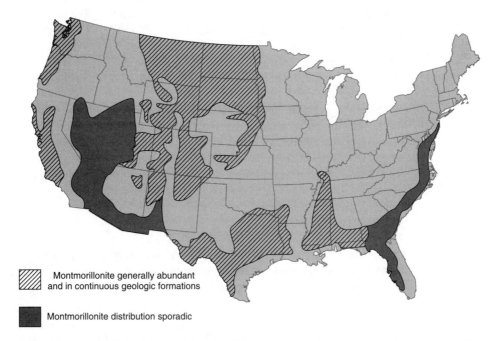

Montmorillonite generally abundant and in continuous geologic formations

Montmorillonite distribution sporadic

Figure 3-11   General distribution of montmorillonite expansive clay across the United States (modified after Tourtelot [302]).

The large volume change property of the expansive clays is typically due to the presence of montmorillonite clay minerals (although significant shrinkage-related problems reported for China occur with clays dominant with illite). Montmorillonites form from the basic igneous rocks (gabbros and basalts, volcanic glass) and some sedimentary rocks (shales containing montmorillonite clay particles but also limestones and marls rich in magnesium). However, climate is an important factor; the montmorillonites are typically found in semiarid tropical and temperate regions. Conditions associated with the formation of montmorillonites are extreme disintegration of the parent material and an alkaline environment, coupled with a semiarid climate, which results in an availability of some but limited water; an absence of water leaching through the material permits magnesium, calcium, iron, and sodium ions—particularly magnesium—to accumulate. (Where leaching has occurred and the majority of ions have been flushed, the clay mineral remaining is typically kaolinite, a "stable" clay.)

The montmorillonite mineral carries a large net negative charge on the surface, and has a high capacity for ion exchange (the *effects* resulting from changing adsorbed cations are greater where the ion exchange capacity is high). Cation exchange occurs outside the structure of the silica–alumina–silica sheet combination, and the resulting most-prevalent ion tends to be the most abundant or available. Where calcium ions are available and replace ions such as sodium but also increase the ion concentration, a lower ion exchange capacity results; an effect is that the tendency for expansion or swelling is reduced.

Relating to construction practice, various methods have been used in the attempt to stabilize expansive clays against swelling and shrinkage. Protecting a soil formation from variations in moisture content responsible for volume changes represents a sensible concept but is often difficult to accomplish or maintain in practical situations.

Chemical stabilization using additives that reduce the inclination to attract or lose moisture can be successful if the additives are adequately mixed or distributed throughout the accumulation of clay particles; lime slurry mixing is a widely used procedure relating to the availability and economy of lime (calcium) materials. A method in use where only a limited area or depth requires protection against soil volume changes is replacement (e.g., excavate the clay material from the zone susceptible to moisture change and replace it with a stable granular soil).

Site investigation that includes obtaining soil samples for testing is important in identifying the presence of expansive clay in an area of proposed construction, so that procedures for preventing problems can be implemented during the design-planning stage of the project. Information on applying soil test results to evaluate a clay's potential for expansion is presented in Section 4.2 and related Figure 4-14.

Dispersive Clays

Fine-grained soils that will deflocculate in still water and erode if exposed to low-velocity water are termed *dispersive clays*.[3] Ordinary clays typically do not erode in the presence of

---

[3]Note that the terms *dispersive clay* and *dispersed structure* for clay refer to different properties. The similarity of terms is unfortunate, but when each term is used in proper context the description will be understood. Simply stated, dispersive clay refers to an easily eroded clay, whereas dispersed structure refers to a positional relationship between particles; a clay having a dispersed structure may or may not be dispersive (easily eroded).

water unless the flow velocity is relatively high (more than about 1 m/sec or 3 to 4 ft/sec). Areas blanketed by natural deposits of dispersive clays characteristically show steep erosion gullies and eroded tunnels, though not always. Embankments constructed with dispersive clays similarly experience the development of gullies and tunnels. Critically, the presence of dispersive clay in earth dams, engineered and otherwise, has been the reason for piping (leakage through eroded tunnels) of such great magnitude that major repair became necessary to keep the structures functional; frequently the piping condition developed very soon after the initial pooling of the reservoir.

A clay's susceptibility to dispersion has been found to be related to the presence of cations (sodium, calcium, magnesium, potassium) in the soil pore water. Generally, repulsive forces between clay particles that act to cause deflocculation decrease as the concentration of ions increases. However, repulsion increases as the quantity of sodium ions increases; the thickness of the diffuse double layer of water adsorbed to the soil particles is relatively great when monovalent sodium ions are present, and the distance-related attractive forces that exist between particles are thus weakened. Usually, in a clay–water system, the repulsive forces and the tendency toward dispersion increase as the concentration of ions increases along the series calcium–magnesium–potassium–sodium–lithium (e.g., a clay high in sodium is more dispersive than one having a high concentration of calcium and little sodium). Figure 3-12 expresses a qualitative relationship between cations present in a clay deposit and the tendency for dispersion; laboratory tests used to indicate dispersive tendencies are discussed in Chapter 4.

The piping that occurs in compacted earth dams that have been constructed of dispersive clay may originate at locations where water enters minuscule cracks, such as those that would develop from settlement or hydraulic fracturing. In the presence of water, the exposed clay particles quickly go into suspension, and a progressive erosion rapidly results. Where the presence of a dispersive clay has not been recognized during construction, dams have experienced piping as the first reservoir filling was underway, or soon after filling has been completed. Relating to the effect of ion concentration, earth dams that have satisfactorily retained reservoirs of saltwater for long periods have been known to develop piping problems quickly when fresh water replaced the saltwater, a result of the dilution of the ion concentration in the dam's clay.

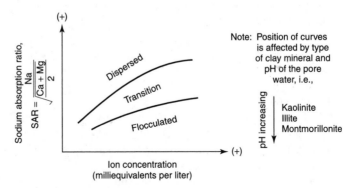

Figure 3-12 Qualitative relationship between ion concentration and dispersivity.

The erodibility of dispersive clay can be reduced through the use of hydrated lime or aluminum sulfate admixtures, 1 to 2 percent by weight. Where core zones of dams must be constructed of dispersive clays, soil transition–filter zones that border the dispersive clay can be designed to control erosion and seal concentrated leaks. The use of geofabrics (geotextiles) can also be investigated as an erosion control.

Laterites

The term *laterite* or description *lateritic soil* refers to a category of residual soil formed from the weathering of igneous rock under conditions of high temperature and high rainfall such as those typically occurring in tropical regions, where the decomposition process results in a soil leached of silica (SiO_2) and calcium carbonate but retaining high concentrations of iron and aluminum sesquioxides (Fe_2O_3, Al_2O_3). An Al_2O_3/SiO_2 ratio of one-half or greater is one criterion used to classify a soil as lateritic. Calcium, magnesium, and potassium contents are very low. Laterites are frequently reddish in color but not always, and all reddish tropical soils are not lateritic. Extensive areas of Asia, South America, and Africa are blanketed with lateritic soils. Concerning their engineering behavior, these soils have been found to possess some unusual properties when compared to soils existing in the temperate regions of the world. Agriculturally, most lateritic soils typically contain only meager concentrations of the nutrients considered necessary for productive farming.

Deposits of lateritic soils may be found in a hard or cemented state, particularly in areas where vegetation is sparse or has been removed. The cementation, attributed to the presence of free iron oxide in the soil, is such that the hardened soil can be quarried and used as a bricklike material for building construction purposes.

Generally, with a tropical residual soil the weathering process is not complete, and material can exist at all stages between fresh rock and soil. However, lateritic soils seem to be typified by a gap grading, existing with a prevalence of gravel-sized and clay materials but having very limited sand- and silt-sized particles. When used as a construction material, these weathered soils often have been found to be unstable, subject to further breakdown during handling and testing or excavation and placement. It appears that the soils from a damp regime are more apt to change physical properties than are soils from an arid region; soil from the damp environment has never been dehydrated but becomes so after exposure to air, an occurrence that has a weakening effect. Further, there is great possibility of significant changes in properties if a laterite is completely dried and then wetted. Reworking a lateritic soil at a given water content frequently alters properties such as plasticity and compaction characteristics. It has become apparent that widely used correlations between soil classification index properties and the subsequent engineering behavior, established primarily from experiences with soils formed in the temperate regions of the world, cannot be relied on to predict behavior of soils formed in tropical areas.

In their native regions, lateritic soils have, of economic necessity, been used as construction material for road bases and embankment structures, and to support foundations. Experience has proved that these materials can serve successfully when protected from percolating or migrating water and the effects of heavy repetitive loadings. Properties can be improved by the use of common admixtures such as cement and lime.

PROBLEMS

3-1. List the soil types included in the coarse-grain category and the fine-grain category and the reason for the division into categories.

3-2. Clay is a soil material that possesses plasticity in the presence of water. What does the term *plastic* mean in relation to clay soils?

3-3. What is the essential reason for the difference in behavior of natural clays and other soil types such as silts and sands?

3-4. Comment on the difference between the shape and size of clay particles compared to other soil types such as silts and sands.

3-5. What are the "building blocks" of most clay minerals? Comment on the comparative length–width–thickness dimensions of a typical clay particle.

3-6. Referring to the attraction that typically exists between water and clay particles, what is adsorbed water?

3-7. Describe what is meant by the dipole nature of a water molecule. How is this related to adsorbed water and plasticity in a clay soil?

3-8. Relating to particles of clay and adsorbed water, what is the diffuse double layer?

3-9. Why does the presence of water in a soil typically have a much greater effect on clays than on sand or gravel materials?

3-10. Expansive (swelling) clays exist in various areas across North America. What clay mineral is typically associated with the swelling condition, and what is the explanation for the expansion? Also, identify some of the problems caused in structures by expanding clays.

3-11. In a general way, the strength of a coarse-grained soil is related to the soil deposit's structure (or particle orientation) and void ratio (or density). What is this relationship?

3-12. With reference to coarse-grained soils, what type of particle structure is susceptible to having large volume changes occur?

3-13. Provide a probable description for the soil deposit that could become a collapsible gravel.

3-14. Briefly describe the phenomenon of liquefaction in soil deposits, indicating type and condition of soil required and occurrences responsible.

3-15. (a) Briefly describe the difference between a flocculent structure and a dispersed structure in clay soils.
   (b) What type of structure is most expected in a clay deposit that has formed in a naturally occurring underwater environment?
   (c) What type of structure would be expected where a clay had been used for a compacted fill on a construction project?

3-16. With regard to the thixotropy phenomenon in clay soils, what is it and what causes it?

3-17. Provide a brief description of dispersive clays and the typical construction-related problems known to develop.

3-18. What are laterites (or lateritic soils), and why are such soils considered in the category of requiring special consideration on construction projects?

CHAPTER 4

Index Properties and Classification Tests, and Soil Classification Systems

As an aid to the engineering profession and the design–construction field, soils have been divided into basic categories or classifications based on certain physical characteristics. But because of the range of characteristics for the different soil variations that exist in nature, classification categories have been, of necessity, relatively broad in scope. All of a soil's properties are not checked to obtain a classification. Consequently, for proper evaluation of a soil's suitability for construction or foundation use, information about its properties in addition to classification is frequently necessary. Those properties that do help to define a soil's engineering qualities and that are used to assist in determining accurate classification are termed *index properties*. The tests necessary to determine index properties are *classification tests*. Index properties include those characteristics that can be determined relatively quickly and easily and that will have bearing on items of engineering importance such as strength or load-supporting ability, tendency to settle or expand, and effect of water and freezing conditions.

Standard Test Procedures. For a test-determined property or identification to be meaningful, the *method* used for testing needs to be standardized in regard to procedure and to the equipment or apparatus unique to the procedure. When soil and rock properties are to be determined for construction projects in the United States, the testing procedures presented in the American Society for Testing and Materials (ASTM) publications[1] and American Society of State Highway and Transportation Officials (AASHTO) publications[2] are probably the most frequently referenced; ASTM standards are considered universal while the AASHTO standards typically relate to transportation projects. Many of the indi-

---

[1]*Annual Book of ASTM Standards,* Volume 04.08, *Soil and Rock* (I) and 04.09, *Soil and Rock* (II). ASTM, West Conshohocken, PA 19428.

[2]*Standard Specifications for Transportation Materials and Methods of Sampling and Testing.* AASHTO, Washington, D.C. 20090.

vidual procedures for testing presented by these two organizations are identical. Typically, a proposed procedure is adopted as a standard by ASTM or AASHTO only after a thorough review and consensus by appropriate researchers, educators, and practitioners. Because of the applicability, ASTM and AASHTO standards are also referenced internationally. Additionally, testing procedures established by some governmental agencies associated with construction work, as well as organizations outside the United States, often carry the designation of the individual agency or organization, but the test methods are identical to the ASTM and AASHTO procedures.

4.1 INDEX PROPERTIES

Index properties refers to those properties of a soil that indicate the type and condition of the soil, and provide a relationship to structural properties, such as the strength and the compressibility or tendency for swelling and permeability.

Generally, for coarse-grained soils, properties of the particles and the relative state of compaction are most significant. For fine-grained soils, the consistency (firm or soft) and plasticity are particularly important. The index properties that provide the desired information for coarse-grained and fine-grained soils defined in Chapter 3 are summarized in Table 4-1.

Table 4-1 Index Properties and Related Classification Tests

Soil Type	Index Property	Classification Test
Coarse-grained	Range of particle sizes and distribution of sizes	Particle-size distribution (mechanical analysis) by sieving, or sedimentation test
	Shape of particles	Visual
	Presence of fine-grained particles	From mechanical analysis (usually from use of a fine-mesh sieve)
	In-place density and relative state of compaction	In situ density determination, and relative density test
	Classification	From mechanical analysis, or visual identification based on grain size
Fine-grained	Consistency (strength and type of structure in the undisturbed state)	Field or laboratory evaluation of unconfined compressive strength or shear strength (cohesion)
	Change in consistency due to remolding	Unconfined compressive strength or cohesion for the remolded soil
	Water content	Water content
	Plasticity	Atterberg limits (liquid limit and plastic limit)
	Classification	From visual identification and Atterberg limits
	Presence and type of clay	Indirectly from determination of plasticity and change in consistency, and/or directly from a clay mineral analysis

It should be recognized that in studies and analyses for construction projects it frequently is not necessary to determine all the index properties for the soil. Properties to be determined relate to the information that is needed and how such information eventually is to be used. For example, a clay mineral analysis requires very specialized equipment and is not performed in studies for foundation designs, unless the conditions are unusual.

For organic soils, it is important to know of the presence and at least the approximate amount of organic material because of its influence on compressibility and strength.

For all soils, the description should include the color. Color may have bearing on the mineralogical composition and is also extremely useful for determining homogeneity of a soil deposit and as an aid for identification and correlation during field construction.

4.2 CLASSIFICATION TESTS

Particle Size Distribution (Mechanical Analysis)

This classification test determines the range of size of particles in the soil and the percentage of particles in each of the sizes between the maximum and the minimum.

Two methods are in common use for obtaining the necessary information. Sieving is generally used for coarse-grained soils, and a sedimentation procedure is used for analyzing fine-grained soils. Sieving is a most direct method for determining particle sizes, but there are practical lower limits to sieve openings that can be used for soils. This lower limit is approximately at the smallest size attributed to sand particles. Information on sieves in common use is shown in Table 4-2.

In the sieve analysis, a series of sieves (screens) having different-sized openings are stacked with the larger sizes over the smaller (Fig. 4-1). The soil sample being tested is dried, clumps are broken, and the sample is passed through the series of sieves by shaking. Larger particles are caught on the upper sieves, and the smaller particles filter through to be caught on one of the smaller underlying sieves. The weight of material retained on each sieve is converted to a percentage of the total sample. The resulting data are conventionally presented as a grain- or particle-size distribution curve plotted on semilog coordinates, where the sieve size opening is on a horizontal *logarithmic* scale, and the percentage (by weight) of the size smaller than a particular sieve opening is on a vertical *arithmetic* scale.

Table 4-2 Common Sieve Types and Mesh Openings

Sieve Size Designation	U.S. Standard		Tyler Standard		British Standard	
	in.	mm	in.	mm	in.	mm
#4	0.187	4.76	0.185	4.70	—	—
#8	0.0937	2.38	0.093	2.362	0.081	2.057
#10	0.0661	1.68	0.065	1.651	0.0661	1.676
#20	0.0331	0.84	0.0328	0.833	—	—
#40	0.0106	0.42	—	—	—	—
#60	0.0098	0.25	0.0097	0.246	0.0099	0.251
#100	0.0059	0.149	0.0058	0.147	0.0060	0.152
#200	0.0029	0.074	0.0029	0.074	0.0030	0.076
#270	0.0021	0.053	0.0021	0.053	—	—
#400	0.0015	0.037	0.0015	0.038	—	—

Figure 4-1 Set of sieves being assembled. Sieves shown are U.S. Standard sieves, typically used in laboratories.

Results may be presented in tabular form also. A detailed procedure for performing the sieve analysis is presented in the *ASTM* (American Society for Testing and Materials) *Testing Manual,* under ASTM Test Designation D-422. A typical presentation is shown in Figure 4-2. Note the "reversal" of the logarithmic scale (size increases from right to left).

Most soil grains are not of an equal dimension in all directions. Hence, the size of a sieve opening will represent neither the largest nor the smallest dimension of a particle, but some intermediate dimension. As an illustration, assume a brick-shaped particle whose length, width, and thickness are different. The dimension that controls whether or not the particle passes through the sieve opening is the intermediate dimension, assuming that the particle is aligned so that the greatest dimension is perpendicular to the sieve opening.

Sieve tests can be performed in a laboratory or in the field (at the area being explored, such as a proposed borrow pit or construction site).

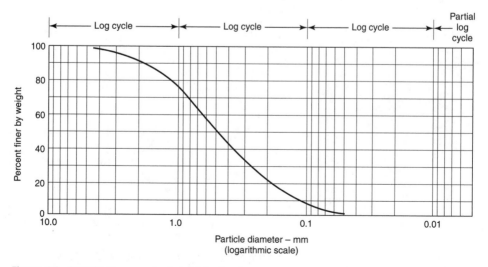

Figure 4-2 Particle-size or grain-size distribution curve.

The appearance of the particle-size distribution plot depends on the range and amounts of the various sizes of particles in the soil sample. These, in turn, have been affected by the soil's origin or the method of deposition. Well-graded soils (a distribution of particles over a relatively large range of sizes) produce a longish straight curve (Fig. 4-3a). A uniform soil (soil having most of the particles of approximately similar size) plots as shown in Figure 4-3b. A gap-graded soil (an absence of intermediate sizes) plots as in Figure 4-3c.

The grain-size plot can provide an indication of a soil's history. A residual deposit has its particle sizes constantly changing with time as the particles continue to break down, and typically produces grain-size curves as shown in Figure 4-4. The curves shown in Figure 4-5a and b represent glacial and glacial–alluvial deposits. River deposits may be well graded, uniform, or gap graded, depending on the water velocity, the volume of suspended solids, and the river area where deposition occurred.

Certain properties of *clean sands* have been related to particle diameters. The *effective size* of a sand is taken as the particle size corresponding to the 10 percent passing size from the grain-size curve, and is indicated as D_{10}. It is this size that is related to permeability and capillarity. It has also been found that the D_{15} particle size can be related to soil permeability. Another relationship, the ratio D_{60}/D_{10}, is termed the *uniformity coefficient, C_u,* and provides a comparative indication of the range of particle sizes in the soil (Fig. 4-6). Sand having a wide range of particle sizes is considered well graded and has C_u values greater than 10. A uniform soil has C_u values less than about 5.

The procedure commonly used for obtaining particle-size distribution information for silts and clays is the sedimentation method. In this method, the soil is placed into solution

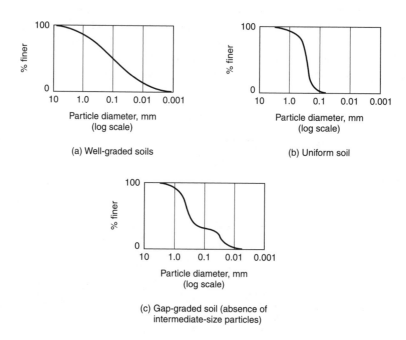

Figure 4-3 Shape of particle-size distribution curve expected for described soil type.

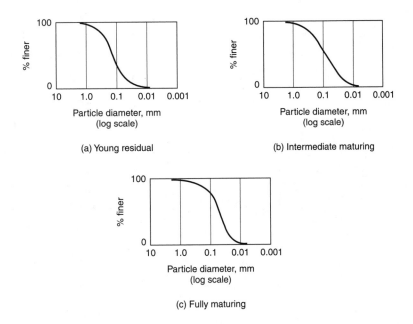

Figure 4-4 Typical particle-size curves for residual soils.

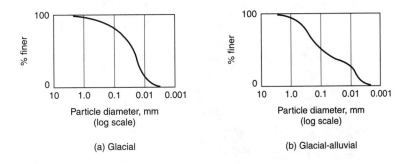

Figure 4-5 Typical particle-size curves for transported soils.

with distilled water, and the soil particles are permitted to settle out of the solution. As settling occurs, the average specific gravity of the solution decreases. Readings of specific gravity or related property by use of a
hydrometer, made at different time intervals, provide an indication of the
weight of soil remaining in solution and also information on the sizes of particles that have settled out of the solution (Fig. 4-7). Most conventionally, the
test data are reduced to provide particle diameters and the percentage
(weight) that is finer than a particular size by using the Stokes equation for
spheres falling freely in a fluid of known properties. This application is not
absolutely correct, since most fine-grained soil particles are not round; in

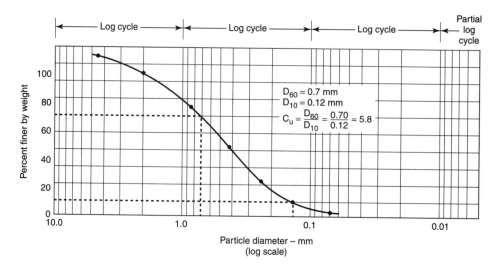

Figure 4-6 Determining uniformity coefficient from particle-size curve.

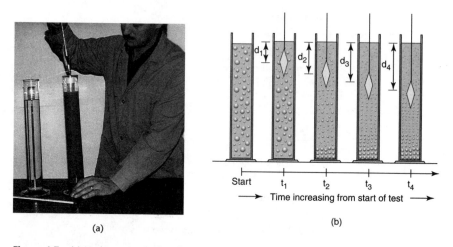

Figure 4-7 (a) Hydrometer used to determine specific gravity of a soil–water solution being inserted into the solution; (b) diagram showing change in hydrometer depth as soil particles settle out of solution.

fact, most clay particles are flat or plate shaped. The method is, therefore, more applicable to silts than to clays. Nevertheless, the method is in wide use, for it is felt to be a practical way to obtain reasonable approximations of the particle-size distribution for fine-grained soils. Resulting effects are not serious, since particle-size distribution is not used for evaluation of the significant engineering properties of fine-grained soils. Details on the method to determine the particle-size distribution by using the sedimentation test (hydrometer method) are described by ASTM Test Designation D-422.

In-Place Density, In-Place Unit Weight

The term *in-place density* refers to the volumetric weight, usually expressed as kilonewtons per cubic meter (kN/m^3), megagrams per cubic meter (Mg/m^3), or pounds per cubic foot (pcf) of a soil in the undisturbed (or in situ) condition or in a compacted fill. (Actually, pcf and kN/m^3 indicate unit weight, whereas Mg/m^3 indicates density. The term *in-place density* has been widely used for years to refer to the field procedure for determining volumetric weight, but with the advent of the SI system, improper use of weight and mass terms has become more noticeable; correct usage is evolving slowly. For reference, the density of water is $1 \, Mg/m^3$, that is, $1,000,000 \, gm/m^3$.) Generally, for the coarse-grained soils, the greater the density, or unit weight, the better the shear strength and the lesser the tendency for compression (settlement). In-place density or unit weight determinations are made of borrow pit soils so as to estimate the volume of shrinkage or swell that will occur as the soil is transported or compacted in place at a fill location. Where compacted earth fills are being constructed, it is standard practice to make in-place density or unit weight determinations of the soil after it is placed to determine whether the compaction effort has been adequate or if more compaction is required.

Of the equipment and methods used for making in-place density or unit weight determinations,[3] the sand cone method and the rubber balloon method have a long history of use (Fig. 4-8). With these methods, a small test area is selected and a volume of compacted soil is dug up and weighed. The sand cone equipment or the balloon equipment is used to determine the volume of the dug hole. Knowing soil weight (or mass) and corresponding volume permits the unit weight (or density) to be calculated.

A modern development for making in-place density and unit weight determinations involves the use of nuclear equipment (Fig. 4-9). Through controlled use of a nuclear material, gamma rays (photons) are emitted into the tested soil. These photons collide with electrons in the soil materials, some being scattered and some being absorbed. The quantity of photons reaching a detection device (part of the test equipment) relates to the soil

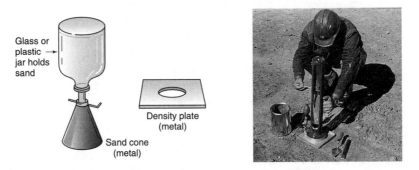

Figure 4-8 Sand-cone and balloon apparatus for determining in-place density. (Photo courtesy of ELE International, Soiltest Products Division)

---

[3]More detailed discussion of the procedures and equipment for performing in-place density testing is included in Chapter 14.

Figure 4-9 (Left) Representative in-place density equipment (left to right): nuclear moisture-density gage, standard sand cone, balloon apparatus, and NYS-DOT sand cone apparatus. (Right) Nuclear moisture–density meter in use. (Courtesy of Troxler Electronic Laboratories, Inc., North Carolina)

density. To determine water content, a neutron-emitting material and detector are used. Compared to the above mentioned methods of determining in-place density, a significant advantage with the nuclear method is the rapid speed with which results are obtained. Detailed information on nuclear moisture–density test equipment is included in Chapter 14.

Figure 4-9 shows a grouping of in-place density or unit weight equipment, for purpose of comparison.

Relative Density

For a granular soil, the shear strength and resistance to compression are related to the density (or unit weight) of the soil; higher shear strength and more resistance to compression are developed by the soil when it is in a dense or compact condition (high density) than when it is in a loose condition (low density). In a dense condition, the void ratio is low; in a loose condition, the void ratio is high. To evaluate the relative condition of a granular soil, the in-place void ratio can be determined and compared to the void ratio when the soil is in its densest condition and when it is in the loosest condition (Fig. 4-10). This comparison is the relative density D_R. Relative density is expressed as a percentage. High values indicate a dense or compact material; low values represent a loose material.

$$D_R\% = \frac{e_{max} - e_0}{e_{max} - e_{min}} \times 100\% \tag{4-1}$$

where e_{max} = void ratio of the soil in its loosest condition
$\quad\quad e_{min}$ = void ratio of the soil in its densest condition
$\quad\quad\;\; e_0$ = void ratio of the soil in the natural condition or condition in question

The maximum density (or minimum void ratio) is determined in the laboratory by compacting the soil in thin layers in a container of known volume and subsequently weighing the soil. The compaction is achieved by applying a vibratory and compressive force simultaneously.

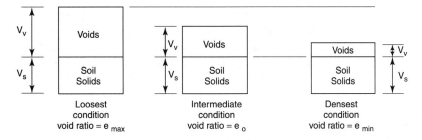

Figure 4-10 Relative conditions of a granular soil.

The compressive force needs to be sufficient to compact the soil without breaking the individual particles of soil. Because of the irregular size and shape of granular particles, it is not possible to obtain a zero volume of void spaces. Practically, there will always be voids in a soil mass.

The minimum density (or maximum void rate) can be determined in the laboratory by carefully letting the soil slowly flow into the test container through a funnel. When this task is carefully performed, the soil will be deposited and remain in a loose condition, from whence the loose density and void ratio can be calculated.

In terms of dry unit weight and dry density (which are more convenient to work with), relative density is

$$D_R\% = \left[\frac{\dfrac{1}{\gamma_{min}} - \dfrac{1}{\gamma_0}}{\dfrac{1}{\gamma_{min}} - \dfrac{1}{\gamma_{max}}} \right] (100\%) \tag{4-2a}$$

$$D_R\% = \left[\frac{\dfrac{1}{\rho_{min}} - \dfrac{1}{\rho_0}}{\dfrac{1}{\rho_{min}} - \dfrac{1}{\rho_{max}}} \right] (100\%) \tag{4-2b}$$

where $\gamma_{min}(\rho_{min})$ = dry unit weight (density) in the loosest condition
$\gamma_{max}(\rho_{max})$ = dry unit weight (density) in the densest condition
$\gamma_0(\rho_0)$ = dry unit weight (density) in the condition in question

Typical relative density values are presented in Table 4-3.

Table 4-3 Representative Values of Relative Density

Descriptive Condition	Relative Density, %	Typical Range of Unit Weight and Density		
		pcf	kN/m³	Mg/m³
Loose	< 35	< 90	< 14	< 1.4
Medium dense	35–65	90–110	14–17	1.4–1.7
Dense	65–85	110–130	17–20	1.7–2.0
Very dense	> 85	> 130	> 20	> 2.0

Illustration 4.1

An undisturbed sample of fine sand is tested in the laboratory and found to have a dry mass of 3.63 kg (dry weight of 8 lb), a total volume of 0.00198 m³ (0.07 ft³), and a specific gravity G_s of 2.70. Other laboratory tests are performed to determine the maximum and minimum density for the sand. At the maximum density, it is determined that the void ratio is 0.35; at the minimum density the void ratio is 0.95. Determine the relative density of the undisturbed sample.

Solution

SI Units

Void ratio of undisturbed sample e_0:

$$V_T = .00198 \text{ m}^3$$

$$V_s = \frac{M_s}{G_s \rho_w} = \frac{3.63 \text{ kg}}{(2.70)(10^3 \text{ kg/m}^3)} = .00134 \text{ m}^3$$

$$V_v = V_T - V_s = .00198 \text{ m}^3 - .00134 \text{ m}^3 = .000636 \text{ m}^3$$

$$e_0 = \frac{V_v}{V_s} = \frac{.636 \times 10^{-3} \text{ m}^3}{1.34 \times 10^{-3} \text{ m}^3} = .475$$

$$D_R\% = \frac{e_{max} - e_0}{e_{max} - e_{min}}(100\%) = \frac{(.95 - .475)}{(.95 - .35)}(100\%) = 79\%$$

$$\text{Dry density} = \frac{M_s}{V_T} = \frac{3.63 \text{ kg}}{.00198 \text{ m}^3} = 1.83 \text{ Mg/m}^3$$

U.S. Customary Units

Void ratio of undisturbed sample e_0:

$$V_T = .07 \text{ ft}^3$$

$$V_s = \frac{W_s}{G_s \gamma_w} = \frac{8\text{lb}}{(2.70)(62.4 \text{ pcf})} = .0474 \text{ ft}^3$$

$$V_v = V_T - V_s = .070 - .0474 = .0226 \text{ ft}^3$$

$$e_0 = \frac{V_v}{V_s} = \frac{.0226 \text{ ft}^3}{.0474 \text{ ft}^3} = .476$$

$$D_R\% = \left(\frac{e_{max} - e_0}{e_{max} - e_{min}}\right)(100\%) = \left(\frac{.95 - .476}{.95 - .35}\right)(100\%) = 79\%$$

$$\text{Dry unit weight of soil} = \frac{W_s}{V_T} = \frac{8.0 \text{ lb}}{.07 \text{ ft}^3} \approx 115 \text{ pcf.}$$

In its natural condition, the soil is probably dense (see Table 4-3).

Water Content

For coarse- and fine-grained soils, water content can have a significant effect on the soils' behavioral properties when used for construction purposes. By definition and as previously described, water content (*w*) is the ratio of the weight of water in a soil to the dry weight of the material. As a result, in laboratory and field work the weight of water in the test sample has to be determined. Conventionally this is done by drying the original wet sample, recording the wet and dry weights, and performing the required calculation.

Various other methods are available for drying soil samples to determine water content, such as by using gas pressure extractors or infrared-based equipment. Another device, the Speedy Moisture Tester, bases its operation on the reaction that occurs between a carbide reagent and soil moisture to determine a soil's water content. The wet soil sample is placed in a sealed container with calcium carbide, and the pressure generated by the vaporized moisture is then related to water content. This method provides a rapid procedure for determining moisture content and is finding increased usage in laboratory and field work.[4]

Consistency of Clays

Consistency refers to the texture and firmness of a soil and is often directly related to the strength. Consistency is conventionally described as *soft, medium stiff* (or *medium firm*), *stiff* (or *firm*), or *hard*. These terms, unfortunately, are relative and have different meanings to different observers. For standardization, it is reasonable and practical to relate consistency to strength. With clays, shear strength is discussed in terms of cohesion and unconfined compressive strength. The unconfined compressive strength is obtained by imposing an axial load to the ends of an unsupported cylindrical sample of clay and determining the load that causes shear failure in the sample. Unconfined compression tests can be performed in the laboratory or in the field. The unconfined compressive strength is, under practical conditions, twice the cohesion (or shear strength) of a clay soil. The cohesion can be determined in a laboratory strength test. For a consistency classification in the field or laboratory, special soil testing equipment such as a vane-shear device or pocket penetrometer provides a means of making quick and easy determinations (Fig. 4-11). A tabulation of strength values for various consistency terms is shown in Table 4-4.

It is well established that the strength of a clay soil is related to its structure. If the original structure is altered because of changes in particle arrangement (from reworking or remolding) or chemical changes, the strength of the altered clay is less than the original strength. *Sensitivity* is the term that provides an indication of remolded strength related to

---

[4]Somewhat unfortunately, the Speedy Moisture reading is expressed as a percentage of the soil's wet weight. Conversion to moisture content by dry weight can be performed using the relationship

$$w\% = \frac{w_{sp}}{1 - w_{sp}} \times 100\%$$

where w_{sp} is the moisture content indicated by the Speedy device, expressed as a decimal.

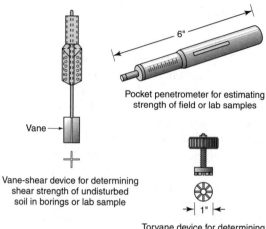

Figure 4-11 Testing equipment for making strength or consistency determinations on cohesive soils.

Table 4-4 Consistency and Strength for Cohesive Soils

Consistency	Shear Strength, t/ft² or Kg/cm² (kN/m² or kPa)	Unconfined Compressive Strength, t/ft² or Kg/cm² (kN/m² or kPa)	Feel or Touch
Soft	< 0.25 (< 24)	< 0.5 (< 48)	Blunt end of pencil-sized object makes deep penetration easily.
Medium (medium stiff or medium firm)	0.25–0.50 (24–48)	0.50–1.0 (48–96)	Blunt end of pencil-sized object makes half-inch penetration with moderate effort.
Stiff (firm)	0.50–1.0 (48–96)	1.0–2.0 (96–190)	Blunt end of pencil-sized object makes moderate penetration (about ¼ in.).
Very stiff (very firm)	1.0–2.0 (96–190)	2.0–4.0 (190–380)	Blunt end of pencil-sized object makes slight indentation; fingernail easily penetrates.
Hard	> 2.0 (> 190)	> 4.0 (> 380)	Blunt end of pencil-sized object makes no indentation; fingernail barely penetrates.

original strength. When remolded, the strength of a clay is affected by the water content. At lower water contents, strength is generally greater. However, sensitivity should be based on comparison of remolded to undisturbed soil strength at an identical water content.

$$\text{Sensitivity } S_t = \frac{\text{Unconfined compressive strength, undisturbed clay}}{\text{Unconfined compressive strength, remolded clay}} \quad (4\text{-}3)$$

For most clays, sensitivities range between 2 and 4. Clays considered *sensitive* have S_t values between 4 and 8. Clays classified as *extrasensitive* have values between 8 and 16. Clays with sensitivity values greater than 16 are classified as *quick* clays. Such clays are very unstable. Normally, clays with a high degree of sensitivity possess a very flocculent structure in the undisturbed condition.

Consistency in the Remolded State and Plasticity

In the remolded state, the consistency of a clay soil varies in proportion to the water content. At higher water content, the soil–water mixture possesses the properties of a liquid; at lesser water contents, the volume of the mixture is decreased and the material exhibits the properties of a plastic; at still lesser water contents, the mixture behaves as a semisolid and finally as a solid.

The water content indicating the division between the liquid and plastic state has been designated the *liquid limit*. The water content at division between the plastic and semisolid state is the *plastic limit*. The water content at the division between the semisolid and the solid state is the *shrinkage limit*.

At water contents above the shrinkage limit, the total volume of the soil–water mixture changes in proportion to change in water content. Below the shrinkage limit, there is little or no change in volume as water content varies (Fig. 4-12).

Definition of the various states of consistency and the establishment of criteria to determine these various states were first formally proposed by A. Atterberg, a Swedish soil scientist, in the early twentieth century. Initially intended for agricultural use, the method has been adapted for engineering use in classifying soils. Although the limit values (water contents) have little direct meaning insofar as engineering properties of soils are concerned, correlations between liquid or plastic limits and engineering properties have been established over the years to aid in evaluating a soil for use as structural fill (dams, embankments, landfills), for highway construction, and for building support.

The liquid limit is taken as the water content at which the soil "flows" (i.e., the condition where a very viscous liquid shears). Special equipment and procedures, defined in ASTM Test Designation D-4318, are required for determining the liquid limit (Fig. 4-13). Relating to the test procedure, the shearing resistance of all soils (free of organic content) is similar when at the liquid limit (near 2 kPa). The plastic limit is the water content when the soil can just be rolled into a $\frac{1}{8}$-inch-diameter thread before crumbling, as described in ASTM Test Designation D-4318.

Various designations have been used to indicate the liquid and plastic limit. Using LL and PL for liquid limit and plastic limit, respectively, provides an easily understood terminology that has little chance of misinterpretation.

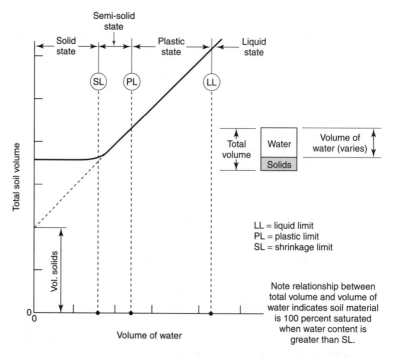

Figure 4-12 Variation of total soil volume with change in water content for a fine-grained soil (for soil in the remolded state).

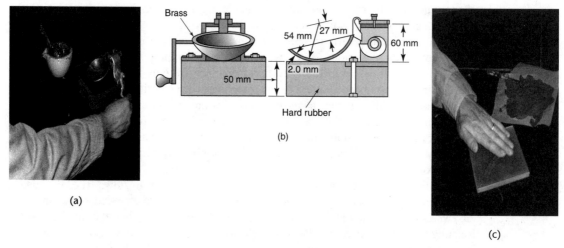

Figure 4-13 (a) Liquid limit trial being performed; (b) details of liquid limit device; (c) soil thread being rolled to determine plastic limit.

The *plasticity index* (PI) is the numerical difference of the liquid and plastic limits, and indicates the range of water content through which the soil remains plastic. For proper evaluation of a soil's plasticity properties, it has been found desirable to use both the liquid limit and the plasticity index values. Engineering soil classification systems use these values as a basis for classifying the fine-grained soils.

For fine-grained soils, determining the natural water content (the water content of a soil in an undisturbed condition in the ground) and relating it to the plastic and liquid limits can provide an indication of the soil's consistency and/or sensitivity potential. One such relationship is the *liquidity index,* LI:

$$LI = \frac{w\% - PL\%}{LL\% - PL\%} = \frac{w\% - PL\%}{PI\%} \tag{4-4}$$

where w is the natural water content of the soil. A value less than 1 indicates that the natural water content is less than the liquid limit. A very low value for the LI, or a value near zero, indicates that the water content is near the plastic limit, where experience has shown that the sensitivity will be low and the cohesive strength relatively high (a stiff or hard consistency). As the natural water content approaches or exceeds the liquid limit, the sensitivity increases. The undisturbed strength still very much depends on the undisturbed structure. Negative values of the LI are possible and normally indicate a desiccated (dried), hard soil.

Presence of Clay Minerals

The presence of even small amounts of certain clay minerals in a soil mass can have significant effect on the properties of the soil. Identifying the type and amount of clay minerals may be necessary in order to predict the soil's behavior or to develop methods for minimizing detrimental effects.

The identification of clay minerals requires special techniques and equipment and trained personnel. Many different techniques are available. Some are useful for identifying only a particular type of clay, whereas other methods are suitable for identifying several types of minerals. These techniques include microscopic examinations, X-ray diffraction, differential thermal analysis, infrared absorption, optical property determination, and electron micrography. Description of these techniques and equipment is beyond the scope of this text; further information is available in publications dealing with clay mineralogy. Even with the techniques available to today's soil scientists, the accurate determination of some clay minerals is not possible. Generally, however, qualitative if not quantitative identifications can be made that are adequate for many engineering problems.

A somewhat indirect method of obtaining information on the type and effect of clay minerals in a soil is to relate plasticity to the quantity of clay-sized particles. It is known that for a given amount of clay mineral the plasticity resulting in a soil will vary for the different types of clays. One relationship is *activity,* defined as

$$\text{Activity} = \frac{\text{Plasticity index, \%}}{\text{Percentage of clay sizes, \%}} \tag{4-5}$$

For this analysis, the percentage of clay sizes is that portion of the soil, by weight, consisting of particle sizes below 0.002 mm. Such information is available from the conventional hydrometer analysis used to determine a particle-size distribution. Activity, therefore, can be determined from standard laboratory tests. Clay materials with kaolinite, a stable clay mineral, will have low activity, whereas those soils with montmorillonite, known to be a type subject to large volume changes depending on available water, will have a high activity value. A relative activity classification is as follows:

Activity	Classification
< 0.75	Inactive clays
0.75–1.25	Normal clays
> 1.25	Active clays

The general relationship between clay's potential for swelling or expansion (high, medium, low) and information about activity, clay particle sizes and plasticity can be indicated diagrammatically as shown in Figure 4-14. The volume expansion associated with the various categories on the diagram is, crudely, as follows:

Category	Potential Volume Expansion
Very high	>10%
High	5–10%
Medium	2–5%
Low	<2%

Testing for Dispersive Clays

On the basis of testing many clay soils for dispersion (dispersive clays as described in Chapter 3), the relationship between the presence of ions and susceptibility to defloccula-tion shown in Figure 4-15 has been proposed. Atomic absorption spectroscopy or flame photometry is used to determine the ion concentrations. Unfortunately, a clay's tendency to disperse is not indicated by conventional soil classification tests such as the Atterberg limits. The atomic absorption spectroscope and flame photometer are not ordinarily found in soil mechanics laboratories; simple *qualitative* tests, using standard laboratory equipment, capable of being performed in the field have been developed and are outlined next. Judgment is involved when evaluating some of the results, but generally the tests are good indicators.

1. *Soil Conservation Service Dispersion Test* (ASTM D4221): This procedure, also known as the *double hydrometer test,* compares the percentage of soil sample particles smaller than 0.005 mm as determined by a conventional hydrometer analysis (where a deflocculating agent is used and the soil sample is mechanically agitated prior to test) and as determined when a similar sample is exposed to a plain distilled water environment without being agitated, whence

$$\% \text{ dispersion} = \frac{\% \text{ finer than 0.005 mm in plain water test}}{\% \text{ finer than 0.005 mm in conventional hydrometer test}}$$

A percentage dispersion greater than 35 percent indicates a dispersive clay soil.

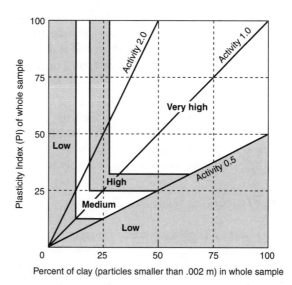

Figure 4-14 Volume change potential classification for clay soils (Van der Merwe, or South African, method). (Modified from NAVFAC DM-7.01, 1986 [311])

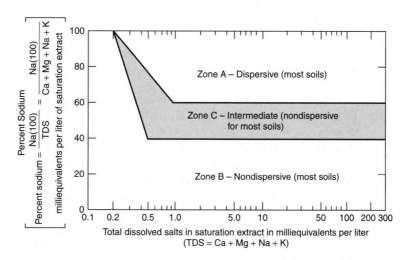

Figure 4-15 Relationship of pore water salts and soil dispersion. (After Sherard, Dunnigan, and Decker [265]).

2. *Crumb Test:* A small sample of soil (10 mm or less in diameter) preserved at the natural water content is placed in a beaker of distilled water. The reaction in terms of a colloidal cloud around the crumb is an indication of the soil's tendency to disperse; no cloud indicates a nondispersive material.

3. *Pinhole Test* (ASTM D4647): The test soil is compacted, and distilled water is then made to flow through a 1-mm-diameter hole in the sample. The test can be set up in the permeameter apparatus used for permeability determinations, using pea gravel filters at the ends of the test sample. With a dispersive clay, the hole erodes and the flowing water discharge is colored, whereas for a nondispersive soil the hole does not enlarge and the water discharge remains clear.

Other Properties

Where a soil is to be used to support building foundations or other types of buried structures (such as pipelines, tunnels, or storage vaults), it is good practice to have pH and soluble sulfate determinations made, so as to obtain information on the corrosion potential of buried metal (piles or piping) or the potential deteriorating effect on concrete foundations.

4.3 CLASSIFICATION SYSTEMS

Soil classification systems have been devised primarily to facilitate the transfer of information between interested parties. In the engineering and construction field, the broad general properties of concern relate to the performance or usefulness for supporting structures and the handling or working qualities of a soil. Because of the wide variation in properties and behavior of soils, classification systems generally group together in relatively broad categories the soils that have similar features or properties that are considered to be of importance. As a result, a classification system is not necessarily an identification system in which all pertinent engineering properties of a material are determined. Because of this, soil classification should not be used as the sole basis for design or construction planning.

Historically, the most widely used method of classifying soils has been through visual identification, the size of soil grains and the plasticity of the soil being used as the basis for indicating the soil type. To a great degree, the refined and more recently developed classification systems that rely on laboratory-determined properties for accurate identification still use these two criteria as the basis for indicating soil type.

The desirable requirements for a satisfactory engineering soil classification system include the following:

1. There should be a limited number of different groupings, so that the system is easy to remember and use. Groupings should be on the basis of only a few similar properties and generally similar behavioral characteristics.

2. The properties and behavioral characteristics should have meaning for the engineering and construction profession. Generally, the properties should, at least crudely, relate to a soil's handling characteristics, shear strength, volume change characteristics, and permeability.

3. Descriptions used for each grouping should be in terms that are easily understood and are in common use for indicating the soil type and its properties. Symbols or shorthand used to describe the grouping should easily relate to the soil type. Coded symbols are not desirable.

4. Classification into any grouping should be possible on the basis of visual identification (generally limited to differentiating between particle sizes, coarse- and fine-grained soils, and plasticity) without special tests or equipment being necessary.

The Unified Soil Classification System satisfies the above requirements of a classification system and is the system that is coming into prevalent use in the engineering and construction fields. This system is shown in Figure 4-16. Classifications are on the basis of coarse- and fine-grained soils, and retain the four common groupings of soil—gravel, sand, silt, and clay. The symbols are easily associated with the classification, being simply the first letter of the soil type (except for silt, which has the designation M, from *mo,* the Swedish word for silt). Refinement in grouping is based upon a coarse soil's being well or poorly graded and a fine-grained soil's being of a high or low plasticity. What at first appear to be a large number of groupings are in fact very logical categories. Experience has proved that previously untrained personnel very quickly learn the system and use it with accuracy. The Unified System includes the use of a plasticity chart for aiding classification of fine-grained soils.

Another system familiar to those in the engineering and construction fields is the AASHTO Classification System (Fig. 4-17), which has been in national use for highway- and transportation-related construction projects. This system classifies soil material into eight groups, with assignment to a particular group based on determination of particle size distribution, liquid limit and plasticity index, and presence of organic material. A plasticity chart aids identification of the fine-grain soil faction. Comparisons of soils within the *same group* are made from a group-index value (a numerical value calculated by use of the empirical formula included in Fig. 4-17). The criteria for the groupings are logical, essentially rating the suitability for use as a pavement base or subgrade material, but laboratory testing is required in order to determine a classification. A generalized comparison of the Unified and AASHTO systems is presented in Table 4-5. (Note: This AASHTO Classification System is designated in both the AASHTO Specifications and the ASTM Standards; the Unified Soil Classification System carries an ASTM Designation but is not included in the AASHTO Specifications.)

The U.S. Department of Agriculture (Soil Conservation Service) textural classification system is a third system commonly encountered by planners, designers, and construction personnel who also work with government soil maps and agriculture-related organizations. Identifying soil as one of the categories shown on the classification chart in Figure 4-18 is done on the basis of texture (that is, percentages of sand, silt, and clay particles) without consideration for other properties such as plasticity. The Department of Agriculture textural classification is determined by following the percentage lines (horizontal for clay, diagonal for silt and sand) to a point of common intersection. To illustrate, a soil sample composed of 45 percent sand particles, 40 percent silt, and 15 percent clay is a *loam.*

The textural classification chart is based only on the percentages of sand, silt, and clay in the sample; the effect of gravel and larger materials is deducted. For example, a soil

Soil Classification Chart

Criteria for Assigning Group Symbols and Group Names Using Laboratory Tests[A]				Soil Classification	
				Group Symbol	Group Name[B]
COARSE-GRAINED SOILS More than 50% retained on No. 200 sieve	Gravels More than 50% of coarse fraction retained on No. 4 sieve	Clean Gravels Less than 5% fines [C]	$C_u \geq 4$ and $1 \leq C_c \leq 3^E$	GW	Well-graded gravel [F]
			$C_u < 4$ and/or $1 > C_c > 3^E$	GP	Poorly graded gravel[F]
		Gravels with Fines More than 12 % fines [C]	Fines classify as ML or MH	GM	Silty gravel [F,G,H]
			Fines classify as CL or CH	GC	Clayey gravel[F,G,H]
	Sands 50% or more of coarse fraction passes No. 4 sieve	Clean Sands Less than 5 % fines [D]	$C_u \geq 6$ and $1 \leq C_c \leq 3^E$	SW	Well-graded sand[I]
			$C_u < 6$ and/or $1 > C_c > 3^E$	SP	Poorly graded sand[I]
		Sands with Fines More than 12 % fines [D]	Fines classify as ML or MH	SM	Silty sand[G,H,I]
			Fines classify as CL or CH	SC	Clayey sand[G,H,I]
FINE-GRAINED SOILS 50% or more pass the No. 200 sieve	Silts and Clays Liquid limit less than 50	inorganic	$PI > 7$ and plots on or above "A" line[J]	CL	Lean clay[K,L,M]
			$PI < 4$ or plots below "A" line[J]	ML	Silt[K,L,M]
		organic	$\dfrac{\text{Liquid limit} - \text{oven dried}}{\text{Liquid limit} - \text{not dried}} < 0.75$	OL	Organic clay[K,L,M,N] Organic silt[K,L,M,O]
	Silts and Clays Liquid limit 50 or more	inorganic	PI plots on or above "A" line	CH	Fat clay[K,L,M]
			PI plots below "A" line	MH	Elastic silt[K,L,M]
		organic	$\dfrac{\text{Liquid limit} - \text{oven dried}}{\text{Liquid limit} - \text{not dried}} < 0.75$	OH	Organic clay[K,L,M,P] Organic silt[K,L,M,Q]
HIGHLY ORGANIC SOILS	Primarily organic matter, dark in color, and organic odor			PT	Peat

[A]Based on the material passing the 3-in. (75-mm) sieve.

[B]If field sample contained cobbles or boulders, or both, add "with cobbles or boulders, or both" to group name.

[C]Gravels with 5 to 12% fines require dual symbols:
GW-GM well-graded gravel with silt
GW-GC well-graded gravel with clay
GP-GM poorly graded gravel with silt
GP-GC poorly graded gravel with clay

[D]Sands with 5 to 12% fines require dual symbols:
SW-SM well-graded sand with silt
SW-SC well-graded sand with clay
SP-SM poorly graded sand with silt
SP-SC poorly graded sand with clay

[E] $Cu = D_{60}/D_{10}$ $Cc = \dfrac{(D_{30})^2}{D_{10} \times D_{60}}$

[F] If soil contains ≥ 15% sand, add "with sand" to group name.

[G] If fines classify as CL-ML, use dual symbol GC-GM, or SC-SM.

[H] If fines are organic, add "with organic fines" to group name.

[I] If soil contains ≥ 15 % gravel, add "with gravel" to group name.

[J] If Atterberg limits plot in hatched area, soil is a CL-ML, silty clay.

[K] If soil contains 15 to 29% plus No. 200, add "with sand" or "with gravel," whichever is predominant.

[L] If soil contains ≥ 30 % plus No. 200, predominantly sand, add "sandy" to group name.

[M] If soil contains ≥ 30% plus No. 200, predominantly gravel, add "gravelly" to group name.

[N] PI ≥ 4 and plots on or above "A" line.

[O] PI < 4 or plots below "A" line.

[P] PI plots on or above "A" line.

[Q] PI plots below "A" line.

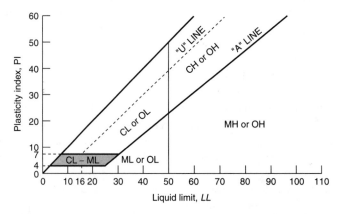

Figure 4-16(b) Plasticity chart for classification of fine-grained soils and fine-grained fraction of coarse-grained soils.

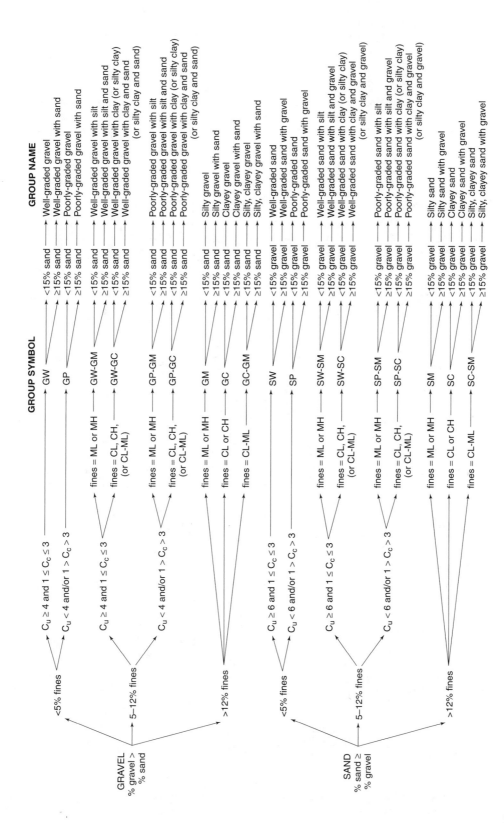

Figure 4-16(c) Flow chart for classifying coarse-grained soils (more than 50 percent retained on No. 200 sieve).

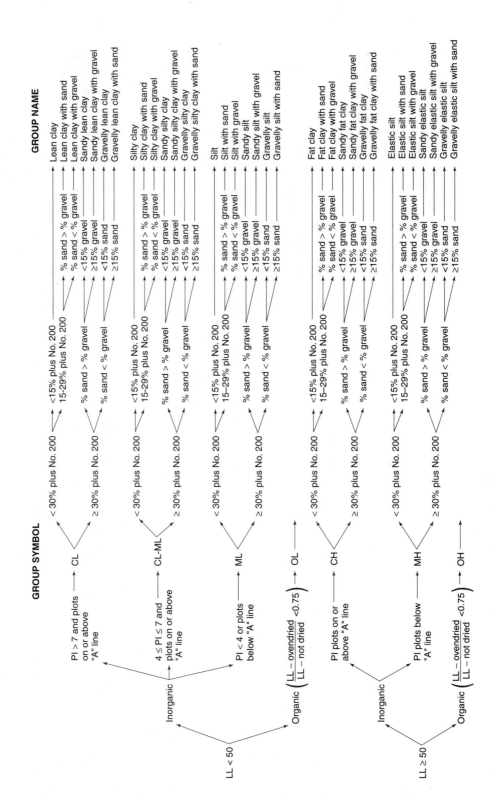

Figure 4-16(d) Flow chart for classifying fine-grained soil (50 percent or more passes No. 200 sieve).

114

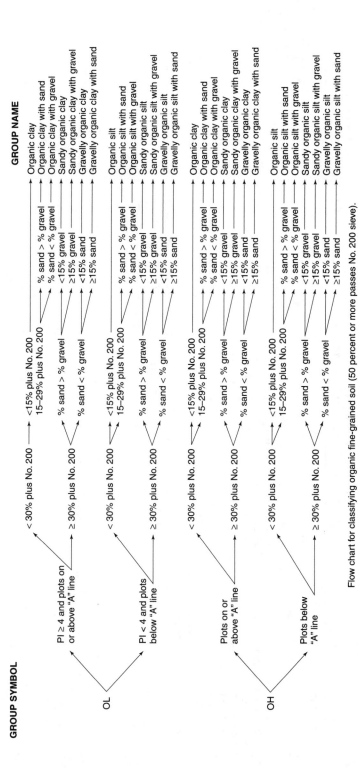

Figure 4-16(e)   Flow chart for classifying organic fine-grained soil (50 percent or more passes No. 200 sieve).

Figure 4-17(a) American Association of State Highway and Transportation Officials Soil Classification System (AASHTO Designation M-45, ASTM Designation D-3282).

Table A Classification of Soils and Soil-Aggregate Mixtures

General Classification	Granular Materials (35 Percent or Less Passing 0.075 mm)			Silt-Clay Materials (More than 35 Percent Passing 0.075 mm)			
Group Classification	A-1	A-3[a]	A-2	A-4	A-5	A-6	A-7
Sieve analysis, percent passing:							
2.00 mm (No. 10)	—	—	—				
0.425 mm (No. 40)	50 max	51 min	—				
0.075 mm (No. 200)	25 max	10 max	35 max	36 min	36 min	36 min	36 min
Characteristics of fraction passing 0.425 mm (No. 40)							
Liquid limit	—	—	[b]	40 max	41 min	40 max	41 min
Plasticity index	6 max	N.P.		10 max	10 max	11 min	11 min
General rating as subgrade	Excellent to good			Fair to poor			

[a]The placing of A-3 before A-2 is necessary in the "left to right elimination process" and does not indicate superiority of A-3 over A-2.
[b]See Table B for values.

Table B Classification of Soils and Soil-Aggregate Mixtures

General Classification	Granular Materials (35 Percent or Less Passing 0.075 mm)							Silt-Clay Materials (More than 35 Percent Passing 0.075 mm)			
	A-1		A-3	A-2				A-4	A-5	A-6	A-7
Group Classification	A-1-a	A-1-b	A-3	A-2-4	A-2-5	A-2-6	A-2-7	A-4	A-5	A-6	A-7-5, A-7-6
Sieve analysis, percent passing:											
2.00 mm (No. 10)	50 max										
0.425 mm (No. 40)	30 max	50 max	51 min								
0.075 mm (No. 200)	15 max	25 max	10 max	35 max	35 max	35 max	35 max	36 min	36 min	36 min	36 min
Characteristics of fraction passing 0.425 mm (No. 40)											
Liquid limit	—		—	40 max	41 min	40 max	41 min	40 max	41 min	40 max	41 min
Plasticity index	6 max		N.P.	10 max	10 max	11 min	11 min	10 max	10 max	11 min	11 min[a]
Usual types of significant constituent materials	Stone fragments, gravel and sand		Fine sand	Silty or clayey gravel and sand				Silty soils		Clayey soils	
General ratings as subgrade	Excellent to Good							Fair to poor			

[a]Plasticity index of A-7-5 subgroup is equal to or less than LL minus 30. Plasticity index of A-7-6 subgroup is greater than LL minus 30 (see plasticity chart).

Process for Classification

(1) Classification is made by using the test limits and group index values.

(2) Use Table A but proceed to Table B if a more detailed classification is desired. Proceed from left to right in Table A or Table B, and identify the correct group by process of elimination; the first group from the left into which the test data fit is the correct classification.

(3) The Group Index value, GI, is calculated from

$$GI = (F-35)[0.2 + 0.005(LL-40)] + 0.01(F-15)(PI-10)$$

where F = percentage of soil finer than 0.074 mm (#200 sieve); base percentage only on material passing a 75 mm (3-inch) sieve, LL, PI = liquid limit, plasticity index.

Assign GI as zero for soils that are non-plastic or where a liquid limit cannot be determined, or where the index calculates to be negative.

For A-2-6 and A-2-7 subgroups, calculate GI using only the second portion (the PI portion) of the formula. Show GI values as whole numbers in parenthesis following the group symbol; e.g., A-2-6(12).

(4) Highly organic soils such as peat or muck may be classified as an A-8 group (base classification on visual identification).

117

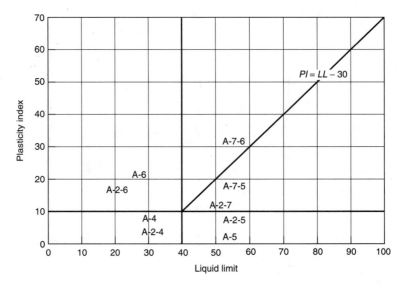

Figure 4-17(b) Plasticity chart: AASHTO soil classification system.

Figure 4-17(c) Description of Classification Groups

GRANULAR MATERIALS - Containing 35 percent or less passing 0.074 mm (#200 sieve).

Group A-1. The typical material of this group is a well-graded mixture of stone fragments or gravel, coarse sand, fine sand and a nonplastic or feebly plastic soil binder. However, this group includes also stone fragments, gravel, coarse sand, volcanic cinders, etc., without soil binder.
Subgroup A-1-a. Includes those materials consisting predominantly of stone fragments or gravel, either with or without a well-graded binder of fine material.
Subgroup A-1-b. Includes those materials consisting predominantly of coarse sand either with or without a well-graded soil binder.
Group A-3. The typical material of this group is fine beach sand or fine desert blow sand without silty or clay fines or with a very small amount of nonplastic silt. The group includes also stream-deposited mixtures of poorly-graded fine sand and limited amounts of course sand and gravel.
Group A-2. This group includes a wide variety of "granular" materials which are borderline between the materials falling in Groups A-1 and A-3 and silt-clay materials of Groups A-4, A-5, A-6, and A-7. It includes all materials containing 35 percent or less passing the 0.074 mm (#200) sieve which cannot be classified as A-1 or A-3, due to fines content or plasticity or both, in excess of the limitations for those groups.

Subgroups A-2-4 and A-2-5. Include various granular materials containing 35 percent or less passing the 0.074 mm (#200) sieve and with a minus 0.425 mm (#40 sieve) portion having the characteristics of the A-4 and A-5 groups. These groups include such materials as gravel and coarse sand with silt contents or plasticity indexes in excess of the limitations of Group A-1, and fine sand with nonplastic silt content in excess of the limitations of Group A-3.

Subgroups A-2-6 and A-2-7. Include materials similar to those described under Subgroups A-2-4 and A-2-5 except that the fine portion contains plastic clay having the characteristics of the A-6 or A-7 group.

NOTE Classification of materials in the various groups applies only to the fraction passing the 75 mm (3-inch) sieve. Therefore, any specifications regarding the use of A-1, A-2, or A-3 materials in construction should state whether boulders retained on the 75 mm (3-inch) are permitted.

SILT-CLAY MATERIALS - Containing more than 35 percent passing the 0.074 mm (#200) sieve.

Group A-4. The typical material of this group is a nonplastic or moderately plastic silty soil usually having 75 percent or more passing the 0.074 mm (#200) sieve. The group includes also mixtures of fine silty soil and up to 64 percent of sand and gravel retained on the 0.074 mm (#200) sieve.

Group A-5. The typical material of this group is similar to that described under Group A-4, except that it is usually of diatomaceous or micaceous character and may be highly elastic as indicated by the high liquid limit.

Group A-6. The typical material of this group is a plastic clay soil usually having 75 percent or more passing the 0.074 mm (#200) sieve. The group includes also mixtures of fine clayey soil and up to 64 percent of sand and gravel retained on the 0.074 mm (#200) sieve. Materials of this group usually have high volume change between the wet and dry states.

Group A-7. The typical material of this group is similar to that described under Group A-6, except that it has the high liquid limits characteristic of the A-5 group and may be elastic as well as subject to high volume change.

Subgroup A-7-5. Includes those materials with moderate plasticity indexes in relation to liquid limit and which may be highly elastic as well as subject to considerable volume change.

Subgroup A-7-6. Includes those materials with high plasticity indexes in relation to liquid limit and which are subject to extremely high volume change.

NOTE Highly organic soils (peat or muck) may be classified in an **A-8 group**. Classification of these materials is based on visual inspection, and is not dependent on percentage passing the 0.074 mm (#200) sieve, liquid limit or plasticity index. The material is composed primarily of partially decayed organic matter, generally has a fibrous texture, dark brown or black color and odor of decay. These organic materials are unsuitable for use in embankments and subgrades. They are highly compressible and have low strength.

Table 4-5(a) Approximate Equivalent Groups of AASHTO and Unified Soil Classification Systems

AASHTO	**Unified**
A-2-6	GC, SC
A-2-7	GC, SC
A-3	SP
A-4	ML, OL
A-5	MH
A-6	CL
A-7-5	CL, OL
A-7-6	CH, OH

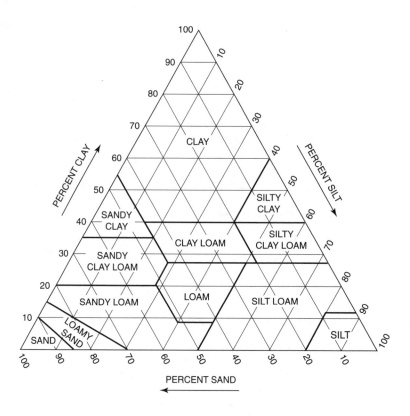

Figure 4-18 U.S. Department of Agriculture chart for textural classification of soil.

sample containing 20 percent gravel sizes, 30 percent sand, 25 percent silt, and 25 percent clay would be modified as follows:

$$\text{percent sand} = \frac{30}{100 - 20}(\%) = 37.5\%$$

$$\text{percent silt} = \frac{25}{100 - 20}(\%) = 31.2\%$$

$$\text{percent clay} = \frac{25}{100 - 20}(\%) = 31.2\%$$

and would be classified as a *clay loam.*

When cross-referencing classification systems, it is important to recognize that the dividing sizes between gravel, sand, silt, and clay particles are not consistent for the three systems discussed in this section (see Table 4-5b).

4.4 ASTM AND AASHTO TEST PROCEDURE DESIGNATIONS

A summary of ASTM and AASHTO test procedure designations for the testing discussed in this and following chapters is presented in Table 4-6. Most commercial, agency, and educational program laboratories perform soil testing in accordance with these procedures.

Table 4-5b   Particle Sizes for Indicated Classification System

Particle Designation	Unified SCS	AASHTO SC	U.S. Dept. of Agriculture Textural Classification System
Gravel	> 4.76 mm	75–2.0 mm	> 2.0 mm
Sand	4.76–0.074 mm	2.0–0.074 mm	2.0–0.05 mm
Silt	< 0.074 mm	< 0.074 mm	0.05–0.002 mm
Clay	< 0.074 mm	< 0.074 mm	< 0.002 mm

Table 4-6   ASTM and AASHTO Test Procedure Designation

Soil property to be determined or test identification name	Related ASTM designations	Related AASHTO designations
1. Particle sizes, distribution of particle sizes	D422	T11, T27, T88
2. Presence of fine-grain particles	D422, D1140	T11
3. Classification, classification systems	D2487, D2488, D3282	M145
4. Shape of particles	D2488	——
5. Water content	D2216, D3017, D4643, D4944, D4959	T93, T217, T239, T265
6. Specific gravity	D854, D5550	T84, T85, T100
7. Plasticity, liquid limit, plastic limit	D4318	T89, T90
8. Presence of clay, type or behavior of clay	D4221, D4647, D4829	——
9. Soil shear strength, consistency	D2166, D2573, D2587, D2850, D3080, D4767, D5315	T208, T223, T234, T236, T296, T297
10. Change in consistency due to remolding	D2166	——
11. In-place density, relative density, compaction	D698, D1556, D1557, D2167, D2922, D2937, D4253, D4254, D5080	T99, T180, T191, T204, T205, T233, T238, T239
12. Permeability, hydraulic conductivity	D2434, D3385, D5126	T215
13. Compressibility, consolidation, soil collapse	D2435, D4186, D4546, D5333	T216

PROBLEMS

4-1. Indicate the difference between index properties of a soil and classification tests.

4-2. (a) What is the purpose of soil classification tests?
(b) What is the importance of soil index properties?

4-3. The following information is obtained from a sieve analysis to determine the range of particle sizes in a granular soil sample.

Sieve Size	Sieve Opening (mm)	Percent Finer by Weight
#4	4.76	96
#10	2.00	80
#20	0.84	51
#40	0.42	38
#60	0.25	25
#100	0.149	12
#200	0.074	5

Present the information as a grain-size curve on semilog coordinates of percent finer versus particle diameters. From the plot, determine the uniformity coefficient C_u.

4-4. Why is the hydrometer method commonly used to obtain data on the sizes of silt and clay particles in soil samples, and not the sieve method?

4-5. The in-place density is determined for a soil at a proposed construction site to plan the foundation design. The in-place density test is performed using rubber-balloon equipment with the following results:
• Mass of soil sample obtained from test hole = 1.45 kg
• Volume of test hole = 0.0008 m³
(a) What is the in-place density for the soil?
(b) For the foundation design, it is necessary to use the property of in-place unit weight; convert the in-place density value to unit weight in kN/m³.

4-6. (a) An in-place density determination (in-place unit weight) is made for the sand in a borrow pit using a balloon-type density apparatus. The damp sample dug from a test hole is found to weigh 8.5 lb. The volume of the test hole is 0.065 ft³. From these data, compute the wet unit weight of the soil.
(b) This soil is found to have a water content of 15 percent. Compute the dry unit weight of the

soil and, by referral to Table 4-3, indicate the probable condition of the natural material.

4-7. A granular soil located in a borrow pit is found to have an in-place dry density of 1895 kg/m³. In the laboratory, values of dry maximum density and minimum density are determined as 2100 kg/m³ and 1440 kg/m³, respectively. From this information, calculate the relative density of the soil in the borrow pit.

4-8. A borrow pit's soil is being used as earth fill at a construction project. The in-place dry density of the borrow pit soil is known to be 1825 kg/m³. The job contract requires the soil at the construction site to be compacted to a dry density of 1975 kg/m³. The project requires 12,000 m³ of compacted soil fill. What volume of soil will the earthmoving contractor be required to obtain from the borrow pit to provide the necessary volume of compacted fill?

4-9. Sand at a borrow pit is determined to have an in-place dry unit weight of 115 pcf. Laboratory tests performed to determine the maximum and minimum unit weights give values of 122 pcf and 102 pcf, respectively. From these data, what is the relative density of the natural soil?

4-10. An undisturbed sample of sand has a dry weight of 4.20 lb and occupies a volume of 0.038 ft³. The soil solids have a specific gravity of 2.75. Laboratory tests performed to determine the maximum and minimum densities indicate void ratios of 0.42 at the maximum density and 0.92 at the minimum density. Compute the relative density of this material.

4-11. A fine-grained soil is found to have a liquid limit of 70 percent and a plastic limit of 38 percent.
(a) What do the percentages for these limit values represent?
(b) What is the plasticity index for this soil?
(c) On the plasticity chart for the Unified Soil Classification System, what soil type is this?
(d) On the plasticity chart for the AASHTO classification system, what soil type is this?

4-12. A silt-clay soil has a plastic limit of 25 and a plasticity index of 30.
(a) If the natural water content of the soil is 35 percent, what is the liquidity index?

(b) What soil type is this, according to the Unified Soil Classification System and the AASHTO system?

4-13. A fine-grained soil is found to have a liquid limit of 90 percent and a plasticity index of 51. The natural water content is 28 percent.
(a) Determine the liquidity index and indicate the probable consistency of the natural soil.
(b) Classify this soil according to the Unified Soil Classification System, and also according to the AASHTO system.

4-14. A saturated 100 cm³ clay sample has a natural water content of 30 percent. It is found that the shrinkage limit occurs when the water content is 19 percent. If the specific gravity of soil solids is 2.70, what will the volume of the sample be when the water content is 15 percent?

4-15. A clay soil is found to have a liquid limit of 75 percent, a plastic limit of 45 percent, and a shrinkage limit of 25 percent. If a sample of this soil has a total volume of 30 cm³ at the liquid limit and a volume of 16.7 cm³ at the shrinkage limit, what is the specific gravity of the soil solids?

4-16. Limit tests performed on a clay indicate a liquid limit of 67 percent and a plastic limit of 32 percent. From a hydrometer analysis to determine particle sizes, it is found that 40 percent of the sample consists of particles smaller than 0.002 mm. From this information, indicate the activity classification for this clay and the probable type of clay mineral.

4-17. Laboratory test results for a sample of clay soil for the purpose of evaluating the potential for volume change (swelling, expansion) are as shown. Rate the volume change potential (high, medium, low) for this soil.
LL = 68%
PL = 24%
Particles smaller than 0.002 mm = 45%

4-18. A soil sample having no particles larger than the No. 4 sieve is classified as a sandy silt, ML, by the Unified Soil Classification System (LL = 25%, PL = 5%; silt particles comprise 70% of the sample, sand 15%, and clay 15%). What is the probable soil classification in the AASHTO system? In the U.S. Dept. of Agriculture textural system? (Do not use Table 4-5.)

4-19. The textural classification used for identifying a soil sample (Dept. of Agriculture classification system) indicates a loam soil. What are the probable identifications by the Unified Soil Classification System and the AASHTO classification system criteria?

4-20. Referring to the USDOA textural classification chart, what percent of the particles in a fine-grained soil (no particles larger than the No. 200 sieve size) must be clay sizes for the soil to be identified as "clay" (no dual name for the classification)?

CHAPTER 5

Site Investigations

Purpose and Methods, Information and Procedures Available[1]

Site investigation refers to the procedure of determining surface and subsurface conditions in an area of proposed construction. Surface and subsurface features may influence *what* can be built and will directly affect the design and construction procedures relating to *how* a structure is built. (Although information on subsurface soil, rock, and groundwater is important, it is also understood that other data will be significant as well.)

Information on surface conditions is necessary for planning construction techniques. Surface topography may affect access to the site with necessary construction equipment, including the ability of equipment to work on and travel across the area. The thickness of vegetation, including the density and height of trees, affects the ease or difficulty associated with preparing a site for construction. Disposal of removed surface material may be a problem, particularly in urban areas. If a condition of surface water develops at times, its presence may hinder construction operations or affect the use of the site after construction. Other factors that could affect construction procedures or postconstruction use of an area include availability of water, availability of electrical power, the proximity to major transportation routes, and environmental protection regulations of various government agencies.

The land drainage pattern, a surface consideration but also partially a subsurface feature, is important, for it might affect construction. A finished grading pattern must be planned so that it does not harm the original area pattern or cause other environmental changes.

---

[1]Author comment: This chapter contains considerable information relating to the various procedures available for determining and evaluating subsurface conditions. Some of the material provides necessary background for the topics presented in later chapters. In that regard, this chapter is properly sequenced. However, because of the broad scope of this chapter, the new reader might benefit by undertaking only a cursory examination of the contents as initial preparation for progressing to subsequent chapters. But it is also recommended that the new reader become familiar with the general content in order to locate material which serves as a reference when the various topics of the later chapters are studied.

Information on subsurface conditions existing at a site is a critical requirement. It is this information that is used to plan and design a structure's foundations and other below-ground work. Construction techniques are planned with the help of data on subsurface conditions. The possible need for dewatering will be revealed by the subsurface investigation. Information necessary to plan and design shoring or bracing of excavations for foundations and pipe trenches is obtained from such explorations. If the construction site is underlain by varying soil conditions, the explorations will be used to indicate "better" areas. For projects where there is flexibility in locating structures, considerable savings in foundation costs may be realized by constructing in the "better soil" areas.

In many locations, information about the area is frequently available in the form of maps showing surface topography plus maps and literature that provide general information on subsurface soil or rock conditions. Aerial photographs can also offer useful data on subsurface conditions.

Though maps and aerial photographs of an area may provide much useful information about soil and rock conditions existing at a site, virtually all major construction projects will have on-site subsurface explorations performed in order to obtain the *detailed* information on soil types and properties necessary for designing foundations and for planning construction activities. Most typically, such information is obtained through the use of borings or test pits or by utilizing geophysical investigative methods. Properly used, these procedures are capable of giving a reliable definition of the type and extent of soil strata underlying an area, and groundwater conditions. Where physical properties of the subsurface soil must be known, soil samples from appropriate depths can be obtained for laboratory testing. Alternatively, in-place testing of soil in its natural location is frequently possible.

These methods of subsurface investigation are also used in exploring for sources of soil fill and for checking material in borrow pit areas.

Importantly, these established procedures also apply to, and currently are in extensive use for, investigating and monitoring subsurface conditions for environmental protection studies.

Frequently, the overall reliability of information obtained from a subsurface investigation is related to the extent of the work performed: The more borings and test pits, the more soil and water sampling and testing, or the more geophysical lines run, the greater the detail. The availability of information reduces the need for interpolating between locations of known conditions.

Accurate preplanning of a subsurface exploration program can be difficult to do properly. The *depth* of borings or test pits considered necessary or adequate for design and planning should relate to the types and properties of earth materials revealed by the investigation as well as to the size and type of structure that is planned. The *number* of borings, test pits, or geophysical lines that will be necessary relates to the variation in conditions. A site underlain by uniform conditions normally requires a less extensive investigation than one underlain by highly variable conditions. It is not unusual to have to plan an exploration program as it progresses, with information as it becomes available determining the need for additional work.

5.1   MAPS AND AERIAL PHOTOGRAPHS AS SOURCES OF INFORMATION

Information on surface and subsurface conditions in an area is useful to the construction and engineering profession and is frequently present in the form of available maps. In the

United States, such sources include U.S. Geological Survey topographic maps, U.S. Department of Agriculture soil conservation maps, and state geologic (rock or soil) maps. Many other areas of the world have similar maps available.

Aerial photographs are also capable of providing considerable information about an area, particularly data on surface conditions and land forms that are not clarified on maps or detectable from ground observations. Aerial photographs frequently serve as the basis for topographic and soil maps. Available sources include the government agencies responsible for developing maps, other municipal governments or agencies that have aerial photographs from construction or tax-mapping projects, and private firms that take the aerial photographs and provide consulting services in photogrammetry.

U.S. Geological Survey Topographic Quadrangle Maps. These maps provide information on surface features and topography. Included on these maps are elevation contours, bodies of water and water courses, indication of areas covered by forests or other vegetation, and man-made features such as buildings, roads, dams and reservoirs, railroads, power lines, airports, harbor development, military bases, and park preserves. Study of contour elevations and indicated land forms on these maps can relate to the geology of subsoil formation and character. These maps are very useful for evaluating the effects that topography and drainage will have on an area. An illustration of a USGS map is shown in Figure 5-1.

U.S. Department of Agriculture Soil Conservation Maps. These maps provide information on surficial soils existing in an area. The information usually represents conditions to a depth of 1.5 m to 2 m (5 to 6 ft.). A coded typing is used to indicate different soil categories on the map. In literature accompanying the maps, various properties and possible uses of the soils are described. Included is the application for construction purposes; AASHTO and Unified Soil Classifications are given. General suitability of the various soils for embankment and highway base course materials, compaction characteristics, and drainage and permeabilities are also provided. An illustration of a soil conservation map is shown in Figure 5-2.

Geologic Maps. These maps indicate the rock types underlying an area. Frequently, geologic maps are accompanied by descriptive papers that complement the actual map. Information on the extent of soil cover overlying the rock and general physical properties of the soil cover and the rock, and data on natural resources such as oil, gas, and commercial minerals are usually included. Developers of geologic maps frequently include state-level governmental agencies. Figure 5-3 provides and illustration of a geologic map.

Aerial Photographs. Air photo interpretation is the method of providing information on subsurface conditions by relating landform development and plant growth to geology. Applied to the building design and construction profession, air photos have particular value in their ability to provide detailed geologic information over a relatively large area and, as a result, can detect conditions that are difficult to observe or evaluate properly from a surface investigation. Illustrations of such special information include the observation of sinkhole cavities in areas underlain by limestone formations, occurrence of an area affected by landslides, past meanderings of existing rivers and probable new locations, presence and extent of glacially deposited land forms (such as drumlins, deltas, kettles), changes in beach areas and offshore deposits (such as sandbars), and land drainage patterns.

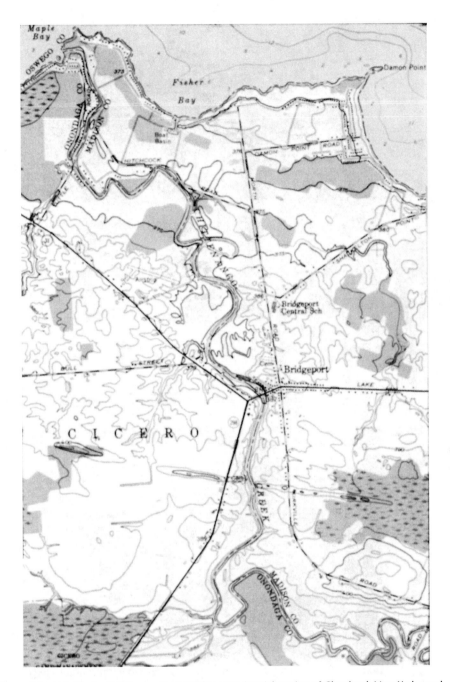

Figure 5-1 U.S. Geological Survey topographic map (partial section of Cleveland–New York quadrangle). (Note: These maps are printed in color.)

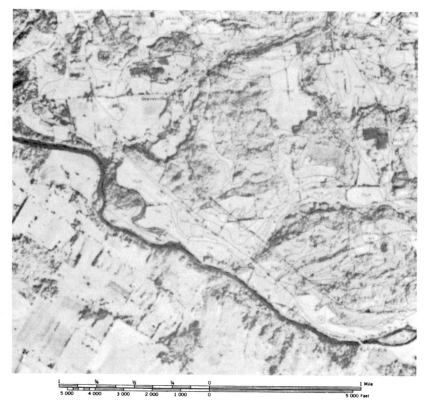

Figure 5-2 U.S. Department of Agriculture soil conservation map showing coded soil categories (partial section from southern Herkimer County, New York)

Not to be overlooked are the benefits of an aerial reconnaissance. A trained observer, or for that matter an experienced designer or constructor, should be capable of detecting items of land formation that can have an effect on design or construction from the view offered by a low and slow flight over a planned construction site. The disadvantage of a flight compared to aerial photographs is that no pictorial record remains for referral in future discussions. Figure 5-4 provides an example of information shown on aerial photographs.

5.2 BORINGS AND TEST PITS

Borings and test pits (observation pits or excavations) serve to provide a visual identification of soil strata underlying an area by literally penetrating into the earth. Subsurface materials are identified in place (test pits) or from samples that are taken at a known depth and brought to the surface for examination (borings). In both borings and test pits, soil samples can be obtained for laboratory testing and analysis. A variety of methods and equipment are in use to obtain soil samples; often a concern is to minimize (hopefully eliminate) disturbance to the soil's natural condition, and use of specialty equipment is necessary.

Boring Methods

With borings, several different techniques are in use for obtaining information on soil conditions. The method utilized depends on the type and extent of information desired from the exploration, the general type of soil conditions in the area being investigated, and the amount of money and time available.

Where only very basic data are necessary, as, for example, if depth to rock (thickness of soil cover) or the thickness of a soft surficial layer (e.g., a marsh deposit) is to be determined, probing by pushing or driving a metal rod until firm resistance or refusal is encountered is a frequently used and economic method. Penetration rods can be handled manually or with mechanical equipment. Usually, manual techniques are practical only where limited depths of probing are necessary.

Hand-operated auger methods (Fig. 5-5) are useful where it is desirable to obtain an indication of various soil types penetrated by the equipment or where it is desired to create an excavation that can give information on the depth to the groundwater table. Information on soil type at a particular depth is usually determined by noting the soil held on the auger. Continuous-flight augers are frequently used to ensure that soil cuttings are carried out of the boring, particularly for deeper borings. When only a limited section of auger is attached to the tip of the rods, the soil does not carry to the surface. Such an occurrence tends to bind the rods and auger in the hole.

Boring contractors and manufacturers have made various refinements to manual penetration and auger equipment that permit small samples of soil to be obtained from a desired depth and brought to the surface for examination or permit the measurement of resistance to penetration offered by the soil at a given depth (Figs. 5-6 and 5-7).

Where the soil investigation is to extend to some depth, mechanical means of drilling borings are utilized. Two techniques are in wide use: the *auger method* and the *wash boring method.* The auger technique (Fig. 5-8) is basically similar to the hand auger method. Continuous-flight augers are used to drill into the earth, and soil cuttings are carried out of the hole by traveling up the flights of the rotating auger. The main difference between manual and mechanical augering is that drilling machinery is utilized to rotate, and simultaneously to mechanically push downward, the auger sections. Because of the drilling machinery, larger-diameter flights than those that are possible with manual methods are used, and the limits to depth are controlled only by the capacity of the equipment.

Subsurface soil types can be detected by noting the soil materials carried to the surface by the augers. In deep drilling, there is the disadvantage of not being able to correlate accurately soil type with depth. In order to better classify soil type with depth, auger samples are used. The procedure consists of augering to a desired depth, withdrawing the augers, inserting an auger sampler to obtain a soil specimen from the bottom of the hole, withdrawing the sampler after cutting the sample (Fig. 5-9), and then reinserting the augers to drill to the next sampling depth. Soil samples obtained with auger methods are disturbed samples suitable for classification and some laboratory tests (e.g., they are acceptable for grain-size and liquid and plastic limit determinations but not for strength or compressibility tests).

Wash boring refers to the method of making a boring by applying an up-and-down chopping and twisting motion to a drill bit (or chopping bit) attached at the end of drill rods while simultaneously having a stream of water under pressure directed through the

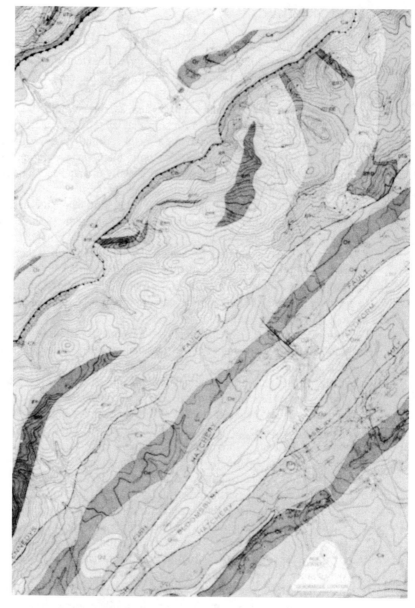

Figure 5-3 (a) U.S. Geological Survey geologic quadrangle map (section from Bloomsbury, New Jersey, quadrangle). (Note: These maps are printed in color.)

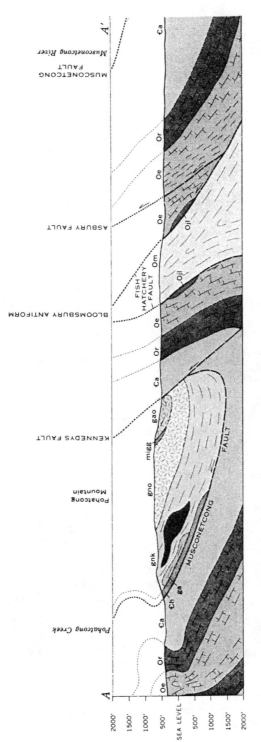

Figure 5-3 (b) Cross section through section line A-A.

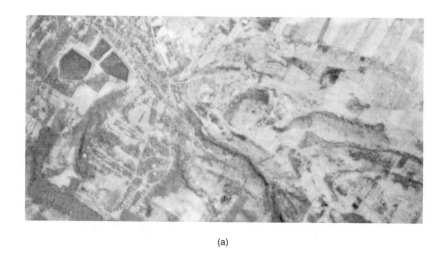

(a)

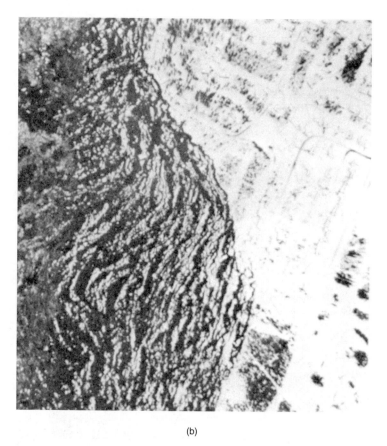

(b)

Figure 5-4 (a) Aerial photograph showing urban development and surrounding land forms: (b) aerial photo showing landslide in Anchorage, Alaska, area resulting from 1964 earthquake.

Figure 5-5 Hand auger with recovered soil sample. (Courtesy of Acker Drill Company)

Figure 5-6 Retractable plug sampler, sequence of operation to recover samples: (1) casing driven with piston extended; (2) piston retracted and casing driven to recover sample; (3) sample recovered and casing withdrawn by hand jack (not shown). Brass sample liners (not shown) are used in bottom length of casing to facilitate recovery and preservation of sample. (Courtesy of Acker Drill Company)

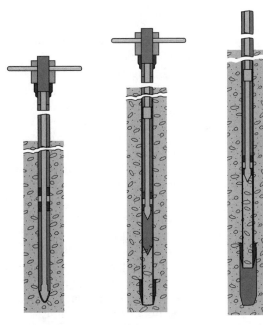

bit to the soil (Figs. 5-10 and 5-11). This combination of chopping plus wash water serves to loosen soil at the bottom of the boring and to flush the soil cuttings to the top of the hole. The method has long been popular for making borings because of the limited equipment necessary. Even small-sized engines for lifting drill rods coupled with small water pumps can be used to make deep borings. An advantage to this is that inexpensive but quite maneuverable and portable drilling equipment is possible. (Portability is a factor of importance in difficult-to-reach drilling sites.) Drilling rods are usually

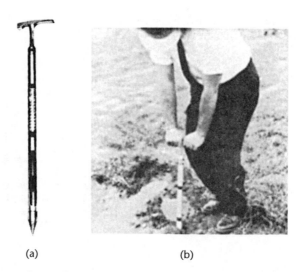

(a)   (b)

Figure 5-7 One type of manual penetrometer: (a) Acker geostick; (b) field use of geostick. (Courtesy of Acker Drill Company)

inserted into and lifted out of the boring by using a tripod or derrick mast to support a rope and pulley lift.

Borings advanced by wash boring methods through cohesionless soils or below water are subject to cave-in. To prevent this, it is typical to drive a protective casing around the boring. The casing consists of heavy wall pipe that is driven into the ground in sections with a drop hammer (drive weight) as the boring is advanced. The casing is withdrawn for reuse at the completion of the boring. An alternate technique for preventing cave-ins consists of using a heavy liquid slurry ("driller's mud") in the boring as the hole is drilled. The slurry, typically a "bentonite" (a montmorillonite clay) and water mix, is dense enough to exert a lateral pressure adequate to keep the walls of the boring from collapsing inward. Soil cuttings are flushed up through the slurry. At the surface, the entire liquid mixture can be caught, filtered, and recirculated back into the boring. Some of the larger and more sophisticated drilling rigs can make and force compressed air into the boring to prevent cave-ins, instead of using casing or slurry.

Some information on soil type and changes in strata can be obtained by examining the soil cuttings washed to the surface. However, such soil is significantly disturbed, and there is the danger that mixing of different soil types can occur before the cuttings reach the surface. Because of the significant soil disturbance and the difficulty in establishing boundaries between strata, the determination of subsurface conditions by using wash borings is no longer an acceptable method.

Soil Sampling

To develop information on subsurface conditions that is considered accurate, it is now established practice to obtain soil samples that are sufficiently undisturbed to permit accurate classification. Such soil samples are recovered from their natural location in the ground by

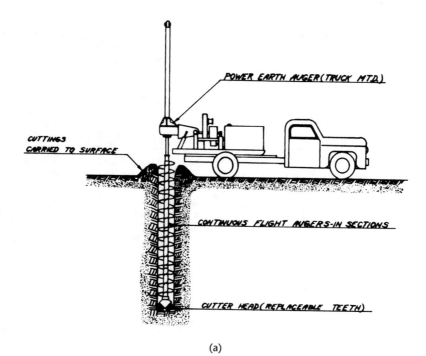

POWER EARTH AUGER (TRUCK MTD.)

CUTTINGS CARRIED TO SURFACE

CONTINUOUS FLIGHT AUGERS-IN SECTIONS

CUTTER HEAD (REPLACEABLE TEETH)

(a)

(b)

Figure 5-8 (a) Soil boring using auger method. (Courtesy of Acker Drill Company) (b) Truck-mounted rotary drill rig showing soil augers in use. (Courtesy of Central Mine Equipment Company)

Figure 5-9 Obtaining soil samples from auger cutting head. (Courtesy of Acker Drill Company)

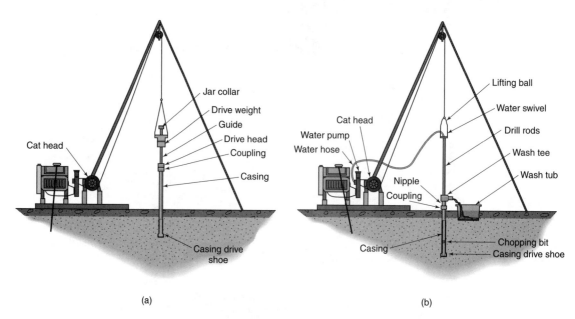

(a)

(b)

Figure 5-10 Typical setup for wash boring: (a) driving casing; (b) chopping and jetting. (Courtesy of Acker Drill Company)

the utilization of special techniques and sampling equipment. Typically, the boring and sampling procedure involves drilling a hole to a desired sampling depth by using wash boring or auger methods and then inserting a soil sampler into undisturbed earth to obtain a soil specimen.

A refinement of the auger method to advance a boring involves the use of hollow-stem augers. The soil sampler fits within the hollow stem of the auger and advances at the bottom of the boring without requiring removal of the augers. Use of this method can considerably speed up the boring and sampling operation.

Figure 5-11   Wash boring rigs in operation.

(a) Land rig (Photo courtesy Acker Drill Company)

(b) Pontoon-barge-mounted rig for working on ice and on water (Photo courtesy Atlantic Testing Laboratories, Ltd., Canton, N.Y.)

To obtain a soil sample, the sampler is advanced by driving with a drop hammer (Fig. 5-12) or by pushing with a hydraulic piston or jack (Fig. 5-13). In its basic form, the soil sampler is a section of metal pipe or tube with a cutting edge at one end and attachments to hold the soil in the sampler as it is brought to the surface. Sampler lengths typically range between 18 inches and 3 feet or approximately 0.5 and 1 m. Samplers come in different diameters and with different wall thicknesses. The choice of sampler type selected relates to the type of soil being encountered and what needs to be done with the recovered specimen. Where in-place properties need to be determined (such as shear strength and compression–consolidation characteristics), "undisturbed" (or only slightly disturbed) samples are required. If samples are for classification or for performing remolded-type laboratory tests (such as grain size and Atterberg limits), "disturbed" samples are permitted. Generally, "undisturbed" soils are obtained by using samplers having a thin wall and a large ratio of inside diameter to wall thickness. Thick-wall samplers and those having a low diameter-to-wall-thickness ratio are considered to provide "disturbed" samples.

One of the most widely known and used soil samplers is the *standard split spoon*. This sampler consists of a longitudinally split tube, or barrel, 2 inches (51 mm) in outside diameter and having a $1\frac{3}{8}$-inch (35 mm) inside diameter (Fig. 5-14). The splitting aspect of

Figure 5-12 Obtaining soil sample by driving with a drop weight. (Courtesy of Acker Drill Company)

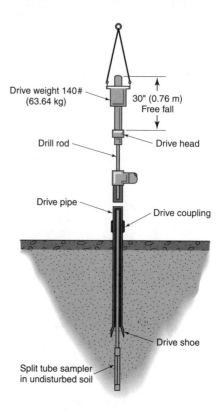

Drive weight 140# (63.64 kg)

30" (0.76 m) Free fall

Drill rod

Drive head

Drive pipe

Drive coupling

Drive shoe

Split tube sampler in undisturbed soil

Figure 5-13 Obtaining soil sample by hydraulic pushing. (Courtesy of Acker Drill Company)

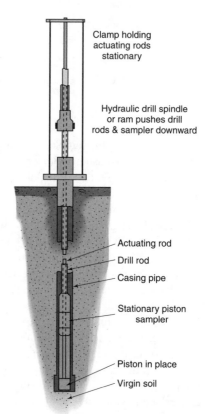

Clamp holding actuating rods stationary

Hydraulic drill spindle or ram pushes drill rods & sampler downward

Actuating rod

Drill rod

Casing pipe

Stationary piston sampler

Piston in place

Virgin soil

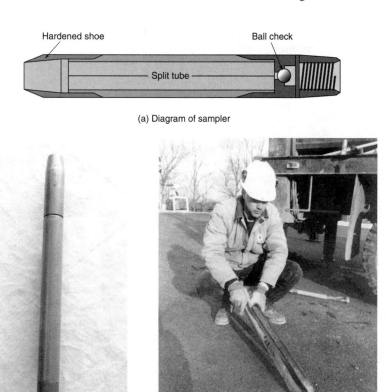

(a) Diagram of sampler

(b) Assembled split-spoon   (c) Sampler opened to expose soil

Figure 5-14 Standard split spoon sampler.

the sampler permits it to be opened so that the soil specimen can easily be examined and then placed in a container for shipping. The wide use and acceptance of this sampler is due to its size (diameter about equal to typical drill rod diameter), its availability and use at the stage of development when investigators and designers started requiring samples for accurate classification, and the comparatively low cost of obtaining this type of sample. The standard split spoon sampler provides samples rated as "disturbed."

Split barrel samplers are also available in diameters larger than the standard split spoon. The large-diameter samplers are considered to provide better samples for laboratory testing and for sampling soils that include gravel-sized particles. Split barrel samplers $3\frac{1}{2}$ inches (89 mm) in diameter for use in 4-inch- (100 mm) diameter borings are included in the typical stock equipment list of many soil-boring contractors. Borings of approximately 6-inch (150 mm) diameter, which permit even larger samples to be obtained, are sometimes used for projects requiring special soil testing or analysis. However, this size is not typical because of the greater expense involved with drilling borings of larger diameter.

Split barrel samplers can be provided with a *liner,* which is a thin metal or plastic tube fitted with the split barrel. The purpose of the liner is to help hold together and protect the sample during handling, shipping, and storage. The liner may consist of a solid length of tube or a series of rings.

The split barrel sampler is a thick-wall sampler and generally is considered to provide disturbed samples (not suitable for shear and consolidation tests). However, in the large-diameter split barrel samplers, disturbance may be limited to just the peripheral zone of the sample.

Samplers for obtaining less disturbed ("undisturbed") specimens are commonly available. Most use a thin-wall tube as the sampler or as a sampler tip or extension (such as the Dames and Moore sampler, Fig. 5-15). Methods to carefully advance the sampler into the natural soil are followed to avoid disturbance. When used by themselves, the thin-wall samplers are typically used only in fine-grained soils (silts and clays). Such samplers are generally not used in coarse-grained soils because of the danger of damaging (buckling or crimping) the thin wall.

Three of the more widely used "undisturbed" samplers are the Shelby Tube, the Piston Sampler, and the Denison Sampler (variations and modifications of this equipment also exist).

The thin-wall Shelby Tube sampler is a seamless metal tube having a limited wall thickness. To obtain the sample, the Shelby Tube is "pushed" (hydraulically pressed) into the earth to obtain the soil specimen. After recovery, the soil is left in the sampling tube for shipping or is hydraulically pushed out of the sampler and placed in an appropriate shipping container (Fig. 5-16).

The Piston Sampler (Fig. 5-17) is an adaptation of the Shelby Tube sampling method. This sampler includes a piston device, which serves to push the thin-wall tube into undisturbed soil from the bottom of the boring.

The Denison Sampler, or Denison Core Barrel (Fig. 5-18), is a double-walled sampler that rotates or cores its way into undisturbed soil. The outer barrel rotates to cut into the soil. The sample is obtained with the inner barrel. Because of the coring feature, it has the advantage of being able to obtain samples in hard or cemented soils, where an otherwise pressed thin wall would be in danger of buckling.

Sample Spacing in Soil Borings

The accuracy of information obtained from a particular soil boring relates to the spacing between samples. Samples can be taken continuously; that is, samples will be obtained for virtually each foot or meter of boring depth. However, it is more usual practice to take samples at selected intervals of depth. In the soil-boring industry it is common to sample at 5-ft (1.5 m) intervals (one sample for each 5-ft section of boring). Thus, if the sampler is driven 18 inches (0.5 m) to recover a sample, the actual distance between samples is $3\frac{1}{2}$ ft (about 1 m). Soil conditions between samples are determined by noting soil cuttings brought to the surface as the boring is made deeper and feeling resistance to the drilling advance. Where it is detected that a change in soil conditions has been encountered, further drilling should cease and an extra soil sample taken at that depth.

The drilling of a soil boring, therefore, actually consists of a repetitive series of steps. The drilling equipment is used to make a hole to the depth where a sample is to be obtained. After sampling, the boring is then drilled to the next point where a sample is required. This procedure continues until a boring of the desired depth has been completed.

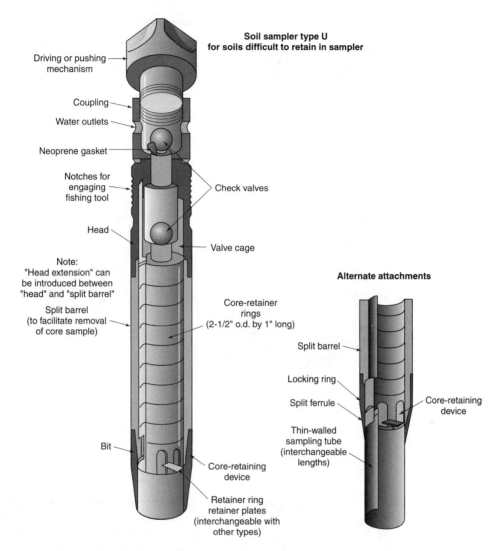

Figure 5-15   Dames and Moore large-diameter split barrel sampler. (Courtesy of Dames and Moore)

Boring Spacing and Depth

The number of borings for a project, or the spacing between borings, is related to the type, size, and weight of structure planned, the extent of the variation in soil conditions that permits "safe" interpolation between borings, the funds available for the boring program, and possibly the demands of a local building code. Generally, these same considerations apply to the depth for borings. A primary purpose of borings is to establish the presence, location, and extent of "good" soil suitable for foundation support and to determine the possible existence and extent of "poor" soil that could have an adverse effect on foundation performance.

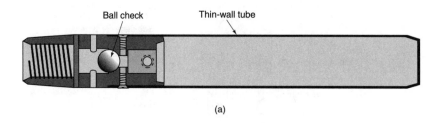

(a)

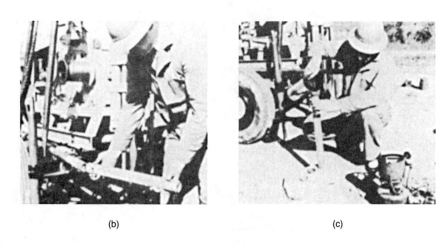

(b)   (c)

Figure 5-16 Thin-wall sampler: (a) schematic diagram; (b) tube being removed from head; (c) sealing thin-wall tube with paraffin for shipping. (Courtesy of Acker Drill Company)

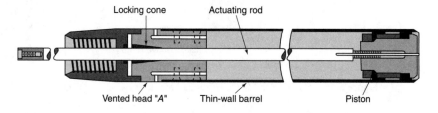

Figure 5-17 Stationary Piston Sampler. (Courtesy of Acker Drill Company)

Figure 5-18 Denison Core Barrel. (Courtesy of Acker Drill Company)

Where poor soils are present, the type of foundation selected would be based on circumventing the location of poor material or would be designed to minimize the effect of its presence.

If an extensive thickness of poor material overlies good soil or rock, deep borings will be required. Where a sufficient thickness of good soil exists close to the ground surface, borings may not have to be deep. Even for this condition, however, it is wise to plan some deep borings to reveal conditions at the greater depths, because a deep material that could have an adverse effect on the structure might exist. Where deep borings are necessary to establish the depth at which good foundation soil can be found, soil sampling in the poor material can be minimized or omitted to help reduce drilling costs.

Because most building projects are unique to some degree, certainly with regard to location, no set of rules can be made for planning a boring program that ensures answering all of a designer's or construction contractor's questions. Where there is much uncertainty about subsurface conditions, a preliminary investigation can be made to obtain general information on the area. From this initial information, a more definite or final exploration program can be planned.

For projects extending over a long horizontal distance, such as a highway or a dam, borings for a preliminary investigation may be on 500- to 1000-ft or 150 to 300 m spacings. The final program may eventually require borings on a 100- to 200-ft or 30 to 70 m spacing. With building projects, borings for a preliminary investigation may be located near the planned boundaries of the building area, with a few borings being drilled at intermediate locations. A final program could include borings at building corners and at important interior locations, with spacings not to exceed 50 to 100 ft or 15 to 30 m, depending on conditions encountered and building code requirements with regard to maximum spacing.

Rock Core Drilling

Rock core drilling relates to the procedure in which rock underlying an area is investigated by coring so as to obtain samples for classification and for determining those properties of rock important to that construction project.

In a structural design, the close presence of an underground rock surface may be desirable, provided that it does not interfere with construction, as the support capabilities of rock are almost always significantly greater than those of soil. The requirement for rock coring develops with the need to establish that an underground rock material encountered in a boring or test pit is actually bedrock and not a large boulder. Rock coring is commonly performed to determine the quality of rock and to check for possible detrimental properties such as cavities, cracks, and weathering or other deterioration that could affect the strength of the formation. Typically, rock cores are necessary to evaluate a rock's ability to support rock anchors, to determine porosity that could affect the flow of underground water or other liquids, or to obtain information for a subsurface rock profile and properties should it be necessary to excavate to depths below the top of the rock.

The method used for most rock drilling is similar to the method of rotary drilling in soil. Machinery used for drilling soil borings is frequently the same equipment as that used for drilling rock (Fig. 5-19). For rock work, special bits and core samplers are necessary, however. Cutting bits are hardened steel or steel alloys, or they are diamond chip bits (Fig. 5-20).

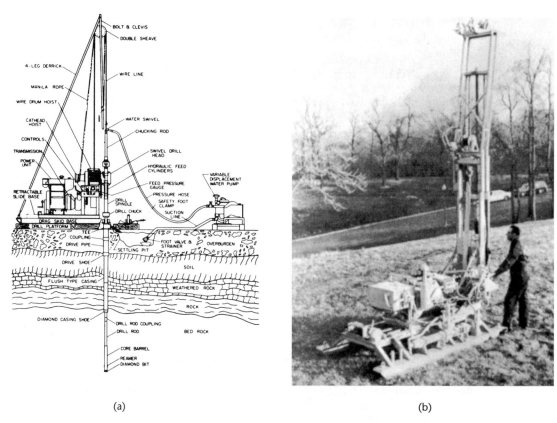

(a)   (b)

Figure 5.19 (a) Diamond core drill setup. (Courtesy of Acker Drill Company) (b) Skid-mounted rotary drill rig. This type is used in locations having difficult access. Machine shown has soil auger and rock core capabilities. (Courtesy of Central Mine Equipment Company)

Diamond Bit and Rock Core Sizes

Bit Designation[a]	Nominal Hole Diameter (in.)	(mm)	Rock Core Diameter (in.)	(mm)
EWG, EWM	$1\frac{1}{2}$	38	$\frac{13}{16}$	21
AWG, AWM	$1\frac{7}{8}$	48	$1\frac{3}{16}$	30
BWG, BWM	$2\frac{3}{8}$	60	$1\frac{5}{8}$	41
NWG, NWM	3	76	$2\frac{1}{8}$	54
$2\frac{3}{4} \times 3\frac{7}{8}$	$3\frac{7}{8}$	99	$2\frac{11}{16}$	68
HWG	$3\frac{7}{8}$	99	3	76

[a]First letter is used to designate approximate size of hole obtained with the indicated size equipment; second letter indicates a group designation for compatible drill rods, casing, etc.; third letter designates type of core barrel for the bit. The W size is considered standard, while the M equipment is recommended for coring poor-quality rock material. The letter W replaces the older X designation (but is synonymous).

Figure 5-20 Standard diamond core bit and summary of common core bit sizes.

To obtain rock core samples, hollow-core barrels are used. Single- and double-tube core barrels (Fig. 5-21) are in wide use. Under rotary action, the core bit advances into the rock. Usually, a circulating supply of water is provided to the cutting edge to help flush rock cuttings and dissipate heat. "Core runs" are made to drill the hole in segments, usually up to 1.5 m or 5 ft, in length. The length of core run is limited by the length of core barrel. At the completion of a core run, the barrel and rock sample are brought to the surface; the rock specimen is removed (Fig. 5-22), and the barrel is reinserted for additional drilling.

Test Pits

In their simplest form, test pits are excavations into the earth that permit visual inspection of the conditions exposed in the walls of the pit. The classification of soil type at various depths is possible, along with the opportunity to learn properties of the soil and determine stratum thicknesses. Where desired, soil specimens for testing can be cut from the walls or bottom of the test pit. Test pits can be made manually or with power equipment; backhoes are frequently utilized (Fig. 5-23).

Test pits provide opportunity for studying subsurface features not possible with borings. Generally, more accurate information as to groundwater elevation is possible (if the pit is deep enough to penetrate to groundwater), as is better information on soil variations or unusual features such as presence of underground springs and fissures or cracks in the soil. When the soil deposit includes gravel or boulders, boring samples may not be large enough to recover and identify these items, but such material will be detected in a test pit.

Figure 5-21   Double-tube rock core barrel. (Courtesy of Acker Drill Company)

Figure 5-22 Rock core samples being placed in core box for shipping. (Courtesy of Acker Drill Company)

Figure 5-23 Test pit excavation with small backhoe.

If the area under investigation represents a proposed borrow pit, bulk samples are easily obtained for laboratory compaction tests or other analysis.

Natural or man-made cuts in soil deposits offer the same advantage for information as a test pit.

Test pits can be a relatively economical way of obtaining information on subsurface conditions. Frequently, many test pit locations can be dug in one workday with a backhoe. The primary shortcoming of test pit exploration is the limited depth possible with commonly available equipment. Unless a large backhoe or similar equipment is obtainable, the test pits will not penetrate beyond depths of 4 to 5 m (12 to 15 ft). However, where shallow foundations can be used, the uppermost 5 m (15 ft) or so of soil are very critical for foundation analysis, and test pits can provide a very useful complement to boring information. When explorations are made for foundation studies, test pits should not be dug at a foundation location, so that the soil eventually required to support the foundation remains undisturbed.

5.3   SOIL RESISTANCE TESTING IN BOREHOLES

Penetration Resistance and the Standard Penetration Test

In situations where soil samples from borings were obtained by using drive methods (the sampler is driven by use of a drop hammer), it became apparent that the number of blows required to advance the sampler (overcoming resistance) was capable of providing a qualitative indication of the in-place properties of the soil. In firm or dense soils, more blows would be required to advance the sampler than in soft or loose soil. Consequently, when samples are taken, it is now common practice to include the number of blows necessary to advance the sampler a distance of 1 foot (0.3 m). For the blow-count information to be meaningful, the weight of the drive hammer used to advance the sampler must be indicated, since a heavy weight will drive a sampler with fewer blows than a light weight.

What has become known as the standard penetration test (SPT) is a soil-sampling procedure that is in wide use and is generally accepted as providing some correlation with in-place properties of a soil. The SPT requires that a 2-in. (51 mm) split spoon sampler be used in conjunction with a 140-lb (63.6 kg) drive weight. The SPT reports the number of blows N to drive the sampler 1 ft (0.3 m) into undisturbed soil by using the 140-lb weight falling 30 in. (0.76 m). In practice, it has become typical to obtain a sample by driving the sampler a distance of 18 in. (0.46 m). The blow count for each 6 in. (152 mm) of penetration is recorded separately, and the standard penetration test result is the number of blows required for the last 12 in. (0.3 m) of driving. Detailed requirements for the taking and care of soil samples using the SPT procedure are presented in ASTM Test Designation D-1586. A correlation between blow count and soil condition is shown in Table 5-1. Further correlations between the SPT and soil properties are presented in Chapters 11 and 13.

Where the SPT is to be related to soil parameters used for foundation design, the *field procedures* utilized for driving the soil sampler require consideration, because the driving energy transmitted to the sampler is affected by the type of drive weight, the dropping procedure, and other factors.

In the United States, two types of drive hammers have long been in common use: the relatively long, thin safety hammer and a short, large-diameter donut hammer. The effective energy (or energy ratio) delivered by the safety hammer is about 60 percent of the theoretical simple energy (the value based on the product of hammer weight and the falling distance). The donut hammer has an energy ratio of about 45 percent. For the same soil, then, the donut hammer would require a greater blow-count and thus would be indicating a higher resistance. In many other countries using equipment with a performance similar to the safety hammer, a proposed standard is to use the energy ratio of 60 percent as the international reference value (213) and identify the blow-count as N_{60}. Soil parameters determined from penetration testing (empirical relationships) would be matched to the 60 percent energy ratio. For example, a field blow count using the donut hammer would be

Table 5-1   Correlation Between Soil Conditions and Standard Penetration Test

Sampler Hammer Soil	Designation*	2" O.D. × $1\frac{3}{8}$" I.D. (51mm × 35 mm) 140 lb, 30" fall (63.64 kg, 0.76 m) N_{60}, Blows/ft (Blows per 0.30 m)
Sand and silt	Loose	0–10
	Medium	11–30
	Dense	31–50
	Very dense	Over 50
Clay	Very soft	0–2
	Soft	3–5
	Medium	6–15
	Stiff	16–25
	Hard	Over 25

*Based on results for drive hammers with 60 percent efficiency.
Source: Acker Drill Company.

modified by a factor of 0.75 to be compared to results from the safety hammer and to relate to the 60 percent standard.

Over the past decade, a newer type of equipment, the automatic hammer, has become common on drilling rigs, a change offering speed and safety. Most of the empirical data relating SPT results to soil properties is based on information developed from the safety hammer and the donut hammer (i.e., relating to N_{60}). Automatic hammers are more efficient than the safety and donut hammers in having the free-fall drive energy transmitted into the string of drill rods for advancing the soil sampler (about 90 percent compared to about 60 percent). Accordingly, for identical soil conditions, the blow count N will be lower where the automatic hammer is used. Job site comparisons suggest that the ratio $N_a = 0.75N_s$ be applied as an approximation to convert data (where N_a is the blow count for the automatic hammer and N_s is the blow-count for the safety hammer). The relationship factor becomes important where SPT results will be correlated to those soil properties associated with predicting soil behavior or used for foundation designs and other studies (for example, if N_a is 20, the value of N_s is estimated to be 27 and studies should be based on the N_s value).

From investigations of cohesionless soil deposits, it is recognized that measured blow-counts will be affected by the soil depth being sampled. For identical soils, the resistance to penetration at a deep location will be greater than the resistance value developed for a shallow location because of the influences of greater overburden and confining pressures and different driving energy losses in the drill rods. Field blow counts, N_{field}, can be corrected for the depth effect by applying a correction factor, C_N, with

$$C_N = 1/\sqrt{\overline{\sigma}_v}$$

when $\overline{\sigma}_v$ is the effective soil overburden pressure expressed in tons per square foot (165). Therefore,

$$N_{corrected} = C_N N_{field}$$

Sand deposits may experience liquefaction during an earthquake or other occurrence that creates seismic shock; liquefaction results in a temporary loss of soil shearing strength. Factors such as the ground acceleration created by shock waves, soil density, and depth are known to have an influence. The susceptibility to liquefaction has been related to the corrected N value (the N value that includes depth and energy ratio corrections). A relatively simple evaluation procedure for delineating sites prone to liquefaction during earthquake, developed from Chinese studies of earthquake-damaged areas, can serve as the basis for determining whether more detailed studies should be undertaken. Sites where the N values are less than N_{critical} (as calculated from the following expression) are indicating susceptibility to liquefaction:

$$N_{\text{critical}} = \overline{N}[1 + .125(Z_s - 3) - 0.05(Z_w - 2) - 0.07(\% \text{ clay})]$$

In this expression, use $\overline{N}$ equal to 6 for an earthquake intensity[2] of VII, 10 for an earthquake intensity of VIII, and 16 for an earthquake intensity of IX. The percent clay refers to the material finer than 0.002 mm. The terms Z_s and Z_w refer to the depth in meters to the sand layer under study and to the groundwater table, respectively.

Generally, corrected N values less than 20 are considered to indicate that the site has a high potential for earthquake damage. Sites where the corrected N values are between 20 and 30 are classified as having the potential for an intermediate degree of damage. Sites where N values are greater than 30 are considered locations where no significant damage from earthquake or seismic shock is expected.

Penetration Resistance and Cone Penetrometers

The cone penetrometer consists of a slender metal rod equipped with a cone-shaped tip. The penetrometer is either pushed or driven into the earth. When pushed, usually by hydraulic jack, the penetrometer is classified as a static cone penetrometer. If driven, usually by blows of a drop hammer, the equipment is referred to as a dynamic cone penetrometer. Electric penetrometers are also available; with this type, the cone tip is advanced by an electrical cell constructed within the penetrometer. In use, the penetrometer's resistance to advancing, q_c, is recorded, where q_c is the resisting force divided by the projected area of cone. The cone penetration test (CPT) resistance is subsequently related to properties of the penetrated soil.

Cone penetrometers offer the advantage that a continuous resistance record is easily obtained for the full depth investigated. The presence of very thin layers can be detected. This is in contrast to the usual practice of taking standard penetration test samples or other

---

[2]Earthquakes are identified in terms of intensity and magnitude. *Magnitude* is an indication of the energy released by an earthquake at the source (the hypocenter). The Richter scale expresses the magnitude of an earthquake on a scale of 1 to 9 (actually, the scale has no upper limit, but no known earthquake has yet succeeded 9). *Intensity* is a measurement of the effects of an earthquake at the earth's surface, using factors such as different degrees of damage to structures. The modified Mercalli scale, ranging from I through XII, is in widespread use for indicating actual or potential damage for a particular area. Though an earthquake will have but a single magnitude value, the intensity value will vary, usually decreasing as distance from the earthquake source increases. Refer to Section 1.4 for further information.

samples at intervals. The method is applicable to cohesionless and cohesive soils. A main disadvantage with many penetrometers is that no soil samples are obtained.

In the past, cone penetrometers have received limited attention in North America. However, a variety of cone penetrometers have been widely used throughout the European countries, and considerable practical experience has been obtained. Consequently, worldwide interest has been developing rapidly, particularly for investigating sites underlain by cohesionless soils. With conventional boring and sampling methods, obtaining undisturbed samples of cohesionless soil below the water table is difficult and unreliable.

Of the various cone penetrometers in use, the Dutch (Delft) cone is one of the most widely known. For illustrative purposes, it is representative of many types of static cone penetrometers. The Dutch cone penetrometer consists of a conical point (Fig. 5-24) having a 36 mm (1.4-in.) diameter and a projected base area of 1000 mm^2 attached to a rod that fits within a larger-diameter rod. For testing the soil at a particular depth, the outer rod (along with the cone and inner rod) is advanced to the desired elevation. Then the outer rod is locked in position and the cone is advanced a set distance, usually 50 mm, by applying a force to the inner rod. The force necessary to advance the cone is determined with a load cell or proving ring and recorded. The rate of advance typically is in the 10 to 20 mm/sec range. Soil properties are correlated with the penetrating force.

The reaction of the force used to advance the cone penetrometer into the earth has to be resisted by the penetrometer rig. If the penetrometer equipment is truck-mounted (Fig. 5-25a), the weight of the vehicle is frequently sufficient to provide the necessary reaction. If lightweight field equipment is used, it is usually necessary to anchor the rig somehow to prevent it from being lifted when the cone is advanced (Fig. 5-25b).

A modification of the Dutch cone penetrometer is the friction cone (Fig. 5-26). A friction sleeve is provided above the cone point. Testing at a particular depth consists of first determining only a cone resistance value, then also measuring the resistance to advancement of the cone plus friction sleeve acting together. The sleeve friction is used to help identify the soil type being penetrated.

The electrical penetrometer includes load sensors that measure axial load at the cone tip and axial load on the friction sleeve. If also equipped with a pressure sensor for measuring soil pore water pressure, the equipment is classified a piezocone penetrometer (Fig. 5-27).

For proper classification of soil, some calibration is generally required. This typically consists of comparison to a conventional boring where samples were recovered. Commonly the ratio of CPT to borings is between 3 to 1 and 10 to 1. Once calibration is established, the penetrometer boring normally progresses at a considerably faster rate than the conventional boring.

Dilatometer and Pressuremeter Tests

Field procedures to investigate soil materials underlying an area and to provide or permit soil properties to be determined are of interest to designers, developers, and contractors because of the advantages of savings in time and, possibly, costs. Penetration tests have been correlated with soil strength and foundation bearing capacity, as discussed in previous sections. Other field procedures, such as those that measure a soil's resistance to

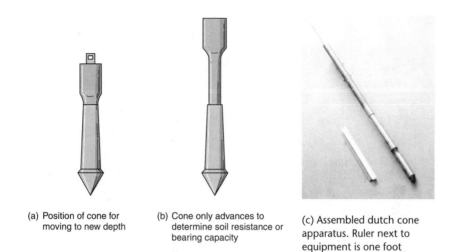

(a) Position of cone for moving to new depth

(b) Cone only advances to determine soil resistance or bearing capacity

(c) Assembled dutch cone apparatus. Ruler next to equipment is one foot (30.5 cm) long, for scale purposes.

Figure 5-24   Dutch cone penetrometer.

(a)

(b)

Figure 5-25   (a) Dutch cone penetrometer conversion unit fitted to conventional truck boring rig. (Courtesy of Soil Mechanics Equipment Company, Maryland) (b) Ten-ton trailer-mounted Dutch cone penetrometer. Sketch of helicoidal anchor indicates method used to anchor rig against lifting. (Courtesy of Soil Mechanics Equipment Company, Maryland)

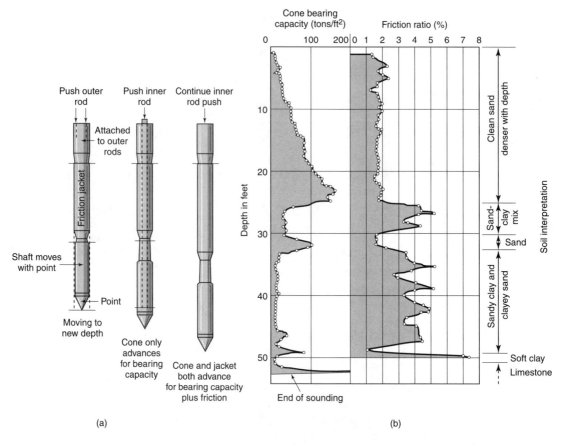

Figure 5-26 (a) Operation of the friction cone; (b) results from a friction cone surrounding [255].

deformation, theoretically are capable of yielding greater detail about soil properties. Two types of field test in this category are the dilatometer test (DMT) and the pressuremeter test (PMT).

The flat-blade *dilatometer* represents a type of penetration equipment that performs a dilation test, whereby resistance to lateral soil deformation is metered. The Marchetti flat-blade dilatometer diagrammed in Figure 5-28 consists of a pointed stainless steel blade that has a thin flexible metal membrane centered on one side of the blade and flush with the surface. The blade is 94 mm wide and 15 mm thick (approximately 4 in. by $\frac{9}{16}$ in.) with a 10° vee-shaped cutting tip. In a field operation, the blade is advanced into the soil using conventional penetrometer methods (by jacking or driving, methods outlined in the sections describing the standard penetration test and the cone penetration test).

A combination gas and electrical line extends from the blade through the hollow drill rods to the dilatometer's ground surface controls. At the test depth, the equipment operator provides and records a gas pressure just to start membrane expansion or "lift-off" (a movement of approximately 0.05 mm is typical); then a second pressure to move the center of

Position   Part

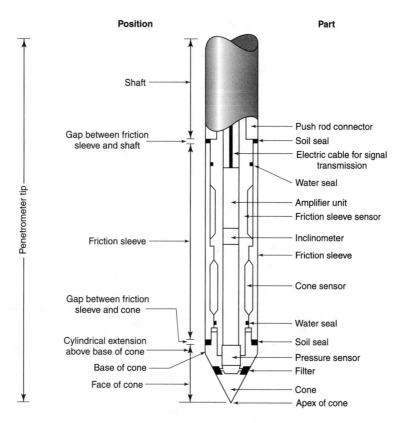

Shaft

Gap between friction
sleeve and shaft

Penetrometer tip

Friction sleeve

Gap between friction
sleeve and cone

Cylindrical extension
above base of cone

Base of cone

Face of cone

Push rod connector
Soil seal
Electric cable for signal
transmission
Water seal
Amplifier unit
Friction sleeve sensor
Inclinometer
Friction sleeve

Cone sensor

Water seal
Soil seal
Pressure sensor
Filter
Cone
Apex of cone

Figure 5-27   Fugro piezocone penetrometer (electrical penetrometer which also measures soil pore water pressure).

Figure 5-28   Sketch of Marchetti flat-blade dilatometer.

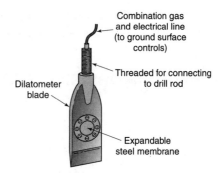

Combination gas
and electrical line
(to ground surface
controls)

Threaded for connecting
to drill rod

Dilatometer
blade

Expandable
steel membrane

the membrane 1.0 mm farther is applied and recorded. An audio signal indicates when the desired membrane movement has occurred. The pair of pressure readings is typically obtained in less than 30 seconds. The gas pressure can then be carefully released so that a reading is obtained when the membrane is returned to the original position; that pressure indicates the in situ water pressure condition (simple hydrostatic or hydrostatic plus excess pore pressure). The blade is then advanced to the next test depth. Because of the speed in obtaining readings, a typical field procedure will test at depth increments of 0.15 to 0.30 m (6 to 12 in.).

There are some disadvantages or limitations to the use of dilatometer equipment; for example, it cannot be used in high-resistance soils and gravelly soils, and no samples are obtained for visual examination or laboratory testing.

The resistance to membrane expansion (i.e., the results of the dilatometer test) relates to the total soil pressure against the blade and is used to indicate in situ soil stress, soil stiffness, and various strength and compressibility parameters. If the difference between an initial dilatometer pressure, p_o, and a final pressure, p_1, is Δp (corrected by calibration factors such as membrane stiffness, etc.) the following dilatometer test indices apply:

Dilatometer modulus, $E_D = 35 \; (\Delta p) = E/(1 - \mu^2)$, where E is the soil modulus of elasticity and μ is the Poisson ratio.

Material index, $I_D = \Delta p/(p_o - u_o)$, where u_o is the pore water pressure at test depth.

Horizontal stress index, $K_D = (p_o - u_o)/\overline{\sigma}_v$, where $\overline{\sigma}_v$ is the vertical stress.

Empirical relationships using these indices to obtain soil properties important for design have evolved from the developers and users of the equipment. Generally, soil type is identified from the I_D value, lateral pressure coefficients and soil strength are related to K_D, and compressibility is related to E_D and K_D [183, 257].

The *pressuremeter* is a device that includes a cylindrical probe that can be pressured and expanded. This equipment differs from the flat-blade dilatometer in that a borehole must be provided before the pressuremeter probe can be inserted (the dilatometer can be advanced to test depth by driving or jacking). Probe pressures are increased in measured increments, and corresponding volume changes are recorded. Correlations between pressure and volume changes are related to soil parameters. Details about the pressuremeter equipment and related foundation design parameters are given in Chapter 13.

5.4 PRESENTATION OF BORING INFORMATION

Boring Logs

Information on subsurface conditions obtained from the boring operation is typically presented in the form of a boring log (boring record). A continuous record of the various soil or rock strata found at the boring is developed. Description or classification of the various soil and rock types encountered and changes in strata and water level data are considered the minimum information that should constitute a log. Any additional information that helps to indicate or define the features of the subsurface material should also appear on the log. Items such as soil consistency and strength or compressibility can be included. "Field" logs typically consist of the minimum information—classification, stratum changes, and water level readings (Fig. 5-29a). A more developed "office" or "lab" log might include laboratory test data presented alongside the boring sample actually tested, so all information pertaining to the various soil or rock types and properties found at a boring location is summarized (Fig. 5-29b).

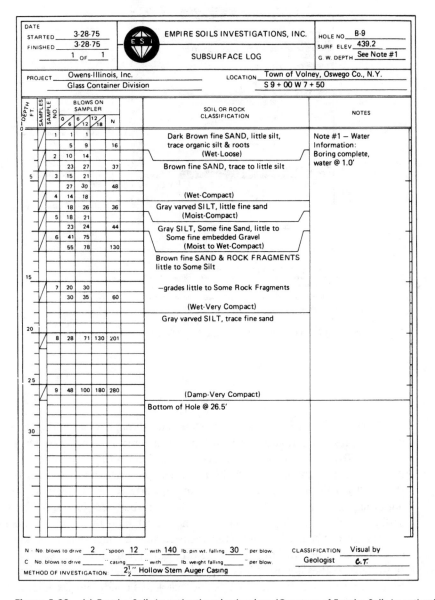

Figure 5-29   (a) Empire Soils Investigations boring log. (Courtesy of Empire Soils Investigations)

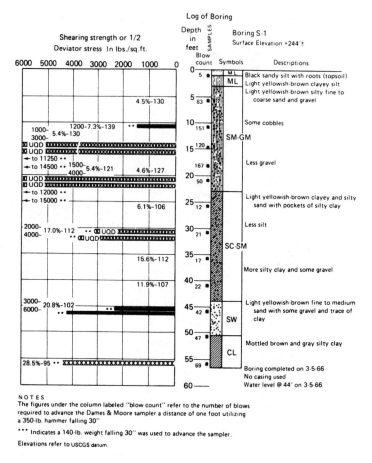

Figure 5-29 (b) Dames and Moore boring log. (Courtesy of Dames and Moore)

Where subsurface conditions are investigated through use of boring or test pits, the exploration program normally consists of making a number of borings or test pits at different locations. To assist in picturing and understanding underground conditions existing at a site, it has become common to use the boring or test pit series to develop a subsurface profile (Fig. 5-30). This profile helps to define the subsurface conditions more clearly, showing uniformity or variation, and can assist in delineating between good and poor areas. If a subsurface profile is developed as data become available (as the borings are completed), the information can be used for indicating the need and general location for additional borings or, conversely, for indicating that some originally planned borings can be omitted. When subsurface profiles are used, it is important to remember that conditions between borings have been estimated by interpolation and that actual conditions are known only at boring and test pit locations.

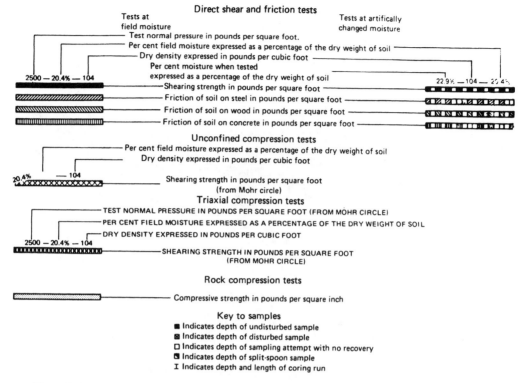

Figure 5-29   (c) Key to test data shown on Dames and Moore log.

Recognizing Limitations of Boring Data

It is accepted that borings are providing information on subsurface conditions only at the actual drilling location and that the standard practice of interpolating between borings to determine conditions does involve some degree of uncertainty. However, it is also important to recognize that there are limitations inherent in the information shown on the typical driller's log. Members of field crews employed by boring contractors are primarily drilling tradespersons. Such individuals typically have limited experience in detailed soil classification and have virtually no familiarity with the importance of subsurface conditions on the features of building design and construction. Important items of information can be innocently passed over by a driller whose major interest is in the rate of drilling progress, for borings typically are contracted for on a footage basis. Consequently, where an exploration is for a final design (and often for preliminary work as well), it is becoming increasingly common to have technically trained personnel assigned to the drilling crew in order to examine and classify recovered soil samples, to direct the depth at which samples should be taken, to select the sequence of drilling borings, and to document other factors relating to

Subsurface section

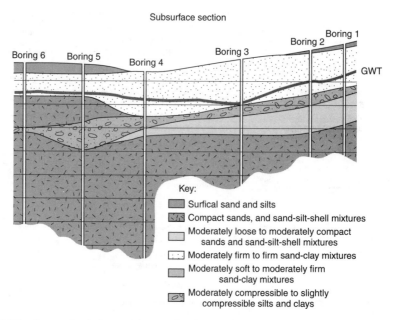

Boring 6   Boring 5   Boring 4   Boring 3   Boring 2   Boring 1

GWT

Key:

☐ Surfical sand and silts

▨ Compact sands, and sand-silt-shell mixtures

☐ Moderately loose to moderately compact
   sands and sand-silt-shell mixtures

☐ Moderately firm to firm sand-clay mixtures

☐ Moderately soft to moderately firm
   sand-clay mixtures

▨ Moderately compressible to slightly
   compressible silts and clays

Figure 5-30   One method of presenting a subsurface profile to summarize boring results.

surface and subsurface conditions that could have an influence on design or construction. Such an individual is typically from a designer's or foundation consultant's office. After the field work is completed, the individual remains an accessible resource person capable of answering questions if issues develop during the design period.

5.5   IN-PLACE TESTING

In-place testing, in the soil and foundation profession, refers to the procedure of determining soil properties or other subsurface conditions at the actual surface or subsurface location. In-place (in situ) soil testing includes determinations of shear strength, permeability, in-place density, plate bearing and settlement, lateral movement, and pore pressure. In-place shear strengths are determined where undisturbed samples for laboratory testing cannot be obtained or where it is desired to eliminate the need to obtain samples. Field permeability determinations offer an advantage over laboratory tests, in that a larger volume of soil is being evaluated in its natural environment. Plate-bearing tests are used to simulate spread footing foundation loading and to obtain load-settlement information on soil that will be supporting shallow foundations or structural slabs (floors, pavements). Lateral movement determinations are of interest where the stability of earth masses and slopes or of embankments or retaining walls is being monitored or if there is a need to know about lateral soil movement caused by foundation or other loading. Underground lateral movements are measured with slope inclination indicator equipment. Methods for determining field permeability are described in Chapter 6. In-place densities are discussed in Chapters

4 and 14. Plate-bearing tests are discussed in Chapter 10. The pressuremeter test (used to determine the volumetric stress–strain properties of a buried soil zone by inserting a probe into a borehole to obtain information directly applicable to foundation design) is discussed in Chapter 13. In-place shear testing, slope inclination equipment, and pore pressure devices are discussed below.

In-Place Shear Tests—Vane Shear

The vane shear test consists of inserting a vane (Fig. 5-31) into the soil and then rotating it by applying a torque. The torque is measured and, for a known-size vane, is easily related to the shearing strength of the soil. In field explorations, the vane testing is typically performed at different depths in a boring as it is being drilled. The soil to be tested should be undisturbed by the boring operation. The vane shear test can be performed on any soil, provided that the vane can be inserted into the soil without causing significant disturbance. Practically, then, the method is useful primarily for fine-grained soils. For typical field or laboratory conditions, where little or no overburden pressure exists, the shear value being determined is primarily cohesion.

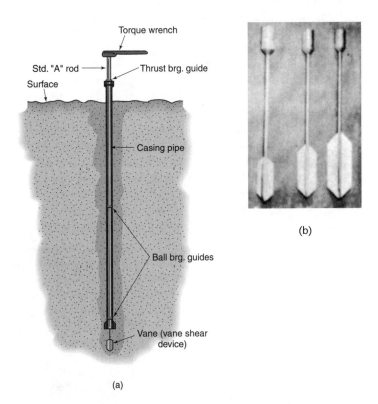

(a)

(b)

Figure 5-31 In-place vane shear test apparatus: (a) schematic diagram of vane shear test procedure; (b) vanes for vane shear testing. (Courtesy of Acker Drill Company)

Slope Inclination Indicator

Slope inclination equipment provides information by indicating the slope or variation from a vertical axis at desired depths in a special casing installed in the ground (Fig. 5-32). A device to determine the variation from a vertical alignment travels in special grooves or tracks on the interior of the casing. The pendulum principle is normally used within the device to establish the vertical axis and for reference. The slope-versus-depth data are recorded. If data collected at different times show a change in slope, this would indicate that lateral movement is occurring. Knowing the slope at different points from top to bottom of the casing permits the lateral position at any depth to be calculated.

Pore Pressure and Piezometer

Pore pressure determination refers to the procedure of measuring the water pressure values developed in the void or pore spaces of an underground soil mass. Usually, the interest is in the "excess" pore water pressure, or the magnitude of pressure greater than a normal hydrostatic pressure resulting from the position of the groundwater table (phreatic surface). Excess pore water pressures can develop in fine-grained soil strata when a new structural loading is placed on the earth above the fine-grained soil, as discussed in Section 10.4, Consolidation. Excess pore pressures provide information on the magnitude of stresses caused in the fine-grained soil by the new loading. On a typical construction project, pore pressure information is utilized to control the rate at which new loading occurs, to keep the stress resulting in the supporting soil from exceeding the strength of that soil.

The *piezometer* is a device to measure pore water pressure. In its simplest form, a piezometer would consist of an open tube or standpipe, with its tip inserted into the soil layer to be checked. Before construction, the water level observed would be the phreatic surface level. As construction loading causes underground stresses to increase, the water level in the open standpipe rises. The rise in elevation multiplied by the unit weight of water gives the magnitude of excess pore pressure. As the stressed soil consolidates, the pore pressure decreases and the water level in the piezometer drops.

Figure 5-32 Slope indicator in use on a field project (equipment shown is a Soiltest slope meter probe). (Courtesy of Soiltest, Inc.)

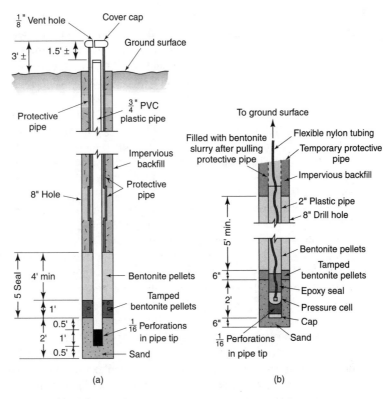

Figure 5-33 Illustrations of piezometer types: (a) open tube type for use in permeable soil; (b) pressure cell for use in impermeable soil.

The simple piezometer just described would be limited to use for soil having a fairly high permeability. With fine-grained soil of low permeability, the time lag for changes in pore water pressures to be measured would be excessive. To indicate changes in pore pressures quickly, modifications (some extensive) have been made to the basic piezometer. The result is that some sophisticated types of equipment have evolved. Most types still include the piezometer unit, which has to be installed to the depth where readings are to apply, and pressure or gage lines, which are carried to a monitor located on the ground surface. Some piezometer types have special conditions for which they are best suited. With some complex units, installation has to be accomplished by specially trained personnel. The diagrams in Figure 5-33 show some piezometer types.

5.6 GROUNDWATER, SOIL WATER, AND SOIL GAS SAMPLING

Obtaining samples of groundwater and soil water is an important phase of any program to monitor or protect a subsurface environment. If a suspected liquid seeps at the ground surface, say from the face of an earth slope or from an artesian condition, or from an open pit, samples can easily be obtained. While any procedure which permits a sampling of ground liquids can be appropriate, monitoring and observation wells are a very commonly used method to obtain water samples that permit checking for the presence of contaminants (Fig. 5-34a). Monitoring and observation wells are typically constructed using one of the procedures for making subsurface borings and installing water supply wells, whereby manual excavation procedures or procedures involving specialty boring or drilling equipment are used (Fig. 5-34b). Additionally, the well may require a protective casing to protect the hole from cave-in. The selection of materials must be such that reactions between the groundwater and a well casing or subsequently used sampling equipment are prevented.

A difference between a monitoring well (or observation well) and a conventional cased-boring or water supply well is the necessity that a perforated casing (perforated with screening or thin slots) be used at the depth where the groundwater sample is to enter into the casing (Fig. 5-34c).

Water samples can be obtained from an open well or open excavation such as a test pit, but the depth that is the source of the contaminated water may not be identified. If a water sample from a particular depth in the well is desired, methods to seal the well below and above the test zone will be necessary. Water samples can be obtained from different depths by using a cluster of individual wells with each extending to the desired depth, or by installing a multiple of different-length casings in a single large-diameter boring (Fig. 5-35).

Water samples may be obtained from locations below the groundwater table or from the vadose zone (above the water table). Procedures and equipment needed for obtaining samples above the water table are different from the procedures and equipment for sampling below the water table. Below the water table, groundwater tends to flow into the well opening because of gravity effects, but above the water table it is common to use suction or vacuum methods to draw water held in soil voids by capillary and adsorption forces.

Wells typically range between 50 mm and 150 mm (2 in. and 6 in.) in diameter. The smaller-diameter wells are preferred where the groundwater flow rate is limited, because a smaller volume of water inflow is necessary to obtain adequate samples. The minimum well size is usually controlled by the size of pumping and sampling equipment planned for the well.

The installation of a monitoring well usually represents an intrusion into previously undisturbed natural earth. The wall area of the well excavation is disturbed because of the smearing disturbance created during the installation, a factor which can later interfere with the flow of liquids into the completed well. The related passage of contaminants into the well could be restricted, an undesirable condition if the presence and concentration is to be determined. The performance of a well, and information obtained from water sampling, can be affected by the skill and care of the well installers.

Following the installation, it is usually necessary to clear the new well of water used for the construction and of the residue sediment accumulated during the installation, to ensure that subsequently obtained water samples are true groundwater samples unmixed with other liquids. A surging and flushing process is used whereby water is forced into the well, then removed; this procedure clears sediment and accumulated water and also helps clear the soil smear along the well wall. Where the well is to be used to obtain groundwater samples over a period of time, as is typical, the top of the well is capped and the surficial soil zone surrounding the well is sealed, to prevent surface water from entering the well and surrounding earth.

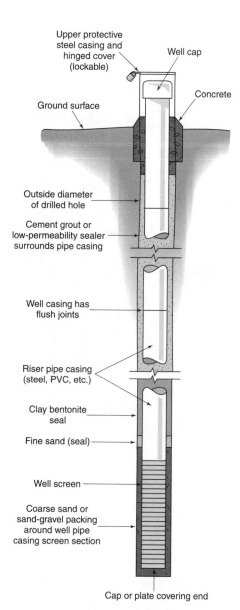

Upper protective steel casing and hinged cover (lockable)

Well cap

Concrete

Ground surface

Outside diameter of drilled hole

Cement grout or low-permeability sealer surrounds pipe casing

Well casing has flush joints

Riser pipe casing (steel, PVC, etc.)

Clay bentonite seal

Fine sand (seal)

Well screen

Coarse sand or sand-gravel packing around well pipe casing screen section

Cap or plate covering end

(a) Schematic – typical monitoring well, screened over one vertical interval (compiled from various sources)

(b) Installation of observation-monitoring well at a gasoline station, to check for leaks from buried storage tanks

(c) Perforated well-casing; small slotted openings in lower sections are not visible in photograph

(d) Removable cover for observation-monitoring well, to permit periodic sampling

Figure 5-34   Observation-monitoring well information.

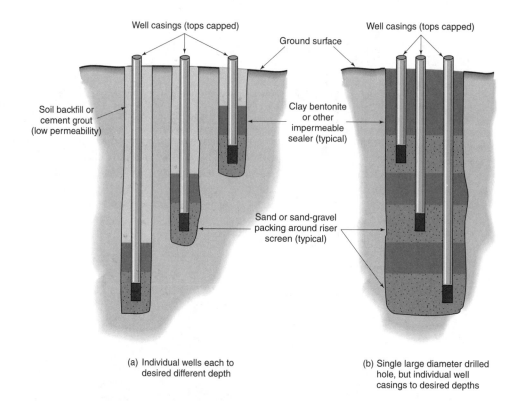

Well casings (tops capped)

Ground surface

Well casings (tops capped)

Soil backfill or
cement grout
(low permeability)

Clay bentonite
or other
impermeable
sealer (typical)

Sand or sand-gravel
packing around riser
screen (typical)

(a) Individual wells each to
desired different depth

(b) Single large diameter drilled
hole, but individual well
casings to desired depths

Figure 5-35   Well cluster to monitor different depths.

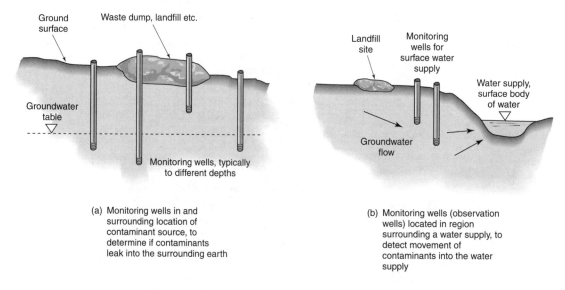

Ground
surface

Waste dump, landfill etc.

Landfill
site

Monitoring
wells for
surface water
supply

Water supply,
surface body
of water

Groundwater
table

Monitoring wells, typically
to different depths

Groundwater
flow

(a) Monitoring wells in and
surrounding location of
contaminant source, to
determine if contaminants
leak into the surrounding earth

(b) Monitoring wells (observation
wells) located in region
surrounding a water supply, to
detect movement of
contaminants into the water
supply

Figure 5-36   Concept of locations for monitoring wells.

The locations selected for monitoring wells usually are adjacent to an area suspected of, or being checked for, escaping contaminants (for example, monitoring the boundary area of a waste dump) or are surrounding a location being protected (such as water supply sources or wells) (Fig. 5-36). Tracers placed into suspected sources of contamination can expedite the process to determine if nearby water well supplies are being affected. Where more than one location is the possible source of groundwater contamination, tracers are most appropriate; a different tracer material is added at each suspected source, and the type of tracer(s) subsequently found at the monitor wells establishes the link to the proper source.

Tracers can be any type of matter—liquid, solid, or gaseous—which can be transferred into the ground at or near a suspected contamination source and flow with characteristics similar to groundwater and the suspected contaminant. The tracer should not be a hazardous material or pose a danger to the environment (e.g., a radioactive material). The methods required to recover samples of the tracer reaching the monitoring points and then to perform analysis to indicate a presence and concentration should be simple. Dyes are commonly used. Water at elevated temperature can be used if flow distance and time are not great. Solid particles, yeast, some types of bacteria, ionic compounds, liquids which retain their identity, and inert gases also can be used. Radionuclides (radioactive isotopes) function well as tracers, but their use is questionable and may be regulated because of the potential dangers.

Water sampling equipment, the equipment used to "grab" a sample of water from a known depth inside the well casing, ranges from very basic, manually operated items to sophisticated, electrical-mechanical-pneumatically operated equipment.

The simplest, bailers, are used for open wells that are screened over one vertical interval (Fig. 5-37a). Bailers are basically open cups or sections of open tubing or piping provided with ball-type check valves. The bailer is lowered to the desired water level depth in the well using a cord line, then withdrawn to the surface. The cup bailer is a container with the top open. Check-valve bailers have a ball-and-seat check valve at the bottom of the bailer tube, or at both top and bottom. With the single, bottom-check-valve-type bailer, the check valve remains open as the bailer is lowered into the water but closes when the bailer is withdrawn because of the downward weight of the water inside the bailer. The double-check-valve bailer (check valve at both bottom and top of the bailer) functions much the same; both check valves are in the open position as the bailer is being lowered but both close as the water sample is being raised to the surface.

The syringe-type sampler illustrated in Figure 5-37b obtains the water sample after being lowered to the desired depth in the well by having a vacuum or negative pressure induced in the container via a tube connected to a pump at the ground surface. Water pressure in the well, in excess of the reduced pressure inside the syringe container, forces water through the syringe needle. The syringe sample is brought up to the surface, where the tip is sealed so the unit can be used to store and transport the water sample.

Wellwater samples can also be obtained by using negative pressure displacement (suction lift) and positive displacement methods. The negative (suction) pressure method, depicted by the schematic diagram of Figure 5-37c, involves use of pump equipment to create a reduced pressure (vacuum pressure) in a container which has a tube extending into the wellwater at the desired sampling depth. Atmospheric and hydrostatic pressure acts on water in the well, then forces the water sample up the tube in the collection container. The height of lift relates to the vacuum pressure which the pump can develop, but typically is limited to about 7 m, or 20 ft. This equipment can also be used to obtain samples of gas that

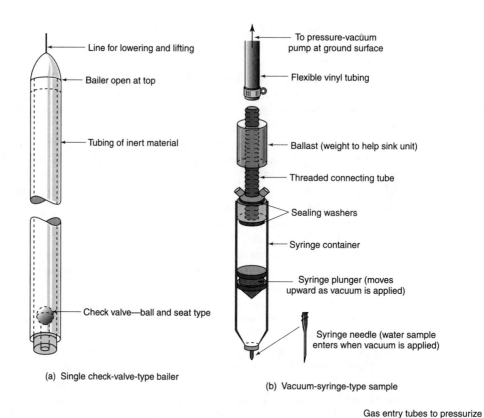

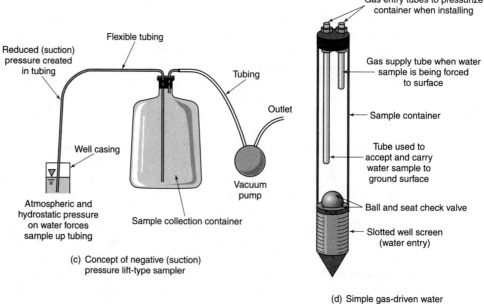

Figure 5-37 Illustrations of common monitor well water samplers.

lies in the well above the water level. The positive displacement methods involve use of conventional-type well-pumping equipment (piston-type, submersible, venturi-jet, etc.), either placed within the well or operating from the ground surface.

Gas-driven water samplers are represented by the simple-in-principle device illustrated in Figure 5-37d. More sophisticated variations exist. To operate the sampler shown, a positive gas pressure is first created in the sampler container using a pressurized supply at the surface, to hold the check valve closed. At the desired sampling depth, the gas pressure is released, and water flows into the sampler. A positive gas pressure is again imposed, a procedure which both causes the check valve to shut and forces the water sample up the discharge tube to be collected at the ground surface.

To obtain samples of soil water in the zone above the water table (the vadose zone), suction equipment is necessary to free water held in the void spaces by capillary and adsorption forces (this soil moisture will not flow into a sampler by the pull of gravity). The concept of a sampler commonly used for the vadose zone, the pressure-vacuum lysimeter, is illustrated in Figure 5-38. This type of sample collector consists of a long container, usually of a rigid plastic material such as PVC, having a cup base made of porous ceramic. In use, the sampler is sealed in place at the desired depth in a boring, so a vacuum will be effective for drawing soil water inward through the porous cup. The lysimeter is provided with two tubes that will extend to the ground surface for connection to a pressure-vacuum pump and a sample holder bottle. To obtain a soil water sample, the discharge tube is closed (or clamped) and the pump is used to create a vacuum in the sampler. The difference between liquid and air pressure in the soil voids and the reduced (vacuum) pressure in the sampler forces a flow through the porous ceramic cup. The collected water sample is then delivered to the surface sample bottle by the procedure of opening (unclamping) the discharge tube and using the pump to deliver air or gas into the sampler at a pressure sufficient to force an upward flow. The method is capable of obtaining soil water samples from locations relatively deep in a monitoring well, but the pressures induced to recover the collected water sample cannot be too great because of the possibility of forcing the water back out through the ceramic cup into the surrounding earth.

Soil gas samples can be obtained using grab sampling techniques or by passive sampling.

Passive sampling involves placing (burying) a sorbent material such as sorbent charcoal in the soil zone being monitored for a period of time (days or weeks, typically). Contaminants diffused through the soil gases are sorbed (trapped) on the charcoal. The retrieved charcoal sampler can then be delivered to the laboratory, where desorption and chemical analysis are performed.

Grab sampling refers to the procedure for obtaining a small volume of the soil gas that is present at the time of sampling. Grab samples are classified as static (the sample is obtained from a more or less immobile body of gas) or dynamic (the sample is obtained from an actively moving volume).

The principle for obtaining grab samples is relatively simple—a small probe is inserted to the depth or the location zone to be tested, then suction equipment such as a hand pump or bellows or a mechanical vacuum pump extracts the desired air-gas sample. The penetrating probe may be as small as a hypodermic needle (for shallow depth penetrations) or of heavy tubing (necessary for deeper penetration). Soil gas samples can also be obtained from borings or wells using the lysimeter apparatus shown in Figure 5-38 as well as by working with soil samples extracted from borings (Fig. 5-39a, b). It is possible to obtain

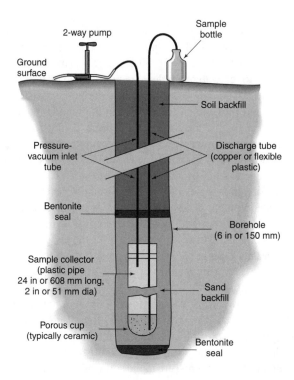

Figure 5-38 Schematic representation of pressure-vacuum lysimeter.

samples of soil gases which reach the ground surface where open observation wells exist (for example, where solid waste fill sites are being monitored). Figure 5-39c shows one type of soil gas sampling probe which can be installed by driving or insertion in a bored hole.

The BAT probe (after Bengt Arne Torstensson, developer) represents another type of apparatus used to obtain samples of soil water/gases in boreholes. The probe type illustrated in Figure 5-40 was developed for use in offshore exploration boreholes drilled in deep waters as undertaken for commercial drilling ventures where the detection of potentially explosive gases in the marine sediments and underlying material is important to planning for the safety of personnel and equipment.

Soil gas analysis methods are particularly effective for determining the presence of volatile organic compounds (VOCs), products commonly escaping from industrial waste fill and solid waste landfills, industrial sites, and so on. Grab samples are often analyzed at the job site in mobile laboratories or with portable instruments (some small enough to be hand held), but the gas samples can also be stored in small glass or stainless steel containers for transportation to a conventional laboratory for testing.

The handheld analyzers typically are limited to indicating gross levels of gases, or total volatiles. One type, the photoionization detector (PID) equipment, is small, reads quickly, and commonly is used to indicate or check for certain components of the volatile compounds (Fig. 5-39d). For identification of individual components, field or fixed laboratory gas chromatography (GC) equipment is necessary, while flame ionization detection (FID) is used to indicate concentrations. The electron capture detector (ECD) is used to

(a) Installation of montoring well at an industrial site. Drilling crew are wearing protective suits

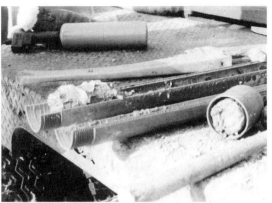

(b) Soil sampling at monitoring well installation, to check for contaminants

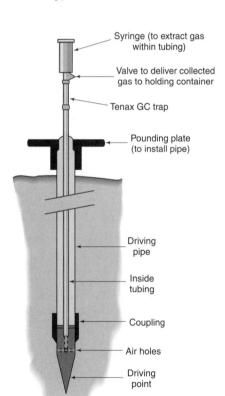

Syringe (to extract gas within tubing)

Valve to deliver collected gas to holding container

Tenax GC trap

Pounding plate (to install pipe)

Driving pipe

Inside tubing

Coupling

Air holes

Driving point

(d) PID instrument to determine presence of soil gas as monitoring well is being installed

(c) Soil gas sampling probe

Figure 5-39   Observation-monitoring well installation and sampling equipment. (Photographs courtesy of Atlantic Testing Laboratories, Ltd., Canton, New York)

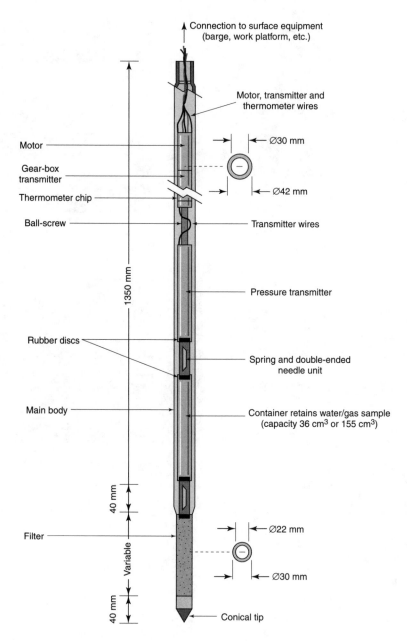

Figure 5-40 Offshore BAT probe [234].

analyze chlorinated hydrocarbons. Field GC and FID information usually is not as detailed as the results from a conventional laboratory.

A practical field procedure available for determining the presence of particular gaseous compounds involves the use of gas detector tubes. A large number of different detector tubes, each for covering a different kind of gas, are available from the marketplace. The detector tube is filled with an appropriate solid reagent. When a measured volume of soil gas is passed through the tube, a color change indicates the presence of a particular compound. The types of detector tubes selected relate to the types of gases that need to be detected or investigated. If the tubes are graduated, the concentration is also indicated.

5.7 GEOPHYSICAL METHODS

The determination of subsurface materials through the use of borings and test pits can be time consuming and expensive. Considerable interpolation between checked locations is normally required to arrive at an areawide indication of conditions. Geophysical methods involve the technique of determining underground materials by measuring some physical property of the material and, through correlations, using the obtained values for identification. Most geophysical methods determine conditions over a sizable distance. Frequently, this is an advantage over the "point" checking accomplished by borings and test pits. Most geophysical measurements can be rapidly obtained. Thus, the methods lend themselves well to the checking of large areas.

In the engineering–construction profession, several types of geophysical investigation have been found useful: the seismic refraction method, the electrical resistivity method, and ground-penetrating radar. Though these methods have proven to be reliable, there are also certain limitations as to the data that can be obtained. Thus, at the present time, subsurface investigations can rely heavily on geophysical methods, but conditions should at least be spot-checked with borings or test pits. Typically, when a thorough investigation is made, a number of borings will be required in order to obtain test samples to make accurate determinations of soil properties such as strength and compressibility. It is these borings that can provide the detail required to check and complement the geophysical data.

Seismic Refraction

When a shock or impact is made at a point on or in the earth, the resulting seismic (shock or sound) waves travel through the surrounding soil and rock at speeds relating to the density and bonding characteristics of the material. In refraction seismology, the *velocity* of seismic waves passing through subsurface soil or rock materials is determined, and the magnitude of the velocity is then utilized to identify the material. A seismograph, the instrument used to make a seismic refraction study, consists of a shock- or impact-inducing mechanism, such as an impact hammer or small explosive, plus a receiver to indicate when the seismic wave reaches a point at a particular known distance from impact as well as a timing instrument for measuring the time for the wave to travel the distance from the point of impact to the point of measurement. In shallow refraction seismology, as used for determining subsurface conditions for construction purposes, the shock impact is created with a sledgehammer hitting a striking plate placed on the ground (Fig. 5-41). The seismic wave

Figure 5-41 Seismic refraction study being performed in field. (Courtesy of ELE International, Soiltest Products Division)

is then picked up by a sensitive geophone. (The geophone is actually a transducer, an electromechanical device that detects vibrations and converts them into electric signals that can be measured.)

The field survey involves obtaining a series of geophone readings at different distances along a straight line directed from the impact point. For geophone spacings close to the strike plate, the vibrations picked up by the geophone will be from those direct waves traveling through the upper layer of earth material (Fig. 5-42). For direct waves traveling through the upper layer, the time to reach the geophone is proportional to the distance from the point of impact.

When the surficial layer is underlain by a harder layer, the seismic waves from the strike plate also progress downward and enter the harder layer. The seismic velocity will be greater in the harder material. Waves traveling through the upper portion of the harder layer transfer energy back into the upper layer through the surface of contact. This energy becomes a refracted wave. For large strike-plate-to-geophone distances, the refracted wave will reach the geophone more quickly than the direct wave, even though the path of travel is longer. This occurs because part of the path is through the harder, high-velocity material (Fig. 5-43).

Seismic velocities for the earth materials in the upper and lower strata would be obtained from plotting the measured values of geophone distance and time on a travel-time graph (both coordinate values use an arithmetic scale); see Figure 5-44. The slope of the first segment of the plot represents the seismic velocity for the material in the upper stratum. The slope of the second segment is the seismic velocity for the deeper layer. To obtain data that properly define the plotted segments, it is recommended that the maximum geophone distance be about five times the depth of investigation.

Figure 5-45a illustrates a soil layer–rock layer subsurface and indicates the type of seismic wave (direct or refracted) picked up at the various geophone positions. The resulting travel-time graph appears in Figure 5-45b.

Figure 5-42   Method of imparting sound waves into a soil.

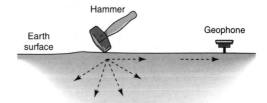

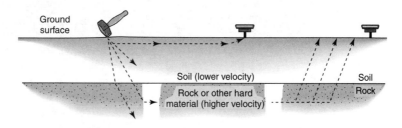

Figure 5-43   Travel of sound waves through different subsurface materials.

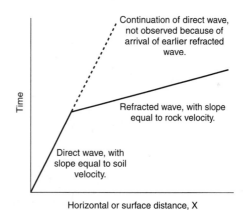

Figure 5-44   Typical travel-time graph for soil overlying rock.

Seismic velocity values representative of different earth materials and conditions are shown in Table 5-2.

Where a two-layer condition exists, the thickness or depth of the upper layer can be determined from

$$H_1 = \frac{X_1}{2}\sqrt{\frac{V_2 - V_1}{V_2 + V_1}} \qquad (5\text{-}1)$$

where H_1 = depth or thickness of the upper layer

   X_1 = distance, taken from the travel-time graph, where the two plotted slopes intersect

V_1, V_2 = seismic velocities in the upper and lower layer, respectively

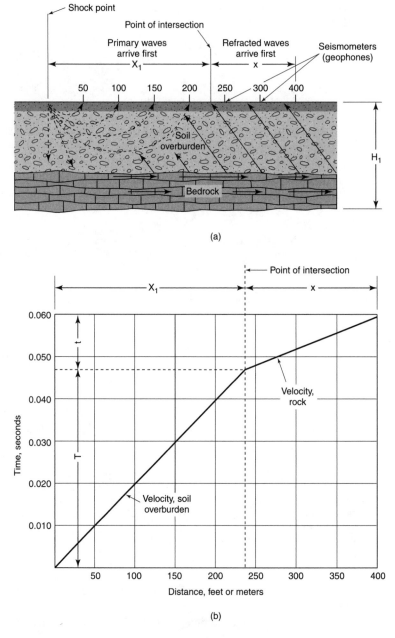

Figure 5-45   Seismic refraction conditions and data: (a) subsurface conditions showing direct and refracted sound waves; (b) travel-time plot for conditions in (a). (Courtesy of Acker Drill Company)

Table 5-2 Representative Seismic Velocity Values (Velocity in ft/sec and m/sec)

Soil—Unconsolidated Material	ft/sec	m/sec
Most unconsolidated materials	Below 3000	Below 900
Soil—normal	800–1500	250–450
—hard-packed	1500–2000	450–600
Water	5000	1500
Loose sand—above water table	800–2000	250–600
—below water table	1500–4000	450–1200
Loose mixed sand and gravel, wet	1500–3500	450–1100
Loose gravel, wet	1500–3000	450–900
Hard clay	2000–4000	600–1200

Rock—Consolidated Material	ft/sec	m/sec
Most hard rocks	Above 8000	Above 2400
Shale—soft	4000–7000	1200–2100
—hard	6000–10,000	1800–3000
Sandstone—soft	5000–7000	1500–2100
—hard	6000–10,000	1800–3000
Limestone—weathered	As low as 4000?	1200?
—hard	8000–18,000	2400–5500
Basalt	8000–13,000	2400–4000
Granite and unweathered gneiss	10,000–20,000	3000–6000
Compacted glacial tills, hardpan, cemented gravels	4000–7000	1200–2100
Frozen soil	4000–7000	1200–2100
Pure Ice	10,000–12,000	3000–3700

Note: Occasionally, formations may yield velocities that lie outside of these ranges.

Source: Soiltest, Inc.

For the condition of three successively harder layers existing in an area, the travel-time graph will show three different slopes. The seismic velocity for each of the materials is the slope of the respective segment of the plot. The thickness of the upper layer can be calculated from Equation 5-1. The thickness of the intermediate layer, H_2, can be determined from

$$H_2 = .85H_1 + \frac{X_2}{2}\sqrt{\frac{V_3 - V_2}{V_3 + V_2}} \tag{5-2}$$

where H_1 = thickness of the upper layer

X_2 = distance from the travel-time graph, where plotted segments 2 and 3 intersect

V_2, V_3 = seismic velocities of layers 2 and 3 as determined from the travel-time graph

There are certain significant limitations to the use of the seismic refraction method for determining subsurface conditions:

1. The method should not be used where a hard layer overlies a softer layer, because there will be no measurable refraction from a deeper soft layer. Refraction seismic test data from such an area would tend to give a single-slope line on the travel-time graph, indicating a deeper layer of uniform material.

2. The method should not be used on an area covered by concrete or asphalt pavement, because these materials will represent a condition of a hard surface over a softer stratum.

3. A frozen surface layer may give results similar to those obtained where a hard layer is over a soft layer because of the velocity increase resulting from the better wave transmission through the more "solid" frozen material.

Further, some topographic and underground features will give seismic data that are difficult to interpret fully and correctly. Such situations include the condition of an irregular or dipping underground rock surface, the condition where discontinuities such as rock faults or earth cuts or banks exist, the condition where layers having gradual changes in their velocity values occur, and the condition of thin layers of varying materials. Because of the possibility for misinterpretation of data with such occurrences, the seismic analysis should be performed by trained personnel, and, as a minimum, spot-checks should be made with borings or test pits.

Where rock exists at or close to the surface at a construction site, predetermining the methods necessary for removal (to lower the surface elevation) or for excavation becomes important for estimating, bidding, and scheduling purposes. Seismic velocity data have been used to determine when the rock material is capable of being ripped with dozer rippers (or other similar equipment) and when drilling and blasting are required. Such seismic rippability information, as developed by one major construction equipment manufacturer, is shown in Figure 5-46.

Electrical Resistivity

Resistivity is a property possessed by all materials. The electrical resistivity method for determining subsurface conditions utilizes the knowledge that in soil and rock materials,

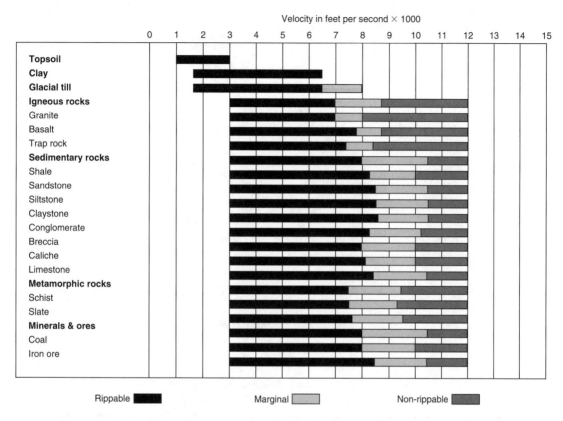

Figure 5-46 Seismic rippability chart developed by Caterpillar Tractor Company for D9 with moulded No. 9 ripper. (Courtesy of Soiltest, Inc.)

the resistivity values differ sufficiently to permit that property to be used for identification purposes.

To determine resistivity at a site, electrical currents are induced into the ground through the use of electrodes. Soil or rock resistivity can then be determined by measuring the change in electrical potential (voltage) between known horizontal distances within the electrical field created by the current electrodes (Fig. 5-47).

A schematic diagram of equipment for resistivity testing, including electrode arrangements (the Wenner configuration), is shown in Figure 5-48. With four electrodes equally spaced along a line, as indicated, the resistivity is calculated from

$$\beta = 2\,\pi S\!\left(\frac{V}{I}\right) \tag{5-3}$$

where β = resistivity of the earth material in ohm-feet

S = electrode spacing

V = difference in potential (volts) between the inner electrodes

I = current flowing between the outer electrodes

Figure 5-47 Path of current flow (solid lines) through a soil in the electrical resistivity study (dotted lines are equipotential surfaces).

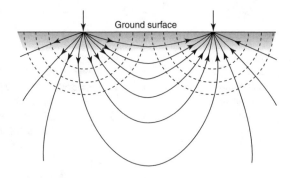

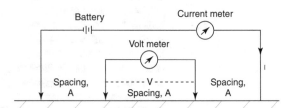

Figure 5-48 Wenner configurations for arrangement of voltage and current electrodes.

The calculated resistivity value is an apparent resistivity. This is a weighted average of all the earth material within the zone created by the electrodes' electric field. The depth of material included in the measurement (depth of penetration) is approximately the same as the spacing between electrodes.

Two different field procedures are in use for obtaining information on subsurface conditions. One method, *electrical profiling,* is well suited for establishing boundaries between different underground materials and has practical application in prospecting for sand and gravel deposits or ore deposits. The second method, *electrical sounding,* can provide information on the variation of subsurface conditions with depth and has practical application in indicating layered conditions and approximate thicknesses. It can also provide information on depth to a water table or water-bearing stratum. Figure 5-49 shows field work for an electrical resistivity study being performed.

In the *electrical profiling* method, an electrode spacing is selected, and this same spacing is used in running different "profile" lines across an area (Fig. 5-50). The information obtained applies for the particular location of the line and is sometimes referred to as an "electrical trench," implying that the subsurface data revealed are similar to the information that would be obtained by examining an open-trench excavation.

The information resulting from a profile line can most simply be plotted on arithmetic coordinates, as shown in Figure 5-51. A change in the plotted curve indicates a change in the underground material. From the series of profile lines, boundaries of areas underlain by different materials can be established on a map of the area (areal map).

In the *electrical sounding* method, a center location for the electrodes is selected and a series of resistivity readings is obtained by systematically increasing the electrode spacing,

Figure 5-49 Electrical resistivity field study being performed. (Courtesy of Soiltest, Inc.)

Figure 5-50 Electrode arrangement in a profiling survey.

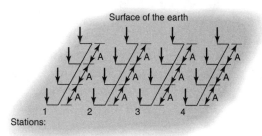

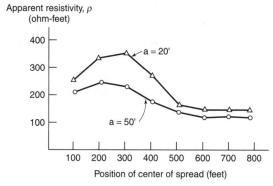

Figure 5-51 Apparent resistivity versus position of the center of electrode spread for two values of electrode separation (Courtesy of Soiltest, Inc.)

Figure 5-52 Representative electrode position during a sequence of sounding measurements.

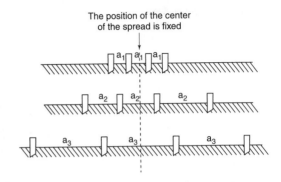

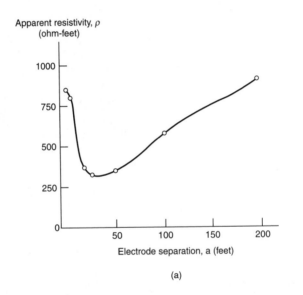

(a)

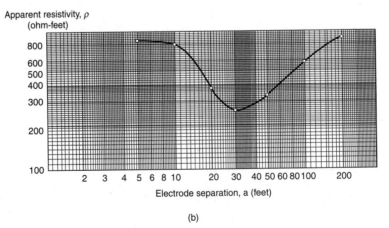

(b)

Figure 5-53 Illustration of two methods for graphing sounding data: (a) Cartesian (arithmetic) coordinates; (b) logarithmic coordinates.

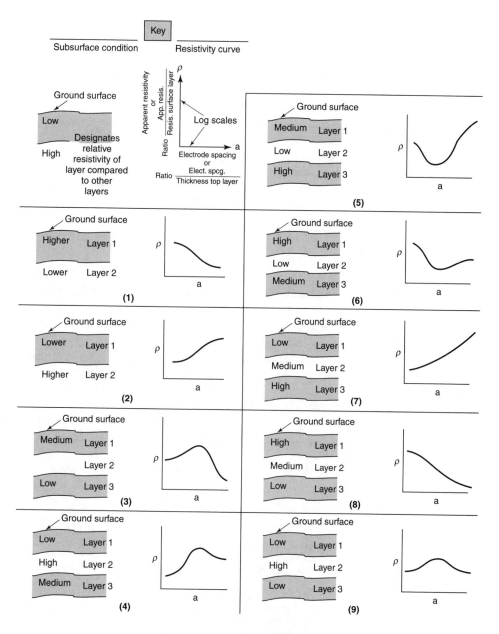

Figure 5-54 Representative resistivity curves for differing subsurface conditions.

as indicated by Figure 5-52. Since the *depth* of information recovered is directly related to electrode spacing, the series of resistivity data obtained from successively increased electrode spacings will indicate changes of resistivity with depth and, hence, information on layering of materials. This method is capable of indicating subsurface variations where a hard layer underlies a soft layer and also the condition of a soft layer underlying a hard layer. Data can be presented on Cartesian (arithmetic-scale) coordinates or logarithmic coordinates, as indicated in Figure 5-53. Logarithmic coordinates are the most popular.

The spacing of electrodes is important. To obtain data that give a reliable indication of conditions, the closest electrode spacing should be no more than half the estimated thickness of the upper stratum. Three feet is the minimum recommended spacing, however. The largest spacing is between 5 and 10 times the total depth of interest. The number of intermediate spacings selected should be adequate to provide sufficient points to plot a well-defined curve.

Representative sounding curves for some typical subsurface conditions appear in Figure 5-54.

Correlation between Resistivity and Earth Materials

In earth materials, resistivity decreases with increasing water content and increasing salt concentration. Increasing degrees of water content and salinity make it easier for an electrical current to flow through the material. Consequently, nonporous materials (holding little water) will have high resistivity values. Such materials include most igneous and metamorphic rock plus some dense sedimentary rock, such as dense limestone and sandstone. In soil materials, clean gravel and sand have a relatively high resistivity value. Silts, clays, and coarse-grained–fine-grained soil mixtures have comparatively low resistivity values. Soil formations in nonglaciated areas typically have lower resistivity values than soils in glacial areas.

Representative values of resistivity for commonly occurring earth materials are presented in Table 5-3.

Table 5-3 Representative Resistivity Values

Types of Materials	Resistivity (ohm-ft)
Wet to moist clayey soils	5–10
Wet to moist silty clay and silty soils	10–50
Moist to dry silty and sandy soils	50–500
Well-fractured to slightly fractured bedrock with moist soil-filled cracks	500–1000
Sand and gravel with silt	1000
Slightly fractured bedrock with dry soil-filled cracks; sand and gravel with layers of silt	1000–8000
Massive bedded and hard bedrock; coarse dry sand and gravel deposits	8000 +

Source: Soiltest, Inc.

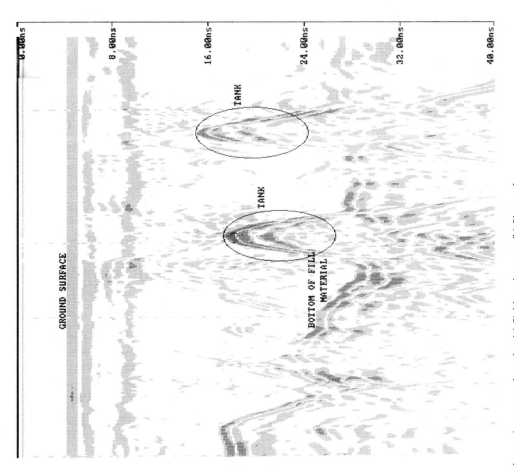

(b)

(a)

Figure 5-55 Ground-penetrating radar equipment and results: (a) Field equipment; (b) Photo of radar wave data on field CRT monitor, indicating presence of buried metal storage tanks (monitor screen is in color; waves shown here in black and white are actually different colors to enhance recognition of different materials). Top of tanks is about 4 ft (1.3 m) deep. (Courtesy of Geophysical Survey Systems, Inc., North Salem, New Hampshire)

183

Thickness of Layers

A method of obtaining a reliable and accurate determination of the depth to, or thickness of, soil layers by using resistivity data is currently not available. Approximation methods exist, but they are cumbersome to handle and not particularly suited for use in field work. With present knowledge, depth or thickness information is best obtained from borings or by seismic methods.

Ground-Penetrating Radar

Ground-penetrating radar is capable of defining the shallow zones of soil and rock materials that underlie an area. The method relies on the penetration and reflection of high-frequency radio waves. For implementation, the equipment is moved across the ground surface along a path whose depth profile is to be outlined (Fig. 5-55a). Radio waves transmitted continuously into the earth then reflect back to the equipment's receiving antenna in accord with the electrical properties of the penetrated material. Notable variations in the reflected signal occur where a change (interface) of material exists (Fig. 5-55b). Typically, the method has been limited to evaluating depths of 10 meters or less. The deeper readings have been obtained in dry, coarse-textured materials such as granular soils. When electrically conductive soils such as wet, fine-grain silt and clay deposits are present, the evaluation is restricted to shallower depths. The most reliable application of the method has been for delineating sharp contrasts in subsurface materials (such as soil to rock), but the method also has been found appropriate for identifying the presence of buried tanks and pipes.

––––––––––––––––––––––––––––– PROBLEMS –––––––––––––––––––––––––––––

5-1. What publicly available maps serve as good references for obtaining information about surface conditions in an area?

5-2. List and describe some of the advantages of learning about land surface features from aerial photographs.

5-3. Suggest a list of procedures that would be practical to implement in order to obtain information on surface and subsurface conditions appropriate for construction planning at a proposed project site that extends over a large area that is presently in a natural, undeveloped state.

5-4. Outline the general procedure used to drill soil borings and obtain soil samples for classification and testing.

5-5. In soil borings, what is the standard penetration test?

5-6. Soil borings are drilled to investigate subsurface conditions at a planned construction site. Soil samples are obtained using the standard penetration test procedure. The drilling equipment utilizes an automatic hammer for driving the soil sampler. Resistance blow counts are $N = 18$, $N = 22$, $N = 23$, and $N = 19$ for samples obtained from the topmost soil stratum. Assume that the blow count data will be applied to preliminary foundation design using foundation bearing capacity information based on a sample drive hammer efficiency of 60 percent. Modify the blow count values obtained with the automatic hammer to relate to the N_{60} data.

5-7. Compare advantages and disadvantages of test pits versus soil borings for obtaining information on subsurface conditions.

5-8. Make a comparison between the static cone penetrometer and standard penetration test methods for determining subsurface soil conditions.

5-9. What are the practical limitations to information provided by soil borings and cone penetration tests?

5-10. Prepare a boring log from the following information, as obtained from a 2½-in. boring where standard split spoon soil samples were taken.

Sample Depth (ft)	Blow Count N	Soil Classification
2–3½	3-4-4	Brown fine to medium sand
7–9½	7-8-7	Brown fine to medium sand
12–13½	8-9-10	Brown fine to medium sand in upper part of sample; red-brown clay in lower part of sample
17–18½	9-12-11	Red-brown clay
22–23½	9-25-27	Gray fine to coarse sand, silt and clay, occasionally gravel (compact glacial till)
27–28½	29-35-39	Gray fine to coarse sand, silt, clay with gravel (compact glacial till)
31–31½	100	Refusal (no sample recovered)

Water level encountered at 21 ft.

5-11. Soil bearings at a site planned for a construction project indicate the presence of a buried sand layer which is below the elevation of the groundwater table. The blow count for the standard penetration test soil samples in the sand layer indicates that the correct N equals 22.

(a) Apply the Chinese criteria for $N_{critical}$ to determine if the sand stratum is vulnerable to liquefaction during an earthquake event, and indicate if the site has high, intermediate, or low potential for having earthquake damage occur. The depth to the groundwater table is 3 m, and the depth to the sand stratum is 5 m. The sand layer includes approximately 5 percent clay-size particles. Evaluate results for the Intensity IX earthquake condition.

(b) Determine the change in $N_{critical}$ if the sand stratum is free of clay content.

5-12. How does obtaining samples of soil groundwater and soil gas relate to diagnosing and monitoring subsurface contamination?

5-13. For groundwater sampling in monitoring wells, what procedures for installing the well are necessary to ensure that the depth indicated for the water sample is accurate?

5-14. What is the main difference in the procedure required for obtaining a sample of groundwater from below the groundwater table and from above the water table in a monitoring well?

5-15. How might samples of soil gas be obtained from a desired depth below the ground surface?

5-16. (a) Briefly describe the principles on which seismic refraction studies for subsurface explorations are based.

(b) Briefly describe the principles on which electrical resistivity studies for subsurface explorations are based.

5-17. List and briefly describe the type of subsurface information that seismic refraction studies can provide and the limitations on information that can be obtained.

5-18. List and briefly describe the type of subsurface information that electrical resistivity studies can provide and the limitations to data that can be obtained.

5-19. A seismic refraction study made for an area provides the following field data:

Distance from Impact Point to Geophone (ft)	Time to Receive Sound Wave (sec)
50	0.025
100	0.05
200	0.10
300	0.11
400	0.12

(a) Graph the time-travel data and determine the seismic velocity for the surface layer and the underlying layer.

(b) Determine the thickness of the upper layer.

(c) Using the seismic velocity information, give the probable earth materials in the two layers.

5-20. List and briefly discuss the major advantages and disadvantages of in-place shear tests, such as those performed by the in-place vane shear.

5-21. Piezometers are installed into a buried stratum of compressible clay to monitor excess pore water pressures during the placement of earth fill in an area where construction of a new highway is taking place. The top of the buried clay layer and the area groundwater table are at a depth of 10 m below the natural ground surface. Initially, the water level reading in the piezometers is at a depth of 10 m. When a 5 m thickness of soil fill is placed over the area, the water level reading in the piezometers rises to an elevation 9 m above the top of the clay layer. What is the stress increase at the level of the clay layer in kPa (i.e., what is the excess pore water pressure that develops)?

PART II

Engineering Properties and Behavior of Soil Deposits

CHAPTER 6

Movement of Water through Soil

Basic Hydrogeology, Subsurface Flow, Permeability, Capillarity

Water is one of the most abundant substances found on this planet. Water can be found in any of the three phases of matter—gas, liquid, or solid—in what are considered normal conditions of nature. Water is a relatively stable material, but it also combines easily with many other substances and permits other solids, liquids, and gases to mix easily into solution. Water has had a profound effect on the development of the planet, and it is essential for many forms of life.

Water is present on the surface of the planet, below the surface, and in the atmosphere. Close to three-fourths of the earth's surface is covered by water. Of the planet's total quantity of water, approximately 97 percent is saltwater, as found in the oceans. The remaining water is classified as "fresh water," even though other materials may be present in solution. Almost three-fourths of the fresh water volume is frozen, locked up in the planet's polar ice caps and glacial ice. Less than one percent of the planet's water is mobile, fresh water, and of that amount more than half is very deep subsurface water (not readily accessible for use at and near the surface). Remarkably, then, much of today's continental topography and plant and animal life, including some marine life, are the result of functioning with approximately one-half of one percent of the planet's supply of water. Nevertheless, the volume of fresh water at, near, and above the earth's surface—that volume influential and accessible to humans—is considerable (see Table 6-1).

Most earth scientists believe that the total volume of water on the planet has been relatively constant from the time the earth reached its present form (i.e., having solid land masses capable of supporting life, the various ocean areas, and an atmosphere). Some of the present-day water supply is lost through the atmosphere when photo-dissociation occurs (the separation of hydrogen and oxygen molecules, with the hydrogens escaping into space), but new water (juvenile water) is also created when gases and liquids escape from the depths of the earth. Surface and near-surface water can be very mobile, however, in form and in location,

Table 6-1 Earth's Water: Estimated Quantity and Distribution

Location and Type	Volume (km3)	Percent of Total
Oceans (saltwater)	1,300,000,000	97+
Ice caps, glaciers (fresh water)	29,000,000	2+
Lakes, rivers, other land surface bodies, fresh	125,000	
Groundwater, fresh (shallow, zone less than 0.8 km deep)	4,000,000	less than 1
Groundwater, fresh (deeper than 0.8 km)	5,000,000	
Atmosphere	13,000	

and the volume in any area is very susceptible to change, even over short periods of time. Natural forces, including the activities of animal and plant life, cause water to move. Humans use large volumes of water in their activities, but as with the effects of other animal and plant life, the usage of such water is actually a "borrowing" or "moving" instead of a "using up."

An important point is that moving water affects the properties and behavior of soil and can influence both construction operations and the performance of completed construction. Since groundwater conditions are frequently encountered on construction projects, those in the construction profession have found it necessary to understand how movement of water through soil can occur and its possible effects. The discussion on permeability (hydraulic conductivity) and capillary action in soil in this chapter relates to type and manner of water movement. In Chapter 7, the discussion on drainage, seepage, and frost heave relates to practical effects of water movement; the discussion on soil and groundwater contamination expands on environmental effects and concerns.

6.1 BASIC HYDROGEOLOGY

Subsurface Flow, Basic Facts

The cycle of changes and movements that surface water and shallow groundwater repeatedly passes through is referred to as the *hydrologic cycle*. The major features of this cycle, indicating typical phases of movement, use, and recycling, are depicted by Figure 6.1. *Hydrogeology* is the study of the groundwater phase of the hydrologic cycle (both surface and subsurface) related to the effects of geophysical features of soil and rock formations.

Precipitation reaching the earth's surface tends to either flow over the ground surface as dictated by gravity or to infiltrate into the ground. Precipitation may infiltrate after it has traveled some distance as surface water. The features of nature that influence the rate of infiltration include the slope of the land, the presence or lack of plant life, and the porosity of the soil or rock. Where soil is present, water seeping into the ground may be prevented from continuing movement by factors of attraction, such as capillarity and adsorption, or may be taken into plant root systems. Infiltrating water that does continue to migrate due to gravity effects is drawn to the underground zone where theoretically all of the pore or void spaces in the soil are filled with water. The surface or upper boundary of this saturated-with-water zone is identified as the *water table* or *phreatic surface*. The water present below the phreatic surface creates a positive hydrostatic pressure so that a total of air plus water pressure is

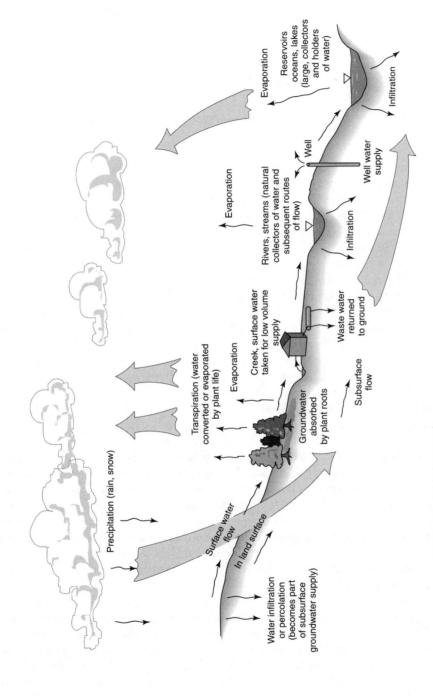

Figure 6-1 Hydrologic cycle.

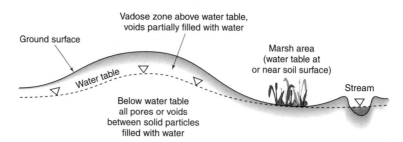

Figure 6-2 The position of the water table generally conforms to the land surface.

greater than atmospheric pressure. The soil zone above the phreatic surface includes air or other gas and is partially saturated with water classified as soil moisture (water held by capillary and adsorption forces); this zone is designated the *vadose zone*. The pore water held in this partially saturated zone above the phreatic surface is at negative pressure so that a total pressure from air plus water is less than atmospheric. In scientific terms, the identification of the water table or phreatic surface is where the groundwater pressure changes from a negative value to a positive value; in other words, where pore water pressure is zero.

In general, the position of the water table tends to reflect the surface topography of the area, being at a higher elevation where the ground surface is high and at a lower elevation where the ground surface is low (Figure 6.2). (Commonly, for some geographical areas, such as glacially effected ones, the relative elevations of a soil surface are also indicative of the boundary of the underlying rock surface).

However, it is also known that the depth or elevation of the groundwater table will vary, even over relatively short horizontal distances. The depth to groundwater can vary seasonally, and from year to year. The presence of groundwater in a soil deposit is affected by factors such as the characteristics of the various soil strata in the soil profile; the depth and type of underlying rock; the elevation of the soil surface related to the surrounding terrain; surface water coming to the area; and the presence of subsurface water in the adjacent soil and rock deposits.

The groundwater in an area may be either relatively stationary or mobile (that is, underground flow is occurring). The case of a stationary volume of groundwater occurs where an equilibrium condition has developed, such as where the phreatic surface is at similar elevation over a large area or when adjacent to a large body of water, or where the top boundary of a subsurface soil or rock layer restricts the entry of water and creates a buried basin in which the collected groundwater is prevented from escaping via movement through the buried soil or rock. (This condition creates what is classified as a *perched water table;* the phreatic surface for the groundwater trapped in the buried basin is positioned or "perched" above the region's true water table.)

In contrast, where strata of soil or rock have porous characteristics that permit the flow of liquid, the groundwater tends to be mobile, moving primarily under the effect of gravity forces. Where the elevation of the groundwater table measured at two different locations is different and the earth material between the two locations is porous (that is, permits the passage of water), flow occurs in the direction from the high elevation toward the lower. Physical conditions responsible for such flow are indicated by the Bernoulli Theorem for a steady condition of laminar flow involving an incompressible liquid such as water. Mathematically, the Bernoulli Theorem expresses the energy head possessed by a body of water at any position as

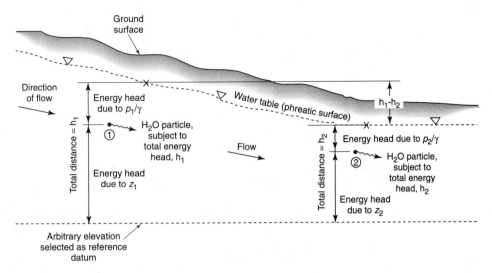

Figure 6-3 Description of conditions for the Bernoulli Theorem of flow.

$$h = \frac{v^2}{2g} + \frac{p}{\gamma} + z \tag{6-1}$$

where h = energy head, based on total or absolute hydraulic head, possessed by a body of
water at an elevated location (units are length, L)

v = flow velocity (L/T)

g = acceleration of gravity (L/T^2)

p = fluid pressure at any point in the body of water (F/L^2)

γ = unit weight of the water (F/L^3)

z = a gravitational position potential head indicating the vertical distance between
an arbitrary reference datum and the point where pressure p is measured (L)

The flow of water between two positions through soil and rock occurs because of the energy head difference between those positions. For flow through soil, where velocity typically is low and can be neglected in computations without incurring significant error, the energy head difference responsible for flow between location 1 and location 2 becomes (Fig. 6-3):

$$h_1 - h_2 = \left(\frac{p_1}{\gamma} + z_1\right) - \left(\frac{p_2}{\gamma} + z_2\right) \tag{6-1b}[1]$$

The pressure difference, Δp, causing flow to continue between the two locations, becomes, approximately,

---

[1]The Bernoulli Theorem was developed from the concept that the total energy (kinetic plus potential energy) of a unit of flowing water remains constant between position 1 and position 2 (i.e., assuming no energy losses occur). For water flowing through a soil or rock formation, however, energy is used in overcoming resistance to flow caused by friction, etc., one reason a steady state of flow having water levels at different elevations can represent a stable condition.

$$\Delta p = \gamma(h_1 - h_2) \text{ (approximately)} \tag{6-2}$$

The underground flow of water through a soil deposit can occur only where the soil material is porous enough to permit continuous passage. Some soil types, such as coarse granular materials, are more porous than other types, such as the fine-grained silt and clay soils (see Section 6.2 on permeability, a property also referred to as *hydraulic conductivity*).

Where a porous stratum of soil (typically a coarse-grained or granular soil) lies below an area's groundwater table or is in contact with an elevated source of water such as a surface body (lake, river, etc.) so that flow in significant quantity can or does occur, that stratum is classified as an *aquifer* or transmitter of water. Aquifers typically are considered to be a good source for obtaining subsurface water (when installing water supply wells, etc.).

Soil strata that restrict or prevent the flow of water, usually because of the fine-grained composition and related absence of large voids necessary to achieve rapid passage, are classified as *aquitards* or preventors of flow. Clay-type soils and, to a lesser extent, silt soil mixtures, function as aquitards.

Aquifers can be classified as unconfined, confined, or artesian (Fig. 6-4). The *unconfined* aquifer typically exists where the porous soil (a granular soil) constitutes the surface stratum. The position for the water table is within the stratum. However, flow through the stratum would not be prevented from reaching or breaking through to the surface should volume and pressure become great enough. The *confined* aquifer condition results where the porous stratum is bounded above and below by an aquitard. The confinement created by the aquitards prevents flow within the aquifer from escaping, unless an outlet exists. The position of the water table may be within the aquifer, or the pressure (energy head) developed by the flow may be great enough to force water entering a well that is tapped into the aquifer to rise above its top boundary. Where the energy head for water flowing within a confined aquifer is sufficient to force water to rise above the top boundary, an *artesian* condition exists. The artesian well, identified by the phenomenon of water flowing from the top of the well without a need for mechanical pumping, actually illustrates the condition where the energy head for flow in a confined aquifer is great enough to raise water to a height above the ground surface.

As described previously, the position of an area's groundwater table can fluctuate seasonally, and from year to year, as influenced by precipitation and weather, the volume of groundwater drawn off by users, etc. It is also recognized that changes in topography, natural or related to man, can be responsible for changes in groundwater, as illustrated by Figure 6-5.

6.2 PERMEABILITY (HYDRAULIC CONDUCTIVITY)

Soil, being a particulate material, has many pore or void spaces existing between the solid grains because of the irregular shape of the individual particles. In a mass of particles that are rounded and roughly equidimensional in shape, such as the gravels, sands, and silts, or are platey or flakelike, such as clays, the pore spaces are interconnected. Fluids (and gases) can travel or flow through the pore spaces in the soil. Thus, soil deposits are porous, and the material is considered a *permeable* material. It should be realized that flow is occurring through the void spaces between particles and not actually *through* the particles themselves (Fig. 6-6).

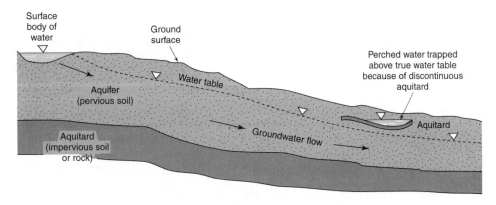

(a) Groundwater flow in
unconfined aquifer

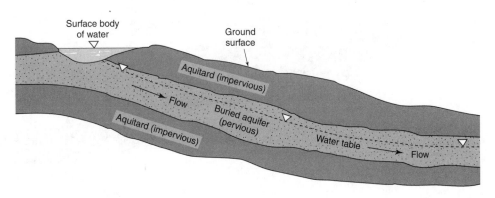

(b) Groundwater flow in
confined aquifer

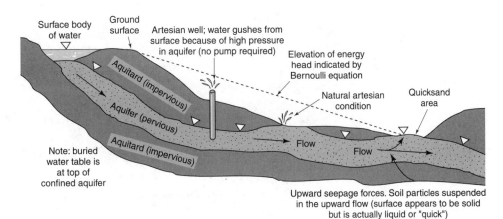

(c) Artesian condition for
groundwater flow

Figure 6-4 Illustrations of groundwater flow in different types of aquifers.

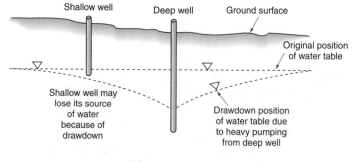

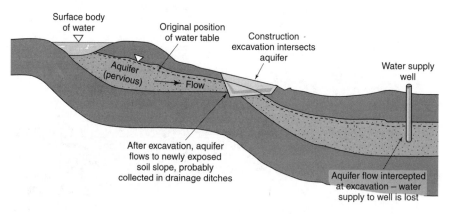

(a) Permanent lowering of water table because of heavy well pumping

(b) Excavation intersects aquifer, cuts off or lowers groundwater flow to well

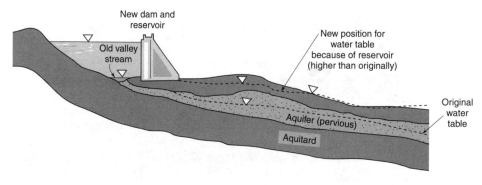

(c) Reservoir responsible for higher water table in area downstream from dam

Figure 6-5 Conditions responsible for changed elevation of groundwater table.

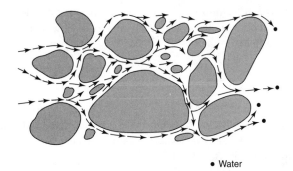

• Water

Figure 6-6   Schematic diagram indicating manner in which water flows through soil.

Factors Affecting Flow

The actual path taken by a fluid particle as it flows through void spaces from one point toward another is a tortuous and erratic one in most soils because of the random arrangement of the soil grains. It is highly probable that the direction of flow and velocity of flow vary considerably.

The factors that can affect the flow of a fluid through soil are known, but the *influence* of all factors has not been clearly established. These factors include:

1. The pressure difference existing between the two points where flow is occurring

2. The density and viscosity of the fluid

3. The size, shape, and number of pore openings

4. The mineralogical, electrochemical, or other pertinent properties of the fluid and the soil particles, which affect the attraction between the two materials

The effects of items 3 and 4 on flow are the most difficult to evaluate, partly because of the tremendous variation that occurs in natural deposits (even in homogeneous soils) and partly because of insufficient knowledge.

In many engineering and construction problems, the concern is primarily with the *quantity* of fluid, usually but not always water, that is flowing through or out of a soil mass. The seepage *velocity* is frequently sufficiently low that no problems result because of this factor, and it may not even require consideration. An exception is where the velocity is great enough to cause movement of the soil particles, or erosion; this problem is discussed in Section 7.1 on flow nets. For the situation where the quantity of flow is to be determined, an *average discharge velocity* is assumed. This is a fictitious velocity compared to the actual velocity of flow. The discharge velocity is simply the volume of fluid flow per unit of time divided by the total area (soil plus voids) measured normal to the direction of flow. If an average *seepage velocity* (average actual velocity of flow) is desired, it can be obtained by dividing the average discharge velocity by n, the porosity of the soil (recall that porosity n equals V_v/V_T).

Experimental studies have shown that fluid flow is affected by the shape and dimensions of the channel through which flow is occurring. To relate these effects to flow velocity, the terms *hydraulic radius* and *shape factor* have been developed. Hydraulic radius, R_H, provides a relation between the cross-sectional area of flow and the channel walls that are in contact with the fluid:

$$R_H = \frac{\text{Area of flow}}{\text{Wetted perimeter}} \qquad (6\text{-}3)$$

The coefficient that reflects the shape factor is C_S. For example, with a circular tube flowing full (Fig. 6-7), the hydraulic radius is one-half the tube radius, and $C_S = \frac{1}{2}$.

To indicate how *permeable* a porous material will be to *any* flowing fluid, the term K, permeability, is used. Mathematically,

$$K = C_S R_H^2 n \qquad (6\text{-}4)$$

where n = porosity of the material. It should be understood that this term applies *only to the material* through which the flow could occur. It reflects the effect of the size, shape, and number of flow channels, and is completely independent of any fluid properties. In the dimensional (units) analysis, the shape factor and porosity are dimensionless terms, and therefore K has a unit of L^2, or area.

For porous materials such as soil deposits, where the void space channels through which fluids (and gases) flow vary significantly in size and are of irregular cross section, the value for K would typically be determined experimentally. The value of K for soils is expressed in *darcys* (one darcy equals 9.88×10^{-13} m^2 or 1.062×10^{-11} ft^2). This term is little used in practical problems for relating the flow of water in soil deposits but does have application when other fluids are studied or comparisons are made.

To indicate the ease or difficulty with which a particular fluid will flow through a permeable material, the properties of the fluid are incorporated with K, the properties of the permeable material, to provide a *coefficient of permeability*, k (the lowercase letter is used for this term), where

$$k = K\frac{\gamma}{\eta_{ad}} = K\frac{\gamma}{\rho\eta_k} \qquad (6\text{-}5)$$

where γ = unit weight of the fluid (e.g., in kN/m^3, pcf)
$\quad \eta_{ad}$ = absolute or dynamic viscosity of the fluid[2]
$\quad \eta_k$ = kinematic viscosity[2]
$\quad \rho$ = density of the fluid

When the values of unit weight, viscosity, and density for a fluid are not known, the information can often be found listed in reference manuals that tabulate physical properties; it is important to recognize that values for unit weight and viscosity usually vary with temperature.

---

[2]Units for absolute or dynamic viscosity are FT/L^2; units for kinematic viscosity are L^2/T. In the SI system, absolute-dynamic viscosity is expressed in Pa·sec (where 1 Pascal = 1N/m^2); in traditional metric units, the absolute-dynamic viscosity is expressed in poises or centipoise (cp) (where 1 poise = 1 g/cm·sec or 0.1 Pa·sec; and in U.S. Customary units, the absolute-dynamic viscosity is expressed in lb·sec/ft^2. For kinematic viscosity, the SI units are cm^2/sec, and the U.S. Customary units are ft^2/sec. For water at 20°C (68°F), absolute viscosity is 1 cp, or 0.001 Pa·sec, or 2.083×10^{-5} lb·sec/ft^2; the kinematic viscosity is 0.0112 cm^2/sec, or 1.076×10^{-5} ft^2/sec. (The SI unit of cm^2/sec for kinematic viscosity is also called a Stoke; that is, 1 Stoke = 1 cm^2/sec.)

Figure 6-7 Hydraulic radius for pipe flowing full.

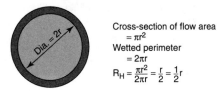

Cross-section of flow area
$= \pi r^2$
Wetted perimeter
$= 2\pi r$
$R_H = \dfrac{\pi r^2}{2\pi r} = \dfrac{r}{2} = \dfrac{1}{2} r$

Historically, the phrase *coefficient of permeability* has been an engineering term. For this same property, geologists, environmentalists, hydrologists, and groundwater specialists use the term *hydraulic conductivity*. This latter term is descriptive of a physical property associated with the transmission of matter or energy, in the family of terms such as *electrical conductivity* and *thermal conductivity*. Geotechnical personnel have become familiar with this dual usage of terms, and it is anticipated that the *hydraulic conductivity* phrase will eventually be used by everyone.

Illustration 6-1

The relationship expressed by Equation 6-5 permits a soil's coefficient of permeability (hydraulic conductivity) to be determined with fluids other than water if the value of K, or k for water, is known. For example, when compared to water (unit weight equal to 62.4 pcf or 9.8×10^3 kN/m³, absolute viscosity at 20°C equal to 1×10^{-3} Pa · sec), the coefficient of permeability (hydraulic conductivity) for a gasoline (assume calculations are performed to study the effects of a spill or underground leak) whose specific gravity is 0.72 and whose absolute viscosity is 0.337×10^{-3} Pa · sec will be approximately two times as great, since for water

$$k_w = \left(K \frac{\gamma}{\eta_{ad}} \right)_w = K \left(\frac{9.8 \times 10^3 \text{ kN/m}^3}{1 \times 10^{-3} \text{ Pa·sec}} \right) = K(9.8 \times 10^9) \text{ m/sec}$$

and for the gasoline

$$k_g = \left(K \frac{\gamma}{\eta_{ad}} \right)_g = K \left(\frac{.72 \times 9.8 \times 10^3 \text{ kN/m}^3}{.337 \text{ Pa·sec}} \right) = K(20.9 \times 10^9) \text{ m/sec}$$

and the ratio

$$k_g / k_w = \frac{K(20.9 \times 10^9)}{K(9.8 \times 10^9)} \cong 2.1$$

Darcy's Law for Flow

In the mid-eighteenth century, H. Darcy performed experiments to study the flow of water through sands. With an arrangement represented by Figure 6-8, it was found that the quantity of water flowing through the soil in a given period was proportional to the soil area normal to the direction of flow and the difference in water levels indicated in the piezometers (open standpipes), and inversely proportional to the length of soil between piezometers through which flow took place. Mathematically,

$$\frac{Q}{t} \propto \frac{\Delta h \times A}{L} = (\text{a constant}) \times \frac{\Delta h \times A}{L}$$

Figure 6-8 Darcy's sand filtration experiment.

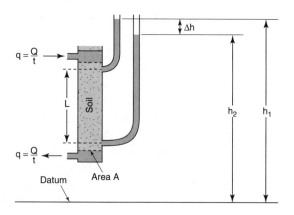

where Q = volume of water flowing through the soil in time t
   t = time period for the volume Q to flow
   $\Delta h = h_1 - h_2$
   A = cross-sectional area of the soil sample
   L = length of soil through which flow occurs, between points h_1 and h_2

The factors A and L relate to the volume occupied by the soil but not to its properties. The value of Δh relates to the pressure acting to force the water to flow through the soil. The constant of proportionality, a factor that indicates if the volume of flow is to be great or small, relates to the ease or difficulty with which the water moves through the soil. This constant of proportionality is k, Darcy's *coefficient of permeability*. This is the same coefficient of permeability indicated in Equation 6-5, but attributed to Darcy, who first established it. Consequently,

$$\frac{Q}{t} = k\frac{(\Delta h)(A)}{L}$$

(6-6)

which is Darcy's law. The ratio of $\Delta h/L$ is termed the *hydraulic gradient, i*, and therefore

$$\frac{Q}{t} = q = kiA$$

(6-7)

where q = volume of flow per unit time.
   The units of k are length per unit time (L/T); i is dimensionless (length divided by length).
   For steady flow, the volume of flow q passing a point is equal to the product of flow velocity v and the cross-sectional area A through which flow occurs:

$$q = Av$$

(6-8)

where units of v are length or distance per unit of time.
   From Equations 6-7 and 6-8, an expression for velocity of flow is obtained. Since

$$q = kiA$$

then

$$v = ki \tag{6-9}$$

This is a theoretical average velocity, and will be lower than an actual average velocity. Its determination is of use in practical problems, however, where only order of magnitude is required.

Laminar and Turbulent Flow

The movement of a fluid through a channel or pore space can be described as laminar or turbulent flow, depending on the path followed by the flowing water particles. Laminar (layered) flow indicates that adjacent paths of water particles are parallel, even when changing direction, and the paths never cross. This is an orderly flow with no mixing. Turbulent flow indicates a disorderly random path for moving water particles, with lines of movement crossing and frequently moving at an angle with or contrary to the general direction of flow. A high degree of mixing occurs.

Velocity has direct bearing on whether a flow is laminar or turbulent.

Darcy's law for fluid flow applies provided that the flow is laminar. In soils, the velocity of flow is affected by the size of the void opening as well as the hydraulic gradient i. Studies show that for soils in the coarse sand and finer range, and frequently for small gravel, laminar flow occurs provided that i is 5 or less. In practical soil mechanics work, Darcy's law thus has a wide range of application.

Effect of Soil Type

The volume of water that can flow through a soil bulk is related more to the size of the void openings than to the number or total volume of voids. This is shown by observing that the values of k for coarse soils are greater than for fine-grained soils (even though void ratios are frequently greater for the fine-grained soils), along with the knowledge that voids in a soil mass can range up to the size of the particles.

This phenomenon of higher permeability in coarse-grained soil can be explained, at least in part, by the manner in which water flows through a conduit. The fluid flow measured at increments of distance extending between the walls of the conduit indicates that the velocity varies from a very low value adjacent to the wall of the conduit (or against the soil particle) to a maximum at the center of the conduit, as shown in Figure 6-9. This variation in flow is caused by the friction developed at the conduit wall and the viscous friction developed in the moving fluid. For fine-grained soil, where void spaces are very small, all lines of flow are physically close to the "wall of the conduit," and therefore only low-velocity flows occur. In clays, flow in already small "flow channels" is further hampered because some of the water in the voids is held, or adsorbed, to the clay particles, reducing the flow area and further restricting flow.

Typical ranges for coefficient of permeability (hydraulic conductivity) for different soil types and resulting drainage characteristics are listed in Table 6-2.

Figure 6-9 Variation of flow velocity across the cross section of a tube.

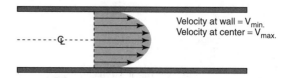

Velocity at wall = $V_{min.}$
Velocity at center = $V_{max.}$

Table 6-2 Typical Ranges of Permeability (Hydraulic Conductivity) for Different Soil Types

Soil Type	Relative Degree of Permeability	k, Coefficient of Permeability or Hydraulic Conductivity (mm/sec)	Drainage Properties
Clean gravel	High	10 to 100	Good
Clean sand, sand and gravel mixtures	Medium	10 to 10^{-2}	Good
Fine sands, silts	Low	10^{-2} to 10^{-4}	Fair–poor
Sand–silt–clay mixtures, glacial tills	Very low	10^{-3} to 10^{-6}	Poor–practically impervious
Homogeneous clays	Very low–practically impermeable	$<10^{-6}$	Practically impervious

Note: To convert, use 1 mm/sec = 0.2 ft/min = 86.4 m/day.

Empirical Relationships

Considerable information on the flow of fluids through porous media has been obtained from studies of flow through tubes and conduits. From such information, attempts have been made to relate permeability to a soil's grain size. Practically, such a relationship appears more possible for sands and silts than for clays, because of the particle size, shape, and overall soil structure. One of the more widely known relationships is

$$k = (D_{10}{}^2)(10^5) \tag{6-10a}$$

where k is given in millimeters per second and D^{10} is the 10 percent particle size, expressed in millimeters, from the grain-size distribution analysis (from the curve resulting from plotting percent finer by weight versus particle diameters).[3] This relationship was developed from the work of Hazen on sands.[4] This expression applies *only* to uniform sands in a relatively loose condition.

---

[3] See Section 4.2 on particle-size distribution curves.

[4] Originally included in "Some Physical Properties of Sands and Gravels with Special Reference to Their Use in Filtration," Allen Hazen, 24th annual report of the State Board of Health of Massachusetts, 1892, and numerous subsequent engineering manuals, including reference [213].

More recent studies to evaluate the permeability and filter properties of sands [266] have determined a relationship between the D_{15} size and the k value for *dense* or *compacted sands*. A close approximation for coefficient of permeability based on the D_{15} size is

$$k = 3.5(D_{15})^2 \qquad\qquad (6\text{-}10b)$$

when D_{15} is expressed in millimeters and the k value is in millimeters per second. The permeability for *loose sands* will be greater than indicated by Equation 6-10b. For sands where the C_u value is less than about 10, the k value for the loose condition can be estimated as 1.5 to 3 times the equation value. When the C_u value is more than 10, the permeability for a loose sand can be considerably greater (5 to 15 times greater) than the value indicated by Equation 6-10b.

Permeability Tests

Much of the available information from studies of flow through uniform porous media is not directly applicable to soils because of the variation in the size and shape of void spaces in a soil deposit. For soil, it has been found to be more practical and accurate to evaluate flow directly through use of laboratory or field tests on the soil in question. Experience has shown that for a given soil, a relationship exists between permeability and void ratio. Generally, a semilog plot (e plotted on an arithmetic scale, k plotted on a logarithmic scale) produces approximately a straight line for most soils. Thus, permeability tests can be performed on a soil at two or three widely different void ratios, and the results can then be plotted. Permeability at intermediate void ratios would be determined by interpolation (Fig. 6-10).

The permeability of a soil deposit is significantly affected by its in-place structure. A loose granular (coarse-grained) soil would have a higher void ratio than a dense soil, and therefore would permit greater flow. Clays are very significantly affected by structure. Even at similar void ratios, a clay with an undisturbed flocculated structure will possess larger void openings than the same clay having a remolded, dispersed (or oriented) structure, with the result that the permeability is many times greater for the flocculated soil (Fig. 6-11).

Extremely important is the significance of stratification. In layered soils, the permeability measured for flow across layers can be greatly different from that for flow parallel to the layering. Fine-grained deposits, such as clay or alternating layers of silt and clay, have a permeability in the direction parallel to bedding of the layers (often, such bedding is approximately horizontal) many times the permeability of cross-bedding flow. Thus, the use of undisturbed test samples, where the sample has retained its original structure and is tested so that flow through the sample is in the correct direction (to correspond to horizontal or vertical flow, as will actually occur in the field), is important if reliable results are to be obtained. This also points out an important consideration regarding use of test data: Data from remolded samples may not apply to field conditions if the natural soil remains undisturbed, or test data from undisturbed samples may not apply if, in the field, the soil is to be disturbed and rehandled (as in placing a compacted fill).

Conditions other than the size and number of voids will affect the quantity of flow through a soil deposit. Trapped air or gases prevent flow, whereas seams, cracks, fissures, and cavities that exist in a soil deposit increase the opportunity for fluid movement. Field investigations need to provide information on the presence of these conditions if reliable

Figure 6-10 Void ratio versus permeability plotted on semilog coordinates.

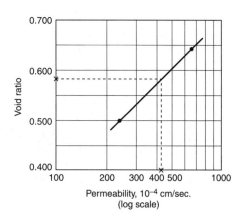

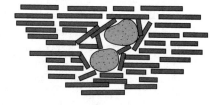

Undisturbed flocculent structure   Remolded or dispersed structure

Figure 6-11 Cohesive soil with flocculent structure will have higher permeability than soil with dispersed structure.

measures of flow through the soil deposit at a construction site are to be obtained. For this reason, field *permeability tests* are felt to provide more accurate information than laboratory tests. Field testing has its disadvantages, however. Costs and time involved are usually greater than for a laboratory test, and the field test provides only information on conditions in the limited proximity of the test location.

In practice, determining permeability from field or laboratory tests or from indirect analytical methods is for *order-of-magnitude* use only. This is realistic when it is realized that subsurface conditions and soil properties most probably vary over even short horizontal distances, that soil conditions and properties are generally not known in all areas of influence at a site, and that, frequently, external factors causing or affecting flow are not accurately known during planning and design.

Laboratory Permeability Tests

Two of the more conventional laboratory permeability tests are the constant-head test and the falling-head test. Schematic diagrams showing each of these methods, and the mathematics to calculate the coefficient of permeability, are shown in Figures 6-12 and 6-13.

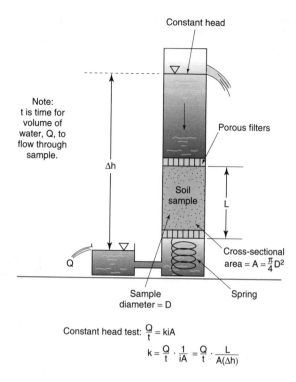

Constant head

Note:
t is time for
volume of
water, Q, to
flow through
sample.

Δh

Porous filters

Soil
sample

L

Q

Cross-sectional
area $= A = \frac{\pi}{4}D^2$

Sample
diameter $= D$

Spring

Constant head test: $\frac{Q}{t} = kiA$

$$k = \frac{Q}{t} \cdot \frac{1}{iA} = \frac{Q}{t} \cdot \frac{L}{A(\Delta h)}$$

Figure 6-12 Constant-head permeability test: Representation of permeameter and definition of related terms.

The constant-head permeability apparatus (permeameter) is in wide use for testing the coarse-grained soils, such as sands, where the volume of flow through the soil will be relatively large. For fine-grained soils, such as silt and clay, the falling-head permeameter is generally used. In the constant-head test, permeability is computed on the basis of fluid that passes through the soil sample. In the falling-head test, permeability is computed on the basis of fluid flowing into the sample. The reason for the distinction is simple. In a fine-grained soil, a very limited volume of fluid will flow through the sample. With the constant-head test, time is required to accumulate the fluid volume necessary to perform computations. Extreme care would be required to prevent leaks in the apparatus and evaporation of discharged water. With the falling-head method, the duration of the test is shortened, and there is no concern about the volume of discharge. Care is required, however, to prevent evaporation of water in the inlet tube.

Darcy's coefficient of permeability is the factor for a condition of steady flow through a soil. In performing laboratory permeability tests, it is essential that volumes be measured only after steady flow has been occurring for some period. It is important to assure that no air or other gases are trapped within the soil to interfere with flow. A vacuum may be required to remove trapped air. In general, the constant-head test is easier to perform and requires less skill and experience than the falling-head test. Care is required during testing of

Figure 6-13 Falling-head permeability test: Representation of permeameter and definition of related terms.

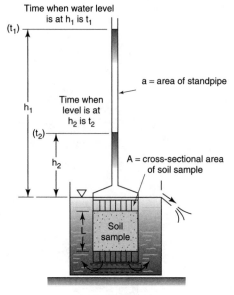

Time when water level is at h_1 is t_1

(t_1)

h_1

Time when level is at h_2 is t_2

(t_2)

h_2

a = area of standpipe

A = cross-sectional area of soil sample

L

Soil sample

Falling head test: $k = \dfrac{L}{(t_2 - t_1)} \cdot \dfrac{a}{A} \cdot \ln \dfrac{h_1}{h_2}$

or

$k = \dfrac{(2.303)L}{(t_2 - t_1)} \cdot \dfrac{a}{A} \cdot \log_{10} \dfrac{h_1}{h_2}$

fine granular soils (such as in the fine sand range) to prevent the particles from being carried along with the discharging water. Details for performing permeability tests are presented in the *ASTM Procedures for Testing Soils* (D2434).

Illustration 6-2

In a laboratory, a constant-head permeability test is performed on a sample of granular soil. The test setup is as indicated in Figure 6-12. The length of the soil sample is 15 cm, and the cross-sectional area is 10 cm². If a 24 ml (or 24 cm³) volume of water passes through the soil sample in a 3-minute period, when Δh is 30 cm, compute the coefficient of permeability.

Solution

Use

$$k = \left(\frac{Q}{t}\right)\left(\frac{L}{A\Delta h}\right)$$

where $Q = 24$ cm³
$t = 3$ minutes
$L = 15$ cm
$A = 10$ cm²
$\Delta h = 30$ cm

$$k = \left(\frac{24 \text{ cm}^3}{3 \text{ min}}\right)\left(\frac{15 \text{ cm}}{10 \text{ cm}^2 \times 30 \text{ cm}}\right) = .4 \text{ cm/min} = .067 \text{ mm/sec}$$

Illustration 6-3

A falling-head permeability test is performed on a silty soil. The test setup is as shown in Figure 6-13. For the test data summarized below, what is the coefficient of permeability for this sample?

Sample length = 8 cm

Cross-sectional area of sample = 10 cm²

Area of standpipe = 1.5 cm²

Height of water in standpipe at start of test period h_1 = 100 cm

Height of water in standpipe at end of test period h_2 = 90 cm

Time for change from h_1 to h_2 = 60 min

Solution

Use

$$k = \frac{(2.303)L}{t_2 - t_1} \left(\frac{a}{A}\right) \log \frac{h_1}{h_2}$$

$$k = \frac{(2.303)(8 \text{ cm})}{60 \text{ min}} \left(\frac{1.5 \text{ cm}^2}{10 \text{ cm}^2}\right) \log \frac{100 \text{ cm}}{90 \text{ cm}}$$

$$= .00212 \text{ cm/min} = 3.5 \times 10^{-4} \text{ mm/sec}$$

Where consolidation tests are performed on soil samples (Chapter 10), the permeability test can be adapted to determine the rate of flow through the consolidation sample, and the coefficient of permeability can be computed from these data. Alternatively, the coefficient can be determined from the consolidation data obtained to calculate the rate of consolidation of the soil (described in Chapter 10).

Field Permeability Tests

Field permeability tests offer the advantage of testing undisturbed soil in the natural location with respect to the ground surface, water table, and other factors that could influence the rate of flow. Various methods for determining permeability are available, depending, among other things, on the soil's being above or below the groundwater table. The methods described herein are of the type where a cased boring is made into the soil that is to be tested. The casing and related equipment necessary are of the type normally used by soil-boring contractors. Whenever possible, it has been found expedient to perform field permeability tests during the investigative stage of planning a project, at the time that the subsurface investigation (soil borings) is being made.

Essentially, the field permeability test involves obtaining a record of the time that it takes for a volume of water to flow out of, or into, the boring casing. A schematic presentation of different conditions and the related equations for calculating the coefficient of permeability are shown in Figure 6-14.

On a practical note, simple field permeability tests of the types depicted in Figure 6-14 are considered appropriate for evaluating granular and silt soils. The low coefficient of permeability associated with clays and silt–clay mixtures would involve a long time period for the field test and require consideration of various additional factors such as evaporation and the like.

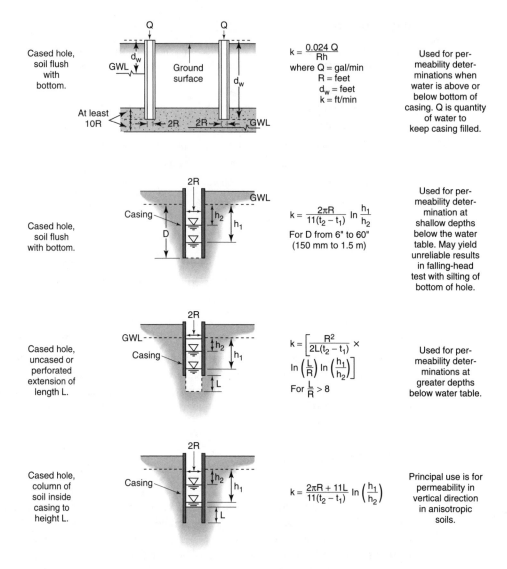

Figure 6-14 Methods for Performing Field Permeability Tests [89]

6.3 CAPILLARITY

The groundwater table (or phreatic surface) is the level to which underground water will rise in an observation well, pit, or other open excavation in the earth. All voids or pores in soil located below the groundwater table would be filled with water (except possibly for small isolated pockets of trapped air or gases). In addition, however, soil voids for a certain height above the water table will also be completely filled with water (full saturation). Even above this zone of full saturation, a condition of partial saturation will exist.

Any water in soil located above the water table is referred to as *soil moisture*. The phenomenon in which water rises above the groundwater table against the pull of gravity but is in contact with the water table as its source is referred to as *capillary rise*. The water associated with capillary rise is *capillary moisture*. The soil region directly above the water table and wetted by capillary moisture is designated the *vadose zone*.

Water in Capillary Tubes

The basic principles of capillary rise in soils can be related to the rise of water in glass capillary tubes (tubes with very small diameters) under laboratory conditions. When the end of a vertical capillary tube is put in contact with a source of water, the water rises up in the tube and remains there. The rise is attributed to the attraction between the water and the glass and to a *surface tension*[5] which develops at the air–water interface at the top of the water column in the capillary tube. This surface tension can be thought of as an infinitely thin but tough film, such as a stretched membrane. (The surface tension phenomenon is one of the reasons that small insects can "walk" on water.) The water is "pulled up" in the capillary tube, to a height regulated by the diameter of the tube, the magnitude of the surface tension, and the density of the water.

The attraction between the water and the capillary tube affects the shape of the air–water interface at the top of the column of water. For water and glass, the shape is concave downward; that is, the water surface is lower at the center of the column than at the walls of the tube. The resulting curved liquid surface is termed the *meniscus* (Fig. 6-15).

The column of water in the capillary tube has risen above the surface of the water supply and against the pull of gravity. For a condition of equilibrium, the effect of the downward pull of gravity on the capillary column of water has to be resisted by the ability of the surface film to adhere to the wall of the capillary tube and hold the column of water. This wall adhesion must equal the surface tension of the fluid. If T_s is the value of surface tension, expressed in units of force per unit length, the vertical loading that can be supported is

$$(\text{Tube circumference}) \times (T_s) \times (\cos \alpha)$$

where α is the angle formed between a tangent to the meniscus and the capillary wall. With water and glass, the meniscus at the wall of the capillary tube is tangent to the wall surface, and the angle α is zero degrees. Therefore, $\cos \alpha$ is 1, and the column weight that is capable of being supported because of surface tension is

---

[5]The kinetic theory explains the surface tension phenomenon (all molecules are in motion and the speed or kinetic energy increases as temperature increases, and molecules attract each other at close range). Within the body of a liquid, a molecule of water will be subject to attractions that are equal in all directions because of the surrounding molecules. At the surface of a body of water (the air–water interface), the water molecules are subject to an unequal force in the direction of the body of water, since in the air (gaseous state) the molecules are widely dispersed and, in total, exert a lesser attractive force. The normal molecular motion of the water molecules is thereby restricted by the unbalanced pull (attraction). The resulting net effect between the natural molecular motion and unbalanced attraction is that the surface zone of the water acts as a stretched membrane (i.e., a membrane in tension). If the attractions internal to the body of water (cohesion) are less than the attractions between the water and a solid (adhesion), then the water surface curves outward (from the body of water) where the water is in contact with the solid, producing a curved meniscus. Since glass materials (and soil particles) are solids that possess a polar surface, the result is an attraction or adhesion between the water molecules and the surface of the glass (or soil particles).

Figure 6-15 Capillary rise and water meniscus in a glass tube.

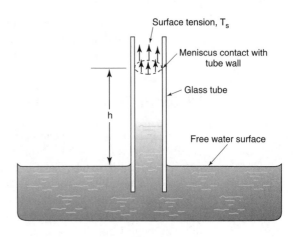

where r is the radius of the capillary tube.

The weight of the column of water in the capillary tube is

$$(\pi r^2) \times (h) \times (\rho) \times (g)$$

where h = height of the column of water

ρ = density of water, mass per volume, taken as 1 gm/cm³, 1 mg/m³, or 1.95 slugs/ft³

g = acceleration of gravity, 9.81 m/sec², 981 cm/sec², or 32.2ft/sec²

If h_c is the maximum height of capillary rise that can occur for prevailing physical conditions (such as for a particular temperature), equilibrium requires that

$$(2\pi r)(T_s) = (\pi r^2 h_c)(\rho)(g) = \pi r^2 h_c \frac{(\text{weight})}{(\text{volume})}\,[6]$$
$$= \pi r^2 h_c \gamma_w$$

where γ_w = unit weight of water, taken as 62.4 pcf, 9.81 kN/m³, or 981 dynes/cm³.[7] The maximum height of capillary rise is then

$$h_c = \frac{(2\pi r)(T_s)}{(\pi r^2)(\rho)(g)} = \frac{2T_s}{(r)(\rho)(g)}$$

---

[6]Since weight equals mass times acceleration of gravity $(W = mg)$, the product of ρg equals $\frac{(\text{mass})}{(\text{volume})} \times g$, or $\frac{\text{weight}}{\text{volume}}$. Also see footnote 2, Chapter 2.

[7]Recall that the dyne is a unit of force; units are $\frac{\text{g-cm}}{\text{sec}^2}$.

$$h_c = \frac{2T_s}{r(\gamma_w)} = \frac{4T_s}{d(\gamma_w)} \tag{6-11}$$

where d = the diameter of the capillary tube.

The value of T_s for water varies according to temperature. At normal room temperatures, T_s is close to 0.005 lb/ft, 0.064 N/m or 73 dynes/cm. In applying the development of capillary rise in tubes to capillary rise in soils, these values for T_s are sufficiently accurate for many practical problems. Thus, the equation for capillary rise can be expressed as

$$h_c \cong \frac{31}{d} \text{ mm} \tag{6-12}$$

provided that d is in millimeters.

Illustration 6-4

Compute the height of capillary rise for water in a tube having a diameter of 0.05 mm.
In SI units
Solution

$$h_c = \frac{4T_s}{d\gamma_w} = \frac{(4)(.064 \text{ N/m})}{(5 \times 10^{-5} \text{ m})(9.81 \text{ kN/m}^3)} = .52 \text{ m}$$

There *are* situations, however, where the temperature effects should be considered. Generally, as temperature increases, the value of T_s decreases, indicating a lessening height of capillary rise under warm conditions or an increasing height of capillary rise for conditions of falling temperatures. At freezing, T_s for water is about 0.067 N/m.

Note from the terms of Equation 6-11 that the height of capillary rise is not affected by a slope or inclination in the direction of the capillary tube, or by variations in the shape and size of the tube at levels below the meniscus (Fig. 6-16). But for water migrating up a capillary tube, a large opening can prevent further upward movement in an otherwise small-diameter tube. The determining factor is the relationship between the size of the opening (tube diameter) and the particular height of its occurrence above the water supply.

In the capillary tube, at the level equal to the free surface of the water in the supply pan, the hydrostatic pressure in the water is zero. Hydrostatic pressures increase below that free water surface according to the relationship

$$p_w = \gamma_w z \tag{6-13}$$

where z = depth below the water surface
p_w = water pressure at depth z

Conversely, hydrostatic pressure measured in the capillary column *above* that free water surface is negative, according to the relationship

$$p_w = -\gamma_w h \tag{6-14}$$

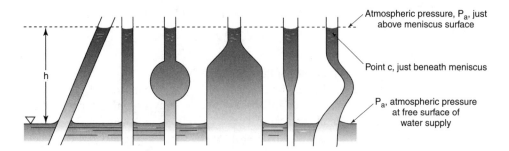

Figure 6-16 Capillary heights of capillary tubes of various shapes are the same if their menisci diameters are the same.

where h is the distance measured upwards, from the free water surface. The negative sign indicates that water pressures in the capillary tube is at less than atmospheric pressure. As a result, water in the capillary column is said to be in *tension*. The maximum negative pressure, or capillary tension, exists at the maximum height of rise h_c. If the pressure at the water surface in the supply pan depicted in Figure 6-16 is atmospheric pressure, p_a, the pressure in the capillary water just beneath the meniscus, point c, is $p_a - h_c \rho g$. By substitution of terms,

$$p_a - h_c \rho g = p_a - \left(\frac{2T_s}{r \rho g}\right)(\rho g) = p_a - \frac{2T_s}{r}$$

or $2T_s/r$ less than the pressure existing just above the meniscus.

As the column of water stands in the capillary tube, "hanging" by the surface tension at the meniscus, the *weight* of the column is transferred to the walls of the capillary tube, creating a compressive force in the walls. An effect of this occurrence within soil deposits is discussed in a later subsection, "Effects of Surface Tension."

Capillary rise is not limited to tube, or enclosed, shapes. If two vertical glass plates are placed so that they touch along one end and, in plan, form a V, a wedge of water will rise up in the V because of the capillary phenomenon (Fig. 6-17). The height of rise relates to the attraction between the water and plates and the physical properties of the water, as in tubes, but also to the angle formed by the V. The significance of this type of capillary rise is discussed in the following subsection, "Capillary Rise in Soil."

Capillary Rise in Soil

In soils, the shapes of void spaces between solid particles are unlike those in capillary tubes. The voids are of irregular and varying shape and size, and interconnect in all directions, not only the vertical. These properties make the accurate prediction of the height of capillary rise in soil almost impossible. However, the features of capillary rise in tubes are applicable to soils insofar as they facilitate an understanding of factors affecting capillarity, and help to establish an order of magnitude for capillary rise in the different types of soils.

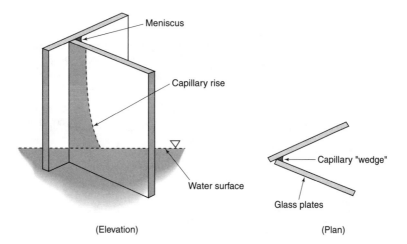

Figure 6-17 Capillary rise in corner formed by glass plates.

Illustration 6-5

Limited laboratory studies indicate that for a certain silt soil, the effective pore size for height of capillary rise is $\frac{1}{5}$ of D_{10}, where D_{10} is the 10 percent particle size from the grain-size distribution curve. If the D_{10} size for such a soil is 0.02 mm, estimate the height of capillary rise.

Solution

$$d = \text{effective capillary diameter} = \frac{1}{5}D_{10} = \frac{1}{5}(.02 \text{ mm}) = .004 \text{ mm}$$

$$h_c = \frac{31}{d} \text{ (from Eq. 6-12)}$$

$$= \frac{31}{.004 \text{ mm}} = 7750 \text{ mm} = 7.75 \text{ m} \cong 25 \text{ ft}$$

Review of Equation 6-11 indicates that even relatively large voids will be filled with capillary water if the soil is close to the source of water, that is, the groundwater table. As the distance from the water table increases, only the smaller voids would be expected to be filled with capillary water. The larger voids represent interference to upward capillary flow and would not be filled. Consequently, the soil closest to the water table but above it is fully saturated as a result of capillarity. Above this zone of full saturation lies a zone of partial saturation due to capillarity (Fig 6-18). The upper limit (or height) to which capillary water rises in the capillary zone is termed *capillary fringe*.

In the upper zone of the capillary fringe, where partial saturation exists, capillary movement occurs through the capillary tube phenomenon but may also occur in the wedges of the capillary V formed where soil particles are in contact (similar to the V formed by vertical plates discussed in the preceding section; see Fig. 6-19).

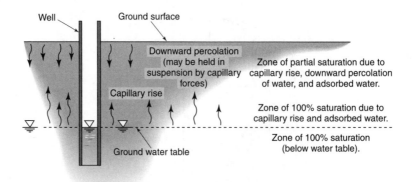

Figure 6-18 Distribution of soil moisture in a soil profile.

Figure 6-19 Wedge of capillary water at point of contact between particles.

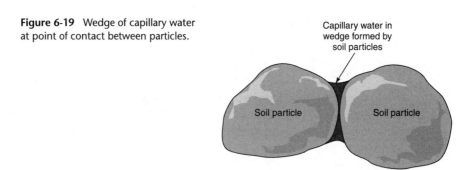

Since the largest void spaces in soil are of the same order of magnitude as the particle sizes, it follows that the height of capillary rise would be greater in fine-grained soils than in coarse-grained soils. For orders of magnitude, relative values of capillary rise for different types of soils are presented in Table 6-3. (Though attempts at establishing a relationship between height of capillary rise and some aspect of soil's grain size or void ratio have been made, the general applicability and accuracy of proposed relationships appear to be limited to providing estimates of order of magnitude.)

Temperature plays an important practical role in capillary rise in soil. The height of rise is greater at lower temperatures than at high temperatures. Demonstrations of this fact have frequently been observed in construction performed in a cool season following a warm season (i.e., in the fall months at locations in the northern hemisphere). Cool temperatures cause an increase in migration of water toward the surface. Frequently the additional moisture increases the difficulty in handling or working with the soil, or creates a muddy surface, which makes the moving of equipment a problem. *Horizontal migration* of water can also occur near a heated building. Water may move from beneath a structure in cold seasons if the presence of the structure affords some protection from cold temperatures. If the building itself contributes to a lowering of the ground temperature, as in a poorly insulated cold-storage warehouse, frozen foods processing plant, or ice skating rink, capillary flow will be induced toward the building.

Table 6-3 Representative Heights of Capillary Rise in Soils

Soil Type	m	ft
Small gravel	0.02–0.1	0.1–0.4
Coarse sand	0.15	0.5
Fine sand	0.3 to 1	1–3
Silt	1 to 10	3–30
Clay	10 to 30	30–90

Time Rate of Capillary Rise

The time necessary for the expected maximum height of capillary rise to occur requires consideration in some practical problems, such as where a structural fill has been placed for highways, buildings, or other purposes. On the basis of typical void sizes, clay and fine silt soils will have significant heights of capillary rise. However, the time period required for the rise to occur may be so great that other influences, such as evaporation and change in groundwater level, also have to be considered.

The term indicating the rate of capillary rise is *capillary conductivity* or *capillary permeability, k_{cap}*. Factors known to affect a soil's capillary conductivity are void sizes, moisture content, and temperature in the soil. Generally, capillary conductivity is greater at higher moisture contents and lower temperatures. Absolute values of capillary conductivity are not available. Though not identical to Darcy's coefficient of permeability, the *relative rates* of capillary conductivity can be thought of in terms of the comparative values for permeability—that is, rapid for coarse soils, low for silts and clays.

Suspended Capillaries

Water attempting to percolate downward through a soil (such as percolating surface water from rain or melting snow or pore water resulting from a formerly higher water table) can be held suspended in the soil voids because of the same surface tension phenomenon responsible for capillary rise. For this, the column of water would have a meniscus at both ends of the suspended column. Each meniscus would be in tension. The maximum length of a column would be controlled by the same factors affecting height of capillary rise.

Elimination of Capillary Water in Soil

For capillary rise to occur, the existence of an air–water interface is required. Capillary water will not exist in soil at a level below the groundwater table. It follows that a condition of capillarity will cease where submergence of a soil zone occurs.

Capillary water can be removed from a soil by evaporation. As a result, capillary water can represent a very mobile category of water where the evaporation is continually replaced by new capillary water.

Effects of Surface Tension

At the location of the meniscus, the surface tension imposes as compressive force onto the soil grains in contact with the meniscus that is equal in magnitude to the weight of the water in the capillary column, as indicated in the earlier discussion for capillary tubes. This effect applies to both water resulting from capillary rise and pore water suspended above a capillary zone. The compressive force is imposed throughout the soil particles in contact with the "held column" of water, and serves to compress or shrink the soil.

In clays, this phenomenon results where the groundwater table drops subsequent to the time the clay deposits were formed. Such "drying" of an area can cause compressive forces within the clay mass of such great magnitude that the affected soil zone becomes firm and strong. This is referred to as drying by desiccation, and the clays are referred to as *desiccated clays*. A land area formed by clay deposits that were originally soft and weak may have a firm crust or upper layer overlying the still soft and weak deeper clays because of desiccation. The thickness and strength of a desiccated zone has frequently been found sufficiently adequate that roads and light building structures could be satisfactorily supported by it.

Typical magnitudes of compressive stress that can result in fine-grained soils are tabulated in Table 6-4.

In partially saturated granular (coarse-grained) soils, the surface tension phenomenon also contributes to the strength of the mass. At partial saturation, all voids are not filled with water. The available water collects in the interstices adjacent to where soil particles touch, forming a wedge of moisture but leaving the center portion of the void filled with air. Thus, an air–water interface, or meniscus, is formed. The surface tension in the meniscus imposes a compressive force onto the soil particles, increasing the friction between particles and thus the shearing strength. The strength gain in granular soil due to partial saturation and the surface tension phenomenon is termed *apparent cohesion*. The strength gain can be quite significant, as the firm condition of wet sandy beach surfaces, which easily support the weight of vehicles, indicates. The menisci and surface tension, along with the apparent cohesion, will disappear when the soil is fully saturated or dries.

Table 6-4 Representative Values of Compressive Stress Resulting from Capillary Forces

Soil Type	psf	Kg/cm^2	kN/m^2
Silt	200–2000	0.1–1.0	10–100
Clay	2000–6000	1.0–3.0	100–300

PROBLEMS

6-1. Of the earth's total volumetric supply of water, approximately how much is fresh water (quantity and percentage)? How much is underground water?

6-2. Define *hydrogeology*.

6-3. Discuss the hydrologic cycle in terms of water on land areas (that is, the source of land-area surface and subsurface water, movement patterns, and change of location).

6-4. What is the groundwater table, and what defines its position?

6-5. Outline practical reasons why the elevation (depth) of the groundwater table in an area may fluctuate seasonally or annually.

6-6. Explain the difference between a soil stratum that is an aquifer and one that is an aquitard.

6-7. Referring to Figure 6-3, if the elevation difference between h_1 and h_2 is 10 m, what is the difference in energy head pressure (approximately) between the groundwater surface at point 1 and point 2?

6-8. Referring to Figure 6-3, if the elevation difference between h_1 and h_2 is 15 ft, what is the energy head pressure difference (approximately) between the groundwater surface at point 1 and point 2?

6-9. Explain how the process of pumping large volumes of water from a well can alter the position of a groundwater table.

6-10. Outline examples of how construction projects can be responsible for altering the elevation of the groundwater table for an area.

6-11. On the basis of general soil types that form an aquifer and an aquitard, indicate a probable range for the coefficient of permeability in aquifers and in aquitards.

6-12. In an area of hilly topography, a section of property has a ground surface that is a long hill extending for a 400 m distance at an average slope of 20 percent. An aquifer is buried beneath the hill as depicted by Figure 6-4c and intersects a large pond at the top of the hill slope and a flowing creek at the base of the hill. For these conditions, determine if a well installed midway up the hill and penetrating to the aquifer will flow as an artesian well; for this problem, assume the ground surface adjacent to the well is 45 m below the top of the hill.

6-13. Compute the value of the hydraulic radius for a circular pipe flowing half-full.

6-14. Calculate the value of the hydraulic radius for flow in an open channel where the bottom of the channel is 3 ft wide, side slopes are 1 on 1 (slope of 1 ft horizontal to 1 ft vertical), and the depth of flow is 2 ft.

6-15. A thick fluid having a high viscosity and a thin fluid having a low viscosity are to be passed through a permeable material. Indicate which fluid will have the greatest resulting coefficient of permeability and briefly explain why. For which fluid (thick or thin) is the quantity of flow through the material expected to be greatest?

6-16. Outline the general relationship between soil particle sizes and size of void spaces between the particles and the corresponding effect on a soil deposit's coefficient of permeability.

6-17. The coefficient of permeability is generally greater for coarse soils (sands and gravels) than for fine-grained soils (silts and clays). What effect does particle size have on permeability?

6-18. The void ratio for clay soils commonly is greater than the void ratio for the coarse-grained soils, but comparisons of the coefficient of permeability indicate considerably lower values for clay. Explain.

6-19. What effect does the presence of adsorbed water in clay have on the coefficient of permeability for this type of soil?

6-20. In the flow of water through soil, what are the conditions necessary in order for Darcy's law to apply?

6-21. (a) Estimate the coefficient of permeability for a uniform sand where a sieve analysis indicates that the D_{10} size is 0.15 mm.
(b) The particle-size distribution curve for a sand indicates that the D_{15} size is 0.19 mm. If the sand deposit is in the dense or compact state, estimate the coefficient of permeability.

6-22. Why might the permeability in fine-grained soil deposits be expected to be greater for horizontal flow than for vertical flow?

6-23. Briefly give reasons why the coefficient of permeability in an undisturbed clay deposit possessing a flocculent structure would be expected to be greater than if the same clay had a dispersed (remolded) structure.

6-24. A constant-head permeability test is performed, and the information below indicates test conditions and results. From the given data, provide a sketch representing the test arrangement, and calculate the value for the coefficient of permeability. On the basis of the computed value for k, indicate the probable soil type.
• Water flows horizontally through the soil sample.

• The height of the reservoir supply source is 2 m above the elevation of the outlet reservoir.
• The soil sample length in the direction of water flow is 200 mm.
• The cross-sectional area of the soil sample is 1950 mm^2.
• The measured volume of flow for the steady-state condition is 500 ml in a 10-minute time period (note 1 ml = 1000 mm^3).

6-25. A constant-head permeability test is performed in a laboratory where the soil sample is 25 cm in length and 6 cm^2 in cross section. The height of water is maintained at 2 ft at the inflow end and 6 in. at the outlet end. The quantity of water flowing through the sample is 200 ml in 2 minutes.

(a) Make a sketch of the described conditions.

(b) What is the coefficient of permeability in millimeters per minute?

6-26. A constant-head permeability test is performed where the hydraulic gradient is 0.75. The cross-sectional area of the sample is 0.25 ft.^{2} The quantity of water flowing through the sample is measured to be 0.004 ft^3/minute.

(a) What is the coefficient of permeability in feet per minute?

(b) What is the coefficient of permeability in millimeters per second?

6-27. How much water will flow through a soil mass in a 5-minute period when the sample length is 150 mm, the cross section is 20 mm by 20 mm, and a constant head of 2 ft is maintained? The soil has a k value of 0.1 mm/sec.

6-28. The coefficient of permeability for a fine-grained soil is determined in a laboratory by use of a falling-head test. Test conditions and results are as indicated below. Determine the coefficient of permeability, then indicate the probable soil classification.

• Length of soil sample = 150 mm
• Cross-sectional area of sample = 1140 mm^2
• Cross-sectional area of standpipe = 200 mm^2
• At the start of the test, the water level in the supply standpipe is 1 m above the top of permeameter.
• One hour after start, the water level in the supply standpipe is 0.95 m above the top of the permeameter.

6-29. A falling-head permeability test is performed on a fine-grained soil. The soil sample has a length of 120 mm and a cross-sectional area of 600 mm^2. The water in the standpipe flowing into the soil is 0.60 m above the top of the sample at the start of the test. It falls 50 mm in 30 minutes. The standpipe has a cross-sectional area of 200 mm^2.

(a) Make a sketch of the described conditions.

(b) What is the coefficient of permeability in millimeters per second?

(c) What is the coefficient of permeability in feet per minute?

(d) On the basis of the calculated value for k, what is the probable soil type?

6-30. A field permeability test is performed by measuring the quantity of water necessary to keep a boring casing (pipe) filled. The distance from the top of the casing to the bottom (in the ground) is 10 ft. The groundwater table is below the bottom of the casing. The casing has an inside diameter of 6 in. In a 10-minute period, 1 gallon of water was used to keep the casing filled. What is the coefficient of permeability for the soil at the bottom of the casing?

6-31. A field permeability test indicates that the coefficient of permeability for a certain soil is 2×10^{-2} mm/sec. Is this a relatively high or low coefficient of permeability? What type of soil would this probably be?

6-32. (a) Outline the practical reasons why field permeability tests to obtain a value for coefficient of permeability might be preferred over the laboratory testing method.

(b) Outline reasons why the coefficient of permeability determined by a laboratory test procedure might be preferred over that from a field test procedure.

6-33. To what height would water rise in a glass capillary tube that is 0.01 mm in diameter?

6-34. What is the water pressure just under the meniscus in a capillary tube where the water has risen to a height of 6 ft?

6-35. A glass capillary tube is 0.001 mm in diameter.

(a) What is the theoretical maximum height of capillary rise for a tube of this size?

(b) What compressive pressure results in the capillary water just under the meniscus?

6-36. In a silt soil, the D_{10} size is 0.01 mm. If the effective pore size for estimating capillary rise is taken as $\frac{1}{5}$ of D_{10}, approximately what height of capillary rise will occur?

6-37. Why is it expected that the maximum height of capillary rise is greater for fine-grained soils than for coarse-grained soils?

6-38. How is the height of maximum capillary rise of water in soil affected by temperature?

6-39. What is the vadose zone in an area's subsurface region, and what is the explanation for the condition?

6-40. The effects of soil capillary water are often cited as the reason why deposits of fine-grained soils (silts and clays) that generally are in relatively weak condition because of high water content can have a drier, firmer surface zone. Explain.

6-41. Explain how capillarity is related to the dried and firm condition frequently observed to exist in the surface zone of fine-grained soil deposits.

6-42. Indicate the ways that capillary water and the effects of capillarity can be removed from a soil.

CHAPTER 7

Movement of Water through Soil

Practical Effects: Seepage, Drainage, Frost Heave, Contamination

The handling of mobile and stationary underground water during construction operations and making provisions so that the effects of its presence will not interfere with the function of completed structures are of vital concern to the construction profession; this chapter's discussion on flow nets and seepage, drainage, and frost heave relates to the practical aspects of controlling groundwater during and after construction. The section on soil and groundwater contamination and problem soil gases discusses factors responsible for the development of such conditions and methods for prevention or accomplishing remediation.

7.1 FLOW NETS AND SEEPAGE

Flow of Subsurface Water

The flow of water beneath the ground surface through all soils except coarse gravel and larger materials occurs as laminar flow; that is, the path of flow will follow a regular pattern, with adjacent paths of water particles all flowing parallel. For this condition, Darcy's law for water traveling through soils can be applied to determine the rate and quantity of flow and the seepage forces that result from this flow. In its most direct form, Darcy's law is

$$q = kiA \tag{7-1}$$

(as developed in Chapter 6, Eq. 6-7), where q is the quantity of flow in a unit time period (or rate of flow), k is the coefficient of permeability for the soil, A is the cross-sectional area of the soil through which flow is occurring (normal to the direction of flow), and i is the hydraulic gradient (the difference in the energy head of water between two points divided by the distance between the same two points), all as previously defined in Chapter 6.

When underground water is flowing over a relatively long distance and within a soil zone having well-defined boundaries, such as that shown in Figure 7-1, the quantity of flow can be determined by using the foregoing expression directly.

Illustration 7-1

A river and a canal run parallel to each other but at different elevations, as indicated by Figure 7-1. If the difference in the water surface elevations is 5 m, the horizontal separation is 200 m, and the thickness of the permeable stratum is 2 m, compute the seepage loss between river and canal, per km of river–canal length. Permeability $k = 1$ m/day.

Solution

$$q = kiA = k\frac{\Delta h}{L}A = (1.0 \text{ m/day})\left(\frac{5 \text{ m}}{200 \text{ m}}\right)(2 \text{ m} \times 1000 \text{ m})$$

$$= 50 \text{ m}^3/\text{day/km of length}$$

The Need for Flow Nets and Flow Net Theory

Where the zones of flow or directions of flow are irregular, where water enters and escapes from a permeable zone of soil by traveling a short distance, or where the flow boundaries are not well defined (a boundary being the separation between where flow does and does not occur), it may be necessary to use flow nets to evaluate flow. Flow nets are a pictorial method of studying the path that moving water follows. Darcy's law can be applied to flow nets to evaluate the effects of flow.[1]

For a condition of laminar flow, the path that the water follows can be represented by *flow lines*. In moving between two points, water tends to travel the shortest distance. If

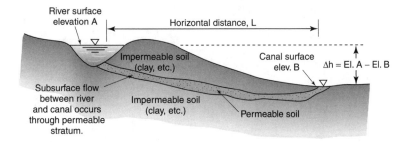

Figure 7-1 Instance in which direct application of Darcy's law can be used to determine underground flow.

---

[1]A scale drawing is necessary where actual quantities of flow are to be calculated or the effects of flow are to be determined.

changes in direction occur, the changes take place along smooth curved paths. A series of flow lines to represent flow through a soil mass would be parallel except where a change in the size of an area through which flow occurs takes place. Figure 7-2 shows flow lines for water seeping through a soil in a simple laboratory model.

Flow of water occurs between two points because of a difference in energy (usually expressed as head of energy or as pressure). Since it is the value of energy difference between two points, energy head can be referenced to an arbitrary datum for convenience. The total head of any point is that resulting from the sum of velocity head, potential head, and pressure head.[2] For water flowing through soil, velocity head is neglected, since it is small compared to potential and pressure head.[3] In Figure 7-3a, piezometers (open standpipes) inserted at points A and B would have the water level rise to the elevations shown. This figure shows the pressure and potential energy heads.

At certain points on different flow lines, the total energy head will be the same, as seen in Figure 7-3a. Lines connecting points of equal total energy head can be drawn, and are termed *equipotential lines* (Fig. 7-3b). It becomes apparent that there is no gradient to cause flow between points on the same equipotential line. On a flow net, equipotential lines must cross flow lines at right angles. (See Appendix A, "Application of the LaPlace Equation to Flow Nets," for mathematical proof.) The flow lines and equipotential lines together are the mesh strands which form the flow net (see box on pages 224–226 for discussion of development of the flow net), and are used to determine the quantities and other effects of flow through soils.

When seepage analyses are made, flow nets can be drawn with as many flow lines as desired. The number of equipotential lines will be determined by the number of flow lines selected. A flow net therefore can be a fine-mesh net or a coarse-mesh net. A general recommendation is to use the fewest flow lines that still permit reasonable depiction of the flow path along the boundaries and within the soil mass. For many problems,

Figure 7-2 Laboratory seepage tank showing flow lines indicating path of water flowing through the soil.

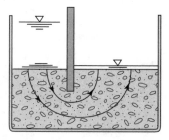

---

[2]Based upon the Bernoulli equation for steady-state laminar flow, total head, $h = v^2/2g + p/\gamma + z$, where v is the velocity of flow, g is the acceleration of gravity, p is pressure, γ is the unit weight of the fluid flowing, and z is the vertical distance between the point where the pressure p is determined and a reference elevation or datum (see Eq. 6-1a).

[3]Because the velocity head can be assumed negligible, the total head difference between two points along a path of flow is equal to the difference in height that water in open tubes would rise—for example, $\Delta h_1 = h_1 - h_2$ as defined in Chapter 6 and shown by Figure 6-3. Thus, flow occurs because of differences in total energy head.

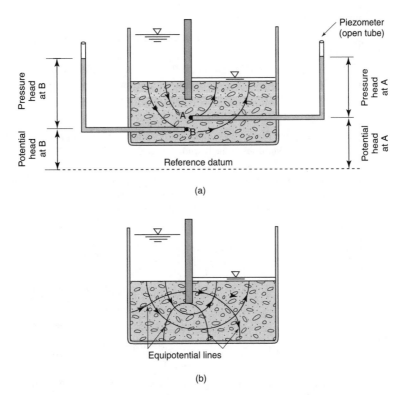

(a)

(b)

Figure 7-3   Points of similar energy head indicated on flow lines, and drawing of the flow net by addition of equipotential lines.

three or four flow channels (a channel being the path between adjacent flow lines) are sufficient.

The expression that results from the flow analysis to calculate the volume of water flowing or seeping through a soil mass (refer to the flow net development discussion in the box on pages 224-226) is:

$$q = kh_w\left(\frac{\mathcal{N}_f}{\mathcal{N}_d}\right)(\text{width}) \qquad \text{[for isotropic soil conditions] (7-2)}$$

where k = coefficient of permeability appropriate to the soil mass

h_w = elevation difference between energy head at upstream and downstream limits of the flow net

$\mathcal{N}_f$ = number of flow channels for the flow net, a channel being the zone between adjacent flow lines

$\mathcal{N}_d$ = number of equipotential drops for the flow net, a drop being the zone between adjacent equipotential lines

width = width of structure or soil zone involved, a distance perpendicular to the cross section or perpendicular to the plane of the flow net

Development of the Flow Net—Isotropic Soil

Figure 1 shows a condition in which flow occurs through a soil mass possessing isotropic properties (applied to fluid flow, the term *isotropic* implies that soil permeability is equal in all directions). Flow occurs because of the difference in total energy head caused by the unequal heights of water, $h_t - h_o$.

1. To begin the flow net, establish the boundaries of the soil mass through which the flow occurs and conditions at the boundaries. In Figure 1, lines a–b and c–e–f–d are drawn to represent flow lines at the boundary. Line c–a is an equipotential line with total head equal to h_t. Line b–d is an equipotential line with total head equal to h_o. These four lines establish the boundaries for flow through the soil in this problem. Note the line that indicates the variation in head between the ends of the soil volume.

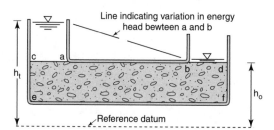

Figure 1

2. Next, sketch in intermediate flow lines parallel to those flow lines along the boundaries. Start by spacing the flow lines an equal distance apart in the sections where it is known that the flow lines will be parallel to each other (where the direction of flow is in a straight line and not curved) (Fig. 2).

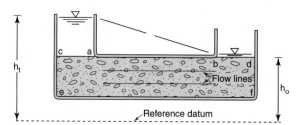

Figure 2

3. After the flow lines have been selected, locate a point of equal total energy head on each flow line. A line connecting these points will be an equipotential line (Fig. 3). Draw additional equipotential lines, selecting a spacing so that the distances y_1, y_2, and y_3 permit the proportion

$$\frac{y_1}{l_1} = \frac{y_2}{l_2} = \frac{y_3}{l_3}$$

to be maintained. Note the corresponding changes in total head at each equipotential line (Fig. 4).

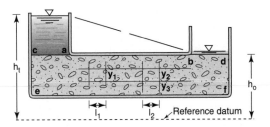

Figure 3

4. Apply Darcy's law, $q = kiA$. Since a volume (of flow) is to be determined, width as well as length and height of soil mass must be considered. For convenience, assume a thickness or width (perpendicular to the page) of unity. Let the flow past y_1 be q_1, the flow past y_2 be q_2, and the flow past y_3 be q_3. Referring to Figure 4, and noting that i (the hydraulic gradient) is the difference in head divided by the distance between the points where the head is measured, or $\Delta h / \Delta L$ (where l_1, l_2, are ΔL), obtain

$$q_1 = (k)\frac{\Delta h_1}{l_1}(y_1)(1)$$

$$q_2 = (k)\frac{\Delta h_2}{l_2}(y_2)(1)$$

$$q_3 = (k)\frac{\Delta h_2}{l_2}(y_3)(1)$$

Since y_1 and y_2 are bounded by the same flow lines, q_1, and q_2 must be equal, because flow is continuous. Also, since the ratios of the sides of each mesh block are equal, y_3 is equal to y_2 and therefore q_3 equals q_2.

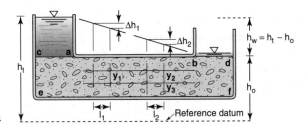

Figure 4

If all the mesh blocks are proportional, the quantity of flow between flow lines is equal, and it follows that Δh_1 equals Δh_2.

In flow net solutions, the mesh blocks are usually drawn as squares, although actually only the proportionality of distances between the two sides of a block must be maintained. By using squares, however, the ratio of the sides (height/length ratio) is 1, and the expression for flow becomes

$$q_1 = k\Delta h$$

where Δh is the total head divided by the number of equipotential line pressure drops $\mathcal{N}_d$, or

$$\Delta h = \frac{h_w}{\mathcal{N}_d}$$

Total flow per unit of width is the summation of the flow in each flow channel, or the sum $q = q_1 + q_2 + q_3 + \ldots$, which is also q_1 times the number of flow channels $\mathcal{N}_f$, or $q_1\mathcal{N}_f$. When analyzing a condition having a known width, total flow q is

$$q = q_1\mathcal{N}_f(\text{width}) = k\Delta h\mathcal{N}_f(\text{width}) = k\frac{h_w}{\mathcal{N}_d}\mathcal{N}_f(\text{width})$$

or

$$q = kh_w\left(\frac{\mathcal{N}_f}{\mathcal{N}_d}\right)(\text{width}) \qquad\qquad \text{[for isotropic soil]}$$

In constructing flow nets for most problems, it is improbable that the figures resulting from the assumed flow lines and equipotential lines will all be squares. The requirement for this condition is that, for each block, the distance across the center of the block between flow lines must equal the distance between the two equipotential lines, and the equipotential lines must cross the flow lines at right angles.

The combination of flow lines and equipotential lines create the mesh (or individual blocks) of the flow net. For most problems, the flow lines and equipotential lines are partially or totally curvilinear, and the mesh resulting will not be uniform throughout the net, but instead will vary in shape and size. The requirement for this situation is that, for each mesh block, the distance across the center of the block between flow lines must equal the distance between the two equipotential lines, and the equipotential lines must cross or meet the flow lines at right angles (refer to Appendix A). Figure 7-4 traces the steps used in drawing a flow net for a practical problem, and shows the computations to determine seepage.

Boundaries for the Flow Net

Figure 7-5 provides illustrations of typical flow net problems. These illustrations can serve as a guide to establish the boundaries of flow and the general pattern of a flow net for many types of problems. For problems like that shown in Figure 7-4, the upper boundary is the flow line following along the base of the structure. The lower boundary is a flow line that, for application to the flow net, is a line following along the surface of the impermeable stratum. The locations and directions of other flow lines within these boundaries are selected by using the methods described earlier.

Figure 7-4 Steps in drawing a flow net: (a) scale drawing of conditions; (b) trial flow lines; (c) trial equipotential lines; (d) final flow net and related seepage computation.

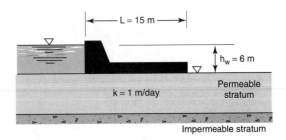

(a)

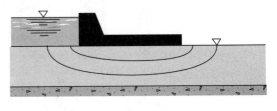

(b)

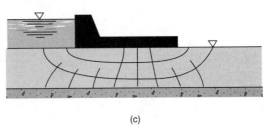

(c)

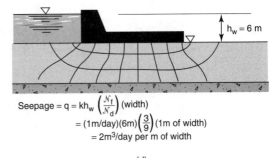

$$\text{Seepage} = q = kh_w \left(\frac{N_f}{N_d} \right) \text{(width)}$$
$$= (1\text{m/day})(6\text{m})\left(\frac{3}{9} \right)(1\text{m of width})$$
$$= 2\text{m}^3/\text{day per m of width}$$

(d)

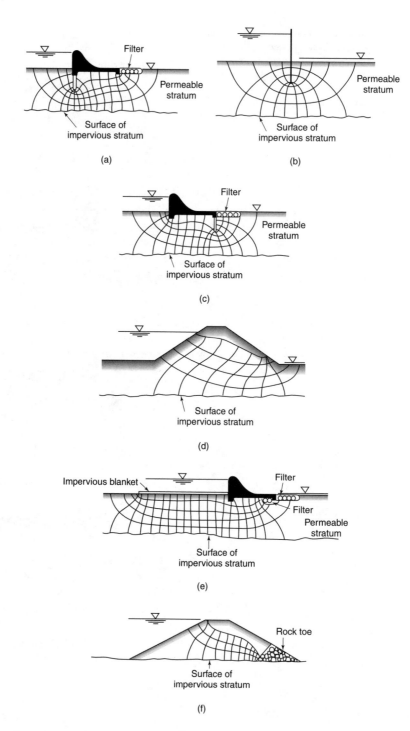

Figure 7-5 Flow nets for representative seepage problems: (a) masonry dam with sheetpile cutoff at heel; (b) sheetpile; (c) masonry dam with sheetpile cutoff and filter at toe; (d) earth dam; (e) masonry dam with upstream impervious blanket and downstream filter; (f) earth dam with downstream rock toe filter.

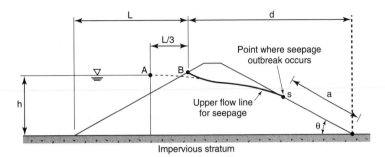

Point A: theoretical beginning point for parabolic shaped upper flow line.
Point B: actual point for water entry on upper flow line.

Figure 7-6 Seepage through an earth dam; definition of terms necessary to locate seepage outbreak.

Earth dams represent the condition in which the upper limit of flow is not defined by a natural boundary.[4] In some cases, such as homogeneous earth dams situated on an impervious stratum (see Fig. 7-6), seepage will break out on the dam's downstream face. The flow line representing the upper limit of seepage moving through the dam is approximately parabolic in shape. The point where the seepage breaks out onto the downstream face can be established from

$$a = \frac{d}{\cos \theta} - \sqrt{\frac{d^2}{\cos^2 \theta} - \frac{h^2}{\sin^2 \theta}}$$

(for slope angles, θ, less than 30°)[5]

where values of a, d, and h are defined in Figure 7-6. The method used to establish the shape and position of the upper line of seepage through the dam is outlined in Figure 7-7. After the upper flow line is established, the remainder of the flow net can be sketched. The quantity of seepage q can be estimated from

$$q \cong k(\sqrt{d^2 + h^2} - d)(\text{width}) \tag{7-3}$$

The effect of seepage forces is discussed in a later section of the chapter.

---

[4]The method for studying seepage through earth dams is presented in Arthur Casagrandes' paper "Seepage through Dams," which appeared in the *Journal of the New England Water Works Association,* June 1937. This paper has been reprinted in *Contributions to Soil Mechanics, 1925-1940,* Boston Society of Civil Engineers, Boston, Mass.

[5]Earth dams typically have slopes flatter than 2 horizontal to 1 vertical—that is, flatter than 30°.

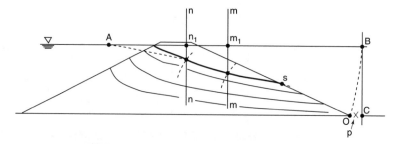

1. Establish point A and swing an arc OB having the radius AO (point O will be the focus of the parabola);

2. Draw vertical line BC (this will be the directrix of the parabola);

3. Locate point p midway between O and C;

4. Draw the vertical line m-m in the general location indicated, and determine the horizontal distance from line BC to line m-m;

5. Using point O as a center, swing an arc with radius m_1 B to intersect line m-m. This intersection establishes a point on the parabolic flow line;

6. Draw the vertical line n-n in the general location indicated, and determine the distance n_1 B;

7. Using point O as a center, swing an arc with radius n_1 B to intersect line n-n. This intersection establishes another point on the parabolic flow line;

8. Continuing the procedure just outlined, establish as many additional points as necessary to sketch the total length of flow line between A and s;

9. Modify the beginning section of the flow line freehand. The freehand line must intersect the upstream face of the dam at a right angle;

10. The parabolic flow line is assumed to follow the downstream face of the dam below point s;

11. Other flow lines are sketched in, assuming a shape generally similar to the upper seepage line.

Figure 7-7 Procedure for locating the upper seepage line for flow through an earth dam.

Flow Nets for Nonisotropic Soils

The method of flow net analysis discussed above applies for the frequently found condition in which the soil permeability in the horizontal and vertical directions if similar. In stratified soil deposits, however, the horizontal and vertical coefficients of permeability may differ; usually the horizontal permeability is greater than the vertical. In such instances, the methods for drawing the flow net need to be modified. Use of a *transformed section* is an easily applied method that accounts for the different rates of permeability. Vertical dimensions are selected in accordance with the scale desired for the drawing. Horizontal dimensions, however, are modified by multiplying all horizontal lengths by the factor $\sqrt{k_v/k_h}$, where k_v and k_h are the vertical and horizontal coefficients of permeability, respectively. A distorted diagram results, with shortened horizontal dimensions, as illustrated in Figure 7-8. The conventional flow net is then drawn on the transformed section using the procedures presented previously.

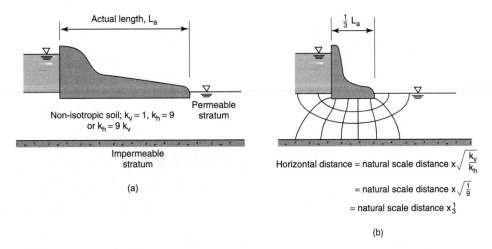

Figure 7-8 Method of drawing the transformed section and flow net for nonisotropic soil conditions: (a) actual section (to scale); (b) transformed section (distorted scale).

For flow through the nonisotropic soil, the seepage equation becomes

$$q = h_w\left(\frac{\mathcal{N}_f}{\mathcal{N}_d}\right)\sqrt{k_v/k_h}(\text{width}) \qquad \text{[for nonisotropic soils]}$$

The values for $\mathcal{N}_f$ and $\mathcal{N}_d$ in this equation are taken directly from the transformed section.

Uplift Forces

For the problem of seepage beneath a structure such as a dam, the uplift force acting on the base of the structure because of this seepage can be evaluated from the flow net. The pressure acting at the upstream and downstream ends of the structure (and at selected in-between points if necessary) is determined and a pressure diagram is made. The total uplift force (per unit of width perpendicular to the drawing) is then the area of the pressure diagram (Fig. 7-9).

Other Seepage Forces

Water seepage beneath a structure and then escaping by flowing upward at the downstream end imposes an upward force on the soil. If the upward force is sufficiently great, soil particles will be carried away, eroded by the escaping water. There follows an illustration of how the flow net can be used to determine if the pressure of the escaping water is great enough to erode the soil. The submerged unit weight soil is

$$\gamma_{sub} = \left(\frac{G_s - 1}{1 + e}\right)\gamma_w \qquad (2\text{-}19)$$

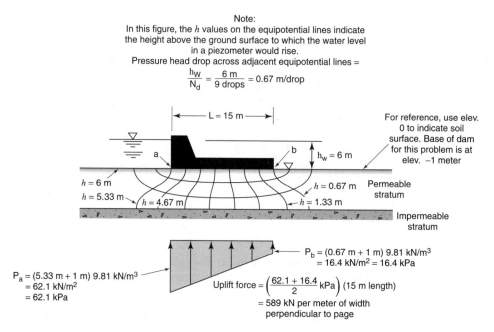

Figure 7-9 Procedure used to compute uplift pressure acting on base of a masonry dam.

If the value of G_s (specific gravity of soil) is about 2.7 and the void ratio is about 0.70 (both reasonable values), the submerged weight of the soil would be about equal to the weight of water. The hydraulic gradient between the last two equipotential lines is the difference in pressure head across these two points divided by the distance between these equipotential lines, or $\Delta h / l$. The pressure difference between the two equipotential lines is Δh times the unit weight of water, γ_w. If the hydraulic gradient is unity, the upward force due to the moving water is equal to the unit weight of water. There then exists an upward pressure of 9.81 kPa or 62.4 psf acting on a soil where gravity effects a downward pressure of 9.81 kPa or 62.4 psf. This is a fully buoyant condition in the soil. A slightly greater uplift pressure would carry away soil particles. For this reason, an escape or exit hydraulic gradient of about unity or greater is usually considered an indication that erosion may occur. To prevent erosion, the pattern of flow could be modified to reduce the pressure of the escape gradient (increasing the length of flow either by making the structure larger or by embedding sheeting beneath the structure are two alternatives), or a coarse material of carefully graded particle sizes that is not susceptible to erosion by the escaping water could be placed beneath the tip of the structure (a soil filter).

Illustration 7-2

A masonry dam having a sheetpiling cutoff at the upstream end is located at a reservoir site, as indicated by the sketch. Draw a flow net for the subsurface flow and compute the seepage. Also calculate the uplift force acting on the base and the escape gradient of the water at the downstream tip of the dam.

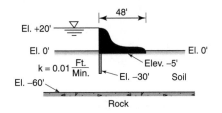

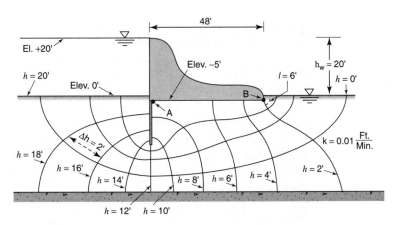

Solution

The number of flow channels, $\mathcal{N}_f = 4$. The number of pressure drops, $\mathcal{N}_d = 10$ (note feet of head acting at each equipotential line, where differences Δh are calculated from

$$\Delta h = \frac{h_w}{\mathcal{N}_d} = \frac{20.0 \text{ ft}}{10} = 2.0 \frac{\text{ft}}{\text{drop}}$$

Seepage

$$q = kh_w \frac{\mathcal{N}_f}{\mathcal{N}_d}(\text{width}), \text{ or}$$

$$q = \left(.01 \frac{\text{ft}}{\text{min}}\right)(20 \text{ ft})\left(\frac{4}{10}\right)(1.0 \text{ ft wide})$$

$$= .8 \frac{\text{ft}^3}{\text{min}} \text{ per foot of width}$$

Uplift Force on Base

$$p_A = (5 \text{ ft} + 7 \text{ ft})\,\gamma_w$$

$$p_B = (5 \text{ ft} + 2 \text{ ft})\,\gamma_w$$

$$\text{Uplift} = \left(\frac{p_A + p_B}{2}\right)(L) = \frac{(12 \text{ ft} + 7 \text{ ft})(62.4 \text{ lb/ft}^3)(48 \text{ ft})}{2}$$

$$= 29{,}200 \text{ lb per foot of width (perpendicular to page)}$$

Escape Gradient at Downstream Tip

Δh between last two equipotential lines = 2.0 ft, and

$$l = 6 \text{ ft}$$

$$i = \frac{\Delta h}{l} = \frac{2}{6} = .33 \text{ (therefore erosion is not expected)}$$

(Figure 7-10 shows a dam where foundation erosion did occur.)

Practical Considerations

The correct drawing of a flow net can be a tedious operation, particularly in the case of nonuniform soil conditions. Fortunately, even crudely drawn flow nets can provide reliable information on quantities of seepage and uplift pressures. Proper evaluation of an escape gradient requires a correct and carefully drawn flow net, however. For flow nets, as well as for all seepage studies, it should be expected that there will be a variation in the coefficient of permeability for the soil in the seepage zone under study. For practical problems, answers obtained from a seepage analysis should be considered as approximate, and are not to be believed as being precise.

Quicksand

The dreaded quicksand condition occurs at locations where a sand or cohesionless silt deposit is subjected to the seepage force caused by an upward flow of groundwater. This condition can occur in depressions of areas where the water table is high (close to the ground surface) or where artesian conditions exist. The upward gradient of the water is sufficient to hold the soil particles in suspension, in effect creating a material with the properties of a heavy liquid. A no-support condition exists even though the soil gives the appearance of being firm ground.

The suction or pull attributed to quicksand is in reality gravity exerting its normal effect but in a heavy liquid environment whose viscous properties also exert "drag" on a thrashing body. Theoretically, a person falling into a quicksand area can float as in water.

Since the quicksand condition is caused by the forces of seepage and not some mystical property of the soil, elimination of the seepage pressure will return the soil to a normal condition capable of providing support.

7.2   DRAINAGE

Among the more common problems in construction work is the need to handle subsurface water encountered during the construction sequence, and to handle subsurface water after construction so that the completed facility is not damaged nor its usefulness impaired.

During construction, lowering of the groundwater table and removal of water from working areas is desirable from the standpoint of better working conditions for workers and equipment. In some cases, dewatering may be necessary to ensure that proper construction

(a) Dam viewed from the downstream side.

(b) Closeup of downstream face and toe of dam.

(c) Foundation cavity approximately 2 feet deep that developed as a result of piping. Light near center of photo is daylight from back (upstream) side of the dam.

Figure 7-10 Small dam that experienced underdam erosion of its foundation due to seepage and piping.

can be achieved. Conditions at some locations require that the structure be protected from erosive effects of flowing groundwater to prevent the loss of foundation stability. Additionally, usable interior portions of structures located below groundwater elevation normally are required to be free from serious seepage or leakage.

Conditions Requiring Drainage

In planing excavations for construction projects that extend below the water table in soils where the permeability is greater than about 1×10^{-5} ft/min or 5×10^{-5} mm/sec, it is generally anticipated that at least some continuous drainage or dewatering procedures will be required if the work area is to be kept dry for construction. Excavations in the more impermeable soil types may remain dry after an initial dewatering, especially if the excavation is to remain open for only a limited time. (For all soil conditions, the actual amount of seepage expected would be affected by the depth below the water table and by the period of time that the excavation is to be open, as well as by the soil properties.)

Dewatering Shallow Excavations

For construction of shallow foundations and for other excavations of limited depth made in coarse soil, open drainage or interceptor ditches can be an expedient and relatively inexpensive method for lowering the groundwater table a slight distance. The interceptor ditch has to penetrate deeper than the elevation of the work area because of the pattern that the underground water surface takes (termed the *drawdown curve*) in the area surrounding the interceptor ditch. Water collecting in such ditches normally has to be pumped out of the ditch for disposal. Since gravity flow is relied on to bring the water to the ditch, the continued inflow is dependent on the water level in the ditch being kept low. With this method, it is common to construct small pits in the ditch, termed *sumps,* for locating the necessary pumps (sump pumps).

Where shallow foundations are to be installed at just about the water table elevation, it has frequently been found possible to obtain sufficient lowering of the water level to permit working in the dry by locating sump pits within, or immediately adjacent to, the foundation excavation.

The drawing down of the water table can also be accomplished by constructing a series of sump pits (Fig. 7-11), or, if greater depth is required, some type of drainage wells around the construction area and pumping the water from these pits or wells.

Subsurface water that flows in an upward direction into an excavation area that is being dewatered imparts a seepage force that tends to loosen the soil, reducing the soil strength. The change in strength should be considered in designing excavation bracing and foundations.

Dewatering Intermediate Depths

Where excavations in coarse-grained and silty soils are to extend more than one meter or a few feet below groundwater level, open ditches or pits typically are not practical, and more advanced methods are used. Discussion of some methods follows.

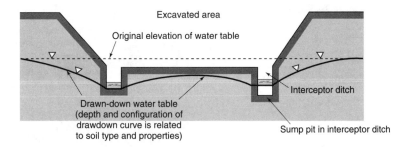

Excavated area

Original elevation of water table

Interceptor ditch

Drawn-down water table
(depth and configuration of
drawdown curve is related
to soil type and properties)

Sump pit in interceptor ditch

Figure 7-11 Shallow drawdown of water table by use of interceptor ditches and sump pit in excavation.

Well Points. For dewatering to intermediate depths (to about 9 m or 30 ft, but more if sufficient area is available for installing the necessary equipment), well-point systems are normally used. Basically, a well point is a closed-end pipe or tube, having perforations along its lower end, that is installed to the desired depth below the water table (Fig. 7-12). Groundwater entering the well point through the perforations is pumped to the surface through a riser pipe connected to the well point (Fig. 7-13). For construction dewatering, the well-point perforations are provided with a protective screen or filter to prevent soil particles from clogging the perforations. Well points are conventionally installed in drilled holes or, most usually, by jetting. In jetting, water is pumped through (into) the riser and discharges through special openings at the tip of the well point, an action that displaces the soil below the tip of the point. This procedure is continued until the desired penetration is achieved.

To dewater an area, a series of well points is installed around the perimeter of the area. The groundwater level within the perimeter will be lowered when the well-point system is put in operation. The spacing of the well points varies according to the soil type and depth

Figure 7-12 A well-point assembly.

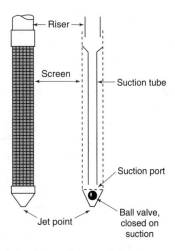

Riser

Screen

Suction tube

Suction port

Jet point

Ball valve,
closed on
suction

External   Cross section

Figure 7-13 Installed well point.

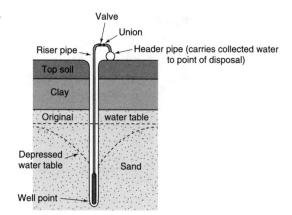

of dewatering. Spacings conventionally vary between 1 to 3 meters (or 3 to 10 feet). Each well-point riser is connected by hose or pipe to a horizontal header or collector pipe, which is connected to a pump. Usually, many risers are connected to a common header. Well-point systems are usually rented from specialty contractors who also install the system. With most contractors, the riser pipes are about 50 mm (2 in.) in diameter, and headers are 150 to 300 mm (6 to 12 in.), depending on the flow to be handled.

Normally, the screen provided with the well point is sufficient to keep medium sands and coarser materials from clogging the perforations. Where finer soils are to be dewatered, it is necessary also to provide a sand filter around the well point. The sand filter has the added benefit of acting to increase the size of the well point, permitting easier and greater flow into the point. To obtain a filter, it is necessary to create a hole having a larger diameter than the well point, prior to or when installing the well point and riser. The sand filter and backfill are then placed around the installed riser. A large-diameter hole can be formed by providing an attachment to the well point so that it can still be jetted to the desired depth, or by installing a casing by jetting or other method, after which the well point and surrounding sand filter are installed and the casing is pulled. For normal requirements, the sand filter material consists of a uniform graded medium to coarse sand.

The horizontal extent to the soil zone surrounding a well point that will show lowering of the groundwater table is affected by the depth of the well point and the permeability of the soil. Generally, the horizontal distance that is affected increases as permeability of the soil increases. The drawdown curve values shown in Table 7-1 can be used to estimate the horizontal distance that will be influenced by well-point dewatering.

With the type of pumping equipment conventionally used for well points, the depth of dewatering that can be achieved by a single line of well points located around the perimeter of an excavation is usually about 6 to 7 m (or about 20 ft). This is due to the limit on the practical lifting, or suction, capacity of the pumping equipment. Lowering the water table through a greater distance may require the use of a two (or more) -stage (multistage) installation. Where a two-stage installation is required, the well points for the first stage of drawdown are located near the extreme perimeter limits of the area that can be excavated, and are put into operation. Well points for the second stage are subsequently located *within* the area that has been excavated, near to the bottom elevation that has been dewatered by

Table 7-1 Approximate Slope of Drawdown Curve Due to a Single Row of Well Points

Soil Being Dewatered	Slope of Drawdown Curve, %[a]
Coarse sand	1–3
Medium sand	2–5
Fine sand	5–20
Silt–clay	20–35

[a]Refer to Figure 7-14 and recall that percentage slope is vertical change (ft or m) per 100 units of horizontal length (ft or m).

the first stage. The second-stage well points then lower the water table to the additional depth necessary to complete the excavation dry (see Figs. 7-14 and 7-15).

Although the principle on which well-point dewatering is based is a simple one, care and skill are required during installation and operation to obtain the intended results. Proper assembly of the piping system is particularly crucial, because if air leaks develop at connections, the efficiency of the pump suction is reduced.

Vacuum Dewatering. Where well points are in use, flow to the well point is by gravity. In the coarser soils (soils with less than 25 percent of particles smaller than 0.05 mm), gravity flow is normally adequate to achieve dewatering. For silts, gravity flow is restricted because of capillary forces that tend to hold the pore water. It has been found, however, that by applying a vacuum to the piping system, satisfactory dewatering of silty soils can be achieved. For maximum efficiency, the vacuum dewatering system requires that the well point and riser be surrounded to within a few feet of the ground surface with filter sand and that the top few feet be sealed or capped with an impervious soil or other material. By having the pumps maintain a vacuum pressure, the hydraulic gradient for flow to the well points is increased. With this system, closer well point spacings are required than for the conventional system.

Where suction pumps are used to draw the collected groundwater from the well points, as is common, the practical maximum height of lift is about 5 to 6 m (15 to 18 ft). Where an excavation is to extend more than this distance below the groundwater table, it will be

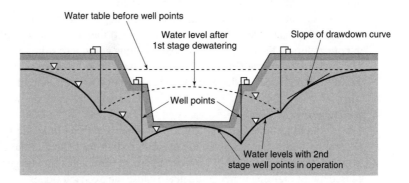

Figure 7-14 Multistage well point operation.

Figure 7-15 Installed two-stage well-point system in operation. (Courtesy of Moretrench American Corporation)

necessary to dewater the area in two or more stages. With multistage dewatering, the well points for the first stage are installed on a perimeter line outside of the actual excavation area required for the construction and put in operation. Excavation proceeds within the perimeter formed by the well points to the depth where groundwater is encountered. Well points for the second stage are next installed within this excavation to further lower the water table for the actual construction area.

Electro-osmosis. Sand and the coarser silt soils usually can be dewatered by using gravity draining or vacuum-assisted well points. In fine silts, clay, or coarse-grained–fine-grained mixtures, the effective permeability may be too low to obtain a successful dewatering by using well-point methods. Drainage of such low-permeability soils may be achieved by electro-osmosis, provided that no more than 25 percent of the soil particles are smaller than 0.002 mm.

The basic principle of electro-osmosis is as follows: If a flow of direct-current electricity is induced through a saturated soil between a positive and negative electrode (anode and cathode, respectively), pore water will migrate toward the negative electrode. If a well or well point is made the cathode, collected water drained from the soil can be removed by pumping as with a conventional well-point system.

Movement of water toward an electrically negative terminal occurs because of the attraction of the cathode for positive ions (cations) that are present in groundwater. Cations in pore water are the result of dissolved minerals going into solution with the groundwater. The cations concentrate around the negatively charged surface of clay particles to satisfy the electrical charge on the particle. These cations, in turn, attract the negative "end" of "dipole" water molecules. As the cations are drawn to the cathode, water molecules held to the cation follow.

In a field installation, it is considered most beneficial to locate the anode close to the excavation perimeter and place the cathode further back, so that the direction of drainage is away from the excavation. This arrangement eliminates the danger that seepage forces caused by the draining water could act to cause the exposed slopes of the excavation to slide or slough in toward the excavation (construction area).

Because of techniques involved, the required special equipment, and the high electrical consumption, drainage by electro-osmosis is expensive compared to gravity drainage methods. Consequently, the method is used only when other methods cannot be applied.

Deep Drainage

If excavations are to extend deep below the groundwater table, or to penetrate through a deep permeable stratum, well points may not be applicable because of the limits to which well points can raise water. For this situation, deep wells and deep well pumps (jet or venturi pumps) can be used. Deep wells are frequently relatively large-diameter (on the order of 600 mm (2 ft) drilled holes. A perforated protective casing is installed, and the deep well pump is then placed inside the casing near the bottom. Coarse filter material is placed between the outside of the casing and the walls of the drilled hole.

Consolidation Drainage

Consolidation refers to the process that occurs in clay deposits when the development of compressive stresses in the soil mass from external loading causes the soil to undergo a reduction in volume (due to a reduction in void spaces) that is simultaneously accompanied by water in the soil pores' being expelled to permit the decrease in voids. Thus, subsurface drainage of fine-grained soils by consolidation can be achieved by imposing a surface load onto an area. In general, a reduction in water content also results in an increase in the shear strength of the clay. Unfortunately, the rate of drainage is quite slow in clay soils because of the low coefficient of permeability.

Sand drains represent one method that has been used with success to accelerate the rate at which consolidation drainage occurs in fine-grained materials. The method is based on reducing the distance that pore water has to travel to escape from the consolidating soil. Sand drains are vertical columns of freely draining sand installed so as to penetrate the soil strata to be consolidated. In clays, the permeability in the horizontal direction is frequently many times the vertical permeability because of the stratified manner in which clay deposits have been formed. The presence of relatively closely spaced sand drains thus reduces the distance that pore water must travel to escape and also permits an easier (horizontal) drainage path. In draining from the clay, water flowing to the sand drains is under pressure as a result of the external loading that is applied to cause the consolidation (usually an earth fill to act as a surcharge). Therefore, the water can flow upward as well as downward in the sand drain. Further details on the mechanics of the drainage that occurs in sand drains and wick drains (a related, more recent development) are presented in Chapter 10.

Summary: Techniques for Subsurface Drainage

A graphical summary relating soil categories and appropriate techniques for accomplishing subsurface drainage is presented in Fig. 7-16.

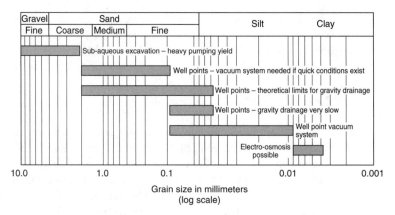

Figure 7-16 Summary: Relationship between method for dewatering soil and range of particle sizes [7].

Drainage after Construction

Preventing groundwater from seeping into a structure or through a structure after it has been built may be necessary in order to obtain proper use of the structure or to protect it from damage.

In some cases, the dewatering methods used during construction can be kept in use to protect the structure. Normally, however, this procedure is not followed; the equipment or methods may interfere with the usefulness of the finished project, may be too expensive for long-term use, or may be impractical in some other way.

When it is known that a usable part of a structure will be located below the groundwater table, it is desirable to build the facility using waterproof design and construction techniques. One of the desirable features for a subaqueous structure is to have all seams and joints provided with water stops and/or to have the structure built with no seams or joints wherever possible. For example, in small conventional buildings, concrete basement walls and floors can be poured monolithically. Admixtures also are available to make concrete more resistant to water penetration; similarly, chemical coatings are available to waterseal the surface of concrete.

Foundation Drains

Where groundwater will be flowing in the vicinity of the structure, provision can be made so that the water will be quickly carried away from the building area, and at worse only a limited height of groundwater buildup against the exterior will occur. With adequate provisions, the large hydrostatic pressures that tend to force seepage entry thus will not develop. A method for achieving this control when the depth below the water table is not too great is the use of foundation drains, conventionally placed around the building exterior at footing level and adjacent to the footing (Fig. 7-17). Such drains should not be lower than the bottom of the footing. Where there is concern over the ability of the exterior drain to handle the expected groundwater, the foundation drains may be located on the interior side of the foundation as well. With a high water table, it may also be necessary to place interceptor drains at some short distance from the building and at an elevation higher than the footings so that the water table is lowered in stags.

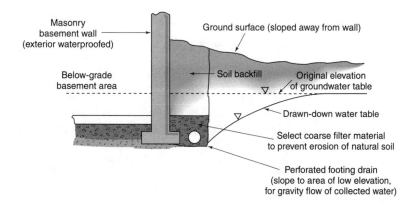

Figure 7-17   Footing drain tile installation for disposing of groundwater against a basement wall.

The installed drain normally consists of pipe or a synthetic composite provided with perforations or installed with open joints so that groundwater can enter into the pipe. Gravel and/or sand filter material, or a geosynthetic fabric, must surround the drain pipe so that soil particles tending to be carried along by the inflowing water are prevented from clogging the drain pipe openings or causing erosion. The topic of geosynthetics and criteria for filter material are presented in subsequent sections of this chapter.

An outlet for collected water is a necessity; preferably disposal will be by gravity flow to a storm drain system or other drainage facility such as a ditch, dry well, or pool located at an area of lower elevation than the building. If disposal by gravity flow is not possible, drainage water will have to be directed to a sump pit or other collector and pumped to a disposal.

Blanket Drains

Where a basement is to be located below groundwater level, a blanket or layer of filter material can be placed beneath the floor slab to provide a highly permeable drainage path for removal of groundwater acting against the bottom of the slab (Fig. 7-18). Providing an escape path serves to reduce uplift pressures and the possibility for seepage to occur through the floor. The blanket connects to a sump where collected water is pumped out, or to drainage pipes where disposal occurs by gravity flow. The filter blanket normally consists of a sand layer placed over the natural subgrade soils and a coarser, small-gravel or crushed rock layer through which most of the horizontal flow is intended to take place.

Interceptor Drains

In paved highways and airfield runways, the design frequently calls for interceptor trench drains located parallel to the shoulder, as illustrated in Figure 7-19. The purpose of the drains is to lower the groundwater table to a level beneath the pavement and to permit easy lateral drainage (escape) for water finding its way into the coarse base material provided

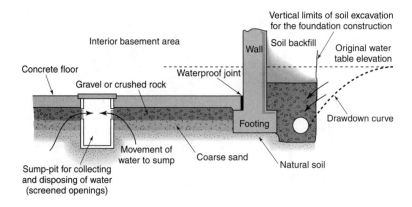

Figure 7-18 Details of a blanket drain beneath a concrete floor.

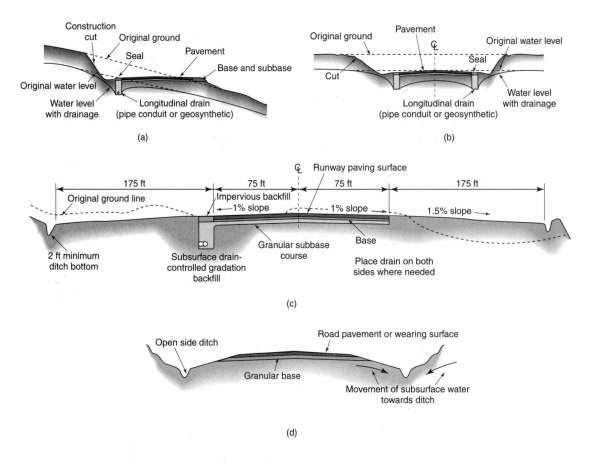

(a)

(b)

(c)

(d)

Figure 7-19 Highway and airfield drains: (a) interceptor drain for highway constructed in a sidehill (after Cedergren [53]); (b) interceptor drain for highway in flat terrain [46]; (c) typical airport runway interceptor drains (Civil Aeronautics Administration); (d) typical open-shoulder ditch for roads.

beneath the pavement (water resulting from upward capillary migration and from infiltration through cracks, joints, or voids in the pavement).

The intent of the drainage facility is to keep the base and subgrade soils dry so that they will maintain high strength and stability. With coarser soils, the intent is probably achieved quite successfully. In fine-grained soils, the degree of achievement is questionable. Most probably, a groundwater table in such soils is not significantly lowered because of the material's low permeability. However, the drainage ditches do serve to prevent excess pore water pressures from building up in the subgrade soils (excess pore water pressures have the effect of weakening the soil). The drains also provide a means for disposal of surface and near-surface water, and will intercept underground flow trying to enter the roadway from the side areas.

Similarly, paved and unpaved roads will benefit somewhat from open drainage ditches located adjacent to shoulder areas (Fig. 7-19d). Such ditches handle surface water, provide a lateral drainage outlet for soil near the level of the road surface that helps prevent the development of excess pore water pressures, intercept surface and near-surface water flowing toward the roadway area from the sides, and provide storage for plowed snow and a flow channel for melting snow so that it does not run onto the roadway.

Currently, the common methods for achieving highway and airfield drainage involve state-of-the-art technology and geosynthetics as shown in Figure 7-22.

Flow through a Structure

In some types of structures, such as earth dams and dikes, it is known that water will be flowing through the structure after it is put in use. The design of the structure, therefore, considers the effect that such flow will have on the stability of the structure. Flow is not prevented but, rather, is permitted and directed so as not to create a stability problem. For instance, in a homogeneous earth dam or dike, flow is typically as indicated in Figure 7-20. If the velocity of the water flowing through the structure is too great, erosion and instability of the downstream side will result. The placement of a toe drain or an underdrain blanket that consists of coarse materials helps direct the flow to those areas that will not erode or lose stability because of the flowing water (Fig. 7-21).

There are basically similar situations encountered in building construction, where localized underground flow occurs through seams or a stratum of pervious soil (frequently referred to as underground springs) and presents a potential seepage problem for the structure. Providing some type of drainage conduits for the flow to pass around or through the structure is frequently technically more desirable and less expensive than attempting to cut off or stop the flow. The main precaution to be observed is that the flow conduit be designed so that seepage forces do not affect the building and that soil erosion does not occur.

Filter Design

Water flowing toward a drainage structure must be able to enter the drain easily and quickly but must not be permitted to carry with it small soil particles that could eventually clog the drain or cause erosion or cavitation in the areas surrounding the drain. Currently, the desirable separation is achieved by use of the geosynthetics discussed in the following section of

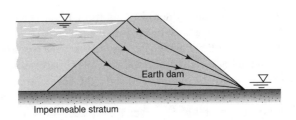

Figure 7-20 Pattern of flow through a homogeneous earth dam.

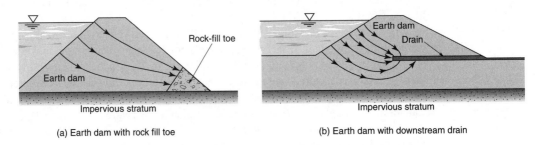

(a) Earth dam with rock fill toe

(b) Earth dam with downstream drain

Figure 7-21 Flow through an earth dam modified by a toe drain or an underdrain.

this chapter. But for the situation where a drainage structure is open to the soil, it is necessary to provide a *drainage filter.* Sand and gravel blends are typically used for such filters.

To permit quick, easy flow into the drain facility, and to "draw" water to the drain, it is necessary to have the permeability of the filter material considerably greater than the permeability of the surrounding natural soil. To satisfy this criterion, the following ratio between particle sizes for the filter material and natural soil has been established:

$$\frac{D_{15} \text{ of filter material}}{d_{15} \text{ of natural soil}} \geq 4$$

To prevent particles of *natural sandy soils* from migrating into the filter material and eventually clogging it, the following ratios of particle sizes should be used as design criteria:

$$\frac{D_{15} \text{ of filter material}}{d_{85} \text{ of natural soil}} < 5$$

$$\frac{D_{50} \text{ of filter material}}{d_{50} \text{ of natural soil}} < 25$$

$$\frac{D_{15} \text{ of filter material}}{d_{15} \text{ of natural soil}} < 20$$

Of these, criteria, the D_{15}/d_{85} ratio is considered to be the most important and indicative of filter performance.

In these ratios, the D_{15}, d_{15}, and so on, values refer to the particle dimensions obtained from a particle-size distribution curve (e.g., the D_{15}, or d_{15} refers to the particle size for which 15 percent of the soil material, by weight, is smaller). In the past, filter design criteria have suggested that natural and filter materials have parallel but transposed particle-size distribution curves, but this feature of filters has been found to be unnecessary and perhaps even undesirable.

Where a sand and gravel filter is to be in contact with a *fine-grained soil,* some additional criteria beyond those just outlined should be applied to the selection of filter materials. For fine clays (a d_{85} range of 0.03–0.10 mm), the D_{15} for the filter should not exceed 0.5 mm. For fine-grained silt (a d_{85} range of 0.03–0.10 mm) with low cohesive strength and low plasticity (liquid limit less than about 30), the D_{15} for the filter should not exceed about 0.3 mm. For sandy silts and clays having noticeable sand content (a d_{85} range of 0.1–0.5 mm), a sand filter material is adequate when the D_{15}/d_{85} ratio is less than 5. The D_{50}/d_{50} and D_{15}/d_{15} ratios should *not* be considered when selecting a sand or a sand–gravel filter that will be in contact with the fine-grained soils described.

To ensure that filter particles will not enter and clog the drain structure, where holes or slots are provided for the entry of water, the following criteria should be followed:

For circular holes:

$$\frac{D_{85} \text{ of filter material}}{\text{Diameter of opening}} > 1.2$$

For slotted openings:

$$\frac{D_{85} \text{ of filter material}}{\text{Width of slot}} > 1.4$$

For proper performance, the filter material must surround the drain, under as well as over. The filter material should also be compacted so that it does not settle. To help minimize any tendency for fine particles to migrate through the filter, and to drain as much water as possible, the holes or slots in the drain structure should be placed facing downward, or as near to downward as possible, so that the flow into the drain facility is *not* downward but instead is lateral or slightly upward.

If the natural soil is fine grained, it may be necessary to provide a filter that consists of two or more layers, each having a different gradation. The material closest to the drain would be coarsest, with the outer material filter that is next to the natural soil being less coarse. For the situation in which two or more filter materials are used, the criteria presented earlier for filter and natural soil would also apply between any two adjacent filter materials.

The filtering function accomplished by selection of soil materials as described in this section can also be achieved using the new technology of geosynthetics described in the section following. However, the use of soil filters is *not* a discontinued practice. Much of the information used to design filter systems with geosynthetics is based on soil filter technology.

Drainage, Filtration, and Separation with Geosynthetics (Synthetic Fabrics)

The successful performance of subsurface construction and earth embankment structures often relates to the proper handling of surface and subsurface water, and to preventing the otherwise detrimental effects of its presence. Usual factors requiring consideration include a means for accomplishing the drainage of subsurface water which might cause problems, a means to achieve filtration where groundwater movement can be tolerated but must occur without migration of soil particles, and a means to maintain separation where differing soil materials in contact must stay zoned.

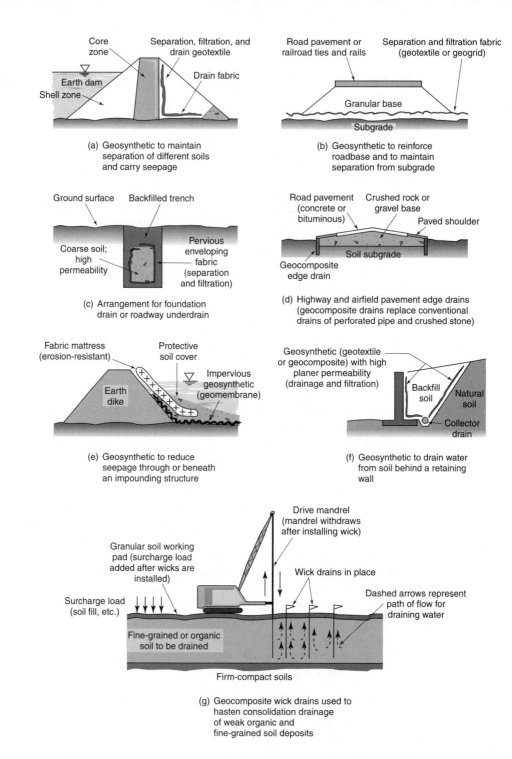

(a) Geosynthetic to maintain
separation of different soils
and carry seepage

(b) Geosynthetic to reinforce
roadbase and to maintain
separation from subgrade

(c) Arrangement for foundation
drain or roadway underdrain

(d) Highway and airfield pavement edge drains
(geocomposite drains replace conventional
drains of perforated pipe and crushed stone)

(e) Geosynthetic to reduce
seepage through or beneath
an impounding structure

(f) Geosynthetic to drain water
from soil behind a retaining
wall

(g) Geocomposite wick drains used to
hasten consolidation drainage
of weak organic and
fine-grained soil deposits

Figure 7-22. Representative uses of geosynthetics for drainage, filtration, and separation. (Photographs courtesy of Hoeschst Celanese Corp., Spartanburg, South Carolina)

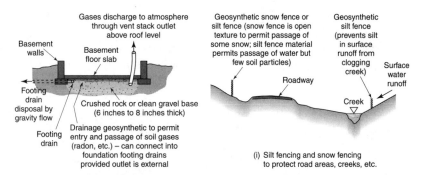

Gases discharge to atmosphere through vent stack outlet above roof level

Basement walls

Basement floor slab

Footing drain disposal by gravity flow

Footing drain

Crushed rock or clean gravel base (6 inches to 8 inches thick)

Drainage geosynthetic to permit entry and passage of soil gases (radon, etc.) – can connect into foundation footing drains provided outlet is external

(h) Soil gas venting for building substructure or slab-on-grade construction

Geosynthetic snow fence or silt fence (snow fence is open texture to permit passage of some snow; silt fence material permits passage of water but few soil particles)

Geosynthetic silt fence (prevents silt in surface runoff from clogging creek)

Surface water runoff

Roadway

Creek

(i) Silt fencing and snow fencing to protect road areas, creeks, etc.

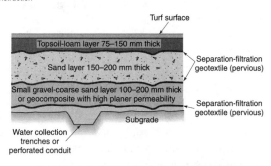

Turf surface

Topsoil-loam layer 75–150 mm thick

Sand layer 150–200 mm thick

Small gravel-coarse sand layer 100–200 mm thick or geocomposite with high planer permeability

Subgrade

Separation-filtration geotextile (pervious)

Separation-filtration geotextile (pervious)

Water collection trenches or perforated conduit

(j) Athletic playing field (section, design for field drainage)

(k) Completed geo-membrane liner in place for waste liquid retention pond

(l) Seaming operation, for joining geotextile sections

(m) Pavement edge drain installation (perforated flexible drain conduit surrounded by small crushed rock, wrapped in filter-separation geotextile)

(n) Edge drain installation completed, waiting to be covered

Figure 7.22 Con't

The subsurface drainage, filtration, and separation functions can be accomplished particularly well by the geosynthetic products when compared to accomplishing these functions in the "conventional" manner with earth materials (see representative uses illustrated in Fig. 7-22). It is expected that in the near future the geosynthetics will be the first choice for all drainage, filtration, and separation designs (except where soil mass is required to aid stability or necessary to provide volume for other necessary functions such as aerobic and anaerobic decompositions). Though the use of the geosynthetics is relatively new, these materials are rapidly becoming popular with designers and construction contractors because of their ability to perform necessary functions while offering practical advantages such as a wide availability of products from the marketplace, the relative ease of shipping and field handling, rapid installation, durability and long life when properly selected, and general environmental safety (will not degrade).

Various details relating to the geosynthetic products are presented in the two insets, "The Role of Geosynthetics in Construction" and "The Materials, Manufacture, and Properties of Geosynthetics Used for Construction Projects." Most of the geosynthetic materials used for filtration and separation are from the *geotextile* grouping described in the insets. Geosynthetics used because of capability of accomplishing the drainage function are from the *geotextile, geonet,* and *geocomposite* groupings. Many of the geosynthetics are cloth-like or sheetlike and available in wide sheets or rolls; application involves spreading the sheet over the earth zone being improved, and large areas can be covered quickly. Since they are manufactured, the synthetics can be produced to possess different properties; for example, geotextiles are available in a range of strengths, in a range of coarse to fine textures, and in a range of permeability.

The concern for separation of soil materials is significant in works such as zoned earth dams, roadways, airfields, and railroads. With projects of this nature, one zone of material is designed to be coarser than an adjacent zone (e.g., an earth dam may consist of a fine-grained soil core with shell zones of coarse material, whereas a roadway base of coarse granular material rests on a subbase or subgrade of finer soil). It is necessary to prevent migration of fine material into the coarse material, or intrusion of coarse material into the fine material. Geotextiles can be used at the boundary between such different soil materials to maintain a separation but still permit the movement and passage of water. Geotextiles and geocomposites similarly have application with foundation drains and underdrains. (It has also been proven that the presence of geotextiles or geogrids incorporated into a structure such as an earth dam or a road pavement foundation will act as reinforcement, helping support and redistribute applied loads because of the material's tensile and shearing strength, as discussed in Chapter 14).

Geosynthetics with low permeability or impermeability (geomembranes and geocomposites) can be used as buried or exposed linings to prevent seepage or infiltration in reservoir and impounding areas. The movement of water through an earth dike can be restricted by installing a geomembrane lining on the dike's upstream side. Seepage beneath a dam can be reduced by placing an impervious geomembrane or geocomposite to act as an upstream blanket. Conversely, geotextiles and geocomposites with high planar permeability (ability to conduct flow within the thickness of the fabric) can be used as a passageway for water (and gases). They have application as protective drains where it is necessary to collect and carry seepage, such as within earth dams and solid waste landfills and behind retaining walls, and as road and airfield pavement drains. Geocomposites with high planar perme-

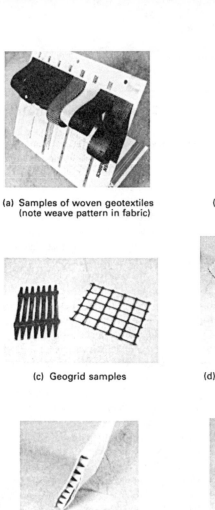
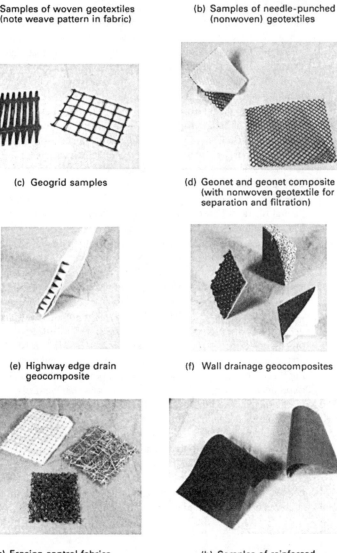

(a) Samples of woven geotextiles
(note weave pattern in fabric)

(b) Samples of needle-punched
(nonwoven) geotextiles

(c) Geogrid samples

(d) Geonet and geonet composite
(with nonwoven geotextile for
separation and filtration)

(e) Highway edge drain
geocomposite

(f) Wall drainage geocomposites

(g) Erosion control fabrics
(synthetic and natural fibers)

(h) Samples of reinforced
synthetic rubber
geomembranes

Figure 7-23 Representative samples of various geosynthetics.

The Role of Geosynthetics in Construction

Of the new engineering materials developed within the last quarter century, the greatest impact on the construction industry has been made by those grouped within the *geosynthetics* classification. The large-scale usage of such synthetics for construction projects is a relatively new development, but use is becoming more common because of the performance experience and confidence gained by designers and contractors. The geosynthetics—made of synthetics instead of natural materials—will perform some function of a construction assemblage that is in contact with the ground (hence the prefix "geo-"). The geosynthetics typically have been produced in the form of fabrics, creating clothlike or sheet materials with which to work. Special assemblages will have other forms, however. The variety of fabric types and the wide range of applications, the general availability, the quality of production, improved methods for field handling and seaming, and the continuing development of specialty products are all factors contributing to the increasing popularity of the construction synthetics.

The basic component materials of the geosynthetics are manufactured by the petrochemicals and plastics industry and consist mainly of polymers such as the polyesters, polypropylenes, polyethylenes, nylon, and the polyvinylchlorides, although other durable materials such as fiberglass and rubber are also used. The synthetics are selected because of their resistance to degradation or decomposition in the presence of moisture and the atmosphere and when buried (although degradation may occur when exposed to some chemicals); fabrics of natural fibers such as cotton and wool will deteriorate when buried and exposed to moisture. The geosynthetics are permitting many important aspects of construction to be accomplished better, faster, or more economically (such as buried drainage facilities, maintaining separation of dissimilar materials, providing ground reinforcement, accomplishing containment of liquids and solids, providing erosion control), and are also permitting some accomplishments that previously were not practical (such as providing reinforcement, containment, sealing, or capping).

Items comprising construction geosynthetics may be grouped into the *geotextiles* (or *geofabrics*), the *geogrids* and *geonets, geomembranes,* and *geocomposites* (see Fig. 7-23). The *geotextiles* consist of synthetic fibers assembled by either a woven, nonwoven, or knitting process to form a porous fabric sheet. The *geogrids* consist of plastic materials where the arrangement of heavy strands outline relatively large and thick gridlike openings in the resulting sheet. The *geonets* consist of plastic or polymer sheets that have been pressed or extruded to produce an open net or weblike form. The *geomembranes* are solid sheets of impervious plastic or synthetic rubber material. The *geocomposites* represent any combination of geotextile, geogrid, geonet, or geomembrane.

The popularity of geosynthetics relates to their being particularly appropriate for functions where immunity to degradation is crucial and the fact that properties such as strength, weight, and flexibility can be controlled or selected because the base materials are selected, sized, and assembled to provide particular features. Geosynthetics are used for construction projects to provide one or more of the following functions (in other words, the function is achieved because of the presence of the geosynthetic):

1. Separation
2. Filtration
3. Drainage
4. Reinforcement
5. Barrier or containment
6. Protection
7. Erosion control

The *separation* function is important where it is necessary to prevent mixing of dissimilar materials that are in contact, such as two different soil types (for example, the coarse granular base for a road pavement and a fine-grained subgrade soil, or different soil materials in a zoned earth dam). In the *filtration* function, the geosynthetic permits the passage of water but prevents the migration of soil particles from one side of the fabric to the other. The *drainage* function relates to providing a passageway within the thickness of the geosynthetic for controlling the movement or flow of liquids and gases, to encourage flow to occur along a desired path, or to direct flow to a selected location for collection, discharge, or escape. The *reinforcement* function involves placing geosynthetics on a surface such as a road subgrade or within a soil fill to improve the ability of the affected soil mass to carry loads by increasing the resistance to tensile and shear stresses. The *barrier* or *containment* function involves use of an impervious geosynthetic for situations where structures require a waterproofing membrane, or to function as a no-leak ground lining for liquid and solid waste disposal sites and the top capping seal. The *protection* function relates to including a protective geosynthetic for strength or resistance to surrounding conditions as part of a geocomposite in a situation where the material used to provide a major function, say drainage, is vulnerable to conditions present in the surrounding environment. The *erosion control* function is concerned with holding soil surfaces in place and preventing erosion. Some geosynthetics permit protective vegetation to grow through the texture of the fabric so that a natural (rooted) resistance to erosion develops (the geosynthetic may be designed to gradually decompose or degrade).

Materials in the geotextiles group are capable of providing any of the functions described above, as long as a proper synthetic and composition are selected. The synthetics used are durable, but the possibility for degradation if exposed to chemicals and sunlight requires consideration. The geogrids have been developed to fulfill the reinforcement function but also may have application for special cases where the separation function is required. The geonets are almost exclusively used for the drainage and erosion control functions. Geomembranes are used for their containment or barrier properties as protective liners and covers.

ability are being used for wick drains, to accelerate the consolidation process for thick deposits of weak compressible soil and in marsh areas. Manufacturers have developed geotextiles and geocomposites for venting radon and other gases that are present in the earth under and surrounding buildings and under waste liquid retention ponds and landfills. It is expected that the development of special products involving the geosynthetics will continue as the opportunity to compete with conventional procedures or materials, or to solve problems differently, is recognized.

Further information on the use of geosynthetics to prevent groundwater contamination appears in Section 7.4, and further information on the use of geosynthetics for reinforcement and erosion control is presented in Chapter 14.

Land Drainage

One of the earliest methods for reclaiming flat expanses of marginal wetland areas for use consisted of constructing a network of deep drainage ditches to lower the water table and obtain a stable surface suitable for agriculture and for supporting light structures. With a network of drainage ditches, drainage is by gravity, and best results are obtained in soils with relatively high coefficients of permeability. In clays or fine silts, where permeability

The Materials, Manufacture, and Properties of Geosynthetics Used for Construction Projects

Most of the geosynthetics used for construction projects are manufactured from polymers such as the polyesters, polypropylenes, polyethylenes, nylons, and polyvinylchlorides (polymers refer to the products obtained by the joining of two or more like molecules to form a more complex molecule whose physical properties are different). Textured fabrics (the *geotextiles*) developed from filaments of polyester or polypropylene are widely used. Some fabrics use bicomponent fibers (fibers of two or more polymer types). Compared to natural fibers, the polymer synthetics offer long-term durability in the presence of elements commonly encountered in construction (e.g., moisture and other types of waste fluids).

The fabric filaments (fibers or strands) may be obtained by an *extrusion* process or by the *slit-film* process. Extruded filaments result when a heated and liquefied polymer is forced through small openings, such as the holes in a spinnerette. The extruded strands cool and solidify and are then ready to be combined to form fabric. In the slit-film process, the polymer is melted and forced (extruded) into a thin sheet, or film. When cooled and solidified, the polymer sheet is cut (slit) into thin strips, which are then ready to be combined to form fabric.

Filaments are assembled into the geotextile fabric using either a woven or a nonwoven procedure.

Woven fabric is obtained by the conventional technique of weaving warp yarn (filaments extending in the long direction of the weave) and filler yarn (filaments in the transverse or cross direction). Properties of the woven fabric are controlled or affected by the spacing of the warp and filler yarns, as well as by the size and properties of the filaments. Generally, strength-related properties of the completed woven fabric (tensile strength, resistance to distortion and tearing, etc.) tend to be bidirectional.

The *nonwoven* procedures include the spun-bonded technique and the dry-laid-nonwoven process.

In the *spun-bonded* process, continuous filaments resulting from extrusion of polymer through a spinnerette are commonly laid down in a circular pattern on a laydown belt to form the fabric web. The web strands can be bonded together using a needle-punching procedure, by heart bonding, or by resin (chemical) bonding. The *needle-punching* procedure involves use of a needle loom containing a number of barbed needles. The barbed needles repeatedly punch through the fabric web, forcing the randomly laid-down strands to interlock mechanically. The needles' barbs are designed so strands are grabbed only on the downward stroke (picture the reverse of fishhook barbs). The resulting fabric texture is soft, with a noticeable thickness.

To *heat-bond* a fabric, the unbonded web is passed through heated rollers in order to cause strand surfaces to melt and fuse together at points of contact. The fuse bonding is retained after the fabric cools. With *resin bonding,* the unbonded web is treated with a liquid-cementing resin that adheres strands at their points of contact when the fabric passes through heated rollers. Excess resin is subsequently flushed from the fabric.

Dry-laid-nonwoven fabrics are manufactured from staple fibers, which are fibers with limited length (typically, fiber lengths range between 1.5 and 6 inches, or 4 to 15 cm). The fibers are laid down on a moving belt by a cross-lapper, to create a crisscross layering of short strands (staple). The staple is bonded by either needle punching, heat, or resin.

The manufactured fabric is wound into rolls for storage and shipping. Roll widths vary according to manufacturer and product, but many products are in the range of 4 to 5 m or 12 to 16 ft wide.

At the construction site, the fabric is unrolled and cut if necessary. Large areas are covered using multiple rolls. The edge and end seams are mechanically sewed, glued, fused, or simply overlapped.

Most *geogrids* are made from high-density polyethylene, polyester, or polypropylene. One method of manufacture involves taking solid sheets of relatively thick or heavy-gauge material, then, by hole-punching and control-stretching the sheet, producing a grid skeleton that has the desired strength strands and size openings. A second method of manufacture involves arranging strips or bundled fiber strands into the desired grid pattern, then bonding the intersections.

At present, most *geonets* are manufactured from polyethylene by an extrusion and stretching process. Heated liquefied polyethylene is extruded through a rotating die called a stenter to create a netlike pattern of closely spaced ribs or strands. The newly formed netted sheet is forced over a cylindrical mandrel which has an increasing diameter, a process which stretches the still-plastic material to achieve the desired-size net strand and net opening. The cooled and stabilized cylindrical-shaped sheet is cut in the direction of the long axis of the mandrel in the desired length, then placed into rolls.

Most of the *geomembranes* are made from modified forms of the synthetic polymers—polyethylene, polyvinylchloride, and polyvinylidine-chloride materials that will result in obtaining virtually impervious sheets. The appropriate polymer resin is blended with additives such as plasticizers, vulcanizers, and other processing aids, then extruded into sheets. Sheets can also be obtained by extruding the liquid onto a cylinder or blowing it into the shape of a long giant bubble; the cylindrical-shaped sheeting is then cut to the desired width and length. Sheets may also be further processed by passing through rollers (calendering) to improve the properties. Geomembranes are also produced by saturating an otherwise porous geotextile with polymer resin to obtain an impervious sheet. Multi-ply geomembranes or geocomposites can be obtained by laminating the single-ply geomembrane with other geosynthetics.

Some generalities exist concerning properties, performance, and suggested uses for synthetic fabrics. (*Note:* Some degree of caution is warranted with generalities; exceptions are to be expected. Furthermore, not all of a fabric's properties are a consideration for use.) Needle-punched fabrics are generally considered more flexible than chemically and heat-bonded fabrics. Nonwoven fabrics are generally considered to have similar strength properties across all directions in the plane of the fabric, whereas woven fabrics are probably strongest in the direction of fabric strands. Fabrics manufactured from continuous filaments generally are considered stronger than fabrics made with staple fibers of similar material. Polyester filaments have good resistance to ultraviolet rays (sunlight), but unprotected polypropylene is subject to deterioration. Polyesters are not expected to be damaged by exposure to diesel fuels, but polypropylenes could be. Polyester fabrics generally retain better flexibility at low temperatures than do polypropylene fabrics. Geotextiles woven from filament fibers and used for the drainage function in cohesionless soils have less tendency to clog than do the nonwoven and slit-film wovens. Geomembranes to be used for liners or capping seals at liquid and solid waste disposal sites should be tested for durability when exposed to the chemicals to be contained.

Description is of representative physical properties commonly used as measures to determine fabric suitability are presented in Table 7-2; testing to determine the properties is usually performed by the manufacturer.

Table 7-2 Summary Tabulation: Commonly Listed Physical Properties of Geosynthetics

Property Designation	ASTM Test Procedures	General Description of Property Test, Purpose, etc.
AOS or EOS (apparent opening size, or equivalent opening size)	D-4751	Property intended to relate ability of fabric to retain or permit passage of solid particles (e.g., soil). Test procedure involves use of a series of equal-diameter glass beads; the smallest-diameter beads for which 95% by weight is retained (or 5% passes) is taken as the equivalent size opening (expressed in mm opening or the equivalent sieve size).
Grab tensile strength and elongation %	D-4632, D-4595	Maximum tensile strength (resistance for one unit of width and an indicated fabric thickness), and percent elongation (stretch) at maximum tensile load.
Trapezoidal tear	D-4533	Tensile type test performed on fabric which is placed in a skewed position (on a bias) in test apparatus so that a progressive tearing or breaking of individual fibers occurs; an indication of resistance to tearing.
Mullen Burst Test	D-3786	Fabric is stretched (distorted) into a hemispheric shape until tearing or bursting occurs.
Puncture resistance	D-3787, D-4833	Resistance to puncture by pointed objects such as sharp stones. Test is performed by pushing an 8.0-mm-dia rod through a stretched fabric (the concept of the "elbow through the sleeve"); indicates resistance to potential damage by stones, pieces of timber and roots, etc.
Permitivity (also referred to as cross-plane permeability)	D-4491	The cross-plane permeability or hydraulic conductivity; indicates ease or difficulty for liquid to flow across (or through) the fabric.
Transmissivity (also referred to as planar permeability)	D-4716	A measure of the liquid volume that can flow within the thickness of the fabric; a significant property for a product intended to facilitate drainage or transmission of liquid.
UV resistance	D-4355	Ability of the fabric to retain original strength property after exposure to ultraviolet (sun) rays.
Soil clogging	Corps of Engineers Test CW-02215	Indication of extent to which fabric will become clogged because of trapped soil particles; clogging can interfere with the cross-plane and the planar permeability.
Fatigue strength	—	Strength of the fabrics after being subject to repetitive loading. Fatigue strength is expected to be less than that for a single application loading, and is affected by the number of load applications.
Seam strength	—	Ability of tensile or shear force to be transmitted across the seam where fabrics are joined.
Creep	—	Measurement of the stretching or deformation that occurs over a period of time under the effect of a constant, long-term load.
Friction (surface friction)	—	An important property for fabrics where sliding along the surface of contact is possible, as when a fabric is placed on a slope. The friction value is determined by applying a normal load to the fabric surface and measuring the force required to move/slide the fabric (a process similar to the basic shear test to determine a coefficient of friction).

is low and capillary forces holding the pore water are high, drainage from gravity effects will be limited. Water can be replaced by normal precipitation as quickly as drainage occurs. However, where the subsurface soils include strata or seams of coarse soils, as frequently exist in coastal and lake areas, predominantly fine-grained soil deposits have been drained. The drainage ditch network, generally laid out in some form of grid pattern, requires outleting to areas of elevation than the draining area.

Effects of Drainage

Water flowing toward a drain also causes seepage forces that act on the soil particles. These forces are capable of moving soil particles, particularly small particles. Unless the migration of particles by a properly functioning filter is prevented, the small particles will be washed into the drain, eventually eroding soil from the area surrounding the drain. This occurrence could cause the drainage structure to collapse because of inadequate support from the surrounding soil. This process is referred to as *internal erosion.*

If an open drainage system (e.g., open ditches) causes seepage forces to act toward an open excavation, there is the possibility that the forces will precipitate a sloughing-in of the soil embankment.

Normally, removing water from soil (particularly a fine-grained soil) will increase the shear strength of the soil, in effect making the soil mass stronger. Lowering of a water table, however, also results in an increased effective vertical pressure acting on the soil mass, because the soil weight changes from a submerged weight to a saturated but unbuoyed weight. The usual result of an additional loading to the area within the zone of drawdown is settlement. The magnitude of the settlement depends on the amount of water table lowering and the strength and compressibility of the soil. If an undeveloped area (one with no buildings, etc.) is dewatered and the surface settles, usually no problems result. If developed areas are in the zone of drawdown, however, settlement of buildings and utilities in the area could occur.

If settlement of structures adjacent to a construction site being dewatered is anticipated, effects can be minimized by the process of *recharging.* Recharging consists of pumping water back into the ground, usually with well points, in the area between where the endangered structures are located and where the dewatering takes place. Careful monitoring of groundwater levels is required for proper overall results.

Soil Percolation Rate

Soil percolation rate refers to the ease or difficulty with which soil accepts fluid, generally water, into it. It is a condition frequently of interest for planning private underground sewage disposal systems, such as the leach fields and leach pits for septic tanks. With a septic tank system, all sewage passes into the septic tank. The solids are retained, but the liquid is passed on for absorption into the earth.

The percolation rate can be related to the coefficient of permeability of a soil: the coarser, permeable soils are capable of accepting larger volumes of water and at a faster rate than fine-grained soils having low permeability. However, *percolation* and *permeability* do *not* refer to the same property of a soil. For example, a coarse soil having a high coefficient of permeability may have a low rate of percolation if the groundwater table is high and the soil is not capable of accepting additional water.

Procedures to check the percolation rate for a soil generally stipulate testing of the area and depth where the septic tank leach field or leach pit is to be located. For the common requirement that leach field drain lines be installed about 0.7 m (or 2 ft) below the ground surface, percolation tests are performed about 0.75 m (or 30 in.) below the surface. A small test hole, usually 150 to 300 mm (or 6 to 12 in.) in diameter, is dug at least 150 mm (or 6 in.) deep. The test hole area is saturated by filling the hole several times with water and letting it drain completely before beginning the actual test. For the test, the hole is filled with water and the time for the water level to drop a specified distance (e.g., from 150 mm down to 125 mm, or from 6 in. deep down to 5 in. deep) is recorded. A fast rate of drop indicates "good" percolation. A slow rate or no drop indicates poor percolation. Areas having poor percolation rates should not be used as leach fields. Details of the applicable methods for performing percolation tests should be obtained from local health department authorities, to ensure that procedures are in accord with their requirements.

7.3 FROST HEAVE IN SOILS

When soil is exposed to a freezing climate for a sufficient period of time, soil temperatures will eventually depress to below freezing. Temperatures will vary within the soil mass, being lowest where contact with the freezing source is made (usually this would be the ground surface in contact with the freezing air temperatures, but the source of a freezing temperature could also be the floor of a frozen storage warehouse or ice rink) and gradually increasing with distance from the freezing source to reach the stable, above-freezing temperature that exists underground.

When freezing temperatures develop in a soil mass, most of the pore water in the soil also is subject to freezing.[6] As water crystallizes, its volume expands approximately 9 percent. In considering normal void ratios and the degree of saturation for soils in general, expansion of a soil material as a result of freezing might be expected to be on the order of 3 or 4 percent of the original volume. This may mean a vertical expansion on the order of 25 to 50 mm (1 to 2 in.) for an area with the climate that exists in the northern half of the United States. Experience indicates, however, that in some soil types, the volume of expansion resulting during freezing periods (a condition termed *frost heave*) is frequently considerably greater than could be expected from expansion of normal pore water. Frost heaves exceeding 0.3 m (1 ft) are not uncommon.

Investigation of heaved soils indicates that pore water indeed has frozen, but additionally, much of the frozen water has been segregated in discontinuous layers, or lenses, throughout the frozen soil, and the volume of the water in the ice lenses is considerably greater than the volume of the original pore water. Further, the magnitude of ground surface heave has been found to approximately equal the combined thickness of ice lenses existing in the soil at the heave location (Fig. 7-24).

---

[6]This explains the "frozen-solid" condition noted for soil ground in cold weather. The frozen water is responsible for the strong bonding between soil particles; the soil particles generally experience no change as "freezing" temperatures occur, for they are already in the frozen or solid state.

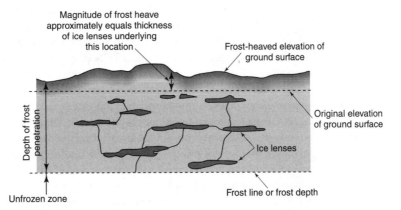

Figure 7-24   Ice lens formation in a frozen soil.

Studies conducted at field sites and in the laboratory to investigate the frost heave phenomenon have been able to identify conditions and occurrences associated with ice lens formation, but all factors involved are not yet understood. Water in soil freezes at temperatures lower than 0°C (32°F) because of ions from dissolved minerals present in the water, soil particle surface forces, and also the existence of negative pore water pressures if the soil is not saturated. The concentration of ions will affect the temperature at which water begins to freeze (nucleate) and, ion concentrations will likely vary across the soil deposit. The temperature for nucleation also varies across the soil void space, being lowest in the diffuse double layer adsorbed to the particle surface and increasing as distance from the particle increases. The thickness to this film of unfrozen water surrounding the soil particle in an otherwise frozen soil is also affected by temperature, decreasing as temperatures decreases. The unfrozen film of water surrounding particle surfaces can be several degrees lower than the conventional freezing temperature for water. Ice, therefore, tends to originate in the zone of soil void space farthest away from the particle surfaces. Additionally, the surface of an ice zone consists of a film of adsorbed water; the ice zone increases by attracting additional water molecules (from adjacent soil voids and particle surfaces) into the adsorbed film. The energy gradients responsible for the attraction of new water to the growing ice zone (the forming ice lens) are complex but probably involve temperature and pressure differences, the phenomenon of capillarity, and osmotic flow.[7] It is probable that migration of water molecules toward an expanding ice lens occurs in a zone within or bordering the films that constitute the diffuse double layer on particle surfaces. But if temperature drops too low, the mobility of this water is restricted. The process of expanding ice lenses will cease when the lens can no longer attract water because the required energy gradient becomes too great or the available sources are diverted to other lenses.

---

[7]Osmotic flow relates to the movement between liquid solutions having different ion concentrations. Simply stated, there is the tendency for the more dilute solution to flow into and mix with the highly concentrated solution. Water in soil normally has some concentration of cations because of dissolved minerals. Applied to the ice lens phenomenon, crystallization of soil pore water may increase the ionic concentration; thus the more dilute free water in the soil would be attracted to the ice lens.

The development of ice lens formations significant enough to cause large heaves requires that a source of water be sufficiently close to the freezing zone of soil so that it can be drawn into this zone. In the normal frost heave occurrence, the source of water is the groundwater table, as illustrated by Figure 7-25 (but other conditions, such as leaking underground pipes, can also be a source). Upward movement from a water table to the freezing zone relates to a potential for migration (i.e., capillary rise). Height of capillary rise is quite limited in clean, coarse-grained soil, and unless a source of water is close to the zone of freezing soil, frost heave problems are normally not expected in such soils. Additionally, sands tend to freeze uniformly throughout the frozen zone and ice lenses do not form. The theoretical height of migratory rise is greatest in the fine-grained silt and clay soils. Practically speaking, however, under typical seasonal conditions of limited cold periods, ice lens growth is affected by the rate of capillary rise in soil. In clays, this rate is so limited that under normal conditions sufficient water cannot be obtained to result in ice lens growth. Silts, however, have a relatively high rate of capillary rise, and water can migrate easily to the zone where ice lenses are forming. Silts, therefore, because of their potential for high capillary rise (distance) coincident with the potential for rapid rise, are conducive to ice lens formation and frost heave. As a result, silts are termed *frost-susceptible* soils. Silt mixtures, such as silty sands and silty clays, are similarly regarded as frost susceptible. Because of the practical limitations on ice lens growth within them, coarse-grained and true clay soils are normally considered as *non-frost-susceptible* soils.

In summary, several simultaneously occurring conditions are required for the notorious frost heave problem to occur:

1. Presence of a frost-susceptible soil

2. Presence of freezing temperatures

3. Presence of a supply or source of water to help form and feed the ice lenses

4. A time period sufficient for ice lenses to form and enlarge

The rate at which freezing temperatures penetrate the frost-susceptible soil has an effect on the total ice lens formation and, consequently, on the frost heave of the ground sur-

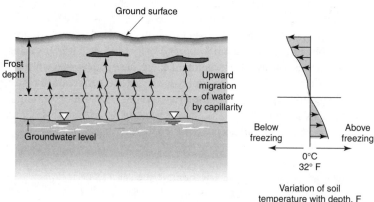

Figure 7-25 Ice lens formation by capillary movement of groundwater into zone of freezing.

face. The maximum effects result when there is a gradual decrease in temperature as opposed to a rapid or flash decrease. The longer and the colder the freezing period, the greater will be the depth of frost penetration and the greater will be the frost heave.

Damages from frost heave result in structures supported on the soils being heaved. Commonly, this includes roadways and pavements and building foundations and exposed ground-level building slabs. The heaving force can be considerable; it has the ability to lift pavements, slabs, and building foundations. Because the magnitude of heave is rarely uniform over even short horizontal distances, cracking of pavements and slabs frequently occurs. In buildings, walls or floors may crack or distort so that doors, windows, and equipment supported on the walls may not operate properly. If heave is great enough, utility pipelines passing through or over the walls may be ruptured because of the movement. If the frozen soil adheres to the exterior side of a masonry foundation wall, the soil heave results in an uplift force on the upper section of wall, which, for light loading conditions, may cause wall mortar joints to crack (refer to Chapter 13, p. 535 for additional information).

Damage is not restricted to that resulting from the heaving. Much of the ice lens volume represents new water introduced to the soil, and when the soil thaws from the surface downward in a warm period, the free water cannot drain through the still-frozen underlying soil. The effect is that the melting ice lenses significantly increase the water content of the soil, and loss of soil strength occurs. Loss of soil bearing and settlement follow. For building foundations not located below frost depths, the cycle of heaving in winter and settling in spring is an annual occurrence that is capable of causing progressive damage.

Similar effects can occur in pavements and unprotected ground-level building slabs. In roadway surfaces, the effects are compounded by the repeated loadings applied to an area by vehicle movement. Under the action of traffic, the roadway surface will be broken up because of loss underlying soil support. This is generally referred to as the spring breakup. Further, the soft soil will be displaced from the road base, causing holes and depressions, which are referred to as potholes or chuck holes.

It should also be realized that frost heave may have its effect where new construction takes place during freezing periods. For example, should the foundation soil for planned structural components such as footings, floor slabs, or pavements be in a frozen and heaved condition at the time of construction, the eventual thawing and settling of the soil can cause the component to experience settlement, distortion, or structural damage.

To prevent the effects of frost heave damage to buildings, it is common practice to construct foundations to a depth at least equal to the depth of frost penetration for the area. Such procedures might be thought to be necessary only in a frost-susceptible situation. If non-frost-susceptible soils exist at a building site or if there is no source of water, locating foundations below frost depth could be unnecessary unless required for other reasons. However, in natural soil deposits, there should also be some suspicion that even "non-frost-susceptible" soils will have some properties that could make them susceptible. For example, many naturally occurring coarse-grained soils include some fines, or include seams or strata of finer, more susceptible material. Clay soils frequently have considerable silt content, or the clay deposit may include fissures and hairline cracks that can act as capillary tubes for the migration of water.

It should also be recognized that during the life of a structure, the depth of the groundwater table is subject to change.

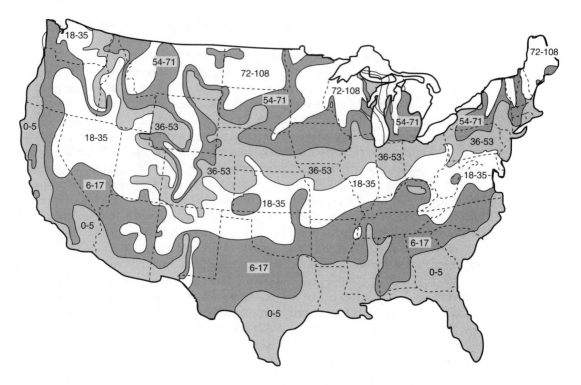

Figure 7-26 Maximum depth of frost penetration in inches (from Frost Action in Roads and Airfields, HRB Publication 211 [103]).

In considering the poor performance and damage that can result if frost heave should affect foundations, most designers and constructions are of the opinion that it is worth the relatively small extra expense to take precaution at the time of construction to install foundations for exterior elements below frost depth. Figure 7-26 shows maximum frost depth data prepared from information obtained across the United States.

For grade-level structures such as building slabs and highway pavements (no deep foundations), other approaches to the control of the frost heave problem are taken. Control usually requires achieving at least one of the following:

1. Removing existing frost-susceptible soils to about frost depth
2. Removing or cutting off the water source that could feed ice lens growth
3. Protecting the susceptible soil from freezing temperatures

For road or airfield pavements, the effort for control or protection has frequently included removing the frost-susceptible soils to frost depth and replacing them with a non-frost-susceptible material or providing drainage facilities to keep the water table low in the vicinity of the pavement. For building slabs, replacing the frost-susceptible soil is frequently a practical solution.

Non-frost-susceptible soils for replacement purposes include coarse-grained soil (sand and sand–gravel mixtures) and clays. Coarse-grained soils are preferred because they are easier to place and compact, are less difficult to work with in inclement weather, and have other desirable properties when in place, such as good drainage. To be non-frost-susceptible, studies and experience indicate that a *well-graded* granular soil should contain not more than 3 percent of particle sizes smaller than 0.02 mm,[8] while *uniform* granular soils can have up to about 10 percent of particle sizes smaller than 0.02 mm. Some ice lens formation may still result with these soils, but such development will generally be of limited extent and tolerable.

Other methods for control of frost heave have included installation of a barrier at about frost depth to prevent upward capillary movement that could feed ice lenses. The barrier can be a very coarse soil of a thickness through which capillary rise cannot occur, an impermeable material such as a densely compacted clay layer, or a membrane or other fabric material through which water cannot penetrate. These methods appear most suitable for protecting limited areas, such as building slabs, or for highway fill sections. The methods are not considered practical for highway construction in areas of cut section or at-grade construction because of the expense of excavating and replacing subgrade soils that would be associated with installation of a barrier. Further, frost-susceptible subgrade soils may be difficult to replace and compact, especially in wet weather.

Successful protection has also been achieved by wrapping pavement subgrade soils within an impermeable membrane after they have been properly compacted, so as to keep the soil at a desired optimum moisture content and to prohibit ice lens formation. A disadvantage of this method of control relates to costs of installing the membrane and placing the soil.

If the soil beneath a slab or pavement can be protected so that it does not freeze, there will be no frost heave problem. In the construction of frozen-storage warehouses, ice rinks, and similar facilities, insulation is placed between the soil subgrade and the floor of the facility, to protect the soil from the cold temperatures prevailing in the structure. Similar methods are adaptable for protecting exposed ground-level slabs for buildings. Various methods that have been used are illustrated in Figure 7-27. For highways, some work has been done with providing insulating material beneath the pavement.

Permafrost

Permanently frozen ground, termed *permafrost,* exists in the northern areas of North America, Europe, and Asia, and across most of Greenland (Fig. 7-28), a result of the severe climate in those areas. It is estimated that one-fifth of the earth's northern hemisphere is underlain by permafrost. The depth of frozen soil is on the order of 300 m (1000 ft) in regions corresponding to Alaska's North Slope, and probably extends to more than 1000 m (3300 ft) at the Arctic Circle.

The surface zone of soil in the permafrost regions thaws in summer (a condition that also occurs where permafrost is located beneath a structure that enables lost heat to penetrate the ground), but the deeper material stays frozen. The zone thawed by a summer climate refreezes the next winter. The soil depth that is subjected to seasonal cycles of thawing and freezing is termed the *active zone.*

---

[8]To retain a perspective on sizes, it is noted that the #200 sieve has openings of 0.074 mm; a #400 sieve has openings of 0.037 mm.

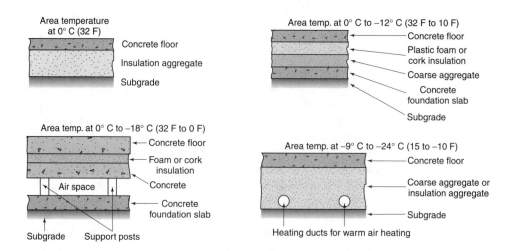

Area temperature at 0° C (32 F)
— Concrete floor
— Insulation aggregate
— Subgrade

Area temp. at 0° C to –12° C (32 F to 10 F)
— Concrete floor
— Plastic foam or cork insulation
— Coarse aggregate
— Concrete foundation slab
— Subgrade

Area temp. at 0° C to –18° C (32 F to 0 F)
— Concrete floor
— Foam or cork insulation
— Concrete
Air space
— Concrete foundation slab
Subgrade   Support posts

Area temp. at –9° C to –24° C (15 to –10 F)
— Concrete floor
— Coarse aggregate or insulation aggregate
— Subgrade
Heating ducts for warm air heating

Figure 7-27 Some methods that have been used to insulate floor systems for refrigerated areas.

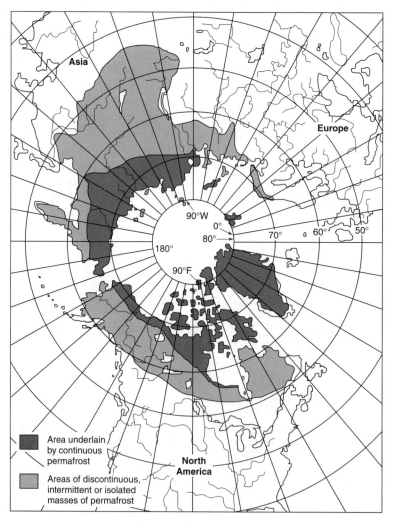

Figure 7-28 Extent of permafrost zones in the Northern Hemisphere [14].

Thawed soil in the active zone is usually weak and compressible because of the trapped water content. Structures built in permafrost areas should have their foundations installed in the permanently frozen earth that exists below the active zone, to protect the structure against seasonal movements that would otherwise result. These foundations must be designed to ensure that heat is not conducted through or by them to the permafrost that provides the support; insulating materials could be necessary along the vertical perimeter of the foundation and at the base. Similarly, to prevent building heat losses from thawing the permafrost, the floors and subsurface walls that would be in contact with the ground must be insulated, artificially cooled, or separated (with small buildings, the first floor can be elevated above the ground surface).

The subarctic plains have a characteristic vegetative cover of mosses and low shrubs underlain by mucklike soil, a combination termed *tundra*. This surficial blanket acts as an insulating material limiting the depth of summer thawing and therefore the active zone. A result of recognizing this beneficial effect of the tundra is that roadways for permafrost areas are now constructed on embankments of gravel, which act as a non-frost-susceptible insulating pad to keep the underlying soil frozen and stable. Equally significant, the presence of tundra has proved to be helpful in keeping sloped and hilly areas stable against slides.

7.4   SOIL AND GROUNDWATER CONTAMINATION

Soil contamination and groundwater contamination can be broadly defined as the inclusion of matter or substances which affect normally expected use. Contamination of soil and groundwater involves degradation due to the presence of physical, biological, chemical, or radiological materials in undesirable concentrations or concentrations considered abnormal. Most natural soil deposits and "normal" groundwater typically include minerals, as well as biological and chemical substances, but in low concentrations considered tolerable. Concentrations considered acceptable for one type of use may be unacceptable for another, however.

Pollution, such as water pollution, refers to the loss of purity through contamination. Polluted water is water unfit for consumption and agricultural and commercial use.

The presence of contaminants can be the result of natural occurrences (natural contamination), but is commonly considered to be a condition associated with human activities (artificial contamination). Most government environmental protection regulations reflect intentions to prevent or control the release of contaminants as related to human activities.

Table 7-3 lists the more well-known types of soil and groundwater contamination sources. Figure 7-29 illustrates similar information.

Most groundwater contaminants are included in one of three broad groups: biological organisms, organic chemicals, and inorganic chemicals. Knowledge of a contaminant general grouping can be helpful in developing a remedial process which achieves decontamination. The source of *biological organisms* is animal, human, and vegetative waste, frequently resulting where heavy concentrations occur such as at inadequate sewage treatment operations, cesspools and septic systems, sewage treatment settling ponds or lagoons, and from poorly controlled land spreading of dried sewage sludge. *Organic chemicals* result from materials containing carbon, hydrogen, and oxygen. Petroleum products and

Table 7-3 Sources of Groundwater Contamination

Category I—Sources designed to discharge substances

Subsurface percolation (e.g., septic tanks and cesspools)

Injection wells:
  Hazardous waste
  Non-hazardous waste (e.g., brine disposal and drainage)
  Nonwaste (e.g., enhanced recovery, artificial recharge, solution mining, and insitu mining)

Land application:
  Wastewater (e.g., spray irrigation)
  Wastewater byproducts (e.g., sludge)
  Hazardous waste
  Non-hazardous waste

Category II—Sources designed to store, treat, and/or dispose of substances; discharge through unplanned release

Landfills:
  Industrial hazardous waste
  Industrial non-hazardous waste
  Municipal sanitary

Open dumps, including illegal dumping (waste)

Residential (or local) disposal (waste)

Surface impoundments:
  Hazardous waste
  Non-hazardous waste

Waste tailings

Waste piles:
  Hazardous waste
  Non-hazardous waste

Materials stockpiles (nonwaste)

Graveyards

Animal burial

Aboveground storage tanks:
  Hazardous waste
  Non-hazardous waste
  Nonwaste

Underground storage tanks:
  Hazardous waste
  Non-hazardous waste
  Nonwaste

Containers:
  Hazardous waste
  Non-hazardous waste
  Nonwaste

Open burning and detonation sites

Radioactive disposal sites

Category III—Sources designed to retain substances during transport or transmission

Pipelines:
  Hazardous waste
  Non-hazardous waste
  Nonwaste

Materials transport and transfer operations:
  Hazardous waste
  Non-hazardous waste
  Nonwaste

Category IV—Sources discharging substances as consequence of other planned activities

Irrigation practices (e.g., return flow)

Pesticide applications

Fertilizer applications

Animal feeding operations

De-icing salts applications

Urban runoff

Percolation of atmospheric pollutants

Mining and mine drainage:
  Surface mine–related
  Underground mine–related

Category V—Sources providing conduit or inducing discharge through altered flow patterns

Production wells:
  Oil (and gas) wells
  Geothermal and heat recovery wells
  Water supply wells

Other wells (nonwaste):
  Monitoring wells
  Exploration wells

Construction excavation

Category VI—Naturally occurring sources whose discharge is created and/or exacerbated by human activity

Groundwater–surface water interactions

Natural leaching

Saltwater intrusion/brackish water upconing (or intrusion of other poor-quality natural water)

Source: Office of Technology Assessment [217]

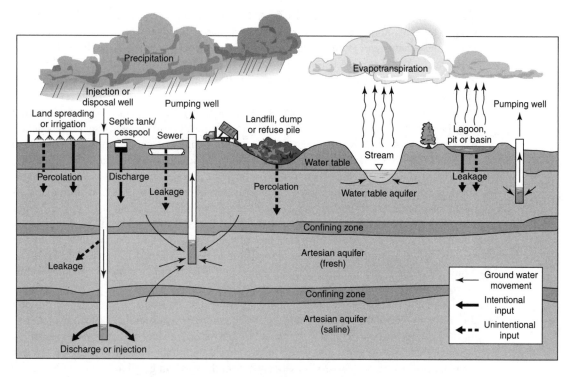

Figure 7-29 How waste disposal practices may contaminate groundwater (from EPA Journal, Vol. 10, No. 6, "Sources of Groundwater Contamination," July-August 1984).

petrochemicals are included in this group, as are the byproducts of decomposing natural and synthetic organics. Commonly, these materials enter the groundwater supply from waste disposal sites, from processing and manufacturing locations, and from accidental spills. *Inorganic chemical* contaminants are of mineral origin, being materials that exist naturally but that become concentrated to dangerous levels at mining and storage operations, or that are a result of accumulated waste products from industrial processing.

Surface and near-surface contaminants can be transported from the source location by natural occurrences such as the movement of water and air, or by intentional and unintentional human activities. Land surface and near-surface contaminants can cause pollution of surface bodies of water (lakes, streams, etc.) via overland flow. Subsurface contaminants are subject to the same forces responsible for the movement of underground water, and for many types of contaminants the water will be the conveying medium (the groundwater transports the contaminant as part of the natural flow). A special case of contaminated subsurface liquid termed *leachate* results when clean surface and subsurface water seeps through landfill or disposal site material and carries along soluble matter responsible for the contamination; leachates are commonly acidic, which increases the capacity for dissolution of toxic materials (usually a benefit) but also the transport of toxic materials (usually a detrimental effect). Solid contaminants are prone to remain at the source location unless transported by a moving medium such as water, air, or humans. Liquid contaminants, however, have different behavioral characteristics, possessing the tendency to flow without requiring the presence of another medium such as groundwater. Liquids having a fluid viscosity and density less than water (for example, gasoline) will be able to move through a soil deposit more easily than water; soils relatively impervious to the movement of water will be pervious to the flow of low-viscosity fluids (refer to Equation 6-5, relating the coefficient of permeability for soil and viscosity of the flowing liquid).

In some instances, the contaminants present at a surface or subsurface source location may pass into solution as water moves through, or particles of matter may be physically transported with the flowing water. Some types of contaminants, however, are subject to chemical reactions and changes, or to biological breakdown, or become absorbed (bonded) to soil particle surfaces; in effect, a filtration occurs. These three factors are often categorized as retardation or attenuation mechanisms. An important result is that some types of contaminants can travel underground long distances without significant change; some types of contaminants will be transported underground but at a retarded rate; some will break down or dilute as part of the transportation process (possibly being sufficiently diluted to become relatively harmless); and some will move only a limited distance. The extent to which the various mechanisms actually function influences the degree to which contaminated groundwater will spread, and affects remedial measures.

The bulk of subsurface contaminant migrations take place in the presence of moving groundwater—both the flowing category and the more or less stationary groundwater condition—due to the process of advection and hydrodynamic dispersion. *Advection* identifies the movement of contaminants suspended or carried within the volume of the groundwater flow. *Hydrodynamic dispersion* is associated with the spreading and dilution that occurs as moving water travels through the irregular orientation of channel paths formed by the interconnected void spaces between soil particles; as flow occurs, portions are diverted or spread out from the main direction of movement. Some osmotic mixing is probably involved (referring to the tendency for liquid solutions that have different ion concentrations to mix). The zone of water carrying contaminants is therefore enlarged along the general direction of the underground flow, although a less concentrated or more diluted volume results. The subsurface zone affected by the spread of contaminated groundwater is referred to as the *contaminant plume,* or simply *plume.* Figure 7-30 illustrates a variety of subsurface soil and groundwater conditions and contaminant plumes that tend to result.

The configuration of a contaminated plume is influenced by an array of factors such as the physical properties of the soil stratum being polluted (the thickness, depth, and slope, the mineral composition of the soil particles, the permeability or hydraulic conductivity); the presence of groundwater; the properties of the contaminant, including the concentration, solubility, and density and flow (viscosity) characteristics; and time.

A liquid contaminant may initially be, or become, mixed with a groundwater supply, but then separates if it is of a significantly lighter or heavier density. For example, gasoline escaping into the earth will accumulate or travel along the top of a groundwater supply and tend to flow faster, because of the lighter density and lower viscosity. Of practical interest, cased water supply wells have continued to provide unpolluted water even with gasoline present in the ground, as long as the intake section of the well casing remained below the level of the gasoline phase.

The migration of contaminants to pollute a water supply is not the full extent of the environmental problem. Contaminants that enter the earth zones important to plant life (either

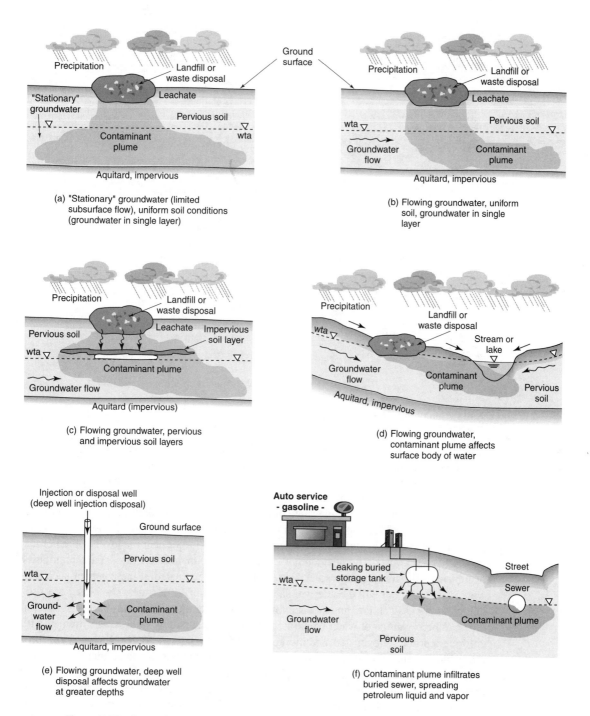

Figure 7-30 Contaminant plume relating to site conditions and contaminant source.

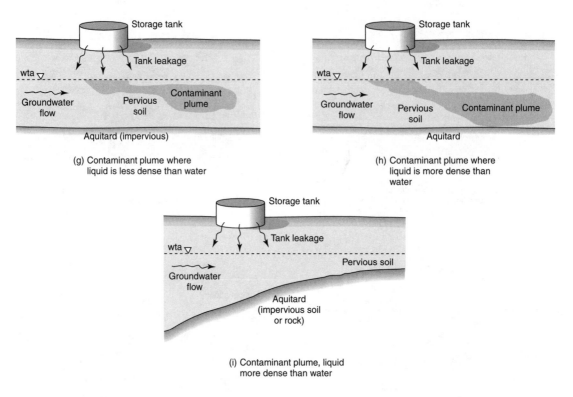

(g) Contaminant plume where
liquid is less dense than water

(h) Contaminant plume where
liquid is more dense than
water

(i) Contaminant plume, liquid
more dense than water

Figure 7-30   (Con't)

a surface, subsurface, or above-surface region) and are then absorbed may have harmful effects on the vegetation and may enter the food chain to harm higher species.

Two situations of groundwater contamination due to factors other than a reaction with waste materials and chemicals are the occurrence of saltwater intrusion into the subsurface supply of fresh water and the presence of trihalomethanes in drinking water. The saltwater intrusion condition develops in seacoast regions when fresh groundwater is removed in sufficient quantity to lower the land area groundwater level to or near the level of the bordering body of seawater. The heavier saltwater displaces, or intrudes into, space previously occupied by the fresh groundwater. Water supply wells situated in the zones of such seawater encroachment will then deliver saltwater instead of fresh water. The presence of trihalomethanes is the rather unusual phenomenon of contamination because of a disinfection effort. When a source of water containing organic matter and bromide salts is treated with chlorine, certain halogenated organic compounds such as chloroform, bromoform, bromodichloromethane, or chlorodibromomethane (the halogen-substituted methanes, or trihalomethanes) are produced. Organic particulate matter and the bromide ions can be found in many natural waters and saltwater. Surface waters tend to contain greater amounts of organic particles in fall and winter because of natural factors such as fallen tree leaves. Trihalomethanes such as chloroform are suspected carcinogens, and desirably the concentration in drinking water supplies will be limited to levels considered tolerable (less than 100 micrograms per liter).

Protecting a subsurface region including the groundwater against pollution involves procedures to monitor for, or detect, the presence of contaminants and, where pollution has occurred, the implementation of remedial processes (that is, become aware of the problem; identify the contaminants and source if possible; then undertake remedial measures). The *monitoring process* conventionally is used where a known source of contaminants exists, such as at a landfill or waste dump site. The air, surface, and subsurface materials surrounding the site are monitored (commonly by analysis of air, soil, vegetation, or groundwater samples) to check for the presence (migration) of contaminants. The typical *detection process* is a protective undertaking for a commodity such as a water supply, and involves checking the air, soil, surface, or groundwater for the arrival of polluting contaminants and the need to provide remediation. The monitoring and detection techniques are similar, using surface and subsurface sensor procedures to indicate change, and sampling-analysis procedures to identify the presence and concentration of contaminants. Commonly, subsurface monitoring or detection programs include the use of wells which permit samples of soil, groundwater, or gases to be obtained for analysis and evaluation (procedures are discussed in Chapter 5). Surface monitoring–detection procedures include physical observation of surface conditions (such as change in vegetation and presence of odors which could indicate escaping soil gases) and air sample analysis. Comparison of aerial photographs taken at different time periods is useful in identifying the changes in surface vegetation that indicate the presence of underground contaminants.

Where a soil or groundwater contamination condition exists, it may be necessary to undertake remediation, or at least containment or other mitigation, for health, safety, environmental protection, or other reasons. Factors influencing the selection of a corrective program relate to the type and extent of contamination, the procedures that could be successful and their cost, the time involved to implement and complete the program, and the hazards associated with implementation. Table 7-4 outlines a variety of corrective actions that have been applied to contamination–pollution problems. It is not unusual to have more than one corrective procedure involved where a complex condition exists or where a large area is affected. Undoubtedly, the refinement of present procedures and the development of new technology will result in more choices being available.

Soil Gas

Recognizing the possibility of a soil gas presence other than normal air has become extremely important for personnel dealing with any aspects of underground construction or environmental safety and protection. Some soil gases are hazardous, being toxic or carcinogenic or presenting the danger of fire or explosion. Significantly, the presence of soil gas in near-surface soils has been found to be a reliable indicator of deeper groundwater and soil contamination, and rapid gas detection procedures are being used that can efficiently check regions for underground pollution and outline affected areas. Methods to obtain and analyze soil gas samples are outlined in Chapter 5.

The gases distributed throughout the voids of a soil deposit would normally be expected to have properties similar to those of the earth's atmosphere. However, displacement by contaminant gases can occur because of pressures or thermal energy generated at a

Table 7-4 Corrective Action Alternatives: Techniques and Descriptions

I. Containment: This category consists of geotechnical methods that act to limit the mobility and prevent the further spreading of contaminants. Contaminants are not actually removed from the subsurface but are contained or isolated from the rest of the environment—e.g., via physical barriers or hydrodynamic pressures. Techniques are applied in relation to either the contaminants or their source.

1. *Slurry wall:* Consists of a material (slurry) barrier wall constructed in place; is usually located below the water table and surrounding a site to limit the horizontal migration of contaminants in the saturated zone; is also used to reduce hydraulic gradients, facilitate withdrawal, or channelize groundwater flow.

2. *Sheet pile:* Consists of a material (e.g., concrete, steel, or wood) barrier wall inserted into place by driving or vibration; is usually located below the water table and around a site to limit the horizontal migration of contaminants in the saturated zone.

3. *Grouting*[a]*:* Consists of a material cutoff injected into voids of water-bearing strata either to cover, bottom-seal, or bind together the subsurface materials at a site.

4. *Geomembrane cutoff:* Involves the insertion of synthetic sheeting into an open trench (combining aspects of both the slurry wall and sheet pile) to form a barrier wall; is used primarily to limit the horizontal migration of contaminants in the saturated zone.

5. *Clay (or other) cutoff*[b]*:* Clay (or other material, e.g., concrete) barrier wall; normally is constructed above the water table and downgradient of a site to limit the horizontal migration of contaminants in the unsaturated zone (which is commonly negligible).

6. *Liner*[c]*:* Consists of a material (e.g., clay or synthetic) barrier constructed or emplaced to isolate (e.g., cover or seal) contaminating sources in order to limit the vertical migration of contaminants; is often a facility design component.

7. *Natural containment:* Involves limitation of contaminant mobility by naturally occurring geochemical, geologic, and/or hydrologic conditions; is evaluated by analytical and/or empirical methods.

8. *Surface sealing*[c]*:* Is used as an infiltration control measure to limit the vertical migration of contaminants by reducing leachate production and/or by recharge.

9. *Diversion ditch*[c]*:* Is used as an infiltration control measure to limit surface runoff into a contamination management area (e.g., a slurry-walled area) by channelizing and diverting surface drainage.

10. *Hydrodynamic control:* Limits the horizontal migration of contaminants in the saturated zone through selective pumping and the subsequent creation of pressure troughs or pressure ridges.

II. Withdrawal: Withdrawal options include methods for either directly removing or facilitating the removal of contaminated groundwater and/or contaminated soils from the subsurface. Techniques are principally applied in direct relation to the contaminants.

1. *Pumping:* Involves the removal of contaminated groundwater by pumping from wells or drains; controls the lateral (and in some cases, vertical) migration of contaminants; can be used for flushing (via artificial recharge).

2. *Gravity drainage:* Involves the removal of groundwater from the subsurface using the force of gravity (e.g., using sumps of French drains) instead of pumps; controls the lateral (and in some cases, vertical) migration of contaminants.

3. *Withdrawal enhancement:* Enhances the ability to withdraw either groundwater or contaminants, typically by increasing contaminant solubility in water (e.g., by injecting steam or heat, bacteria or nutrients, or surfactants).

4. *Gas venting:* Removes gases associated with contamination (e.g., methane and petroleum-related products).

5. *Excavation:* Involves the direct removal of contaminated soil and/or groundwater resulting from source leakage.

III. Treatment: This category includes physical and chemical/biological treatment methods for detoxifying contaminants found in groundwater. These methods presume that contaminants have already been withdrawn from the subsurface (e.g., via withdrawal methods) in the form of contaminated groundwater or contaminated soils. Treatment can be applied at the source, at the site of contamination (e.g., in on-site treatment units), prior to the distribution of groundwater for use (e.g., in municipal wastewater treatment facilities), and at the point of end use (e.g., at the tap).

A. Physical treatment:

1. *Skimming:* Involves the removal of floating contaminants (e.g., oil, grease, and hydrocarbons) in a multilayer solution.

2. *Filtration:* Involves the physical retention and subsequent removal of contaminants present as suspended solids.

3. *Ultrafiltration:* Involves the physical filtration, through semipermeable membranes, of suspended and dissolved metals, emulsified hydrocarbons, and substances of high molecular weight.

4. *Reverse osmosis:* Involves the osmotic filtration, through semipermeable membranes, of contaminants (e.g., metals and radioactive wastes) present as dissolved solids; operates at high pressures (up to 1,500 psig).

5. *Air stripping:* Uses air injection to facilitate the volatilization and removal to the atmosphere of contaminants (e.g., volatile organics and hydrogen sulfide) that are present in water as dissolved solids.

6. *Steam stripping:* Involves the fractional distillation of volatile organics or gases by heating.

Table 7-4   (*continued*)

B. Chemical/biological treatment:

7. *Precipitation/clarification/coagulation:* Removes contaminants (e.g., suspended and colloidal solids, phosphates, and heavy metals) through the use of chemical additives such as coagulants and coagulant aids.

8. *Ion exchange:* Removes selected ions (primarily inorganic) via the exchange of ions between an insoluble solid salt ("ion exchange") and a solution containing the ion(s) to be removed.

9. *Adsorption:* Removes contaminants (primarily organics) via their tendency to condense, concentrate, or adhere on the surface of another substance (e.g., granular activated carbon and synthetic resins) with which they come into contact.

10. *Electrodialysis:* Separates and removes positive or negative ions under the action of an electrical field.

11. *Chemical transformation:* Involves oxidation-reduction reactions for the chemical conversion of contaminants to less toxic substances (e.g., by ozone treatment, hydrogen peroxide treatment, ultraviolet photolysis, and chlorination).

12. *Biological transformation:* Involves the transformation and removal by microorganisms of dissolved and colloidal biodegradable contaminants; includes both aerobic and anaerobic processes.

13. *Incineration:* Involves the high-temperature transformation of contaminants into constituent components; many types of thermal destruction systems are included.

IV. *In situ rehabilitation:* In situ rehabilitation techniques are directed at immobilizing or otherwise detoxifying contaminants in place.

1. *Biological degradation:* Involves either stimulating the growth of native microflora or injecting specific organisms to consume or otherwise alter contaminants.

2. *Chemical degradation:* Involves the injection of specific chemicals that react with or otherwise alter contaminants.

3. *Water table adjustment:* Involves either the isolation of the contaminated zone (and creation of a detoxifying unsaturated environment) by lowering the water table or the artificial inducement of increased flushing action by raising the water table.

4. *Rehabilitation via natural processes:* Involves the natural degradation, dispersion, or detoxification of contaminated groundwater; is evaluated by analytical and/or empirical methods.

V. *Management options:* Management options are usually applied to prevent further contamination or to protect potential exposure points from contaminated groundwater. These methods thus focus on sources and exposure points rather than on contaminants per se. The methods also tend to be institutionally based rather than technology based.

1. *Limiting/terminating aquifer use:* Limits access or exposure of receptors to contaminated groundwater.

2. *Development of alternative water supply:* Involves the substitution of alternative supplies (e.g., surface water diversions and/or storage, desalination, and new wells) for contaminated groundwater.

3. *Purchase of alternative water supply:* Includes bottled water and water imports.

4. *Source removal[d]:* Involves the physical removal of the source of contamination and includes measures to eliminate, remove, or otherwise terminate source activities; could also include modification of a source's features (e.g., operations, location, or product) to reduce, eliminate, or otherwise prevent contamination.

5. *Monitoring:* Involves an action evaluation program with a "wait and see" orientation.

6. *Health advisories:* Involves the issuance of notifications about groundwater contamination to potential receptors.

7. *Accepting increased risk:* Involves the decision to accept increased risk; is usually a "no action" alternative.

---

[a]Can be considered a form of chemical immobilization if injected directly into the plume of contamination.

[b]Physical barriers located above the water table will not affect the horizontal migration of contaminants in the saturated zone.

[c]Most often used in the context of either "source removal" or the prevention of recharge to the groundwater system, rather than as a containment option per se.

[d]Modification of a source's features is often an important element of corrective action in the context of preventing future groundwater contamination (i.e., reducing the need for future corrective action).

Source: Office of Technology Assessment [217].

buried source or because of heavier density. Buried sources of gas producers can be present in an area because of natural occurrences or because of human activities.

The types of gases found in a soil deposit relate to the types of buried materials responsible for their formation, and the possibilities for such gas-producing reactions are numerous. But conditions responsible for the development of soil gas volumes that possess the capacity for fire or explosion are most associated with the presence of buried organic materials and organic chemicals. Seeping mature natural petroleum–gas deposits and juvenile gases generated within still-decomposing masses of buried vegetation (e.g., in peat or marsh areas) represent the typical natural occurrence. Waste dumps, areas of buried vegetation resulting from land clearance operations, and garbage-trash landfills, rich with organic throwaway materials, represent the more active gas producers associated with human activities. Radioactive gas such as radon can result from natural sources of uranium and from processed or waste materials containing uranium, radium, or polonium.

On the elemental level, decomposition of buried organic matter occurs by reactions classified as oxidation procedures and reduction procedures. *Oxidation,* the removal of electrons from atoms in the organic molecule, occurs where oxygen is present. *Reduction,* the addition or gain of electrons, occurs where oxygen is lacking but organic matter is plentiful. Where the decomposition process involves oxidation and reduction (the electrons gained equal those being removed or, in other words, being transferred so as to change the involved molecules), the process is termed a *redox reaction.* With organic matter, oxidations tend to occur before reductions if oxygen is available. With buried organic materials, the early stages of decomposition usually occur by oxidation because of the presence of trapped air and oxygen in groundwater. As the available oxygen is consumed, the reduction process prevails.

Oxidation processes include the breakdown of organic matter to carbon dioxide, ammonia to nitrate, methane to carbon dioxide, and sulfide to sulfate. Reduction processes occurring in the presence of anaerobic acid–forming microorganisms include the change of nitrate to nitrogen gas,[9] nitrate to nitrite, nitrite to ammonia, sulfate to sulfide (such as hydrogen sulfide, the "rotten egg" odorous gas), carbon dioxide to methane, and nitrogen gas to ammonia.

Methane is a volatile, explosive gas. Additionally, other gases can be carried along with methane (for example, vinyl chloride, a known carcinogen which results from degradation of the solvent trichlorethylene, as well as benzene and xylene, have been found to be included when methane was detected). Hydrogen sulfide gas possesses an obnoxious odor and is toxic in high concentrations. Otherwise harmless gases such as carbon dioxide can, in large volumes, cause asphyxiation. Nitrates and sulfates can pollute groundwater.[10] Nitrate is considered toxic to infants. Ammonia is a pungent gas that can affect respiration. Recognizing the possibility of problem gases, deep construction work such as tunneling and even shallow rock excavation (particularly in sedimentary formations) includes the use of gas detection (sniffer) equipment, while solid waste landfill projects will build into the protective containment system a means for venting gases that are expected to be generated.

---

[9]The loss of nitrates from soil, termed *denitrification,* is undesirable in the eyes of the agricultural community, where nitrate is considered to be a plant nutrient.

[10]Denitrification often is part of the treatment process in municipal wastewater (sewage) treatment plants.

Radon is a radioactive gas whose presence is related to rock, soil, and waste material whose composition includes uranium, radium, or polonium. Radium and polonium, products of decay (the progeny) from naturally occurring uranium, in turn decays into radon. Radon gas is invisible, odorless, and tasteless, and is soluble in water. As radon gas decays into solid particles of lead and bismuth, it emits small amounts of radiation. Naturally occurring radon gas exists in low concentration in many areas of the earth without causing harm. On the basis of information obtained from experiences with highly radioactive materials, however, medical and health specialists fear that inhaling air that includes high concentrations of radon for extended periods could cause the type of serious health problems associated with excessive exposure to radiation and carcinogens. Where waterborne, the dissolved radon apparently represents a direct hazard only after it separates and becomes airborne.

Unusual concentrations of radon have been encountered in and near uranium mines, at locations where uranium or phosphate tailings (i.e., waste materials and byproducts of mining operations) exist, at disposal sites for manufacturing processes that included uranium–radium materials, in some geologic areas underlain by extensive deposits of granite and sillimanite (aluminum silicates), and from granular soil deposits resulting from glacially transported granite debris. In open space at such locations, health may not be affected because dangerous concentrations do not accumulate. But significant hazard to health is anticipated where high concentrations accumulate indoors (in houses and at places of work) when a building is constructed over uranium-bearing rock, soil, or waste material and where radioactive materials have been used inadvertently to manufacture material used for the building construction (e.g., masonry building blocks made with radioactive phosphate slag). A concentration of airborne radon exceeding about 4 picocuries per liter of air (pCi/l) typically is taken as the condition requiring a remedial action, but some agencies recommend 2 pCi/l for the threshold.[11] The radon gas works its way into the building through construction joints or other openings in contact with the earth. Domestic water will be the source of radioactive material if the surface or subsurface supply has absorbed radon; for example, the radon becomes airborne when the water is sprayed. A concentration of 10,000 pCi per liter of water is taken as the threshold level requiring a remedial action.

If the presence of radon in limited concentrations is known or suspected at the site of a *planned* building, measures for having the radon vented harmlessly to the atmosphere (instead of becoming dispersed or trapped within the building) can be incorporated into the building design and construction phases (Fig. 7-22h). When radon in limited concentration is detected from groundwater or subsoil conditions at the location of an *existing* building, adequate mitigation methods consisting of gas interception and venting procedures often can be achieved at moderate expense. If a planned-for development is known as a location of high radon concentrations, the fill or soil or rock responsible can be removed (and possibly replaced with an unoffending material) if practical. Where the problem source cannot

[11]The *curie* (Ci) is the unit to indicate radioactivity. Radioactivity is the release of particles or radiation from the nucleus of an unstable atom as it is in the process of transforming to a stable atom. The release can be alpha particles, beta particles, or gamma radiation. One curie represents 3.7×10^{10} disintegrations per second, that is, the release of 3.7×10^{10} alpha particles or beta particles per second, or 3.7×10^{10} photons per second of gamma radiation. One picocurie represents 1×10^{-12} Ci, or (3.7×10^{10}) $(1 \times 10^{-12}) \times (60) = 2.2$ disintegrations per minute.

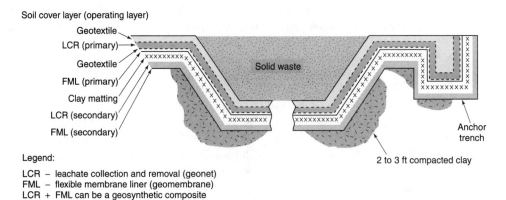

Soil cover layer (operating layer)

Geotextile
LCR (primary)
Geotextile
FML (primary)
Clay matting
LCR (secondary)
FML (secondary)

Solid waste

Anchor trench

2 to 3 ft compacted clay

Legend:

LCR – leachate collection and removal (geonet)
FML – flexible membrane liner (geomembrane)
LCR + FML can be a geosynthetic composite

(a) Double-liner system for solid waste landfills (geotextile, drainage composite and geomembrane with clay underlayer constitute each liner)

Note: Amendment to the 1976 Federal Government Resource Conservation and Recovery Act (1984) requires use of a double liner for hazardous waste disposal facilities

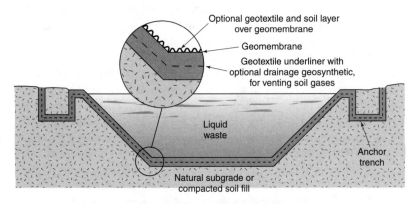

Optional geotextile and soil layer over geomembrane

Geomembrane

Geotextile underliner with optional drainage geosynthetic, for venting soil gases

Liquid waste

Anchor trench

Natural subgrade or compacted soil fill

(b) Single-liner system for liquid waste retention ponds (geomembrane for containment overlies geotextile with high planar permeability for gas venting; optional design includes soil layer plus geotextile above geomembrane)

Figure 7-31   Use of geosynthetics to protect against ground contamination (continued next page).

be identified or worked, realistic concerns relating to liability and health issues would probably dictate that the site be abandoned.

Use of Geosynthetics to Prevent Ground Contamination

Though some soil and groundwater contamination occurs because of natural phenomena, the waste disposal locations associated with human activities—the solid waste (trash) landfills, sewage treatment and industrial waste liquid retention ponds, and mining waste

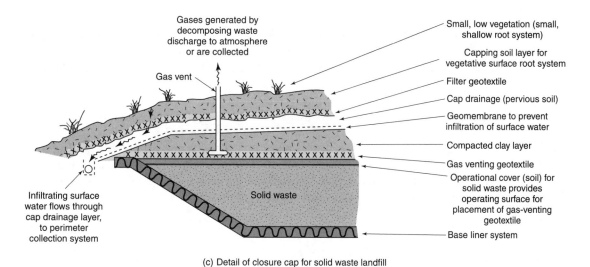

(c) Detail of closure cap for solid waste landfill

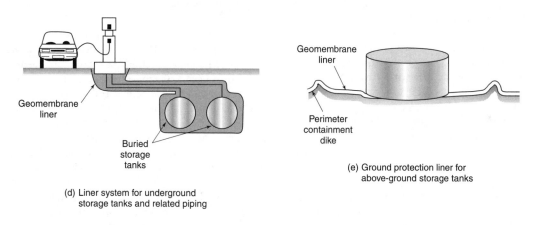

(d) Liner system for underground
storage tanks and related piping

(e) Ground protection liner for
above-ground storage tanks

Figure 7-31   (Con't)

dumps—are potentially the greatest problem because of the large volume and localized heavy concentrations. Recent efforts toward protecting the environment have dealt with preventing contaminants present at such waste disposal locations from escaping into the surrounding earth. Practical reasons involve factors such as ready identification of the contaminant source and of the property owner, which may be a municipality. In the case of a new facility, the disposal site can be properly prepared before being put into operation, or, for an older, in-use site, a modification can be accomplished to limit contaminant migration. The relatively recent availability of the geosynthetics discussed earlier in the chapter have improved on the opportunity to achieve protective measures that are reliable and practical. The use of geosynthetics for environmental protection against ground contamination almost always involves containment.

Where possible, such as where planning identifies a waste disposal site before it is put into service, the geosynthetics are used to create a bottom seal for the ground surface across the basin area so the waste material will be isolated from the natural surroundings. When a top cover seal is eventually added, the waste material ends up being enveloped within the protective geosynthetic system. As is typical for present state-of-the-art systems, the protective geosynthetics will be multilayered and include zones where filtered liquids and gases can be collected for safe disposal. Representative designs for solid waste disposal and waste liquid retention ponds are shown in Figure 7-31a, b, and c.

Liquid chemical and petroleum storage tanks also have been a source of ground contamination (leaks, spills). As protection against the possibility of leaking, above-ground and buried tanks can be multishell, or installed where a geosynthetic envelope is also provided to ensure separation and confinement in the event of tank failure (Fig. 7-31d and e).

PROBLEMS

7-1. A canal is located in an area where the alignment parallels that of a nearby river. The difference in the water surface elevation of the canal and the river is 10 m. A 3-m-thick buried stratum of granular soil (permeable) extending between the canal and river is bounded by layers of clay (impermeable). The horizontal distance between the two bodies of water is 400 m. The coefficient of permeability for the granular soil is 0.75 m per day.
(a) Make a sketch of the described conditions.
(b) Calculate the volume of seepage that occurs from the canal to the river, per kilometer of canal length.

7-2. A river is located in an area where its route parallels that of a nearby canal. The elevation of the water surface in the river is +375 ft. In the canal, the water surface is at elevation +345 ft. A stratum of coarse-grained soil 10 ft thick is sandwiched between relatively impermeable fine-grained soils, and extends between the canal and the river. The distance between the river and the canal is 500 ft, and the coefficient of permeability of the buried coarse stratum is 0.01 ft per hour.
(a) Make a sketch of the described conditions.
(b) Calculate the seepage loss from canal to river, in gallons per day per mile of river canal length.

7-3. What are flow nets and why are they used?

7-4. Flow lines and equipotential lines constitute flow nets used to study the subsurface movement of fluids. Explain the physical factor that the flow lines represent and what the equipotential lines indicate.

7-5. Are flow nets to study subsurface flow appropriate only to coarse-grained soil with a high coefficient of permeability? Explain.

7-6. A concrete dam is 30 m long from upstream side to downstream end. The dam is situated on a 15-m-thick stratum of soil that has a coefficient of permeability equal to 0.5 m per day. Firm, hard clay and rock underlies the described soil stratum. The base of the dam is installed at a level 2 m below the natural soil surface. Draw the flow net for seepage occurring beneath the dam, and compute the seepage in cubic meters per day per meter of dam width (perpendicular to the paper), for the condition where the reservoir water against the upstream face of the dam is 18 m above the soil surface and the water level at the downstream end is at the soil surface.

7-7. A concrete gravity dam is 100 ft long from upstream to downstream end. It is constructed on a stratum of sand-silt soil 40 ft thick. The coefficient of permeability for this soil is 0.1 ft per day. Rock underlies the soil stratum. The bottom of the dam is 6 ft below the soil surface.
(a) For the condition where the height of water against the upstream face of the dam is 45 ft above the soil surface and the water level at the downstream end is just at the soil surface, draw the flow net for seepage beneath the dam and calculate the quantity of seepage in gallons per day per foot of width of dam.
(b) For the condition where the height of water against the upstream face of the dam is 35 ft

above the soil surface and the water level at the downstream end is 5 ft above the soil surface, draw the flow net for seepage beneath the dam and compute the seepage in gallons per day per foot of width of dam.

7-8. The vertically upward flow of groundwater (an artesian condition) occurs through a sand deposit where the void ratio is 0.60 and the specific gravity of soil particles is 2.65. What hydraulic gradient is necessary for a quicksand condition to develop?

7-9. The hydraulic gradient for an occurrence where there is a vertically upward flow of water through a sand mass is 0.95. If the specific gravity of the soil particles is 2.75 and the void ratio is 0.65, determine if a quicksand or erosion condition could develop.

7-10. Explain why a pumped well results in a cone-shaped, depressed or drawn-down water table in the region surrounding the well location.

7-11. Why is the vacuum well-point dewatering system expected to withdraw a greater quantity of groundwater than the conventional well-point dewatering system?

7-12. Well points are used to dewater an excavation that extends 5 m below the surface of the water table. The soil being dewatered is a fine sand. At what distance (approximately) outside of the well-point system is the original groundwater table unchanged (not drawn down)?

7-13. Well points are to be used to dewater a large trench excavation that extends 35 ft below the surface of the water table. Assume that the bottom width of the trench excavation is 70 ft.

(a) Develop a simple sketch of the necessary two-stage well-point system. Use separate diagrams to show the progress of the necessary stages of installation.

(b) Assume that half the necessary lowering of the water table elevation is achieved with the outer row of well points. If the soil being dewatered is a fine to medium sand, approximately at what distance beyond the well point location is the original groundwater table not affected by well-point pumping?

7-14. Indicate the intended functions for foundation drains and the major requirements for design and installation so as to achieve a properly operating system.

7-15. What are the intended functions and limitations of interceptor drains as conventionally used for highway and airfield construction?

7-16. What are the primary reasons for having limitations on the sizes of particles used to construct drainage filters?

7-17. Relating to interceptor drains for highway and airfield pavements:

(a) Outline the traditional method for installing pipe drains.

(b) Outline techniques and materials for installations using geosynthetics.

7-18. Relating to the use of geosynthetics on construction projects, briefly describe what is meant by (a) the separation function, (b) the filtration function, (c) the drainage function, (d) the reinforcement function, and (e) the containment function.

7-19. Why is it important that the geotextiles (geofabrics) consist of synthetics and not natural fibers?

7-20. Relating to geotextiles (geofabrics), what is a significant difference in appearance and texture between the woven and nonwoven materials?

7-21. Briefly describe the conditions necessary and the resulting sequence of occurrences that typically take place in a frost heaving situation. Specify the soil type considered most susceptible to large frost heaves.

7-22. Why is it expected that a period of rapidly developed extreme cold (weather) will not produce severe soil frost heave?

7-23. For extensive freezing-related soil heave to occur, an important contributing natural feature must continue. What is this feature and how does it occur? (*Hint:* Consider the extent of ice lens formations.)

7-24. What methods are commonly used to prevent the occurrence of, or to offset the effects of, frost heave?

7-25. In regions of permafrost, why is it expected that the thickness of the frozen soil zone tends to be relatively constant?

7-26. For construction work in permafrost regions, what type of foundation construction and protection is necessary to maintain structural stability?

7-27. (a) Briefly define *groundwater contamination*.

(b) What is leachate?

7-28. Briefly describe how the direction of groundwater flow and the density properties of a contaminating liquid affect the configuration of a contaminant plume in an aquifer.

7-29. Relating to contaminant materials buried or accumulated underground, list the categories (general procedures) available to correct or control problems that tend to result, and briefly describe the procedures involved.

7-30. List the types of soil gases expected to be generated at a solid waste landfill site and the typical problem associated with each.

7-31. Briefly describe what radon is and the reasons for being concerned about its presence.

7-32. Indicate briefly, using words or sketches, how geosynthetics have been used to prevent groundwater contamination.

CHAPTER 8

Combined Stresses in Soil Masses

Stress at a Point and Mohr's Circle

When a body or mass is subjected to external loading, various combinations of *internal* normal and shear stresses are developed at the different points within the body or mass. Generally, data concerning internal stress conditions are used to determine deformations (in soils work, deformation is most frequently associated with settlement) and to check for the possibility of a material (soil) failure occurring because the strength is exceeded. In performing a stress study, it is convenient to use methods from engineering mechanics for analyzing stress at a point. With these methods, stresses acting on any plane passing through the point can be determined. The combinations of normal and shear stress that develop will vary, depending on the plane being analyzed (Figure 8-1) and the magnitude of the external loading. The "stress at a point" analysis is also applicable for determining the weakest plane or potential plane of failure in a material, not always easily evident, and for indicating the magnitude of the stresses that act on this plane. This type of analysis has particular application for soil and foundation studies, since stability failures in soil masses are the result of the shear strength of the soil being exceeded.

8.1 STRESS AT A POINT: ANALYTICAL DEVELOPMENT

The stresses acting on any plane passed through a point within a material consist of a normal stress (compression or tension) and a shearing stress. Depending on the type of external loading causing the stress condition, it is possible for the shear *or* the normal stress, or both, to be zero on some planes. In soil problems, most external loadings are compression. (The downward weight of a structure supported by a soil mass would be a compressive loading.) For a situation where the loading is compressive, normal stresses that develop on any plane would almost always have a value other than zero; some shear stress would act on all planes with the exception of two planes, where it will be zero (discussed in the following paragraphs).

Figure 8-1 Concept for analysis of stress at a point where external loading causes different combinations of normal and shear stress within the material.

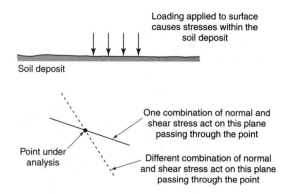

Loading applied to surface causes stresses within the soil deposit

Soil deposit

One combination of normal and shear stress act on this plane passing through the point

Point under analysis

Different combination of normal and shear stress act on this plane passing through the point

If the combination of normal and shearing stresses acting on any two mutually perpendicular planes (orthogonal planes) is known, the combination of stresses acting on any other plane through the same point can be determined.

In analyzing stress at a point, it is convenient to assume an incremental element that represents the stress conditions at the point and to show the known stresses acting on it, as indicated in Figure 8-2a. For equilibrium, the sum of *forces* (not stresses) acting in any direction must equate to zero, and the rotational moments about any axis caused by forces similarly must equal zero. To satisfy this latter requirement, shear stresses acting on orthogonal planes must be equal in magnitude.

At the same point, but on a differently oriented element (actually a different plane passing through the identical location), the combination of normal and shearing stresses that act will be different (Fig. 8-2b). As will be shown, however, there is a relationship between the normal and shear stresses acting on all planes (or orientation of elements) that pass through the same point.

Consider all the planes that can be passed through a point. On one particular plane, not yet defined, the shear stress will be zero, whereas the normal stress will be the maximum possible value of all the normal stresses acting on the various planes through that point. On a plane perpendicular to the plane just referred to, the shear stress will also be zero, but the normal stress that acts will be the least, or the minimum, of all the normal stresses acting on different planes through the point. These *maximum* and *minimum normal stresses* are called *principal stresses.* The planes on which they act are *principal planes.* The shear stress on a principal plane is *always* zero.

In many practical soil problems, the principal stresses act in the vertical and horizontal directions (or on horizontal and vertical planes) and are easily calculated. For instance, where the incremental element represents a point within a soil mass where the ground surface is horizontal, the vertical stress is due to the weight of the soil overburden at that point. This would be the effective unit weight of soil multiplied by the depth of the point below the ground surface. The horizontal stress is proportionate to the vertical stress. Thus, the magnitudes of the principal stresses are known, as are the orientations of planes on which the stresses act.

If the major and minor principal stresses are σ_1 and σ_3, respectively, the magnitude of a normal stress σ_n and shear stress τ_n on any other plane can be determined. Referring to the stressed element in Figure 8-3a and letting the area on the cut plane be unity (Fig. 8-3b), it is apparent that the area of the plane on which σ_1 acts becomes $(1) \times (\cos \theta)$, and the

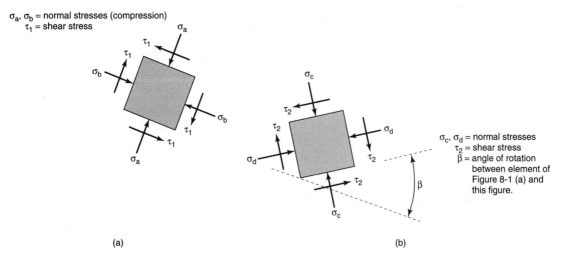

Figure 8-2 Basic representation for stress-at-a-point analysis, indicating stress combinations acting on an incremental element. (a) Incremental element with representative stresses assumed for analysis of "stress at a point"; (b) method of rotating incremental element to determine stresses on different planes of study.

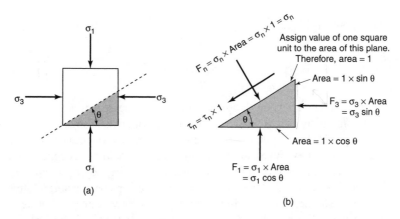

Figure 8-3 Basic step in determining stresses on a random plane through a point in terms of known principal stresses.

area on which σ_3 acts becomes $(1) \times (\sin \theta)$. The total normal *force* on the cut plane is $(\sigma_n) \times (1)$. The total force on the vertical surface is $(\sigma_3) \times (\sin \theta)$, and the total force on the horizontal surface is $(\sigma_1) \times (\cos \theta)$.

To determine σ_n and τ_n in terms of σ_1 and σ_3, the forces acting on the horizontal and vertical planes are resolved into components parallel and perpendicular to the cut plane (Fig. 8-4). Summing forces parallel to σ_n (normal to the cut plane) gives

$$\sigma_n = \sigma_1 \cos \theta \cos \theta + \sigma_3 \sin \theta \sin \theta$$

$$= \sigma_1 \cos^2 \theta + \sigma_3 \sin^2 \theta$$

Figure 8-4 Resolution of principal stresses into components parallel and perpendicular to a random plane. This step is in preparation for determining stresses on the plane in terms of principal stresses.

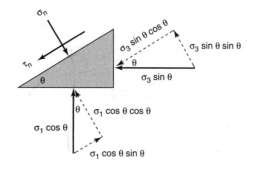

and by trigonometric identity,

$$\sigma_n = \sigma_1 \cos^2 \theta + \sigma_3 \sin^2 \theta = \frac{\sigma_1 + \sigma_3}{2} + \frac{\sigma_1 - \sigma_3}{2} \cos 2\theta \qquad (8\text{-}1)$$

Summing forces parallel to the cut plane in the direction of τ_n gives

$$\tau_n = \sigma_1 \cos \theta \sin \theta - \sigma_3 \sin \theta \cos \theta = (\sigma_1 - \sigma_3) \sin \theta \cos \theta$$

and by trigonometric identity,

$$\tau_n = (\sigma_1 - \sigma_3) \sin \theta \cos \theta = \left(\frac{\sigma_1 - \sigma_3}{2}\right) \sin 2\theta \qquad (8\text{-}2)$$

When applying these equations, it is necessary to use some method that will prevent compressive stress from being confused with tensile stress; for geotechnical problems, it is typical to identify compressive stress as a positive value and tensile stress as a negative value.

From examination of Equation 8-2 it should be recognized that the maximum shear stress will occur on a plane that is 45° from the major principal plane ($\theta = 45°$) and will have a magnitude equal to $\frac{1}{2}(\sigma_1 - \sigma_3)$. On this plane, the normal stress is *always* $\frac{1}{2}(\sigma_1 + \sigma_3)$.

The previous developments apply to a two-dimensional stress analysis (stresses in a plane). In actuality, for soil problems, the study of stress at a point involves a three-dimensional analysis. An incremental element with three principal stresses acting—σ_1, σ_2, and σ_3—is shown in Figure 8-5. For many practical conditions of loading, the intermediate stress, σ_2, is equal to σ_1, or to σ_3. For some situations, however, σ_2 may be a value intermediate between σ_1 and σ_3. As far as practical effects of the intermediate stress are concerned, it appears that there is some influence on the strength and stress–strain properties of the material, but its effect is not clearly understood. To keep a proper perspective, however, it is pointed out that the methods available for making determinations of the stresses within a soil mass that result from external loading are not highly refined. As a result, for most practical problems, the degree of accuracy does not appear to be significantly diminished by neglecting the effect of the intermediate stress and working with the simpler two-dimensional condition (σ_1 and σ_3 only).

Figure 8-5 Three-dimensional incremental element showing principal stresses in mutually perpendicular directions.

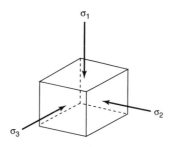

8.2 MOHR'S CIRCLE

The varying values of normal stress and shear stress corresponding to differing values of θ can be determined by using Equations 8-1 and 8-2. If the combination of normal and shear stress resulting from each value of θ is plotted as a point on a coordinate system where the horizontal axis represents normal stress and the vertical axis represents shear stress (Fig. 8-6a), the locus of many plotted points will form a circle (Fig. 8-6b).

This fact can be used to advantage. By working with simple properties of a circle, a graphical or pictorial method of solving for normal and shear stresses on any plane is easily developed once the stresses on orthogonal planes are known. The method simplifies the calculation necessary for determining stresses and eliminates the need to work with cumbersome equations. The method is referred to as the Mohr's circle for determining stresses, after Otto Mohr (1835–1918), who is credited as the developer of the method.

In using Mohr's circle, a sign convention is required. For soil problems, compressive stresses are conventionally assumed to be positive, and shearing stresses that provide a clockwise couple are also considered positive (Fig. 8-7).

For a stress combination like that shown in Figure 8.8a, σ_1 and σ_3 are the major and minor principal stresses. There is no shear stress acting on these major and minor principal planes. To construct the Mohr's circle for this combination (Fig. 8-8b), locate σ_1 (or 2.5 kPa) and σ_3 (or 0.50 kPa) on the *horizontal axis* of the coordinate system (since the shear is zero). Next, establish the location (or value) for the center of the Mohr's circle, knowing that the diameter of the circle has a value equal to σ_1 minus σ_3. (This numerical difference between σ_1 and σ_3 is called the *deviator stress*.) For this illustration, the deviator stress is 2.50 kPa minus 0.50 kPa, or 2.0 kPa. The radius of the circle is then 1.0 kPa, and the center of the circle will plot at 1.5 kPa, which is the value of σ_3 plus the radius. After the center of the circle and the diameter are established, the circle itself can be constructed; thereafter, the stress combination on *any* plane can be determined.

In working with the Mohr's circle, it is convenient to reference the plane under study to the major principal plane whenever possible. Remember that the major principal stress acts on the major principal plane; the angular measurement to the plane in question is made on the Mohr's circle by starting from the point representing the major principal stress. If the angle to be measured is formed by two radii, the angle is measured at the center of the circle. Because of the properties of a circle, the central angle on the circle must be twice the value of the angle θ measured on the original element (Fig. 8-9). The direction of

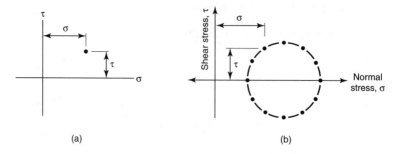

(a)   (b)

Figure 8-6 Method of representing the combination of normal and shear stresses acting on any plane through an incremental element on shear stress–normal stress coordinates: (a) single point represents stress combination on one plane; (b) stress combinations occurring on different planes through an incremental element provide points that form a circle.

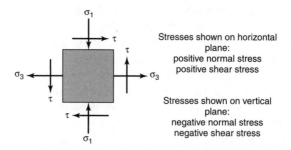

Stresses shown on horizontal plane:
positive normal stress
positive shear stress

Stresses shown on vertical plane:
negative normal stress
negative shear stress

Figure 8-7 Sign convention assigned to stresses for the Mohr's circle analysis.

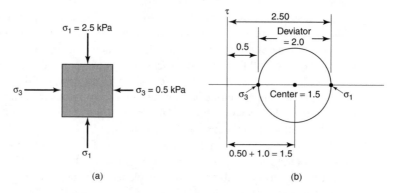

(a)   (b)

Figure 8-8 Representation of principal stresses acting at a point and the related Mohr's circle plot: (a) stresses acting on incremental element; (b) Mohr's circle plot.

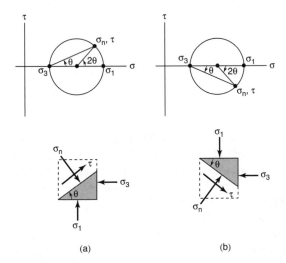

Figure 8-9   Relation of positions of planes on the incremental element to points on the Mohr's circle: (a) reference angle measured counterclockwise; (b) reference angle measured clockwise.

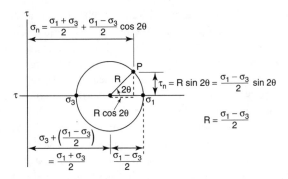

Figure 8-10   Establishment of the general equations for normal stress and shear stress from the Mohr's circle.

measurement from the principal plane to the cut plane on the element, clockwise or counterclockwise, is the same direction to be used on the Mohr's circle (Fig. 8-9). By reference to the Mohr's circle diagram in Figure 8-10, the expressions for shear and normal stress for any point (representing any plane) can be reaffirmed. The normal stress coordinate is the value of the center of the Mohr's circle plus or minus the horizontal projection of the radius. As shown in Figure 8-10, the normal stress would then be

$$\sigma_n = \frac{\sigma_1 + \sigma_3}{2} + \frac{\sigma_1 - \sigma_3}{2} \cos 2\theta \qquad (8\text{-}1)$$

which agrees with Equation 8-1. The shear stress coordinate is the vertical projection of the radius, or

$$\tau_n = \frac{\sigma_1 - \sigma_3}{2} \sin 2\theta \qquad (8\text{-}2)$$

which agrees with Equation 8-2.

As an illustration, to determine the magnitude of the normal and shear stress acting on a plane 45° from the major principal plane (Fig. 8-11), the angle measured at the center of the Mohr's circle would have to be twice 45°, or 90°. Note that, on the element, the angle θ is 45° measured counterclockwise, regardless of whether the top or bottom of the original element is used as the reference plane. On the Mohr's circle the central angle, 2θ or 90°, is also measured counterclockwise. If the principal stress equations developed previously are used, or by applying simple mathematics if the Mohr's circle is used pictorially, or by scaling if the graphical method is used, the value of the normal stress is found to be 1.5 kPa and the shear stress is 1.0 kPa.

From examination of the Mohr's circle, it should be evident that the maximum shear stress (in this case 1.0 kPa) always acts on the plane that is 45° from the major principal plane. Further, the maximum shear stress always has a magnitude equal to the radius of the Mohr's circle.

As a second illustration, with the same principal stresses used previously, assume that it is desired to find the stresses acting on a plane that is 60° clockwise from the major principal plane (Fig. 8-12). On the Mohr's circle, the central angle to be measured is 120°. Mathematically, from a pictorial Mohr's circle, the shear stress τ_n equals $(R) \times (\sin 60°)$, or $(1.0\,\text{kPa}) \times (0.866)$, which is 0.866 kPa. The normal stress σ_n is $1.5\,\text{kPa} - (R) \times (\cos 60°)$, or 1.0 kPa, as shown in Figure 8-12. The 1.5 kPa value locates the center of the circle. On the desired plane, the stresses are as shown in Figure 8-13.

If the stress combination is to be determined on a plane measured with reference to the minor principal plane, the related angular measurement on the Mohr's circle is made from the minor principal stress, σ_3, as shown in Figure 8-14. The direction of the angle will be the same on the Mohr's circle as on the incremental element.

The Mohr's circle pictorial or graphical method can similarly be used for determining principal stresses if the stress conditions for any two orthogonal planes other than principal planes are known. For an element where the normal shear stresses are as shown in Figure

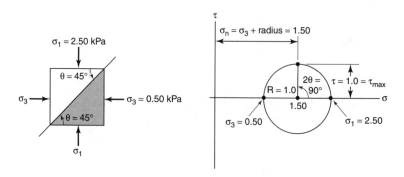

Figure 8-11 Use of the Mohr's circle to determine the shear and normal stress acting on a plane 45° from the major principal plane.

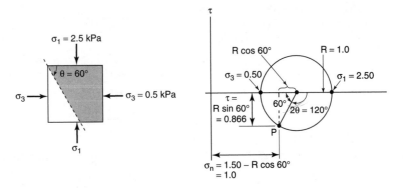

Figure 8-12 Use of the Mohr's circle to determine stresses on a plane 60° clockwise from the major principal plane.

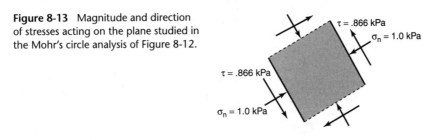

Figure 8-13 Magnitude and direction of stresses acting on the plane studied in the Mohr's circle analysis of Figure 8-12.

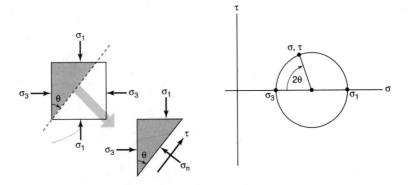

Figure 8-14 Method of locating the stress–combination point on the Mohr's circle when the minor principal plane is used as a reference.

Figure 8-15 An incremental element with stresses that are not principal stresses acting on principal planes.

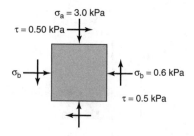

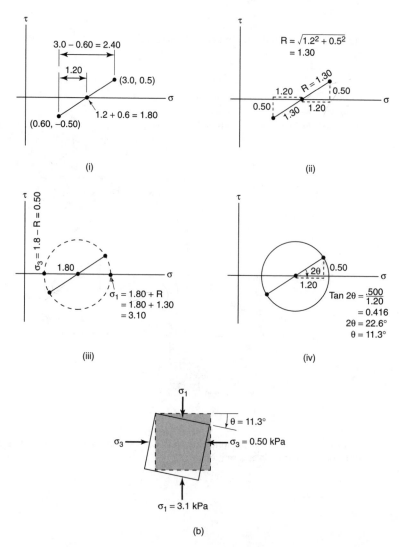

Figure 8-16 Procedure for applying the Mohr's circle analysis to solve for principal stresses.

8-15 (note that the shear stresses on the mutually perpendicular planes are equal in magnitude), the Mohr's circle construction proceeds as follows (Fig. 8-16):

1. On the Mohr's circle coordinates, establish the two points corresponding to σ_a, τ, and σ_b, τ.

2. Connect these two points, thereby establishing the diameter of the Mohr's circle (Fig. 8-16a), and determine the coordinate for the center of the circle.

3. Calculate the value for the radius of the Mohr's circle (Fig. 8-16b).

4. The value of σ_1 (the major principal stress) is the circle radius added to the value established for the circle center. The value of σ_3 (the minor principal stress) is the radius subtracted from the coordinate for the circle center (Fig. 8-16c).

5. The angle 2θ on the Mohr's circle is obtained from simple geometry, as shown in Figure 8-16d.

6. The orientations of the principal planes with respect to the original element are shown in Figure 8-16e.

7. Since the complete Mohr's circle is established, stresses on any other plane can be determined by using the methods covered previously.

Illustration 8-1

The major and minor principal stresses acting at a point are 50 Pa compression and 10 Pa compression, respectively. Draw the Mohr's circle for this stress combination, and determine the magnitude of shear and normal stress on one of the planes where shear is a maximum.

Solution

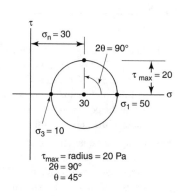

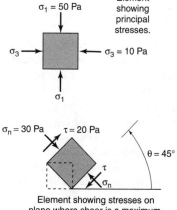

Element showing stresses on plane where shear is a maximum, and the orientation relating to principal planes.

Illustration 8-2

At one point in a soil mass, the minor principal stress is 1000 psf. What is the maximum value possible for the major principal stress at this point if the shear stress cannot exceed 2000 psf?

Solution

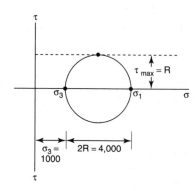

The maximum shear stress at a point is the value of the radius of the Mohr's circle; that is,

$$Radius = 2000 \text{ psf}$$

Since the major principal stress $\sigma_1 = \sigma_3 + 2R$,

$$\sigma_1 = 1000 + 2(2000) = 5000 \text{ psf}$$

PROBLEMS

8-1. In terms of the stress-at-a-point analysis, define principal stress. Indicate how it is different from other normal stresses (tensile or compressive) that also act at the point.

8-2. At a point in a stressed material, the major principal stress is 50 kPa compression and the minor principal stress is 20 kPa compression. Use Equations 8-1 and 8-2 to determine the maximum shear stress that acts through the point and the normal stress on the plane of this maximum shear.

8-3. At a point in a stressed material, the major principal stress is 44 kPa tension and the minor principal stress is 18 kPa tension. Using Equations 8-1 and 8-2, determine the maximum shear stress that acts through the point and the normal stress on the plane of maximum shear.

8-4. At a point in a stressed material, the major principal stress is 7.2 ksi compression and the minor principal stress is 3.6 ksi compression. Using Equations 8-1 and 8-2, determine the maximum shear stress that acts at the point and the normal stress that acts on the plane of the maximum shear stress.

8-5. Using the equations for determination of stress at a point, calculate the value for the maximum shear stress that develops at a point if the major principal stress is 36 kPa compression and the minor principal stress is 14 kPa tension. Also indicate the value of normal stress acting on the plane of maximum shear.

8-6. Using the equations for determination of stress at a point, calculate the value for the maximum shear stress that develops if the major principal stress is 7.2 ksi compression and the minor principal stress is 3.6 ksi tension.

8-7. At a point in a stressed material, the major principal stress is 120 kPa compression and the minor principal stress is 40 kPa compression.
  (a) Draw the Mohr's circle for this stress combination.
  (b) Determine the maximum shear stress. Also indicate the normal stress on the plane where shear is a maximum.
  (c) Determine the value of normal and shear stress acting on the plane 30° counterclockwise (ccw) from the major principal plane.

8-8. At a point in a stressed material, the major principal stress is 5000 psf compression and the minor principal stress is 2000 psf compression.
  (a) Draw the Mohr's circle for this stress combination.
  (b) Find the maximum shear stress, and also indicate the value of normal stress acting on this same plane.
  (c) Determine the values of the normal and shear stresses acting on a plane that is 60° ccw from the major principal plane.

8-9. At a point in a stressed material, the major principal stress is 180 kPa compression and the minor principal stress is 40 kPa tension.
  (a) Determine the Mohr's circle for this stress combination.
  (b) What is the maximum shear stress that acts at the point, and what is the value for the normal stress acting on the plane of maximum shear?
  (c) Determine the value of the normal and shear stress acting on the plane 22.5° ccw from the major principal plane.

8-10. At a point in a stressed material, the major principal stress is 80 psi compression and the minor principal stress is 20 psi compression.
  (a) Draw the Mohr's circle for this stress combination.
  (b) What is the maximum shear stress acting at the point, and what value of normal stress acts on this same plane?
  (c) Determine the value of the normal and shear stress acting on a plane that is 60° clockwise (cw) from the minor principal plane.

8-11. At a point in a stressed material, the major principal stress is 66 kPa compression and the minor principal stress is 34 kPa tension.
  (a) Use the Mohr's circle to determine the shear and normal stress on a plane 45° ccw from the major principal plane.

  (b) Determine the stress combination acting on the plane 75° ccw from the major principal plane.

8-12. The major and minor principal stresses acting at a point in a stressed material are 80 psi compression and 20 psi tension.
  (a) Draw the Mohr's circle for this stress combination.
  (b) What is the maximum shear stress acting at the point, and what value of normal stress acts on the plane?
  (c) Determine the value of the shear and normal stresses acting on a plane that is 30° ccw from the major principal plane.

8-13. At a point in a stressed material, the stress combination acting on a plane is a 90 kPa compressive stress and a 30 kPa shear stress, while on a perpendicular plane, the stress combination is 30 kPa compression and 30 kPa shear. Use the Mohr's circle for this stress analysis.
  (a) Determine the values of the major and minor principal stresses.
  (b) Determine the angle between the major principal plane and the plane where the 90 kPa compressive stress acts.

8-14. At a point in a stressed material, the stress combination acting on a plane is 60 kPa compression and 20 kPa shear, while on an orthogonal plane, the stress combination is 40 kPa tension and 20 kPa shear. Use the Mohr's circle for this stress analysis.
  (a) Determine the values for the major and minor principal stresses.
  (b) Determine the value of the maximum shear stress and the value of the normal stress on the plane of maximum shear.

8-15. At a point, the normal and shear stresses acting on one plane are 6000 psf compression and 2000 psf, respectively. On a perpendicular plane, the normal and shear stresses are 2000 psf compression and 2000 psf, respectively.
  (a) Draw the Mohr's circle.
  (b) Determine the value of the principal stresses.
  (c) Find the angle between the plane on which the 8000 psi compressive stress acts and the major principal plane.
  (d) What is the maximum shear stress acting at the point?

8-16. The normal and shear stresses acting on one plane passing through a point in a soil mass are 120 psi compression and 25 psi, respectively. On an orthogonal plane, the respective stresses are 40 psi compression and 25 psi.

(a) Draw the Mohr's circle for this stress condition.

(b) What are the principal stresses?

(c) Determine the angle between the plane on which the 120 psi stress acts and the major principal plane.

(d) What is the maximum shear stress acting at the point?

8-17. The major principal stress and deviator stress at a point are 100 kPa and 60 kPa, respectively. What is the value of the maximum shear stress and of the normal stress acting on the same plane? (Deviator stress is the algebraic difference between the values for the principal stresses.)

8-18. For a minor principal stress of 50 kPa, what is the maximum value possible for the major principal stress if the maximum shear stress is not to exceed 40 kPa?

8-19. At a location where the ground surface is horizontal, the vertical stress at one point in the underlying soil deposit is 190 kPa and the horizontal stress is 76 kPa. If these stresses represent principal stresses, what is the maximum shear stress passing through the point? What is the value of the normal stress on the plane of maximum shear?

8-20. In a soil where the ground surface is horizontal, the vertical stress acting at a point is 2000 psf and the horizontal stress is 1000 psf. If these stresses represent the principal stresses, what is the maximum shear stress acting at the point?

8-21. A level ground surface exists in an area where soil in the underlying stratum has a unit weight of 19 kN/m³. The vertical pressure at any depth represents the major principal stress. The lateral pressure at a corresponding depth is 0.45 times the vertical pressure and represents the minor principal stress.

(a) Determine the principal stresses acting at a depth 4 m below the surface, and draw the Mohr's circle for the condition.

(b) What is the maximum shear stress acting at this depth?

8-22. Assume a soil mass with a level ground surface and a material having a unit weight of 120 pcf. The vertical pressure at any depth represents the major principal stress at that particular point. The lateral pressure acting at any point is one-half the vertical pressure and represents the minor principal stress.

(a) Determine the principal stresses acting at a depth of 10 ft below the ground surface, and draw the Mohr's circle for these conditions.

(b) What is the value of the maximum shear stress at this depth?

8-23. In an area where the ground surface is level, the underlying soil has a unit weight of 16 kN/m³. Determine the principal stresses acting at a depth of 5 m if the lateral pressure is one-half the vertical.

8-24. The soil in a deposit underlying an area where a level ground surface exists has a saturated unit weight of 21 kN/m³. The groundwater table is at the soil surface.

(a) Determine the principal stresses for the total stress condition at a depth of 5 m below the soil surface if the lateral pressure is one-half the vertical pressure.

(b) Determine the principal stresses for the effective stress condition at the same 5 m depth.

(c) Determine the value of the maximum shear stress for the total stress and the effective stress condition.

8-25. A soil deposit exists with a level ground surface. The saturated unit weight of the soil is 125 pcf, and the water table is at the ground surface.

(a) Determine the principal stresses acting at a depth 15 ft below the ground surface that results from the effective (submerged) soil weight if the lateral pressure is 0.45 times the vertical pressure.

(b) What is the magnitude of the maximum shear stress at this depth?

8-26. A soil deposit exists in an area where the ground surface is level and the groundwater table is 3 m below the surface. The saturated unit weight for the soil is 22 kN/m³, and the wet weight for the soil above the water table is 18 kN/m³. The lateral pressure is 0.5 times the vertical pressure. The vertical and lateral pressures represent principal stresses. Determine the value for the major and minor principal stresses at a depth of 10 m (determine values for both the total stress and effective stress condition).

CHAPTER 9

Subsurface Stresses

At a point within a soil mass, stresses will be developed as a result of the soil lying above the point and by any structural or other loading imposed onto that soil mass. The magnitude of the subsurface stress at a point is affected by the groundwater table if it extends to an elevation above the point.

In most foundation design problems, the safe bearing capacity of the soil (the ability to support structural load) and the settlement (the soil volume change resulting from loading) that will develop under a given intensity of structural loading are major items of concern. For such analysis, the significant stresses are considered to be those acting in the vertical direction. In the design of vertical structures such as retaining walls, sheeting for braced excavations and waterfront structures, and some types of pile foundations, the soil stresses acting in the horizontal or lateral direction are the most significant.

9.1 STRESSES CAUSED BY THE SOIL MASS

Vertical Stresses

In a soil mass having a horizontal surface, the vertical stress caused by the soil at a point below the surface is equal to the weight of the soil lying directly above the point. Vertical stress thus increases as the depth of the soil overburden increases. The vertical stress can be calculated as the weight of a "column" of soil extending above a unit area (Fig. 9-1a). For a homogeneous soil having a wet unit volumetric weight of γ_t (normally expressed as kilonewtons per cubic meter or pounds per cubic foot), the stress σ_v (normally in kilonewtons per square meter or kilopascals, or pounds per square foot) at a depth Z m or ft below the ground surface is

$$\sigma_v = \gamma_t Z \tag{9-1}$$

If the soil mass is made up of strata of different soil types and the unit weights of the soil in each stratum are different, the vertical stress at a depth Z will be equal to the total weight of the different segments of the soil "column," as indicated by Figure 9-1b.

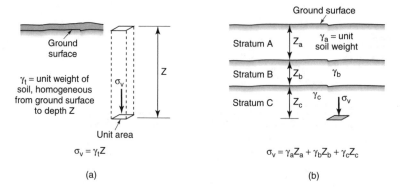

Figure 9-1 Vertical subsurface stress resulting from the soil mass.

Effect of Groundwater Table

When a soil exists below the groundwater table, the submerged soil particles are subject to a buoyant force resulting from the hydrostatic water pressure, the same phenomenon that acts on any submerged solid. The submerged weight of the soil, γ_{sub}, is termed the *effective soil weight,* and the subsurface stress that results is termed the *effective stress.* Effective stress represents the actual intergranular pressure that occurs between soil particles. This effective stress is the stress that influences shear strength of the soil and volume changes or settlements.

If a condition exists where the *water table is at the ground surface* and the soil mass is homogeneous, the effective stress $\overline{\sigma}_v$ at a depth Z is

$$\overline{\sigma}_v = \gamma_{sub}Z \tag{9-2}$$

If the total weight of soil as it exists above the water table is γ_t (the total soil weight before the buoyant effects of submergence are considered), the effective stress is

$$\overline{\sigma}_v = \gamma_t Z - \gamma_w Z \tag{9-3}$$

where γ_w is the unit weight of water. (It is conventional to assume 62.4 pcf or 9.81 kN/m³ for γ_w.) The last term of this equation is the total water pressure at the depth Z. Total water pressure at a point is termed the *neutral stress, u,* for it acts equally in all directions. Neutral stress refers to any water pressure that develops at a point as caused by hydrostatic conditions. It is important to recognize that the neutral stress acts to *reduce* the intergranular stress that develops between soil particles. This condition frequently has an adverse effect on the strength of a soil. The effective stress for conditions just described can be expressed as

$$\overline{\sigma}_v = \gamma_t Z - u \tag{9-4}$$

To compute the effective stress for a condition where the *groundwater table lies below the ground surface,* and for the condition where strata of soils of different types and weights exist, two different approaches are possible. One approach involves determining the total soil pressure (disregarding buoyancy effects) and then subtracting the hydrostatic pressure (the neutral pressure) at the point being analyzed. The neutral pressure u is the unit weight of water γ_w, multiplied by the depth below the water table. A second approach is to determine directly the effective stress of the column of soil above the point by using the effective or submerged weight of all soil in the "column." Above the water table, the effective soil weight is the total soil weight, including the pore water; below the water table, the effective soil weight is the submerged or buoyant weight (Fig. 9-2a).

Where the soil surface is below water (such as in oceans and lakes), the effective stress should be computed by using the submerged or effective soil weight multiplied by the depth measured from the soil surface (Fig. 9-2b).

A summary of the analytical procedures for computing subsurface stresses resulting from various soil and water table conditions, and numerical illustrations, are presented in Figure 9-3.

Horizontal (Lateral) Stresses

The magnitude of vertical stress is relatively simple to determine when the ground surface is level. When this condition does exist, it is also convenient to indicate horizontal (lateral) stresses that exist in a soil mass in terms of the soil vertical stress. The ratio of lateral stress to vertical stress, K, is termed the *coefficient of lateral earth pressure.* Mathematically,

$$K = \frac{\text{Horizontal soil pressure } \overline{\sigma}_h}{\text{Vertical soil pressure } \overline{\sigma}_v} \tag{9-5}$$

The use of effective pressures for determining the lateral pressure coefficient accounts for the influence of submergence where the soil deposit is wholly or partially below water or the water table (i.e., submerged unit weights are used when appropriate to determine $\overline{\sigma}_v$).

In a horizontal, uniform soil mass of infinite extent, the lateral movement of the soil at any depth is not possible, because the confining pressure is equal in all horizontal directions. Thus a state of static equilibrium exists and the soil is in the *at-rest condition.* The coefficient of lateral pressure for the at-rest condition is indicated by K_o.

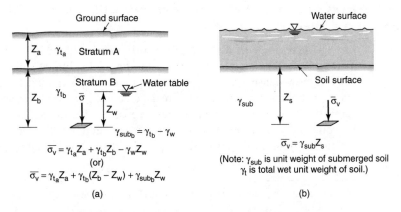

Figure 9-2 Subsurface stress below water.

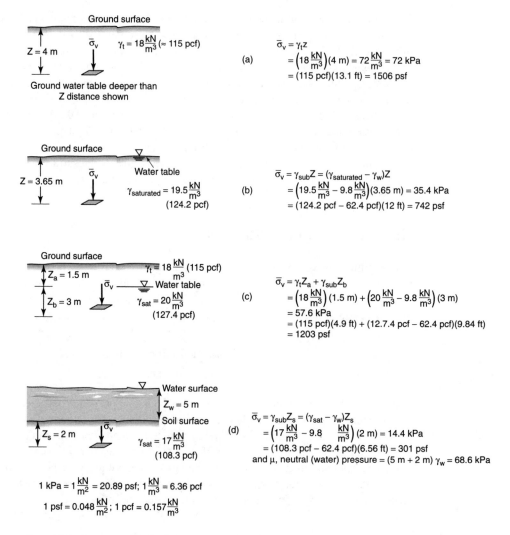

Figure 9-3 Summary: Method to compute subsurface stress in a soil mass.

The magnitude of K_o for a given soil mass is affected by the soil deposit's stress history. Soils that have been subjected to heavy loading at some time in their history, such as now-dense granular (sand or gravel) deposits and hard, overconsolidated[1] clays, would have had to develop resistance to high lateral stress in order to maintain stability. Deposits that have not been exposed to heavy loading, such as loose granular soils and soft, normally consolidated or underconsolidated clays, would not have developed high lateral strength. Typically, then, dense granular soils and hard clays

---

[1]Clays that at some past time were subject to loading greater than the weight of all currently existing overlying soil.

Table 9-1   Typical Values of K_o

Soil Type	K_o
Granular, loose	0.5–0.6
Granular, dense	0.3–0.5
Clay, soft	0.9–1.1 (undrained)
Clay, hard	0.8–0.9 (undrained)

end up having lower values of K_o than do loose granular soils and soft clays. Such typical values are presented in Table 9-1. The topic of lateral pressures is discussed in greater detail in Chapter 16.

Illustration 9-1

The unit weight of the soil in a uniform deposit of loose sand is 100 pcf. Determine the horizontal stress that acts within the soil mass at a depth of 10 ft.

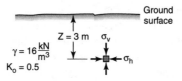

Solution

$$\sigma_v = \sigma Z = (16 \text{ kN/m}^3)(3 \text{ m}) = 48 \text{ kN/m}^2 = 48 \text{ kPa}$$

$$\sigma_h = K_o\sigma_v = (.5)(48 \text{ kPa}) = 24 \text{ kPa}$$

where K_o is obtained from Table 9-1.

In soils below the water table, determination of the *total* lateral pressure requires that the hydrostatic pressure due to the water be added to the effective at-rest soil pressure computed by using a value from Table 9-1 and the effective soil weight.

Illustration 9-2

A concrete basement wall for a structure extends below the groundwater table. For conditions indicated by the sketch, calculate the total lateral pressure acting against the wall at a point 8 ft below the ground surface.

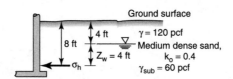

Solution

$$\text{Soil pressure, } \overline{\sigma}_h, \text{ at 8 ft} = K_o \gamma Z$$

$$= (.4)[(4 \text{ ft} \times 120 \text{ pcf}) + (4 \text{ ft} \times 60 \text{ pcf})]$$

$$= 288 \text{ psf}$$

$$\text{Total lateral pressure} = K_o \gamma Z + \gamma_w Z_w = 288 \text{ psf} + (4 \text{ ft})(62.4 \text{ pcf})$$

$$= 538 \text{ psf}$$

9.2 STRESS WITHIN THE SOIL MASS RESULTING FROM VERTICAL SURFACE LOADING

Uniform Homogeneous Soils

When a vertical loading from a structure or other body is applied at the surface of a soil mass, new stresses are created within the mass. Because of shearing resistance developed within the soil, loading transferred to the soil mass will be spread laterally with increasing depth from the point or area of application (Fig. 9-4). With increasing depth, the area over which new stresses develop will increase but the magnitude of the stresses will decrease. For an equilibrium condition, the sum of the new vertical stresses developed in the soil mass on any horizontal plane must equal the weight or force of the surface loading.

The manner in which the stresses become distributed throughout the soil mass is affected by the properties of the soil, including modulus of elasticity and Poisson's ratio and any stratification of different soil types.

Boussinesq Stress Distribution

One of the methods in common use for calculating stresses that result in a soil mass from a surface loading is based upon the work of Boussinesq, a nineteenth-century French mathematician. Boussinesq assumed a homogeneous, isotropic material (properties the same in all directions) of semi-infinite extent (unlimited depth) and developed equations for the stress distribution resulting from a point load. Adapted to soil masses, the described conditions are as indicated by Figure 9-5. The vertical stress increase $\Delta\sigma_v$ resulting at a

Figure 9-4 Variation of vertical stress at different depths Z.

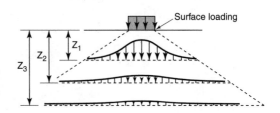

Figure 9-5 Definition of terms applicable to Boussinesq and Westergaard equations.

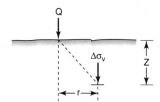

depth Z and a distance r, measured horizontally from where a point loading Q is applied, becomes

$$\Delta\sigma_v = \frac{3Q}{2\pi}\frac{Z^3}{(r^2+Z^2)^{5/2}} = \frac{Q}{Z^2}\frac{3}{2\pi\left[1+\left(\dfrac{r}{Z}\right)^2\right]^{5/2}} \tag{9-6a}$$

This equation indicates that as the depth increases the stress decreases. Similarly, the stress decreases as the horizontal distance from the point of loading is increased. For a given depth, the intensity of stress is greatest directly beneath the point of load application.

With the Boussinesq equation for vertical stress, the modulus of elasticity and Poisson's ratio are not required, which indicates that the stress is independent of these properties as long as the material is homogeneous and isotropic.

The suitability of using the Boussinesq equation for determining subsurface stresses in a foundation analysis depends on how closely the actual soil conditions and properties resemble the theory's original assumptions. For practical problems, conditions of a homogeneous and isotropic material are commonly assumed for homogeneous clay deposits, for man-made fills where the soil fill has been placed and compacted in thin layers, and for limited thicknesses of uniform granular soil deposits.

Westergaard Stress Distribution

Some sedimentary soil deposits consist of alternating thin layers of sandy soil (coarse, relatively incompressible material) and fine-grained silt–clay soils (compressible material)— for example, stratified deposits such as laminated clays. For such conditions, the Westergaard equations provide a better means of evaluating the subsurface stresses. In his development, Westergaard assumed that thin layers of a homogeneous and anisotropic material were sandwiched between closely spaced, infinitely thin sheets of rigid material that would permit compression but no lateral deformation. For the case where Poisson's ratio is zero, the equation for subsurface stress resulting from a concentrated point loading reduces to

$$\Delta\sigma_v = \frac{Q}{Z^2\pi\left[1+2\left(\dfrac{r}{Z}\right)^2\right]^{3/2}} \tag{9-7a}$$

The terms Q, r, and Z are as defined for the Boussinesq equation and as shown in Figure 9-5.

Computational Aids

With both the Boussinesq and Westergaard equations, the subsurface stress resulting from a given point load will be related to the Z and r distances, or the r/Z ratio. If the Boussinesq equation is written in terms of a stress influence factor I_B, which is related to r/Z as follows:

$$\Delta\sigma_v = \frac{Q}{Z^2} \frac{3}{2\pi\left[1 + \left(\dfrac{r}{Z}\right)^2\right]^{5/2}} = \frac{Q}{Z^2}I_B \qquad (9\text{-}6b)$$

calculations for values of I_B for different r/Z ratios can be computed and presented as shown in Figure 9-6. The effort to determine a subsurface stress then becomes greatly simplified. Similarly, the Westergaard equation can be written in terms of an influence factor I_w as follows:

$$\Delta\sigma_v = \frac{Q}{Z^2\pi\left[1 + 2\left(\dfrac{r}{Z}\right)^2\right]^{3/2}} = \frac{Q}{Z^2}I_w \qquad (9\text{-}7b)$$

and values of I_w versus r/Z ratios can be developed as presented in Figure 9-6.

Illustration 9-3

For Boussinesq conditions, what subsurface stress will result at a point 3 m below where a 45 kN point load is applied?

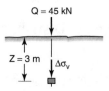

Figure 9-6 Values of I_B and I_w for calculating vertical stress resulting from surface load Q.

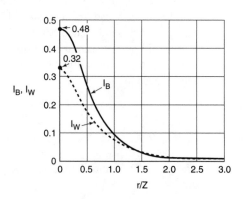

Solution

$$r = 0 \text{ ft}, \quad Z = 3 \text{ m}, \quad r/Z = 0$$

For $r/Z = 0$, obtain $I_B = 0.48$ (from Fig. 9-6) and $\Delta\sigma_v = \dfrac{Q}{Z^2}I_B$

$$= \frac{45 \text{ kN}}{(3 \text{ m} \times 3 \text{ m})}(.48) = 5 \text{ kN/m}^2 = 5 \text{ kPa}.$$

Illustration 9-4

For the Westergaard conditions, what subsurface stress will result 10 ft below and 10 ft horizontally from where a 10,000-lb concentrated load is applied?

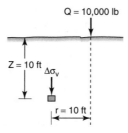

Solution

$$r = 10 \text{ ft}, \quad Z = 10 \text{ ft}, \quad r/Z = 1.0$$

For $r/Z = 1.0$, obtain $I_w = 0.065$ (from Fig. 9-6) and $\Delta\sigma_v = \dfrac{Q}{Z^2}I_w$

$$= \frac{10,000 \text{ lb}}{(10 \text{ ft} \times 10 \text{ ft})}(.065) = 6.5 \text{ psf}.$$

Application for Foundation Loading

In construction practice, the condition of a concentrated point loading is rarely encountered. More commonly, building loads are supported on foundations that cover a finite area (square, rectangular, or round footing), or the structure itself directly imposes loading over a finite area (earth structures such as dams and dikes). The subsurface stresses that result from loading acting over an area can be determined by integration of Equation 9-6 or 9-7, where the loading on infinitely small increments of the foundation area can be assumed as point loads. The subsurface stress at a point is the summation of the effects resulting from all of the applied point loadings. Such integrations have been accomplished for uniform loads acting on square, rectangular, strip, and circular areas, and also for uniformly varying loads such as those developing from the weight of an earth structure having a sloped cross section (dams and dikes, for instance). Results are available in the form of charts, tables, and graphs in order to provide generalized solutions. In these presentations, the subsurface stresses are expressed as a percentage of the foundation loading intensity. Measurements of the depth and horizontal distance at which a subsurface stress acts are expressed in terms of the dimensions of the loaded foundation area. Subsurface stress conditions indicated by Boussinesq and Westergaard equations for commonly occurring foundation loadings are presented in Figures 9-7 and 9-8. The following illustration problems serve to demonstrate use of these figures.

The stress curves in Figures 9-7 and 9-8 show that the stresses beneath the center of a loaded foundation area will be greater than stresses beneath the edge of the foundation un-

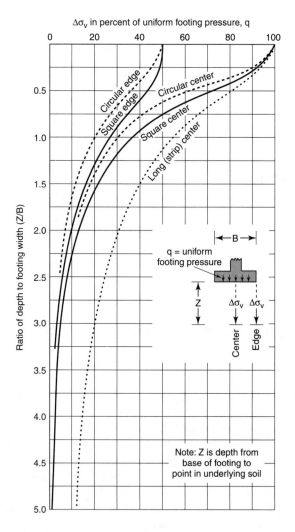

Figure 9-7 Variation of vertical stress beneath a foundation: Boussinesq analysis.

til a depth of about twice the foundation width is reached. Below this level, the stresses beneath the center and edge become practically equal. Consequently, in making determinations of the stresses resulting from a foundation loading, it is suitable to assume a concentrated point loading, if convenient, where the subsurface depth is greater than twice the foundation width.

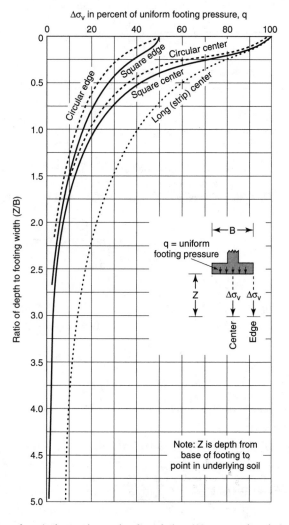

Figure 9-8   Variation of vertical stress beneath a foundation: Westergaard analysis.

Illustration 9-5

A foundation supported on the surface of a uniform, homogeneous soil is 2 m square and carries a loading of 500 kN. What subsurface stress increase occurs beneath the center of the foundation at a depth of 2 m?

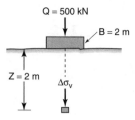

Solution

For uniform homogeneous soil, use Boussinesq conditions.

$$q = \frac{Q}{A} = \frac{500 \text{ kN}}{(2 \text{ m} \times 2 \text{ m})} = 125 \text{ kPa at base of foundation}$$

From Figure 9-7, $\Delta\sigma_v$ in terms of q is 34% for $\dfrac{\text{Depth}}{\text{Width}} = \dfrac{2 \text{ m}}{2 \text{ m}} = 1.0$.
Therefore

$$\Delta\sigma_v = .34(125 \text{ kPa}) = 42.5 \text{ kPa}$$

Illustration 9-6

A circular storage tank is supported on soils that satisfy the Westergaard assumptions. What subsurface stress increase develops 10 ft beneath the edge of the tank? The tank is 20 ft in diameter, and the stored fluid material causes a pressure of 1000 psf at the tank base.

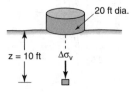

Solution

$$\text{Surface load } q = 1000 \text{ psf}$$

$$\text{Ratio of depth/width} = \frac{10 \text{ ft}}{20 \text{ ft}} = .5$$

From Figure 9-8, $\Delta\sigma_v = 23\%$ of q from "circular edge" curve. Therefore $\Delta\sigma_v = .23(1000 \text{ psf}) = 230 \text{ psf}$.

Sixty-degree Approximation

A method in wide use for making rough estimates of subsurface stresses resulting from a loaded foundation area is the so-called 60° approximation. In this method, it is assumed that the subsurface stresses spread out uniformly with depth, the stressed area increasing at a slope of 1 m or ft horizontally for each 2 m or ft of depth as measured from the edges of the

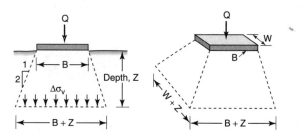

Figure 9-9   Method for approximating vertical stress increase resulting beneath a loaded foundation (60° approximation).

foundation. These assumed conditions are sketched in Figure 9-9. At a given depth, the subsurface stress $\Delta\sigma_v$ is assumed to be uniform over the area stressed; the method differs from the Boussinesq and Westergaard theories in this respect. The stress at a depth Z then becomes

$$\Delta\sigma_v = \frac{Q}{(B + Z)(W + Z)} \tag{9-8}$$

This approximation method is incorrect in representing subsurface stresses as being uniform across a plane area. In a homogeneous soil, the computed subsurface stresses directly beneath the foundation will be less than indicated by the Boussinesq analysis, whereas at distances beyond the edges of the foundation the computed stress will be greater. This method's best application may be estimating stress conditions in deep layers below a foundation and for determining an order of magnitude in a preliminary analysis.

Layered Soils Having Different Properties

The condition of soil deposits having uniform properties over a great vertical distance is not always found in practice. There frequently is encountered the situation where, within the zone (depth) that will be affected by structural loads, two or more layers of significantly different soils exist. For the condition of two different strata, the possibilities of relative properties include a firm upper layer overlying a soft layer, and the reverse.

For the condition where a firm layer overlies a soft layer (a dense sand above a compressible clay, or a surface zone of firm desiccated clay overlying the still-saturated, softer and compressible lower zone of the clay), the firm layer tends to bridge over the soft layer, spreading out the area over which stresses are transferred into the softer layer. This results in creating lower stresses in the soft material than would result for a homogeneous soil.

With a soft layer overlying a firm layer, the stresses reaching the deeper layer in the area directly beneath the foundation will be greater than those indicated for a uniform homogeneous deposit.

A diagram for determining how the variation in subsurface stress beneath a uniformly loaded circular area is affected by a two-layer soil mass appears in Figure 9-10. This diagram is for the special case where the thickness of the upper soil layer is equal to the radius of the loaded area.

The curves of Figure 9-10 can be used for approximating stress conditions beneath a square area as well as the circular area by assuming foundation dimensions giving the

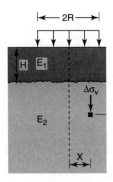

E = modulus of elasticity for soil

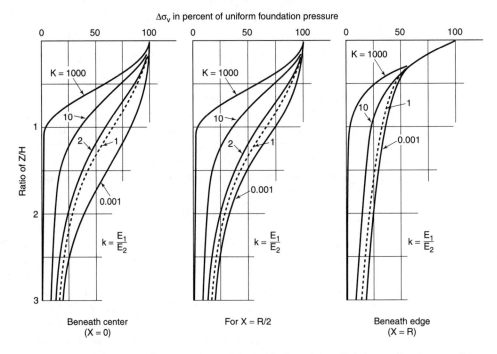

$\Delta\sigma_v$ in percent of uniform foundation pressure

Figure 9-10 Variation of vertical stress beneath a circular foundation. Two-layer subsurface condition for condition where $R = H$.

same area as the circular foundation. The illustration presented below indicates the use of Figure 9-10.

Illustration 9-7

A circular storage tank 10 m in diameter is located in an area where soil conditions are as shown by the sketch. If the tank pressure at the soil surface is 15 kPa, what stress results beneath the center of the tank at a depth of 7.5 m?

Solution

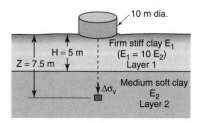

$$H = \text{radius} = 5 \text{ m}$$

$$\text{Ratio of } \frac{Z}{H} = \frac{7.5 \text{ m}}{5 \text{ m}} = 1.5$$

$$k = \frac{E_1}{E_2} = 10$$

From Figure 9-10, $\Delta\sigma_v$ is 20 percent of surface load q. Therefore, $\Delta\sigma_v = 0.20(15 \text{ kPa}) = 3 \text{ kPa}$.

Effect of Foundation Installation Below Finished Grade

In the practice of building design and construction, foundations are seldom placed directly on the ground surface. More usually, foundations are placed at some depth below the ground surface as protection against frost and seasonal soil volume changes, erosion, or other factors. In other words, some excavation is performed before the foundation can be constructed. The subsurface stresses developed in the soil below the foundation are determined by using the previously described methods, and the depth for computing the stress increase is measured from the base of the foundation.

The total stress at a point in the soil below the foundation is the sum of the stress caused by the soil overburden plus the stress due to the foundation loading. In many cases, a foundation excavation is backfilled to the ground surface or interior floor level after the foundation is constructed. For practical purposes, then, the stress caused by the soil remains equal to the original soil overburden.

If an excavation is made and not backfilled, the effect is to reduce the subsurface stresses. The reduction in stress can be computed by assuming that the excavation is a "foundation" whose loading is the weight of the excavated soil. The size of the "foundation" is the plan area of the excavation. The subsurface stress is determined as for a real foundation but is handled as a negative stress to be subtracted from the stress caused by the actual foundation that is ultimately installed in the excavation.

Illustration 9-8

A heavy materials storage building 30 ft by 30 ft is constructed with a basement that extends 6 ft below the ground surface. A foundation for an interior supporting column is constructed at floor level in the middle of the basement. The column footing is 4 ft square and carries a loading of 50 kips. What is the net stress change in the soil 5 ft below the middle of the basement floor? (For this example, neglect the weight of the concrete footing and floor.)

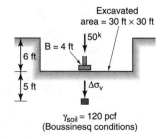

Excavated
area = 30 ft × 30 ft

50^k

B = 4 ft

6 ft

5 ft

$\Delta\sigma_v$

γ_{soil} = 120 pcf
(Boussinesq conditions)

Solution

A "negative" foundation loading results from the basement excavation. Negative load is 6 ft deep × 120 pcf = 720 psf. For a foundation 30 ft × 30 ft placed at a depth 5 ft below the floor,

$$\frac{\text{Depth}}{\text{Width}} = \frac{5 \text{ ft}}{30 \text{ ft}} = .17$$

and the stress decrease $\Delta\sigma_v$ is 95% × 720 psf = −685 psf (percentage as determined from Fig. 9-7). Stress increase from column footing, where

$$q = \frac{50 \text{ kips}}{4 \text{ ft} \times 4 \text{ ft}} = 3.14 \text{ ksf} = 3,140 \text{ psf}$$

is + $\Delta\sigma_v$ = (25%)(3,140 psf) = +780 psf, for the ratio of

$$\frac{\text{Depth}}{\text{Width}} = \frac{5 \text{ ft}}{4 \text{ ft}} = 1.25 \text{ (ref. Fig. 9-7)}.$$

Therefore, actual stress increase,

$$\Delta\sigma_{v_{net}} = +780 - 685 = 95 \text{ psf}$$

Effect of Changing the Surface Grade

On large construction sites, the surface grade may be changed to improve the topography and to obtain a surface more suitable for the needs of the project. If the surface grade is lowered, this results in a reduction in the subsurface stresses. Where earth fill is placed to raise the surface level, an increase in the subsurface stresses results. When a uniform thickness of material is placed or removed over a large area, the subsurface stress increase or decrease is the product of the unit weight of the soil added or removed and the thickness involved. This subsurface stress change is constant with depth; it does not diminish with increasing depth, as for foundation loadings.

Where the thickness of fill or soil removal is variable, as is frequently the case, a limited area can be analyzed and the average thickness of soil added or removed can then be used to calculate the subsurface stress change for the area.

Illustration 9-9

A compacted fill for a shopping center development is placed over an area where soil conditions are as shown by the sketch. What stress increase results in the middle of the clay layer from the weight of the fill?

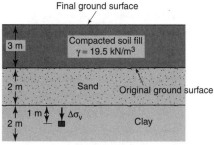

Solution

Since this is an areal fill (fill placed over a large area), the stress increase in the clay layer is

$$\Delta\sigma_v = (\gamma_{\text{fill}})(\text{height of fill})$$
$$= 19.5 \text{ kN/m}^3 \times 3 \text{ m} = 58.5 \text{ kPa}$$

Subsurface stresses caused by foundations subsequently installed in fill or cut areas are computed by using the methods already presented. It is important to remember that the final total subsurface stress must reflect the changes caused by filling or lowering the surface of the site as well as the foundation loading.

Illustration 9-10

Assume that a building foundation is constructed at a location where an 8 ft soil fill has been placed. If the footing is 5 ft square (5 ft by 5 ft) and carries a loading of 75 kips, what net stress results in the middle of the clay layer beneath the center of the footing for conditions indicated by the sketch? (Consider effects of fill *and* foundation loading.)

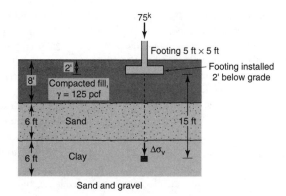

Solution

$$q_{footing} = \frac{75 \text{ kips}}{5 \text{ ft} \times 5 \text{ ft}} = 3 \text{ ksf} = 3000 \text{ psf}$$

$$\frac{\text{Depth}}{\text{Width}} = \frac{15 \text{ ft}}{5 \text{ ft}} = 3$$

Estimate stress increase by either Boussinesq or Westergaard, $\Delta\sigma_v \approx 5\% \times q$ (percentage determined from Fig. 9-7 or Fig. 9-8), therefore

$$\Delta\sigma_v = .05(3,000 \text{ psf}) = 150 \text{ psf from foundation}$$

Stress increase from fill $= 125 \text{ pcf} \times 8 \text{ ft} = 1000 \text{ psf}$

Total stress increase $= 150 + 1000 \text{ psf} = 1150 \text{ psf}$

PROBLEMS

9-1. Relating to soil deposits, provide a definition of subsurface effective stress and total stress, and explain the relationship between the two.

9-2. At a planned construction site, subsurface sampling indicates that the soil wet unit weight is 19.5 kN/m³.
 (a) Determine the effective vertical stress at a depth of 4 m if the water table is deep.
 (b) Determine the effective vertical stress if the groundwater table is at the ground surface.
 (c) Determine the effective vertical stress at the 4 m depth if the groundwater table is 2 m below the surface.

9-3. At a planned construction site, subsurface sampling indicates that the wet unit weight of the soil is 123 pcf.
 (a) Determine the effective vertical stress at the 12-ft depth if the water table is deep.
 (b) Determine the effective vertical stress and the neutral stress at the 12-ft depth if the water table rises to within 6 ft of the ground surface.
 (c) Determine the effective vertical stress at the 12-ft depth if the water table is at the ground surface. ,

9-4. At a lake location, the soil surface is 6 m under the water surface. Samples from the soil deposit underlying the lake indicate a wet unit weight of 18.8 kN/m³. Determine the effective vertical stress in the deposit at a depth 12 m below the soil surface. Also indicate the total vertical stress at the same point.

9-5. At an offshore location, the soil surface is 30 ft below the water surface. Weighing of soil samples obtained in the subsurface investigation indicates the saturated unit weight to be 118 pcf. Determine the effective vertical pressure at a depth 55 ft below the water surface.

9-6. Estimate the lateral earth pressure developed at a depth 5 m below the soil surface of a loose sand deposit (wet unit weight of soil is 17.5 kN/m³). Also estimate the lateral pressure for the same depth if the sand is in a dense condition (unit weight is 20.5 kN/m³). What would be the total lateral pressure for the loose sand condition described if the water table were at the ground surface?

9-7. (a) Determine the lateral earth pressure at a depth 20 ft (6.1 m) below the ground surface in a loose sand deposit. The wet unit weight of the sand is 115 pcf (19 kN/m³). What would the total lateral pressure be if the water table rose to the ground surface (in psf or kPa)?
 (b) What difference in lateral pressure is expected when the soil is in a dense condition (unit weight is 130 pcf or 20.4 kN/m³)?

9-8. A basement wall for a commercial building extends 3.1 m below the ground surface. Assume the soil unit weight is 18.5 kN/m³. For a point 3 m below the ground surface, compare the lateral soil pressure acting against the wall for the condition (a) when the soil is a sand–gravel material and (b) when the soil is clay.

9-9. A deep basement for a building is constructed in dense granular soil whose unit weight is 130 pcf (20.4 kN/m³). For a point 10 ft (3 m) below the ground surface, determine the increase in total lateral pressure that results by having the water table change from a location below the basement level to the ground surface (in psf and kPa).

9-10. A concrete tank structure is constructed below ground level for a large commercial swimming pool in an area where dense granular soil exists. The diving board end of the tank-pool is 6 m deep (below the ground surface). For the design, it is assumed that the groundwater table could rise close to the ground surface at various times. For the condition where the pool is empty, determine the total lateral pressures (due to soil and to water) acting against the exterior side of the end wall at points 2 m, 4 m, and 6m below the ground surface for the high water table condition.

9-11. A 200 kN concentrated (point) load acts on the surface of a soil mass. Determine the vertical stress 3 m below the ground surface at locations directly beneath the point of load application, 3 meters horizontally from the load, and 6 meters horizontally from the load.
(a) Assume the Boussinesq conditions apply.
(b) Assume the Westergaard conditions apply.

9-12. A 20-kip concentrated (point) load acts on a surface of a soil mass. Determine the vertical stress 10 ft below the ground surface at locations directly beneath the load, 10 ft horizontally from the load, and 20 ft horizontally from the load, for
(a) Boussinesq conditions
(b) Westergaard conditions

9-13. A 30 kN concentrated load is imposed at the surface of a granular soil deposit. Determine the subsurface vertical stress created at depths of 5 m and 10 m directly beneath the loading and at a horizontal distance 5 m from the line of application (5 m and 10 m depths) for
(a) Boussinesq conditions
(b) Westergaard conditions

9-14. A square footing foundation, 3 m by 3 m, and positioned on the ground surface of a soil deposit,

supports a column load of 1350 kN. Determine the vertical stress resulting from the foundation loading at a depth 3 m below the base of the footing for locations beneath the center and beneath the edge, assuming
(a) Boussinesq conditions apply.
(b) Westergaard conditions apply.

9-15. A square foundation, 4 ft by 4 ft, located on the surface of a soil mass, supports a column load of 96 kips. Determine the vertical stress resulting from the foundation loading at a point 6 ft below the ground surface for locations beneath the center of the footing and beneath the edge of the footing, assuming
(a) Boussinesq conditions apply.
(b) Westergaard conditions apply.

9-16. A 300 kN loading is supported on a 1.5 m by 1.5 m square foundation constructed on the ground surface of a homogeneous soil deposit (soil unit weight is 19 kN/m³). Determine the values of the preconstruction and the postconstruction vertical stress imposed on the soil at a point 2 m below the center of the foundation area (assume the Westergaard conditions apply).

9-17. A long strip-footing foundation will be used to support a masonry bearing wall for a commercial building. The footing will be 1.5 m wide and 45 m long. The wall loading imposed onto the foundation will be 200 kN per meter of wall length. Determine the vertical stress increase for points 1 m, 2 m, and 4 m below the footing, assuming
(a) Boussinesq conditions apply.
(b) Westergaard conditions apply.

9-18. A 4-ft-wide long (strip) footing carries a wall loading of 20,000 lb per foot of wall length. What vertical stress increase results below the center of the footing at depths of 4 ft, 8 ft, and 12 ft, assuming
(a) Boussinesq conditions apply?
(b) Westergaard conditions apply?

9-19. A 2 m by 2 m square foundation will be used to support an item of machinery weighing 800 kN. The foundation will be placed on the soil surface. Determine the values of preconstruction and postconstruction vertical stress in the soil zone 1.5 m below the base of the foundation, under the center, and under the edge.
(a) Assume Boussinesq conditions apply.
(b) Assume Westergaard conditions apply.

9-20. A 1-m-wide surface-supported strip footing carries a loading of 100 kN per meter of wall length. Determine the total vertical stress acting at a depth of 1 m below the center of the foundation width (note that the total stress will be the sum of the original vertical stress due to the soil mass plus the increase due to the foundation loading). Assume the Boussinesq conditions apply. Use a soil unit weight equal to 18 kN/m³.

9-21. A steel storage tank 20 m in diameter will be used to hold a liquid petroleum product. When filled, the tank causes a bearing pressure of 150 kPa. The soil underlying the tank has a unit weight of 18.5 kN/m³. For a depth 10 m below the base of the tank, compute the effective vertical stress when the tank is empty and when full, for points beneath the center and beneath the edge.
(a) Assume the Boussinesq conditions apply.
(b) Assume the Westergaard conditions apply.

9-22. An oil storage tank 50 ft in diameter imposes a maximum loading of 2500 psf onto the ground surface where the tank is supported. The soil underlying the tank has a unit weight of 125 pcf. For a point 25 ft below the ground surface, compute the effective vertical stress when the tank is empty and when the tank is full, at locations beneath the center and the edge of the tank. Assume that the Boussinesq conditions apply.

9-23. Compare the stress increase occurring 2 m below the center of a 3 m by 3 m square foundation imposing a bearing pressure of 145 kPa (145 kN/m²) when
(a) the Boussinesq stress distribution is assumed.
(b) the 60° approximation is assumed.

9-24. Compare the stress increase resulting 8 ft below the center of a 10-ft square foundation imposing 500 psf when the 60° approximation is assumed and when the Boussinesq conditions are assumed.

9-25. A 5-m-diameter tank supported on the surface of a soil deposit imposes a bearing pressure of 225 kPa (225 kN/m²). For a point 4 m below the tank base, compare the vertical stress increase due to the tank loading when
(a) the Westergaard conditions are assumed.
(b) the 60° approximation is assumed.

9-26. A circular foundation 12 ft in diameter imposes a pressure of 8000 psf onto the soil. At the 12-ft depth, determine the vertical stress increase beneath the center and the edge of the loaded area, assuming
(a) the Westergaard conditions apply.
(b) the 60° approximation

9-27. An 8-m-diameter storage tank is located on the ground surface of an area where a 4 m thickness of desiccated (firm) fine-grained soil is underlain by a very thick layer of similar fine-grained soil that retains a high water content and, as a result, is relatively weak and compressible. The modulus of elasticity for the firm upper zone is 10 times the modulus for the underlying layer.
(a) Determine the increase in vertical stress beneath the center of the tank at a depth of 6 m.
(b) Compare the result from (a) to the value of stress increase if homogeneous conditions (Westergaard) are assumed.

9-28. A 30-ft-diameter storage tank is supported on the ground surface at a site where a 15-ft-thick layer of dense sand overlies a very thick clay layer. The modulus of elasticity for the upper layer is ten times the modulus for the lower layer. The tank imposes a pressure of 3 ksf.
(a) Determine the increase in vertical stress below the center of the tank and at a depth of 30 ft.
(b) Determine the stress increase at the 30-ft depth if subsurface conditions were homogeneous (assume Boussinesq), and compare the results with (a).

9-29. The site for a commercial building project has the surface elevation lowered 2 m to improve area grading for building access, parking, and drainage. Foundations for interior building columns will be constructed at the level of the new ground surface. Each foundation is 1.5 m by 1.5 m square and imposes a bearing pressure of 200 kPa. What net stress increase results in the soil 1.25 m below the center of the foundation?

9-30. For a construction project, the ground surface at a building location is lowered 5 ft. A 4-ft square footing (4 ft by 4 ft) imposing a bearing pressure of 4000 psf is then constructed at the level of the new surface. What net stress increase results in the soil mass 4 ft below the center of the foundation? Assume that the soil unit weight is 120 pcf.

9-31. A 3 m thickness of compacted earth fill is placed across an area as preparation for constructing an industrial building. An item of heavy industrial equipment will be placed at a location within the building and supported on a foundation installed at the surface of the compacted soil fill. The equipment has a total weight of 5000 kN. The supporting foundation will be 5 m by 5 m. What net stress increase results beneath the center of the foundation in the original soil mass 2 m below the natural ground surface. Use 19 kN/m² for the soil unit weight.

9-32. For a construction project, 10 ft of compacted earth fill is placed in preparation for erecting a building. At one of the building's interior locations, a machine foundation 20 ft square will be installed at floor level to support equipment weighing 2000 kips. What net stress increase results below the center of the machine in the original soil mass, 5 ft under the natural ground surface? Assume a unit weight of 125 pcf for the compacted fill.

CHAPTER 10

Settlement

Soil Compression, Volume Distortion, Consolidation

The weight of any structure on the earth will result in stresses being imposed on the soils below the level of the base or foundation of that structure. The deformations that develop in the soil because of these stresses cause dimensional changes in the soil volume, with the result that the structure undergoes settlement. The extent of foundation settlement that will actually occur is related to the bearing pressures (stresses) imposed on the soils and the stress–strain properties of the soil.

10.1 BASIC CONSIDERATIONS

For the typical situation, loadings imposed by a structure at foundation level act in a downward vertical direction. The resulting stresses developed within the soil mass beneath the structure act vertically but in other directions as well. When the stresses that develop are well below the ultimate strength of the soil (a usual condition of foundation designs), settlement estimates of practical accuracy can be obtained by considering only the vertical stresses.

Load-induced settlement of foundations is due to volume distortion in the foundation soil caused by shear strains (volume distortions occurring without change in volume) and by soil compression resulting from decrease in soil void ratio (soil particles are forced into a closer arrangement). Volume distortion tends to occur upon application of loading and results in an almost immediate initial settlement. Primary soil compression occurs at the rate in which foundation stresses are transferred onto the soil particles (or the soil skeleton). Primary compression occurs almost instantly in coarse-grained soil, but is a time-related condition in fine-grained soil deposits having a high degree of saturation because of the lag period associated with pore water escaping (or being squeezed) from the soil void spaces; such time-related compression is termed *consolidation compression*. An additional time-related compression, termed *secondary compression,* is volume decrease attributed to creep or rearrangement of soil particle positions when a condition of constant stress exists over an extended period.

Total foundation settlement results from the combination of immediate settlement plus primary (or consolidation) compression settlement plus secondary compression settlement, or

$$S_t = S_{vd} + S_{pc} + S_{sc} \tag{10-1}$$

where S_t = total foundation settlement
S_{vd} = settlement due to volume distortion
S_{pc} = settlement due to primary compression
S_{sc} = settlement due to secondary compression

For foundations bearing on coarse-grained soils free of organic content (sands, sand mixed with silt or gravel), most of an expected total settlement occurs upon application of load. The settlement due to volume distortion and primary compression is immediate; the time-related transfer of foundation stress onto the soil skeleton—that is, primary (consolidation) compression—occurs rapidly even in fully saturated coarse-grained deposits because of the high coefficient of permeability (or hydraulic conductivity) and rapid draining characteristics. Secondary compression or creep effects typically are small.

For foundations supported on fully saturated and almost-saturated fine-grained soil deposits free of organic content (the soils possessing cohesion such as clays and silt–clay mixtures), some volume distortion (immediate) settlement occurs, but primary (consolidation) compression settlement which extends over a period of time is expected to be the cause of most of the total settlement. A major exception is where foundations bear on firm soil (high shear strength) and the foundation pressures do not exceed the preconsolidation stress for the soil (preconsolidation pressure is discussed in Section 10.2); for this situation, immediate settlements may be on the order of half the total settlement. Long-term secondary compression settlement also takes place with fine-grained soil deposits, but typically is limited for soils free of organic matter.

Where foundations bear on soils with significant organic content, the immediate settlement, primary (consolidation) settlement, and secondary compression settlement can each be significant. Typically, soils containing organic material are expected to undergo greater total settlement that nonorganic soil deposits.

Primary (consolidation) compression and secondary compression may begin when foundation loads are first imposed onto a soil deposit, occurring concurrently with the processes responsible for immediate or initial settlement.

For determining volume distortion settlements, calculation methods based on elastic theory and knowledge of soil properties are in use. Primary and secondary compression properties of fine-grained soils, relating to volume changes occurring in the soil deposit as void spaces between particles decrease when the particles rearrange to develop resistance to new (foundation) loading, are typically determined from laboratory compression tests on undisturbed soil samples. Laboratory compression tests are seldom performed on cohesionless soils, relating to the practical difficulty associated with obtaining truly undisturbed samples of granular soil from the construction site. Also, because the rate of settlement for granular soils is rapid (so that settlements often occur during the construction period), there may not be the problems of drawn-out postconstruction settlement as occur where the fine-grained cohesive soils are present; an exception for coarse-grained soil deposits is where much of the anticipated loading is from postconstruction activity, such as live loads and repetitive loads. Where compression characteristics of a granular soil are needed, the information is usually obtained indirectly, such as from an in-place density determination, relative density testing, or correlations from the results of procedures utilized in subsurface investigations such as penetration resistance or soil sampling in borings.

Table 10-1 provides a summary of the relationship between foundation settlements and soil conditions.

In practice, determination of the settlement characteristics of a soil deposit is frequently attempted by one of, or combination of, several methods:

1. Working backwards from observations of the behavior of structures in the area near the planned new building site.

2. Prior to construction, performing large-scale field load tests at the actual building site.

3. Performing laboratory tests (such as compression tests) on soil samples obtained from borings or test pits made at the planned construction site.

4. Estimating compressibility or volume change characteristics on the basis of index property tests performed on soil samples from borings or test pits. (Such tests typically include classification, moisture density determinations, liquid limit and plastic limit determinations, and relative density determinations.)

5. Using data from field exploration results (such as standard penetration tests used to obtain boring samples and cone penetrometer tests) that have been correlated empirically to design values for allowable foundation bearing pressures and foundation settlements.

6. Performing in situ tests in the field (in drilled boreholes, etc.) that relate to soil compressibility or expected settlement for a planned foundation.

Methods 1 and 2 have inherent shortcomings. With method 1, it can be difficult to accurately ascertain the actual extent of settlement that an existing structure has experienced, and, perhaps more importantly, soil conditions may not be identical at different sites. However, checking the behavior or performance of existing structures in an area is good practice, provided that the procedures discussed hereafter are also followed. For method 2, field load tests seldom truly represent the loading conditions that will be developed by a structure, and the resulting settlement or compression data may be seriously misleading. Data from field load tests may be grossly in error when cohesive soil strata underlie the test area (relating to the time lag for consolidation settlement). Full-scale load tests are relatively expensive to perform.

Evaluation of a soil's compressibility characteristics can be accurately determined from laboratory tests if the tested samples represent the natural or undisturbed condition (method 3). With soil borings or other form of subsurface investigation being an accepted necessity for construction projects, it becomes practical to also obtain soil samples for purposes of performing laboratory tests. Compression tests can provide accurate information (method 3), but index property tests (method 4) may also provide adequate information (particularly if great accuracy is not required) and are quicker and less costly to perform.

Correlating foundation settlement to soil conditions data obtained from cone penetrometer results and standard penetration tests in soil borings (method 5) represents a practical approach to designing foundations, since borings and penetrometer tests are typically included in the subsurface investigation phase of planning a project. This procedure also represents a necessary approach for conditions where undisturbed soil samples suitable for strength-deformation tests cannot be obtained (i.e., it is considered virtually impossible to

Table 10-1 Summary: Type and Rate of Foundation Settlement that Occurs, Related to Soil Classification

Soil Type	S_{wd}	S_{pc}	S_{sc}	Additional Comments
Granular or coarse-grained soils (sands, sands with limited silt)	Yes. Occurs quickly (upon application of load or soon after)	Yes. Occurs quickly (upon application of load or soon after)	Yes. Magnitude typically is small, but foundations subjected to repetitive or fluctuating loads are expected to experience greater time-related settlement than foundations supporting static loading.	Most of the expected total settlement occurs quickly upon application of loading. For many types of structures, settlement may be almost completed during the construction period. But structures subjected to repetitive or fluctuating loads will experience postconstruction settlement. In overconsolidated or prestressed soil deposits, total settlement expected is considerably less than where normally compressed soil exists.
Cohesive or fine-grained soils (clay, silt–clay mixtures)	Yes. Occurs quickly but for typical soils is expected to be a small part of total settlement.	Yes. Occurs gradually, relating to consolidation properties of the soil deposit.	Yes. Typically expected to be a small part of total settlement.	In overconsolidated soils, settlement expected is considerably less than for normally consolidated and underconsolidated soil deposits.
Organic soils (soils including high organic content, swamp deposits, peats, etc.)	Probably yes. Expected to occur quickly.	Probably yes. Expected to occur quickly.	Yes. Can be considerable and extend over very long period.	Highly organic and highly fibrous materials typically have very low shear strength and very low bearing capacity.

obtain samples of cohesionless soil without affecting the soil particle structure which influences settlement properties).

In situ field testing procedures to directly determine the volume change characteristics of a soil zone, such as accomplished with the pressuremeter apparatus (discussed in Chapter 13) and flat-blade dilatometer (Chapter 5), are not yet in widespread use.

Procedures for calculating the settlement of foundations supported on sand deposits and those supported on clay deposits are discussed separately in the following sections. If the subsurface conditions consist of a cohesionless soil stratum underlain by a cohesive layer, it may be necessary to calculate the settlement of the granular layer and of the clay layer separately then sum these separate values to obtain the total expected settlement.

10.2 SETTLEMENT OF FOUNDATIONS ON SAND

The settlement of foundations supported on sand relates to volume distortion and primary compression occurring in the stressed soil zones, and is expected to occur rapidly after application of loading (almost instantaneously). For construction where the major loading is from dead load due to the weight or mass of the structure, such settlement commonly is completed within the timeframe of the construction period, with the result that postconstruction settlement tends to be very limited. Conversely, the structures whose function involves significant live loading will experience postconstruction settlement. Limited documentation [43] further indicates that settlement attributed to the secondary compression or creep properties of the sand deposit should be anticipated where the live loading is fluctuating or variable, such as from wind effects and vibration.

Schmertmann Method

Immediate settlement of cohesionless soil will be due to the combined effects of volume distortion and primary compression. The process of determining soil properties related to these effects through the use of laboratory testing has been hampered by the practical difficulty associated with obtaining undisturbed samples; important physical relationships such as soil particle orientations, effect of past compressive loadings and degree of compaction, and influence of in-ground confinement are altered by the sample-taking procedure. However, equations from elastic theory and model studies indicate distributions of vertical strains in the soil zone stressed by a foundation loading that are similar (Fig. 10-1) and provide the basis for obtaining settlement information that is in agreement with field observations.

The Schmertmann method [256,258], utilizing empirical relationships between subsurface investigation data and soil properties to approximate the pattern for vertical strains in the stressed soil zone, offers a procedure to calculate settlement resulting from the combined effects of volume distortion and compression in sand deposits that have not been precompressed (previously loaded). Strain influence factors for the Schmertmann method are shown in Figure 10-2.

From elastic theory, the expression for vertical strain, ϵ_v, at a point in the stressed soil zone is

$$\epsilon_v = \frac{\Delta q}{E_s} I_v \tag{10-2}$$

Figure 10-1 Vertical strain in sands below foundations (representative results: theory and model studies).

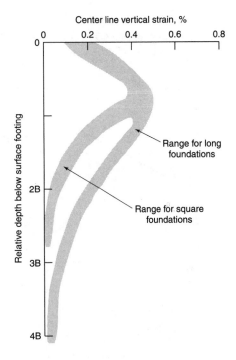

where Δq = net foundation bearing pressure imposed onto soil

E_s = sand modulus of elasticity (also referred to as Young's Modulus)

I_v = strain influence factor for the soil zone beneath the foundation

The value for settlement is obtained by summing the strains occurring throughout the depth being stressed by the foundation loading. Since the values of Δq, E_s, and I_v are expected to vary with depth below the base of the foundation area, the calculation procedure to determine settlement involves subdividing the stressed zone into finite sublayers and assigning a representative or average value of Δq, E_s, and I_v to each sublayer. The settlement for each sublayer is calculated and the results summed to obtain total settlement, or $\Delta s_1 + \Delta s_2 + \Delta s_3 + \ldots$, where Δs values are the settlements of the respective sublayers.

The settlement for any sublayer Δs_n is determined as the product of the average strain value within the layer, $\varepsilon_{v_{avg}}$, and the related sublayer thickness, Δz_n, or

$$\Delta S_n = (\varepsilon_{v_{avg}})(\Delta z_n) = \Delta q(I_v/E_s)(\Delta z_n) \tag{10-3}$$

The strain influence factors, I_v, can be approximated from the diagram of Figure 10-2. A correction factor C_1 is applied to compensate for the effect of foundation depth (or embedment), where

$$C_1 = 1 - .5(\sigma_{vo}/\Delta q) \tag{10-4}$$

where $\Delta\sigma_{vo}$ = soil overburden pressure at the level of the foundation base (corresponds to q_{vo})

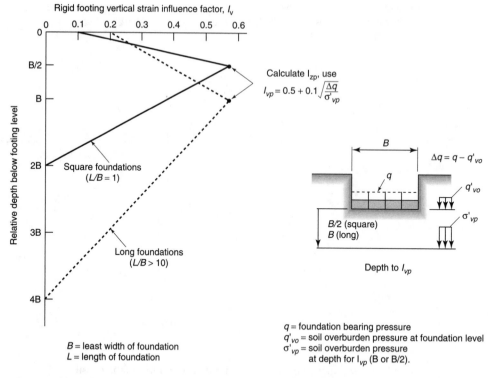

Figure 10-2 Strain influence factors for use with Schmertmann method to estimate settlement of foundations on sand.

The minimum value assigned to C_1 is 0.5. A second correction factor C_2 can be applied to consider effects of creep and other long-term compression factors for sand deposits, which include thin zones of fine-grained soil (but a factor of unity is suggested where it is known that the sand deposit is free of thin fine-grained soil layers). Where appropriate, C_2 is determined from

$$C_2 = 1 + 0.2 \, (\log 10t)$$

where t = elapsed time in years

Total settlement S_t then becomes

$$S_t = C_1 C_2 \sum (\varepsilon_{v_{avg}})(\Delta z_n)$$

$$= C_1 C_2 \sum (\Delta q)(I_v / E_s)(\Delta z_n) \tag{10-5}$$

Relating to the impractical aspects associated with obtaining values of E_s through the testing of soil samples, correlations have been developed to relate this soil modulus to cone penetrometer resistance results, q_c, determined as part of the subsurface investigation for a construction site. Values suggested for E_s extend between $1.5q_c$ and $4q_c$; values from the lower range are appropriate where sand and silt mixtures underlie square-shaped foundations, while values from the higher range are applicable where dense sand or sand–gravel mixtures underlie long foundations.

Subsurface investigations frequently consist of drilled borings where the soil sampling involves using the standard penetration test (SPT) procedure and where the blow count N associated with obtaining each sample is recorded. The SPT method (a dynamic method involving use of a drop hammer procedure) is not considered an appropriate method for determining values of the soil modulus E_s. But for sites where cone penetrometer data are not available, crude correlations between the SPT blow count N and the cone penetrometer resistance q_c provide estimations of E_s for use in settlement studies. Results considered representative of the correlations between q_c and N are shown in Figure 10-3.

A summary of the relationships between q_c, the SPT blow count N, and the soil modulus E_s is presented in Table 10-2.

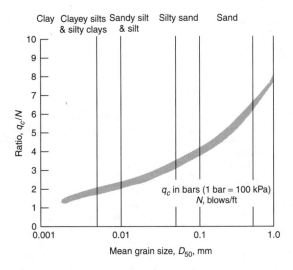

Figure 10-3   Ratio q_c/N related to mean grain size. (Range based on standard penetration test N, resulting from typical field procedure having approximately 60 percent efficiency.) [245]

Table 10-2   Values of E_s Related to Soil Type

| | Approximate Value for E_s (kgf/cm², 0.1 MPa, ton/ft² | |
Soil Type	In terms of N	In terms of q_c
Sand–silt mixture	$4N$	$1.5q_c$
Fine to medium sands,	$7N–10N$	$2q_c–3q_c$
fine–medium–coarse sands	(relating to density and compactness)	(relating to density and compactness)
Sand–gravel mixtures	$12N$	$4q_c$

Illustration 10-1

Using the Schmertmann method, calculate the settlement for a foundation being supported on a sand stratum, given the following conditions:

Square foundation, 3 m by 3 m; foundation bearing pressure $q = 165$ kPa; foundation installation depth $D_f = 1.5$ m; soil unit weight $\gamma = 17.5$ kN/m³

Solution
Net foundation pressure Δq:

$$\Delta q = 165 \text{ kPa} - (17.5 \text{ kN/m}^3 \times 1.5 \text{ m}) = 138.75 \text{ kPa}$$

Vertical pressure at depth $B/2$ below foundation σ_{vp} (prior to foundation construction)

$$\sigma_{vp} = (17.5 \text{ kN/m}^3 \times 1.5 \text{ m}) + \left(17.5 \text{ kN/m}^3 \times \frac{B}{2}\right) = 52.5 \text{ kN/m}^2 = 52 \text{ kPa}$$

Maximum strain influence factor I_{vp} (refer to Fig. 10-2):

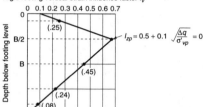

Rigid footing vertical strain influence factor I_v

$$I_{vp} = .5 + .1\sqrt{\Delta q/\sigma_{vp}} = .5 + .1\sqrt{138.75/52.5} = .66$$

$$C_1 \text{ factor} = 1 - .5\left(\frac{\sigma_{vo}}{\Delta q}\right) = 1 - .5\left(\frac{17.5 \times 1.5}{138.75}\right) = .905$$

$$C_2 \text{ factor} = 1.0$$

The value for $(I_v/E_s)\Delta z$ is obtained using the tabulation procedure following. Table values shown are for a site where cone penetrometer resistance data are available.

Layer Identification and Soil Type (below fdtn level)	Layer Thickness Δz (m) (below fdtn level)	Distance to Center of Layer (m) (below fdtn level)	q_c (or N)[a] (kgf/cm²)	I_v[b]	E_s (MPa)	$(I_v/E_s)\Delta z$ (m/MPa)
(1) f-c sand	1.0	0.5 (or 0.17B)	30 ≈ 3 MPa	0.25	2.5q_c = 7.5	0.033
(2) f-m sand	1.5	1.75 (or 0.58B)	25 ≈ 2.5 MPa	0.66	2.5q_c = 6.25	0.158
(3) f-m sand	1.0	3.0 (or 1B)	27.5 ≈ 2.75 MPa	0.45	2.5q_c = 6.88	0.065
(4) f-c sand	1.5	4.25 (or 1.42B)	35 ≈ 3.5 MPa	0.24	2.5q_c = 8.75	0.041
(5) f-c sand	1.0	5.5 (or 1.83B)	30 ≈ 3.0 MPa	0.08	2.5q_c = 7.5	0.011
						$\Sigma=0.31$

[a]Use average value for the layer thickness.
[b]Use I_v values to depth equal to 2B for square foundation (ref. Fig. 10-2).

Solution
$$\text{Settlement, } S_t = (C_1 C_2)\Delta q \sum (I_v/E_s)\Delta z$$

$$= (.905 \times 1)(138.75 \text{ kPa})(0.31 \text{ m/MPa})$$

$$= 0.039 \text{ m} = 39 \text{ mm}$$

The Schmertmann procedure applies to normally-loaded sand deposits and will overestimate foundation settlement if the sand has been precompressed or densified by construction activity such as compaction. Where geologic or other information indicates pre-compressed sands, the settlement to expect can be estimated as roughly half of the value calculated by the Schmertmann method.

Approximations for Estimating Settlement of Foundation on Sand

The planning associated with construction projects may require that estimates of foundation settlements be obtained before subsurface conditions and soil properties are fully identified.

Burland Data.   The following information permits a probable order-of-magnitude settlement to be calculated. The data represented by Figure 10-4, developed from a large assortment of case study observations [42], can be used to perform rapid calculations to indicate a probable magnitude of settlement. If the soil conditions are identified and uniform, the ratio of settlement to applied pressure can be taken from the range between one-half and three-fourths of the upper limit value indicated by Figure 10-4.

Illustration 10-2

The preliminary design for a structure indicates that a foundation having a width of 3 m would impose a soil bearing pressure of 250 kPa (or 250 kN/m²). Site data from maps and preliminary field studies indicate that the construction area is underlain by a uniform deposit of dense sand. Estimate the foundation settlement using Figure 10-4.

Solution
For B equal to 3 m, select an s/q ratio equal to 0.04 (from Figure 10-4). For $s/q = 0.04$, obtain settlement:

$$S = .05\,(q) = .05(250 \text{ kN/m}^2)\left(\frac{3}{4}\right) =$$

$$\approx 10 \text{ mm}\left(\text{or less than } \frac{1}{2} \text{ in.}\right)$$

Subgrade Reaction Method.   In uniform, cohesionless soil deposits, the soil rigidity (or modulus of elasticity) is expected to increase with depth because the overburden pressure and confining pressure, which affect the rigidity, increase with depth. For the condition where the increase in rigidity is uniform with depth, the following expression can be used to estimate settlements, provided that the *depth* of the foundation below the ground surface is less than the width of the foundation and that the foundation width is not greater than about 6 m (20 ft):

$$\text{Settlement, } S = \frac{4qB^2}{K_v(B + 1)^2} \tag{10-6}$$

where $B = $ width of foundation in meters or feet
$\quad q = $ pressure imposed by the foundation in kN/m², kPa, or kips/ft²
$\quad K_v = $ modulus of vertical subgrade reaction for 0.3 m square plate (1 ft square) bearing on the ground surface, in kN/m³ or kips/ft³

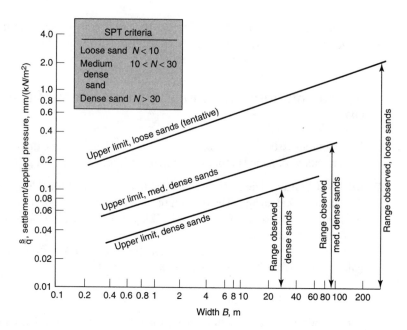

Figure 10-4 Data for estimating settlement of foundations on sands. Ranges of s/q related to foundation width as derived from case study observations (after Burland [42]).

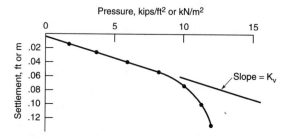

Figure 10-5 Plot of load settlement data from a plate-bearing test, to determine modulus of vertical subgrade reaction.

The modulus of vertical subgrade reaction K_v can be determined directly from a plate-bearing test performed in the field at the planned location for the structure. In this test, loads are applied to the plate in increments, and the plate settlement is recorded for each load. The results are plotted as shown in Figure 10-5.

The tangent to the initial straight-line portion of the plot is K_v, the modulus of vertical subgrade reaction. Mathematically,

$$K_v = \frac{\text{Pressure}}{\text{Settlement}} = \frac{\text{kN}}{\text{m}^3} \text{ or } \frac{\text{kips}}{\text{ft}^3}$$

Field plate-bearing tests are time consuming and costly. For purposes of estimating settlements, representative values of K_v that have been correlated with other soil properties can be used; see Table 10-3.

Table 10-3 Typical Values of K_v for Differing Sand Densities

Condition	Relative Density, %	Representative Values of Dry Unit Weight		Values of K_v[a]	
		pcf	kN/m³	kips/ft³	kN/m³
Loose	<35	<90	<14	100	15×10^3
Medium dense	35–65	90–110	14–17	150–300	$(25$–$50) \times 10^3$
Dense	65–85	110–130	17–20	350–550	$(55$–$85) \times 10^3$
Very dense	>85	>130	>20	600–700	$(95$–$110) \times 10^3$

[a]For the condition where the groundwater table is at a depth greater than 1.5 B. If the water table is at the base of the foundation, use $\frac{1}{2}K_v$. Interpolate for intermediate locations of the water table.

Illustration 10-3

A foundation footing is to support a total column loading of 2250 kN. Half of this will be live load. The building site is underlain by a thick stratum of sand. Tests on soil samples obtained from the site indicate dry unit weights for the sand between 15 and 17 kN/m³. The design calls for the footing to be located 1 m below the ground floor level of the building. Preliminary design data indicate that a soil bearing pressure of 250 kN/m² can be used. Estimate the foundation settlement.

Solution

Assume a square footing shape. The required footing area becomes (2250 kN)/(250 kN/m²) = 9 m² and for a square footing, $B = \sqrt{9 \text{ m}^2} = 3$ m. From Table 10-3, estimate a value for K_v; select $K_v = 45 \times 10^3$ kN/m³. Estimated settlement, S, is:

$$S = \frac{4qB^2}{K_v(B + 1)^2} = \frac{(4)(250 \text{ kN/m}^2)(3 \text{ m} \times 3 \text{ m})}{(45 \times 10^3 \text{ kN/m}^3)(4 \text{ m} \times 4 \text{ m})} = .012 \text{ m} = 12 \text{ mm}$$

The expected postconstruction settlement (settlement due to live loading only) would be in the ratio of the live load to the total load. Since half of the total load is live load, the postconstruction settlement will be about ½ (12 mm) = 6 m (or less than ¼ in.).

10.3 SETTLEMENT OF FOUNDATIONS ON CLAY SOILS

The total settlement expected for foundations bearing on clay soils is the sum of volume distortion (immediate) settlement plus primary compression (consolidation) settlement plus secondary compression settlement, or $S_t = S_{vd} + S_{pc} + S_{sc}$.

Volume Distortion Settlement for Clays

The settlement due to soil volume distortion, S_{vd}, is typically a small proportion of the total settlement where foundations are supported on cohesive soil deposits. However, for firm, stiff clays (soil possessing high shear strength), the initial settlement may approach half the total settlement if the foundation pressures do not exceed the clay's preconsolidation stress (but total settlement for this situation is usually small). Expressions from the linear theory of elasticity are in use for estimating the magnitude of volume distortion settlement; Equation 10-7 below is based on a foundation bearing on a cohesive soil deposit possessing homogeneous and isotropic properties, and of infinite horizontal extent. The expression provides reliable results for static loading conditions when the foundation stresses are low, well within the limits of the ultimate soil bearing capacity (a condition satisfied where the factor of safety applied to the ultimate soil bearing capacity described in Chapter 13 is on the order of 3, a typical value).

$$S_{vd} = C_s q B \left(\frac{1 - v^2}{E_u} \right) \tag{10-7}$$

where C_s = shape and foundation rigidity factor

q = magnitude of equivalent distributed load acting on the foundation area (total load/foundation area)

B = plan width or diameter of the foundation

E_u = undrained elastic modulus clay (Young's Modulus or modulus of elasticity)

v = Poisson's ratio for the applied stress range (assume 0.5 for saturated clays, slightly less for partially saturated)

The shape factors, C_s, are based on the design function of the foundation, rigid or flexible, because foundation bearing pressure distributions vary, as shown by Figure 10-6. Tables 10-4 and 10-5 present values, based on rigid or flexible category, shape, and position beneath the footing, where the settlement calculation applies.

The elastic modulus, E_u, can be determined from the results of undrained triaxial compression tests performed on undisturbed soil samples, as indicated by Figure 10-7 (also refer to Chapter 11). Values for E_u frequently lie in the range between 500 c_u and 1500 c_u, where c_u is the soil cohesion shear strength as determined from undrained tests. The lower range is for clays of high plasticity and where foundation loads are large, while the higher range is for clays of low plasticity and where foundation loading is low. An estimated value for E_u, based on the consistency of the soil, can be selected from Table 10-6.

Illustration 10-4

Calculate the immediate settlement (due to volume distortion) expected beneath the center of a rigid rectangular-shaped foundation bearing on a deep clay deposit for the following conditions:

Foundation length = 6 m

Foundation width = 3 m

L/B ratio = 2

Foundation bearing intensity, q = 200 kPa

Clay cohesion (shear strength), c = 90 kPa

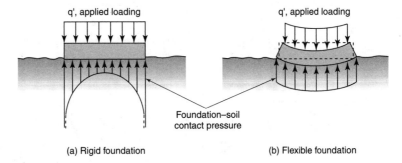

Figure 10-6 Variation of foundation–soil contact pressures for rigid and flexible foundations on cohesive soil, and related settlement pattern.

(a) Rigid foundation

(b) Flexible foundation

q', applied loading

q', applied loading

Foundation–soil contact pressure

Table 10-4 Values of C_s for Foundations on Clay Soil of Infinite Depth[a]

Shape	Center	Corner	Edge at Middle of Long Side	Average
Flexible foundation:				
Circular	1.00	—	0.64	0.85
Square	1.12	0.56	0.76	0.95
Rectangular:				
$L/B = 2$	1.53	0.76	1.12	1.30
$L/B = 5$	2.10	1.05	1.68	1.82
$L/B = 10$	2.56	1.28	2.10	2.24
Rigid foundation:				
Circular	0.79	—	0.79	0.79
Square	0.82	0.82	0.82	0.82
Rectangular:				
$L/B = 2$	1.12	1.12	1.12	1.12
$L/B = 5$	1.60	1.60	1.60	1.60
$L/B = 10$	2.00	2.00	2.00	2.00

[a]Soil depth extends greater than 10B.

Table 10-5 Values of C_s for Foundation on Clay Soil of Limited Depth (D) above a Rigid Substratum (Rock)

C_s Under Center of Rigid Circular Foundation	Depth to Width Ratio (D/B)	C_s, Under Corner of Flexible Rectangular Foundation[a]				
		$L/B = 1$	$L/B = 2$	$L/B = 5$	$L/B = 10$	$L/B = \infty$
0.35	1	0.15	0.12	0.10	0.04	0.04
0.54	2	0.29	0.29	0.27	0.26	0.26
0.69	5	0.44	0.52	0.55	0.54	0.52
0.74	10	0.48	0.64	0.76	0.77	0.73
0.79	∞	0.56	0.76	1.05	1.28	—

[a]To determine C_s for center of a foundation area, divide foundation shape into four equal subrectangles, then assign the B dimension based on the size of one of the subrectangles. Multiply the selected value of C_s by 4 to use with Equation 10-7.

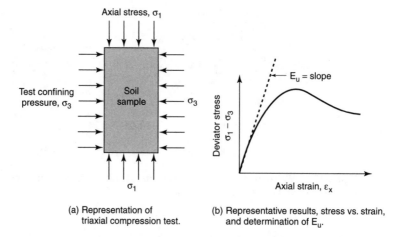

Figure 10-7 Method used to determine E_u from triaxial compression tests.

Table 10-6 Values of E_u Related to Soil Consistency

Clay Consistency	Cohesion (Shear Strength)	Values for E_u MPa	ksf
Soft	<24 kPa (<500 psf)	2.5–15	50–300
Medium to stiff	25–100 kPa (500 to 2000 psf)	15–50	300–1000
Very stiff to hard	>100 kPa (>2000 psf)	50–200	1000–4000

Solution

Estimated value for elastic modulus, $E_u = 45$ MPa

Use Poisson's ratio, $v = 0.5$

Shape and foundation rigidity factor, $C_s = 1.12$

$$S_{vd} = C_s qB\left(\frac{1 - v^2}{E_u}\right)$$

$$= (1.12)(200 \text{ kPa})(3 \text{ m})\left(\frac{1 - .5^2}{45 \text{ MPa}}\right)$$

$$= (1.12)(200 \text{ kPa})(3 \text{ m})\left(\frac{.75}{45 \times 10^3 \text{ kPa}}\right) = .011\text{m} = 11\text{mm}$$

Primary Compression Settlement for Clays

In primary compression, decreases in soil volume are attributed principally to decreases in void spaces between soil particles as the particles rearrange so as to develop resistance to the new external loading; practically, there is no decrease in the actual volume of soil particles. The term *compressibility* is used to indicate one-dimensional volume changes that occur in a soil deposit when compressive loading is applied and the particles rearrange as a reaction. Although applicable to all soil types, the term is more typically associated with the study of fine-grained, cohesive soils (clay, clay–silt mixtures). The compression properties of a fine-grained soil can be determined directly by performing a laboratory compression test, frequently called a *consolidation test.* In this test, an undistorted sample is fit into the ring of a consolidometer apparatus so that the sample is confined against lateral displacement, and compressive loading is imposed on the soil. [Basically, the consolidometer apparatus consists of a heavy brass metal ring to hold the soil sample being tested, and includes porous stone discs to cover the top surface and bottom surface of the sample to enable pore water to escape from the soil as compression is taking place (ref. Fig. 10-8); an extensometer or micrometer dial gage is used to measure changes in soil sample thickness.] For known magnitudes of load, the amount of compression and also the time required for compression to occur are recorded. The test is usually performed by imposing a series of increasing compressive loadings and determining time-rate-of-compression data for each increment of loading. The entire body of data permits the compressive stress–strain characteristics of the soil to be determined (discussed more fully later in this chapter).

The speed or rate of time that is required for volume changes to occur differs significantly for the coarse-grained (cohesionless) soils and the fine-grained (cohesive) soils. The cohesionless soils experience compression relatively quickly—frequently instantaneously—after loading is imposed. Conversely, the fully saturated and almost fully saturated clay soils (common conditions for many naturally occurring clay deposits) generally require a significant period before full compression under an applied loading results. Relating compression with the *time period* necessary for the compression to occur (that is, the time rate of compression) includes the process termed *consolidation.* Consolidation and its effects are discussed in Section 10.4.

Presentation and Analysis of Laboratory Compression Test Data. Compression test data are presented in any of several ways, usually depending on the preferences and experiences of the individual using the data. Because compression is due to changes in the void spaces in the soil, methods in common use frequently indicate compression as a change in the void ratio. Using arithmetic coordinates, a typical test result for change in void ratio versus the increase in loading pressure is as shown in Figure 10-9. The slope of the curve at any point is a_v, the *coefficient of compressibility.* Mathematically, $a_v = de/d\overline{\sigma}_v$. The dimensional units for a_v are square feet per pound (or kip or ton) or square meter per kilonewton. Because of the constantly changing slope of the curve, it is somewhat difficult to use a_v in a mathematical analysis, as is desired in order to make settlement calculations. When semilog coordinates are used and the void ratio is plotted versus the logarithm of pressure, the data will plot approximately as a straight line (or, as described later, a series of straight lines) (Fig. 10-10). In this form, the test data are more adaptable to analytical

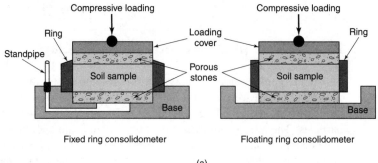

Fixed ring consolidometer   Floating ring consolidometer

(a)

(b)

Figure 10-8 (a) Schematic representation of conventional types of laboratory consolidometers; (b) typical laboratory consolidation equipment.

Figure 10-9 Presentation of compression test data on arithmetic coordinates.

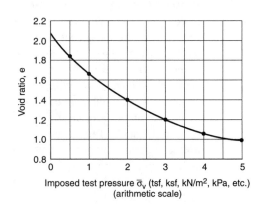

use. For all plots, the imposed compressive test pressure is the intergranular or effective pressure (stress), $\overline{\sigma}_v$.

Another method of presentation shows unit change in sample thickness (or strain) versus the logarithm of pressure (Fig. 10-11). This method gives essentially the same results as the method of Figure 10-10, but it has the advantage of minimizing the amount of work associated with reducing the test data. It may also make for simpler settlement calculations. With this method, no conversion to void ratio is necessary. The compression data for the sample are converted to strain by dividing total compression by the original sample thickness, and strain is then plotted against the logarithm of pressure.

If, during the compression test, the pressures on a sample are increased to a certain magnitude, unloaded to a lesser value, and then reloaded and increased to magnitudes greater than previously, results like those shown in Figure 10-12 are obtained. Note that the soil does not expand to its original volume when pressure is removed. Some of the volume change due to external loading is permanent. Soil, therefore, is not an elastic material. Upon reloading, the resulting slope of the compression curve is less steep than the original slope. These factors of soil behavior have significant effect on the settlement of structures.

Consider a soil sample obtained from a site where conditions are as shown in Figure 10-13a. The ground surface overlying the sample has never been above the existing surface, and there never was extra external loading action on the area. For this condition, the maximum vertical pressure ever imposed on the sample being considered is the current weight of overlying soil, $\overline{\sigma}_{v_0}$. The result of a compression test performed on the sample is as shown in Figure 10-13b. For laboratory loading less than $\overline{\sigma}_{v_0}$, the slope of the compression curve is less steep than it is for loads greater than $\overline{\sigma}_{v_0}$ since, to the soil, values less than $\overline{\sigma}_{v_0}$ are in the "reloading range." The slope of the curve at loading greater than $\overline{\sigma}_{v_0}$ is termed the *virgin compression curve,* and the slope of the curve is the *compression index, C_c*. The slope of the curve at values less than $\overline{\sigma}_{v_0}$ is the recompression slope C_r. Mathematically, the slope of the curve, for either C_c or C_r, is $\Delta e / \Delta \log \overline{\sigma}_v$. C_c and C_r are terms without dimensional units (that is, they will be indicated as numerical values that have no dimensional units).

From examination of Figures 10-12 and 10-13, it should be concluded that a change in the slope of the compression curve results when the previous maximum pressure ever imposed onto the soil is exceeded. If the ground surface had at some time in past history been

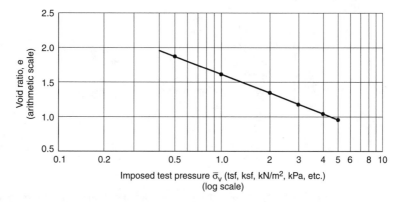

Figure 10-10   Presentation of compression test data on semilog coordinates.

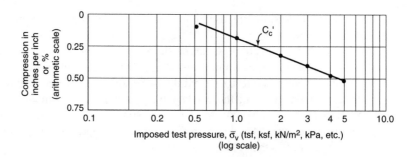

Figure 10-11 Method of indicating soil compression as strain instead of void ratio.

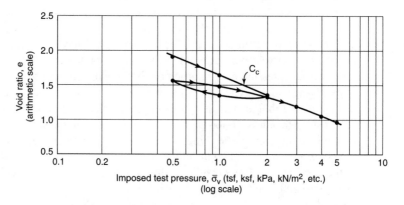

Figure 10-12 Results of loading–unloading–reloading cycle applied to a soil.

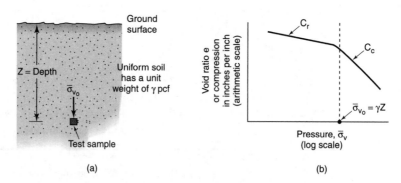

Figure 10-13 Description of conditions applying to compression test sample: (a) location of soil sample obtained for compression test; (b) result of compression test.

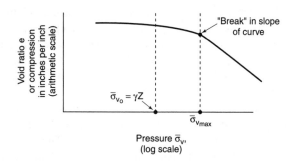

Figure 10-14   Compression test results where past pressure on soil has exceeded existing overburden pressure.

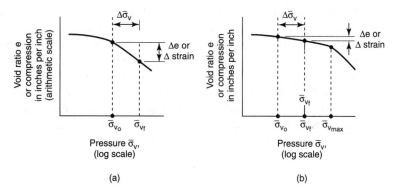

Figure 10-15   Comparison of compression that occurs for (a) normally consolidated and (b) overconsolidated soils at loading above the overburden pressure.

above the existing surface and eroded away, or if the weight of a glacier had been imposed on the area at some time in history, $\overline{\sigma}_{v_0}$ (the existing overburden) is not the maximum pressure that has been imposed on the soil sample. The greatest pressure that previously existed on the soil is $\overline{\sigma}_{v_{max}}$, which would be the total pressure that developed from the existing soil overburden *and* the weight of the eroded soil or glacier. Compression test results for this occurrence would be as shown in Figure 10-14.

The effect of this soil property on the settlement of structures is shown by a comparison of conditions from Figures 10-13b and 10-14. Assume that the weight of the structure would cause an additional stress equal to $\Delta\overline{\sigma}_v$ to act on the sample being studied. The total soil compression that will occur within the soil deposit, and therefore the settlement that the structure experiences, is related to the compression occurring in the test sample. Comparative results are shown by Figure 10-15a and b. It is seen that, for the same magnitude of structural loading, soil conditions of Figure 10-15a cause more compression and, therefore, more building settlement than do the conditions of Figure 10-15b. This indicates that the stress history of a soil may be more significant than other soil properties insofar as settlement of structures in concerned.

Soil deposits whose condition is represented by Figures 10-13b and 10-15a are termed *normally loaded* or *normally consolidated,* meaning that the present overburden pressure is the greatest pressure that has ever been imposed on the soils. Soils whose condition is represented by Figures 10-14 and 10-15b are termed *precompressed, preconsolidated,* or *overconsolidated,* meaning that at some time in past history there were imposed pressures greater than those that currently exist.

If the virgin curve of a laboratory compression test begins at a loading less than $\overline{\sigma}_{v_0}$, the soil is termed *underconsolidated,* which means that the soil has not fully adjusted or stabilized under the current overburden pressures. (This would represent a "new" soil deposit).

For reliable results, undisturbed soil samples are required for the compression test. If the soil sample to be tested is disturbed because of the soil boring techniques used to obtain the sample or during transportation and handling, the test data may be in error. Typical test results occurring from disturbed samples are shown in Figure 10-16.

Estimating Compressibility from Index Properties. Laboratory compression tests on fine-grained soils normally require days to complete (about 2 weeks is the period conventionally used). It is frequently necessary to obtain information about the compressibility of a soil in as short a time as possible. Or, in the interest of job economy, it may be desirable to limit the number of compression tests but still to evaluate the compressibility characteristics of many soil boring samples. For these situations, it is possible to use correlations that have been established between compression properties and some more easily or more quickly determined properties of soils. These correlations, which permit *estimates* of the compressibility of silts and clays to be made, are

$$C_c = .54(e_0 - .35) \tag{10-8a}$$

where e_0 is the in-place void ratio,

$$C_c = .0054(2.6w - 35) \tag{10-8b}$$

where w is the in-place water content, and

$$C_c = .009(\text{LL} - 10) \tag{10-8c}$$

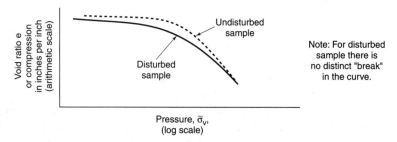

Figure 10-16 Typical compression test curve shapes for disturbed and undisturbed test samples.

where LL is the liquid limit. Equation 10-8c should be applied only for normally consolidated clays. In a normally consolidated clay, the natural water content is about at the liquid limit. If a clay has its natural water content significantly less than the liquid limit, it is preconsolidated.

The shear strength of a clay soil can also be used to indicate if the soil is normally consolidated or overconsolidated. The magnitude of the maximum pressure imposed on a soil is from 3 to 4 times its shear strength. By determining the shear strength and then $\overline{\sigma}_{v_{max}}$, the magnitude of $\overline{\sigma}_{v_{max}}$ can be compared with the overburden pressure that acted on the sample. Close agreement would indicate a normally consolidated soil.

Settlement Due to Primary Compression of Clay.   Assume that a building is to be constructed at a location where soil conditions are as indicated by the sketch of Figure 10-17a. The results of a laboratory compression test performed on a sample obtained from the center of the clay stratum are shown in Figure 10-17b. The value of $\overline{\sigma}_{v_0}$, the overburden pressure, is equal to $y\gamma_{sand} + (H_0/2)\gamma_{clay}$, where γ_{sand} and γ_{clay} are the unit weights (in pcf or kN/m³) of the sand and clay soils, respectively, and the y and H distances are in units (m, ft) compatible with the soil unit weight values.

The foundation load imposed by the new structure results in an additional pressure, $\Delta\sigma_v$, acting on the clay. Figure 10-18 uses the phase diagram method to illustrate changes that the clay sample experiences under the increased loading. The change in volume is equal to the change in void ratio. The relationship of the change to the original volume is

$$\frac{\Delta V}{V_T} \text{ or } \frac{\Delta e}{1 + e_0}$$

If the clay sample represents the *average* volume change that occurs throughout the clay stratum of Figure 10-17a, the following proportion will apply:

$$\frac{S_{pc}}{H_0} = \frac{\Delta e}{1 + e_0}$$

where H_0 is the original thickness of the clay layer and S_{pc} is the compression that this layer will experience. The settlement of a foundation supported above the clay layer will be equal to the compression that occurs, or

$$\text{Settlement} = S_{pc} = H_0\left(\frac{\Delta e}{1 + e_0}\right) \tag{10-9}$$

As previously noted, the slope of the curve from the compression test for loadings greater than any previous maximum overburden pressure is C_c, where

$$C_c = \frac{\Delta e}{\Delta \log \overline{\sigma}_v}$$

Rearranging gives

$$\Delta e = C_c(\Delta \log \overline{\sigma}_v)$$

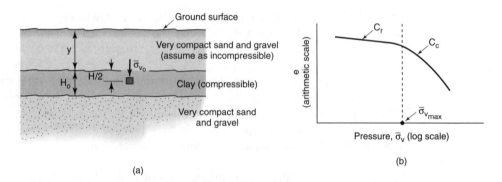

(a)

(b)

Figure 10-17 Description of conditions applying to compression test sample: (a) subsurface profile indicating source of compression test sample; (b) compression test results.

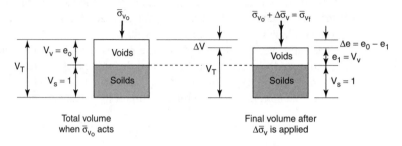

Total volume when $\bar{\sigma}_{v_0}$ acts

Final volume after $\Delta\bar{\sigma}_v$ is applied

Figure 10-18 Phase diagram illustrating change in soil volume that occurs with increase in loading.

Substituting in the equation for settlement gives

$$S_{pc} = \frac{H_0}{1 + e_0} C_c (\Delta \log \bar{\sigma}_v) \tag{10-10a}$$

or

$$S_{pc} = \frac{H_0}{1 + e_0} C_c (\log \bar{\sigma}_{v_f} - \log \bar{\sigma}_{v_0}) = \frac{H_0}{1 + e_0} (C_c) \log (\bar{\sigma}_{v_f} / \bar{\sigma}_{v_0}) \tag{10-10b}$$

In this equation, $\bar{\sigma}_{v_0}$ is the overburden soil pressure, and $\bar{\sigma}_{v_f}$ is the sum of the overburden soil pressure $\bar{\sigma}_{v_0}$ and the pressure caused by the weight of the structure $\Delta\bar{\sigma}_v$.

If the compression test data are presented in the form shown by Figure 10-11, the settlement equation is

$$S_{pc} = H_0 \times \text{(change in percentage of compression for the load change } \Delta\bar{\sigma}_v)$$

or

$$S_{pc} = H_0 C_c' (\log \bar{\sigma}_{v_f} - \log \bar{\sigma}_{v_0}) \tag{10-11}$$

where C_c' is the slope of the compression versus $\log \bar{\sigma}_v$ plot (see Fig. 10-11).

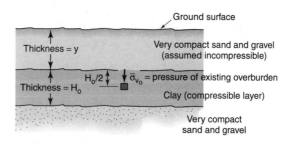

Figure 10-19   Soil conditions for problem of settlement due to compression in a buried clay layer.

Illustration 10-5

Referring to the conditions shown by Figure 10-19, assume that y is 3.66 m (12 ft), H_0 is 2.44 m (8 ft), and γ_{sand} and γ_{clay} are 21.2 kN/m³ (135 pcf) and 17.27 kN/m³ (100 pcf), respectively. The clay stratum is normally consolidated (e.g., $\bar{\sigma}_{v_0} = \bar{\sigma}_{v_{max}}$; e_0 is 1.20 and the compression index C_c is 0.20 (both dimensionless values). The weight of the structure causes a stress of 28.8 kN/m² (600 psf) at mid-height of the clay layer. [Therefore, $\Delta\bar{\sigma}_v = 28.8$ kN/m² (600 psf).] (*Note:* Values of stress increase, $\Delta\bar{\sigma}_v$, are computed by using methods described in Chapter 9.)

Solution
The settlement would be

$$S_{pc} = \frac{H_0}{1 + e_0} C_c (\log \bar{\sigma}_{v_f} - \log \bar{\sigma}_{v_0})$$

(a) Problem worked in SI units:
Since

$$\bar{\sigma}_{v_0} = (3.66 \text{ m} \times 21.2 \text{ kN/m}^3) + \left(\frac{2.44 \text{ m}}{2} \times 17.27 \frac{\text{kN}}{\text{m}^3}\right) = 98.9 \frac{\text{kN}}{\text{m}^2}$$

and

$$\bar{\sigma}_{v_f} = \bar{\sigma}_{v_0} + \Delta\bar{\sigma}_v = (98.9 + 28.8) \frac{\text{kN}}{\text{m}^2} = 127.7 \frac{\text{kN}}{\text{m}^2}$$

therefore

$$S_{pc} = \frac{(2.44 \text{ m})}{(2.20)}(.20)(\log 127.7 - \log 98.9)^1 = .0246 \text{ m} = 25 \text{ mm}$$

If the groundwater table is at the soil surface, the effective soil overburden pressure, $\bar{\sigma}_{v_0}$, is due to the submerged (or effective) weight of the soil. If the unit weights of 21.2kN/m³ and 17.27 kN/m³ represent saturated unit weights, the submerged unit weights will be (using $\gamma_w = 9.81$ kN/m³),

---

[1]From mathematics, recall that the process of subtracting the logarithms of numerical values is actually accomplishing a division of those values. Hence, the *units* for the involved terms, such as psf or kN/m², will cancel, and the result of the logarithmic subtraction is dimensionless (without units).

For the sand:

$$\gamma_{sub} = \gamma_{sat} - \gamma_w = 21.2 \text{ kN/m}^3 - 9.81 \text{ kN/m}^3 = 11.39 \text{ kN/m}^3$$

For the clay:

$$\gamma_{sub} = \gamma_{sat} - \gamma_w = 17.27 \text{ kN/m}^3 - 9.81 \text{ kN/m}^3 = 7.46 \text{ kN/m}^3$$

If the soils are not fully saturated, it is generally sufficiently accurate to assume that a submerged effective soil weight is about half its weight when not submerged. The stress increase $\Delta\bar{\sigma}_v$ is not affected by submergence. Using the values calculated, the overburden pressure $\bar{\sigma}_{v_0}$ is then

$$\sigma_{v_0} = (3.66 \text{ m})\left(21.2 \frac{\text{kN}}{\text{m}^3} - 9.81 \frac{\text{kN}}{\text{m}^3}\right) + \left(\frac{2.44}{2} \text{ m}\right)\left(17.27 \frac{\text{kN}}{\text{m}^3} - 9.81 \frac{\text{kN}}{\text{m}^3}\right)$$

$$= (41.69 + 9.10) \frac{\text{kN}}{\text{m}^2} = 50.79 \frac{\text{kN}}{\text{m}^2}$$

$$\bar{\sigma}_{v_f} = \bar{\sigma}_{v_0} + \Delta\bar{\sigma}_v = (50.79 + 28.8) \text{ kN/m}^2 = 79.59 \text{ kN/m}^2$$

and

$$S_{pc} = \frac{(2.44 \text{ m})}{(2.20)}(.20)(\log 79.59 - \log 50.79) = .043 \text{ m} = 43 \text{ mm}$$

Thus, the effect of a high water table is to cause more settlement.

(b) Problem worked in U.S. customary units:

Since

$$\bar{\sigma}_{v_0} = (12 \text{ ft} \times 135 \text{ pcf}) + (4 \text{ ft} \times 110 \text{ pcf}) = 2060 \text{ psf}$$

and

$$\bar{\sigma}_{v_f} = \bar{\sigma}_{v_0} + \Delta\bar{\sigma}_v = (2060 + 600) \text{ psf} = 2660 \text{ psf}$$

therefore

$$S_{pc} = \frac{8 \text{ ft}}{1 + 1.20}(.20)(\log 2660 - \log 2060)^1$$

$$= \frac{(8 \text{ ft})(.20)}{2.20}(3.426 - 3.315) = .0806 \text{ ft} = 1 \text{ in. } (\pm)$$

If the goundwater table is at the soil surface, the effective soil overburden pressure, $\bar{\sigma}_{v_0}$, is due to the submerged (or effective) weight of the soil. If the unit weights of 135 pcf and 110 pcf represent saturated unit weights, the submerged unit weights will be

For the sand:

$$\gamma_{sub} = \gamma_{sat} - \gamma_w = 135 \text{ pcf} - 62.4 \text{ pcf} = 72.6 \text{ pcf}$$

For the clay:

$$\gamma_{sub} = \gamma_{sat} - \gamma_w = 110 \text{ pcf} - 62.4 \text{ pcf} = 47.6 \text{ pcf}$$

If the soils are not fully saturated, it is generally sufficiently accurate to assume that a submerged effective soil weight is about half its weight when not submerged. Using the values calculated, we obtain

$$\overline{\sigma}_{v_0} = (72.6 \text{ pcf} \times 12 \text{ ft}) + (47.6 \text{ pcf} \times 4 \text{ ft}) + 1062 \text{ psf}$$

and

$$\overline{\sigma}_{v_f} = \overline{\sigma}_{v_0} + \Delta\overline{\sigma}_v = 1062 \text{ pcf} + 600 \text{ pcf} = 1662 \text{ psf}$$

(Note that $\Delta\overline{\sigma}_v$ is not affected by the water table or submergence.)

$$S_{pc} = \frac{H_0}{1 + e_0} C_c (\log \overline{\sigma}_{v_f} - \log \overline{\sigma}_{v_0})$$

$$= \frac{(8 \text{ ft})(.20)}{2.20}(3.222 - 3.028) = .14 \text{ ft} = 1.7 \text{ in. } (\pm)$$

If the soil is an overconsolidated soil, the equation for settlement calculations is modified to:

$$S_{pc} = \frac{H_0}{1 + e_0} C_r (\log \overline{\sigma}_{v_f} - \log \overline{\sigma}_{v_0}) \tag{10-12}$$

provided that $\overline{\sigma}_{v_{max}} > \overline{\sigma}_{v_f}$. (Recall that $\overline{\sigma}_{v_{max}}$ is the pressure where the slope of the compression test plot changes from C_r to C_c.)[2] For an overconsolidated soil where $\overline{\sigma}_{v_f}$ is greater than $\overline{\sigma}_{v_{max}}$, the settlement equation should be used as:

$$S_{pc} = \frac{H_0}{1 + e_0} C_r (\log \overline{\sigma}_{v_{max}} - \log \overline{\sigma}_{v_0})$$

$$+ \frac{H_0}{1 + e_0} C_c (\log \overline{\sigma}_{v_f} - \log \overline{\sigma}_{v_{max}})$$

or

$$S_{pc} = \frac{H_0}{1 + e_0}[C_r(\log \overline{\sigma}_{v_{max}} - \log \overline{\sigma}_{v_0}) + C_c(\log \overline{\sigma}_{v_f} - \log \overline{\sigma}_{v_{max}})] \tag{10-13}$$

---

[2]To obtain a value of C_r from the laboratory compression test for use in this equation, the recommended practice is as follows: Impose loading on the sample in increments to above the precompression stress, unload or rebound to at least the existing overburden, then reload the sample, and use the last recompression slope for determining C_r.

Illustration 10-6

Assume that a buried stratum of clay 1.83 m (6 ft) thick will be subjected to a stress increase of 33.6 kPa (700 psf) at the center of the layer. The magnitude of the preconstruction soil overburden pressure is 48 kPa (1000 psf) at the center of the layer. A laboratory compression test indicates that the clay is overconsolidated, with $\bar\sigma_{v_{max}}$ equal to 72 kPa (1500 psf). The value of C_c is 0.30, and the value of C_r is 0.05. What change in thickness results in the clay layer due to the stated conditions?

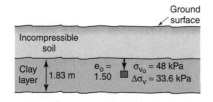
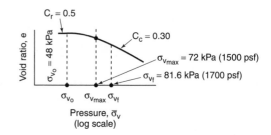

Solution

The change in thickness of the clay layer, from Equation 10-13, is:

$$S_{pc} = \frac{H_0}{1 + e_0} [C_r(\log \bar\sigma_{v_{max}} - \log \bar\sigma_{v_0}) + C_c(\log \bar\sigma_{v_f} - \log \bar\sigma_{v_{max}})]$$

(a) Problem worked in SI units:

$$S_{pc} = \frac{1.83 \text{ m}}{2.50} [(.05)(\log 72 - \log 48) + (.30)(\log 81.6 - \log 72)]$$

$$= .0184 \text{ m} = 18.4 \text{ mm}$$

(b) Problem worked in U.S. customary units:

$$S_{pc} = \frac{72 \text{ in.}}{2.50} [(.05)(\log 1500 - \log 1000) + (.30)(\log 1700 - \log 1500)]$$

$$= .73 \text{ in.}$$

If the thickness of the compressible soil layer is great, it should be divided into thinner layers for purposes of making the settlement calculations. The increment of settlement for each layer should be handled by using the methods discussed. The value of $\bar\sigma_{v_0}$ is the overburden pressure at midheight of each layer, and $\Delta\bar\sigma_v$ is the increase in pressure from the weight of the structure, also calculated for the midpoint of each layer. The total settlement will be the sum of increments occurring in all the layers. The depth of soil that is considered to be affected by foundation loading is limited to where $\Delta\bar\sigma_v$ is approximately one-tenth of the soil overburden pressure.

Illustration 10-7

A settlement calculation is to be made for a structure that is planned for a site underlain by a thick deposit of normally consolidated clay soil. Soil properties are as indicated on the sketch. What building settlement is expected?

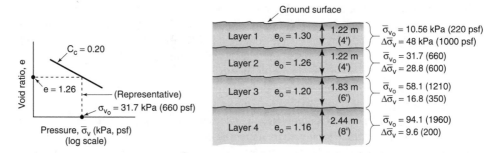

Solution

For this condition, the settlement calculation should be performed by dividing the thick clay deposit into sublayers and calculating the compression that occurs in each sublayer. The overburden pressure is calculated for the center of each layer, and the stress increase due to the structure is also calculated for the center of each layer. Results of these computations are shown on the sketch. The related settlement calculations follow.

(a) Problem worked in SI units:

$$S_1 = \frac{1.22\ m}{2.30}(.20)\log\left(\frac{58.56}{10.56}\right) = .0789\ m = 79\ mm$$

$$S_2 = \frac{1.22\ m}{2.26}(.20)\log\left(\frac{60.48}{31.68}\right) = .0303\ m = 30\ mm$$

$$S_3 = \frac{1.83\ m}{2.20}(.20)\log\left(\frac{74.9}{58.1}\right) = .0184\ m = 18.4\ mm$$

$$S_4 = \frac{2.44\ m}{2.16}(.20)\log\left(\frac{103.7}{94.1}\right) = .0095\ m = \underline{\underline{9.5\ mm}}$$

$$\text{Total settlement} = 137\ mm$$

(b) Problem worked in U.S. customary units:

$$S_1 = \frac{48\ in.}{2.30}(.20)(\log 1220 - \log 220) = 3.11\ in.$$

$$S_2 = \frac{48\ in.}{2.26}(.20)(\log 1260 - \log 660) = 1.19\ in.$$

$$S_3 = \frac{72\ in.}{2.20}(.20)(\log 1560 - \log 1210) = .72\ in.$$

$$S_4 = \frac{96\ in.}{2.16}(.20)(\log 2160 - \log 1960) = \underline{\underline{.38\ in.}}$$

$$S,\ \text{total settlement} = \underline{\underline{5.4\ in.}}$$

Secondary Compression of Clay Soil

Primary consolidation compression, discussed in the preceding section, refers to the process where soil particles in the stressed zone of saturated soil are rearranging into a more compact or tighter configuration as void water is squeezed (expelled). The effective stress acting on the soil skeleton (the soil particles) is gradually changed from the initial (before-load) condition to the final condition (where the full magnitude of the construction loading is imposed onto the soil particles), while simultaneously, the excess void water (pore water) pressure gradually diminishes to zero (i.e., theoretical full consolidation has occurred). *Secondary compression* is the additional compression that occurs at a constant value of effective stress after excess pore water has been dissipated. The full nature of secondary compression is not yet completely understood but does relate to the slowly occurring adjustment of the particle arrangement under the effect of newer imposed loads. The process typically continues slowly over a long period of time. For most inorganic deposits, secondary compression is usually small compared to consolidation compression. On a practical basis, the procedure to analyze the effects of secondary compression in a soil layer subject to usual types of construction loading is complicated because the process of secondary compression can be occurring simultaneously with the process of consolidation compression. (In section 10.5 it is more fully explained that in a consolidating soil layer the degree of consolidation varies across the depth or thickness of the layer until full consolidation for the entire layer is reached; the soil zones adjacent to the drainage surface for the escaping water consolidate faster than do the soil zones farthest from the drainage surface.)

A compression index for secondary compression, C_α, can be determined from the plot of void ratio versus logarithm of time (data obtained from the laboratory consolidation test) and the relationship

$$C_\alpha = \frac{\Delta e}{\Delta \log t} = \frac{\Delta e}{\log t_2 - \log t_1} \tag{10-14a}$$

The value for Δe is the change in test sample void ratio that occurs over the time period between t_1 and t_2 that follows the time when full theoretical consolidation (i.e., full primary compression) has been reached. The value of C_α should be determined for the consolidation test stress increment range comparable to that which will develop at the construction project.

Typically, the values of C_α are small compared to the value of C_c (the compression index for consolidation compression). Test-determined values of C_α and C_c imply that a relatively limited range of the ratio C_α/C_c exists for naturally occurring soil deposits. For soft inorganic clays and highly organic, plastic silt–clays, the range is

$$C_\alpha/C_c = .03 \text{ to } .06 \tag{10-14b}$$

Values from the lower end of this range are expected for the inorganic clays, while the higher values are expected for the organic, highly plastic silt–clay soils. These values to not apply to the highly organic deposits such as peat.

After a value of C_a has been obtained, secondary compression settlement can be estimated from the following expression:

$$S_{sc} = C_\alpha \left(\frac{H_0}{1 + e_0} \right) (\log t_f - \log t_p)$$

$$= C_\alpha\left(\frac{H_0}{1 + e_0}\right)\log (t_f/t_p) \tag{10-15}$$

where t_p is the time at the end of consolidation compression and t_f is the time at the end of the time period (a final time) over which the settlement occurs (or is to be determined for).

Illustration 10-8

The following data are applicable to settlement studies for a construction project: The construction site is underlain by a buried stratum of normally consolidated, inorganic clay 5 m thick. Laboratory tests indicate $C_c = 0.20$ and $e_0 = 1.15$. Calculations indicate that consolidation settlement (primary compression) will be 30 mm for the expected construction loading. Full consolidation settlement (primary compression settlement) will require approximately 8 years.

Estimate the total settlement to be expected over a 20-year time span, considering the effects of secondary compression.

Solution

Estimate $C_\alpha = .04C_c = .04(.20) = .008$

$$S_{sc} = C_\alpha\left(\frac{H_0}{1 + e_0}\right)\log (t_f/t_p) = .008\left(\frac{5\text{ m}}{1 + 1.15}\right)\log \frac{20}{8} = .0074\text{ m} = 7.4\text{ mm}$$

Total settlement $= S_{pc} + S_{sc} = 30\text{ mm} + 7.4\text{ mm} = 37.4\text{ mm}$

10.4 SETTLEMENT RESULTING FROM EARTH FILL

It has been indicated that volume changes in subsurface soils should be expected as a result of any new loading imposed on the soil. Filling of construction sites with compacted earth so as to raise the grade of a low area or make an uneven site level is now a relatively common undertaking in the construction industry. Areal settlements due to fills should always be considered. And where structures are to be supported on the compacted earth fill, it should be recognized that the total settlement for the structure could well be the sum of the volume changes occurring in the natural soil from the weight of the fill plus the volume changes in the fill and natural soil caused by the weight of the structure itself. Only if the completed fill is left in place a sufficient time so that full compression of the natural soils occurs before the structure is built can the effect of settlement due to fill weight be neglected. Settlement of granular soils will still, of course, occur almost instantaneously. However, compression of fine-grained soils does usually require a time period.

Areal settlement due to the placement of fill can be computed by using methods already presented. However, the increase in stress $\Delta\sigma_v$ caused by the weight of the fill is assumed to remain constant throughout the full depth of the natural or original soil. This is in contrast to the condition that occurs beneath a foundation area, where the value $\Delta\sigma_v$ decreases with increasing depth below the foundation.

The settlement of fill can be assumed to result from *compression* that occurs in the original soils underlying the fill. For cohesive soils, the necessary value of C_c or C_r is determined or estimated from laboratory tests. For granular soils, the expression for soil compression is also used. Appropriate values of C_c can be estimated from index properties, or from the following:

For loose uniform sands (D_R from 25% to 40%),

$$C_c = .05 \text{ to } .06 \tag{10-16a}$$

For dense or compact sands (D_R from 60% to 80%).

$$C_c = .02 \text{ to } .03 \tag{10-16b}$$

10.5 CONSOLIDATION

The compression of clay deposits usually occurs gradually when new loading is applied (in contrast to coarse-grained deposits where compression related to loading typically is rapid). The causes for this are related to the low permeability of these fine-grained soils and the condition of their common occurrence in nature. Many clay deposits were formed under-water, an exception being the clays in residual and volcanic soil deposits. In the underwater deposit, all void spaces were occupied with water. Even with the passage of time, these sedimentary materials remained fully or almost fully saturated. In soil, volume decreases are due to reductions in the void spaces between the soil particles as the particles rearrange in order to support the additional pressure resulting from external loadings. For compression to occur in a saturated soil, water in the voids must be expelled to permit the decrease in void spaces to occur. This behavior may be pictured as similar to that of a saturated sponge having the water squeezed out of it by a pressing device.

When external load is applied to a saturated clay, the water in the voids (or pores) becomes subject to an immediate increase in pressure equal to the subsurface stress caused by that loading. This is the same principle that is in effect for any hydraulic loading system, where an external force results in pressure developing in the fluid throughout the system. The soil's trapped pore water then starts to flow toward regions of lesser pressure, moving away from the area where loading is occurring or toward bordering soil layers where a lesser pore water pressure exists. But because of the low permeability of fine-grained soil, the movement or escape of this pore water takes place slowly. Only as water escapes from the stressed zone can soil particle rearrangement and decrease in void spaces take place. The result is that, for saturated clay soils, compression is a gradual occurrence. As water escapes, the pressure caused by the external loading is transferred from this water to the soil particles. The process of load transfer to the soil as pore water escapes is the *consolidation* process.

The rate of consolidation for a stratum of clay soil is affected by several factors, including:

1. The permeability of the soil
2. The extent or thickness of the compressible soil and the distance that pore water in the soil must travel to escape from the zone where pressures due to foundation loading exist

3. The in-place void ratio of the compressible soil

4. The ratio of new loading to the original loading

5. The compression properties of the soil

The consolidation properties of a soil (not the soil stratum) are dependent on items 1, 3, 4, and 5. The effect of these factors can be grouped together to obtain a property termed the *coefficient of consolidation, c_v*, a value which indicates how rapidly or slowly the process of consolidation takes place. The coefficient of consolidation is

$$c_v = \frac{k(1 + e)}{a_v \gamma_w} \left(\text{Note: Units for } c_v \text{ are } \frac{\text{length}^2}{\text{time}}. \right) \qquad (10\text{-}17)$$

where e = void ratio (dimensionless value)

$\quad k$ = coefficient of permeability (mm/sec, cm/min, m/day, ft/min, ft/day, etc.)

$\quad a_v$ = coefficient of compressibility (m²/kN, ft²/lb) (ref. Fig. 10-9)

$\quad \gamma_w$ = unit weight of water (kN/m³ or lb/ft³)

Values for k, e, and a_v can be determined separately and substituted into the equation to determine c_v. It is common practice, however, to determine c_v directly from a laboratory consolidation test. The consolidation test is a compression test but with the provision that time readings also are obtained (amount of compression versus time, for each applied loading increment). Appendix B "Laboratory Procedure to Determine Coefficient of Consolidation" provides details of the procedures used to compute values for c_v from laboratory consolidation test results.

As previously identified, the consolidation process requires two interrelated occurrences: the soil particles will be shifting to a denser configuration so that the particle arrangement (that is, the soil structure or soil skeleton) develops the additional resistance necessary to support the increased loading while simultaneously, pore water in the void spaces is escaping so that the required decrease in voids can occur. Referring to Figure 10-20 and 10-21, note that initially the stress increase due to structural loading $\Delta\sigma_v$ is imposed fully on the trapped pore water. This means that, at first, none of the new loading is carried by the soil particles. Only as water is being squeezed from the voids does the new loading start transferring to the soil particles; the rate at which pore water escapes from soil voids and the newly imposed construction related loads are shifted to the soil particles is affected by the soil coefficient of permeability.

Consolidation theory indicates that the end of the consolidation process has been reached when the water pressure in the soil voids drops to the hydrostatic pressure normal for the location so that the draining of pore water ceases. Full consolidation (100 percent) is assumed to have occurred when the full loading of the structure is carried by the soil particles (no excess pressure exists in the pore water). Values of consolidation less than 100 percent indicate that, for the applied loading, soil compression (and settlement of the structure overlying the consolidation stratum) is still in process.

If the term $U_{avg}\%$ represents the average percent consolidation throughout the thickness of a compressing soil layer, $U_{avg}\%$ and excess pore water pressure are related by the expression

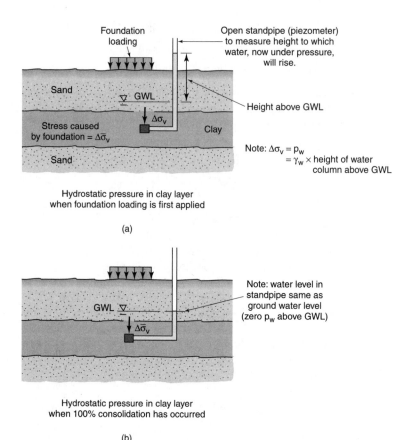

Figure 10-20 Change in pore water pressure occurring as clay soil consolidates: (a) hydrostatic pressure in clay layer when foundation loading is first applied; (b) hydrostatic pressure in clay layer when 100% consolidation has occurred.

$$U_{avg}\% = \left(1 - \frac{u_t}{u_i}\right)(100\%) \qquad (10\text{-}18)$$

where u_t = average excess pore water pressure in the consolidating soil layer, corresponding to the time when the percent consolidation is being determined

u_i = initial excess pore water pressure in the consolidating soil layer

Until the time the consolidation process is completed, excess pore water pressures in the consolidating soil vary in magnitude across the thickness of the compressing layer, being lower near the boundaries of the layer and greatest at the locations farthest from the boundaries because of the time lag associated with travel distance (similar to the manner in which a heated metal plate cools from the surface inward) (Fig. 10-22).

A time factor T is used to relate the rate at which excess pore water pressures dissipate to the period required for an average percentage consolidation to occur (Fig. 10-23). T_v is used for the condition where the escaping pore water travels vertically (i.e., vertical

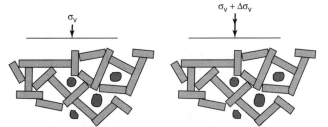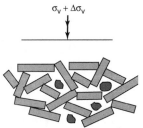

σ_v

$\sigma_v + \Delta\sigma_v$

$\sigma_v + \Delta\sigma_v$

Fine-grained soil deposit; soil deposit is 100% saturated. Before construction, a compressive stress due to the soil overburden exists (is imposed onto the soil particles); $\sigma_{vo} = \gamma z$. Water pressure in soil void spaces is normal hydrostatic pressure.

Construction loading causes a stress increase in the soil deposit. Due to effect of the new stress $\Delta\sigma_v$, particles attempt to shift into a more compact or tighter arrangement (so as to develop more compressive resistance as necessary for stability). However, water occupies the void spaces and prevents the soil particles from moving into those void spaces.

The compressive stress increase, $\Delta\sigma_v$, therefore is imposed onto the water trapped in the void spaces, and not onto the soil particles. The result is that the void water is under an "excess" pressure (pressure greater than normal hydrostatic pressure expected from the position of the groundwater table).

Because void water is subject to excess pressure created by $\Delta\sigma_v$, the water gradually flows to locations of lesser water pressure (toward location of "normal" water pressure). The rate of flow depends on soil permeability. As void water pressure is dissipated, the soil particles gradually shift and the $\Delta\sigma_v$ stress is slowly transferred from the void water over to the soil particles (that is, onto the "soil skeleton" for the soil deposit). The process of stress being slowly transferred onto the soil skeleton as void water is being expelled or squeezed from the stressed region of the soil deposit, and related gradual compression of the soil deposit, is "consolidation."

Figure 10-21 Sequence of foundation loading being transferred onto soil particles (the soil skeleton) by the consolidation process.

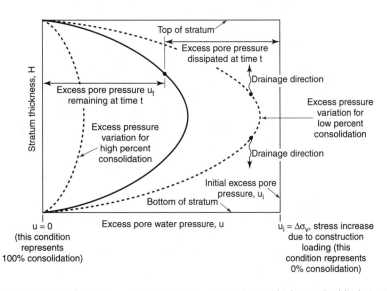

Figure 10-22 Variation of excess pore water pressure across stratum thickness, doubly drained stratum.

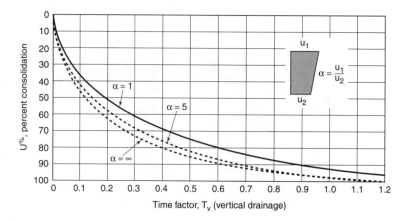

Figure 10-23 Variation of time factor T_v with percentage of consolidation U.

drainage) (Fig. 10-24), while T_h will apply to the condition where horizontal drainage of pore water takes place. The solid curve of Figure 10-23 applies to the condition of vertical double drainage and also single drainage of the consolidating layer for the occurrence where the initial excess pore pressure u_i is uniform throughout the full thickness of the layer (note from Fig. 10-24 that for double drainage of a layer, the drainage distance is half the layer thickness, while for single drainage, the drainage distance is the full layer thickness). The situation of uniform initial excess pore pressure can be assumed where the construction loading responsible for the development of excess pore pressure is of large areal extent (such as results from placement of an earth fill over a large area or from the loading of a structure that occupies a large area). For the case where construction loading acts over a limited area (such as with individual foundations for a building), the increase in subsurface stress due to the structural loading diminishes with depth; since the initial excess pore water pressure is expected to equal the increase in construction-caused subsurface stress, the excess pore water pressure will therefore vary across the thickness of the consolidating layer (greatest at the top, least at the bottom). If the ratio of the initial pore pressure at the top of the layer to the pore pressure at the bottom of the layer is α (i.e., $\alpha = U_{i\,top}/U_{i\,bot}$), the dashed line curves on Figure 10-23 can be used to select a time factor T_v appropriate for the expected variation in excess pore pressure distribution.

For given field conditions, where the distance that trapped pore water must travel to escape from the compressible layer (H_{dr} in Fig. 10-24) and c_v are known, the time period t required for a given percentage of consolidation to occur is

$$t = \frac{T_v H_{dr}^{\,2}}{c_v} \tag{10-19}$$

where T_v = time factor for consolidation due to vertical drainage (a dimensionless value)
H_{dr} = longest drainage distance for escaping pore water
c_v = coefficient of consolidation

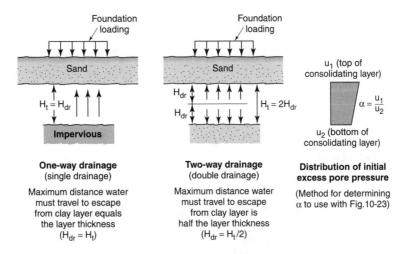

Figure 10-24 Vertical drainage conditions in consolidation theory.

Compression of a clay layer resulting because trapped pore water under pressure has escaped is consolidation compression or primary compression. Some additional long-term compression from what is considered secondary compression also takes place as soil particles continue to adjust under the applied loading. However, for inorganic soil deposits, the effects of secondary compression are expected to be small; where the calculated primary compression settlements are small, the effects of secondary compression commonly are ignored. The time rate for secondary compression settlement is estimated from the plot of consolidation test results for the appropriate load increment, using the ratio of time for full compression to time for the theoretical 100 percent consolidation compression.

Illustration 10-9

Referring to Illustration 10-5 and Figure 10-19; assume that laboratory consolidation test data for the clay indicate that, for the range of loading applied to the soil, c_v is 0.016 m² per month. (Recall that the settlement computed in Ill. 10-5 is 25 mm ±). (a) How long will it take for half of the estimated settlement to occur? (b) How much settlement will occur in one year?

Solution
(a) Half the estimated settlement is a U of 50 percent. Referring to Figure 10-23, assuming the solid curve applies, for $U = 50\%$, obtain the value $T_v = 0.20$.

$$t = \frac{T_v H_{dr}^2}{c_v} = \frac{(.2)\left(\dfrac{2.44 \text{ m}}{2}\right)^2}{(.016 \text{ m}^2/\text{month})} = 18.6 \text{ months}$$

(Note that for this problem, H_{dr} is one-half the total thickness of the clay layer.)

(b)
$$t = 1 \text{ yr} = 12 \text{ months}$$

$$T_v = \frac{tc_v}{H_{dr}^2} = \frac{(12 \text{ months})(.016 \text{ m}^2/\text{month})}{(1.22 \text{ m})^2} = .13$$

From Figure 10-23, for $T_v = 0.13$, we obtain $U = 42\%$, and therefore settlement, S_{pc}, is

$$S_{pc} = .42 \times 25 \text{ mm (from Ill. 10-5)} \cong 11 \text{ mm}$$

10.6 SURCHARGING

The situation may exist where, because of compressible or weak subsoils, a proposed structure would undergo settlements that would exceed tolerable limits. Alternatively, stresses resulting in the weak soil from the weight of the structure could cause a weak soil to fail. If the site is to be used, the problem soil somehow must be circumvented (by removing it if possible, or by using pile foundations penetrating to below the poor soil) or improved so that the strength is adequate and the compressibility reduced. For construction projects extending over large areas, such as highways, improvement of poor subsoils is always given consideration because of possible cost savings.

The improvement of poor soils may possibly be accomplished by using a surcharge program. Surcharging of a soil, in its simplest intent, merely involves imposing an external loading for a long enough duration to cause desirable changes in the soil before the structure planned for the area is erected and supported on the soil.

Under the weight of a surcharge load, a poor soil will compress and increase in strength. The surcharge is made to the desired weight or loading and continued for a sufficient period to achieve results satisfactory for the need of the planned structure. Generally, this might include having the surcharge cause the magnitude of soil compression (settlement) that the planned structure would have caused, so that after the surcharge is removed and the structure is built, its settlement will be negligible, or having the surcharge cause a sufficient increase in the strength of the poor soil so that it can safely support the weight of the planned structure.

Any material that will impose pressure onto the subsoils that require improvement can be used for the surcharge loading. Most conventionally, the material used is soil (borrow or fill). Soil is used because of its general availability and low cost, its ease of handling, and the absence of problems with deterioration.

Surcharging for improvement is most commonly applied for cohesive and organic soils—soil types that require a time period for compression (consolidation) and strength gain to occur. Surcharging programs *are* applicable for improving loose, granular soil deposits, but other methods, frequently quicker and more economical, can be used for these soils (methods that do not have application with cohesive soils).[3]

Surcharges for improving the strength of weak soils have to be applied slowly or in increments so that the weak soils are not overstressed (failed) *before* they have time to improve. With cohesive soils, the compression–consolidation process involves a readjust-

---

[3]See vibro-compaction and dynamic compaction methods discussed in Chapter 14.

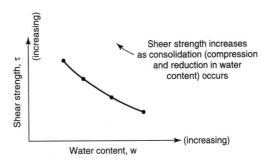

Figure 10-25 Change in soil strength and water content as consolidation under surcharge occurs.

ment of soil particles as water is squeezed from the voids of the soil. A gain in strength occurs because of the consolidation, but the strength gain results only as rapidly as water is squeezed from the soil (an implication from the strength curve shown in Fig. 10-25). Consequently, a properly performed surcharge program requires field monitoring to establish a safe rate for increasing the surcharge and to determine how long it should remain in place. Monitoring normally includes instruments such as piezometers (see Chapter 5) to measure excess pore water pressures developed within the weak soil, to learn of the rate at which consolidation is occurring, and to indicate when additional increments of the surcharge can be safely added. Settlement plates located in or on the surface of the original soil are used to monitor the rate and amount of settlement caused by a surcharge. For accurate records, any reference benchmark to measure settlement must be located outside of the area that is settling.

In some cases, the structure itself can be used to apply a surcharge load in increments. An example of this is with ground-level storage tanks, such as those used to hold petroleum or other fluids. For many situations, the tolerable settlement of storage tanks is relatively large, provided that large *differential* settlements that might cause the tank to rupture do not occur. A fluid, possibly the material to be stored but more frequently water, is used to partially fill the tank. The tank is left partially filled until the subsoil achieves the necessary increase in strength. Then additional liquid is pumped in, and the next waiting period is allowed to elapse. This procedure continues until the subsoils supporting the tank reach the desired degree of stability. The tank can then be put into permanent use. Permanent piping connections should not be made until all, or most of, the settlement has occurred.

With this method, a laboratory test program should be performed before field operations are begun to ensure that the weak subsurface soils will be able to reach the strength required to support the tank and to obtain an estimate of the time required to complete the consolidation process. A laboratory program involves consolidating samples of the weak subsoil under the same vertical pressures that the tank will impose onto the soil and then performing shear tests to determine if adequate shear strength (Chapter 11) is being obtained.

Acceleration of the Soil Improvement Process

Use of a surcharge program requires that the surcharge be applied and left in place for long enough to obtain the desired compression or strength gain. Depending on the properties and thickness of the poor soils, a surcharge may remain in place for months or longer before it accomplishes its purpose and can be removed. Construction cannot begin until the surcharge is removed, of course. If time is an important factor, means of accelerating the consolidation period can be undertaken.

One method to reduce the time for a desired amount of compression (or settlement) to result is to use an excessive surcharge, provided that the subsurface soils have sufficient strength. A given amount of settlement will occur more quickly under a heavy surcharge than under a light surcharge.

Illustration 10-10

Assume that an area is underlain by a stratum of compressible soil 5.3 m thick. The weight of the proposed structure, say an earth embankment for a new highway, will result in an increase of 28.2 kPa pressure to the compressible soil. The settlement under this loading would be 0.30 m. The compressible layer is doubly drained, and the soil has a c_v value of 0.4 m²/month. Calculations indicate that it would take approximately 22 months for the settlement to occur under the weight of the proposed embankment. How long would be required for a surcharge imposing a pressure of 56.4 kPa to cause the 0.30 m of settlement?

Solution

The first step of the solution requires that the total settlement to be expected from the 56.4 kPa surcharge be computed. Using Equation 10-10, we find that the settlement would be 0.51 m. To obtain but 0.30 m of settlement, about 60 percent consolidation (0.3 m/0.51 m × 100%) must occur. The T_v value for 60 percent consolidation is 0.3, and the time required is

$$t = \frac{T_v H_{dr}^2}{c_v} = \frac{(.3)(2.65 \text{ m} \times 2.65 \text{ m})}{0.4 \text{ m}^2/\text{month}} \cong 5.3 \text{ months}$$

Horizontal Drainage

When a surcharge is used, another method for accelerating the consolidation process is to shorten the length of the drainage path for the pore water escaping from the consolidating soil. This can be accomplished by providing vertical drains or drain wells at spacings closer than the drainage distance for vertical flow so that horizontal drainage flow will occur. Frequently, too, the horizontal coefficient of permeability for cohesive soil deposits is many times greater than the coefficient of permeability for the vertical direction (because of the manner of deposition), which further increases the speed of consolidation.

Sand Drains. Vertical drain wells can be holes that are drilled into the compressible soil and filled with sand (frequently called *sand drains*). Since the sand is very permeable material, pore water under pressure because of the surcharge will flow to the drain wells and be forced upward (or downward, too, if the well extends to a deeper permeable stratum underlying the compressible soil). After flowing upward to the surface, the water is carried away from the surcharge area (Fig. 10-26a).

If the drain wells are located as indicated in Figure 10-26b, the zone of drainage can be assumed to be circular. The expression for the time period necessary to achieve consolidation when this condition exists is

Figure 10-26 Cross section and plan of typical sand drain installation: (a) cross section of installed drain wells and surcharge; (b) drain well pattern [23].

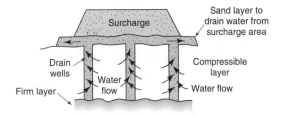

(a) Cross section of installed drain wells and surcharge

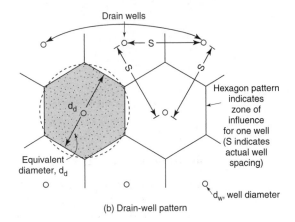

(b) Drain-well pattern

$$t = \frac{T_h(\gamma_w a_v)}{k_h(1 + e_0)} = \frac{T_h d_d^2}{c_h} \qquad (10\text{-}20)$$

where t = elapsed time

T_h = time factor for horizontal drainage (dimensionless)

c_h = horizontal coefficient of consolidation (*note*: units are length^2/time)

d_d = diameter of the zone draining to a drain well (ft or m)

The value of T_h varies according to the ratio of the draining area to the escape (drain well) area. If d_d is the diameter of the draining area and d_w is the drain well diameter, n can be the ratio of d_d to d_w. Curves of T_h, similar to those for T_v used for vertical consolidation, are available for different ratios of n (Fig. 10-27).

The drain well spacing affects the time for consolidation more than does the drain well size. Quicker results will be obtained by using a small drain well spacing. However, there is the practical problem of constructing the wells close together without collapsing or otherwise damaging them. There is also the potential problem that the sand drains may shear along their length if the compressible soils experience large settlement or lateral movement during compression, with the result that water flowing from the consolidating soils to the drain well cannot escape to the surface, thus invalidating the drain well system. Because of the specialized techniques and equipment required for a sand drain installation, it is generally not economical to use the method for improving small projects.

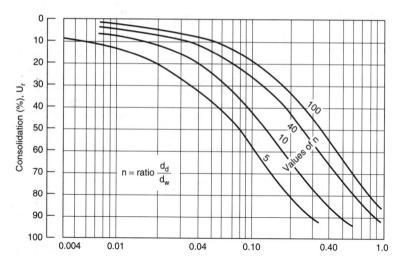

Figure 10-27 Variation of time factor T_h with percentage of consolidation U for sand drain installations [23].

Illustration 10-11

The settlement problem of Illustration 10-10 examined the effects of using a surcharge to improve the subsurface soils for a planned highway embankment. The surcharge program involves the placement and removal of an amount of soil greater than that required for the embankment, with a related expense. It is desired also to study the effect of using a sand drain installation, on the possibility that it may be less costly than surcharging (because less surcharge is required) and that faster results may be provided.

Laboratory testing of samples indicates that the coefficient of permeability for horizontal flow ranges from 5 to 15 times the coefficient for vertical flow; the coefficient of consolidation for horizontal drainage would be similarly affected. For a preliminary analysis, use a value of c_h that is 7 times the value of c_v. Since c_v was 0.4 m²/month, c_h will become about 2.8 m²/month.

It is planned to install sand drains that are 0.7 m in diameter and on a center-to-center spacing of 3.5 m in a pattern as indicated in Figure 10-26b.

Settlement calculations indicate that a 37.6 kPa surcharge would cause a settlement of 0.34 m. Related to this, the 0.30 m settlement expected from the necessary embankment fill represents 0.3 m/0.34 m or 89 percent consolidation.

The ratio of d_d to d_w is 3.5/0.7, or 5. With reference to Figure 10-27, the value of T_h for 89 percent consolidation and n equal to 5 gives a time factor of 0.27. Substitution of appropriate values into Equation 10-20 gives

$$t = \frac{T_h(d_d)^2}{c_h} = \frac{(.27)(3.5 \text{ m} \times 3.5 \text{ m})}{2.8 \text{ m}^2/\text{month}} \cong 1.2 \text{ months}$$

Wick Drains. A practical development of the 1970s was the prefabricated band drain consisting of synthetic fabrics and including a permeable core. The typical band drain has a small cross section (e.g., 4 in. wide by 0.25 in. thick, or 100 mm by 7 mm), is extremely flexible, and can be used on a job site while being unwound from a storage or dispensing

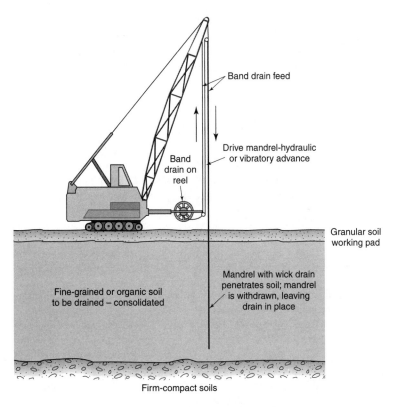

Band drain feed

Band drain on reel

Drive mandrel-hydraulic or vibratory advance

Granular soil working pad

Mandrel with wick drain penetrates soil; mandrel is withdrawn, leaving drain in place

Fine-grained or organic soil to be drained – consolidated

Firm-compact soils

Figure 10-28 Typical wick drain installation equipment and procedure.

reel. The fabric drain can be rapidly installed with mobile crane or mast-equipped rigs by using a needle-punching technique (a nondrilling, soil displacement procedure; see Fig. 10-28). A long, thin mandrel (shaft) on the installation rig is vibrated or hydraulically pushed into the soil to be drained. The fabric drain, attached at the tip of the mandrel, is pulled into the earth as the mandrel advances. At the desired penetration, the drain is disengaged from the mandrel and also cut from the band reel at the ground surface. The mandrel is withdrawn, but the fabric drain remains in place. This vertical drain or *wick* provides the escape path for excess pore water in the penetrated soil. The rate of drain installation can be on the order of 1 m or 3 ft per second. Installation depths up to about 30 m or 100 ft have been used. In the typical application, the wick drains are installed at closely spaced intervals (on the order of 1.5 m or 5 ft or less) in a staggered pattern similar to that indicated in Figure 10-26b. A surcharge material is applied to the area after the drains have been installed to create an increase in subsurface stress and pore water pressure, as for the sand drain installation. As with sand drains, a granular soil layer is placed on the site surface under or as part of the surcharge, to provide a drainage pad for water wicked upward. The close spacing of drains is conducive to rapid consolidation because of the small lateral distance that squeezed pore water must travel to reach a drain. The fabric drains offer some distinct advantages over sand drains, such as lower materials cost, easier and more rapid installation, and less danger of damage to the drain as the treated soil consolidates and settles. Accordingly, wick drains generally have supplanted the use of sand drains. The theory previously outlined for sand drains can be applied where analytical studies involving wick drains are necessary. As an additional item of information, installed columns of gravel-size material are still being selected for projects, for reinforcing primarily (to increase the shear strength or bearing capability of a weak soil formation) but to simultaneously function as paths of drainage for the consolidation process.

Whether or not the use of drains is more economical than a surcharge program without drains (where a heavier surcharge or longer time period would be involved) depends on the relative cost of the drain installation and price of embankment fill. Factors to make a drain installation more attractive would include the possibility for the surcharge to be adequately compacted during placement so that it can be left in place as the structural embankment when the desired settlement has been reached.

PROBLEMS

10-1. The settlement of foundations is typically the result of three separate occurrences that take place in the soil zones providing support. List these factors and give a brief description of each.

10-2. Indicate the typical time period associated with foundation settlement for the condition where granular soil provides the support and for the condition where saturated clay provides the support. Discuss the reasons for the difference.

10-3. The major portion of settlement typically resulting for a foundation bearing on a sand deposit is different than for a foundation bearing on a saturated clay deposit. Indicate and explain the cause for the major portion of settlement for the two soil categories.

10-4. Compare the relative degree of settlement typically expected where a foundation bears on a soil deposit with significant organic content to the settlement expected where a foundation bears on soil free of organic material.

10-5. For structures supported on foundations that bear on sandy soils, it is presumed that the major portion of expected settlement occurs upon application of loading; the practical effect is that total settlement can take place by the end of the time period required to complete the construction. However, there is a major exception to this statement; indicate this condition.

10-6. Provide reasons why field plate-loading tests performed at a planned construction site may give er-roneous settlement information if the area is underlain by cohesive soils. (*Hint:* Field loading tests are frequently performed with bearing plates smaller in area than the foundation footings will be. Also consider the effects of consolidation.)

10-7. Use the Schmertmann method to estimate the settlement for a foundation bearing on a sand deposit and where the following conditions apply:
- Square foundation, 2 m by 2 m, installed at a depth of 1 m
- Foundation bearing pressure, $q = 200$ kPa
- Soil unit weight, $\gamma = 18$ kN/m³, deep water table

Use the tabulation procedure as shown in Illustration 10-1. Analyze for conditions extending to a depth $2B$ below foundation level, using four 1-m-thick sublayers. Cone penetration resistance values (in MPa), q_c are 3.0, 3.5, 5.0, and 3.5, respectively, for the four sublayers. Assume that the C_2 factor equals 1.

10-8. Use the Schmertmann method to estimate the settlement of a long strip-footing foundation bearing on a sand deposit where the following conditions apply:
- Footing width = 1.5 m, installed at a depth of 1 m
- Foundation bearing pressure, $q = 150$ kPa
- Soil unit weight, $\gamma = 18.5$ kN/m³, deep water table

Use the tabulation procedure as shown in Illustration 10-1. Analyze for conditions extending to a depth $4B$ below foundation level, using sublayers 1 m, 1 m, 2 m, and 2 m thick. Subsurface samples have been obtained using the standard penetration test (SPT), where N values of 24, 22, 29, and 36 can be used as the respective average values for the sublayers. Assume $E_s = 10N$ (Table 10-2), $C_2 = 1$.

10-9. Use the Schmertmann method to estimate the settlement of a foundation bearing on a sand deposit and where the following conditions apply:
- Square foundation, 4.5 ft by 4.5 ft, installed at a depth of 3 ft
- Foundation bearing pressure, $q = 4$ ksf = 2 tsf
- Soil unit weight, $\gamma = 118$ pcf, deep water table

Use the tabulation procedure as shown in Illustration 10-1. Analyze for conditions extending to a depth $2B$ below foundation level, using three sublayers each 3 ft thick. Standard penetration test (SPT) values for the three layers are $N_{avg} = 36, 27$, and 32. Assume $E_s = 9N$ and $C_2 = 1$. The groundwater table is at the elevation of the base of the foundation (use submerged unit weight for soil zone below foundation level when computing σ_{vp}).

10-10. Referring to the foundation settlement information summarized by the Burland studies (Fig. 10-4), estimate the settlements expected for the following foundations, assuming that the supporting sand deposit is medium-dense to loose:
(a) Square foundation 2 m by 2 m, $q = 200$ kPa
(b) Long foundation 1.5 m wide, $q = 150$ kPa
(c) Square foundation 4.5 ft by 4.5 ft, $q = 4$ ksf

10-11. A plate-bearing test is performed to determine a value for the modulus of subgrade reaction at a construction site underlain by granular soil. The load versus settlement data graphs as a straight line for the initial range of applied loads. The graphical plot indicates a settlement of 1.4 cm when the loading is 450 kN/m². What is the value for the modulus of subgrade reaction?

10-12. A plate-bearing test is performed on soil near the surface of a granular stratum in order to determine the modulus of vertical subgrade reaction. The load versus settlement information appears as a straight line on a graphical plot. At a point where the load is 5 tons/ft², the indicated settlement is 0.5 in. What is the value for the modulus of vertical subgrade reaction?

10-13. A square footing 3 m by 3 m supports a building column load that results in a foundation bearing pressure of 200 kPa. The deep sand stratum underlying the foundation has an average unit weight of 17.0 kN/m³ and is classified as medium-dense.

Use the subgrade reaction method to estimate the foundation settlement.

10-14. A square footing 5 ft by 5 ft supported on a sandy stratum is designed to carry a column loading of 250,000 lb. The unit weight of the sands underlying the footing location vary between 120 and 125 pcf. Estimate the foundation settlement to be expected using the subgrade reaction method. (Neglect the weight of the footing.)

10-15. A square foundation is to be supported on a thick clay soil deposit. Calculate the foundation settlement due to volume distortions occurring in the supporting soil zone, given the following conditions:
- Square foundation (flexible category), 2 m by 2 m
- Total loading on foundation is 700 kN
- Clay shear strength cohesion, $c = 100$ kPa; saturated clay

10-16. A square footing foundation (flexible category), 5 ft by 5 ft, imposes a total loading of 75 kips. The foundation bears on a thick stratum of saturated clay where the shear strength c is 1500 psf. Estimate settlement due to volume distortions occurring in the clay.

10-17. A long strip-footing foundation 1.5 m wide will be used to support an exterior bearing wall for a commercial building. The foundation will be installed near the soil surface and impose a loading of 200 kN per meter of wall length. The soil is saturated clay, with cohesion c (shear strength) equal to 50 kPa.
(a) Calculate the settlement under the center of the foundation due to volume distortion in the soil, assuming a flexible foundation.
(b) Calculate the volume distortion settlement, assuming the foundation is rigid.

10-18. Calculate the settlement under the center of a flexible foundation 4 m by 4 m due to volume distortions occurring in a saturated clay stratum but where rock exists at a depth 8 m below the foundation. The clay shear strength c is 60 kPa. The total foundation loading imposed onto the soil is 2400 kN.

10-19. Calculate the settlement for the center of a flexible foundation due to volume distortions in the underlying clay stratum. The clay soil has a cohesion shear strength c equal to 1000 psf. The foundation is 6 ft by 6 ft and imposes a total loading of 120 kips. The clay stratum is 12 ft thick; rock underlies the clay.

10-20. Calculate the soil volume distortion settlement under the center of a 2-m-wide strip foundation (flexible type) which carries a loading of 400 kN per meter of wall length. The foundation bears on a saturated clay stratum (cohesion c is 75 kPa) that is 10 m deep. Rock underlies the clay.

10-21. Calculate the settlement due to volume distortion in a clay where a rectangular-shape (flexible category) foundation 6 ft by 12 ft imposes a total loading of 220 kips. The layer of saturated clay (cohesion c is 1500 psf) extends 12 ft deep and is underlain by rock.

10-22. Indicate advantages related to performing laboratory compression tests on clay samples obtained from strata underlying a building site for determining soil compression data.

10-23. Briefly review the manner in which laboratory compression tests are conventionally performed so as to obtain load deformation data for a soil.

10-24. In the conventional laboratory compression test, what is the cause of soil volume decrease?

10-25. In a laboratory compression test, the void ratio in a soil sample changes from 1.46 to 1.32 when the compressive loading increases from 100 kN/m² to 200 kN/m². Calculate the value for the coefficient of compressibility and the compression index for this load range.

10-26. In a laboratory compression test, the void ratio changes from 1.55 to 1.36 as loading increases through the range from 2000 to 4000 psf.
  (a) What is the value of the coefficient of compressibility for these conditions?
  (b) What is the compression index for this loading range?

10-27. Referring to occurrences observed in a typical field or laboratory compression test, explain why a soil mass does not expand to its original volume when loading is released.

10-28. Explain the difference between soil compression and consolidation.

10-29. Briefly indicate the difference between a normally consolidated and an overconsolidated soil deposit.

10-30. A cohesive soil sample obtained from a known normally consolidated clay deposit is found to have a liquid limit of 80 percent. Approximately what would be the compression index for this soil?

10-31. At a planned construction site, a 2-m-thick stratum of normally consolidated clay underlies a surface layer of compact granular soil 3 m deep. The unit weight for the compact granular soil is 20.2 kN/m³. The clay material has a unit weight of 17.6 kN/m³. The groundwater table is very deep. Laboratory testing of the clay indicates an in-place void ratio of 1.35 and a compression index of 0.42. The building planned for the site will create a stress increase of 24.5 kN/m² at the center of the clay layer.
  (a) Assume that the foundations for the building will be situated near the surface of the upper compact granular soil layer. Determine the foundation settlement due to primary compression occurring in the clay layer because of the stress increase.
  (b) Calculate the settlement to be expected if the groundwater table were at the soil surface, and compare to (a).

10-32. For the foundation loading and soil conditions described in problem 10-31, determine the compression occurring in the clay layer if the clay is overconsolidated and
  (a) the preconsolidation pressure is 120 kPa.
  (b) the preconsolidation pressure is 90 kPa.

10-33. A 7-ft layer of clay is buried beneath a 10-ft stratum of very compact granular soil. Compact sand underlies the clay. The layer of granular soil is composed of material having a unit weight of 130 psf. The clay unit weight is 105 pcf. A laboratory compression test on a sample of the clay indicates a compression index of 0.40 and a natural void ratio of 1.30. A planned building loading will cause a 550-psf stress increase at the middle of the clay layer.
  (a) What amount of primary compression occurs in the clay layer for the indicated conditions?
  (b) How much primary compression of the clay layer would result if the groundwater table was at the ground surface (all other conditions remain the same)?
  (c) How much clay layer compression would occur if the clay was an overconsolidated material, the past maximum pressure was 2000 psf, and the C_r value was 0.10? Assume a deep water table.

10-34. Calculate the foundation settlement due to primary compression occurring in a buried clay layer where the following conditions exist:
  • The groundwater table is very deep.
  • The thickness of buried clay layer is 6 ft.

- An 8-ft-thick layer of compact sand overlies the clay layer and extends to the ground surface.
- A deep zone of compact sand underlies the clay layer.
- The unit weight of the compact sand is 135 pcf, and the unit weight for the clay is 100 pcf.
- The foundation size and loading causes a stress increase of 700 psf at the center of the clay layer.
- Laboratory testing of the clay indicates the clay is normally consolidated, the soil void ratio is 1.28, and the compression index is 0.33.

10-35. A building foundation 3 m by 3 m will impose 1500 kN total loading onto the surface of a thick stratum of normally consolidated clay soil. Determine the foundation settlement due to primary compression in the clay, using layers that are 1 m, 1 m, 2 m, and 2 m thick, respectively, from foundation level downward. For simplification, use a soil unit weight of 17.5 kN/m³ constant with depth, an in-place void ratio of 1.20, and a compression index equal to 0.38 for each layer.

10-36. A planned construction site is underlain by a thick deposit of normally consolidated clay soil. A building foundation 6 ft square will be located on the ground surface and carry a total loading of 180,000 lb. Determine the foundation settlement by analyzing the volume changes due to primary compression in layers that are 2, 4, and 6 ft thick, respectively, from the foundation level downward. For simplification, assume a soil unit weight of 115 pcf constant with depth, an in-place void ratio of 1.05, and a compression index of 0.35 for each layer analyzed.

10-37. Estimate the settlement due to effects of primary compression plus secondary compression in a clay stratum 3 m thick where the compression index C_c is 0.33 and the value for e_0 is 1.25. Settlement due to primary compression (or consolidation) is computed to be 45 mm and will require approximately 6 years to complete. Calculate the settlement expected after a 15-year time period by summing the primary compression and secondary compression (select a ratio for C_α/C_c from the lower values of the typical range).

10-38. Approximate the settlement expected after a 25-year period from the effects of secondary compression in a clay layer 10 ft thick where the primary compression settlement is calculated to be 2 in. and will require a period of about 12 years

for completion. The compression index for the clay is 0.28, and e_0 is 1.15. (Refer to Eq. 10-15.)

10-39. Describe the events that take place when a saturated clay soil undergoes consolidation (the process of load transferring onto the soil skeleton, etc.).

10-40. What is the physical meaning of the coefficient of consolidation, c_v?

10-41. At a planned construction site, a 2-m-thick buried clay layer lies beneath a surficial stratum of free-draining granular soil. Free-draining granular soil also underlies the clay layer. Double drainage from the clay layer can therefore occur when construction loads cause consolidation. The coefficient of consolidation for the clay is 0.001 m²/day. Settlement calculations indicate that the clay layer will eventually compress 4 cm (primary compression or consolidation) due to the effect of building loads.
  (a) How long a time period is required for 90 percent of the estimated settlement to occur?
  (b) How much settlement occurs in the first 12 months?
  (c) What time period is required for a settlement of 2 cm?

10-42. A buried clay layer 10 ft thick is sandwiched between strata of free-draining granular soil so that double drainage during consolidation can occur. Calculations indicate that, because of a planned building loading, an ultimate settlement of 2 in. is expected as a result of primary compression in the clay layer. The coefficient of consolidation for the clay is 0.01 ft²/day.
  (a) How long will it take for 90 percent of the estimated settlement to take place?
  (b) How much settlement will occur in one year?
  (c) How long will it take for 1 in. of settlement to take place?

10-43. At a building site, a 3-m-thick clay layer is overlain by a stratum of sand and bounded on the bottom by rock. Calculations indicate that foundation settlement due to primary compression (consolidation) will be 40 mm. The coefficient of consolidation for the clay is 0.055 m²/month.
  (a) Determine the time period for 90 percent of the primary compression settlement to occur.
  (b) Indicate the settlement expected one year after construction.

10-44. A buried saturated clay layer 8 ft thick is bounded on top by a layer of granular soil having a high coefficient of permeability and along the bottom by impervious rock. Calculations for a foundation indicate that 2.5 in. of settlement will occur due to primary compression in the clay. The coefficient of consolidation (vertical drainage) for the clay is 0.02 ft²/day.

(a) Determine the period of time for 90 percent of the primary compression settlement to occur.

(b) Indicate the settlement (inches) that will occur in the first year after construction.

10-45. A surcharge and sand drain installation is proposed for an airport project to accelerate the consolidation of a thick deposit of fine-grained soil underlying the construction area. The plan is to install sand drains 1.5 ft in diameter at a staggered spacing so that each sand drain will handle a plan area 15 ft in diameter. Laboratory tests indicate that the coefficient of consolidation for horizontal drainage of the soil is 0.5 ft²/day. Computations indicate that the surcharge load will cause an eventual settlement of 12 in. How long a period will be required for 90 percent of this settlement to take place?

10-46. A surcharge and wick drain installation is planned as part of a surcharge program for a highway project undertaken to accelerate the consolidation of a fine-grained soil deposit underlying the construction area. The wick drains will be approximately 10 mm by 15 mm in cross section. Wicks will be installed in a hexagonal pattern so that each drains an area having an equivalent diameter of 1.25 m (ref. Fig. 10-26); assume an equivalent n ratio to be about 20 for the conditions described. Laboratory tests indicate that the coefficient of consolidation for horizontal drainage is 2.5 m²/month. Calculations indicate that the surcharge loading will cause an eventual total consolidation settlement of 1 m. How long a period will be required for 90 percent of this settlement to occur?

CHAPTER 11

Shear Strength Theory

The ability of a soil deposit to support an imposed loading or of a soil mass to support itself is governed by the shear strength of the soil. As a result, the shearing strength of the soil becomes of primary importance in foundation design, highway and airfield design, slope stability problems, and lateral earth pressure problems that deal with forces exerted on underground walls, retaining walls, bulkheads, and excavation bracing.[1]

In the study of the shear strength of soils, it is common to consider the two major categories of soil types—cohesionless and cohesive—separately. Overall, the factors that can affect the shearing strength of both soil types are the same. Practically, however, the factors that have the *most* influence on the shear strength that is or will be developed by each soil type are different.

The shearing strength and related deformations (or stress–strain relationship) of a foundation or construction soil is conventionally studied in the laboratory by testing soil samples obtained from the construction site, using established testing procedures. Additionally, field test methods have been developed for determining the shear strength of soil in its natural location, for reasons of expediency and economy and sometimes necessity, if samples for testing cannot be obtained. Typically, field determinations of shear strength are quick procedures, so that many soil samples (from different borings or test pits and different depths) can be checked easily and economically. Commonly with field tests, strength values only (no deformation data) are obtained.

11.1 LABORATORY TESTS

The most widely used laboratory tests for studying the shear strength and related deformations of soils include the direct shear test (single and double shear), the triaxial compression

---

[1]The shearing strength of soil deposits (which we recognize as consisting of an accumulation of discrete particles, possibly with some degree of interparticle bonding present) almost always refers to the resistance along a plane that passes between or along particle surfaces (but not through the particle). The mineral compositions of soil particles produce materials that have relatively high resistance to compressive and shear forces and that rarely are fractured or sheared when a soil mass "shears." A plane of resistance passing between or along particle surfaces is weaker than a plane of resistance through the solid particles.

test, and the unconfined compression test. For cohesive soils, vane shear tests can be used in the laboratory and the field to determine the cohesive strength, or cohesion, of soils (cohesion is related to shear strength).

Currently, the most preferred type of strength test is the triaxial test. The unconfined compression test is a type of triaxial test that is appropriate only for cohesive soils. The use of the direct shear test has decreased since the development of the triaxial test, but it still represents the basic approach for studying the stress–strain characteristics of a soil during shearing and possesses much merit as a learning tool.

Direct Shear Test

The direct shear apparatus for performing single shear is essentially a rectangular or circular box having separated lower and upper halves (Fig. 11-1a). After the sample to be tested is placed in the apparatus, a normal (compressive) loading is applied to compress the soil. The upper half of the apparatus is then moved laterally by a recorded shearing force, forcing the sample to shear across the plane between the two halves of the apparatus. The normal force is kept constant during the test. The shearing force starts at zero and increases until the sample fails (is sheared). Usually, a record of the magnitude of the shearing force and the resulting lateral movement is kept so that shearing resistance stress versus shearing displacement can be computed and plotted graphically. Changes in sample thickness that occur during the shearing process are also recorded so that volume change versus shearing stress or shearing strain can be studied. Typically, results of shearing stress versus shearing strain are as shown in Figure 11-2. The initial portion of the diagram is curvilinear (constantly changing slope), which continues until the maximum shear is reached, after which continuing deformation occurs with no increase in loading—that is, no increase in resistance. Failure is considered to have occurred at the maximum shearing value.

In direct shear testing, shearing can be accomplished by either controlling the rate of strain or the rate of stress. For the *strain-controlled* test, the shearing deformation (lateral movement) occurs at a controlled rate, usually continuously and at a constant speed. With this type of test, the shearing force necessary to overcome the resistance within the soil is automatically developed.

In the *stress-controlled* test, the magnitude of the shearing force is the controlled variable. The force (and therefore the stress) is increased at either a uniform rate or in established increments. For each increment of shearing force, it is applied and held constant until the shearing deformation ceases.

The strain-controlled shear test appears to be the most widely used of the two methods, probably because a mechanically operated strain-controlled apparatus is the simplest to devise.

Double shear testing is similar to the single shear test, except that two parallel surfaces are sheared. This type of test is usually performed on soil samples obtained from test borings that have utilized special tube equipment designed to fit directly into a shearing apparatus, so that a minimum of sample handling is required. As an illustration of one method, the soil-sampling apparatus used to extract samples from test borings is such that the soil is obtained in a tube whose interior lining consists of a series of rings. In the laboratory, three rings of soil are taken for the shear test. The shearing takes place as indicated in Figure 11-3. The stress–strain data from the double shear test are similar to the data from the single shear test.

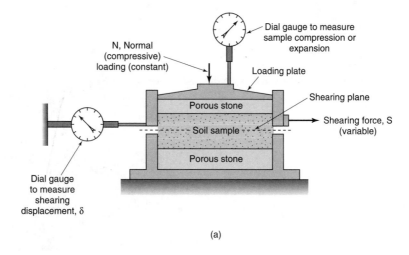

(a)

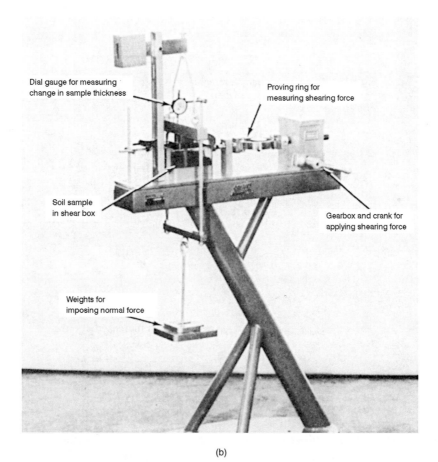

(b)

Figure 11-1 (a) Schematic diagram of direct shear apparatus. (b) Laboratory direct shear equipment (manually operated) (Courtesy of Soiltest, Inc.)

Figure 11-1 (c) State-of-the-art electronic direct shear apparatus (Courtesy of ELE International, Soiltest Products Division)

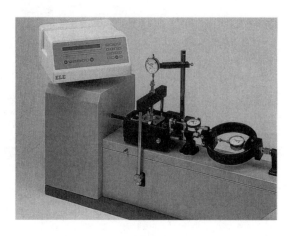

Figure 11-2 Typical plotted representation of shearing stress versus shearing displacement data.

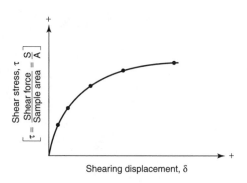

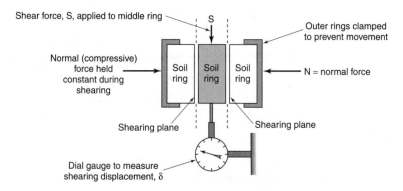

Figure 11-3 Schematic diagram of double-ring shear.

A variation of the direct shear test is a torsional test, where one section of the tested sample is twisted relative to the other. The torsional resistance developed on the failure plane is related to the shear strength of the soil. This test is not in common use. Similarly, some past studies have been made of using an extrusion-type test for cohesive soils, but such tests are not in common use.

Triaxial Compression Test

The triaxial test is currently the most popular test for determining the shearing strength of soils. Though not as simple a test as the direct shear test, it has several advantages that are of practical importance: (1) the loading conditions in the triaxial test can be made to simulate more accurately the loading conditions that the soil was (or will be) subjected to in its natural state; and (2) failure is not forced to occur across a predetermined plane, as occurs for the direct shear test.

In the triaxial test, a cylindrical soil sample (as conventionally obtained in soil borings) is wrapped in an impervious membrane for protection and placed in a chamber where an all-around, or confining, pressure can be applied. The confining pressure is usually applied through water or air introduced into the sealed chamber. The sample sits on a fixed pedestal, and a cap attached to a vertical piston rests on the top of the sample. In testing, the confining pressure is applied all around and to the top of the sample and, usually, held constant. An axial (vertical) load is subsequently applied to the sample through the piston, which passes through the top of the chamber. The axial load is steadily increased until failure of the sample occurs. Figure 11-4a shows the triaxial testing arrangement schematically.

Analysis of a three-dimensional element within the tested triaxial specimen shows that a lateral normal stress equal to the confining pressure p_c acts on orthogonal vertical planes, and a vertical stress acts on the horizontal plane. The vertical stress is equal to the sum of the confining pressure p_c plus that pressure increase resulting from the axial loading, Δp (Fig. 11-5).

In practical problems, the confining pressure is frequently selected to be equal to the confining or lateral pressure that acted on the soil in its natural location within the soil mass, and the axial pressure that eventually causes the sample to fail is assumed to be representative of the maximum vertical pressure that can be imposed on the soil at the depth at which the confining pressure acts. By relating test pressures to subsurface stresses, the confining pressure represents σ_3, the minor principal stress, while the *total* axial stress represents σ_1, the major principal stress (principal stress as defined in Chapter 8).

In the triaxial test setup, porous discs or stones are placed at the top surface and bottom surface of the sample so that pore water in the sample can be drained, if and when desired. The pedestal and loading cap are provided with drainage tubes for the water to escape and from which the pore water pressure can be determined. If the problem under study requires, water can also be introduced into the sample to determine the effect of increased water content on the strength.

The numerical difference between the total axial pressure σ_1 and the confining pressure σ_3 in the triaxial test is termed the *deviator stress*. When stress–strain data are graphed, it is the deviator stress that is conventionally plotted against strain. Typical qualitative results are shown in Figure 11-6.

For axially loaded test specimens, the related slope of the stress–strain curve is the modulus of elasticity E. The modulus of elasticity is an indication of the stiffness or

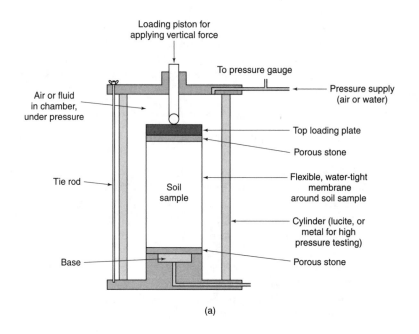

Loading piston for
applying vertical force

To pressure gauge

Pressure supply
(air or water)

Air or fluid
in chamber,
under pressure

Top loading plate

Porous stone

Tie rod

Soil
sample

Flexible, water-tight
membrane
around soil sample

Cylinder (lucite, or
metal for high
pressure testing)

Base

Porous stone

(a)

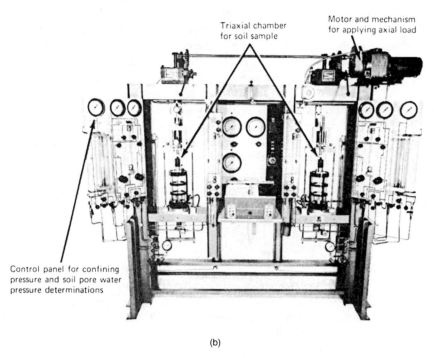

Triaxial chamber
for soil sample

Motor and mechanism
for applying axial load

Control panel for confining
pressure and soil pore water
pressure determinations

(b)

Figure 11-4 (a) Schematic diagram of triaxial compression test apparatus; (b) laboratory triaxial equipment (dual-chamber, motorized). (Courtesy of Soiltest, Inc.)

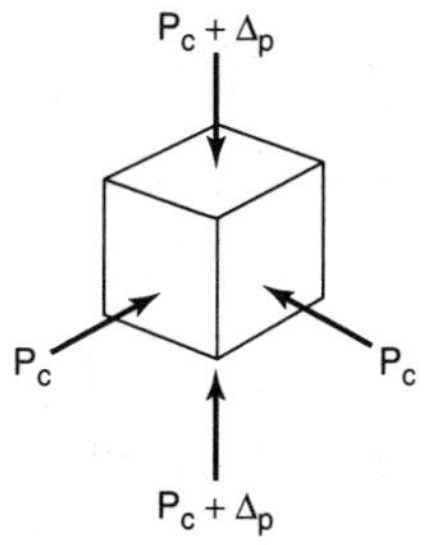

Figure 11-5 Stress combination acting on incremental element of soil subjected to triaxial testing.

Figure 11-6 Representative triaxial test data: (a) typical result for loose sands and normally consolidated clays; (b) typical result for dense sands and overconsolidated clays; (c) indication of common terms used to relate stress and strain.

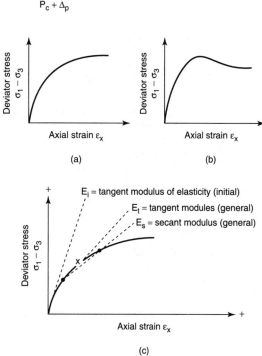

(a)

(b)

E_i = tangent modulus of elasticity (initial)

E_t = tangent modules (general)

E_s = secant modulus (general)

(c)

resistance to deformation of the material; the higher the value of E, the stiffer or stronger the material. A *tangent modulus* is the slope of a line drawn tangent to a point on the curve. A *secant modulus* is the slope of a line connecting any two points on the stress–strain curve. By reference to Figure 11-6, it should be clear that a tangent modulus or a secant modulus will not be constant for all parts of the curve. Both the tangent modulus and secant modulus are used in problem analysis.

Unconfined Compression Test

The unconfined compression test (Fig. 11-7) is a triaxial compression test but a type where the all-around confining pressure is zero. The axial force represents the only source of external pressure imposed onto the soil. Because it is necessary that the soil sample under test be capable of standing in the testing apparatus under its own internal strength, the test is limited to use for soils possessing cohesion. Test results are generally similar to conventional triaxial test results.

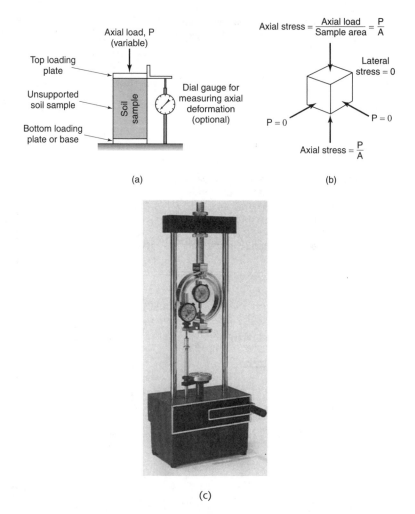

Figure 11-7 Representation of unconfined compression test: (a) test arrangement; (b) stresses acting on incremental element; (c) unconfined compression apparatus. (Photo courtesy of ELE International, Soiltest Products Division)

Vane Shear Test

Vane shear tests are coming into wider use for determining the in situ strength of cohesive soils. The vane shear apparatus consists of thin-bladed vanes that can be pushed into the soil with a minimum of disturbance. A torque applied to rotate the vanes is related to the shear strength of the soil. A vane shear apparatus, as pictured in Figure 11-8, can be attached to a long vertical rod and inserted into borings, so that the in-place strength of the soil can be determined without removing the soil from its natural state in the ground. Similar equipment is also used in the laboratory. It is specifically useful for determining the cohesive strength of soft or sensitive clays that could possibly be affected by the handling that the conventional laboratory shear tests require.

Figure 11-8 Vane of vane shear apparatus, for determining in-place shear or cohesive strength (also ref. Fig. 5-31).

Torque

Figure 11-9 Torvane for determining shear or cohesion of soil in field or laboratory.

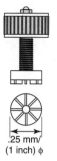

.25 mm/
(1 inch) ϕ

An adaptation of the vane shear apparatus is the torvane (Fig. 11-9), used for determining the cohesive strength of samples in the field or laboratory.

11.2   SHEAR TEST RESULTS PLOTTED ON MOHR'S CIRCLE COORDINATES

Results of the direct shear and triaxial tests can be plotted graphically (or semigraphically) on a shear stress versus normal stress coordinate system. They are the same coordinates used for Mohr's circle plots (Chapter 8). The stress combination that represents the condition at the maximum shearing strength or the ultimate shearing strength (at failure) is generally used for making the plot.

For the direct shear test (Fig. 11-10a), the results of *each* test would plot as a point, as indicated in Figure 11-10b.

With triaxial test results, the values of σ_3 and σ_1, or the principal stresses, are known, and plotting these points permits the Mohr's circle for these stress conditions to be drawn (Fig. 11-11b).

If identical homogeneous soil samples were tested under identical conditions in a direct shear test *and* a triaxial test, the same strength information, for practical purposes, would be obtained. Where the results from a direct shear test are superimposed on the results from a triaxial test, the plotted point of τ versus σ appears as shown in Figure 11-12a or b.

Strength or Failure Envelope

If a series of direct shear tests were performed on different samples of homogeneous soil and the normal loading were different for each test, the results plotted on τ_f versus σ coordinates could be as shown in Figure 11-13. A curve passing through the plotted points

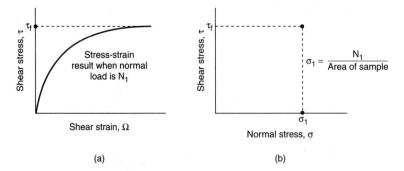

(a)

(b)

Figure 11-10 Direct shear results for one test: (a) stress–strain plot; (b) result of test on normal stress–shear stress coordinates.

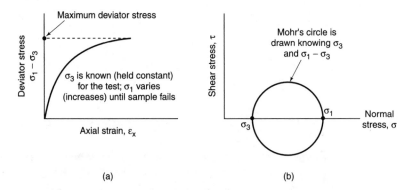

(a)

(b)

Figure 11-11 Triaxial results for one test: (a) stress–strain plot; (b) result of test on normal stress–shear stress coordinates.

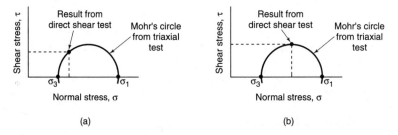

(a)

(b)

Figure 11-12 Result of direct shear and triaxial test plotted on same coordinates: (a) general location of τ versus σ point from direct shear test on cohesionless soil; (b) general location of τ versus σ point from direct shear test on cohesive soil.

establishes what is referred to as the *strength* or *failure envelope curve*. This envelope indicates the limiting shear strength that a material will develop for a given normal stress. Since the strength or failure envelope is a boundary established from test results, it should be evident that it is physically impossible to achieve a stress combination whose coordinates plot above the boundary.

The same information can be obtained from a series of triaxial tests performed on a homogeneous soil. In each test the confining pressure σ_3 is different. This consequently effects the maximum axial stress σ_1 that develops at failure. The Mohr's circle test results would be as shown in Figure 11-14.

A curve that is tangent to all the Mohr's circles establishes the strength or failure envelope curve. The *point of tangency* to each circle establishes the combination of shear and normal stresses that act on the failure plane in the tested soil sample. These should be the same combinations of shear and normal stresses obtained from direct shear tests.

From reference to the Mohr's circle plot, it should be understood that the plane of failure does not necessarily occur on the plane where the shear stress is a maximum. Rather, it is the *combination* of shear and normal stress developing within the material that is critical. The combination that *is* critical relates to the slope or angle of the failure envelope curve. If the strength (failure) envelope is horizontal (parallel to the normal stress axis), the *maximum* shear stress indicated by the Mohr's circle is the shear stress on the failure plane. Where the envelope curve makes an angle with the horizontal axis, the shear stress on the failure plane is *less* than the maximum shear stress represented by a Mohr's circle. In soils, the condition of both the horizontal and the sloped strength–failure envelope exists. This is discussed further in Section 11.3.

Quite frequently in soil analysis, the Mohr's circle plot requires use of only the upper half of the circle, the lower half being considered redundant because it is a mirror image of the upper half. However, the entire circle is best shown when strength–failure envelopes are

Figure 11-13 Results of direct shear tests on samples of a homogeneous soil.

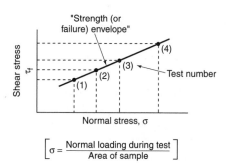

$$\left[\sigma = \frac{\text{Normal loading during test}}{\text{Area of sample}}\right]$$

Figure 11-14 Results of triaxial test on samples of a homogeneous soil.

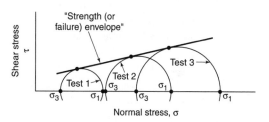

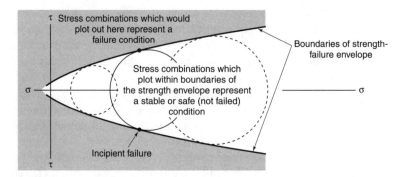

Figure 11-15 Concept of strength–failure envelope showing combination of stress representing failure and stable conditions.

discussed so as to establish clearly the concept of boundaries for the nonfail and fail combination of stresses (see Fig. 11-15).

For strength conditions indicated by the Mohr's circle of Figure 11-15, it should be seen that an incipient failure situation develops on two different planes simultaneously. But when it is desired only to learn the combination of principal stresses that will cause failure or the magnitude of shear and normal stresses acting on the failure planes, as is usual, using only half of the Mohr's circle is sufficient.

11.3 SHEARING STRENGTH

Basic understanding of the shearing strength of cohesive and cohesionless soils can be obtained by reference to the results from triaxial or direct shear tests. A discussion relating to the more simple direct shear test is easiest to follow, and is presented first. The behavior described, and the explanation of the factors affecting the behavior, apply to triaxial conditions as well as to direct shear conditions. A discussion of triaxial testing and related results is presented later in this chapter.

Shearing Strength of Cohesionless Soil

The results of a direct shear test performed on a dry cohesionless soil (e.g., a dry sand), where the normal load is held constant during the shearing process, are presented on stress–strain coordinates. Typical results are indicated in Figure 11-2.

The actual configuration of the stress–strain curve will be affected by the size and shape of the soil particles and by the density of the sample at the beginning of the test, as well as by the magnitude of the normal loading. A sample that is initially loose will develop a stress–strain curve as indicated in Figure 11-16a, whereas an initially dense sample will present a curve as shown in Figure 11-16b. The ultimate strength for samples whose only difference is the initial density will be the same. The difference in the behavior is explained by the factors that contribute to the shearing strength of dry cohesionless soil. Resistance to movement across the failure plane, and hence the shearing *strength,* is developed from friction that occurs between particle surfaces under the applied normal loading and to interlocking

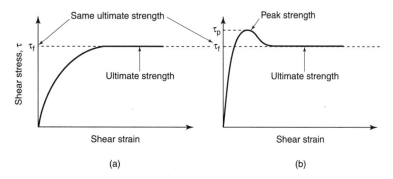

Figure 11-16 Comparison of shear test results as affected by soil density: (a) cohesionless soil, sample initially loose; (b) cohesionless soil, sample initially dense.

between particles. The extra shearing resistance for the initially dense sample (difference between peak strength and ultimate strength shown in Fig. 11-16b) is attributed to a greater degree of interlocking. This greater interlocking is overcome as shearing displacements increase, indicating that the initially dense sample loosens during shearing. This is verified by recording the thickness (or volume) of the sample during testing and observing that a thickness increase occurs. In other words, the void ratio of the initially dense sample increases during shearing. Conversely, the thickness of the initially loose sample decreases during testing, indicating a decrease in the void ratio. At the ultimate strength, after shearing movement, the final void ratio will be similar for the initially dense and loose materials, regardless of the initial void ratio. This final void ratio is referred to as the *critical void ratio.* Soils whose natural void ratios are above the critical void ratio will attempt to decrease in volume during shearing, whereas the reverse is true for soils whose natural void ratio is below the critical ratio (Fig. 11-17).

If a series of shearing tests are performed on identical samples, but the normal loading that acts during shearing is different, the ultimate shearing resistances (and peak resistances if the soil is initially dense) will also be different. Greater shearing resistances, or shearing strengths, are obtained at higher normal loadings. (Fig. 11-18).

If the results of the series of direct shear tests are plotted on shear stress versus normal stress coordinates, as for a Mohr's circle analysis, a strength–failure envelope curve for the soil is obtained (Fig. 11-19).

Through the range of normal stresses usually encountered in foundation problems, the plotted points of τ versus σ establish what is basically a straight line. The angle that the strength–failure envelope creates with a horizontal line is indicated as ϕ, the *angle of internal friction* for the soil. For dry cohesionless soils, the strength–failure envelope obtained from ultimate shear strength values is assumed to pass through the origin of coordinates. Thus, the relation of shearing strength to normal stress can be calculated from

$$\tan \phi = \frac{\tau}{\sigma}$$

or

$$\tau = \sigma \tan \phi \qquad (11\text{-}1a)$$

Figure 11-17 Effect of initial density on change in void ratio during shearing.

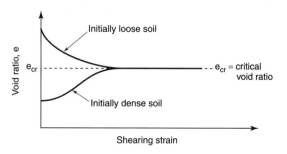

Figure 11-18 Effect of normal loading on shear strength of dry cohesionless soil.

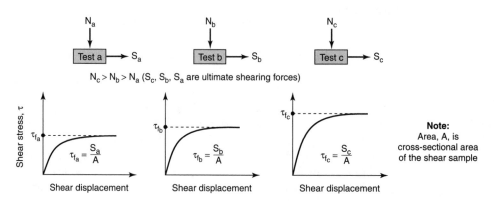

$N_c > N_b > N_a$ (S_c, S_b, S_a are ultimate shearing forces)

$\tau_{f_a} = \dfrac{S_a}{A}$

$\tau_{f_b} = \dfrac{S_b}{A}$

$\tau_{f_c} = \dfrac{S_c}{A}$

Note:
Area, A, is cross-sectional area of the shear sample

Figure 11-19 Summary representing maximum shear strength versus applied normal stress for dry cohesionless soil.

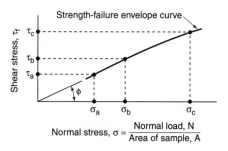

Normal stress, $\sigma = \dfrac{\text{Normal load, N}}{\text{Area of sample, A}}$

Table 11-1 Representative Values of ϕ for Cohesionless Soils

| | Angle ϕ (degrees) | |
Soil Type	Ultimate	Peak
Sand and gravel mixture	33–36	40–50
Well-graded sand	32–35	40–50
Fine to medium sand	29–32	32–35
Silty sand	27–32	30–33
Silt (nonplastic)	26–30	30–35

If the peak values of shear strength are plotted, values of φ somewhat greater than obtained from the ultimate shear strength values will be obtained. Representative values for φ are presented in Table 11-1.

The value of φ (ultimate or peak) selected for use in practical soil or foundation problems should be related to the soil strains that are expected. If soil deformation will be limited, using the peak value for φ would be justified. Where deformations might be relatively great, ultimate values of φ should be used.

Illustration 11-1

A sample of dry sand is tested in direct shear. A normal load equivalent to 96 kPa is imposed for the test. The shearing force applied to fail the sample is increased until shearing does occur. The shear stress at failure is 65 kPa. What is the angle of internal friction φ for the sand?

Solution

$$\tan \phi = \frac{65 \text{ kPa}}{96 \text{ kPa}} = .677$$

$$\phi = 34° \pm$$

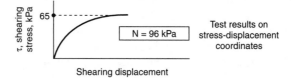

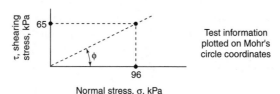

Illustration 11-2

A dry cohesionless soil is tested in a triaxial test to determine the angle of internal friction φ. A confining pressure equal to 1000 psf is used. The sample fails when the axial load causes a stress of 3200 psf. What is the value of φ?

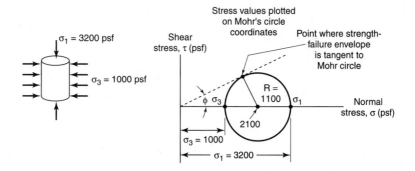

Solution

$$\text{Radius of circle} = \frac{3200 - 1000}{2} = 1100 \text{ (psf)}$$

$$\text{Center of circle} = 1000 + 1100 = 2100 \text{ (psf)}$$

From the Mohr's circle plot,

$$\sin \phi = \frac{1100 \text{ psf}}{2100 \text{ psf}} = .525$$

$$\phi = 31.5° \pm$$

If moisture is present in the soil samples during shearing, the subsequent plotting of ultimate shearing strength versus the applied normal stress does not present a failure envelope curve that passes through the origin. Rather, at zero normal stress, the failure envelope intersects the shear stress coordinate, as shown in Figure 11-20. This indicated "no load" shearing strength is referred to as *apparent cohesion,* a shear strength value attributed to factors other than friction developed from the normal stress. With damp cohesionless soils, the extra strength is due to compressive forces exerted on soil particles as a result of surface tensions where water menisci have formed between soil particles (as in capillary water; see Chapter 6). The extra shearing strength exists as long as the soil retains some moisture. The extra strength would be lost if the soil were to dry out or to become saturated or submerged. For the saturated and submerged case, all voids are filled with water, and all water menisci are lost. For these reasons, the extra shear strength attributed to apparent cohesion generally is neglected in foundation studies.

Values of ϕ are not significantly affected when the soil is below a groundwater table or otherwise submerged. However, any hydrostatic pressure u that acts on a plane under analysis reduces the intergranular compressive stress (and will affect the soil shear strength). The resulting *effective stress,* $\overline{\sigma}$ (equal to $\sigma_t - u$; see Eq. 9-4) is utilized to obtain a preferred *general* expression for the shear strength of a cohesionless soil, whereby

$$\tau = (\sigma_t - u) \tan \phi = \overline{\sigma} \tan \phi \qquad (11\text{-}1b)$$

where σ_t is the total stress acting at a point. When u is zero, $\overline{\sigma}$ equals σ_t and Equation 11-1a, derived for the case of a dry cohesionless soil, results.

Figure 11-20 Results of shear tests on moist cohesionless soils indicate an "apparent cohesion" value.

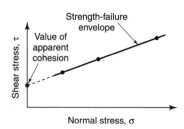

Illustration 11-3

(a) Samples taken from a uniform deposit of granular soil are found to have a unit weight of 19.6 kN/m³ and an angle of internal friction of 35°. What is the shearing strength of the soil on a horizontal plane at a point 4 m below the ground surface?

Solution

At a depth of 4 m, the soil overburden pressure, or normal stress, is

$$(19.6 \text{ kN/m}^3)(4 \text{ m}) = 78.4 \text{ kN/m}^2 = \sigma \text{ on horizontal plane}$$

The shearing resistance that can be developed is

$$\tau = \sigma \tan \phi = (78.4 \text{ kN/m}^2)(\tan 35°)$$
$$= 54.9 \text{ kN/m}^2$$

(b) A proposed structure will cause the vertical stress to increase by 60 kN/m² at the 4 m depth. Assume that the weight of the structure also causes the shearing stress to increase to 52 kN/m² on a horizontal plane at this depth. Does this shearing stress exceed the shearing strength of the soil?

Solution

The total vertical pressure due to the structure and soil overburden is

$$60 \text{ kN/m}^2 + 78.4 \text{ kN/m}^2 \text{ overburden pressure} = 138.4 \text{ kN/m}^2$$

The shearing strength that can be developed by the soil at this depth is

$$\tau = (138.4 \text{ kN/m}^2)(\tan 35°) = 96.9 \text{ kN/m}^2$$

This would indicate that the shear strength of the soil is greater than the imposed shear stress; therefore, a shear failure does not occur (96.9 kN/m² > 52 kN/m²).

If the water table rose to the ground surface, the effective soil overburden pressure would be reduced to about

$$\left(\frac{1}{2} \times 19.6 \text{ kN/m}^3\right)(4 \text{ m}) = 39.2 \text{ kN/m}^2$$

[This value represents the effective vertical stress, i.e., $\bar{\sigma} = \sigma_t - u = \gamma_{sub}Z = (\gamma_t - \gamma_w)Z$.]
    The total vertical stress would be

$$39.2 \text{ kN/m}^2 + 60 \text{ kN/m}^2 = 99.2 \text{ kN/m}^2$$

The shear strength available is

$$\tau = (99.2 \text{ kN/m}^2)(\tan 35°) = 69.46 \text{ kN/m}^2$$

This is still greater than the shear stress resulting from the loading conditions; that is, 69.46 kN/m² > 52 kN/m².

The preceding discussion has indicated that the shearing strength that can develop is directly proportional to the effective normal stress that acts on the plane under analysis. For most practical problems, the effective stress in cohesionless soil is calculated by using the effective overburden weight plus any stress increase created by structural loading. However, if the soil is saturated, or nearly so, there is the possibility that the stress resulting from loads newly applied onto the soil mass will not result in a related increase to the effective stress acting within the soil. This is the situation of an excess hydrostatic pressure condition developing (pore water in the soil voids becomes subject to a pressure greater than the normal hydrostatic, as discussed in Section 10-4.). With this situation, there is the potential danger that the shearing stress resulting from the external loading may increase faster than the soil shearing strength (which is controlled by the effective stress), and a shear failure may occur. Fortunately, cohesionless soils have relatively high rates of permeability that under most conditions permit rapid drainage of pore water when new loadings and stresses are imposed onto the soil mass. The danger of excess hydrostatic, or excess pore pressure, conditions in cohesionless soils and possible shear failures are generally limited to the situation where loose saturated material is exposed to vibratory, instantaneous, or shock loading, such as from explosives, earthquakes, and traveling trains. The occurrence of loss of strength under these conditions is called *liquefaction,* for the soil momentarily liquefies and tends to behave as a dense fluid (refer to Chapter 3). Generally, for sandy soils, the number of vibrations or shock waves required for liquefaction to occur increases as the soil density increases, but the presence of fine-grained material decreases the susceptibility to liquefaction. The soils found to be most susceptible to liquefaction are the saturated and loose fine to medium sands having a uniform particle-size range.

Approximating Values of ϕ from Boring Data

Soil borings for subsurface investigations are frequently sized so that soil samples can be obtained with a "standard" two-inch- or 51-mm-diameter split-spoon soil sampler. Soil-boring and sampling procedures are discussed in Chapter 5. With the split-spoon sampler, the blow count required to drive the sampler into undisturbed soil is recorded for all samples obtained. The samples recovered with this equipment are considered to be disturbed and are unsuitable for performing strength tests. Obtaining undisturbed samples of cohesionless soil is generally difficult under many existing subsurface conditions and may require the use of large-diameter borings and special soil samplers. Because of the widespread usage of the split-spoon sampler (partially because of the relative economy), correlations between the blow count N from the standard penetration test and the angle of internal friction have been developed. These correlations are presented in Table 11-2. Due to the generalized nature of the correlation, any application of such data to final foundation designs should be made with caution.

Shearing Strength of Clay Soils

The shearing strength that a clay deposit possesses is related to the type of clay mineral and the water content but, very importantly, also to the effective stress or consolidation pressure to which the soil has been subjected in its past (the soil's stress history). The change that is possible in a clay's shear strength (such as if loading conditions change because of the

Table 11-2 Approximate Relationship between N and ϕ for Cohesionless Soil

Value of N[a]	Relative Condition of Soil	Approximate Value of ϕ
10	Loose	30° ±
20	Medium-dense	32° ±
30	Medium-dense to dense	35° ±
40	Dense	38° ±
50	Dense to very dense	40° ±
60	Very dense	42° ±

[a]In the so-called standard penetration test, N is the number of blows required to drive a standard 2-in.-outside-diameter (51 mm) split-barrel soil sampler 12 inches (.3 m) into undisturbed soil with a 140-lb weight falling 30 inches (or 63.5 kg with a 0.76 m drop). Values in this table refer to soil sampling procedures where the efficiency of the drop hammer is approximately 60%. For further information on the standard penetration test, refer to Chapter 5.

weight of a new structure) is affected by these just-mentioned factors but also by the pore water drainage that can occur as shearing deformations tend to occur. Consideration of drainage is of practical importance, because many clays in their natural condition are close to full saturation, and the low permeability of these soils tends to inhibit changes in pore water content that try to occur during shearing.[2]

The significance of effective stress and drainage during shearing can be explained by reference to a series of shearing tests made on fully saturated clays. Practically, the discussion also applies to nearly saturated clays.

Assume that the soil for all the test samples is obtained from one location in a homogeneous, isotropic, normally consolidated clay deposit. All properties of all samples to be tested are identical (Fig. 11-21). Samples are to be tested in direct shear.

In each series of direct shear tests, each test sample will have a different normal load applied during shearing to determine the effect. The normal loads are greater than the effective overburden (or consolidation) pressure.

For the first series of tests, no drainage of pore water is permitted, either when the normal load is applied or during shearing. This "no drainage" control can be achieved by having the sample wrapped in a thin, tight, impermeable membrane when it is placed in the shearing apparatus. The normal load is applied just prior to beginning the shearing process. Shearing is then completed relatively quickly. These test conditions are referred to as *unconsolidated–undrained* (U–U) shear tests, because the samples are not permitted the time or the drainage necessary to consolidate to the pressure exerted by the normal loading, and no drainage or volume change is permitted during the shearing process. For this condition, excess pore water pressures develop in the samples during shearing. Results for this series of tests are shown in Figure 11-22. For the U–U conditions, the shearing strengths for all samples are the same, since the effective

---

[2]Strictly speaking, stress history and drainage factors can also affect the shearing properties of a cohesionless soil, although shape and size distribution of particles have very considerable influence. A granular soil that had been subjected to high effective stresses or high overburden pressures in its past would be dense and capable of developing a peak strength, as shown in Figure 11-16b. The prevention of pore water drainage during shearing of a saturated soil would lessen the shearing strength, but under most practical conditions there are no restrictions on drainage because of the relatively high permeability.

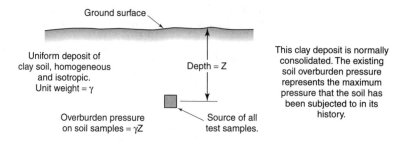

Figure 11-21 Source of samples for analysis of shear strength of clay soils.

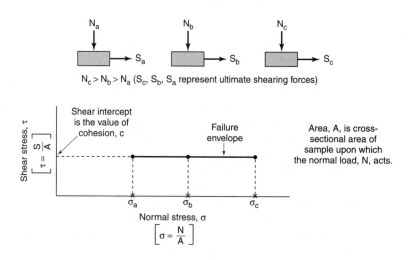

Figure 11-22 Shear strength results for unconsolidated–undrained (U–U) tests.

stress within each sample has not changed from the condition that existed when the soil was in the ground. Even though different normal loads are applied during shearing, consolidation to the normal load pressure can only occur if pore water is permitted to drain from the sample. Since drainage and consolidation cannot occur, the internal effective stresses are the same for each of the samples, and therefore, the shearing strengths will remain the same.

For the second series of tests, different normal loads are applied to each test sample and full consolidation to the respective new normal load is permitted. After full consolidation has taken place so that the effective stress acting is equal to the applied normal loading, the samples are wrapped in the impermeable membrane (so that no further drainage or volume change can occur) and are placed in the shearing apparatus. Shearing is performed quickly. These test conditions are referred to as *consolidated–undrained* (C–U), indicating that the samples are consolidated to their respective normal load prior to shearing but that during shearing no drainage or volume change is permitted. For this condition, excess pore water pressures develop in the sample during shearing, but they are not as great as those that occur in the U–U test. Results for such a series of tests are shown in Figure 11-23. The increase in shearing strength (compared to the U–U conditions) is the result of the increased effective pressure to which the test samples have been consolidated.

Figure 11-23 Shear strength results for consolidated–undrained (C–U) tests.

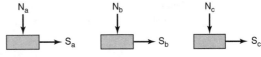

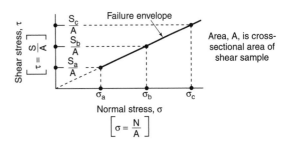

For the third series of tests, the samples are fully consolidated to the normal loads that will be applied during the shearing test. However, in this series, shearing will take place very slowly, and drainage and volume changes *are permitted* during the shearing process. These test conditions are referred to as *consolidated–drained* (C–D) conditions. No excess pore water pressures develop in the soil during shearing. Results for such a series of tests are shown in Figure 11-24.

The results from the three series of tests are presented together in Figure 11-25. A difference in the C–D and C–U curves results because of a normally consolidated soil's tendency to undergo volume decrease during shearing. If volume change is prevented, as in the undrained test, slight excess pore pressures are developed, with a subsequent reduction in the effective stress and shear strength.

As for tests on cohesionless soils, the slope of the τ versus σ curve is designated by the angle ϕ.

The stress represented by point A in Figure 11-25 is approximately the value of the effective stress $\overline{\sigma}_{v_c}$ that acted on the soil in its natural subsurface location. If the soil has been normally consolidated, $\overline{\sigma}_{v_c}$ is equal to the soil overburden pressure ($\overline{\sigma}_{v_c} = \gamma Z$; see Fig. 11-21).

If C–U or C–D tests were performed and the applied normal loading was less than $\overline{\sigma}_{v_c}$, the plotted shear strengths would fall close to the values for the failure envelope curve obtained for the U–U test series.

The strength envelope curves of Figures 11-22, 11-23, 11-24, and 11-25 indicate the total stress condition existing on the shear failure plane (i.e., the total normal stress and related soil shearing resistance). Recall that the total normal stress represents the combined influence of effective stress $\overline{\sigma}$ and neutral stress u (the latter term indicating stress resulting from pore water pressure), or

$$\sigma_t = \overline{\sigma} + u \qquad \text{(from Eq. 9-4)}$$

A rearrangement gives

$$\overline{\sigma} = \sigma_t - u$$

Figure 11-24 Shear strength results for consolidated–drained (C–D) tests.

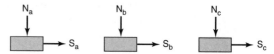

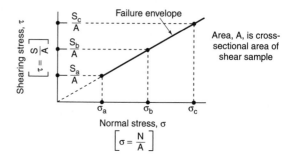

$$N_c > N_b > N_a \ (S_a, S_b, S_c \text{ represent ultimate shearing forces})$$

$$\left[\sigma = \frac{N}{A} \right]$$

Figure 11-25 Qualitative comparison of shear strength results for U–U-, C–U, and C–D tests.

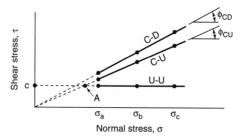

Values on the strength envelope from the consolidated–drained (C–D) test series indicate effective stress as well as total stress, since no excess pore water pressure existed during the shearing process. For any C–U- or U–U-type test, the magnitude of excess pore water pressure existing in the soil at shear failure is indicated from the relationship between the C–D and C–U envelopes, or the C–D and U–U envelopes; at points of similar shear strength, the difference in the normal pressures shown on the C–D and C–U curves, or the C–D and U–U curves, represents the excess pore water pressure (Fig. 11-26).

Illustration 11-4

To determine the strength properties of a clay soil, a series of consolidated–undrained direct shear tests is performed. For the first test, the normal pressure is 36 kPa and the sample fails when the shear stress is 21.6 kPa. The second sample is tested under a normal load of 72 kPa, and failure occurs when the shear stress is 24 kPa. A third sample is tested under a normal loading of 120 kPa, and failure occurs when the shear stress is 38.4 kPa. From these data, estimate the cohesion of the soil in the in situ condition and the value of ϕ_{CU}.

Solution
A problem of this type frequently is best solved graphically.

1. Locate test data on a scaled plot.

2. Draw a sloped failure envelope line to estimate ϕ_{CU}.

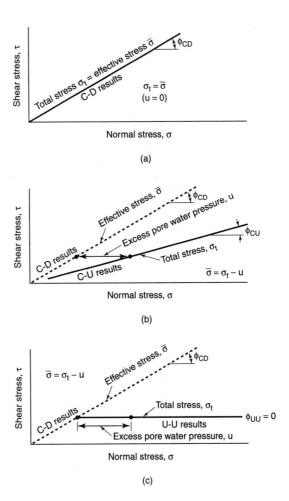

Figure 11-26   Relationship between effective stress, total stress, and pore water pressure: (a) C–D shear test results; (b) C–U shear test results; (c) U–U shear test results.

3. For a test having a low normal (compressive) pressure, assume that the pressure is less than $\overline{\sigma}_{v_c}$, and draw a horizontal failure envelope line through it to obtain the value of in situ cohesion.

4. The intersection of sloped and horizontal failure envelope lines gives a normal pressure value that is approximately the value of $\overline{\sigma}_{v_c}$ (the maximum past overburden stress).

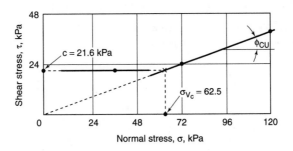

From the scaled plot, $c = 21.6$ kPa, $\phi_{CU} = 17° \pm$

In many practical soils problems involving clays, applied structural loads result in excess pore pressures being developed. Some drainage does occur, but it is restricted because of the low permeability of the soil. Consequently, the actual shearing conditions that are developed fall somewhere between the U–U and the C–D case. In many soil and foundation design problems, the U–U strength is used where the initial period of loading is critical, since it is realized that the shear strength will become greater (increase) and conditions safer with time as excess pore water drainage occurs.

These shear test results also indicate that it is possible to improve the shear strength of clay soils by consolidation, provided that time is available for permitting the necessary pore water drainage to take place. In effect, consolidation results in decreasing the water content of the clay and obtaining a related increase in shear strength, as indicated by Figure 11-27.

Shear Strength of Clay Related to Triaxial Testing

Although the direct shear test is a convenient reference for introducing the factors affecting the shearing strength of clay soils, the triaxial compression test is the method most typically used in the laboratory for determining soil shearing strength properties. Recall from the early sections of this chapter that the triaxial compression test subjects the soil sample to a three-dimensional state of compression (i.e., compressive stresses act in the three orthogonal directions), thereby causing shear stresses to develop within the sample (Fig. 11-28; also ref. Fig. 11-4).

The following discussion relates to clay soil as defined by Figure 11-21; that is, the clay soil is fully saturated and the deposit from which the test sample is obtained is a *normally consolidated* deposit (i.e., the existing soil overburden and lateral pressures represent the maximum pressures the soil has been subjected to in its history as part of the deposit).

The strength-failure envelope for a soil can be determined by performing the triaxial test on separate but identical samples; each test uses a different confining pressure (Fig. 11-29). As explained in Chapter 8, it has been found adequate to work with only the σ_3 and σ_1 values on the Mohr's circle coordinates. In Figure 11-29, the value of σ_3 is the all-around confining pressure or stress acting on the soil sample at the start of the test, and σ_1 is the *measured* vertical (axial) stress that subsequently causes the soil sample to shear. If $\sigma_{3c} > \sigma_{3b} > \sigma_{3a}$, it will be found that $\sigma_{1c} > \sigma_{1b} > \sigma_{1a}$.

Figure 11-27 Change in strength for a cohesive soil as water content changes.

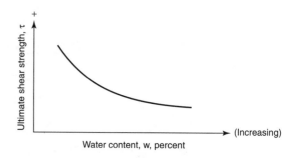

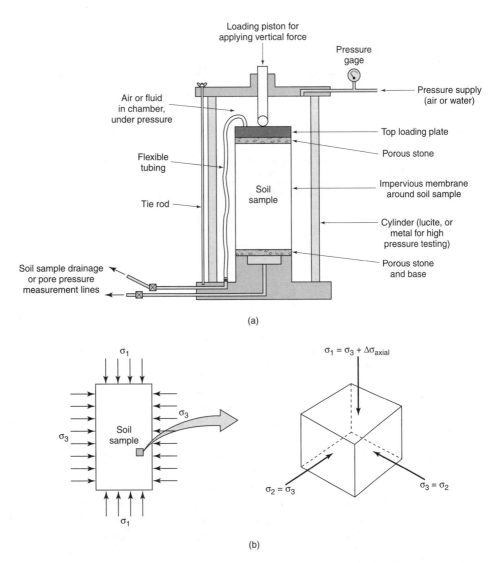

Figure 11-28 Triaxial compression test representations: (a) schematic diagram of test apparatus; (b) state of stress on an incremental element in the soil sample.

The typical triaxial equipment depicted by Figure 11-28a includes tubing lines connected to the porous stones positioned on the top and bottom of the soil sample; these lines can be utilized to drain pore (void) water from the sample or to measure or control or vary the pore water pressure existing within the soil sample during the test procedure.

In the following descriptions, the magnitudes of the σ_3 and σ_1 test pressures are *greater* than the σ_3 and σ_1 stresses that acted on the soil sample in situ (in the ground).

For the consolidated–drained (C–D) test, the confining pressure is imposed onto the test sample and the porous stone drain lines are left open to enable soil pore water to escape

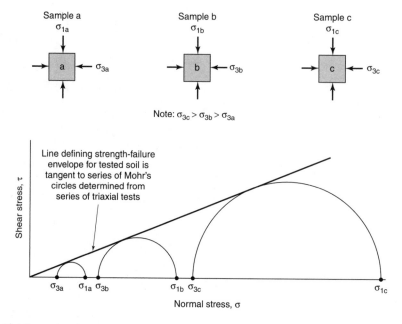

Figure 11-29 Results of a series of triaxial tests performed on identical soil samples, as used for determining the strength envelope for the soil. (Representative of results for cohesive soils subject to consolidated-type tests and for cohesionless soils.)

from the soil as the confining pressure squeezes the sample (i.e., consolidates it). After the sample is fully consolidated by the confining pressure (identified as the condition when equilibrium for the imposed loading is reached, whereby no further pore water drains), the axial force is slowly applied. The drain lines remain open as the axial force increases, permitting additional drainage of sample pore water to occur: Since additional compression (i.e., volume decrease) of the sample tends to occur because of the increasing (new) vertical stress, the continuous escape of pore water must be permitted if the fully consolidated (no excess pore water pressure) condition is to be satisfied. The axial and confining pressure acting when the soil sample shears provide the values to plot one Mohr's circle for the C–D case. A series of C–D triaxial tests on identical soil samples provide the data for determining the *effective stress,* $\overline{\sigma}$, envelope curve, as indicated in Figure 11-30.

In the consolidated–undrained (C–U) test, the soil sample in the triaxial apparatus is fully consolidated to the all-around confining pressure, σ_3, by permitting the pore pressure drain lines to remain open (as for the C–D test). Then the drain lines are closed and the axial load is imposed. Pressure gages on the drain lines permit measurements of the pore water pressure developing in the sample as the axial load is applied. The increasing axial force acting on the soil sample tries to cause sample compression (i.e., further consolidation), but since the pore water is now prevented from escaping because the drain lines are closed, an "excess pore water pressure" develops. The pore water pressure continues to increase in relation to the increase in axial loading. When the soil sample shears, the axial load, the internal pore water pressure, and the confining pressure are known (measured and recorded). The axial and confining pressures indicated by the triaxial gages represent the

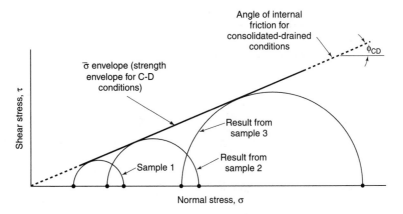

Figure 11-30 Triaxial results from consolidated–drained (C–D) tests and related strength envelope.

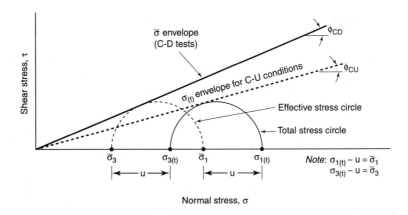

Figure 11-31 Results from consolidated–undrained (C–U) triaxial test and related strength envelope.

total axial and confining stresses acting on the soil sample. A total stress minus the measured pore water pressure provides the value of interparticle, or effective, stress acting within the soil sample at the time of failure[3]:

$$\overline{\sigma} = \sigma_t - u$$

The test results are presented on the Mohr's circle coordinates, as shown by Figure 11-31. As for the C–D test, a series of tests performed on identical samples but at different confining pressures permits a strength envelope to be obtained.

---

[3]This equation helps clarify the phrase "excess pore water pressure." That term refers to the magnitude of pressure developed in the void space water that is above the magnitude of the interparticle pressure that acts. The occurrence is similar to the undergraduate strength of materials problem, where a structural member composed of two different materials carries a compressive load; the portion of the total load carried by each material relates to the stress–strain properties or rigidity of the material (the least deformable or more rigid material supports more than its share of the total load).

In the performance of the unconsolidated–undrained (U–U) test, the soil sample is assembled in the triaxial apparatus and the pore water drain lines are closed. The confining pressure, σ_3, is applied and quickly followed by application of the axial load. The imposed pressures act to compress the soil sample (i.e., force the soil particles closer together), but because the pore water drains are closed, the trapped pore water prevents a decrease in sample volume from occurring. The trapped pore water has prevented the sample from consolidating. The pressure gages on the pore water drain lines will show increasing pore water pressure as the axial force increases. The values indicated by the triaxial gages for the confining pressure, σ_3, and axial stress, σ_1, represent the total pressure or stress internal to the soil sample. The effective stress, or intergranular pressure, when the sample shears is determined by deducting the pore water pressure from the total pressure (as for the C–U test), as shown in Figure 11-32a, or

$$\bar{\sigma} = \sigma_t - u$$

A series of tests on identical samples, but at differing confining pressures, σ_3, provide the data to obtain the strength envelope for the U–U case as shown in Figure 11-32b. Note that the Mohr's circles will be the same for all samples but will be positioned horizontally in accord with the σ_3 value used to perform the test. Even though the σ_3 and σ_1 values that cause the sample to shear are different for each of the tests, the internal *shear strength* of the soil remains the same because the internal conditions remained the same (the presence and quantity of pore water, the sample void ratio, and the particles' spacing will not change when trapped pore water prevents a volume decrease).

Somewhat different occurrences result when the confining pressure applied in the triaxial test is *less than* the confining pressure that existed when the soil was in situ (for this occurrence, the soil is considered overconsolidated in regard to the test conditions). When the test confining pressures *exceed* the in situ pressures, the sample volume tends to *reduce* during testing; conversely, the soil sample tends to *expand* during testing when the confining pressure is *less than* the in situ pressure. If an overconsolidated sample is prevented from taking in water during testing as the sample attempts to expand (the triaxial drain lines are closed), a negative pore water pressure develops in the soil void spaces (a condition similar to the effects of capillary tension) and the relationship between total and effective stress becomes

$$\bar{\sigma} = \sigma_t - (-u) = \sigma_t + u$$

The results plotted on the Mohr's circle coordinates are shown in Figure 11-33.

In the natural (in situ) condition, overconsolidated clay soils will tend to undergo expansion during some types of construction activity, such as excavation work, and negative pore water pressures will develop within the soil. A high soil shear strength results, as indicated by the Mohr's circle for the initial strength condition (Fig. 11-33). However, natural water, which is or becomes present, will tend to be drawn into the clay voids. Eventually, as the negative pore pressures vanish, the soil shear strength decreases toward the value indicated by the effective stress envelope (Fig. 11-34).

Where the U–U strength is required to be known for analysis of a soil or foundation study, it is convenient to determine the shear strength from unconfined compression tests. As explained earlier in this chapter, the unconfined compression test is a triaxial test where

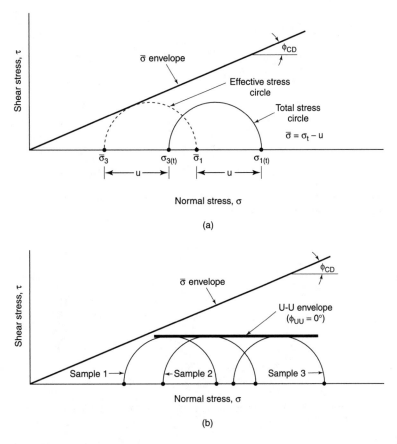

Figure 11-32 Unconsolidated–undrained (U–U) triaxial test data: (a) relationship between total stress, effective stress, and pore water pressure; (b) results of a series of tests on identical soil samples provide Mohr's circle of equal size but transposed laterally.

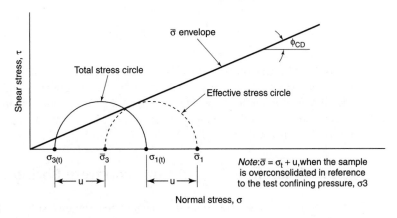

Figure 11-33 Data from U–U triaxial test on overconsolidated clay soil.

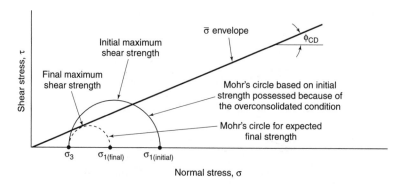

Figure 11-34 Mohr's circles illustrating expected change in strength for overconsolidated clay when soil water content increases.

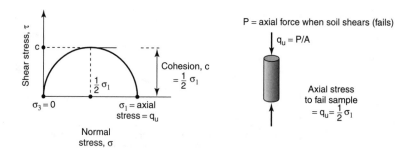

Figure 11-35 Unconfined compression test data used to determine cohesion c.

the confining pressure acting on the sample is zero. This test is quick and easy to perform. The axial stress required to fail the tested soil is designated q_u. The result of an unconfined compression test can be plotted on τ versus σ coordinates, as shown in Figure 11-35. The cohesion, or shear strength, is simply one-half the unconfined axial stress:

$$c = \frac{1}{2} q_u \tag{11-2}$$

Some clay deposits have developed cracks or fissures (sometimes called *slickensides*). This condition may be the result of desiccation, seismic effects, or other factors. If tested in unconfined compression, shear failure along the slickensides could give a low and misleading indication of the strength of soil confined in the natural deposit. For such material, a triaxial test is preferred.

Approximating Values of Cohesion from Boring Data

Because of the widespread use of the standard penetration test for making soil borings, approximate relationships between values of cohesion and N have been developed. Values are

Table 11-3 Approximate Relationship Between N and Cohesion of Clays

Value of N*	Relative Condition of Soil	Approximate Value of Cohesion, c	
		psf	kN/m² (kPa)
2–4	Soft	250–500	12–24
4–8	Medium	500–1000	24–48
8–15	Stiff	1000–2000	48–96
15–30	Very Stiff	2000–4000	96–190
> 30	Hard	>4000	>190

*Values refer to sampling procedures where the efficiency of the drive hammer is approximately 60 percent; refer Table 11-2.

presented in Table 11-3. Due to the approximate nature of the values, the data should be applied with caution.

Shear Strength of Mixed Soils

Mixtures of clay and granular soils in nature are not unusual. Material that is predominantly clay, in which all granular particles are surrounded by clay materials, will behave essentially as a clay. Mixtures that are predominantly granular soil with limited clay will present a sloped failure envelope curve on a τ versus σ plot, but the intercept is on the τ axis, as shown in Figure 11-36. The relationship of shear strength to normal stress can be expressed as

$$\tau = c + \overline{\sigma} \tan \phi \tag{11-3}$$

For such soils, the limitations on permeability and drainage should not be overlooked when considering new loadings that are applied to the soil mass. Excess pore pressures may develop and only part of the newly applied stress may represent effective stress, with a subsequent lag in the development of shear strength.

Position of Failure Plane Related to Angle ϕ

Homogeneous soils stressed to failure during a triaxial test or unconfined compression test typically develop a distinct plane of failure, as indicated in Figure 11-37a. Applying the analysis for stress at a point (Fig. 11-37b and c), a practical relationship between the angle of internal friction, ϕ, and the position of the failure plane is obtained:

$$\phi + 90° + (180 - 2\theta) = 180°, \text{ the sum of the interior}$$
$$\text{angles in a triangle}$$

$$\phi + 90° = 2\theta$$

$$\theta = \frac{90° + \phi}{2}$$

$$\theta = \left(45 + \frac{\phi}{2}\right) \text{deg} \tag{11-4}$$

Figure 11-36 Typical plot of τ versus σ for mixed soils.

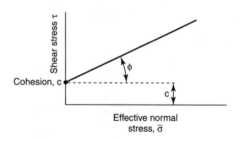

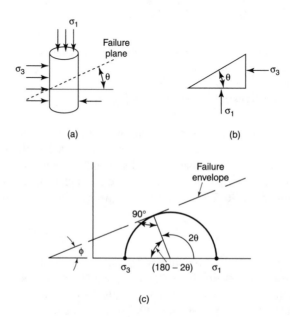

Figure 11-37 Mohr's circle analysis to relate ϕ and position of failure plane: (a) failure plane on tested sample; (b) incremental element at failure plane; (c) Mohr's circle.

11.4 STRESS PATHS

Stress paths refer to the series of progressive changes in shear and normal stress that develop within a soil mass as a result of construction loading being applied (such as resulting from loading due to structures and earth fill embankments that cause stress increases) or load being reduced (e.g., excavations).

Section 11.3 explained the method of relating shear and normal stresses on the σ-τ Mohr's circle coordinates (or σ-τ space). An alternative method for indicating the relationship is to use a p and q coordinate system, and indicate combinations of shear and normal stress in the related p-q space (Fig. 11-38).

The values of p and q represent the coordinates for the maximum shear stress on the Mohr's circle and the related normal stress. The p and q coordinates, using total stress conditions, are identified as

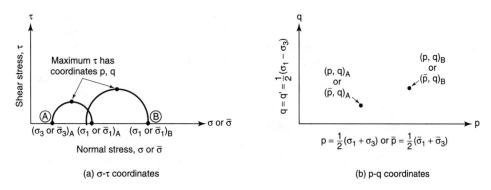

Figure 11-38 Comparison of results of triaxial tests plotted on the σ-τ coordinates and p-q coordinates.

$$p = \frac{1}{2}(\sigma_1 + \sigma_3)$$

$$q = \frac{1}{2}(\sigma_1 - \sigma_3)$$

In terms of effective stress conditions, the coordinate values p and q are identified as

$$\bar{p} = \frac{1}{2}(\bar{\sigma}_1 + \bar{\sigma}_3)$$

$$\bar{q} = \frac{1}{2}(\bar{\sigma}_1 - \bar{\sigma}_3)$$

Since total stress is represented as σ_1 and σ_3, and effective stress is total stress less pore water pressure u (or neutral stress), the values for effective stresses $\bar{\sigma}_1$ and $\bar{\sigma}_3$ become

$$\bar{\sigma}_1 = \sigma_1 - u$$

$$\bar{\sigma}_3 = \sigma_3 - u$$

and

$$\bar{q} = \frac{1}{2}(\bar{\sigma}_1 - \bar{\sigma}_3) = \frac{1}{2}[(\sigma_1 - u) - (\sigma_3 - u)] = \frac{1}{2}(\sigma_1 - \sigma_3)$$

so therefore $\bar{q} = q$ (i.e., since the q term represents shear stress, its value is not affected by pore water pressure).

Correspondingly,

$$\bar{p} = \frac{1}{2}(\bar{\sigma}_1 + \bar{\sigma}_3) = \frac{1}{2}[(\sigma_1 - u) + (\sigma_3 - u)] = \frac{1}{2}[\sigma_1 + \sigma_3 - 2u] = p - u$$

The stress path analysis has practical application for the condition where construction loads are responsible for the development of excess pore water pressures in the soil foundation zone, such as with clay and silt–clay soils having a low coefficient of permeability. A concern is that shear stresses developing in the soil as a result of foundation loads will exceed the shear strength possessed by the soil. The problem can be studied using the σ-τ coordinate system to compare construction-induced stresses to the soil strength, but the necessary comparisons become cumbersome for the typical field condition where construction-related loadings are changing (e.g., increasing from structural loads, decreasing for excavations) while, simultaneously, excess pore water pressures are changing (e.g., increasing as structural loads increase but then decreasing as drainage occurs). The p-q stress path method allows easier comparisons between construction-induced shear stresses and soil shear strength.

Stress Path for Tests in Consolidated–Drained Conditions

Understanding of the p-q stress path may be most easily grasped by referring to a triaxial test performed on a normally consolidated clay under consolidated–drained conditions (C–D) and where the applied test stresses exceed the in situ stress condition for the soil sample. As part of the sample preparation, the soil is fully consolidated in the triaxial chamber to an all-around confining pressure σ_3. Since the sample is consolidated to the confining pressure σ_3, the total stress and effective stress are similar, or $\sigma_3 = \bar{\sigma}_3$. The confining pressure acts in the vertical (axial) direction as well as horizontal, so $\bar{\sigma}_3 = \bar{\sigma}_1$. The test will be performed slowly so that drainage of pore water in the sample can occur as the axial load increases (σ_1 increases); referring to Figure 11-28b, $\bar{\sigma}_1 = \bar{\sigma}_3 + \Delta\sigma_{axial}$.
Therefore, at the start of the test,

$$\bar{p} = \frac{1}{2}(\bar{\sigma}_1 + \bar{\sigma}_3) = \frac{1}{2}(\bar{\sigma}_3 + \bar{\sigma}_3) = \bar{\sigma}_3$$

$$\bar{q} = \frac{1}{2}(\bar{\sigma}_1 - \bar{\sigma}_3) = \frac{1}{2}(\bar{\sigma}_3 - \bar{\sigma}_3) = 0$$

As the test progresses, $\bar{\sigma}_1$, (i.e., $\bar{\sigma}_3 + \Delta\sigma_{axial}$), increases, and

$$\bar{p} = \frac{1}{2}(\bar{\sigma}_3 + \Delta\sigma_{axial} + \bar{\sigma}_3) = \bar{\sigma}_3 + \frac{\Delta\sigma_{axial}}{2}$$

$$\bar{q} = \frac{1}{2}(\bar{\sigma}_3 + \Delta\sigma_{axial} - \bar{\sigma}_3) = \frac{\Delta\sigma_{axial}}{2}$$

The stress path for axial load increasing until the sample fails in shear is shown in Figure 11-39. Shear failure of a soil occurs when the p-q stress combination reaches the k'_f strength (or failure) envelope line; this k'_f line defines the limiting strength for the soil (a strength envelope is similarly used for the Mohr's circle stress analysis).
The relationship between the angles ϕ and α defining the strength (failure) envelope lines shown on Figure 11-39 can be determined, recalling that

$$\sin\phi_{CD} = \frac{\dfrac{1}{2}(\bar{\sigma}_1 - \bar{\sigma}_3)}{\dfrac{1}{2}(\bar{\sigma}_1 + \bar{\sigma}_3)}$$

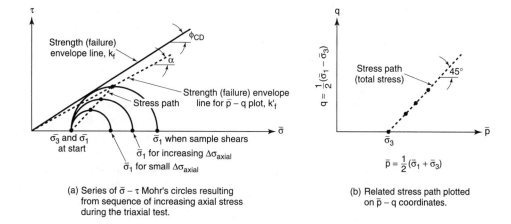

(a) Series of $\bar{\sigma} - \tau$ Mohr's circles resulting from sequence of increasing axial stress during the triaxial test.

(b) Related stress path plotted on $\bar{p} - q$ coordinates.

Figure 11-39   Results of consolidated–drained (C–D) triaxial test relating $\bar{\sigma}$-τ and $\bar{p}$-q values to develop the stress path for isotropically consolidated soil sample, $\bar{\sigma}_1 = \bar{\sigma}_3$ at start.

and noting that

$$\tan \alpha = \frac{\frac{1}{2}(\bar{\sigma}_1 - \bar{\sigma}_3)}{\frac{1}{2}(\bar{\sigma}_1 + \bar{\sigma}_3)}$$

Both expressions are the same, and therefore

$$\sin \phi_{CD} = \tan \alpha$$

The stress path indicated for the C–D triaxial test is also representative of effects when load is applied to granular soil where the high coefficient of permeability permits rapid drainage of pore water so that excess pore water pressures from the presence of groundwater do not develop as the construction loads are imposed.

Stress Path for Tests in Consolidated–Undrained Conditions

The stress path for soil subjected to triaxial compression under consolidated–undrained (C–U) conditions is significantly different than the stress path for soil under drained conditions. On a practical basis, the C–U condition is representative of the effects of construction loading applied to a normally consolidated clay deposit, where the low soil coefficient of permeability retards drainage so that excess pore water pressures develop.

Consider a soil sample of a saturated clay, normally consolidated, in a triaxial compression test under C–U conditions. The clay sample is fully consolidated in the triaxial chamber to the test value all-around confining pressure. When the sample is fully consolidated, the total lateral stress equals the confining pressure σ_3. Since excess pore water has been allowed to drain from the soil sample and the full triaxial chamber pressure is imposed onto the soil skeleton, $\overline{\sigma}_3$ equals σ_3. At this stage , the axial stress (vertical stress) also equals the chamber confining pressure so that $\sigma_1 = \overline{\sigma}_1 = \overline{\sigma}_3$. To fail the soil in the shear, axial loading will be increased while lateral pressure is held constant. The triaxial equipment drain lines are closed so that no additional pore water can drain from the sample. Pore water pressures are measured independently from measurement of the applied loading (direct pore pressure can be determined by use of a pore pressure needle inserted into the sample or from a pore water supply line connected at the base of the sample).

The applied axial load results in a stress increase, $\Delta\sigma_1$. But because the equipment drain lines are closed to prevent drainage during the axial loading procedure, an excess pore water pressure, Δu, is created. Because of the volume strains that occur during application of the axial load, the excess pore pressure value that develops is different than the stress increase $\Delta\sigma_1$. The relationship can be expressed as $\Delta u = A(\Delta\sigma_1)$ where the A factor is the Skempton pore pressure coefficient. The excess pore water pressure is exerted equally in all directions, and the resulting effective stresses become

$$\overline{\sigma}_1 = \sigma_3 + \Delta\sigma_1 - \Delta u$$

$$\overline{\sigma}_3 = \sigma_3 - \Delta u$$

where $\Delta\sigma_1$ = increase in axial stress

σ_3 = all-around confining pressure (stress) at start of test, held constant

Δu = excess pore water pressure resulting from application of $\Delta\sigma_1$

The concept of Δu on total and effective stress conditions in a foundation soil is illustrated by Figure 11-40. If the ratio $\Delta u/\Delta\sigma_1$ were to remain constant for the range of applied loading, the effective stress path would follow a sloping straight line. However, in soil the rate of volume strains changes as applied stresses increase; the result is that the A term or

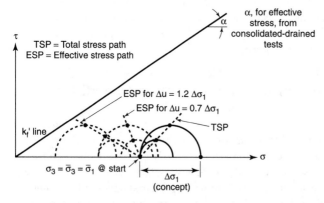

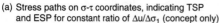

(a) Stress paths on σ-τ coordinates, indicating TSP and ESP for constant ratio of $\Delta u/\Delta\sigma_1$ (concept only)

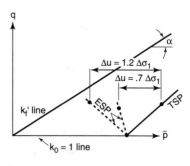

(b) Related stress paths on $\overline{p}$ – q coordinates.

Figure 11-40 Stress paths for consolidated–undrained triaxial tests for condition of constant $\Delta u/\Delta\sigma_1$ ratio (concept of effect of Δu only) for isotropically consolidated soil samples, where $\overline{\sigma}_1 = \overline{\sigma}_3$ at start.

Figure 11-41 Stress paths (ESP and TSP) for consolidated–undrained triaxial test where ratio $\Delta u/\Delta\sigma_1$ increases (for isotropically consolidated soil sample, $\bar\sigma_1 = \bar\sigma_3$ at start).

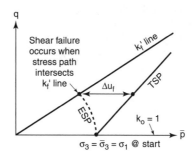

related $\Delta u/\Delta\sigma_1$ are not constant values but increase with increases in applied stress. The value for A_f or $\Delta u/\Delta\sigma_1$ when the sample reaches the failure shear stress has been found to range between 0.5 and 1.0 for normally consolidated clays. The A_f value ranges from near zero to 0.5 for overconsolidated clays and is between 1.0 and 1.25 for sensitive clays with highly flocculent structures. The effective stress path curves representative of normally consolidated clays are as illustrated in Figure 11-41.

Applications

On a practical basis, the concern is when the effective stress path resulting because of construction loading extends to intersect the strength envelope line k_f' for the soil, a condition indicating the start of shear failure.

Continuing, as drainage of pore water in the foundation soil occurs, the excess pore water pressures dissipate. The effective stress path curve (ESP) gradually shifts toward the total stress path curve (TSP), a stress condition where the foundation soil possesses greater shear strength (i.e., the soil can resist greater foundation-induced shear stress before a failure condition is reached).

The triaxial test conditions described in the preceding paragraphs and Figure 11-41 represent the situation where isotropic stress conditions exist in the foundation soil (that is, $\sigma_h = \sigma_v$, and the lateral pressure coefficient K_o value of 1 is located on the horizontal axis of the σ-τ and p-q coordinate systems. For a soil deposit where anisotropic stress conditions exist and K_o is less than 1 (as is common in normally consolidated and underconsolidated deposits), the influence can be shown as a sloping K_o line on the p-q coordinates. The start point for the stress path is from the sloping K_o line instead of from the horizontal axis (Fig. 11-42).

Where a saturated, normally consolidated clay is excavated, the vertical load on the remaining foundation soil zone decreases, resulting in progressive decreases in p and q values. The in-place soil tends to expand because of the stress relief. With a clay that was originally saturated, the expanding void spaces result in a partially saturated clay and the development of negative pore pressures, $-\Delta u$ (sometimes referred to as *suction*). The total and effective stress paths for the excavation condition extend downward, instead of upward as occurs when construction load increases (Fig. 11-43). The effective stress path is to the right of the total stress path, indicating that the soil shear strength is greatest at the time of excavation but then decreases with time as the negative pore pressures diminish.

Figure 11-42 Stress paths for consolidated–undrained triaxial test (for anisotropically consolidated soil sample, $\bar{\sigma}_1 > \bar{\sigma}_3$). Note stress path begins at K_o line ($K_o = \bar{\sigma}_3/\bar{\sigma}_1 = 0.6$ for this diagram).

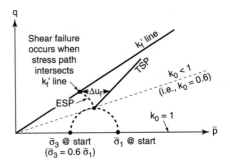

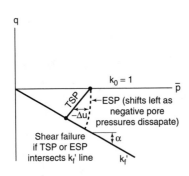

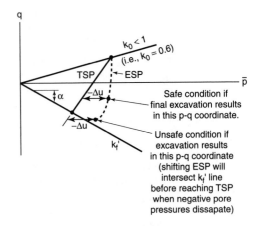

(a) Stress paths for isotropic stress conditions: $\sigma_v = \sigma_h$ and $k_0 = 1$.

(b) Stress paths for anisotropic stress conditions: $\sigma_v > \sigma_h$ and $k_0 < 1$.

Figure 11-43 Stress paths (ESP and TSP) for rapid excavation in normally consolidated clay (representing consolidated–undrained conditions).

PROBLEMS

11-1. (a) A dry sand sample is tested in direct shear. The test procedure includes having a normal (compressive) stress of 200 kPa imposed while the sample undergoes shearing. The sample fails when the shear stress reaches 135 kPa. From this data, determine the angle of internal friction ϕ for the soil.

(b) A second sample of the same sand is also to be tested in direct shear, but the applied normal (compressive) stress will be 145 kPa. What shear stress is expected to cause the sample to fail?

11-2. A sample of dry sand is tested in direct shear. The shear box holding the sample has a circular cross section with a diameter of 50 mm. The normal (compressive) load imposed is 200 N. The sample shears when the shear force is 130 N.

(a) Determine the test normal stress and shear stress at failure.

(b) Determine the angle of internal friction ϕ for this soil.

(c) What is the probable condition of the tested sample (dense, loose, etc.)?

11-3. (a) A sample of dry sand is tested in direct shear. Under an applied normal stress of 4800 psf, the sample falls when the shear stress reaches 3100 psf. What is the angle of internal friction for this soil?

(b) A second sample of the same sand material is also to be tested in direct shear, but the applied compressive loading will be 3500 psf instead of 4800 psf. What shear stress is expected to fail the sample?

11-4. Samples of damp sand are tested in direct shear (samples are tested at the in situ water content). The procedure involves two separate tests, with different normal loads for each test. In test one, using a normal loading which causes a compressive stress equal to 50 kPa, the sample is failed at a shear stress equal to 33 kPa. For test two, the normal stress is 80 kPa and the sample shears at a stress of 51 kPa. Use this test data to determine the angle of internal friction and the value of apparent cohesion for this sand.

11-5. Determine the values of apparent cohesion and angle of internal friction for a damp sand using the results of direct shear tests performed on two identical samples of the soil. In test one, the sample shears at a stress of 71 kPa when the compressive (normal) stress is 95 kPa. In test two, the sample shears at a stress of 104 kPa when the normal stress is 150 kPa.

11-6. A sand sample is subjected to direct shear testing at its normal (in situ) water content. Two tests are performed. For one of the tests, the sample shears at a stress of 3000 psf when the normal stress is 4000 psf. In the second test, the sample shears at a stress of 4000 psf when the normal stress is 6000 psf. From these data, determine the value of apparent cohesion and the corresponding angle of internal friction.

11-7. A sample of dry sand in relatively loose condition is subject to a triaxial test. The sample fails when the confining stress (minor principal stress) is 50 kPa and the axial stress (major principal stress) is 170 kPa. What is the angle of internal friction for the soil?

11-8. A dry sand is known to have an angle of internal friction equal to 36°. A triaxial test is planned, where the confining pressure will be 40 kPa. What maximum axial stress (major principal stress) should be predicted?

11-9. A dry sand sample in a triaxial test failed when the confining stress (minor principal stress) was 1000 psf and the axial stress (major principal stress) was 4000 psf. What is the angle of internal friction for this soil?

11-10. A dry sand is known to have an angle of internal friction equal to 35°. What is the maximum major principal stress that the soil can withstand when the minor principal stress is 20 psf?

11-11. It is known that the angle of internal friction for the soil comprising a granular deposit is 37°. At one depth in the deposit, the lateral pressure is 45 kPa, and this is considered the value of the minor principal stress. Use the Mohr's circle analysis to determine the maximum vertical pressure (major principal stress) that can be applied (i.e., the vertical pressure for incipient shear).

11-12. The soil in a dry granular deposit has an angle of internal friction equal to 35°. At a point in the soil mass where the lateral pressure (minor principal stress) is 1000 psf, what maximum vertical pressure (major principal stress) can be imposed?

11-13. Design studies for a planned building indicate that the foundation loading will cause the principal stresses in the soil at one point below the building area to increase to 195 kPa and 70 kPa (major and minor principal stress values, respectively). If the soil is a dry sand having an angle of internal friction equal to 34°, determine if the indicated stresses would cause shear failure to occur.

11-14. Computations indicate that a planned building loading will cause the principal stresses in the soil at a point beneath the building to increase to total values of 4500 psf and 1500 psf (major and minor principal stresses, respectively). If the soil is a dry sand with an angle of internal friction equal to 35°, will the indicated stresses cause a shear failure at the point?

11-15. A sample of dry sand is tested in direct shear and a separate sample of the same soil is tested in a triaxial compression test. In the triaxial test, the confining pressure (or minor principal stress value) is held at 15 kPa, and the sample shears (fails) when the axial pressure reaches 51 kPa. What shear strength is expected for the sample tested in direct shear when the test normal (compressive) stress is 125 kPa?

11-16. Samples of a dry sand are to be tested in a direct shear test and a triaxial test. In the triaxial test, the sample fails when the major and minor principal stresses are 140 psi and 40 psi, respectively. What shear resistance is to be expected in the direct shear test when the normal load stress equals 5000 psf?

11-17. A triaxial test performed on a clay sample, under unconsolidated–undrained conditions, reaches a maximum (failure) axial stress (major principal stress) of 106 kPa when the confining stress (minor principal stress) is 40 kPa. What is the value of c (cohesion) for this sample? If another sample of the same soil is tested in an unconfined compression test, what axial load is expected to cause failure?

11-18. A clay soil is subjected to a triaxial test under unconsolidated–undrained conditions. At failure, the major and minor principal stresses are 3000 psf and 1000 psf, respectively.
(a) What is the cohesion for this soil?
(b) If this soil were subjected to an unconfined compression test, what axial load would result in failure?

11-19. A clay soil has a cohesion of 800 psf. How could the strength of this soil be increased?

11-20. A deposit of homogeneous mixed soils (soil including both coarse-grained and fine-grained sizes) possesses shear strength due to both cohesion and internal friction ($c = 22$ kPa, $\phi = 28°$). What shear strength is expected on a plane where the normal stress is 70 kPa?

11-21. A mixed soil is found to possess a unit cohesion of 500 psf and an angle of internal friction of 30°. What shear strength is expected at a point in the soil deposit where the normal stress is 1500 psf?

11-22. Separate triaxial compression tests are performed on identical soil samples. The soil is a saturated mixed soil. Both tests follow the procedure for consolidated–undrained conditions. At failure, the respective minor and major principal stress values are 20 kPa and 46 kPa for sample one and 40 kPa and 80 kPa for sample two. From these data, determine the value for ϕ_{CU} and the cohesion intercept c on the Mohr's circle shear stress coordinate. (This problem can be solved graphically using the Mohr's circle.)

11-23. It is desired to obtain information about the in situ strength properties, c and ϕ, for a saturated clay soil, and also the preconsolidation pressure. Direct shear tests under consolidated–undrained conditions are performed on three identical samples of the clay. For sample one, the applied normal pressure is 30 kPa and the sample fails when the shear stress is 19 kPa; for sample two, the ap-

plied normal stress is 70 kPa and the sample shears at 28 kPa; for sample three, the applied normal pressure is 120 kPa and the sample shears at 45 kPa. From these data, estimate the values of the in situ cohesion c and ϕ_{CU}, and the preconsolidation stress σ_{vc}.

11-24. A triaxial test is performed on a normally consolidated clay. The sample is further consolidated by the test all-around confining pressure prior to application of the axial load. The axial force is then applied very slowly. The equipment pore pressure lines remain open so that drainage of soil pore water can occur during the test. (The described conditions outline the procedure for a consolidated–drained-type test.) The sample fails (shears) when the σ_3 pressure is 21 kPa and the σ_1 value is 61 kPa. From this data, determine the angle of internal friction for the effective stress strength-failure envelope (that is, determine ϕ_{CD}).

11-25. A triaxial test is performed on a clay sample as described in the preceding problem, but the sample fails when the axial stress (major principal stress) is 30 psi and the confining stress (minor principal stress) is 10 psi. Determine the angle of internal friction for the effective stress strength envelope, ϕ_{CD}.

11-26. A normally consolidated clay sample is subject to a triaxial test where pore water pressure measurements are made. A consolidated–undrained type of test is performed. The sample fails (shears) when the total all-around confining pressure is 45 kPa and the total axial pressure is 97 kPa. At failure, the recorded pore water pressure is 20 kPa. Determine the angle of internal friction ϕ_{CD} for the effective stress strength envelope and also the value for ϕ_{cu}.

11-27. Provide a brief description of the term *stress path* as related to stress conditions in a soil mass.

11-28. The results of consolidated–drained triaxial compression tests on samples of saturated clay indicate that ϕ_{CD} is 24° for this soil. Use this information to determine the slope of the K_f' line to use for a stress path analysis.

11-29. A consolidated–undrained triaxial compression test is performed on a sample of saturated clay. The sample has been isotropically consolidated so that at the start of the test, $\overline{\sigma}_1 = \overline{\sigma}_3 = 25$ kPa. The slope of the K_f' line is 30°. Use the stress path method to determine the value of the axial stress

($\overline{\sigma}_1$ at start plus $\Delta\sigma_v$) when the effective stress path intersects the K_f' line to cause shearing (failure) of the sample. Assume that the Skempton pore pressure coefficient at failure, A_f, is 1.0. (*Hint:* Refer to Fig. 11-40.)

11-30. A consolidated–undrained triaxial compression test is performed on a sample of saturated clay. The clay is anisotropically consolidated, and $\overline{\sigma}_3 = 0.6\,\overline{\sigma}_1$. The slope of the K_f' line is 28°. At the start of the test, $\overline{\sigma}_3$ is 24 kPa and $\overline{\sigma}_1$ is 40 kPa. Use the stress path method to determine the value of axial pressure when the effective stress path intersects the K_f' line and the sample shears. Assume that the Skempton pore pressure coefficient at failure A_f is 1.

11-31. Stress path diagrams for applications of increasing loading as illustrated in Figure 11-41 indicate that a clay soil shear strength will increase as consolidation progresses and pore water pressures dissipate (that is, the effective stress eventually equals the total stress). Excavation and related effects on clay soil represent a different condition, however. Referring to Figure 11-43, briefly describe in general terms what concerns exist relating to construction safety when excavations are undertaken in clay deposits.

Application of Soil Mechanics Theories

CHAPTER 12

Foundations
Introductory Concepts

For many of humankind's structures, it is the earth underlying the structure that provides the ultimate support. The soil at a building location automatically becomes a material of construction affecting the structure's stability. Typically, soil is a material weaker than the other common materials of construction, such as steel, concrete, and wood. To carry a given loading satisfactorily, a greater area or volume of soil is necessarily involved. In order for loads carried by steel, concrete, or wood structural members to be imparted to the soil, load transfer devices—the structural foundations—are required. The major purpose of the structural foundation is the proper transmission of building loading to the earth in such a way that the supporting soil is not overstressed and does not undergo deformations that would cause serious building settlement. The type of structural foundation utilized is closely related to the properties of the supporting soils. A structural foundation performs properly only if the supporting soil behaves properly. Consequently, it is important to recognize that building support is actually being provided by a soil–foundation system, a combination that cannot be separated. Designers and constructors are aware of this relationship, but it has become common practice to consider the structural foundation separately, primarily because it is a cost item that is built or installed, while the supporting soil is usually the natural earth that "is there."

Since the soil–foundation system is responsible for providing support for the lifetime of a structure, it is important that all forces that may act over that time period are considered. For a building to endure, its foundations should be designed for the worst conditions that may develop. Typically, the foundation design always includes the effect of the structure's dead plus live loads. It is important also to consider load effects that may result from environmental factors such as wind, ice, frost, heat, water, earthquake, and explosive blasts.

12.1 GENERAL TYPES OF FOUNDATIONS—FOUNDATION CATEGORIES

The various types of structural foundations can be grouped into two broad categories: shallow foundations and deep foundations. Generally, the classification indicates the depth of the foundation installation and the depth of the soil providing most of the support. Spread footing and mat (or raft) foundations usually fall within the shallow foundations category. Deep foundation types include piles, piers, and caissons. The floating foundation, a special category of foundation, is actually not a different type, but it does represent a special application of soil mechanics principles to a combination mat–caisson foundation.

Spread Footings

Spread footing foundations are typically of plain concrete or reinforced concrete, although masonry and timber have also been used. The spread footing foundation is basically a pad used to "spread out" building column and wall loads over a sufficiently large soil area. Spread footings are constructed as close to the ground surface as the building design permits (considering requirements such as basements and the need to resist lateral forces) and as controlled by local conditions (considering factors such as frost penetration, soil shrinkage and expansion, the possibility of soil erosion, and building code stipulations). Footings for permanent structures are rarely located directly on the ground surface. To be classified as a spread footing, the foundation does not have to be at a shallow depth; spread footings will be located deep in the ground if soil conditions or the building design requires. Spread footing foundations for building columns, walls, and equipment bases commonly have the shapes of squares, rectangles, trapezoids, or long strips (Fig. 12-1). Usually, the shape and dimensions for a footing result from having the structural loading positioned so that, theoretically, a *uniform* bearing pressure on the soil beneath the foundation is achieved. For the support of walls and single columns, the loading is usually centered on the footing. For foundations supporting two or more column loads, or machinery, the positioning of the loading or weight often makes a rectangular or trapezoidal shape necessary.

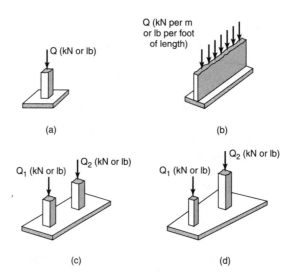

Figure 12-1 Types of shallow spread footing foundations: (a) square spread footing to support column loading; (b) long (strip) footing to support wall loading; (c) rectangular footing for two columns (combined footing) or machine base; (d) trapezoidal footing for two columns (combined footing) or machine base.

Mat (or Raft) Foundations

The mat (or raft) foundation can be considered a large footing extending over a great area, frequently an entire buildings area (Fig. 12-2). All vertical structural loadings from columns and walls are supported on the common foundation. Typically, the mat is utilized for conditions where a preliminary design indicates that individual column footings would be undesirably close together or try to overlap. The mat is frequently utilized as a method to reduce or distribute building loads in order to reduce differential settlement between adjacent areas. To function properly, the mat structure will be more rigid and thick than the individual spread footing.

Pile and Pier Foundations

Piles and piers are foundation types intended to transmit structural loads through upper zones of poor soil to a depth where the earth is capable of providing the desired support (Fig. 12-3a and b). In this respect, where loadings developed at one level are transmitted to a lower level, piles and piers are similar to structural columns. Though considered as long, slender structural members, such foundations typically obtain adequate lateral support from the embedding soil along their length so usually there is no concern about buckling under axial load, as with conventional columns. These deep foundation types are also utilized in situations where it is necessary to provide resistance to uplift or where there is concern about possible loss of ground or erosion due to flowing water or other causes.

In years past, the pile or pier category was indicative of the method used to install the foundation or of its size. Piles were slender foundation units, usually driven into place. Piers, typically larger in area than piles, were units formed in place by excavating an opening to the desired depth and pouring concrete. Often, such foundations were large enough to permit a man to enter and inspect the exposed earth. Currently, a clear distinction between pile- and pier-type foundations is not always present because of changes and innovations in construction or installation techniques. For example, some types of cast-in-place piles are constructed by using the basic methods historically attributed to pier foundations. As a result, the developing practice is to classify all deep, slender foundation units simply as pile-type foundations, with terms such as *driven, bored,* or *drilled* and *precast* or *cast-in-place* to indicate the method of installation and construction.

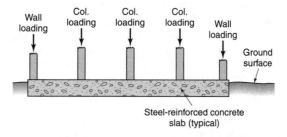

Figure 12-2   Mat foundation—for soils having low bearing capacity or where soil conditions are variable and erratic—is used to obtain low bearing pressure and reduce differential settlement. The mat foundation may also consist of a grid arrangement of closely spaced and interconnected grade beams capped with a floor slab.

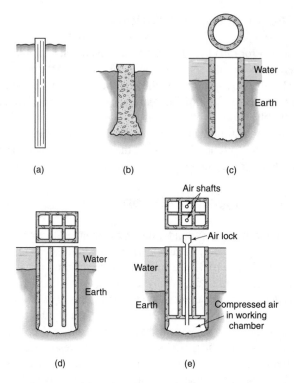

Figure 12-3 Representative types of deep foundations: (a) slender driven, drilled, or cast-in-place pile; (b) drilled or cast-in-place pier with enlarged base; (c) pile-type open caisson; (d) box-type open caisson; (e) pneumatic caisson.

Caissons

A caisson is a structural box or chamber that is sunk in place or built in place by systematically excavating below the bottom of the unit, which thereby descends to the final depth. Open caissons may be box type or pile type. Usually, the top and bottom are open during installation. When in place, the bottom may be sealed with concrete if necessary to keep out water, or the bottom may be socketed into rock to obtain a high bearing capacity. Pneumatic caissons have the top and sides sealed and use compressed air to keep soil and water from entering the lower working chamber, where excavation to advance the caisson is occurring. Representative types of caissons are illustrated in Figure 12-3c, d and e.

Floating Foundations

The floating foundation is a special type of foundation construction that is useful under proper conditions. Particularly, it has application in locations where deep deposits of compressible cohesive soils exist and the use of piles is impractical. The floating foundation concept requires that a building's substructure (the below-ground structure) be assembled as a combination mat and caisson to create a rigid box, as shown in Figure 12-4. This foun-

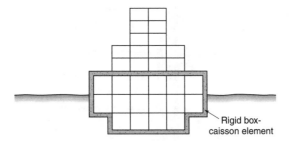

Figure 12-4   Rigid box caisson foundation utilizing floating foundation concept.

dation is installed at a depth so selected that the total weight of the soil excavated for the rigid box equals the total weight of the planned building. In theory, the soil below the structure is therefore not subjected to any change in loading. For such an occurrence, there would be no settlement. Usually, however, some settlement does occur because soils at the bottom of a deep excavation commonly expand somewhat after excavation because of stress relief, then recompress during and following construction.

12.2   PILE FOUNDATION TYPES AND INSTALLATION PROCEDURES

Pile Drivers for Driven Piles

Driven piles are installed from the ground surface by hammering a ready-made unit or hollow shell, usually with special pile-driving equipment (Fig. 12-5). Most driving is done with an impact type of hammer; a moving weight falls or is forced against the top of the pile (pile driving has been described as a brutal method for installing foundations). General categories of modern pile hammers include the single- and double-acting units (steam, compressed air, or hydraulic) and diesel hammers, illustrated in Figure 12-6. The term *hammer* refers to the entire driving unit; usually, the moving weight that strikes the pile is the *ram.*

Single-acting hammers use steam or compressed air to raise the hammer ram to a ready-for-driving position. The ram is then released to enable it to drop on the top of the pile. A *double-acting hammer* uses steam or compressed air to raise the ram to a ready-for-driving position and also to accelerate the ram's downward thrust. *Differential-acting hammers,* another category, are similar in operation to the double-acting hammers. Double-acting hammers actuated by hydraulic pressures are also available.

Diesel hammers are self-contained, self-activated units. The ram is located within an enclosed cylinder. Initially, the ram is mechanically raised to the top of the cylinder and released for its fall. A fuel mixture injected into the cylinder ahead of the falling ram is compressed due to the piston effect. Near the bottom of the stroke, the fuel is detonated, and the force of the resulting explosion and the ram impact is delivered to the pile, driving it. Within the cylinder, the force of the explosion raises the ram up to the top of the cylinder, where it is ready to begin another cycle.

Single-acting hammers need to rely on the weight of a heavy ram for driving piles. Double-acting and diesel hammers can develop high driving energy that is equal to or greater than the energy of the falling ram in a single-acting hammer, but they achieve this

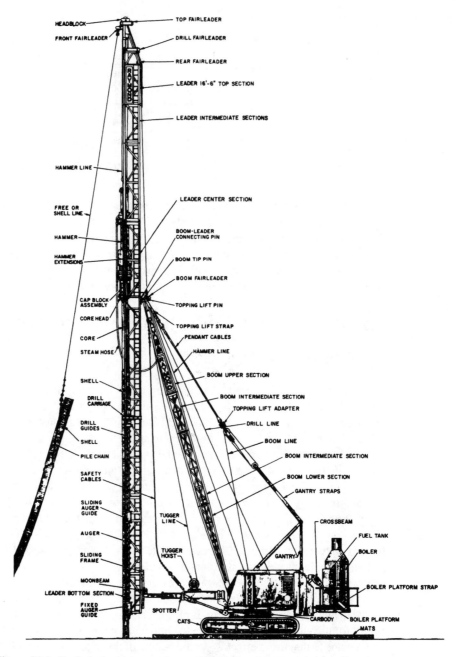

Figure 12-5 Typical pile-driving equipment. (Courtesy of Raymond International, Inc.)

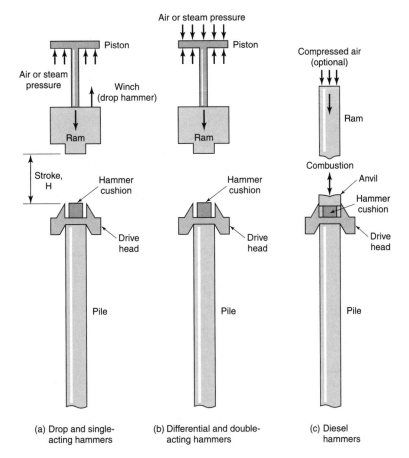

Figure 12-6   Schematic diagram of ram-type pile-driving hammers [220].

through greater impact velocity. One advantage of the double-acting and diesel hammers over the single-acting hammer is the greater operating speed; a greater number of blows will be delivered per unit of time. When long or heavy piles are driven, a hammer that has a heavy ram should be used. For effective driving, the weight of the hammer ram needs to be approximately the same weight as the pile or greater. Equipment manufacturers provide data on the important characteristics of their hammers, including the weight of the ram and the stroke, hammer efficiency, driving energy per hammer blow, and the number of blows per minute (Table 12-1).

Vibratory drivers rely on a principle different from the conventional hammers for installing a pile. Vibratory units are typically more contained and compact than the falling-ram type of hammer. Basically, the vibratory driver (Fig. 12-7) consists of a pair of counter-rotating weights that are synchronized so that the lateral components of thrust always counteract, or cancel, each other's effects. The vertical components of thrust are additive, however, and create up-and-down pulsations, or vibrations. For an installation, the driver is clamped to the pile. The pile is then vibrated into the earth under its own weight.

Table 12-1 Data on Representative Pile-driving Hammers

Impact Hammers		Weight: Ram, Piston, or Moving Parts		Blows per Minute	Rated Energy per Blow	
Make	Model (type)	kips	kN		k-ft	kN-m
Vulcan Iron Works, Florida	08 (single-acting)	8	35.6	50	26	35.3
	040 (single-acting)	40	177.9	60	120	162.7
	060 (single-acting)	60	266.9	62	180	244.1
	3100 (single-acting)	100	448.8	58	300	406.8
	140 C (differential-acting)	14	62.3	103	36	48.8
	200 C (differential-acting)	20	89	98	50.2	68.1
	400 C (differential-acting)	40	178	100	40	177.9
	N-33 (diesel)	3	13.3	50–60	24.6	33.4
McKiernan-Terry, New Jersey	S-8 (single-acting)	8	35.6	55	26	35.3
	S-20 (single-acting)	20	89	60	60	81.4
	C-5 (double-acting)	5	22.2	110	16	21.7
	DE-20 (diesel)	2	8.9	40–50	To 18	To 24
	DE-70B (diesel)	7	31	40–50	To 63	To 85
Raymond International, Texas	8/0 (single-acting)	25	111	35	81	110
	150 C (differential-acting)	15	66.7	95–105	48	66
Link Belt, Iowa	520 (diesel)	5.07	22.6	80–84	26.3	35.7
	440 (diesel)	4.0	17.8	86–90	18.2	24.7
Kobe Diesel, Japan	K-25 (diesel)	5.51	24.5	40–60	50.7	68.8
	K150 (diesel)	33.1	147.2	45–60	280	380
Mitsubishi International, Japan	M-145 (diesel)	2.97	13.2	42–60	12–26	16–35
	MB70 (diesel)	15.84	70.5	38–60	63–141	86–191

Vibratory Hammers		Total Weight		Available HP	Force at Operating Frequency	
Make	Model	kips	kN		kips/cps	kN/hertz
Bodine, USA	B	22	99	1000	63/100 to 175/100	280/100 to 780/100
Foster, France	250	11.2	50	100	101/17	450/17
Menck, Germany	MVB 44–30	8.6	38	100	97/—	432/—
Uraga, Japan	VHD-3	15	67	120	574/20	129/20

Vibratory drivers may be low-frequency type (operating range between about 10 and 30 Hz or cps) or resonant type (capable of an operating frequency up to about 150 Hz or cps). The intent of a resonant driver is to create a condition of resonance between the driver, pile, and zone of affected soil so that penetration occurs very rapidly. Vibratory drivers are considered to be most effective when installing piles in sand and silty sand soils.

An item of importance in pile installation is the factor of stresses created in the pile by the driving. For wood piles and concrete piles, excessive driving when high resistance is met (a high number of blows per inch of penetration) may result in structural damage or breakage in the pile. It is helpful for the foundation designer to impose an *upper limit* to the

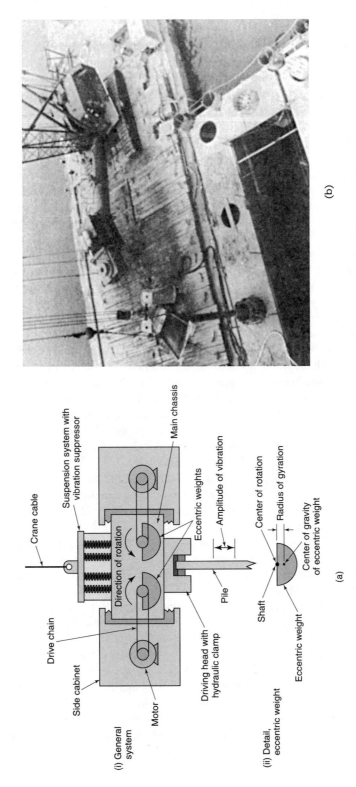

Figure 12-7 (a) Schematic diagram of vibratory pile driver; (b) vibratory hammer utilized to install steel-casing cast-in-place concrete piles for a ship channel. Photo shows concrete being tremied into a casing. The vibratory hammer is shown hanging from the crane boom.

415

number of blows that can be safely imposed, for reference by field personnel. Breakage while driving can be difficult to detect but is often indicated by a drop in blow count following a high blow count. Metal tips or end covers can be installed on piles to help protect them in hard driving and to aid penetration, as shown in Figure 12-8.

Pile Types and Materials

The common types of piles that are *installed by driving* include timber piles, steel H (Fig. 12-9) or pipe, reinforced concrete, and prestressed concrete. Generally, with these types, the pile unit is complete and ready to be installed when it is delivered to the job site.

The *driven steel shell* (Fig. 12-10) is a different category of pile installed by driving. Often, with this type, a hollow steel shell is installed to the desired depth by driving on a steel mandrel or steel core that fits inside the shell (Fig. 12-11). Much of the force of driving thus acts at the tip of the pile, so, in effect, the pile is partially being pulled into the ground. At the desired penetration, the mandrel or core is withdrawn and plain or reinforced concrete is placed in the shell. Corrugated or fluted steel is conventionally used for the shell so that it will possess adequate strength to withstand the stresses of driving. This type of pile is usually considered a cast-in-place unit.

Timber piles (Fig. 12-12) can be treated or untreated. "Treated" refers to the procedure by which the wood has been impregnated with a protective material or preservative. Pressure impregnation is a typical treatment for piles that are to be exposed to moisture but not permanently submerged in water. Untreated wood permanently below water will not decay and is considered to have unlimited life unless threatened by other effects. Preservative–protection treatment is required to protect wood piles from marine borers (if for a marine environment), from wood-infesting insects such as termites, or from decay (wet rot) if the pile is embedded in soil above the water table.

Cast-in-place piles include the steel-shell and concrete type, described previously; the shell-less type, which is formed by excavating to a desired depth and then filling the opening with concrete; and the type in which a concrete unit is formed in the ground without prior excavation.

Where pre-excavation is performed for a *shell-less pile* before the concrete is placed, the excavation may be unlined or may be provided with a temporary lining, such as a steel shell, whose purpose is to keep soil from caving into the open excavation and to seal off water from the soil walls. Excavation for this type of pile is typically performed by augering or by applying a wash boring technique similar to that utilized when borings are drilled (Chapter 5). With either of these methods, a bentonite clay slurry may be used in the excavation while it is being drilled (Chapter 5) instead of a temporary steel lining to keep the soil walls from caving in. The bentonite slurry is a heavy liquid whose purpose is to exert a lateral pressure sufficient to hold the soil walls in place. Concrete for the foundation unit is tremied (placed under water through a large-diameter flexible hose or tube in order to prevent contamination) to the bottom of the excavation. The pile is formed from the base up, and the slurry is displaced from the excavation. The shell-less, cast-in-place pile is often referred to as a *bored-pile* or *drilled-shaft foundation*. Representative types are illustrated in Figure 12-13.

Figure 12-8 Piles being provided with protective points. (Courtesy of Associated Pile and Fitting Corporation)

(a)   (b)

Figure 12-9 (a) H-piles being driven. Photo shows battered piles. (Courtesy of Associated Pile and Fitting Corporation) (b) Installation of pipe piles. (Courtesy of MKT Corporation)

In cohesive soils, the base of the excavation may be enlarged to provide a greater bearing area. The excavation for the shaft section and the belled or underreamed base is often performed by special auger-type drilling equipment, but can be accomplished manually. This type of foundation is often referred to as a *drilled (drilled and belled) pier, drilled (drilled and belled) caisson,* or *drilled-shaft foundation* (Fig. 12-14).

A different type of cast-in-place pile, the *augered* type, is constructed by augering into the earth without removing soil, then pumping concrete to the bottom of the drilled zone through a hollow stem in the auger and forming the pile from the bottom up as the auger withdraws and removes soil (Fig. 12-15).

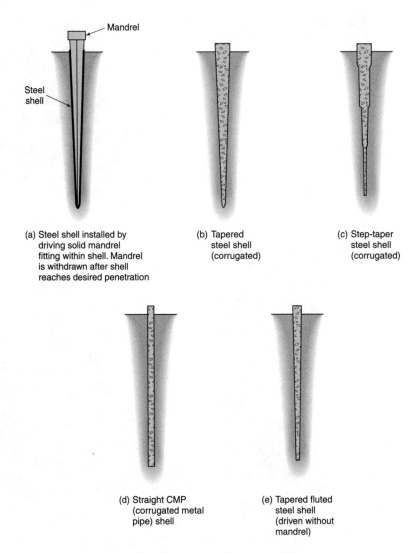

Figure 12-10 Types of driven steel-shell, cast-in-place concrete piles.

Still another category of cast-in-place foundation is the *bulb-type uncased concrete pile* (Fig. 12-16). To construct this type, a mass of very dry concrete is placed in a steel casing standing on the ground surface and formed into a plug by a falling heavy ram. High friction develops between the concrete and the inside casing wall, and continued blows of the ram force the concrete plug and casing into the ground. At the depth where it is desired to form a base for the pile, the casing is locked to the driving rig, and continued blows of the ram force the concrete plug from the bottom of the casing into the earth. An enlarged base forms as the driven concrete plug compacts the soil around this level. Additional dry concrete is then added and rammed while the casing is slowly lifted. A continuous pile shaft is formed as the rammed concrete continues to be driven from the bottom of the withdrawing casing.

Figure 12-11 Installation of step-taper steel-shell cast-in-place concrete piles. (Courtesy of Raymond International, Inc.)

Figure 12-12 Timber piles being installed with single-acting hammer. (Courtesy of Associated Pile and Fitting Corporation)

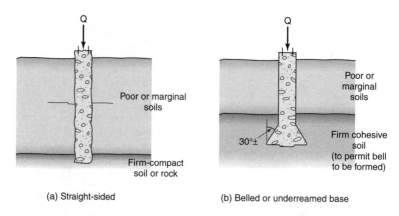

Figure 12-13 Typical drilled-shaft foundations: (a) straight-sided; (b) belled or underreamed base.

(a)   (b)

Figure 12-14 (a) Truck-mounted rig for constructing drilled-shaft foundations. (Courtesy of Association of Drilled Shaft Contractors, Inc.) (b) Crane-mounted equipment for constructing drilled-shaft foundations. Photo shows underream tool for forming enlarged base. (Courtesy of Association of Drilled Shaft Contractors, Inc.)

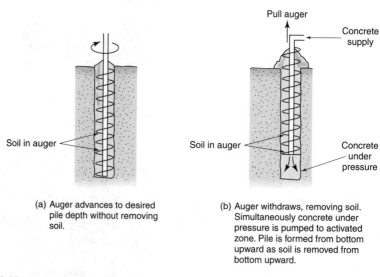

Figure 12-15 Augered cast-in-place pile.

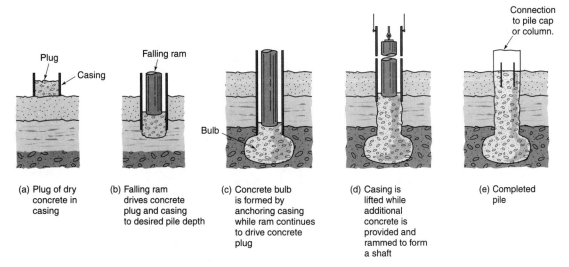

Figure 12-16   Procedure to form bulb-type uncased concrete pile.

The term *composite piles* usually refers to piles that have different materials for their different sections of length. The most typical combinations include timber or steel H pile or pipe pile for the lower section and cased cast-in-place concrete for the upper section (Fig. 12-17). Composite piles are normally used for reasons of economy (timber is less costly than steel or concrete) or to take advantage of certain structural features that one material possesses (steel will fare better than timber or concrete in hard driving; timber or cast-in-place concrete piles may not be available in the required length). Where composite piles are utilized, the splice joint between sections should be as strong as the pile materials, particularly if the pile will be subject to uplift forces. *Composite pile* is also the term applied to a pile cross section of more than one material, such as a concrete-filled pipe pile.

Representative Pile Load Capacities and Available Lengths

The pile type and cross section to be selected for a project will be influenced by soil conditions, required pile length, required structural capability of the pile, and consideration of the pile installation method. Information on the maximum lengths generally available for the various pile types is summarized in Table 12-2. Representative structural load ranges for various pile types are shown in Table 12-3.

Methods to Aid Pile Installation

The installation of driven piles can be aided by the use of spudding or predrilling. These techniques are frequently utilized in situations where obstacles that could damage the driven pile are buried in the soil to be penetrated, where compact or hard soil must be penetrated, or where driving vibrations may affect nearby structures. Another form of installation assistance is jetting, a procedure that eliminates some driving and hastens pile penetration.

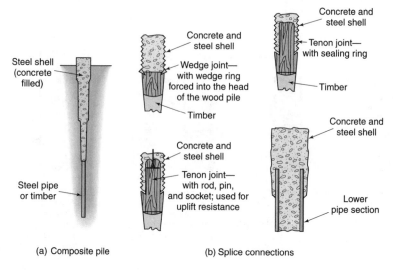

(a) Composite pile   (b) Splice connections

Figure 12-17 Composite pile showing splice connections: (a) typical makeup of composite pile; (b) typical splice connections.

Table 12-2 Available Lengths of Various Pile Types

Pile Type	Comments, Available Maximum Length
Timber	Depends on wood (tree) type. Lengths in the 15 to 18 m (50- to 60-ft) range are usually available in most areas; lengths to about 25 m (75 ft) are available but in limited quantity; lengths up to the 30 m (100-ft) range are available, but supply is very limited.
Steel H and pipe	Unlimited length; "short" sections are driven and additional sections are field-welded to obtain a desired total length.
Steel shell, cast-in-place	Typically to between 30 and 40 m (100 and 125 ft), depending on shell type and manufacturer-contractor.
Precast concrete	Solid, small-cross-section piles usually extend to the 15 to 18 m (50- to 60-ft) length, depending on cross-section shape, dimensions, and manufacturer. Large-diameter cylinder piles can extend to about 60 m (200 ft) long.
Drilled-shaft, cast-in-place concrete	Usually in the 15 to 25 m (50- to 75-ft) range, depending on contractor equipment.
Bulb-type, cast-in-place concrete	Up to about 30 m (100 ft).
Composite	Related to available lengths of material in the different sections. If steel and thin-shell cast-in-place concrete are used, the length can be unlimited; if timber and thin-shell cast-in-place concrete are used, lengths can be on the order of 45 m (150 ft).

Table 12-3   Typical Capacities for Various Pile Types

Pile Type	Typical Design Load Range	
	tons	kN
Wood	15–30	100–300
Concrete, cast-in-place, steel shell and uncased	30–75	250–700
Concrete, reinforced or prestressed (lower range for smaller cross sections, upper range for larger cross sections)	30–200	250–2,000
Concrete, bulb-type	75–1,000	600–9,000
Composite: wood and concrete	30–60	250–600
Steel pipe (lower range for small diameter and wall thickness, upper range for larger diameter and heavy wall thickness)	30–100	250–1,000
Steel HP (lower range for light sections, upper range for heavier sections)	40–200	400–1,500

Spudding refers to the procedure of driving a steel H or similar section into the earth to break up obstacles before installing the pile. Beyond the depth where the obstacles exist, the spud is withdrawn. The pile is then installed in the hole and driven to its final depth.

Predrilling consists of drilling a hole, approximately the diameter of the pile, through very hard soils to eliminate the danger of pile damage that might result if driving were attempted (Fig. 12-18). Predrilling is frequently utilized on projects where driving effects must be minimized, as protection against possible damage from driving vibrations to nearby facilities. The procedure is also utilized when piles are installed in clay soils to prevent ground heave, which can result from pile driving. If ground heave occurs, previously driven piles may also heave up and have to be reseated.

Jetting is the technique of using a powerful stream of water directed below the tip of a long pile penetrating sandy soil to wash ahead of the pile to assist it in advancing through the sand. The jetting nozzle may be temporarily attached directly to the pile. As sand is flushed from below the pile tip, the pile settles into the created void or is easily driven if the driving hammer is activated. Jetting causes the sand that eventually surrounds the pile to be loose. Typically, the jetting is stopped and the pile is driven for the last segment of the desired penetration to develop high end friction and tip bearing. For short piles, jetting is also used to wash a hole at a pile location, before driving, to make installation easier.

Other Installation Considerations

Pile foundations may be required to resist lateral forces instead of or in addition to vertical loads. Driven piles and some types of formed-in-place piles can be installed at an angle to the vertical to develop high resistance to lateral forces. Such piles are referred to as *batter piles.* Most vertical piles are capable of resisting lateral forces also, but usually of only small magnitude.

Deep foundations may be required at locations where available headroom is limited. Driven and formed-in-place piles can be installed where space is limited, but special equipment is necessary. If headroom is very limited, driven-type piles can be installed by working with very short sections and jacking. Sections are jacked and added to until the desired penetration is reached.

Figure 12-18 Predrilling equipment for installing steel-shell piles in clay soils. (Courtesy of Raymond International, Inc.)

The augered-excavation cast-in-place piles can offer the advantage of relatively quiet, vibration-free installation compared to driven piles. Such factors may be important if construction is in a highly developed area where there is concern about noise or the effect of driving vibrations on nearby structures.

Comparative installed costs for different foundation types will vary, depending on soil conditions to be penetrated, required pile length, desired load capacity, geographic area and labor costs, availability of pile materials, site accessibility and site conditions, and contractor availability. Although drilled-shaft or cast-in-place piles are often considered more economical than driven piles where short lengths are required, and vice versa, each project should be considered unique and costs should be determined accordingly.

12.3 RELATING SOIL CONDITIONS AND FOUNDATION TYPES

A structural foundation serves as the intermediary element to transmit forces from a building's superstructure to the supporting soil. It is necessary to know soil conditions and soil properties underlying an area as well as the magnitude and type of building loading before selection of a proper foundation type can be made. After an appropriate foundation has been decided upon, each unit is sized for proper carrying capacity. Individual foundations should be analyzed, as is any other structural member, to ensure that the element itself possesses adequate internal strength.

As a guide for developing an understanding of the intereffect of soil conditions and a required type of foundation, illustrations of different subsurface conditions and related foundation considerations are presented in Figure 12-19. For the design comments, it was assumed that a multistory commercial structure, such as an office building, was to be supported.

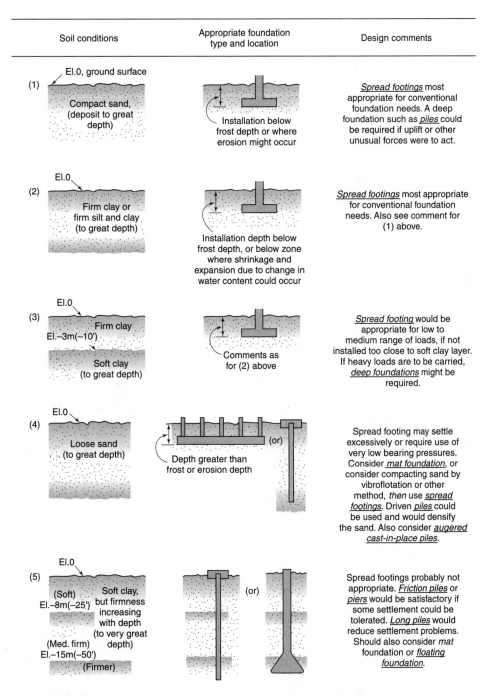

Soil conditions	Appropriate foundation type and location	Design comments
(1) El.0, ground surface Compact sand, (deposit to great depth)	Installation below frost depth or where erosion might occur	*Spread footings* most appropriate for conventional foundation needs. A deep foundation such as *piles* could be required if uplift or other unusual forces were to act.
(2) El.0 Firm clay or firm silt and clay (to great depth)	Installation depth below frost depth, or below zone where shrinkage and expansion due to change in water content could occur	*Spread footings* most appropriate for conventional foundation needs. Also see comment for (1) above.
(3) El.0 Firm clay El.−3m(−10') Soft clay (to great depth)	Comments as for (2) above	*Spread footing* would be appropriate for low to medium range of loads, if not installed too close to soft clay layer. If heavy loads are to be carried, *deep foundations* might be required.
(4) El.0 Loose sand (to great depth)	(or) Depth greater than frost or erosion depth	Spread footing may settle excessively or require use of very low bearing pressures. Consider *mat foundation*, or consider compacting sand by vibroflotation or other method, *then* use *spread footings*. Driven *piles* could be used and would densify the sand. Also consider *augered cast-in-place piles*.
(5) El.0 (Soft) Soft clay, El.−8m(−25') but firmness increasing with depth (to very great (Med. firm) depth) El.−15m(−50') (Firmer)	(or)	Spread footings probably not appropriate. *Friction piles* or *piers* would be satisfactory if some settlement could be tolerated. *Long piles* would reduce settlement problems. Should also consider *mat foundation* or *floating foundation*.

Figure 12-19 Illustrations relating soil conditions and appropriate foundation types.

Soil conditions	Appropriate foundation type and location	Design comments

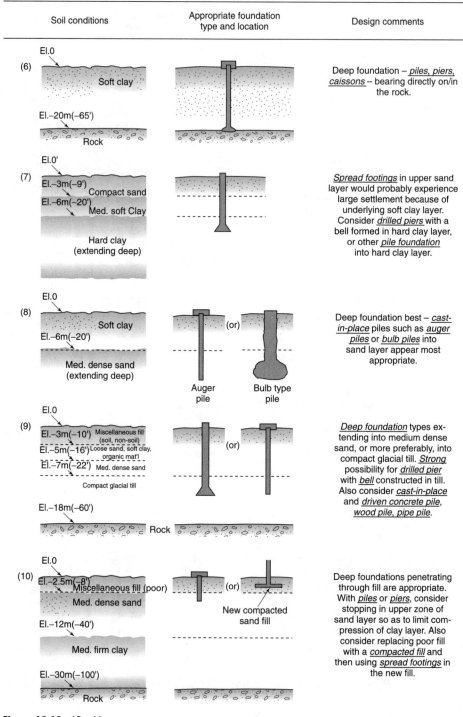

(6)

El.0

Soft clay

El.−20m(−65')

Rock

Deep foundation – *piles, piers, caissons* – bearing directly on/in the rock.

(7)

El.0'

El.−3m(−9')
Compact sand
El.−6m(−20')
Med. soft Clay

Hard clay
(extending deep)

Spread footings in upper sand layer would probably experience large settlement because of underlying soft clay layer. Consider *drilled piers* with a bell formed in hard clay layer, or other *pile foundation* into hard clay layer.

(8)

El.0

Soft clay
El.−6m(−20')

Med. dense sand
(extending deep)

Auger pile

Bulb type pile

(or)

Deep foundation best – *cast-in-place* piles such as *auger piles* or *bulb piles* into sand layer appear most appropriate.

(9)

El.0

El.−3m(−10') Miscellaneous fill
(soil, non-soil)
El.−5m(−16') Loose sand, soft clay, organic mat'l
El.−7m(−22') Med. dense sand

Compact glacial till

El.−18m(−60')

Rock

(or)

Deep foundation types extending into medium dense sand, or more preferably, into compact glacial till. *Strong* possibility for *drilled pier* with *bell* constructed in till. Also consider *cast-in-place* and *driven concrete pile, wood pile, pipe pile*.

(10)

El.0

El.−2.5m(−8')
Miscellaneous fill (poor)

Med. dense sand

El.−12m(−40')

Med. firm clay

El.−30m(−100')

Rock

(or)

New compacted sand fill

Deep foundations penetrating through fill are appropriate. With *piles* or *piers*, consider stopping in upper zone of sand layer so as to limit compression of clay layer. Also consider replacing poor fill with a *compacted fill* and then using *spread footings* in the new fill.

Figure 12-19 (Con't)

426

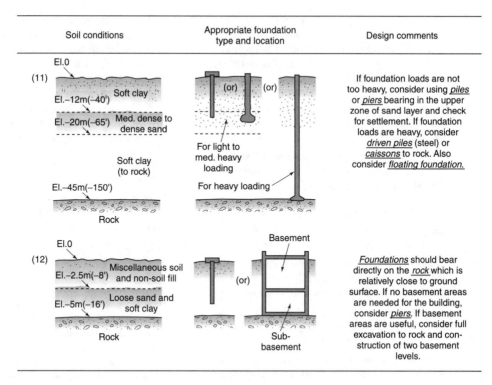

Soil conditions	Appropriate foundation type and location	Design comments

(11) — soft clay / med. dense to dense sand / soft clay (to rock) / rock

El.0
El.−12m(−40') Soft clay
El.−20m(−65') Med. dense to dense sand
Soft clay (to rock)
El.−45m(−150')
Rock

For light to med. heavy loading
For heavy loading

If foundation loads are not too heavy, consider using _piles_ or _piers_ bearing in the upper zone of sand layer and check for settlement. If foundation loads are heavy, consider _driven piles_ (steel) or _caissons_ to rock. Also consider _floating foundation._

(12)
El.0
El.−2.5m(−8') Miscellaneous soil and non-soil fill
El.−5m(−16') Loose sand and soft clay
Rock

Basement
(or)
Sub-basement

Foundations should bear directly on the _rock_ which is relatively close to ground surface. If no basement areas are needed for the building, consider _piers_. If basement areas are useful, consider full excavation to rock and construction of two basement levels.

Figure 12-19 (Con't)

PROBLEMS

12-1. Why are foundations, as a distinct structural element, required for buildings?

12-2. Typically, a foundation design will be most influenced by the properties of the supporting soil and the magnitude of structural loading to be carried. Of these two factors, soil properties represent the primary concern. Explain.

12-3. List the various types of foundations in use for construction projects such as buildings, including type of materials commonly chosen and general definition of the depth where soil support is achieved.

12-4. (a) What factors determine whether a foundation type is in the shallow or deep foundation category?

(b) Piles and piers are considered deep foundation units. Why are deep foundation units typically long, slender members?

12-5. Generally, shallow foundations, such as poured concrete spread footings, offer definite cost advantages over deep foundation types, such as piles. Assume that you are making a presentation to a nontechnical audience; outline reasons for the usually significant cost differential between shallow and deep foundations.

12-6. Why are site factors such as a high water table or the potential for frost heave or soil shrinkage and swelling considerations important to the design of shallow foundations?

12-7. Provide a brief description of the mat-type foundation, and list the practical reasons why a mat foundation type could be selected for a building.

12-8. Outline the scientific reasoning which supports the concept of the floating foundation.

12-9. A planned subterranean vault for heavy power transformer equipment is required in an area where a thick deposit of weak soils exists. The vault is 8 m by 12 m in plan area and is 4 m high. The soil unit weight is 15 kN/m³. The approximate weight of the vault structure plus included equipment is 5500 kN. It is proposed to apply the floating foundation concept to the design–installation of the vault, to eliminate the expense of deep foundation support. Does the proposal appear reasonable?

12-10. A multistory building with a plan area of 200 ft by 100 ft will impose an estimated total loading of 20,000 kips onto the supporting earth. If the floating foundation concept is to be utilized for this structure and the average unit weight of soil underlying the site is 120 pcf, what volume of soil excavation (and subsurface structure) is required?

12-11. Compile a summary list of the common types of pile foundations, including materials used, available lengths, and common range of load applications.

12-12. List the general types of pile drive hammers in use, and provide a brief description of the basic operating principle for each.

12-13. Structural designers commonly mention that steel piles are a good choice for sites where long piles are required and the depth to bearing will be variable and somewhat unpredictable. What are the probable reasons for this belief?

12-14. Indicate the difference between pile and pier types of foundations.

12-15. Outline the difference between the caisson type of foundation and the pier type of foundation.

12-16. Relating to the installation of driven piles, provide a brief explanation of spudding, predrilling, and jetting (indicate the general purpose for the procedure, and the differences in the methods).

12-17. The selection of pile foundations for any project requires that the designer and contractor consider various practical issues to ensure an adequate rate of progress (completed installations) and proper (quality) installations (no damage, etc.). For driven wood piles, driven steel-shell and concrete-filled piles, and augered-type cast-in-place piles, list some of these practical issues that probably would have influence on the type chosen (such as transporting longer piles to the job site, a possible problem in an urban setting, and on-site storage).

12-18. Soil conditions at a proposed construction site make it necessary for a driven pile foundation system to be utilized to support a planned building. The designer has the following design–cost estimate data:

(a) Wood piles, 25-ton capacity, required length 35 ft, $15 per lineal foot in place.

(b) Steel-shell, cast-in-place concrete piles, 50-ton capacity, required length 45 ft, $22 per lineal foot in place.

(c) Steel HP piles, 95-ton capacity, required length 75 ft, $27 per lineal foot in place.

Of these types of piles, which appears to be the most economical for supporting the typical column loadings of 350 tons and 550 tons?

12-19. A piling contractor has been awarded the pile foundation contract for a large warehousing structure. The contractor's fee is based on the following conditions and terms:

- Building will require 120 separate pile foundation units, with the design indicating that the foundation units will consist of pile groups varying between 5 and 9 piles per group.
- Design estimate calls for a total of 1100 piles, lengths to vary between 40 and 50 ft, with an average length of 46 ft expected (the estimated total footage for this project is therefore 50,600 linear feet, and this total serves as the basis for calculating final contractor installation costs).
- Mobilization and demobilization for 4 pile rigs is $45,000.
- Contract base payment for piles, total estimated linear foot basis, is $26.50 per foot.
- Overage, for total linear footage installed in excess of the estimated total of 50,600 ft, will be $32 per foot.
- Underage, if installed linear footage is less than the estimated total of 50,600 ft, will be $20 per foot payment reduction for each foot less than the estimated total.
- Installation period for the pile foundations is 30 days; contractor will be penalized $2,000 per day for each day beyond 30 days; and contractor bonus will be $1,000 per day if the installation is completed in less than 30 days.

Compute the contract final (total) fee if

(a) 48,000 linear feet are actually installed and the contractor finishes four days ahead of the 30-day schedule.

(b) 55,000 linear feet are installed, and the contractor finishes in 30 days.

12-20. A brief description of subsurface conditions at different building locations follows. Assume a commercial–industrial building is planned. Column loads to be imposed on foundations will vary, and could be up to 2000 kN. For each location, list the type(s) of foundation which, for preliminary planning, probably could be considered suitable.

(a) Thick stratum of glacially deposited sand–silt–clay mixture, relatively firm–compact, is underlain by rock; water table is deep.

(b) Soil deposit across the area consists of widely varying layers of fine-grained soil and sand soil possessing low shear strength.

Rock underlies the area at depths ranging from 10 m to 15 m below the soil surface; groundwater table is at a depth of 3 m.

(c) Thin surface layer of predominantly sand soil (2 to 3 m thick) is underlain by a 3- to 4-m-thick stratum of weak, compressible fine-grained soil. A thick stratum of firm clay underlies the weak compressible layer. The groundwater table is at a depth of 10 m.

(d) A surface layer of weak organic soil ranging from 2 to 4 m thick is underlain by a thick stratum of relatively dense–compact sandy soil with limited silt and gravel content. The area groundwater table corresponds to the surface of the sand stratum, or slightly below.

CHAPTER 13

Foundations

Design Considerations and Methods

Building foundations need to be capable of carrying an imposed loading without undergoing movement that causes structural damage or affects the facility's planned usage. These considerations require that the soil responsible for supporting a foundation not be stressed beyond its strength limits. Simultaneously, the deformations resulting within this soil because of loading and action of natural forces cannot be excessive. The pressure that a foundation unit can impose onto the supporting earth mass without causing overstressing (or shear failure) is the soil's *bearing capacity*. Deformations occurring because of foundation loading usually cause settlement, but lateral movements may also be of concern. It is important also to consider the possibility of foundation movements due to natural phenomena, such as soil expansion and shrinkage if moisture changes or freezing occurs. The magnitude and type of loading (static, live, or repetitive), the foundation performance requirements (how much settlement is permissible), and properties of the supporting soil all have influence on the type and size of foundation that will be necessary and its resulting behavior.

Methods in widespread use for determining a soil's bearing capacity include the application of bearing capacity equations, the utilization of penetration resistance data obtained during soil explorations, and the practice of relating the soil type to a presumptive bearing capacity recommended by building codes. Permissible bearing capacities determined by the equation method or from building code tables typically do not consider effects of soil compressibility and the possible influence of poorer soil layers underlying the bearing layer. Consequently, settlement determinations and other results of soil deformations must be analyzed separately (e.g., by using the methods discussed in Chapter 10). Foundation design criteria developed from boring–penetration resistance data often relate a foundation bearing pressure to settlement. Design data in this form are convenient to use, but the methods available do not cover all foundation and soil types and may be less precise than the analytical methods for determining bearing capacity and settlement.

An additional method in use for determining the permissible or safe design loading (applied mostly for piles, less so for footing-type foundations) is the field load test performed on an in-place foundation unit. Load tests relate carrying capacity and settlement together, which is an advantage. Disadvantages include the cost and time involved. Further, test results require care in their evaluation, for it is known that a load test on a small shallow foundation may not be representative of the behavior of a large shallow foundation. With piles, the load test results on a single unit may not be indicative of the behavior of a loaded pile group.

Of the common procedures for foundation design, the analytical method using soil mechanics principles (e.g., the use of bearing capacity equations with settlement analysis) and the use of penetration resistance data are preferred. Properly applied, these methods consider the effects of foundation type and size as well as the properties of soil to the depth that will have significant effect on the foundation performance. Load test data will provide reliable design information if properly related to the results of a subsurface exploration and a final foundation design. The use of presumptive bearing capacities is discouraged because of the heavy reliance on the soil-type description, with little correlation to the soil's actual physical properties and no consideration of the possible existence of weak soil strata underlying the foundation bearing level.

The procedures for foundation design that follow represent currently accepted methods. These methods generally have a history of being conservative. As in the past, foundation design procedures are subject to improvement for greater precision as new techniques are developed to better determine soil properties and behavior or as new analytical tools evolve.[1] The foundations profession is still considered a "state of the art" profession. Scientific methods and principles are utilized, but exact answers are not always expected. Final decisions concerning best foundation type, design criteria, expected behavior, and methods for construction and field control are greatly influenced by experience and intuition.

SHALLOW FOUNDATIONS—DESIGN AND INSTALLATION CRITERIA

13.1   BASIC CONCEPTS—LONG (STRIP) FOOTINGS

The ultimate soil-bearing capacity for foundations (the loading that will cause a shear failure in the supporting soils) is related to the properties of the soil, including the past stress history and the proximity of the groundwater table; it is also affected by the characteristics of the foundation, including size, depth, shape, and the method of construction or installation.

For the case of an increasing load being imposed onto a shallow, horizontal strip footing resting on a homogeneous soil, a characteristic load-settlement curve is shown in Figure 13-1. Information important for predicting foundation behavior and for developing criteria appropriate for foundation design has resulted from studies in which influential soil characteristics have been correlated with the settlement data (e.g., the soil descriptions have identified the materials as dense or firm, exhibiting brittle-type stress–strain behavior, or as loose or soft, exhibiting plastic properties). The soil zones involved in developing resistance to foundation loading (therefore, the soil zones responsible for bearing capacity) have been

---

[1]For example, the pressuremeter method discussed later in this chapter represents a procedure that is new to most of the world's building designers but that offers the promise of applying field-determined pressure-resistance data directly to foundation design and behavior.

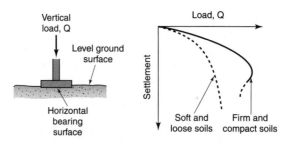

Figure 13-1 Typical load-settlement relationship for shallow foundations.

identified in a qualitative manner. The three principal modes of soil failure, established by the patterns of the shearing zones, are defined as *general shear failure, local shear failure,* and *punching shear failure.* Figure 13-2 illustrates the differences in the three modes and the foundation load-settlement curve typical to each mode.

The *general shear failure* (Fig. 13-2a), expected for soils possessing brittle-type stress–strain characteristics, is identified by a well-defined wedge beneath the foundation and slip surfaces extending diagonally from the side edges of the footing downward through the soil, then upward to the ground surface. The ground surface adjacent to the footing bulges up-

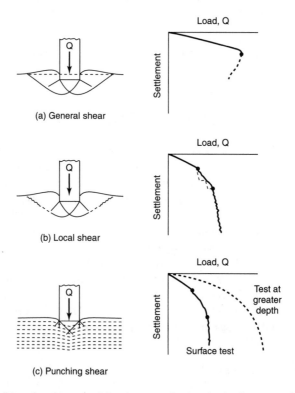

Figure 13-2 Conditions for ultimate soil-bearing capacity: involved soil zones and related foundation settlements [317].

ward. Soil displacement is accompanied by tilting of the foundation (unless the foundation is restrained). The load-settlement curve for the general shear case indicates that failure is abrupt.

The *punching shear failure* (Fig. 13-2c), occurring in soil possessing the stress–strain characteristics of a very plastic material, involves poorly defined shearing planes. Significant compression of a wedge-shaped soil zone beneath the foundation is accompanied by the occurrence of vertical shear beneath the edges of the foundation. The soil zones beyond the edges of the foundation are little affected, and no significant degree of surface bulging occurs. Aside from large settlement, failure is not clearly recognized.

The *local shear failure* (Fig. 13-2b) involves some of the characteristics of both the general shear and the punching shear failure modes. As for general shear, the well-defined wedge and slip surfaces are formed *beneath* the foundation, but the slip surfaces fade into the soil mass beyond the edges of the foundation and do not carry upward to the ground surface. Slight bulging of the ground surface adjacent to the foundation does occur. As with punching shear, significant vertical compression of soil directly beneath the foundation takes place. The local shear condition represents a transitional mode between the general shear and the punching shear failures and is expected for soils possessing somewhat plastic stress–strain characteristics.

The load-settlement curves of Figure 13-2 are qualitative in nature. Investigators of the bearing capacity problem have found that relatively large settlements are required in order to have foundations reach the "failure" load (viz., 3 to 15% of the foundation width for very shallow installations and up to 25 percent of the foundation width for deeper installations).

For application to shallow foundation design, it is commonly considered that the general shear case applies to dense granular soil and to firmer saturated cohesive soils subject to undrained loading (the U-U and C-U shearing conditions apply). The punching shear case is considered appropriate for compressible soil, such as sands having a low to medium relative density, and for cohesive soils subject to slow loading (the C-D shearing conditions apply).

13.2 BEARING CAPACITY EQUATIONS

Historically, over the past one hundred years, a number of investigators have undertaken studies relating to foundation bearing capacity, typically applying the classical theories of elasticity and plasticity to soil behavior to develop equations appropriate for foundation design.[2] Behavior described by the classical theory of plasticity has been widely used to obtain a solution for the case of a general shear failure. The original theoretical concepts for analyzing conditions considered applicable to foundation performance using the theory of plasticity are credited to Prandtl [231] and Reissner [239]. Prandtl studied the effect of a long, narrow metal tool bearing against the surface of a smooth metal mass that possessed cohesion and internal friction but no weight. The results of Prandtl's work were extended by Reissner to include the condition where the bearing area is located below the surface of the resisting material and a surcharge weight acts on a plane that is level with the bearing area. Terzaghi [293] applied the developments of Prandtl and Reissner to soil foundation problems, extending the theory to consider rough foundation surfaces bearing on materials that possess weight.

Conditions for relating the classical theory of plasticity to the case of a general shear failure are indicated by Figure 13-3. The arrangement shown establishes criteria for

---

[2]Jumikis (1962) [146] provides a summary of the works of early investigators.

developing the ultimate bearing capacity for a long strip foundation; because of the infinite foundation length, the analysis proceeds as for a two-dimensional or plane-strain problem. The theory assumes that the (soil) material in zones I, II, and III possesses the stress–strain characteristics of a rigid plastic body (viz., the material shows an infinite initial modulus of elasticity extending to the point of shear failure, followed by a zero modulus; see Fig. 13-3b). Applied to the soil mass providing support for the foundation, the theory assumes that no deformations occur prior to the point of shear failure but that plastic flow occurs at constant stress after shearing failure. It is also assumed that the plastic deformations are small and the geometric shapes of the failure zones remain essentially constant. The use of an equivalent surcharge to substitute for the soil mass above the level of the foundation, along with some estimations, simplifies the analysis, but the effect is to provide conservative results. When subject to a foundation loading near to the ultimate, zone I behaves as an active zone that pushes the radial zone II sideways and the passive zone III laterally upwards. Boundaries AC and DE shown on Figure 13-3c are essentially straight lines; the shape of section CD varies from circular (when the soil angle of internal friction ϕ is zero degrees) to a curve intermediate between a logarithmic spiral and a circle (when ϕ is greater than zero degrees).

Terzaghi developed a general bearing capacity equation for strip footings that combined the effects of soil cohesion and internal friction, foundation size, soil weight, and surcharge effects in order to simplify the calculations necessary for foundation design. His equation utilized the concept of dimensionless bearing capacity factors whose values are a function of the shear possessed by the supporting soils. Through ensuing years, the ultimate bearing capacity for shallow and deep foundations has continued to be studied in the quest for refined definition of foundation soil behavior and a generalized bearing capacity equation that agrees well with failure conditions occurring in model and large-scale foundation

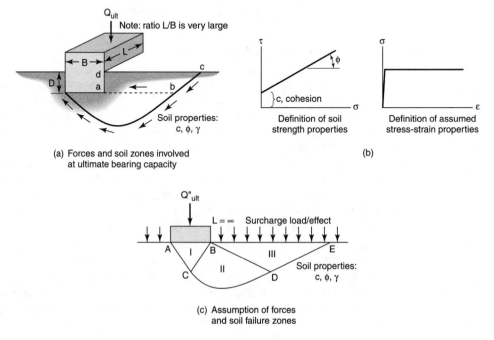

(a) Forces and soil zones involved at ultimate bearing capacity

(b)

(c) Assumption of forces and soil failure zones

Figure 13-3 Definition of conditions for developing the ultimate bearing capacity equation.

tests.[3] Modifications to early concepts have emerged from such studies, but the general form of the Terzaghi bearing capacity equation has been retained because of its practicality.

The ultimate bearing capacity equation, shown as Equation 13-1a and developed from the conditions outlined in Figure 13-3c for long footings, has been found to provide good agreement with ultimate loads observed for model and large-scale foundations. This equation shows the Terzaghi format, but the current dimensionless bearing capacity factors are the result of better definition of the soil failure zones.

$$\frac{Q_{\text{ult}}}{(B \times L)} \approx \frac{Q''_{\text{ult}}}{(B \times 1)} = q_{\text{ult}} = cN_c + \frac{1}{2}B\,\gamma_1 N_\gamma + \gamma_2 D_f N_q \qquad (13\text{-}1a)$$

where Q_{ult} = total loading acting at the base of the strip foundation whose width is B and length is L (kips, kN, etc.)

Q''_{ult} = load acting at base of foundation for unit length in L direction (kN/m, k/ft, etc.)

q_{ult} = ultimate gross bearing capacity or soil bearing pressure (kN/m², ksf, etc.)

c = cohesion of the soil below foundation level (kN/m², kPa, ksf, etc.)

D_f = depth of footing below lowest adjacent soil surface (m, ft, etc.)

γ_1 = effective unit weight of soil below foundation level (kN/m³, pcf, kcf, etc.)

γ_2 = effective unit weight of soil above foundation level (kN/m³, pcf, kcf, etc.)

N_c, N_γ, N_q = soil-bearing capacity factors, dimensionless terms, whose values relate to the angle of internal friction, ϕ, whereby

$$N_q = \tan^2\left(45 + \phi/2\right)(e)^{\pi\tan\phi}$$

where e = base of natural logarithims, 2.71828

$$N_c = (N_q - 1)\cot\phi \qquad \text{(for } \phi > 0°)$$
$$= 5.14 \qquad\qquad \text{(for } \phi = 0°)^{[4]}$$
$$N_\gamma = 2(N_q + 1)\tan\phi$$

The relationship between the bearing capacity factors, N_q, N_c, and N_γ and the angle of internal friction is presented in Figure 13-4.

The first and third terms in this equation represent the Prandtl–Reissner solution for the condition of a weightless soil. The second term represents the separate solution to a determination of the ultimate bearing capacity for a cohesionless soil with weight but no overburden effects (a necessary consideration because the Prandtl–Reissner terms do not include a shearing resistance from the effect of soil weight and internal friction, ϕ, in the failure zones). This method of superposition (i.e., a simple combination of three terms) to produce Equation 13-1a is not a theoretically correct procedure, but errors are on the safe side and small for most practical applications.

Equation 13-1a implies that the bearing capacity for foundations on soil possessing internal friction, ϕ, increases indefinitely with increases in foundation width, as indicated by the solid lines of Figure 13-5. However, physical studies have indicated that the *rate* of increase for bearing capacity decreases somewhat as foundation width increases, and there may be an up-

---

[3]Meyerhof [190], Hansen [119], Vesic [317], DeBeer [74].
[4]A rigorous mathematical solution for the value of N_c when ϕ equals zero is presented by Jumikis [147, pp. 626–27].

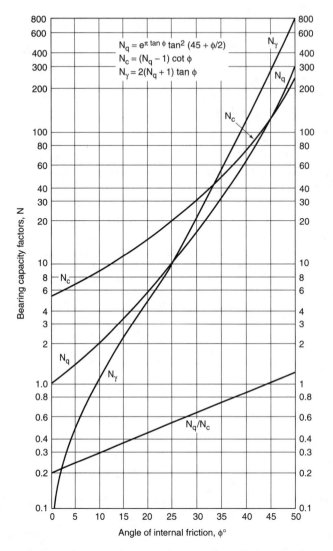

Figure 13-4 Bearing capacity factors for the bearing capacity equation.

per limit to a soil's bearing capacity (see dashed lines of Fig. 13-5). Foundation widths beyond the value that develops the upper limit would not contribute to the bearing capacity. Fortunately, the width corresponding to such an upper limit on bearing capacity apparently is relatively large; as a result, the condition is not a serious restriction for commonly used sizes of shallow foundations. This "size effect" can be compensated for during design by using the ϕ value from soil testing appropriate to the general range of the bearing capacity pressure and by applying the rigidity index reduction factors discussed in the following paragraphs. Nevertheless, the effect of using higher bearing pressures on larger foundations requires careful consideration where settlement is important. Because of the manner in which stresses from foundation loadings are transferred into the earth (Chapter 9), a large foundation tends to settle more than a small foundation when both impose the same intensity of bearing pressure, because the loaded area and

Figure 13-5   Implications of the general bearing capacity equation: Allowable pressures increase with increases in footing width and depth (solid lines). Note that the actual rate of bearing capacity increase *decreases* as width becomes greater (dashed lines).

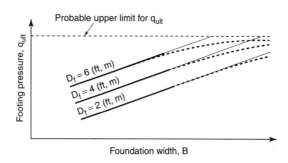

total load Q are greater. Using a greater bearing pressure for a large footing further increases the tendency for greater settlement. Consequently, if lightly loaded and heavily loaded foundations for a structure are designed only on the basis of allowable bearing capacity, there is a strong possibility of differential settlement between the large and small footings.

The assumption that the soil responsible for developing a foundation bearing capacity behaves as a rigid material is satisfied for the case of general shear but is not appropriate for punching shear and local shear. Comparison of the relative load-settlement curves of Figure 13-2 indicates that, for the punching and local shear cases, the ultimate load is less and the settlement is greater than for the condition of the general shear failure. Methods applied in the past to compensate for lower bearing capacity design values where a foundation was to be supported on soil that is compressible and that would undergo significant volumetric changes (i.e., where the local or punching shear condition was expected) typically utilized reduced values for the angle of internal friction, ϕ, and the cohesion, c, in the general bearing equation. Results could be overly conservative, however. Vesic [317] has proposed the use of a rigidity index, I_r, which provides a relative measure of soil deformability, whereby

$$I_r = \frac{G}{c + \bar{\sigma}_v \tan \phi} = \frac{E}{2(1 + v)(c + \bar{\sigma}_v \tan \phi)} \tag{13-2}$$

where G = soil shearing modulus

c, ϕ = soil cohesion and angle of internal friction

$\bar{\sigma}_v$ = effective soil overburden pressure at a depth corresponding to $B/2$ below the foundation base

v = Poisson's ratio for the bearing soil (see Table 13-1)

E = soil modulus of elasticity[5] = $\dfrac{1}{m_v}\left[\dfrac{(1 + v)(1 - 2v)}{(1 - v)}\right]$ (also see Table 13-1)

$\dfrac{1}{m_v}$ = constrained modulus

$m_v = \dfrac{\Delta\varepsilon_v}{\Delta\bar{\sigma}_v} = \dfrac{a_v}{1 + e_0} = \dfrac{.435\, C_c}{(1 + e_0)\bar{\sigma}_{v_a}}$ where $\bar{\sigma}_{v_a}$ is the average of the initial and final stress, determined at a depth of $B/2$.

(*Note:* Values of E and m_v determined from field and laboratory tests are considered to include the effects of elastic and plastic deformations.)

---

[5]Soil modulus of elasticity can also be obtained as the slope of the deviator stress versus strain curve, as taken from the triaxial test.

Table 13-1 Range of Values: Modulus of Elasticity and Poisson's Ratio

Soil Type	Modulus of Elasticity, E	
	ksf	kN/m²
Sand, loose	200–500	9,000–25,000
Sand, dense	1,000–1,700	45,000–80,000
Sand, silty	150–450	7,000–21,000
Sand and gravel, loose	1,000–3,000	45,000–145,000
Sand and gravel, dense	2,000–4,000	90,000–180,000
Silt	50–400	2,400–20,000
Loess	300–1,100	15,000–50,000
Clay, soft	10–100	500–5,000
Clay, medium	100–200	4,000–10,000
Clay, firm	150–400	7,000–20,000
Clay, sandy	550–850	25,000–40,000

Soil Type	Poisson's Ratio, v
Sand, loose	0.10–0.30
Sand, dense	0.30–0.40
Silt	0.30–0.40
Loess	0.20–0.40
Clay, saturated	0.40–0.50
Clay, partially saturated	0.30–0.40
Clay, with sand and silt	0.20–0.40

Note: Values for Poisson's ratio lie within a narrow range and for analytical studies usually are estimated. The modulus of elasticity varies widely, being affected by stress history, water content, density, and grain–size distributions; with a given soil, values will vary according to the applied load range.

Using a critical value of rigidity index $I_{r_{\text{critical}}}$ as a reference value, the actual rigidity index for a site is compared to the critical index, whence

$$I_{r_{\text{critical}}} = \frac{1}{2} \exp\left[\left(3.30 - 0.45\frac{B}{L}\right)\cot\left(45 - \frac{\phi}{2}\right)\right] \tag{13-3}[6]$$

Values of the critical rigidity index $I_{r_{\text{critical}}}$ appropriate for a strip foundation and a square foundation are presented in Figure 13-6. Values for rectangular shapes (for $0 < B/L < 1$) can be interpolated between the two limit curves or calculated from Equation 13-3.

Where the actual rigidity index is *greater than* the critical index, the general shear failure mode applies and Equation 13-1a is suitable. Where the actual rigidity index is *less than* the critical index, the local or punching shear failure mode applies and the ultimate bearing capacity will be less than indicated by Equation 13-1a. For this case, reduction factors r_c, r_γ, and r_q are applied to modify the bearing capacity factors N_c, N_γ, and N_q, resulting in equation 13-1b. Reduction factors are obtained from Figure 13-7.

---

[6]Note: e^x can also be written as $\exp(x)$; that is, $e^x = \exp(x)$

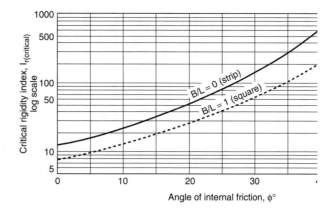

Figure 13-6   Relationship of critical rigidity index, $I_{r_{critical}}$, and soil angle of internal friction, ϕ.

$$q_{ult} = c(N_c r_c) + \frac{1}{2}B\,\gamma_1(N_\gamma r_\gamma) + \gamma_2 D_f(N_q r_q) \qquad (13\text{-}1b)$$

Equations 13-1a and 13-1b should not be used directly for foundation design. These equations serve as the basis for designs but require modification because of the effect of various practical factors. Such factors and the form of the equations suited to foundation design are discussed in the following sections. (For designs, see Eq. 13-6.)

Additional Considerations, Bearing Capacity Equation

The ultimate bearing capacity of a long strip footing (Eq. 13-1b) indicates the gross or total bearing pressure that can be imposed at the level of the base of the footing. For the usual condition of a foundation installed below the ground surface, the gross bearing pressure includes the pressure developed from the foundation loading *and* the weight of the buried footing and soil backfill directly above the footing. For practical design purposes, the net bearing pressure, q_{net}, then becomes

$$q_{net} = q_{ult} - \gamma_2 D_f$$
$$= cN_c r_c + \frac{1}{2}B\,\gamma_1 N_\gamma r_\gamma + \gamma_2 D_f(N_q - 1)r_q \qquad (13\text{-}4)$$

where $\gamma_2 D_f$ is the pressure of the soil overburden adjacent to the level of the foundation base. Where the overburden is not of constant height, the lesser value for $\gamma_2 D_f$ should be used (Fig. 13-8).

The values for soil unit weight to be used in the bearing capacity equation (γ_1 and γ_2) are effective unit weights. When the soil in the yield zone is below the water table, the submerged unit weight must be used. For the condition where the water table is at the base of

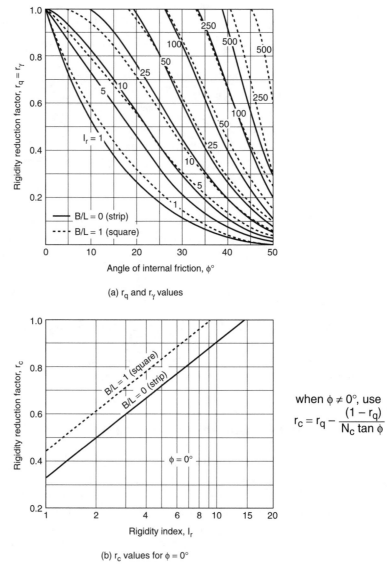

(a) r_q and r_γ values

(b) r_c values for $\phi = 0°$

Figure 13-7 Rigidity reduction factors when actual soil rigidity index is less than $I_{r_{critical}}$.

a footing, the submerged (or buoyant) soil unit weight applies for the γ_1 term. Since the depth of the failure zone is considered to be approximately equal to the width of the footing, the full unit weight of the soil (not the submerged unit weight) is used when groundwater is lower than a distance B below the footing. For in-between depths of the water table, an interpolated value of effective soil unit weight (between $\gamma_{wet\ soil}$ and γ_{sub}) can be used. Continuing, if the water table extends above footing level, a submerged soil unit weight should also be used for the γ_2 value.

Figure 13-8   Illustration of unequal surcharges encountered in foundation design.

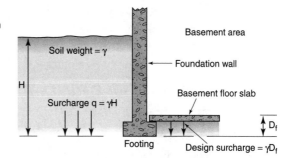

A high water table results in a lowered bearing capacity because of the requirement to use a reduced effective soil weight (the submerged weight) to determine the safe capacity. Further, situations arise in which a high water table forces additional factors to be taken into account. In the case where a building's basement will be below the level of an expected groundwater table, unequal pressures acting on the structure's sidewalls and floor as a result of this groundwater need to be considered in the design (Fig. 13-9). The possible requirement for dewatering during construction (Chapter 7) should also be studied.

Equations 13-1a, 13-1b, and 13-4 provide ultimate values of bearing capacity. Designers conservatively assume that some variation in soil conditions will exist over an area to be occupied by construction or could develop because of environmental factors, such as a fluctuating water table or drying, and they apply the resulting soil properties determined from explored locations with some caution. Also, as discussed earlier, ultimate loadings result in foundation settlements that are large and typically greater than desired or tolerable. As a result, when the bearing capacity equation is used for foundation design, it is conven-

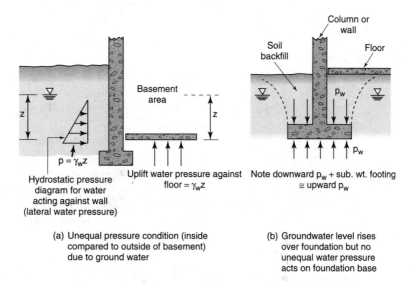

Figure 13-9   Effects of high groundwater on foundation and basement design.

tional practice to apply a factor of safety of 2.5 or 3 to the value of q_{net} (Equation 13-4) to obtain q_{design}:

$$q_{design} = \frac{q_{net}}{2.5} \text{ or } \frac{q_{net}}{3} \qquad (13\text{-}5)$$

Application for Design—The Extended Bearing Capacity Equation

Equations 13-1a, 13-1b, and 13-4 apply for the condition of an infinitely long strip foundation with the base horizontal, with a vertical load centrically applied, where the ground surface surrounding the foundation is horizontal; also recall that the shearing strength of the soil above the plane level with the foundation base was neglected. In practice, these conditions are not always present: Foundations have finite dimensions (the commonly used shapes include square, rectangular, and circular as well as the long strip); the soil above the level of the footing may have significant shear strength; the loading on a horizontal foundation may be inclined or eccentrically positioned; the foundation base may be inclined; or the soil surface may be sloped.

The general bearing capacity equation developed for the infinite strip can be extended to consider variations from the basic assumptions by applying modification factors that account for the effect of each variation. The *extended bearing capacity equation* [119] is written

$$q_{net} = cN_c(r_c s_c d_c i_c b_c) + \frac{1}{2}B\gamma_1 N_\gamma(r_\gamma s_\gamma d_\gamma i_\gamma b_\gamma)$$
$$+ \gamma_2 D_f(N_q - 1)(r_q s_q d_q i_q b_q) \qquad (13\text{-}6)$$

where r_c, r_γ, r_q = reduction factors to compensate for a possible punching–local shear condition
s_c, s_γ, s_q = shape factors for square, rectangular, and circular foundations
d_c, d_γ, d_q = depth factors
i_c, i_γ, i_q = load inclination factors
$b_c, b_\gamma, b_q,$ = base tilt factors

Suitable modification factors to use with Equation 13-6 are presented in the following sections. Where a modification factor does not apply for the foundation condition under study, its value for Equation 13-6 is considered to be unity (that is, neglected; factors that do not apply simply are disregarded in the equation).

In summary, Equation 13-6 should be utilized for all practical applications when calculating the value of q_{net} *that is substituted into Equation 13-5 to obtain* q_{design}.

The illustrations that follow demonstrate application of the theory that has been developed to this point for long strip footings supporting vertical load. Procedures to modify the extended bearing capacity equation, for use with other shapes of footings and load conditions, are discussed in the sections following these illustrations.

Illustration 13-1 _____

A strip footing for a masonry bearing wall is 1.1 m wide and is supported in soil that has the properties indicated by the sketch. What design loading can be imposed onto the foundation per meter of length? Use a factor of safety equal to 3 with the general bearing capacity equation (Equations 13-4 and 13-5). For this problem, ignore the depth factor terms.

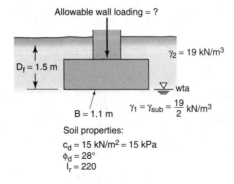

Allowable wall loading = ?

Soil properties:
$c_d = 15$ kN/m^2 = 15 kPa
$\phi_d = 28°$
$I_r = 220$

Solution

From bearing capacity factor chart, obtain $N_c = 26$, $N_\gamma = 16$, $N_q = 15$. From Figure 13-6, $I_{r_{critical}} \cong 140$ for $B/L = 0$ and $\phi = 28°$. Since $I_r > I_{r_{critical}}$, $r_c = r_\gamma = r_q = 1$. Therefore,

$$q_{net} = cN_c(1) + \frac{1}{2}B\,\gamma_1 N_\gamma(1) + \gamma_2 D_f(N_q - 1)(1)$$

$$= (15 \text{ kPa})(26) + \frac{1}{2}(1.1\text{m})\left(19\frac{\text{kN}}{\text{m}^3} \times \frac{1}{2}\right)(16)$$

$$+ (19\text{kN/m}^3)(1.5 \text{ m})(15 - 1) = 872.6 \text{ kPa}$$

$$q_{design} = \frac{872.6 \text{ kPa}}{3} = 291 \text{ kPa}$$

Allowable wall loading = (291 kPa)(1.1 m wide)(1 m long) = 320 kN per meter of wall length.

Illustration 13-2 _____

A strip foundation similar to that used in Illustration 13-1 is located in an area where medium-soft clay soil exists. Properties are indicated on the sketch. Using the general bearing capacity equation, with a factor of safety equal to 3, calculate the wall load that can be carried by the footing (Equations 13-4 and 13-5). For this problem, ignore the depth factor terms.

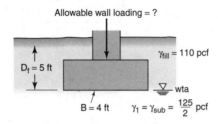

Allowable wall loading = ?

Soil properties:
$c_d = 150$ psf
$\phi_d = 22°$
E (from triaxial test) = 750 psi = 108 ksf
Poisson's ratio = 0.35 (estimated on basis of soil type)

Solution

Bearing capacity factors: $N_c = 17$, $N_\gamma = 7$, $N_q = 8$. Compute

$$I_r = \frac{E}{(2)(1 + v)(c + \bar{\sigma}_v \tan \phi)} = \frac{108 \text{ ksf}}{(2)(1 + .35)(.150 + .675 \tan 22°) \text{ ksf}} = 48$$

where $\bar{\sigma}_v$ at depth of $B/2$ beneath footing $= (.110 \text{ kcf} \times 5 \text{ ft}) + \left(\frac{.125}{2} \text{ kcf} \times 2 \text{ ft}\right) = 675$ psf $= .675$ ksf.

From Figure 13-6, obtain $I_{r_{\text{critical}}} \cong 60$, and since $I_r < I_{r_{\text{critical}}}$, use rigidity reduction factors from Figure 13-7. For $I_r = 48$, obtain $r_q = r_\gamma = 0.92$, $r_c = 0.91$.

Calculate

$$q_{\text{net}} = cN_cr_c + \frac{1}{2}B\gamma_1N_\gamma r_\gamma + \gamma_2 D_f(N_q - 1)r_\gamma$$

$$= (.15 \text{ ksf})(17)(.91) + \frac{1}{2}(4 \text{ ft})\left(\frac{.125}{2} \text{ kcf}\right)(7)(.92)$$

$$+ (.110 \text{ kcf})(5 \text{ ft})(8 - 1)(.92) = 6.67 \text{ ksf}$$

Therefore

$$q_{\text{design}} = \frac{6.67 \text{ ksf}}{3} = 2.22 \text{ ksf}$$

Allowable wall loading $= (2.22 \text{ ksf})(4 \text{ ft wide} \times 1 \text{ ft long}) = 8.9 \text{ k/ft of wall length.}$

Illustration 13-3

A wall footing is to be constructed on a clay soil, as indicated by the sketch. The footing is to support a wall that imposes a loading of 135 kN per meter of wall length. What footing width should be provided to have a factor of safety of 3 for a bearing pressure computed from the general bearing capacity equation (Equations 13-4 and 13-5). Ignore the depth factor terms for this problem.

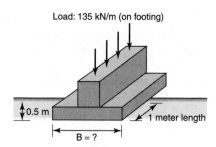

Load: 135 kN/m (on footing)

0.5 m

1 meter length

B = ?

Soil properties:
$\gamma = 17.3$ kN/m^3
$c_u = 72$ kN/m^2 $(\phi = 0°)$
$E = 5500$ kN/m^2
$v = 0.35$
$I_r = 28$

Solution

From Figure 13-6, $I_{r_{\text{critical}}} = 13$ for $\phi = 0°$ and curve for $B/L = 0$. Since $I_r > I_{r_{\text{critical}}}$, we have $r_c = 1$. For $\phi = 0°$, $N_c = 5.14$; N_γ and N_q are negligible.

Calculate

$$q_{ult} = cN_cr_c = (72 \text{ kN/m}^2)(5.14)(1.0) = 370 \text{ kN/m}^2$$

$$q_{design} = (370 \text{ kN/m}^2)\left(\frac{1}{3}\right) = 123 \text{ kN/m}^2$$

Allowable footing load $= (123 \text{ kN/m}^2)(B \text{ m}) = 123B \text{ kN/m}$ of wall length.

Required footing width, $B = \dfrac{135 \text{ kN/m}}{123 \text{ kN/m}^2} = 1.1 \text{ m}$.

Modification Factors for Square, Rectangular, and Circular Footings.   For practical use, the yield zones occurring in the soil beneath an infinitely long footing can be adequately analyzed in a two-dimensional study. Where foundations of a finite length are involved, such as with square, rectangular, and circular shapes, the yield zones are three-dimensional and Equations 13-1a, 13-1b, and 13-4 do not directly apply. Mathematical solutions have been managed for the case of a circular footing, and use of a shape factor permits the format of the general bearing capacity equation to be used. Mathematical solutions have not yet been obtained for square and rectangular shapes because of the complexity of the problem, but shape factors are available, as developed from empirical data, to include with the general bearing capacity equation. The shape factors appropriate for square, rectangular, and circular foundations are as follows:

$$s_c = 1 + (B/L)(N_q/N_c) \tag{13-7a}$$

$$s_\gamma = 1 - .4(B/L) \tag{13-7b}$$

$$s_q = 1 + (B/L)\tan\phi \tag{13-7c}$$

For the square and circular foundation shapes, the B/L ratio equals unity. Note that for a strip foundation, the B/L value approaches zero and the shape factors become 1.

Illustration 13-4

A square foundation for a column is to carry a loading of 1500 kN. Determine the footing dimensions when bearing on the surface of a sand having properties indicated on the sketch. Use a safety factor of 3 for the *extended bearing capacity equation*.

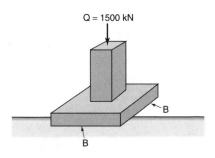

Q = 1500 kN

B

B

Soil properties: sand
  $\gamma = 20 \text{ kN/m}^3$
  $\phi_d = 36°$ (c = 0)
  $N_\gamma = 50$
  $I_r = 450,\ I_{r(critical)} = 150$

Solution

For a square foundation on the surface of sand (cohesion, $c = 0$), with no inclination or tilt,

$$q_{net} = \frac{1}{2} B \gamma_1 N_\gamma r_\gamma s_\gamma$$

where $s_\gamma = 1 - .4(B/L) = .6$. Since $I_r > I_{r_{critical}}$, $r_\gamma = 1$. Obtain

$$q_{net} = \frac{1}{2}(B \text{ ft})(20 \text{ kN/m}^3)(50)(1.0)(.6) = 300B \text{ kN/m}^2$$

$$q_{design} = (300B \text{ kN/m}^2)\left(\frac{1}{3}\right) = 100B \text{ kN/m}^2$$

with $Q_{design} = 1500$ kN, we also need to satisfy the requirement that

$$q_{design} = \frac{Q}{A} = \frac{1500 \text{ kN}}{B^2}$$

Therefore

$$q_{design} = 100 \, B \text{ kN/m}^2 = 1500/(B \text{ m})^2$$

$$\text{or } B^3 = 15 \text{ m}^3$$

$$\text{so } B = 2.47 \text{ m}$$

Illustration 13-5

In an industrial building, two columns are close together and will be supported on a common rectangular foundation. The soil is a clay having a cohesion of 1700 psf. For the conditions indicated by the sketch, what footing dimensions should be provided? Use a safety factor of 3 with the extended bearing capacity equation.

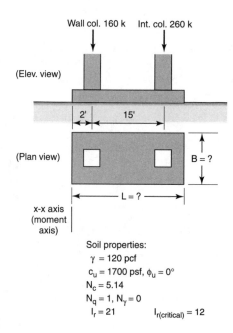

x-x axis
(moment axis)

Soil properties:
$\gamma = 120$ pcf
$c_u = 1700$ psf, $\phi_u = 0°$
$N_c = 5.14$
$N_q = 1$, $N_\gamma = 0$
$I_r = 21$ $I_{r(critical)} = 12$

Solution

For the rectangular foundation on clay, $q_{net} = cN_c r_c s_c$. Assume $L/B \cong 5$; therefore,

$$s_c = 1 + (B/L)(N_q/N_c) = 1 + (.2)(.2) = 1.04$$

Since $I_r > I_{r_{critical}}$, obtain $r_c = 1.0$.

$$q_{net} = (1.7 \text{ ksf})(5.14)(1.0)(1.04) = 9.09 \text{ ksf}$$

$$q_{design} = (9.09 \text{ ksf})\left(\frac{1}{3}\right) = 3.03 \text{ ksf}$$

but use 3.0 ksf. Two unknowns require two independent equations; use

$$(i) \ \Sigma F_v = 0 \qquad (ii) \Sigma M_{xx} = 0$$

For $\Sigma F_v = 0$: Downward column loading = upward reaction on foundation base.

$$\text{Use} \quad (160 + 260) \text{ k} = (3\text{ksf})(BL) \text{ sf} \qquad \text{(weight of foundation is neglected)}$$

$$BL = 140$$

$$B = 140/L$$

For $\Sigma M_{xx} = 0$: Moment due to column loading = moment of soil reaction on foundation.

$$(160 \text{ k})(2 \text{ ft}) + (260 \text{ k})(17 \text{ ft}) = (3 \text{ ksf})(BL \text{ sf})(L/2 \text{ ft})$$

$$BL^2 = 3150 \text{ ft}^3$$

Since $B = 140/L$, get $BL^2 = (140/L)(L^2) = 3150 \text{ ft}^3$.
Obtain $L = 22.5 \text{ ft}$ and therefore $B = 6.25 \text{ ft}$.

Modification Factors for Footing Depth. The development of the general bearing capacity for a long strip foundation does not consider the shearing resistance of the failure plane passing through the soil zone above the level of the foundation base. If the soil in this zone is weak or compressible or consists of a poorly placed backfill, the omission is considered proper. However, if this upper soil zone possesses significant shearing strength, because of natural conditions or from compaction, the ultimate value for bearing capacity would be increased. For this case, depth factors (d_c, d_γ, d_q) can be applied to the extended bearing capacity equation, whereby

$$d_\gamma = 1 \tag{13-8a}$$

$$d_q = 1 + 2 \tan \phi (1 - \sin \phi)^2 \frac{\tan^{-1}(D_f/B)}{57.3} \tag{13-8b}$$

$$d_c = d_q - \frac{1 - d_q}{N_c \tan \phi} \quad \text{(when } \phi > 0°) \tag{13-8c}$$

$$d_c = 1 + \frac{\tan^{-1}(D_f/B)}{(3)(57.3)} \quad \text{(when } \phi = 0°) \tag{13-8d}$$

Modification Factors for Eccentric and Inclined Loads. If the loading imposed onto a horizontal foundation does not act at the center, or if the loading is not vertical, the pattern of the soil yield zone beneath the foundation is different from the pattern that develops beneath a level footing carrying a vertical load.

For a design, eccentricity should be limited to one-sixth of the foundation dimension to prevent the condition of theoretical uplift occurring under part of the foundation (a condition that causes rapidly increasing and potentially dangerous changes in bearing imposed onto the soil beneath the loaded side of the foundation). The effect of eccentric placement of foundation loading can be conveniently and conservatively considered by substitution of a reduced foundation width (B') or length (L') in the bearing capacity equation, using

$$B' = B - 2e_B \qquad (13\text{-}9a)$$

or

$$L' = L - 2e_L \qquad (13\text{-}9b)$$

where e_B and e_L are the eccentricity distances with reference to the center of the foundation area (Fig. 13-10).

The inclined loading can be handled by the inclusion of inclination factors (i_c, i_γ, i_q) with the extended bearing capacity equation, whereby

$$i_q = \left(1 - \frac{Q_h}{Q + BLc \cot \phi}\right)^m \qquad (13\text{-}10a)$$

$$i_\gamma = \left(1 - \frac{Q_h}{Q + BLc \cot \phi}\right)^{m+1} \qquad (13\text{-}10b)$$

$$i_c = i_q - \left(\frac{1 - i_q}{N_c \tan \phi}\right) \quad \text{(when } \phi > 0°) \qquad (13\text{-}10c)$$

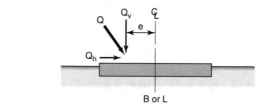

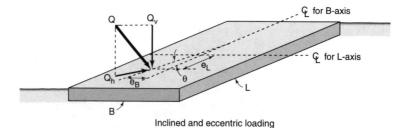

Inclined and eccentric loading

Figure 13-10 Definition of conditions for eccentric and inclined loading.

$$i_c = 1 - \frac{mQ_h}{cN_c BL} \qquad \text{(when } \phi = 0°) \qquad \text{(13-10d)}$$

Recommended values of m are

$$m_L = (2 + L/B)/(1 + L/B) \qquad \begin{array}{l}\text{if the angle of} \\ \text{inclination is in the} \\ \text{plane of the } L\text{-axis}\end{array} \qquad \text{(13-10e)}$$

$$m_B = (2 + B/L)/(1 + B/L) \qquad \begin{array}{l}\text{if the angle of} \\ \text{inclination is in the} \\ \text{plane of the } B\text{-axis}\end{array} \qquad \text{(13-10f)}$$

Should the inclined load act in a plane that makes an angle θ with the L-axis of the foundation, the value for m becomes

$$m_\theta = m_L \cos^2 \theta + m_B \sin^2 \theta \qquad \text{(13-10g)}$$

When an inclined loading is also eccentric, the modified values B' and L' should be used in Equations 13-10a, b, c, and d. The actual values of B and L are still used to determine the m factors, however.

The horizontal component of an inclined load acts to slide the foundation laterally. If the shearing or frictional resistance developed under the base is not adequate, sliding occurs. The maximum horizontal force that can be resisted along the base of the footing is

$$Q_{h(max)} = Q_v \tan \delta + A' c_a \qquad \text{(13-11)}$$

where A' is the effective bearing area of the footing ($B'L'$), δ is the angle of friction between the foundation material and soil, and c_a is the adhesion between the foundation and soil (for clay, a value related to the undrained shear strength).

Modification Factors for Inclined Foundation Base. Where the base of a foundation is inclined or tilted from the horizontal and an applied load acts normal to the base, the pattern of the soil yield zone beneath the foundation is different from the pattern that develops beneath the level footing carrying a vertical load. For this condition, base tilt factors, b_c, b_γ, b_q, developed originally for the strip footing foundation but considered also appropriate for other foundation shapes, can be applied to the general bearing capacity equation:

$$b_\gamma = \left(1 - \frac{\alpha}{57.3} \tan \phi \right)^2 \qquad \text{(13-12a)}$$

$$b_q = b_\gamma \qquad \text{(13-12b)}$$

$$b_c = b_q - \frac{(1 - b_q)}{N_c \tan \phi} \qquad \text{(when } \phi > 0°) \qquad \text{(13-12c)}$$

$$b_c = 1 - .0067 \alpha \qquad \text{(when } \phi = 0°) \qquad \text{(13-12d)}$$

In these expressions, α represents the angle of the base inclination, in degrees, with respect to the horizontal.

Footings on Slopes

The bearing capacity for footings on slopes is less than that for footings on level ground. The subsurface soil zones that provide resistance to foundation loading on the downhill side of the slope are smaller than they are where the ground is level. The safe loading is related to the soil failure zone involved. For footings of limited size, the soil failure zone is usually confined to the region near the vicinity of the foundation (regarding footings on level ground, see Fig. 13-2). For large footings, the failure zone may include all or most of the slope.

For continuous-strip footings, the ultimate bearing capacity is evaluated from

$$q_{ult} = cN_{cs} + \frac{1}{2}\gamma BN_{\gamma s} \tag{13-13}$$

where N_{cs}, $N_{\gamma s}$ are bearing capacity factors for footings on slopes. Values for the bearing capacity factors are shown in Figure 13-11. For cohesive soil, N_{cs} is related to a slope stability factor, M_s, as defined in Figure 13-11. The bearing capacity factor curves can also be used to indicate the footing setback distance where the slope inclination is no longer influential. For an adequate setback, the bearing capacity factors for level ground are used.

Illustration 13-6

A bearing wall for a warehouse building is to be located close to a slope, as indicated by the sketch. For the given conditions, what size strip footing should be provided?

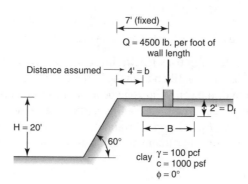

Solution

The allowable bearing pressure relates to the footing setback and the resulting b/B ratio, as well as the M_s factor. A trial-and-error procedure can be undertaken; it is initially assumed that $D_f/B = 0.5$ and $b/B = 1$. Since B is less than H, refer to the curves for $M_s = 0$. From Figure 13-11c, for $M_s = 0$, $b/B = 1$, and $\beta = 60°$, obtain $N_{cs} = 5.2$ (average of $N_{cs} = 4.9$ for D/B of 0 and 5.5 for D/B of 1).

$$q_{ult} = cN_{cs} = 5.2c = 5.2(1000 \text{ psf}) = 5200 \text{ psf}$$

$$q_{design} = \frac{5200 \text{ psf}}{\text{safety factor}} = \frac{5200 \text{ psf}}{3 \text{ (assumed)}} \cong 1700 \text{ psf}$$

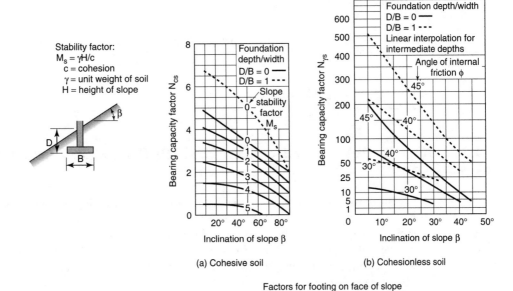

(a) Cohesive soil

(b) Cohesionless soil

Factors for footing on face of slope

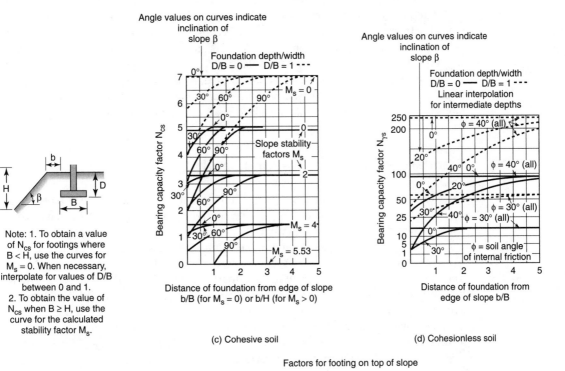

Note: 1. To obtain a value of N_{cs} for footings where $B < H$, use the curves for $M_s = 0$. When necessary, interpolate for values of D/B between 0 and 1.
2. To obtain the value of N_{cs} when $B \geq H$, use the curve for the calculated stability factor M_s.

(c) Cohesive soil

(d) Cohesionless soil

Factors for footing on top of slope

Figure 13-11 Bearing capacity factors for long footings on face of slope: (a) cohesive soil, (b) cohesionless soil. Bearing capacity factors for footings on top of slope; (c) cohesive soil, (d) cohesionless soil.

Required footing width, B, is

$$B = \frac{Q}{(q_{design})(1 \text{ ft length})}$$

$$= \frac{4500 \text{ lb}}{(1700 \text{ psf})(1 \text{ ft})} \cong 2.7 \text{ ft}$$

Use $B = 3$ ft (conservative).

13.3 FOOTING DESIGN DATA FROM PENETRATION RESISTANCE IN BOREHOLES

Obtaining foundation design criteria directly from the soil borings or soundings normally performed to explore conditions underlying a site is desirable from the view of cost and time savings. At locations where it is difficult to recover soil samples suitable for determining soil properties (a requirement when an analytical study for a foundation design is to be performed), obtaining foundation design criteria based directly on exploration data may be a necessity.

Typically, it is not a great problem to recover samples of soil possessing cohesion from borings, although care is required to minimize disturbance. On the other hand, obtaining undisturbed samples of cohesionless soil can be difficult, particularly when explorations extend below the groundwater table. Partly because of necessity, empirical relationships between the standard penetration test (SPT) (Chapter 5) and performance of spread footings supported on sands have evolved. Relationships have also been developed for the cone penetrometer (Chapter 5) and foundations on sand. Associated to practical worth, correlations between penetration resistance and the shear strength for clay soils also have been established. Various relationships applicable to sands and clays are described in the following sections.

Standard Penetration Test and Sand

Early information relating SPT results (the blow count N) in sand to spread footing size, bearing pressure, and settlement was presented by Terzaghi and Peck in 1948[7] and has been widely referred to for foundation design. This empirically determined design method was intended to provide foundations whose maximum settlement would not exceed one inch (25 mm), with the expectation that the greatest differential settlement between different footings would not exceed one-half to three-fourths inch, or 12 to 20 mm, a tolerable range for most structures. These early recommendations were based on limited data and were deliberately conservative.

Subsequent studies, which included performance evaluations of foundations designed in accord with the original recommendations, have indicated that modification of the original design criteria is necessary. Greater allowable bearing pressures can be permitted.

---

[7]Also in their subsequent publications through 1968, including [296].

Importantly, the value of N was found to be related to the soil overburden pressure (or depth) where the soil sample and blow count were taken. For a given soil density, the blow count is higher at deeper sampling depths because of the greater confinement and lateral pressure existing at deep, compared to shallow, locations. In contrast, blow count values obtained close to the ground surface are lower than should be indicated because of lack of overburden pressure and confinement. Thus, when SPT values are used for foundation design, the field blow counts need to be corrected to reflect the effect of sample depth on actual soil properties (Fig 13-12). The stress history of a sand deposit, including effects of compaction rolling or the presence of a high water table, also will affect a footing's performance, and these factors become a design consideration.

Empirically developed foundation design methods are typically based on the averaged or representative performance of case studies, but they will be conservative. Practically, deviations will always exist between individual cases and the design recommendations that evolve. Often, empirical design methods are recommended only for preliminary studies. Realistically, designers frequently need also to apply the procedures to final designs because of lack of further information or methods. When such methods are applied to future designs, it is necessary to expect a range in the predicted performance accuracy (e.g., an expected maximum settlement of one inch could end up being somewhat lesser or greater). Figure 13-13 presents design criteria for spread footings located at the shallow installation depths normally associated with basementless buildings (1 to 2 m, or 3 to 5 ft, below grade). The bearing pressures are intended to be values that produce a maximum settlement on the order of one inch or 25 mm. The maximum differential settlement that results between all foundation units on the site is expected to be less than three-fourths inch or 20 mm.

The N values in Figure 13-13 used for design are corrected values, which consider effects of depth or overburden pressure on the field SPT value of N. Figure 13-12 provides a factor C_N that should be applied to the field blow count in order to arrive at a modified N value (i.e., $N_{corr} = C_N N_{field}$). The discussed blow count values N refer to results from the field procedures which deliver approximately 60 percent of the drive energy to the soil

Figure 13-12 Correction factor for field SPT values [176].

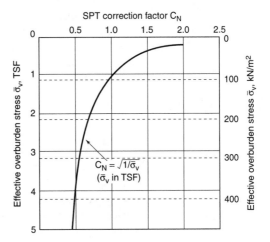

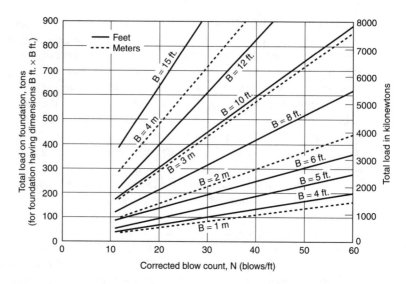

Figure 13-13 Chart for spread footings on sand. Corrected standard penetration test results related to foundation size and loading for approximately 25 mm or one-inch settlement. Water table more than depth *B* below footing. (Data presentation after F. H. Collopy). *Note:* SPT values shown are referenced to a soil sampling procedure delivering approximately 60% drive energy (ref. p. 147).

sampler; soil sampling with some types of boring equipment will deliver greater energy (such as when the automatic hammer is used), resulting in a need to use a modified *N* count (ref. article 5-3).

The *N* values in Figure 13-13 apply for dry or saturated sands. Where a condition of partial saturation exists, the sands possess some "apparent cohesion" because of the presence and effects of the air–water meniscus. Typically, this creates a greater strength. If the meniscus effect could disappear at some future time (through evaporation or submergence), a slight reduction should be applied to the *N* value used for design.

The procedure for designing foundations from Figure 13-13 involves working initially with the most heavily loaded units. For each boring location, the average corrected *N* value (see Ill. 13-7) is determined for the soil that lies within the zone between the bottom of the footing and a depth of about 1.5*B*. A value for *B* is based upon approximating a bearing pressure value from the design curves of Figure 13-13. Since soil conditions are expected to vary somewhat at most construction sites, the *N* value selected for design is the lowest N_{avg}.

Illustration 13-7

SPT results from a soil boring located adjacent to a planned foundation for a proposed warehouse are shown below. If it is assumed that little site grading will be performed, and if spread footings for the project are to be founded 4 ft below surface grade, what foundation size should be provided to support a 400-kip column load? Assume that a one-inch settlement is tolerable.

	Boring Log Summary	
SPT Sample Depth (ft)	**Blow Count, N_{field}**	**Soil Classification**
1	9	Fine to medium sand, trace silt
4	10	Fine to medium sand
8	15	Fine to medium sand
12	22	Fine to coarse sand
16	19	Fine to medium sand
20	29	Fine to coarse sand
25	33	Fine to coarse sand
30	27	Fine to coarse sand
	Driller's Note: Water table encountered at 25 ft.	

Solution

A solution requires, first, that the field SPT values be corrected. A tabular arrangement, as shown below, is practical. Correction factors are from Figure 13-12.

SPT Depth (ft)	N_{field}	Correction factor	N_{corr}	N_{avg} to depth shown
1	9	2.5	22	—
4	10	2.04	20	21
8	15	1.44	21	21
12	22	1.18	26	22
16	19	1.02	19	21
20	29	0.91	26	22
25	33	0.82	27	23
30	27	0.78	21	23

For a foundation to support 200 tons (400 kips) when the corrected N_{avg} value (to a depth between about 10 and 15 ft) is between 21 and 22 blows, Figure 13-13 indicates that a footing approximately 7.5 ft square ($7\frac{1}{2}$ by $7\frac{1}{2}$ ft) is required.

The design bearing pressures shown in Figure 13-13 presume the water table to be greater than a depth B below the bottom of the footing. If the water table is within the zone close to the bottom of the footing, a bearing pressure value obtained from the design curves should be reduced to keep settlement from exceeding the one-inch or 25 mm limit. For the condition where the water table is at the base of the foundation, a one-third reduction in the bearing pressure value should be applied. A linear interpolation can be assumed for water table depths intermediate between the foundation level and a distance B below it.

A modification to the blow count value obtained in the field should also be applied to the condition where fine sands and silty fine sands are submerged. This modification can be applied before correcting for the effects of overburden pressure. Letting N' be the modified field value, the following relationship is recommended:

$$N' = .6N_{field} \qquad \text{(for submerged fine sand and silty sand)} \qquad (13\text{-}14)$$

The bearing pressure–settlement values indicated by Figure 13-13 do not apply to narrow foundations. Where narrow footings are installed on cohesionless soil, bearing capacity values on the basis of shear failure of the supporting soil will control design. Angles of internal friction for the sand to permit analytical design can be estimated from the SPT blow count (Chapter 11, Table 11-2).

Sites showing a blow count value of less than 10 are indicating loose sand. The data of Figure 13-13 should not be applied to such conditions. Consideration should be given to improving the density of the sands by vibro-compaction, vibro-flotation or Terra-probe, blasting, or other methods before construction. Where an entire area is improved, foundation design and performance will benefit, but other aspects of construction, such as building slabs, roads, parking areas, and buried utilities, are also affected. Foundation design should be made on the basis of the improved soil properties.

Different foundations usually carry different magnitudes of loading (e.g., interior column foundations generally are more heavily loaded than wall column footings). Foundations carrying different loads can have their allowable bearing pressures determined individually from the curves in Figure 13-13, using the lowest corrected N_{avg} or an N value more appropriate for the actual footing area. Frequently, however, a bearing pressure is selected for the worst condition (lowest N value and heaviest foundation loading), and this pressure is used for proportioning all footing units. Although it is usually a safe procedure, this method tends to increase the range of differential settlement that results between light and heavily loaded foundations.

On building sites underlain by sand, many designers take the precaution of having the surface of the finish-graded area compacted by vibratory equipment to increase the density of the upper zone of soil and achieve a better degree of uniformity across the site before foundation construction takes place. Where sand fill projects are concerned, vibro-compactors are typically selected because of their effectiveness and efficiency. Densifying sand with vibratory methods causes high lateral pressures to develop in the compacted zone, a condition that produces high resistance to further volume changes and settlement. For the condition where the soil zone providing the foundation support has been compacted by vibratory methods, an increase in the bearing pressure values obtained from Figure 13-13 can be made. Limited indications are that a one-fourth increase will be conservative.

Figure 13-13 is intended to apply only to sand. Where a subsoil contains gravel, the SPT blow count can be erroneously high. For gravelly soils, a large-diameter soil sampler should be used to recover samples for examination. Better, test pits can be excavated and the compactness of soil that will lie beneath footing level can be checked visually or with density testing equipment. Where the density of deep soil zones need to be known, nuclear density meters capable of extending to deep depths can be used in the borings.

If a maximum settlement greater than one inch or 25 mm can be tolerated, or if it is desired to restrict the maximum settlement to less than one inch or 25 mm, bearing pressure values presented in Figure 13-13 can be increased or reduced in proportion to the allowable settlement, within reasonable limits (e.g., for half of one-inch or 12 mm settlement, assume half the bearing pressure).

The information of Figure 13-13 can also be utilized for designing mat foundations. Although mat foundations typically impose comparatively low bearing pressures in accord with the purpose of their design, allowable pressures from the design can be doubled without expecting differential settlement of the mat to exceed three-fourths inch or 20 mm.

The Standard Penetration Test and Cohesive Soil

Shallow foundations supported on clay are not designed directly from SPT blow count data. A major reason for this situation has been the difficulty in relating blow count values to a cohesive soil's stress history accurately enough to permit reliable foundation settlement predictions. Since practical procedures exist for separately determining the properties of a clay (vane shear tests can be performed in borings on the in-place soil, or laboratory strength and compressibility tests can be performed on undisturbed samples recovered from borings), some of the usual reasons for converting boring results directly into foundation design data do not apply. However, blow counts from the standard penetration test are useful for estimating the value of cohesion, c (the unconfined shear strength):

$$c = \frac{N}{4} \text{ ksf} \quad \text{(for clay)} \tag{13-15a}$$

$$c = \frac{N}{5} \text{ ksf} \quad \text{(for silty clay)} \tag{13-15b}$$

The N count indicated in these relationships are for SPT samples obtained using procedures that deliver approximately 60 percent drive energy to the soil sampler; blow counts obtained with other equipment delivering different drive energy, such as the automatic hammer, should be modified before the estimate of cohesion, c, is calculated. The value of cohesion c can then be substituted into Equation 13-6 to estimate the ultimate bearing capacity for the clay.

Static Cone Penetrometer and Sand

Static cone penetrometers are used extensively throughout Europe in the performance of investigations to determine subsoil stratification and related soil properties. In recent times, investigations utilizing penetrometer methods have received considerable attention in virtually all areas of the world. Interest has been generated by the developing awareness that direct relationships between penetrometer results and foundation design have been formulated.

For studying the relationship between static cone penetration results and foundation performance, much attention has been directed to the Delft (Dutch) cone penetrometer (Chapter 5). This apparatus has a 60-degree cone point with a base area of 1000 mm². Many other static penetrometers in use have similar cone dimensions. To evaluate the soil in a boring or sounding, the cone is advanced a short distance to determine the point resistance q_c (kg/cm² or tons/ft²) of the undisturbed soil. European experience indicates that for spread footing foundations located at shallow depths (on the order of 1 m or 3 ft), the allowable bearing pressure, q_{allow}, is about one-tenth of the point resistance obtained with the Dutch cone, or

$$q_{allow} \cong \frac{q_c}{10} \quad \text{(in kg/cm}^2 \text{ or ton/ft}^2\text{)} \tag{13-16}$$

More detailed design information, reflecting the effect of footing width and depth, has been developed, as shown in Figure 13-14.

Easily usable, refined relationships between static cone penetration, foundation size, bearing pressure, and *settlement* have not been developed, although the issue continues to receive attention. For *order-of-magnitude* information, it is practical to use correlations between cone penetration resistance and the standard penetration test. For clean sands and sands with little silt or gravel, the ratio of q_c to N usually ranges between 3.5 and 5. A reasonable, practical approximation is

$$q_c = 4N \quad \text{(approximately)} \tag{13-17}$$

where q_c is in kg/cm² or ton/ft² and N is the blow count from the SPT (N in terms of N_{60}).

With this value, the information of Figure 13-13 relating shallow foundation design to settlement (an anticipated maximum settlement of 25 mm or one inch) has been replotted in terms of the static cone resistance q_c (Fig. 13-15). In practical terms, the positioning of the bearing pressure curves are not highly sensitive to slight variations in the q_c-to-N ratio. Hence, the approximations of Figure 13-15 are applicable for the sand categories indicated earlier. The conditions assumed for developing Figure 13-13 would also apply to Figure 13-15 (i.e., the foundation embedment is on the order of 1 to 1.5 m, or 3 to 5 feet, below grade, the value of q_c is representative of the soil zone below the base of the footing to a depth extending approximately $1.5B$ further, and the water table is deep). Reductions in bearing pressures recommended for the case of a high water table also apply to Figure 13-15.

Figure 13-14 Relationship among static cone resistance, footing size and depth, and allowable bearing pressure [254].

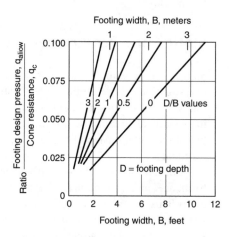

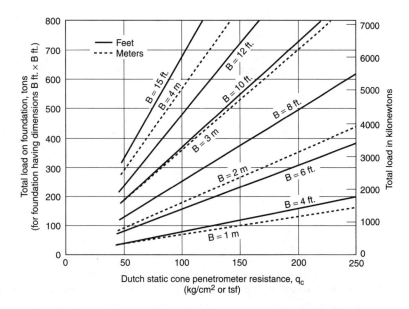

Figure 13-15   Chart for spread footings on sand. Dutch static cone resistance related to foundation size and loading for approximately one-inch settlement. Water table more than depth B below footing. (Data presentation after F. H. Collopy)

Static Cone Penetrometer and Cohesive Soil

In practical terms, good relationships have been established between static cone values of q_c and the cohesion and stress history of clay. The bearing capacity for the clay is subsequently calculated from the theoretical bearing capacity equations (Eq. 13-6). For the Delft (Dutch) cone:

$$c = \text{between } \frac{q_c}{15} \text{ and } \frac{q_c}{18} \tag{13-18a}$$

(for normally consolidated clay, $q_c < 20$)

$$c = \text{between } \frac{q_c}{10} \text{ and } \frac{q_c}{14} \tag{13-18b}$$

(for soft clays where a local shear failure is expected)

$$c = \text{between } \frac{q_c}{22} \text{ and } \frac{q_c}{26} \tag{13-18c}$$

(for overconsolidated clays, $q_c > 25$)

Settlement estimates can be obtained by using the methods described in Chapter 10. The value for C_c, the compression index, to apply when Equation 10-10b (repeated below) is used can be determined from Table 13-2.

Table 13-2 Correlation of Cone Resistance, Compressibility Index, and Water Content

Point Resistance, q_c (tsf, kg/cm²)	Water content, w (%)	Compression Index, C_c
$q_c > 12$	$w < 30$	$C_c < 0.2$
$q_c < 12$	$w < 25$	$C_c < 0.2$
	$25 < w < 40$	$0.2 < C_c < 0.3$
	$40 < w < 100$	$0.3 < C_c < 0.7$
$q_c < 7$	$100 < w < 130$	$0.7 < C_c < 1$
	$w > 130$	$C_c > 1$

Source: [215].

$$S = \frac{H_0}{1 + e_0} C_c(\log \overline{\sigma}_{v_f} - \log \overline{\sigma}_{v_0}) \qquad (10\text{-}10b)$$

If soil samples are not recovered from borings or test pits, an appropriate value of e_0 can be obtained by estimating the water content from data in Table 13-2 while assuming full saturation and selecting a representative value for G_s.

If the coefficient of volume compressibility, m_v, is defined as

$$m_v = \frac{\Delta e}{\Delta \sigma_v(1 + e_0)} \qquad (13\text{-}19)$$

then Equation 10-10b can be rewritten as

$$S = H_0(\Delta \sigma_v)(m_v) = \frac{H_0}{\eta q_c} \qquad (13\text{-}20)$$

where $\Delta \sigma_v$, represents the stress increase in the clay layer, as for Equation 10-10b, and η is obtained from Table 13-3.

The Pressuremeter and Foundation Design

The pressuremeter is a device developed to determine in situ properties of earth materials. As used for evaluating soil deposits, the pressuremeter is simply a cylindrical apparatus whose volume can be increased by expanding in the lateral direction only. To test a soil zone, the pressuremeter cylinder, identified as the *probe,* is lowered to the desired position in a boring hole (borehole), and an internal pressure is applied that causes the probe to expand laterally into the surrounding soil. Pressure is increased in measured increments, and related changes in volumes are recorded. The test continues until the soil is considered to be failed, a condition that is assumed where large increases in volume (indicative of large volumetric strain) occur under small increments of pressure.

Credit for the concept and development of the pressuremeter types that are presently in use to obtain information applicable to foundation design is given to Louis Menard of France. In the mid-1950s, Menard conceived the idea that the pressure required to expand a balloonlike device that had been inserted into a soil mass would also be indicat-

Table 13-3 Values of the η Coefficient for Clay and Silt Soils

	(q_c in tsf or kg/cm²)	
	$q_c < 7$	$3 < \eta < 8$
CL—low-plasticity clay	$7 < q_c < 20$	$2 < \eta < 5$
	$q_c > 20$	$1 < \eta < 2.5$
ML—low-plasticity silt	$q_c < 20$	$3 < \eta < 6$
	$q_c > 20$	$1 < \eta < 2$
CH—very plastic clay	$q_c < 20$	$2 < \eta < 6$
MH–OH—very plastic clay	$q_c > 20$	$1 < \eta < 2$

Source: [218].

ing that material's resistance to load-carrying elements such as structural foundations. To date, usage of the pressuremeter has been limited but interest and application has been increasing as knowledge of the equipment and the related foundation design procedure spreads.

The major component of Menard's prototype pressuremeter apparatus was an in-line, three-cell probe that, positioned in the earth, would provide pressure-related, soil-volume-change information. The center cell was the test or measuring cell; the end cells were guard cells. During the test, each cell would be independently inflated to the same pressure. This arrangement protects the center cell from end effects that would develop in a single-cell device and also permits the test cell to expand only radially outward from the probe's longitudinal axis. Though refinements and modifications have been made to the original design, present pressuremeters retain the feature of a protected center test cell. The effect on the soil zone surrounding the expanding center cell is that of a probe having an infinite length, where compression of the soil can be attributed to deformations resulting from plane strain.[8] This latter condition permits available solutions to the heretofore theoretical problem of expansion of a cylindrical cavity in an infinite medium to be applied to the pressuremeter.

The Menard-type pressuremeter consists of the probe, which is inserted into a borehole, a surface-stationed control unit that controls and monitors the probe's pressure and volume changes, and tubing that connects the control unit and probe (Fig. 13-16). Water typically is used in the test cell to measure the volume changes that occur as pressure increases. Containerized compressed gas is used to pressurize the test and guard cells.

To provide visualization of equipment size, one of the most widely used probes, described as the BX probe, has a diameter of 58 mm, a length of 42 cm, and a test cell volume, V_c, of 535 cm³. The BX designation is taken from the diameter of a borehole made with a 60 mm, or BX-size, diamond bit.

The field test procedure consists of inserting the probe into a drilled borehole to the depth to be tested. Pressures are applied in increments until the soil "fails." Usually, failure is considered to have been reached when the total expanded volume of the test zone reaches twice the volume of the original cavity. Each increment of pressure is held for a period, typically one minute, and the related volume-change readings are noted. Subsequently, a plot

---

[8]All displacements that result from deformation are parallel to the same plane.

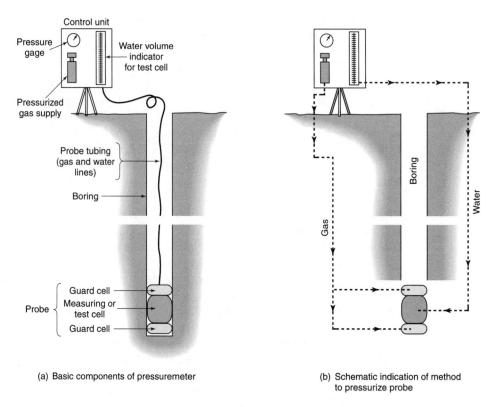

(a) Basic components of pressuremeter

(b) Schematic indication of method to pressurize probe

Figure 13-16 Illustration of Menard-type pressuremeter.

of pressure versus volume (or change in volume) is made to obtain parameters necessary for foundation design. A typical pressuremeter curve is shown in Figure 13-17.

The limit pressure, p_l, indicated by the pressuremeter curve of Figure 13-17b, is taken as the cell pressure that has caused the volume of the tested borehole cavity to be doubled. The ultimate bearing capacity, q_{ult} for the soil zone supporting a shallow foundation, based on theoretical considerations with empirically determined modifications, has been related to the limit pressure as follows:

$$q_{ult} = \overline{\sigma}_v + k_{bc}(p_l - \sigma_h) \qquad (13\text{-}21)$$

where q_{ult} = ultimate bearing capacity of the soil, in kN/m²

$\overline{\sigma}_v$ = effective vertical stress at the planned foundation depth, in kN/m²

σ_h = at-rest horizontal soil pressure at the planned foundation depth, or $K\overline{\sigma}_v$, in kN/m²

k_{bc} = bearing capacity factor for the proposed foundation shape and embedment and soil type (see Table 13-4)

p_l = limit pressure (see Fig. 13-17b)

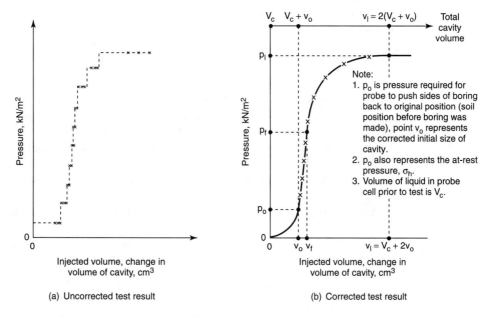

Figure 13-17 Pressuremeter test results.

Table 13-4 Values of k_{bc} for Pressuremeter Bearing Capacity Equation

	Values of k_{bc}, Strip Foundations								
Soil Type →	Sand			Silt			Clay		
↓ p_l (kN/m²)　D/B ratio →	0	2	4	0	2	4	0	2	4
500	0.8	1.5	1.8	0.8	1.4	1.6	0.8	1.3	1.5
1000	0.8	1.7	2.2	0.8	1.45	1.75	0.8	1.4	1.7
3000	0.8	2.0	2.6	0.8	1.65	2.0	0.8	1.5	1.8
6000	0.8	2.3	3.0						
Values of k_{bc}, Square and Round Foundations									
500	0.8	2.2	3.0	0.8	2.1	2.45	0.8	1.9	
1000	0.8	3.0	4.0	0.8	2.25	2.75	0.8	2.2	2.5
3000	0.8	3.7	4.8	0.8	2.45	3.1	0.8	2.4	2.9
6000	0.8	3.9	5.4						

The settlement, S, of a shallow foundation situated in homogeneous soil can be estimated from

$$S = \frac{q_{des}}{9E_m}\left[2B_0\left(\lambda_d\frac{B}{B_0}\right)^\alpha + \alpha\lambda_c B\right]F_d \qquad (13\text{-}22)$$

where q_{des} = foundation design bearing pressure minus the vertical pressure of the soil overburden adjacent to the foundation base

B_0 = a reference dimension, equal to 0.6 m or 2 ft

B = width or diameter of the foundation, provided that B is equal to or greater than B_0

α = rheological or creep deformation factor, which depends on the soil type and the ratio $E_m/(p_l - \sigma_h)$ (see Table 13-5)

λ_c, λ_d = shape factors that are based on length-to-width ratio of the foundation (see Table 13-6)

E_m = Menard pressuremeter modulus, equal to $2.66G_m$, where G_m is the Menard shear modulus for the pressuremeter curve between

v_0 and v_f; i.e., $G_m = \left(V_c + \dfrac{v_0 + v_f}{2}\right)\dfrac{\Delta p}{\Delta V}$,

or, also;

$E_m = 2.66\left(V_c + \dfrac{v_0 + v_f}{2}\right)\left(\dfrac{p_f - p_0}{v_f - v_0}\right)$, where the respective pressure and volume values are obtained from the pressuremeter curve (Fig. 13-17); in kN/m^2

F_d = depth factor, equal to 1 if the foundation depth is greater than B, equal to 1.20 for a foundation at the ground surface, equal to 1.10 for a foundation depth equal to one-half B.

Table 13-5 Pressuremeter Settlement Equation, α Factors

Soil Type	$E_m/(p_l - \sigma_h)$	α
Clay, normally consolidated	9–16	2/3
Clay, overconsolidated	>16	1
Silt, normally consolidated	8–14	1/2
Silt, overconsolidated	>14	2/3
Sand	7–12	1/3
Sand, compact	>12	1/2

Table 13-6 Pressuremeter Settlement Equation, λ Shape Factors

L/B Ratio:	Circular	Square	2	5	20
λ_c	1	1.10	1.20	1.40	1.50
λ_d	1	1.12	1.53	2.14	2.65

13.4 SEISMIC BEARING CAPACITY OF SPREAD FOOTING FOUNDATIONS

The foundation bearing capacity of soil will be affected by the ground motion accelerations and velocities that occur during earthquake activity. For spread footing foundations, the seismic-induced ground accelerations and velocities affect the behavior of the soil zones providing support for the foundation, as well as create forces that act on the foundation element. As shown on Figure 13-18, the active and passive soil zones responsible for foundation bearing capacity under the conditions of static loading will become shallower and smaller during seismic activity as inertial lateral and vertical forces caused by the event act to increase the forces responsible for movement of those zones. (These additional earthquake forces will be the product of the weight of the soil in the active zone and in the passive zone, and an acceleration coefficient, k_h or k_v; the k term represents the ratio of ground acceleration, a_h or a_v, to the acceleration of gravity, g, or $k_h = a_h/g$ and $k_v = a_v/g$.) The result is that a foundation bearing pressure that causes large movement (failure) in the supporting soil zones is lower for the seismic condition than for the static loading condition (i.e., the seismic bearing capacity becomes less than the static bearing capacity). The reduction in bearing capacity explains why foundation failures involving large settlements have occurred during earthquakes, even in areas where firm soil existed and foundation stability was expected to be maintained during ground shaking. Such occurrences are an indication that seismic bearing capacity should be determined for structures located in areas where a probability of significant earthquake activity exists.

In reporting the effect of earthquake ground motion on shallow strip foundations,. Richards, et al. [241], also provides a method to estimate soil bearing capacity under earthquake conditions; the seismic bearing capacity can then be compared with static bearing capacity. The analytical procedure considers the effect of ground motion both in the foundation soil and on the structural foundation but includes several simplifying and practical assumptions (Fig. 13-19). Comparisons of actual foundation behavior and settlement calculated using the analytical method indicate general agreement. The form of the bearing capacity equation for spread footing foundations used for static load conditions (Eq. 13-1a) has been retained but rewritten for seismic bearing capacity

$$q_{\text{ult}_E} = cN_{CE} + \frac{1}{2}B\,\gamma_1 N_{\gamma E} + \gamma_2 D_f N_{qE} \qquad (13\text{-}23)$$

where N_{CE}, $N_{\gamma E}$, and N_{qE} are bearing capacity factors for earthquake conditions. The earthquake bearing capacity factors are influenced by the strength and weight properties

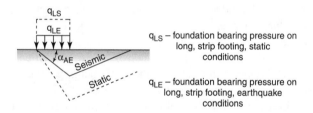

Figure 13-18 Comparison of foundation failure zones for static and seismic conditions. (After Richards, et al. [241]).

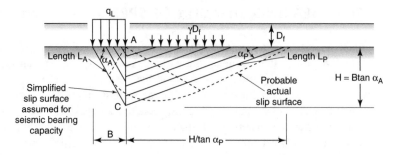

(a) Simplified slip zone foundation wedges assumed for seismic analysis.

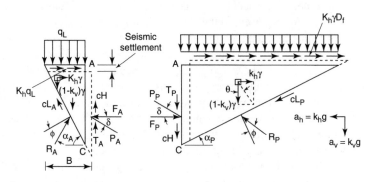

(b) Seismic related forces acting on slip zone wedges.

Figure 13-19   Summary of conditions assumed to determine seismic bearing capacity for strip foundations. (after Richards et al. [241]).

of the soil in the foundation bearing zones (as for the static load bearing capacity) but also by the magnitude of an earthquake's horizontal and vertical ground acceleration factors. Figure 13-20 presents the ratio of static bearing capacity factors to earthquake bearing capacity factors for a range of horizontal and vertical ground acceleration coefficients k_h and k_v.

During a seismic event, foundation settlement occurs if the imposed bearing pressure exceeds the allowable earthquake bearing pressure (the soil zones responsible for foundation support undergo displacement). The allowable bearing pressure for a particular earthquake is affected by the acceleration ratio $[k_h/(1-k_v)]$ as indicated by Figure 13-20. For a foundation design based on static bearing capacity parameters, the acceleration ratio value which causes the earthquake bearing capacity to be reached is termed the *critical acceleration ratio,* or $[k_h/(1-k_v)]^*$. Where a foundation is supported on cohesionless soil (soil cohesion $c = 0$) and when k_v is zero, the critical acceleration ratio then becomes $k_h.^*$ The relationship between the static bearing capacity factor of safety and k_h^* for a range of D_f/B ratios is shown in Figure 13-21. (The analytical studies described in Ref. 241 imply that the values of k_v and cohesion c have limited effect on the seismic bearing capacity values.)

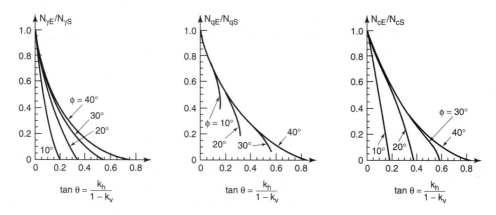

Figure 13-20   Ratios of static to seismic bearing capacity factors. (After Richards, et al. [241])

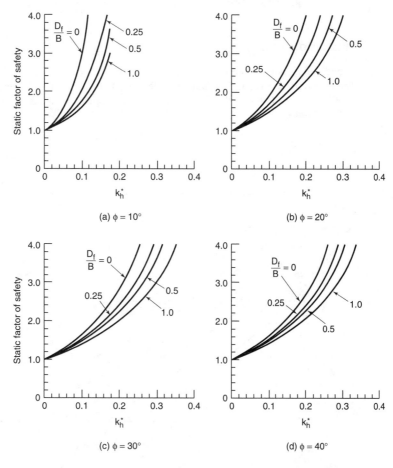

Figure 13-21   Critical acceleration k_h^* for incipient foundation settlement ($c = 0$, $k_v = 0$). (After Richards, et al. [241]).

When the imposed foundation bearing pressure exceeds the value of the earthquake bearing capacity, the settlement expected for an identified earthquake condition k_h can be estimated from

$$\text{Earthquake settlement, } S_E = .174 \frac{V^2}{Ag} \left(\frac{k_h{}^*}{A}\right)^{-4} \tan \alpha_{AE} \qquad (13\text{-}24)$$

where V = ground motion peak velocity (ft/sec or m/sec), ref. Fig. 1-27(b)
$\quad A$ = ground acceleration coefficient (dimensionless), ref. Fig. 1-27(c)
$\quad g$ = acceleration of gravity (9.81 m/sec² or 32.2 ft/sec²)
$\tan \alpha_{AE}$ = angle between the plane of the rupture surface with respect to the horizontal for the active soil zone formed under the foundation during the earthquake condition, as used for relating allowable soil bearing pressure to the resistance developed in the active and passive soil zones in the bearing capacity derivation, ref. Fig. 13-22

Illustration 13-8

A long wall foundation bears on a dense cohesionless soil with properties $\phi = 40°$ and $\gamma = 20\,\text{kN/m}^3$. The D_f/B ratio is 0.5. The foundation was designed to have a factor of safety (FS) of 3 for the static bearing capacity. Estimate the foundation settlement for the earthquake event causing $A = 0.35$ and $V = 0.27$ m/sec.

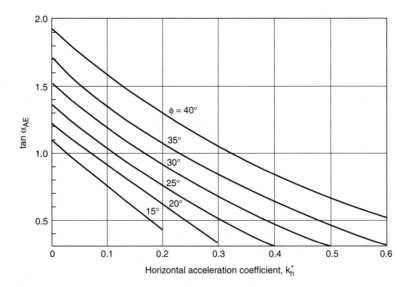

Figure 13-22 Relationship between $k_h{}^*$, soil friction angle ϕ, and angle α_{AE} defining foundation failure zone. (After Richards, et al. [241]).

Solution

For FS $= 3$ and $D_f/B = 0.5$, obtain $k_h* = 0.26$ (from Fig. 13-21) and $\tan \alpha_{AE} = 1.2$ (from Fig. 13-22).

$$S_E = .174 \frac{(.27 \text{ m/sec})^2}{(.35)(9.81 \text{ m/sec}^2)} \left(\frac{.26}{.35}\right)^{-4} \tan \alpha_{AE}$$

$$= .174 \frac{(.27 \text{ m/sec})(.27 \text{ m/sec})}{(.35)(9.81 \text{ m/sec}^2)} \left[\frac{1}{(.26/.35)^4}\right](1.2)$$

$$= .0146 \text{ m} = 14.6 \text{ mm}$$

13.5   PRESUMPTIVE BEARING PRESSURES

Allowable foundation bearing values that are related to the visual classification of a site's bearing soil, a type of design information that may be included as part of the building codes in effect for cities and suburban areas, are termed *presumptive bearing pressures* (i.e., assumed safe bearing pressures). Table 13-7 illustrates presumptive bearing values.

Presumptive bearing pressures predate the development of soil mechanics. Their origin lies in the experiences accumulated by the early builders in an area. Information of this type easily became part of the early building codes, for the developing cities needed reference criteria for new construction that presumably would ensure safe design.

Table 13-7   Presumptive Unit Soil Bearing Values

Class	Material	Allowable Bearing Value[1] (tons/ft^2)
1	Massive crystalline bed rocks, such as granite, gneiss, trap rock, etc.; in sound condition	100
2	Foliated rocks, such as schist and slate: in sound condition	40
3	Sedimentary rocks, such as hard shales, silt-stones, or sandstones: in sound condition	15
4	Exceptionally compacted gravels or sands	10
5	Gravel and sand–gravel mixtures; compact	6
6	Gravel, loose; coarse sand; compact	4
7	Coarse sand, loose; sand–gravel mixtures, loose; fine sand, compact; coarse sand, wet (confined)	3
8	Fine sand, loose; fine sand, wet (confined)	2
9	Stiff clay	4
10	Medium-stiff clay	2
11	Soft clay	1
12	Fill, organic material, or silt	(^{2})

[1]Presumptive bearing values apply to loading at the surface or where permanent lateral support for the bearing soil is not provided.

[2]Except where, in the opinion of the enforcement officer, the bearing value is adequate for light frame structures, fill material, organic material, and silt shall be deemed to be without presumptive bearing value. The bearing value of such material may be fixed on the basis of tests or other satisfactory evidence.

Source: (Excerpted from New York State Building Construction Code, 1977; for illustration only; information currently not in use.)

Presumptive bearing pressures can be appropriate for design when utilized in the area for which the values were developed and for the type of construction that provided the original experience data. Unfortunately, presumptive bearing pressures frequently no longer reflect local geologic conditions and experiences but are values extracted from other codes or references. Importantly, methods and materials of construction that affect the behavior of commercial–industrial buildings likely have changed significantly since the origin of the presumptive bearing values for places such as older cities; for example, compared to the older style of low masonry construction, modern structures are taller, many with steel frame construction, and the performance demands of foundations have been changed correspondingly. Since presumptive bearing pressures do not consider the important influence of factors such as foundation size and embedment, position of the groundwater table, soil density or consistency, and soil stress history, unsafe or uneconomical foundation designs can occur. Though building codes must be followed in the area of their jurisdiction, it is fortunate that many modern codes also permit foundation designs made on the basis of engineering studies that include a site investigation and soil testing to supersede the presumptive values.

13.6 FOUNDATION WALLS

Foundation walls are the components of a substructure (the below-ground section of a structure) that typically are required to resist lateral soil and water pressures as well as provide support for vertical loads. Foundation walls commonly define a subsurface space that is part of a structure, such as a basement, cellar, cellar crawl space, or other underground or underwater chamber.

Conventionally, foundation walls are constructed of concrete, masonry, or stone materials because of their durability for the common conditions of exposure (water or moisture, chemicals and minerals in the earth, etc.). Other materials such as wood, metal, reinforced earth, and sand bags also have been used to construct, or function as, foundations walls.

Foundation walls are often, but not always, supported on the long, strip-category footings discussed earlier in this chapter (the strip footing creates a wide base section for the wall, so as to limit the vertical bearing pressure imposed on the soil). But thick-section walls may be without a separate footing if the downward loading is of limited magnitude.

Foundation walls can suffer structural damage if soil settlement or a soil bearing capacity failure occurs. Such walls will also experience structural damage where the wall does not develop adequate resistance to the lateral soil and water pressures (forces) or other forces that act against the exterior side. Structural problems due to inadequate resistance against lateral forces may be the more prevalent of the noted categories of problems (methods to determine lateral pressures against a wall are presented in Chapter 16). The deformation or movement that a wall experiences and the resulting cracking pattern often is indicative of the type of soil-related problems responsible for the wall movement (Table 13-8 and Figure 13-23).

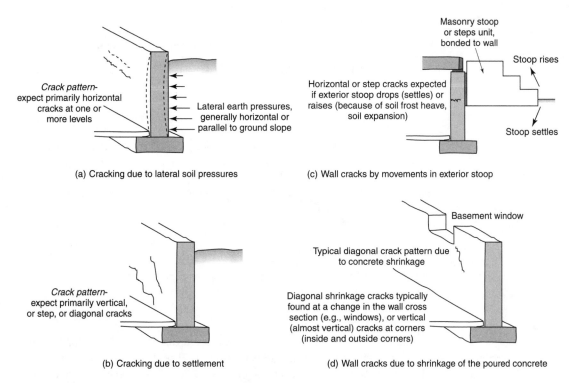

(a) Cracking due to lateral soil pressures

Crack pattern-
expect primarily horizontal
cracks at one or
more levels

Lateral earth pressures,
generally horizontal or
parallel to ground slope

(c) Wall cracks by movements in exterior stoop

Masonry stoop
or steps unit,
bonded to wall

Stoop rises

Horizontal or step cracks expected
if exterior stoop drops (settles) or
raises (because of soil frost heave,
soil expansion)

Stoop settles

(b) Cracking due to settlement

Crack pattern-
expect primarily vertical,
or step, or diagonal cracks

(d) Wall cracks due to shrinkage of the poured concrete

Basement window

Typical diagonal crack pattern due
to concrete shrinkage

Diagonal shrinkage cracks typically
found at a change in the wall cross
section (e.g., windows), or vertical
(almost vertical) cracks at corners
(inside and outside corners)

Figure 13-23   Foundation wall cracks and typical causes.

a. For settlement cracks, consider raking then patching if no further settlement expected, but recommend also patch exterior side to reduce seepage entry.

b. For cracking due to lateral forces (the horizontal crack condition) patch cracks, but also consider bracing with buttressing pilasters on interior side (expected to be successful if inward deflection is not excessive) as a relatively inexpensive mitigation (see sketch).

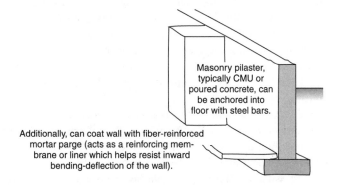

Masonry pilaster,
typically CMU or
poured concrete, can
be anchored into
floor with steel bars.

Additionally, can coat wall with fiber-reinforced
mortar parge (acts as a reinforcing membrane or liner which helps resist inward
bending-deflection of the wall).

Figure 13-24   Some of the methods available for repairing cracked foundation walls.

Table 13-8 Summary: Typical Foundation Wall Problems and Related Causes

Type of Wall Problem	Typical or Possible Causes
Primarily horizontal cracks, noticeable inward movement or bulging	Lateral soil and/or water pressures acting against exterior side
	Settlement or heave of concrete slab or a stoop–steps unit attached high on the exterior side
	Careless backfilling of soil against exterior side during the construction period
	Frost heave of frozen soil that has adhered to the wall exterior during the ground-freezing occurrence (masonry walls carrying light structural loads are most susceptible)
Primarily vertical or diagonal cracks at random locations	Settlement or expansion/heave of foundation soils underlying the wall
Primarily vertical and diagonal cracks in a concrete wall at corners and at changes in a wall cross section such as a basement window	Tensile stresses developing in the concrete as shrinkage during curing occurs.

Soil compression and structural settlement can occur due to foundation loading. But also, walls placed on soil fill will settle and crack if the fill settles (compresses) after construction. (To provide adequate support for a foundation, soil fill should be properly placed and compacted such as by spreading and compacting in thin layers; refer to methods discussed in Chapter 14.) Additionally, the weight of a great thickness of soil fill placed in an area tends to cause the natural underlying soil to compress, a condition that also results in the fill settling. (Such a fill area can be provided with an adequate waiting time period before construction begins, or surcharged, so as to have the soil compression precede the construction period.) Foundation wall cracks may also occur where large-size rigid piping (such as cast iron sewer pipe or steel water supply pipe) passes through the foundation wall and the supporting soil on the exterior side settles. (Such rigid conduits typically are installed in a trench excavation which may not be provided with adequate bedding support under the conduit, or the trench is not backfilled properly.)

Ground freezing and soil frost heave can be responsible for cracking in masonry foundation walls. If the frozen soil bonds to the exterior side of the wall, soil heave results in an uplift force on the upper section of the wall, which, for lightweight structures, can cause wall mortar joints to crack; if the soil surface is restrained against vertical movement, the ice lens–soil expansion may increase lateral pressures against the wall sufficient to cause cracking.

Cracked foundation walls often can be adequately repaired using basic crack patching methods, providing the cracking and wall movements are limited. For cracking due to lateral pressures against the wall exterior side, a structural bracing or reinforcing is typically required where the wall movement is more than minor; installing buttressing pilasters on the accessible interior side of a wall is one of the practical procedures in use (Fig. 13-24). Where wall movement and cracking is excessive, reconstruction may be necessary to achieve proper function.

DEEP FOUNDATIONS—CAPACITY AND INSTALLATION CRITERIA

13.7   PILE AND PIER FOUNDATIONS

The load-carrying capacity of deep foundations such as piles and piers is typically evaluated by analytical methods, by in-place field load tests on an installed unit, or through the application of pile-driving formulas. The analytical method applies the principles of soil mechanics and requires a knowledge of subsurface conditions and soil properties, as well as information on pile type, dimensions, and installation method. Field load tests usually consist of imposing loadings in increments onto the in-place pile and obtaining settlement data for each increment; the foundation design capacity is frequently taken as the value where a predetermined tolerable settlement is reached or where the load-settlement plot is no longer proportional. Load test results are considered to provide reliable information on the capacity of an installed pile (but long-term settlement remains to be determined analytically). Historically, pile driving formulae have been developed from a concept that relates dynamic energy imparted by the pile-driving hammer to the axial capacity of the pile. (Typically, the driving record for each pile on a project is documented to have information on embedded length, resistance to penetration, etc.) In general, such driving formulae are not considered appropriate for determining the true capacity of a driven pile because the procedure does not properly evaluate effects of soil and pile behavior that occurs during installation. However, use of a driving formula for field control can be appropriate if load test results are used to modify the basic formula in order to account for the effects of pile-soil behavior, and also to relate the driving record for each production pile to the tested pile. An alternative dynamic method for predicting pile capacity is based on relating pile driving data and the currently accepted wave equation analysis (which utilizes wave transmission theory to represent drive hammer forces moving through the pile length and the associated pile-soil reactions); instrumentation attached to a test pile for the installation is used to determine the magnitude and velocity of force moving through the pile as a result of the impact from the pile driver. The measured data is input to a computer analysis of the wave equation to calculate the pile's axial capacity. The relationship between pile capacity and driving record can then be used to establish the driving criteria for production piles.

Statical Analysis

For a pile or pier having adequate structural strength to carry an intended loading, the total downward capacity, Q_{total}, will be based on soil conditions (Fig. 13-25). The ultimate capacity of a pile is due to soil resistance developed by friction or adhesion between the soil and pile shaft, $Q_{friction}$, and the end bearing at the tip of the pile, Q_{tip}:

$$Q_{total} = Q_{friction} + Q_{tip} \tag{13-25a}$$

$$Q_{total} = fA_{surface} + q_{tip}A_{tip} \tag{13-25b}$$

where $A_{surface}$ = effecive surface area of pile in contact with soil along the embedded shaft length

f = average unit skin friction or adhesion between soil and pile surface

q_{tip} = bearing pressure of soil at pile tip

A_{tip} = pile tip bearing area

Figure 13-25 Basic concepts—pile capacity related to soil support.

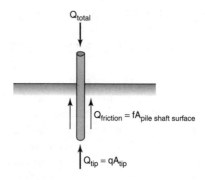

Methods in use to compute $Q_{friction}$ and Q_{tip} relate to the type of soil penetrated. The analyses differ for clay soil and for sand soil, as detailed below. Where strata of soil possessing different properties are penetrated or the pile cross section and surface area vary along its length, the skin friction can be calculated by using segments of pile length and the appropriate soil value and pile area.

Statical Method—Driven Piles in Sand

The ultimate capacity of a single pile driven in sand is the sum of the point resistance and the skin friction acting along the shaft of the pile:

$$Q_{total} = fA_{surface} + q_{tip}A_{tip} \tag{13-25b}$$

The unit skin friction for a straight-sided pile is a function of the soil pressure acting normal to the pile surface and the coefficient of friction between the soil and pile material (Fig. 13-26). Thus, skin friction on a pile shaft is determined in the manner used in engineering mechanics to compute frictional resistance developing between two solid materials with movement impending.

Soil pressure acting normal to a vertical pile surface is assumed to be directly related to an effective vertical soil pressure (overburden pressure) and a lateral pressure coefficient K.

$$\overline{\sigma}_h = K\overline{\sigma}_v \tag{9-5}$$

where $\overline{\sigma}_h$ = horizontal soil pressure acting at any depth Z in a soil mass
$\overline{\sigma}_v$ = effective vertical pressure (overburden) acting at the same depth Z within a soil mass
K = lateral pressure coefficient expressing the ratio of $\overline{\sigma}_h$ to $\overline{\sigma}_v$

The unit value of skin friction f acting at any depth in a sand mass theoretically becomes

$$f_{sand} = \overline{\sigma}_h \tan \delta = K\overline{\sigma}_v \tan \delta \tag{13-26}$$

where $\tan \delta$ = coefficient of friction between sand and the pile surface.

Figure 13-26   Solid-to-solid friction developing against pile shaft.

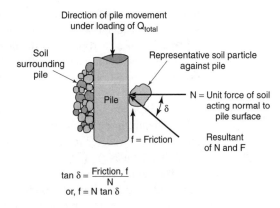

$$\tan \delta = \frac{\text{Friction, f}}{\text{N}}$$
$$\text{or, f} = \text{N} \tan \delta$$

Representative values for the coefficient of friction between silica sand and the pile materials are shown in Table 13-9. The total skin friction acting along the embedded length of pile is then

$$Q_{\text{friction}} = f_{\text{sand}} A_{\text{surface}} = (K \overline{\sigma}_v \tan \delta) A_{\text{surface}} \tag{13-27}$$

In cohesionless deposits, the vertical (overburden) stress, σ_v, is conventionally assumed to increase with depth, according to Equation 9-1:

$$\sigma_v = \gamma_{\text{soil}} Z \tag{9-1}$$

where γ_{soil} = unit weight of soil
$\quad\quad Z$ = depth below ground surface

However, in evaluating skin friction that develops with *driven piles,* evidence indicates that the *influence* of the effective overburden stress of soil adjacent to the pile does *not* continue to increase without limit, as implied by Equations 9-1 and 13-26. Adjacent to a pile, the effective vertical stress of influence, $\overline{\overline{\sigma}}_v$, increases only until a certain distance of penetration, termed the critical depth D_c, is reached. Below this depth, the influential effective vertical pressure remains essentially constant, or increases at a low rate. (Evidence

Table 13-9   Representative Values for Coefficient of Friction Between Sand and Pile Materials

Material	Tan δ
Concrete	0.45
Wood	0.4
Steel (smooth)	0.2
Steel (rough, rusted)	0.4
Steel (corrugated)	Use tan ϕ of sand

Note: Reference 58 suggests δ equal to 0.8 times the peak soil friction angle, ϕ_{peak}.

indicates that driving a pile in sand results in a thin zone of loose sand around the pile wall, which is encircled by a zone of densified sand. By arching, the cylindrical zone of loose and dense sand prevents the development of high lateral earth pressure against the pile [64].) The point where the critical depth is reached is influenced by the initial condition of the sand (loose or compact) and the dimension of the pile. Field and model tests suggest that the critical depth ranges from about ten pile diameters for loose sands to about twenty pile diameters for dense, compact sands. Figure 13-27 provides information on critical depth for different conditions that can be conservatively used for pile design: The indicated depths refer to soil above the water table; for submerged deposits, the critical depth is 1.6 times greater.

The value of K that acts at a particular depth in a soil mass after the pile has been driven is affected by the initial condition of the sand (such as relative density), the shape and size of the pile, and the method of installation. Piles are driven into undisturbed ground or installed in predrilled holes or with the assistance of jetting. The latter two procedures are commonly utilized where it is anticipated that a pile cannot be driven to a desired depth without damage because of the presence of hard material or hazardous objects in upper zones of soil. Figure 13-28 relates installation conditions with values of K that are commonly assumed in pile capacity analysis.

Driven piles causing large soil displacement (e.g., cylindrical piles) develop greater unit skin friction than piles with small volume displacement (e.g., H-piles). This is the effect of a greater K value resulting from the pile installation. Similarly, tapered piles develop greater capacity along the shaft than do straight piles because of the greater change in properties of the soil surrounding the pile and the greater inclination in the angle of the soil force resultant acting against the pile shaft.

Where information on soil conditions and properties as well as on the type and shape of the pile to be driven is available, the previously detailed methods should be applied to

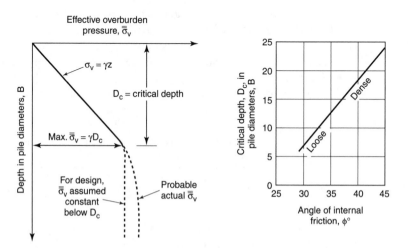

Figure 13-27 Variation of influential effective overburden stress in sands adjacent to driven straight-sided piles [198, 313, 315, 64].

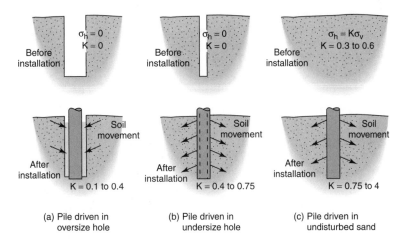

Figure 13-28 Values of lateral pressure coefficient K used for design of piles in sand.

compute the skin friction. Because the condition of the sand surrounding the pile has significant influence on the skin friction that develops, it is possible to establish a probable range of skin friction values acting along the shaft of a pile based on the relative density of the sand. Such information is shown in Figure 13-29 for driven piles having straight sides and limited diameter. Such data can be used to obtain *preliminary estimates* of skin friction during the absence of detailed information.

For computing the skin friction that develops along the shaft of an embedded pile, it is convenient to work with a diagram indicating that the variation of $\overline{\overline{\sigma}}_v$, with depth (see Ill. 13-9). Multiplying the area of this diagram by K and $\tan \delta$ provides the total skin friction acting on an incremental strip along the embedded length of pile. The product of this increment of skin friction and the pile perimeter (circumference of the pile cross section) completes the numerical integration that gives the total skin friction.

The end bearing pressure, q_{tip}, at a pile tip can be computed from the bearing capacity equation for deep foundations in cohesionless soil:

$$q_{\text{tip}} = .4\,\gamma B N_\gamma + \overline{\overline{\sigma}}_v N_q \qquad (13\text{-}28a)$$

where $N\gamma$, N_q = bearing capacity factors for deep foundations
$\overline{\overline{\sigma}}_v$ = effective vertical pressure of influence acting at pile tip depth
B = pile tip diameter or width
γ = unit weight of soil in zone of pile tip

This bearing capacity equation has the same form as the bearing capacity for shallow foundations. With most driven piles, the first term of the equation is small compared to the second term because of the limited B dimension. Thus, for many design problems,

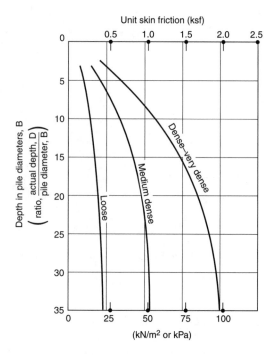

Figure 13-29 Range of values: Skin friction versus depth for straight-sided piles driven in sand. (after Coyle and Castello [64])

$$q_{\text{tip}} = \overline{\overline{\sigma}}_v N_q \tag{13-28b}$$

and the total end bearing becomes, for practical purposes,

$$Q_{\text{tip}} = (\overline{\overline{\sigma}}_v N_q) A_{\text{tip}} \tag{13-28c}$$

If the pile is a large-diameter pile, the tip dimension will be significant, and the end bearing should be determined from Equation 13-28a.

The value of the bearing capacity factor, N_q, is related to the angle of internal friction of the sand in the vicinity of the pile tip (several pile diameters above and below the tip) and the ratio of pile depth to pile width. Determining the value of N_q from field studies is not easily accomplished because of the difficulty of separating it from other influential factors. Values of N_q presented by various investigators have been developed primarily from theoretical analysis. Results range widely (Fig. 13-30) because of the assumptions made in defining the shear zones near the pile tip. Values of N_q obtained from the zone between the Berezontzev and Terzaghi curves (Fig. 13-30) are believed to be most applicable for commonly encountered soil conditions. The magnitude of effective vertical pressure, $\overline{\overline{\sigma}}_v$, that develops at the pile tip is limited below the critical depth. For preliminary design purposes, it can be conservatively assumed that the value of $\overline{\overline{\sigma}}_v$ in Equations 13-28a, b, and c is equal to the effective overburden existing at the critical depth (see Fig. 13-27). The angle of in-

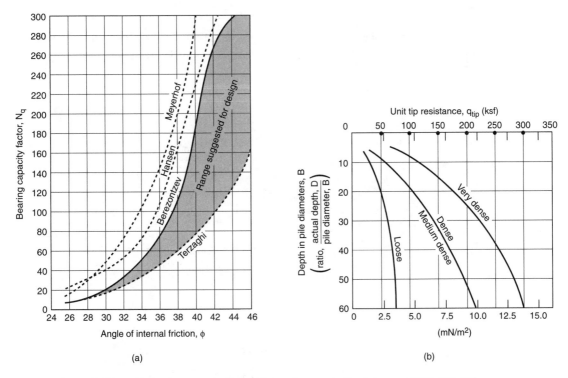

Figure 13-30a (a) Bearing capacity factor N_q for piles penetrating into sand [189, 215, 64]; (b) Range of values: ultimate unit tip resistance for piles driven in sand (after Coyle and Castillo [64]).

ternal friction for soil in the zone of the pile tip is determined from laboratory tests on re-covered samples or from correlations with penetration resistance in borings (e.g., the SPT test) or from soundings. If Dutch cone data are available, the end bearing for a pile is di-rectly related to the cone resistance, q_c:

$$Q_{tip} = \frac{q_c}{2} A_{tip} \qquad (13\text{-}29)$$

Figure 13-30b presents direct values for unit tip resistance, q_{tip}. The curves have been developed from regression analysis on a number of pile load test results [64] and at this time should be considered tentative. Direct values for q_{tip} are highly practical and improve the de-signer's ability to predict a safe pile loading; the term combines the influence of $\overline{\sigma}_v$ and sand density ϕ or the N_q factor on the driven pile, values difficult to determine with accuracy.

If the pile analysis requires use of Equation 13-28a, the value of N_γ for a deep founda-tion can be conservatively taken as twice the N_γ value used for shallow foundations.

Values of N_q shown in Figure 13-30a assume that the soil above the pile tip is com-parable to the soil below the pile tip. If the pile penetrates a compact layer only slightly and loose material exists above the compact soil, an N_q value for a shallow foundation will be more appropriate than a value from Figure 13-30a.

Illustration 13-9

Prestressed concrete piles are planned for use as the foundation for a waterfront structure. Soil conditions are as indicated by the sketch. What is the approximate axial capacity for a single 0.3-m-square pile driven to a depth of 10 m?

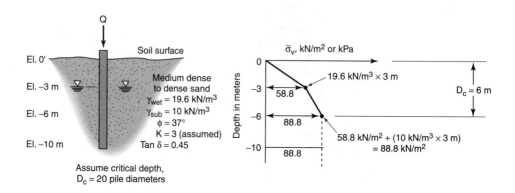

Solution

Total skin friction = (area of $\bar{\bar{\sigma}}_v$ diagram)(K)(tan δ)(pile circumference)

$$= \left[\left(58.8 \text{ kN/m}^2 \times \frac{3 \text{ m}}{2}\right) + \left(\frac{88.8 \text{ kN/m}^2 + 58.8 \text{ kN/m}^2}{2}\right)(3 \text{ m})\right.$$

$$\left. + (88.8 \text{ kN/m}^2 \times 4 \text{ m})\right](3 \times .45)(4 \times .3 \text{ m})$$

$$= 1077 \text{ kN}$$

End bearing $= \bar{\bar{\sigma}}_v N_q A_{\text{tip}} = (88.8 \text{ kN/m}^2)(80)(.3 \text{ m} \times .3 \text{ m})$

$$= 640 \text{ kN}$$

$$Q_{\text{total}} = 1077 \text{ kN} + 640 \text{ kN} = 1717 \text{ kN; then}$$

$$Q_{\text{design}} = \frac{1717 \text{ kN}}{\text{FS}}$$

Statical Method—Driven Piles in Clay

The ultimate supporting capacity of a driven pile in clay is

$$Q_{\text{total}} = Q_{\text{friction}} + Q_{\text{tip}} \tag{13-25a}$$

As driven piles penetrate into a saturated clay, the soil in the vicinity of the pile is remolded because of displacement and disturbance. The remolded shear strength of a clay is almost always less than the original (cohesion), but typically the strength improves with time (thixotropy). The rate of strength gain will be related to the consolidation characteristics of the clay, since dissipation of excess pore water pressures developed because of the pile installation must occur.

The skin friction or adhesion, f, that develops between the soil and pile shaft has been related to the undisturbed cohesive strength of the clay.

$$f_{\text{clay}} = \alpha c \qquad (13\text{-}30a)$$

where f_{clay} = unit adhesion or skin friction developed between clay and pile shaft
$\quad\quad\ c$ = cohesive strength of undisturbed clay
$\quad\quad\ \alpha$ = factor that relates adhesion to cohesion (or the friction ratio)

For *normally consolidated* clays in the soft-to-firm range (cohesion less than 50 kPa or 1 ksf), the value of α can be taken as unity; therefore,

$$f = c \quad \text{(for normally consolidated clays with } c \leqq 50 \text{ kPa or } 1.0 \text{ ksf)} \qquad (13\text{-}30b)$$

Equation 13-30b may also be valid for normally consolidated clays having cohesive strengths greater than 50 kPa (1 ksf), but current data are too limited for confirmation.

Various investigators have reported values of α for clays that are *overconsolidated*. Results vary but fall within the band shown in Figure 13-31.

The magnitude of Q_{friction} for clay to use in Equations 13-25a and b then becomes

$$Q_{\text{friction}} = fA_{\text{surface}} = \alpha c A_{\text{surface}} \qquad (13\text{-}31)$$

It is recognized that the lateral pressure existing in the remolded soil adjacent to the pile shaft has an influence on the dissipation of excess pore water pressures in this zone. If

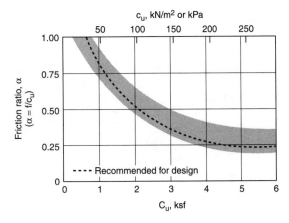

Figure 13-31   Correlation of friction factor α and clay cohesion c_u.

sufficient pore water drainage can occur, the final strength of the clay conceivably may be greater than the undisturbed strength. Because of the phenomenon of strength gain with time, load tests on piles in clay should be performed some weeks after installation to obtain a reliable indication of actual capacity.

Illustration 13-10

A 0.3-m-square prestressed concrete pile is to be driven at a site where soil conditions are as indicated by the sketch. If a penetration of 14 m is assumed, approximately what total skin friction is expected to develop along the embedded length of the pile?

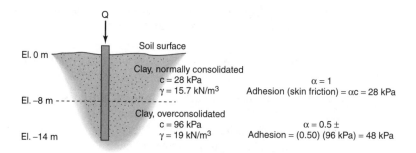

Solution

For depths from 0 to 8 m: $\alpha = 1$; therefore, adhesion = skin friction = 28kPa.

For depths from 8 to 14 m: $\alpha = .5 \pm$ and adhesion = αc; therefore, adhesion = $(.50)(96\ kPa)$ = 48 kPa

$$Q_{friction} = (28\ kN/m^2)(.3\ m \times 4\ sides)(8\ m) + (48\ kN/m^2)(.3\ m \times 4)(6\ m)$$

$$= 269\ kN + 345\ kN = 614\ kN$$

Note: A value of Q_{total} would include Q_{tip} but then would be reduced by the factor of safety to yield Q_{design}.

Calculating the value of skin friction developing along the shaft of a pile by use of the method in Illustration 13-10 has proved to be fairly reliable, though not precise, for piles, provided that the embedded length has not been too great (less than about 25 m or 75 ft). For very long piles, the method has been found to be too conservative in the prediction of the frictional capacity. Reasons for the discrepancy are not yet clear. Disturbance and remolding may not be as severe at greater depths as at shallower depths, with the result that the α factors are too low. Or, as recent evidence implies, the manner in which skin friction in clay occurs may be basically similar to the development of friction in sands, where resistance is related to an effective lateral pressure.

In recognition of the probable relationship between effective overburden pressures and skin friction, an empirical method for determining the frictional capacity of long cylindrical steel piles in normally consolidated clay had been proposed (the lambda method [318]). The method has been extended to other pile types [162]. The method considers the theoretical maximum passive pressures that would act in the soil zone adjacent to the pile, $\overline{\sigma}_v + 2c$ (Fig. 13-32), and applies a variable lambda factor λ to account for a limit to the influence on the effective overburden $\overline{\sigma}_v$ and the adhesion between the pile shaft and surrounding soil. The unit skin friction is taken as

$$f = \lambda(\overline{\sigma}_v + 2c_u) \tag{13-32a}$$

where λ = a coeficient that evaluates the effect of soil–pile adhesion plus the influence of the soil overburden

c_u = undrained shear strength (cohesion) of the clay

The total friction capacity for a pile is then

$$Q_{\text{friction}} = \lambda(\overline{\sigma}_{v_m} + 2c_m)A_{\text{surface}} \tag{13-32b}$$

where $\overline{\sigma}_{v_m}$ = mean effective vertical pressure for the embedded length

c_m = mean undrained shear strength for depth of pile (Fig. 13-33)

A_{surface} = surface area of pile shaft

Figure 13-32   Mohr's circle presentation indicating relation between vertical pressure and maximum horizontal pressure.

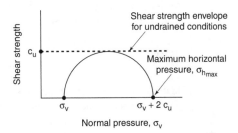

Figure 13-33   Mean vertical pressure and mean shear strength used in designing long piles in clay. [188,318]

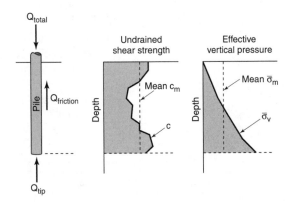

Figure 13-34 Relationship between λ coefficient and pile length for normally consolidated clays. (After Kraft, Focht, and Amerasinghe [162]).

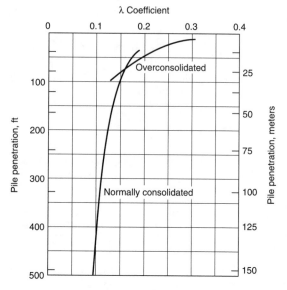

Note: Use normally consolidated curve for $c_m/\bar{\sigma}_{v_m} < 0.4$

Use overconsolidated curve for $c_m/\bar{\sigma}_{v_m} > 0.4$

The curves relating design values for λ to normally consolidated soil deposits and over-consolidated soils (Fig. 13-34) have resulted from regression analysis on a number of pile load test results. The curves are considered appropriate for all driven pile types.

Illustration 13-11

A pipe pile with an outside diameter of 18 in. and a flat end plate is driven 200 ft into a normally consolidated clay deposit having the properties shown. The water level is above the ground surface. Applying the lambda (λ) method, calculate the skin friction expected to develop along the shaft of the pile.

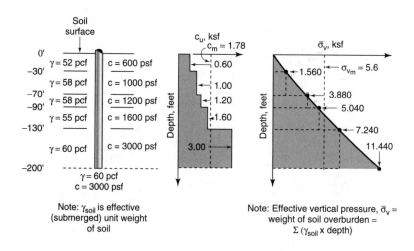

Depth (feet)	Cohesion, c (psf)	Sub. unit wt. (pcf)
0 to 30	600	52
30 to 70	1000	58
70 to 90	1200	58
90 to 130	1600	55
130 to 200	3000	60

Solution

$$c_m = \frac{\text{area of } c_u \text{ diagram}}{\text{pile depth}}$$

$$= \frac{\begin{bmatrix} (.60 \text{ ksf} \times 30 \text{ ft}) + (1.0 \text{ ksf} \times 40 \text{ ft}) + (1.20 \text{ ksf} \times 20 \text{ ft}) \\ + (1.60 \text{ ksf} \times 40 \text{ ft}) + (3.00 \text{ ksf} \times 70 \text{ ft}) \end{bmatrix}}{200 \text{ ft}}$$

$$= \frac{365 \text{ k/ft}}{200 \text{ ft}} = 1.78 \text{ ksf}$$

$$\bar{\sigma}_{v_m} = \frac{\text{area of } \bar{\sigma} \text{ diagram}}{\text{pile depth}} = \frac{1120 \text{ k/ft}}{200 \text{ ft}} = 5.6 \text{ ksf}$$

$$Q_{\text{friction}} = \lambda(\bar{\sigma}_{v_m} + 2c_m)A_{\text{shaft}}$$

$$= (.12)[5.6 \text{ ksf} + (2)(1.78 \text{ ksf})] \times \left(200 \text{ ft} \times 3.14 \times \frac{18}{12} \text{ ft}\right)$$

$$= 1035 \text{ kips}$$

The values of *end bearing* for a pile in clay can be computed from

$$q_{\text{tip}} = cN_c \tag{13-33}$$

where c = cohesion of the clay in the zone surrounding the pile tip
  N_c = bearing capacity factor for deep foundations, ranging between 6 and 10, depending on the stiffness of the clay. A value of 9 is conventionally used.

The foregoing equation assumes that the lower section of pile penetrates at least five diameters into the clay whose cohesion is c.

If the subsurface investigation includes Dutch cone data, the end bearing for a pile can be directly related to the cone resistance, q_c in kg/cm² or tons/ft². For homogeneous conditions, where the pile penetrates at least eight diameters into the clay whose q_c is used in the design computation and no soft soil underlies the pile tip, the end bearing can be taken as

$$Q_{\text{tip}} = A_{\text{tip}}q_{\text{design}} = A_{\text{tip}}\frac{(q_c)}{1.5} \tag{13-34}$$

Factor of Safety

The value of Q_{total} obtained by using the analytical procedures discussed for piles in sand and clay represents an estimate of the maximum load that can be applied to a pile. Whenever possible, predicted capacities should be verified by a field load test before a final design loading is selected. Where load tests have not been performed, it is usual practice to apply a factor of safety of 2 during the analysis to determine the downward design load:

$$Q_{\text{design}} = \frac{Q_{\text{total}}}{2} \qquad \text{(for downward loading)} \qquad \text{(13-35a)}$$

Piles subject to uplift develop resistance to pullout only from the skin friction developed along the embedded length. End bearing does not apply, but the weight of the pile can be included in uplift resistance. For design purposes, it is common to apply a factor of safety of *at least* 2. If nonhomogeneous conditions exist or soil properties are not accurately known, a larger theoretical factor of safety is warranted. Tapered piles are commonly assigned a larger factor of safety than are straight-sided piles. Generally, a larger safety factor is applied to a design uplift capacity than to the downward capacity. The reason frequently stated is that an overrated capacity for downward loading probably results only in a greater-than-estimated settlement, whereas a pile whose uplift is overrated may experience significant upward movement with serious effects to the supported structure. The strength of the pile-to-pile cap connection becomes critical where uplift forces are imposed, because tensile failure at this location will negate any potential pullout resistance that a pile possesses.

$$Q_{\text{design}} = \frac{Q_{\text{friction}} + W_{\text{pile}}}{\text{FS}} \qquad \text{(for uplift)} \qquad \text{(13-35b)}$$

where W_{pile} = weight of pile
 FS = factor of safety, 2 or greater

Negative Skin Friction

When the properties of the soil through which a pile penetrates are evaluated, it is important to watch for and recognize the condition that will create *negative friction,* or *downdrag.* The phenomenon occurs when a soil layer surrounding a portion of the pile shaft settles more than the pile. This condition can develop where a soft or loose soil stratum located anywhere above the pile tip is subjected to new compressive loading. As an illustration, new loading to existing soil results when fill is placed on a construction site to raise the area. If the soft or loose layer settles after the pile has been installed, the skin friction-adhesion developing in this zone is in the direction of the soil pulling downward on the pile (Fig. 13-35). Extra loading is thus imposed onto the pile. It is necessary to subtract negative skin friction values from the total load that the pile can support in order to know what building load can be carried. Values of negative skin friction are computed in the same manner as those of positive skin friction.

Where it is anticipated that an upper zone of soil would impose undesirable or intolerable downdrag on a pile, a common practice is to provide the pile with a protective

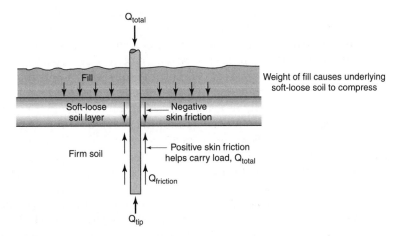

Figure 13-35 Action of negative skin friction on pile shaft.

sleeve or coating for the section that is embedded in the settling soil. Skin friction for this section of pile is eliminated, and downdrag is prevented.

Other Design Considerations

The development of skin friction along the shaft of a pile does require some shear strain in the soil adjacent to the pile. If the tip and end section of a pile are embedded in a very firm or compact material, high end bearing can be developed with relatively little downward movement of the pile. Only limited deformation may occur in the upper zone of soil surrounding the pile shaft, and the maximum skin friction may not develop. The percentage of reduction in skin friction that should be attributed to comparatively soft or loose strata in contact with the pile shaft is generally a judgment factor, influenced by a comparison of relative values of cohesion or density in the various soil zones. Conversely, if a pile should penetrate through firm or dense soils and terminate in a soft or loose material, the maximum computed tip resistance may not be able to develop unless significant downward movement occurs.

Statical Method—Bored Piers and Piles

Bored piers and piles refer to the type of deep foundations that are constructed by drilling into the earth and subsequently placing concrete in the excavation, usually directly against the soil, to form the foundation unit. The concrete is often reinforced with steel. This type of foundation can be straight-sided for its full depth, or it may be constructed with a belled or underreamed base if in cohesive soil, in order to increase the tip bearing. Bored foundations have gone under a variety of names in the past, such as drilled piers, piles, or caissons, and cast-in-place or cast-in-situ piles. When provided with an enlarged base, the foundation may be referred to as a drilled-and-belled type. The designation currently receiving general acceptance is *drilled-shaft foundation.*

Drilled-shaft foundations are constructed by utilizing mechanical auger drill equipment to excavate the hole in the earth. When subsurface conditions permit, such as a sufficiently strong cohesive soil and no groundwater, the hole can be drilled dry, after which the cast-in-place concrete comes in direct contact with the soil forming the walls of the excavation. If cohesionless soils are penetrated or the water table is encountered, a bentonite slurry may be circulated into the hole as it is being drilled to prevent soil cave-in and to assist in flushing soil cuttings to the surface. (This method is similar to the procedure used for drilling uncased soil borings.) Reusable protective casing may also be utilized with the bentonite slurry if a particularly bad groundwater condition is encountered. Concrete for the finished foundation is placed at the bottom of the excavation and worked upward in a continuous pour so as to force displacement of the bentonite slurry. If a casing has been used, it is pulled as the concrete is poured, but in such a manner as to prevent the soil walls from falling into the excavation and mixing with the concrete. This is accomplished by pouring high-slump concrete inside the casing and keeping the height of concrete above the bottom of the withdrawing casing at all times. The earlier presence of a bentonite slurry leaves a film on the soil surrounding the excavation, an effect that will reduce the skin friction between the soil and the hardened concrete.

Design criteria for drilled-shaft foundations *in clay* have evolved from instrumented studies performed on full-scale foundations. For foundations embedded in relatively homogeneous clay, the ultimate capacity is due to the resistance provided from the end bearing and skin friction. In calculating the foundation depth that effectively provides skin friction, the lower 1.5 m or 5-ft section and the belled section (if provided) are neglected because of disturbance and loss of strength caused by construction. There is also a strong possibility of disturbance and loss of strength occurring in the surface zone of soil. Skin friction should be neglected in such a zone. A depth of 1.5 m or 5 ft can be assumed unless more accurate information is available.

$$Q_{\text{total}} = Q_{\text{base}} + Q_{\text{friction}} \qquad (13\text{-}25a)$$

becomes $\qquad\qquad\qquad\qquad\qquad\qquad\qquad\qquad\qquad\qquad\qquad\qquad (13\text{-}36a)$

$$Q_{\text{total}} = cN_c A_{\text{base}} + fA'_{\text{shaft}}$$
(for drilled shaft foundations)

where c = cohesion of soil at base of foundation
N_c = bearing capacity factor
A_{base} = bearing area of foundation base
f = unit skin friction developing between clay and concrete shaft
A'_{shaft} = surface area of shaft that is effective in developing skin friction

The skin friction that develops along the shaft is related to the cohesion of the surrounding clay and the field procedure for drilling the foundation. Values appropriate for design are shown in Table 13-10.

If the base of the foundation is in very firm clay but the shaft is not, the supporting capacity of the unit is due primarily to end bearing. If the movement of the foundation base

Table 13-10   Skin Friction Values for Drilled Shaft Foundations in Clay

Foundation Type and Drilling Method Utilized	Skin Friction, f	Upper Limit on Skin Friction	
		ksf	kPa
Straight shaft, excavation drilled dry	$0.5c$	1.8	86
Straight shaft, drilled with slurry	$0.3c$	0.8	38
Belled, drilled dry	$0.3c$	0.8	38
Belled, drilled with slurry	$0.15c$	0.5	24

Note: c is soil cohesion determined from triaxial testing, not in situ vane shear tests.

is very limited, the soil adjacent to the shaft does not experience the shear strain necessary for significant skin friction to develop; thus,

$$Q_{total} = Q_{base} = cN_cA_{base} \qquad (13\text{-}36b)$$

(for drilled shaft foundations bearing on very firm clay)

An N_c value of 9 is assigned for all the conditions described previously.

When the methods presented above are applied to a foundation design, it is recommended that a factor of safety of 3 be applied to the bearing capacity of the base of the unit. Thus,

$$Q_{design} = \frac{1}{3}cN_cA_{base} + fA'_s = 3cA_{base} + fA'_{shaft} \qquad (13\text{-}37)$$

(for drilled shaft foundations in clay)

Illustration 13-12

A 1-m-diameter drilled-shaft foundation is constructed with a 2-m-diameter base in soil conditions shown by the sketch. The excavation is drilled dry. For the foundation length shown, what maximum axial load (design load) should be planned?

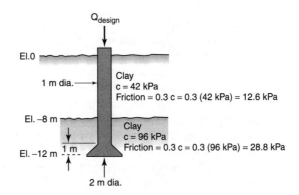

Solution

$$Q_{tip} = 9cA_{tip} = (9)(96 \text{ kN/m}^2)\left(\frac{\pi}{4} \times 2 \text{ m} \times 2 \text{ m}\right)$$

$$= 2713 \text{ kN}$$

$$Q_{skin \, friction} = fA'_{shaft} = (12.6 \text{ kN/m}^2)(\pi \times 1 \text{ m})(8 \text{ m} - 1.5 \text{ m})$$

$$+ (28.8 \text{ kN/m}^2)(\pi \times 1 \text{ m})(4 \text{ m} - 1 \text{ m} - 1.5 \text{ m}) = 393 \text{ kN}$$

$$Q_{design} = \frac{1}{3}Q_{tip} + Q_{friction} = (2713/3 \text{ kN}) + 393 \text{ kN} = 1297 \text{ kN}$$

The capacity of drilled-shaft foundations *in sand* can be analyzed by applying the procedures discussed for driven piles in sand:

$$Q_{total} = Q_{tip} + Q_{friction} \tag{13-25a}$$

$$= \overline{\overline{\sigma}}_v N_q A_{tip} + \Sigma \, (K\overline{\overline{\sigma}}_v \tan \delta)(\Delta A_{shaft})$$

$$= \overline{\overline{\sigma}}_v N_q A_{tip} + (\text{area of } K\overline{\overline{\sigma}}_v \tan \delta \text{ diagram})(\text{shaft circumference}) \tag{13-36c}$$

where $\overline{\overline{\sigma}}_v$ = effective vertical pressure considering the limits imposed by the concept of critical depth

$\tan \delta$ = coefficient of friction between sand and concrete

K = lateral pressure coefficient for a drilled foundation in sand

The value of K for drilled-shaft or bored foundations ranges between about 0.3 for loose sand and 0.75 for compact sand. The value of $\tan \delta$ can be taken as equal to $\tan \phi$ for the sand when the excavation has been drilled dry, because of the roughness of concrete against soil. If a slurry has been used when drilling the excavation, some reduction should be applied. The design load should include a factor of safety at least equal to that for driven piles.

Arrangement for Piles in a Group

Pile analysis discussed previously refers to the capacity of an individual unit. When used for foundation support, driven piles will almost always be used in a grouping. This requirement results from the desire to ensure that the imposed structural load (e.g., column or wall load) falls within the support area provided by the foundation. Usually, driven piles are not used singularly beneath a column or wall because of the tendency for the pile to wander laterally during driving. If a single pile were to be used as the foundation, a designer could not be certain that the pile would be centered beneath the foundation. If an unplanned eccentric loading results, the connection between pile and column may be inadequate, or the pile may fail structurally because of bending stresses created within it. As a result, piles for walls are commonly installed in an arrangement staggered to both sides of the wall centerline. For a column or isolated load, a minimum of three piles is used in a triangular pattern, even for light loads. Where more than three piles are required in order to obtain adequate capacity, the pile arrangement is symmetrical about the point or area of load application. Representative patterns are illustrated in Figure 13-36. Column and wall loads are usually transferred to the pile group through a pile cap. The pile cap is typically a reinforced concrete slab structurally tied to the pile butts (top) to help the group act as a unit (Fig. 13-37).

The requirement for group arrangement of driven piles does not necessarily apply to bored piles (drilled-shaft foundations). Drilled shafts can be located and installed quite accurately. For building construction, it is common to use a single, large-diameter drilled-shaft foundation to support a column. For light loads, a single unit can usually be more properly matched with soil conditions to provide necessary capacity than can a driven pile foundation, where the three-pile minimum usually provides excessive carrying capacity. Economy can result. Additional cost savings will be realized with a single-pile foundation where the need for a pile cap is eliminated.

Group Capacity

The capacity of a pile group is not necessarily the capacity of the individual pile multiplied by the number of piles in the group. Soil disturbance because of pile installation, and overlap of stresses between adjacent piles, may cause the group capacity to be less than the sum of individual capacities. Conversely, soil between individual piles may become "locked-in" because of strong adhesion or densification from driving, and the group may tend to behave as an equivalent single large pile. Densification and improvement of soil surrounding the group can also occur. These latter factors tend to provide a group capacity greater than the sum of the individual piles. The capacity of the

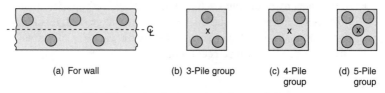

(a) For wall   (b) 3-Pile group   (c) 4-Pile group   (d) 5-Pile group

Note: Pile spacings for groups typically range between $2\frac{1}{2}$ and 4 pile diameters, center to center.

Figure 13-36   Representative pile patterns for wall and column foundations.

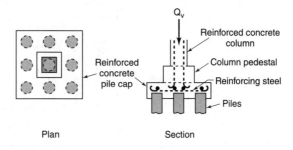

Plan   Section

Figure 13-37   Pile cap of type provided for reinforced concrete column.

equivalent large pile is analyzed by determining the skin friction around the embedded perimeter of the group and calculating the end bearing by assuming a tip area formed by this same perimeter (Fig. 13-38).

To determine the *design load* for a pile group, compare the sum of the individual pile capacities with the capacity of the single large equivalent (group) pile. With a proper factor of safety, the least capacity will be the design load. The capacity of the equivalent (group) pile is affected by soil type and properties and pile spacing. Generally, there is a greater tendency for the group to act as a large single unit when pile spacings are close and the soil is firm or compact.

For driven piles embedded in cohesionless soil, the capacity of the large equivalent (group) pile will almost always be greater than the sum of the individual pile capacities. This is an indication that the soil in the immediate proximity of each individual pile controls the group capacity. Consequently, for design, the group capacity is typically based on the single-pile load and the number of piles in the group.

For bored piles in sand, the group capacity will be less than the sum of the individual pile capacities, in part because of limited densification of the soil zone surrounding the pile group (compared to the driven pile group) and because of the overlapping of stresses between adjacent units. The capacity of a bored-pile group is typically about two-thirds the capacity of the sum of the individual pile capacities.

The capacity of closely spaced piles embedded in clay is often limited by the behavior of the group acting as a single large unit. In determining the capacity of the single equivalent large pile, the skin friction and end bearing are calculated by using the general methods discussed for the individual pile. The unit value of perimeter skin friction is intermediate between adhesion or remolded shear strength and the undrained shear strength. For pile spacings greater than about three diameters, the group capacity is, for practical purposes, equal to the capacity of the piles acting individually. The capacity should be reduced by one-third if the pile cap for the group does not rest on the ground; this modification accounts for the condition that group load is controlled by the capacity of individual piles as influenced by stress overlap between adjacent units.

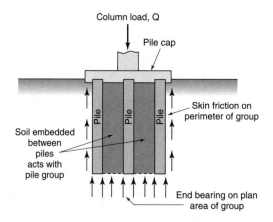

Figure 13-38 Skin friction and end bearing acting on the single equivalent pile formed by a pile group.

Illustration 13-13

A nine-pile group composed of 12-in. diameter pipe piles is embedded 50 ft deep in soil conditions indicated by the sketch. A $2\frac{1}{2}$-diameter spacing is used. What design capacity should be estimated for the group?

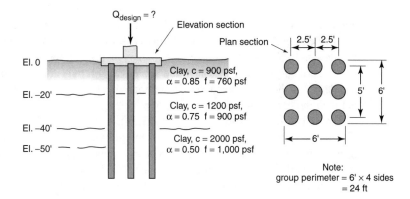

Solution

Capacity, Piles Acting Individually

$$Q_{\text{friction}} = (.85 \times 900 \text{ psf})(3.14 \times 20 \text{ ft}) + (.75 \times 1200 \text{ psf}) \times (3.14 \times 20 \text{ ft})$$

$$+ (.50 \times 2000 \text{ psf})(3.14 \times 10 \text{ ft})$$

$$= 136,000 \text{ lb} = 136 \text{ k}$$

$$Q_{\text{tip}} = 9cA_{\text{tip}} = (9 \times 2000 \text{ psf})\left(\frac{\pi}{4} \times 1 \text{ ft}^2\right) = 14,000 \text{ lb} = 14 \text{ k}$$

$$Q_{\text{total}} = 136 \text{ k} + 14 \text{ k} = 150 \text{ k/pile}$$

Group total = 150 k/pile × 9 piles = 1350 k

Capacity, Group Action

$$Q_{\text{friction}} = \left(\frac{900 \text{ psf} + 760 \text{ psf}}{2}\right)(24 \text{ ft} \times 20 \text{ ft}) + \left(\frac{1200 \text{ psf} + 900 \text{ psf}}{2}\right)(24 \text{ ft} \times 20 \text{ ft})$$

$$+ \left(\frac{2000 \text{ psf} + 1000 \text{ psf}}{2}\right)(24 \text{ ft} \times 10 \text{ ft}) = 1,260,000 \text{ lb} = 1260 \text{ k}$$

$$Q_{\text{tip}} = 9cA_{\text{tip}} = (9)(2 \text{ ksf})(6 \text{ ft} \times 6 \text{ ft}) = 650 \text{ k}$$

Group total = 1260 k + 650 k = 1910 k

Use least group total for design. Therefore,

$$Q_{\text{design}} = \frac{1350 \text{ k}}{\text{FS} = 2} = 675 \text{ k}$$

Pile Group Settlement

The vertical movement that occurs at the pile cap level is the result of compressive short-ening within the pile from the loading plus the settlement occurring in the soil supporting the pile. Analytical methods available to predict the settlement of pile groups provide only approximations. A widely used procedure assumes that the pile group acts as a single large, deep foundation, such as a pier or mat. Where the piles are embedded in a uniform soil (fric-tion plus end bearing piles), the total load is assumed to act at a depth equal to two-thirds the pile length. Conventional settlement analysis procedures assuming the Boussinesq or Westergaard stress distribution are then applied to compute compression of the soil beneath the pile tip. If the piles have their tip section embedded in a stratum firmer or more com-pact than the overlying soil (end bearing piles), the total load is assumed to act at a depth corresponding to pile tip elevation (Fig. 13-39).

Compressive properties of soil below the pile tip are determined from laboratory tests on recovered soil samples (typical for cohesive soils) or from empirical correlations devel-oped from soil exploration penetration tests (typical for cohesionless soil). For piles em-bedded in sand, the following expressions for approximating the settlement can be applied:

$$S = \frac{q\sqrt{B}}{N} \text{ (approximately)} \tag{13-38a}$$

where S = settlement, in inches

    q = net foundation bearing pressure, in tons/ft^2

    B = width of pile group, in feet

    N = blow count from standard penetration test, corrected for depth and hammer energy

The N value depth applies to the sand in the zone most subject to compression by the pile load. A depth distance of one group width (or B feet) is recommended. If static cone pene-tration data (e.g., Dutch cone) are available for the soil below pile tip level, the settlement can be approximated from

$$S = \frac{qB}{q_c} \text{ (approximately)} \tag{13-38b}$$

where q_c = static cone resistance, in same units as q (the foundation load) and B (the pile group width).

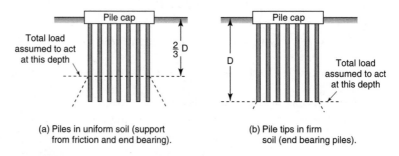

(a) Piles in uniform soil (support from friction and end bearing).

(b) Pile tips in firm soil (end bearing piles).

Figure 13-39 Assumed conditions for estimating settlement of pile groups.

If load testing is performed on a single pile, caution is in order when one is extrapolating settlement results to a group. Group settlements typically range from two to more than ten times the settlement of the single pile. The greater settlement ratio (i.e., group settlement to single-pile settlement) occurs with the larger pile groups. For piles installed in sand deposits, settlement data obtained from a field load test on a single pile, S_{single}, can be used to estimate the settlement for a pile group, S_{group}, by the following [311]:

$$S_{group} = S_{single} \sqrt{\bar{B}/B} \qquad (13\text{-}38c)$$

where B = diameter of tested pile
$\quad\quad \bar{B}$ = smallest dimension of the pile group

Pile Load Tests

Applied to construction projects, on-site load tests are performed on test piles installed during the design stage to check estimated capacities (as predetermined by analytical or other methods) and to help deliver criteria for the foundation installation contract, or as a check on contract piles installed by the builder to verify the carrying capacity (proof testing).

Both cohesive and cohesionless soils will have their properties altered by the installation of a driven pile. In clays, the disturbance causes remolding and loss of strength. With time, much of the original strength will be regained. Consequently, to obtain load test results that are practical for design, testing of piles in clay should be performed only after a lapse of at least several weeks. The effect of installing driven piles in sand is to create a temporary condition where extra resistance is developed. Shortly after the installation, however, the extra resistance is lost (stress relaxation). To obtain test results applicable for design, a period of at least several days should lapse before testing begins.

Loading applied to test piles frequently is obtained by hydraulically jacking against a supported weight box or platform (Fig. 13-40a) or against a reaction girder secured to anchor piles (Fig. 13-40b). If the jack is outfitted with a pressure gage, the magnitude of test load imposed onto the pile is read directly. However, some designers prefer that a proving ring or pressure capsule be used for measuring the load, feeling that it is more accurate.

Measurement for pile movement is related to a fixed reference mark. The support for reference marks needs to be located outside the soil zone that could be affected by pile movements.

Several different methods for performing pile load tests are in use. Probably the most common is the *slow-maintained load (slow-ML) test* (ASTM D-1143). With this procedure, the test load is applied in eight equal increments until twice the intended design load is reached. Time–settlement data are obtained for each load increment. Each increment is maintained until the rate of settlement becomes less than 0.25 mm (0.01 in.) per hour, or for 2 hours, whichever occurs first. The final load (double the design load) is maintained for 24 hours. Unloading also occurs in increments.

Another common procedure is the *constant-rate-of-penetration test.* In this method, the load is increased on the pile as necessary to force settlement (penetration) at a predetermined rate. A rate of 0.5 mm (0.02 in.) per minute is typical. The force required to achieve penetration is recorded, thus giving load–settlement data similar to the slow-ML test. This test is considerably faster than the slow-ML test.

Other general test methods include *cyclic loading,* in which each increment of load is repeatedly applied and removed (or reduced), and the *quick-maintained test,* in which the

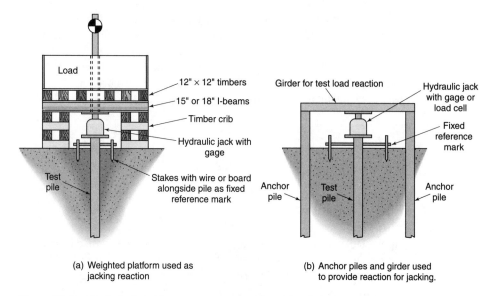

Figure 13-40 Typical pile load test arrangements: (a) weighted platform used as jacking reaction; (b) anchor piles and girder used to provide reaction. (ASTM D-1163).

load increments are imposed for short periods of time. Pile movement for all loads, or cycles of loads, is recorded.

The load–settlement data are used to determine the design load for the pile. Frequently, the design load is obtained from a "failure load" indicated by the load test with a factor of safety applied. The definition of failure load is arbitrary; it may be taken when a predetermined amount of settlement has occurred (e.g., one-tenth of the pile diameter or a certain number of inches) or where the slope of the load–settlement plot is no longer proportional. A method in use for the slow-ML test is to plot both load and settlement values on logarithmic coordinates. The results typically plot as two straight lines (Fig. 13-41). The intersection of the two lines is referred to as the failure load. This is the failure load for design purposes and is not the actual failure or ultimate load.

The design factor of safety used for contract piles should relate to the extent of information known about subsurface conditions, and to the number and comparative results of pile load tests performed at different locations on the site. If soil conditions, including properties, are uniform and load tests at different areas compare well, a relatively low factor of safety could be justified (i.e., 1.5 to 2.0). If subsurface conditions and load test results are variable, a greater factor of safety is usually necessary (2 to 3) as protection against possible unexpected poorer soil conditions and lesser pile capacity at untested locations.

Load testing by use of an expanding load cell positioned at the base of a drilled shaft concrete pile prior to placing the concrete (such as the Osterberg Load cell) represents a variation of the conventional downward-load pile test. In basic form, the load cell for pile testing is a hydraulic or pneumatic jack, with controls at the ground surface. When pressurized, the cell expands upward and downward. The soil under the base acts as the lower support reaction to jack against, and the installed pile acts as the upper reaction. The downward force imposed by the load cell represents the end bearing resistance that develops at the base of the pile while

Figure 13-41   One method of plotting pile load test data to determine pile design capacity.

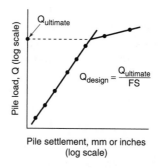

the upward force is opposed by the frictional resistance developed along the shaft of the pile (i.e., capacity due to skin friction). The total downward capacity of the pile indicated by the test procedure is then twice the value of the expansive force developed by the load cell (assuming that shaft friction for upward movement and downward movement are equal).

Pile-Driving Formulas

The axial capacity of piles driven in certain types of soil can be related to the resistance against penetration developed during driving. Relating the load capacity that a pile will develop to the driving resistance is appropriate for piles penetrating soils that will not develop high pore water pressures during the installation. Such soils include free-draining sands and hard clays. In saturated fine-grained soils, high pore water pressures develop because of soil displacement and vibration caused by the driving. The strength of the soil surrounding the pile is affected. A predicted capacity based on such soils' resistance to pile penetration is different from the capacity that develops after the excess pore pressures dissipate.

Using driving data to determine pile capacity requires knowledge of the effect that a hammer blow has on the pile and supporting soil. As the top of a pile is struck with the pile hammer, the force of the blow causes a stress wave to be transmitted through the length of the pile. Some of the force of the transmitted wave is absorbed by the soil surrounding the pile shaft, while some is imparted to the soil at the pile tip. What has become known as the *wave equation method* applies wave transmission theory to determine the carrying capacity developed by a pile and the maximum stresses that result within the pile during driving. The basic wave equation method presented in 1962 (ref. 249) has been modified as a result of subsequent in-depth research programs, most sponsored by the (USA) Federal Highway Administration. New computer programs resulting from such studies are WEAP[9], WEAP86[10], and TTI.[11] Currently these are considered to best model actual pile driving installations.

---

[9]"Wave Equation Analysis of Pile Driving—WEAP Program," Volumes 1 through 4, FHWA IP-76-13.1 through IP-76-14.4, April 1976 and March 1981. G.C. Goble and F. Rausche.

[10]"Wave Equation Analysis of Pile Foundations—WEAP 86," Volumes 1 through 4, FHWA Contract DTFH 61-84-C-00100, March 1986. G.C. Goble and F. Rausche.

[11]"Pile Driving Analysis Wave Equation Users Manuals TTI Program," Volumes 1 through 4, FHWA IP-76-13.1 through IP-76-13.4, April 1986. T.T. Hirsch, L. Carr, and L. L. Lowery, Jr.

In a wave equation analysis, the method assumes that the pile and its behavior when embedded in soil can be represented by a series of individual spring-connected weights and spring-damping resistances (Fig. 13-42).The various weight values, W, correspond to the weight of incremental sections of pile. The spring constants, K, relate to the elasticity of the pile. The spring damping, R, represents the frictional resistance of soil surrounding the shaft of the pile and the soil resistance at the pile tip. Spring damping along the shaft of the pile accounts for a gradual diminishing of the longitudinal force (from the hammer blow) that travels along the length of the pile. Spring damping at the pile tip is necessary to account for the force that remains within the pile to be transmitted at the tip.

To obtain a solution for the wave equation, it is necessary to know approximate pile length, pile weight, cross section, elastic properties, and the pile hammer characteristics, including efficiency, ram weight, and impact velocity, to have data on the pile cap and capblock, and to assign values for soil damping and the spring constants. Determining the effect of a stress wave traveling through the pile is a dynamics problem. However, if the effect of a pile hammer blow at one particular instant in time is selected (the reaction of each weight and spring to the forces acting on overlying weights is determined), the analysis can be han-

Figure 13-42 Method of representing pile for wave equation analysis [272].

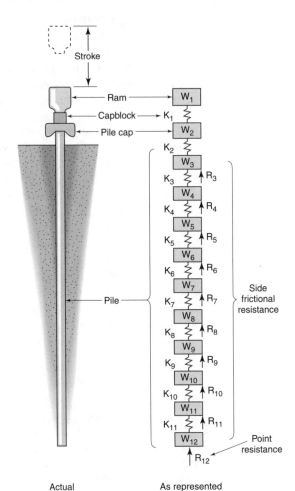

Figure 13-43   Relation between pile capacity and driving resistance indicated by wave equation.

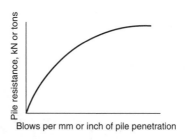

dled as a statics problem. By analyzing changing conditions for successive small increments of time, the effects of the force wave traveling through the pile to the tip will be simulated. This analysis requires a numerical integration, a task conveniently undertaken by computer. Results obtained will be only for a particular pile driven by a specified pile hammer. Separate analyses are required for different conditions.

For field use, it is convenient to have the pile capacity expressed in terms of mm or inches of penetration per hammer blow in order to know when driving can cease. Representative results for a solution are shown in Figure 13-43. Under normal circumstances, proper evaluation of factors such as the equivalent spring constant and soil damping values requires driving data for the pile under study. This means that results for field control are not always available before the start of a project. However, values based on experiences are evolving and current information can produce close estimates of actual capacity.

Practical application of the wave equation to produce information on pile capacity at the *time of installation* has resulted from sponsored research studies conducted through Case Western Reserve University (1975).[12] The procedure is now referred to as the Case Method (of pile analysis). Information required to implement the Case Method include data on the pile driving hammer system, measurement of the force transmitted by the drive hammer into the pile, and the pile velocity that results from the hammer blow. Values of force and velocity are calculated from measurements (obtained by electronic instruments) of strain and the related acceleration that occurs when the pile is struck by the drive hammer. Strain is determined by (strain) transducers attached to the side of the pile (but near the top) and acceleration is measured by separate motion sensors (accelerometers) also attached to the side of the pile. Force developed in the pile is determined from the relation between strain, modulus of elasticity of the pile material and cross-sectional area of the pile. Acceleration is converted to velocity by integration over an assigned period of time. Specialty equipment has been developed to measure the necessary pile strain and acceleration resulting from a pile hammer blow, then instantly process the information into a field computer to obtain solution of the wave equation (the feedback) which indicates the axial capacity of the embedded pile (Figure 13-44). Pile capacity can be determined either at the time of installation, as is often desired, or at a later time. Testing after installation is appropriate if the properties of the soil surrounding the pile are expected to improve after the effects of disturbance due to driving dissipate, and, for cases where piles in-place from a previous structure could be utilized for a planned new structure but the capacity is uncertain. The re-strike hammer blow for such after-installation-testing can be from any falling weight of known values.

[12]"Bearing Capacity of Piles from Dynamic Measurement—Final Report." Goble, G. G., Likens, G. E., and Rausche, F. Department of Civil Engineering, Case Western Reserve University, Cleveland, Ohio, 1975.

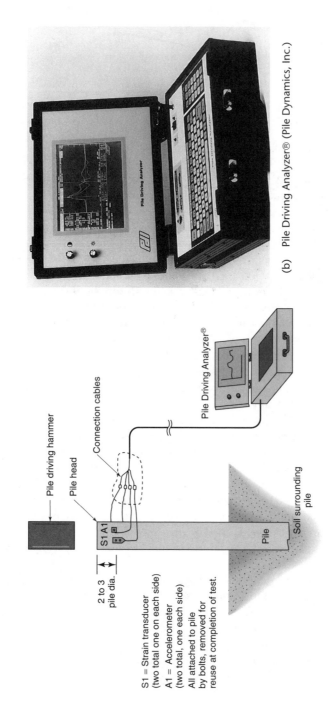

(a) Schematic of equipment setup for Pile Driving Analyzer® test procedure to determine pile capacity.

S1 = Strain transducer
(two total one on each side)

A1 = Accelerometer
(two total, one each side)

All attached to pile by bolts, removed for reuse at completion of test.

(b) Pile Driving Analyzer® (Pile Dynamics, Inc.)

Figure 13-44 Equipment and arrangement for field test to determine pile capacity through application of wave equation. (Courtesy Pile Dynamics, Inc., and Goble Rausche Likins & Assoc., Cleveland, Ohio).

Wave transmission theory shows that the maximum force that can pass through a pile is related to the modulus of elasticity and mass density of the pile material and the cross-sectional area, in a term referred to as *pile impedance:*

$$\text{Pile impedance} = \rho v A \tag{13-39}$$

where impedance = ability of a pile to pass a longitudinal (axial) force by wave transmission resulting from a pile hammer blow

ρ = mass density of the pile material, γ_{pile}/g (where γ_{pile} is the pile material's unit weight; g is the acceleration of gravity)

v = velocity of longitudinal wave propagation, $\sqrt{E/\rho}$ (where E is the modulus of elasticity for the pile material)

A = cross-sectional area of the pile

If it is assumed that pile driving occurs with a hammer capable of delivering adequate energy, greater capacity is possible with piles having greater impedance values. Other comparative effects of pile impedance are shown by Figure 13-45. The relative position of the curves indicates that high pile capacities cannot be achieved when low-energy driving hammers or piles having low impedance values are used.

Wave transmission theory also explains other occurrences noted in pile driving. If the soil at the pile tip is very hard or compact, some of the stress wave may be reflected back along the pile. As a result, tensile stress (as well as compressive stress) may develop within the pile during driving. If a pile material is weak in tension (e.g., poorly reinforced concrete), tension cracks may develop near the top of the pile. If the pile penetrates through firm soil into soft material, tensile stresses may develop near the tip and cause damage to the end section of the pile.

The pile foundations for a project may be designed by engineering personnel, but the installations are handled by specialty contractors who frequently receive the job as the result of a bidding procedure. Pile foundation contractors typically possess a variety of driving equipment. If the contractor is required to provide a description of the proposed driving system *before* beginning the field work, the designer can introduce this information along with soil parameters (assumed or developed from boring data) into the wave equation to evaluate the adequacy and effects of the proposed driving procedure (e.g., Can the hammer efficiently drive the piles to the desired depth? Will the piles be overstressed during driving? etc.). If problems are indicated, the proposed driving equipment can be changed before it is brought to the construction site.

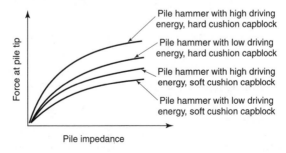

Figure 13-45   Relationship among pile capacity, pile impedance, and equipment used for pile driving [220].

Expressions relating pile capacity to driving resistance have also been developed from work-energy theory and are simpler to apply than the wave equation. Such pile-driving formulas have been in use since at least the mid-nineteenth century. A basic assumption for these driving formulas is that the kinetic energy delivered by the pile-driving hammer equates to the work done on the pile (work defined as the product of force and distance), or

$$(W_h)(H)(\text{Eff.}) = (\text{pile resistance})(\text{pile penetration per hammer blow}) \quad (13\text{-}45)$$

where W_h = weight of pile hammer ram
H = height of ram fall
Eff. = efficiency of pile hammer

For application, these formulas need to include corrections to account for energy losses during driving that are caused by factors such as elastic compression of the pile, soil, capblock, and cushion, and by heat generation. Variations in the many pile formulas that have been proposed result because of the different methods for handling these energy losses.

The assumption of work-energy theory to develop pile-driving formulas does not properly consider the effect of impact on a long member such as a pile; therefore, such formulas will not be theoretically correct. However, some expressions that include empirical energy loss correction factors have shown a reliability for predicting the axial capacity of piles, particularly those driven in cohesionless soil. Some of these formulas are lengthy and complicated (e.g., the Hiley formula [50]) and are not reported herein. Load test studies on piles in sand indicate that the simple-to-use Danish formula is as reliable as any of the more complicated formulas [4].[13]

$$Q_{\text{ult}} = \frac{W_h H(\text{Eff.})}{s + \dfrac{1}{2}s_0} \qquad (\text{Danish formula}) \qquad (13\text{-}46a)$$

where Q_{ult} = axial capacity of the pile
s = average penetraion of the pile from the last few driving blows (pile "set")
s_0 = elastic compression of the pile

$$= \left[\frac{(2)(\text{Eff.})(W_h HL)}{AE}\right]^{1/2}$$

where L = length of pile
A = cross-sectional area of pile
E = modulus of elasticity for pile material

Where the formula is used, a factor of safety of 3 should be applied ($Q_{\text{design}} = \frac{1}{3}Q_{\text{ult}}$).

During pile installation, it is convenient to know when a desired capacity has been obtained and driving can cease. The pile formula above can be rearranged to indicate the final set, usually inches per blow, that provides the design capacity:

---

[13]Some test data are available to indicate that the Danish formula can also apply for piles embedded in cohesive soil [102]; at the present level of experience, use of this driving formula should be limited to projects where pile load tests are performed to correlate load-carrying capability.

$$s = \frac{W_h H \times \text{Eff.}}{3Q_{\text{design}}} - \frac{1}{2}s_0 \qquad (13\text{-}46b)$$

If the set is to be in inches per blow, all lengths should be expressed in inches. Similarly, the units for the weight of ram and pile capacity have to be the same.

Illustration 13-14

Steel pipe piles intended to provide a design capacity of 50 tons are used at a construction site underlain by cohesionless soils. Preliminary analysis indicates that the required pile length will be between 45 and 50 ft. If the Danish formula will be used as a field control during driving to indicate when the desired capacity is obtained, what should be the set under the last driving blows? Pile installation will be accomplished with a Vulcan Model 0 hammer having a rated driving energy of 24,375 ft-lb.

Solution

Assume a pile length of 50 ft, pile cross section of 10 in.^{2}, and E_{steel} of 29×10^6 psi.

$$\text{Set } s = \frac{W_h H \times \text{Eff.}}{3Q_{\text{design}}} - \frac{1}{2}s_0$$

where $s_0 = \sqrt{\dfrac{2 \times \text{Eff.} \times W_h HL}{AE}}$

$$= \left[\frac{(2)(24{,}375 \text{ ft-lb})(50 \text{ ft})(144 \text{ in.}^2/\text{ft}^2)}{(10 \text{ in.}^2)(29 \times 10^6 \text{ lb/in.}^2)} \right]^{1/2} = 1.1 \text{ in.}$$

Therefore,

$$\text{Set } s = \frac{(24{,}375 \text{ ft-kip})(12 \text{ in./ft})}{(3 \times 50 \text{ tons} \times 2 \text{ k/ton})} - \frac{1}{2}(1.1 \text{ in.})$$

$$= .98 \text{ in.} - .55 \text{ in.} = .43 \text{ in.} \cong \frac{1}{2} \text{ in.}$$

One of the most widely known pile-driving formulas is the *Engineering News* (ENR) formula, presented in the late nineteenth century for determining the capacity of timber piles. The formula is not reported herein, because it does not have application for existing pile-driving methods. The use of the ENR formula should be discouraged. Though it has a theoretical factor of safety of 6, various load test programs have indicated that the actual factor of safety can range from less than 1 (representing pile failure at less than the expected capacity) to about 20.

13.8 SUPERVISION OF FOUNDATION CONSTRUCTION

Since foundations frequently are covered or completely buried soon after their construction, there is limited opportunity for noting and correcting improper work or detecting conditions not in accord with design assumptions. To minimize the risk of problems resulting from improper or inadequate foundation construction, the practice of having such work guided (or supervised) by personnel with soil mechanics (geotechnical) training is increasing.

The responsibilities of such personnel would include matching soil and groundwater conditions observed in excavations with the conditions indicated by the borings used for design. For projects where shallow foundations are being installed, the proper location, size, and depth should be verified. The bottoms of excavations should be examined prior to concrete placement to ensure that disturbed or muddy soil is not left in place; compaction or removal can be ordered if such conditions are observed.

Where drilled foundations are being installed, the bottom of the excavation and soil exposed in the walls should be examined whenever conditions permit. Depth to, or bearing in, a proper soil stratum should be verified. Proper location, dimensions, and installation of required reinforcing steel should also be verified.

With driven-pile foundations, driving records (the continuous record of blow count versus depth) should be kept for all piles. The size and length of piles should be verified. Detailed information on the equipment used for driving the piles should be obtained.

Personnel doing supervision work should keep a written record (daily logs or reports) of activities, observations, and decisions regarding the foundation work. Copies frequently are expected to be forwarded to the designer's office. Written records can serve as a diary of the construction progress, and frequently serve as a reference if disputes develop.

PROBLEMS

13-1. List and briefly describe the (generalized) methods or procedures for determining the design bearing capacity for shallow foundations. Include the advantages and disadvantages perceived for each of the methods.

13-2. The procedure of designing spread footing foundations on the basis of field exploration data (standard penetration test, cone penetrometer results, etc.) offers some practical benefits. Indicate the advantages of this method compared to basing foundation design on plate-bearing tests, presumptive bearing values, and the use of bearing capacity equations.

13-3. (a) Determine the wall loading that can be carried by a long footing 1 m wide and located 1.5 m below the ground surface. Strength tests indicate that the soil has a cohesion of 58 kN/m^2 and an angle of internal friction of 20°. The total soil unit weight is 18 kN/m^3. The soil above the footing is well compacted, and the depth factors will apply. The soil rigidity index is greater than the critical index. The water table is expected to rise to the ground surface. Use a factor of safety equal to 3 with the extended general bearing capacity equation.

(b) Recalculate the design bearing capacity assuming the depth factors do not apply. What percentage change in bearing capacity does this represent (compared to the answer in part a)?

(c) Recalculate the bearing capacity for the conditions of part a but where the water table is very deep.

13-4. (a) A long footing, 3 ft wide, is to be installed 4 ft below grade. Drained shear tests indicate that the supporting soil has a cohesive strength of 900 psf and the angle of internal friction is 20°. The soil rigidity index is greater than the critical index. The soil unit weight of 115 pcf and the water table is deep. Calculate the safe bearing capacity, applying a factor of safety of 3 to the extended bearing capacity equation. Assume that the backfill is not well compacted and the depth factors do not apply.

(b) Determine the design bearing capacity if the depth factors do apply (all other conditions of part a are the same). What percent change in the design bearing capacity results from including the depth factors?

(c) Recalculate the bearing capacity for the conditions of part a but where the groundwater table is at the soil surface.

13-5. Relating to spread footing foundation design, calculate the value for the soil rigidity index, I_r, given the indicated conditions, to use with the extended bearing capacity equation:

(a) Soil modulus of elasticity is 25,000 kPa, Poisson's ratio is 0.42, soil unit weight is 18 kN/m³, shear strength values are cohesion = 36 kPa, and the angle of internal friction is 22°. Find I_r for a point 3 m below the soil surface.

(b) The modulus of elasticity is to be estimated from the results of a consolidation test where C_c is 0.35 over the pressure range between 50 and 100 kN/m² and the value of e_0 is 1.40. Poisson's ratio is 0.30. In place, the soil has a unit weight of 17 kN/m³, a cohesion equal to 60 kN/m², and an angle of internal friction of 15°. Find I_r for a point 4 m below the soil surface.

13-6. For the conditions described, calculate the soil rigidity index I_r:

(a) Soil modulus of elasticity is 150 ksf, Poisson's ratio is 0.40, soil unit weight is 115 pcf, cohesion is 600 psf, and angle of internal friction is 28°. Find I_r for a point 10 ft below the soil surface.

(b) Modulus of elasticity is 18 ksf, Poisson's ratio is 0.45, soil cohesion is 1200 psf (undrained shear), and soil unit weight is 105 pcf. Find I_r for a point 8 ft below the soil surface.

(c) Modulus of elasticity is 200 ksf, Poisson's ratio is 0.30, angle of internal friction is 35°, and soil unit weight is 122 pcf. Find I_r for a point 10 ft below the soil surface.

13-7. A long strip footing 1.25 m wide is used to support the perimeter masonry bearing walls for a warehouse building. The soil underlying the building area is a clay where the undrained shear strength is 40 kPa (use zero for the angle of internal friction), and the soil unit weight is 16 kN/m³. The rigidity index for the clay is 9. The foundation is constructed 1 m below the ground surface, and the soil backfill is well compacted. Applying the extended bearing capacity equation with a factor of safety of 3, calculate the maximum load to be carried by the foundation, per meter of length.

13-8. A long strip footing 3.5 ft wide is to be used to support a masonry bearing wall for a building. The subsurface investigation indicates that the soil at the site is a clay with a unit weight of 110 pcf and a cohesion, c, of 1200 psf. The angle of internal friction is zero. The clay has a rigidity index of 10. The foundation is constructed 3 ft be-

low the ground surface, and the soil backfill is well compacted. Using a factor of safety of 3, what maximum load should the foundation be designed to support per foot of wall length?

13-9. A square footing is to be designed to carry a column load of 1000 kN (including the weight of the foundation). The footing will be installed at a shallow depth on clay soil whose undrained shear strength (cohesion) is 96 kPa. Assume that the angle of internal friction is zero. The water table is deep. The soil rigidity index value is 30. If a factor of safety of 3 is used with the extended general bearing capacity equation, what size footing should be provided?

13-10. A square footing foundation is required to carry a column load of 100 kN (including the weight of the foundation). The footing will be installed close to the ground surface on a uniform clay soil having an undrained shear strength (cohesion) equal to 45 kPa. Assume that the angle of internal friction is zero. The water table is deep. The soil rigidity index is 9. Apply the extended bearing capacity equation with a factor of safety of 3 to determine the required size foundation.

13-11. An interior column for a manufacturing building will be supported on the site's natural clay soil, at a shallow depth below the concrete floor slab. The column imposes a loading of 475 kips and will be supported on a square spread footing foundation. The clay soil properties are 117 pcf unit weight and 2200 psf cohesion value, c (undrained shear strength). The soil rigidity index is 38. The groundwater table is very deep. What size footing should be provided, as determined by the extended bearing capacity equation, using a factor of safety of 3?

13-12. An 800 kN column load will be supported on a square spread footing foundation bearing on the surface of a cohesionless soil deposit. The soil has a unit weight of 20 kN/m³ and an angle of internal friction of 34°. The rigidity index for the affected soil zone is 75. Assume that the groundwater table is very deep. Determine the required size footing, using the extended bearing capacity equation with a factor of safety of 3.

13-13. A square footing is to be provided in order to support a 350-kip column loading. If the footing is to be located on the surface of a cohesionless soil deposit, what size foundation is required? The soil has a unit weight of 125 pcf, and the angle of internal friction is 36°. The rigidity index for the soil zone near the ground surface is 100. It is expected that the water table could rise to the ground surface. For the design, apply a factor of safety of 3 to the extended bearing capacity equation.

13-14. An existing building is to be renovated. One of the existing footings is 1.5 m by 1.5 m in area and 1 m below the adjacent soil surface. The structural changes for the building would require this footing to support a loading of 2000 kN. Determine if the footing size and soil bearing capacity are adequate. Soil investigation indicates that the foundation soils are medium-dense to dense sands, with a rigidity index value of 140. The soil angle of internal friction is 34°, and the unit weight is 19 kN/m³. The soil surrounding the foundation is well compacted, and the depth factors should be applied. Base the calculations on use of the extended bearing capacity equation, applying a factor of safety between 2.5 and 3.

13-15. A footing for an existing building is 6 ft square (6 ft × 6 ft) and 3 ft below the surface. A proposed building alteration would require this footing to carry a loading of 400 kips. Is the footing of adequate size? Soil investigation indicates that the foundation soils have an angle of internal friction of 35° and a unit weight of 120 pcf. The soil rigidity index value is 150. The soil above the base of the footing is well compacted, and the foundation depth factors can be applied. Use a factor of safety of 3 with the extended general bearing capacity equation.

13-16. Design a square footing to carry a column load of 750 kN. The footing is to be installed 1 m below the ground surface on sand whose angle of internal friction is 33° and whose unit weight is 18 kN/m³. The soil rigidity index value is 125. Assume that the backfill soil is not well compacted and that depth factors do not apply. The water table is very deep. Use a factor of safety of 3 with the extended bearing capacity equation.

13-17. Changes at a manufacturing facility will result in replacement equipment being installed on an existing concrete pad (separate foundation). The pad is 2 m by 4 m in plan area and is supported near floor slab level directly on the surface of soil for which drained shear tests indicate a cohesion of 25 kPa and an angle of internal friction of 22°. A representative soil unit weight is 17 kN/m³. The soil rigidity

index is 30. The groundwater table is very deep. Determine the maximum weight that the pad foundation should carry, applying the extended bearing capacity equation with a factor of safety of 3.

13-18. What loading could a rectangular foundation, 4 ft by 8 ft, support if it is located on the surface of a soil for which drained shear tests indicate a cohesion of 500 psf and an angle of internal friction of 20°? The soil unit weight is 110 pcf. The soil rigidity index value is 25. The water table is deep. Use a factor of safety of 3 with the extended general bearing capacity equation.

13-19. Outdoor storage containers for a manufacturing facility impose a total loading of 5500 kN when full. The storage containers will be supported on a concrete pad foundation constructed near the soil surface. Space constraints require the foundation to have a length twice the width. The soil at the site is a clay; undrained shear tests indicate that the cohesion c is 105 kPa (the angle of internal friction is zero). The soil rigidity index is 60. Determine the required foundation size based on the extended bearing capacity equation and a factor of safety of 3.

13-20. A heavy machine is to be provided with a temporary concrete pad foundation that needs to have a length twice the width. The foundation will be installed close to the ground surface on a clay soil; undrained shear tests indicate that the cohesion is 2500 psf. The soil rigidity index is 75. The machine weighs 1200 kips. To have a factor of safety of 3, what size foundation should be constructed?

13-21. Determine the plan dimensions for a rectangular-shaped footing that will support two building columns (an exterior or wall column and an interior column). The interior column design loading is 3000 kN, and the wall column loading is 1900 kN. The columns are 6 m apart (center-to-center distance). The footing for the wall column cannot extend more than 1 m beyond the centerline of the column. The foundation will be installed near the soil surface. Preliminary foundation analysis indicates that a foundation bearing pressure of 175 kPa can be used for the footing design without exceeding the tolerable settlement or the soil bearing capacity. Assume that the column loads include the weight of the footing.

13-22. Two building columns (a wall column and an interior column) are to be supported on a common rectangular (combined) footing. The wall column imposes a loading of 120 tons; the interior column carries 200 tons. The columns are 16 ft apart. The end of the footing near the wall cannot project more than 3 ft beyond the centerline of the wall column. Using an allowable soil bearing pressure of 4000 psf, give the length and width of foundation that should be provided.

13-23. Two adjacently positioned building columns for an industrial building (an exterior wall column and an interior column) will be supported on a common trapezoid-shaped (combined) footing foundation. The columns are located 3 m apart, center to center. Because of space constraints, the total length of the foundation is limited to 5 m. The end section of the wall footing can project 0.75 m outward beyond the center of the wall column. The wall column imposes a load of 600 kN; the interior column, 950 kN. The allowable design soil bearing pressure is 165 kN/m². Determine the plan dimensions of the necessary trapezoidal shape, using the maximum allowable length of 5 m. (*Hint:* Theoretically, to achieve a uniform soil bearing pressure, the position for the resultant of the column loads and the centroid of the foundation coincide.)

13-24. A long footing, 1 m wide, has its base inclined 10° from the horizontal. The footing bears on the surface of a soil for which drained shear tests show a cohesion of 24 kPa and an angle of internal friction of 25°. The soil rigidity index value is 50. What load acting normal to the foundation base can be imposed if a factor of safety of 3 is used with the extended general bearing capacity equation?

13-25. A strip footing foundation will be constructed in a sloped area where an inclination of 40° exists. This hillside is a clay soil whose soil unit weight is 17 kN/m³ and whose soil cohesion shear strength c is 65 kPa. The width of the footing will be less than the height of the slope, and therefore the slope stability factor is zero when using the bearing capacity factor curves. Assume that the D/B ratio is close to zero. What width footing should be provided to support a wall loading of 155 kN per meter of length? Use a factor of safety of 3.

13-26. A long footing is to be constructed in the slope of a clay hillside that makes an inclination of 30° with the horizontal. The cohesion of the clay is

1500 psf, and the unit weight is 105 pcf. If the slope stability factor is zero and the D/B ratio is close to zero, what width footing is required to support a wall loading of 8000 lb per foot of length? Use a factor of safety of 3.

13-27. A strip footing foundation 1 m wide is constructed in the slope of a hill whose surface inclination is 20°. The soil in the hill is a dense sand, with a unit weight of 19 kN/m³ and an angle of internal friction of 40°. Assume that the D/B ratio is about 1. Applying a factor of safety of 3, determine the wall loading the foundation can carry, per meter of footing length.

13-28. A long footing, 3 ft wide, is located in the slope of a hill whose inclination is 30°. The soil of the hill is sand that has an angle of internal friction of 35° and a unit weight of 115 pcf. If a D/B ratio of 1 is assumed, what wall loading can the footing carry? Use a factor of safety of 3.

13-29. A long footing 2 m wide is located in the level area at the top of a long, 45° slope where clay soils exist. The slope height is greater than the footing width. The soil unit weight is 17.5 kN/m³, and the cohesion shear strength is 65 kPa. The D/B ratio is 1, and the b/B ratio also is 1. What loading can be imposed onto the footing per meter of length, (using a factor of safety of 3 with the bearing capacity equation appropriate for footings along the top of slopes)?

13-30. A 4-ft-wide long footing is located at the top of a slope whose inclination is 30°. The soil of the hill is sand with an assumed angle of internal friction of 30°. If the D/B ratio is 1 and the b/B ratio is also 1, what loading can be imposed on the footing per foot of length? Use a factor of safety of 3.

13-31. Soil borings indicate that a building site is underlain by clean, fine to coarse sand. The following standard penetration test information is obtained from a boring log: At 2-ft depth, N is 12; at 6-ft depth, N is 14; at 10-ft depth, N is 18; at 14-ft depth, N is 20; at 20-ft depth, N is 25; at 25-ft depth, N is 26. These N values refer to N_{60}. No groundwater was encountered. It is planned to construct a square footing 4 ft deep to carry a 300-kip column load near to this boring location. If a 1-in. settlement is tolerable, what size foundation should be provided? What size footing should be provided if the groundwater table could rise to footing level?

13-32. Static cone penetrometer resistance data obtained from a building site underlain by sand show an average q_c value of 100 kg/cm². From this information, estimate the size of square footing necessary to support a column load of 250 kips. Assume that the footing will be placed approximately 4 ft deep and that a 1-in. settlement is tolerable.

13-33. Static cone penetrometer resistance data indicate that a building site is underlain by an overconsolidated clay. The average q_c value in the upper 20-ft soil zone is 75 kg/cm². Using this information, estimate the cohesion of the soil and a safe bearing capacity for shallow square footings.

13-34. A long strip footing foundation 1 m wide is to be designed for a region where earthquakes could occur. The soil underlying the area has a unit weight equal to 17 kN/m³, an angle of internal friction equal to 20°, and a cohesion value c equal to 25 kPa. Assume that the footing will be installed at a depth of 1 m below the adjacent soil surface, that the soil rigidity index is greater than the critical index, and that the depth factors do not apply. For the condition where k_h equals 0.20 and k_v is half of k_h, compare the value for static ultimate bearing capacity (Eq. 13-1a) with the value for seismic bearing capacity (Eq. 13-23 and Fig. 13-20). For this problem, do not use a factor of safety when making the comparison.

13-35. A strip footing foundation supports a bearing wall for an industrial building. The foundation bears on a cohesionless soil where the unit weight is 18 kN/m³ and the angle of internal friction is 30°. The footing depth-to-width ratio is 1. Estimate the foundation settlement (in mm) expected if the original design for static conditions included a factor of safety of 3 but an earthquake event results in a ground motion peak velocity of 0.23 m/sec and the ground acceleration coefficient A is 0.3.

13-36. Foundation walls for the basements of different commercial buildings have developed cracks. For the conditions described, indicate the probable cause of the wall crack condition.

(a) Walls are concrete block construction; cracks exist along several horizontal mortar joints; wall deflects (bulges) inward near midheight.

(b) Walls are concrete block construction; diagonal cracking, or step-cracking, exists where two walls meet and form a corner.

(c) Walls are poured concrete construction; diagonal cracks exist, all starting at the corners of various window openings.

13-37. A steel pipe pile having a diameter of 0.35 m is driven 15 m into a loose sand with a unit weight of 16.5 kN/m³ and an angle of internal friction of 33°. Compute the design axial downward loading, using the statical analysis with a factor of safety of 2.

13-38. To support a heavy column loading, a nine-pile group consisting of steel pipe piles 0.35 m in diameter is installed in a loose sand where the unit weight is 16.5 kN/m³ and the angle of internal friction is 33°. The piles are embedded 15 m and the pile spacing is 3 diameters. Determine the group capacity for the nine piles, assuming that the sand between the piles is densified by the driving operation to the extent that the piles and the interlocked soil act as a unit. Use the statical analysis for a group, with a factor of safety of 2. Then, compare this group capacity with the capacity based upon a single pile load (refer to Problem 13-37) times the number of piles.

13-39. A 14-in.-diameter pipe piles is driven 50 ft into a dry loose sand whose unit weight is 105 pcf. The angle of internal friction is 33°. Compute the safe axial downward loading, using the statical analysis. Apply a factor of safety of 2.

13-40. The 14-in. diameter pipe piles installed as described in the preceding problem will be used for a four-pile group where the pile spacing is 3 diameters. Determine the group capacity for the four piles, assuming that the soil between the piles is densified by the driving operation to the extent that the piles and interlocked soil act as a unit. Then, compare this group capacity with the capacity based upon single pile capacity (85 kips) times the number of piles. For both values, apply a factor of safety of 2.

13-41. Determine the allowable downward axial capacity for the pile and conditions described. A prestressed concrete pile with a square cross section 0.4 m by 0.4 m will be installed in sandy soil for a waterfront building. The piles will be embedded 12 m, being jetted into place for the first 9 m but driven the last 3 m. Assume the condition of a water table at the soil surface. The saturated unit weight for the soil is 19 kN/m³, and the angle of internal friction is 35°. Based on the conditions described, assume that the critical depth for computing effective soil stresses is 15 pile diameters. To calculate skin friction, use a lateral pressure coefficient for a pile driven in an oversized hole for the jetted depth and the coefficient for a pile driver in undisturbed sand for the deeper driven depth (refer Fig. 13-28). Calculate tip bearing using the bearing capacity equation then compare to the value obtained from the Coyle–Castello curves for medium-dense sand. Apply a factor of safety of two to the computed capacity.

13-42. A 12-in.-square prestressed concrete pile is to be installed in sandy soil for a marine structure. The pile will be 40 ft long. It will be jetted 30 ft and driven the last 10 ft. The water table is at the ground surface. The saturated soil weight is 120 pcf. The angle of internal friction is 35°. For these conditions, assume that the critical depth for computing effective soil stresses is 20 pile diameters. Calculate the safe (design) axial downward loading, using the statical analysis and a factor of safety of 2.

13-43. Steel pipe piles are to be used as friction piles in an area where a deep deposit of normally consolidated clay exists. The soil unit weight is 17 kN/m^3, with shear strength cohesion equal to 45 kPa. Determine the design axial capacity due to skin friction only for a pile diameter of 0.30 m and an embedded length of 10 m. Apply a factor of safety of 2.

13-44. Timber piles are to be used as friction piles at a site underlain by a deep deposit of normally consolidated clay soil whose cohesion is 1000 psf and whose unit weight is 105 pcf. For a pile embedded 30 ft, having a tip diameter of 8 in. and a 14-in. butt diameter, calculate the design axial capacity, assuming skin friction only. Use a factor of safety of 2.

13-45. A 0.3 m steel H-pile (HP) is driven 15 m into clay soil where the following conditions exist: From the ground surface to a depth of 10 m, the clay is normally consolidated, with a unit weight equal to 16 kN/m^3 and shear strength cohesion equal to 40 kPa; below 10 m, the clay is overconsolidated, with a unit weight equal to 18 kN/m^3 and cohesion equal to 100 kPa. Determine the design axial capacity of this pile, using a factor of safety of 2. Compute the shaft capacity, assuming that skin friction is developed on the surface of the rectangle (0.3 m by 0.3 m) that encloses the pile.

13-46. A 12-in. steel H-pile is driven 60 ft into clay soil as follows: From the ground surface to a depth of 30 ft, the cohesion is 800 psf and the unit weight is 105 pcf; below 30 ft, the cohesion is 2000 psf and the unit weight is 115 pcf. Assume that the upper clay layer is normally consolidated and the deeper clay is overconsolidated. Determine the axial capacity of the pile, using a factor of safety of 2. Compute the shaft capacity assuming that skin friction is developed on the surface of the rectangle (12 in. square) that encloses the pile.

13-47. A square prestressed concrete pile with a cross section 0.3 m by 0.3 m is driven 15 m into overconsolidated clays with properties as defined: From the ground surface to a depth of 7 m, the unit weight is 17 kN/m^3 and the shear strength cohesion is 70 kPa; below 7 m, the soil unit weight is 18.5 kN/m^3 and cohesion is 105 kPa. Calculate both the design downward axial capacity and upward axial capacity, using a factor of safety of 2.

13-48. A 14-in.-diameter steel pile is driven 50 ft into overconsolidated clays having the following properties: From zero to 20 ft depth the cohesion is 1500 psf and the soil unit weight is 110 pcf; below 20 ft, the cohesion is 2500 psf and the soil unit weight is 120 pcf. Calculate the downward and upward axial capacity of the pile, applying a factor of safety of 2.

13-49. A four-pile group consists of prestressed concrete piles having a cross section 0.4 m by 0.4 m. The piles are embedded (driven) 12 m into an overconsolidated clay where the shear strength cohesion is 100 kPa and the soil unit weight is 18 kN/m^3. The pile spacing is $2\frac{1}{2}$ diameters. Find the capacity of the pile group, applying a factor of safety of 2. (*Hint:* When analyzing the shaft capacity of the group, use soil-to-pile friction where the pile and soil are in contact, but soil cohesion as shear strength for the soil embedded between the piles; ref. Ill. 13-13.)

13-50. A four-pile group consists of 12-in.-square prestressed concrete piles 35 ft long. The piles are driven into an overconsolidated clay whose cohesion is 2000 psf and whose soil unit weight is 115 pcf. The pile spacing is $2\frac{1}{2}$ diameters. The water table is at the ground surface. Using a factor of safety of 2, find the capacity of the pile group.

13-51. Use the lambda method to calculate the axial capacity of a long steel pipe pile installed for an offshore project. The pile has a 0.45-m outside diameter and penetrates 60 m into normally consolidated clay. Soil conditions are as follows: From zero to 15 m depth, the soil cohesion is 45 kPa and the saturated unit weight is 16.5 kN/m³; between depths of 15 m and 35 m, the soil cohesion is 65 kPa and the saturated unit weight is 18 kN/m³; below 35 m, the soil cohesion is 110 kPa and the saturated unit weight is 19 kN/m³. The water surface is above the soil surface. Apply a factor of safety of 2.

13-52. For an offshore project, a steel pipe pile with an outside diameter of 18 in. penetrates 150 ft through normally consolidated clays. Soil conditions are as follows: From zero to 50 ft depth, the cohesion is 800 psf and the saturated soil unit weight is 105 pcf; from 50 to 100 ft depth the cohesion is 1500 psf and the saturated unit weight is 115 pcf; below 100 ft, the cohesion is 2400 psf and the saturated unit weight is 125 pcf. The water surface is above the soil surface. Using the lambda method, determine the axial capacity of the pile in kips, using a factor of safety of 2.

13-53. A drilled-shaft foundation having a total length of 12 m is constructed in an area underlain by clay soil having the following properties: From the ground surface to a depth of 7 m, the soil cohesion is 45 kPa and the unit weight is 16.5 kN/m³; below 7 m, the soil cohesion is 110 kPa and the unit weight is 19 kN/m³. This foundation unit will have a belled (under-reamed) bottom; the shaft section has a 1 m diameter, and the base of the bottom section has a 2 m diameter. The bell forms a 60° angle with the horizontal plane. Compute the design downward capacity if
(a) the excavation is drilled dry.
(b) it is necessary to use a bentonite slurry during drilling of the excavation.

13-54. A drilled-shaft foundation is to be constructed in clay soil having the following properties: From zero to 30 ft depth, the cohesion is 1000 psf and the soil unit weight is 105 pcf; below 30 ft, the cohesion is 2500 psf and the unit weight is 115 pcf. The foundation is to extend to a total depth of 36 ft. A 3-ft-diameter shaft will be used with a belled (under-reamed) bottom 6 ft in diameter. Assume

that the bell forms a 60° angle with the horizontal plane. Compute the design downward capacity if
(a) the excavation is drilled dry.
(b) it is necessary to use a bentonite slurry during drilling.

13-55. A 12-in.-square prestressed concrete pile is driven 50 ft. Under the last blow of a Vulcan Model 010 hammer, the pile penetrates 0.25 in. Using the Danish pile-driving formula, what safe design load is indicated? Is this size hammer proper for this type and length pile?

13-56. The Danish pile-driving formula is to be used for field control of a pile installation. Steel pipe piles penetrating approximately 40 ft are to provide a design capacity of 60 tons. Each pipe has a net cross-sectional area of 20 in.² What should be the set under the last blow of an MKT-11B3 hammer to obtain the desired axial capacity?

13-57. The following data are obtained from a slow-ML pile load test. Plot the load and settlement information on logarithmic coordinates and, from the plot, determine the "failure load."

Load (tons)	Settlement (in.)
25	0.05
50	0.15
75	0.26
100	0.40
125	0.60
150	0.80
175	0.95
200	1.55
225	2.45

13-58. Outline the practical reasons associated with having foundation installations (the construction of foundations) supervised by personnel familiar with the factors used to develop the foundation design. (*Hint:* Consider the relationship between soil properties, transfer of load from foundation into the supporting soil, etc.)

CHAPTER 14

Earth Moving, Compaction, and Stabilization

In the natural location and condition, soil provides the foundation support for many of humankind's structures. But soil is also extensively used as a basic material of construction, as witnessed by the existence of earth structures such as dams, dikes, and embankments for roads and airfields. For situations where the natural topography needs to be changed to make the area more suitable for building development, soil is the material most used for filling low locations. The desirability of utilizing soil as a building material stems from its general availability, its durability, and its comparatively low cost.

When soil is used for construction purposes, it is typical for it to be placed in layers to develop a final elevation and shape. Each layer is compacted before being covered with a subsequent layer. Properly placed and compacted, the resulting soil mass has strength and support capabilities that are as good as or better than many natural soil formations. In the case of earth structures such as dams, the compacted earth is capable of supporting itself and the forces to which it is subjected. With earth fills, it is possible to support buildings, highways, and parking areas on the compacted soil mass. Such soil is referred to as a *compacted earth fill* or a *structural earth fill*.

Whether soil is used as the foundation material to support buildings, roads, or other structures or is used to build the structure itself (e.g., an earth dam), it is desirable that the in-place material possess certain properties. The soil should have adequate strength, be relatively incompressible so that future settlement is not significant, be stable against volume change as water content or other factors vary, be durable and safe against deterioration, and possess proper permeability. These desirable features can be achieved with a compacted fill by proper selection of the fill soil type and by proper placement. The important properties of a fill could be checked independently, but the more desirable characteristics, such as high strength, low compressibility, and stability, are normally associated with high-density (or unit weight) values and hence will result from good compaction.

Virtually any type of soil can be used for structural fill, provided that it does not contain organic or foreign material that would decompose or otherwise undergo change after it is in place. Granular soils are considered the easiest to work with on a construction site.

This material is capable of developing high strength, with little volume change expected after compaction. Permeabilities are high, which can be an advantage or a disadvantage. Generally, compacted silts are stable, are capable of developing fairly good strength, and have limited tendency for volume change. Silty soils can be difficult to compact if wet or if work is performed in wet periods. Permeability is low. Properly compacted clay soils will develop relatively high strengths. Their stability against shrinkage and expansion is related to the type of clay mineral; for example, the montmorillonite clays would have a greater tendency for volume change than the more stable kaolinite clays. Compacted clays have a very low permeability, a factor that can be used to advantage where movement of water needs to be restricted. Clay soils cannot be properly compacted when wet. Working with clay soil is difficult under wet conditions.

14.1 FIELD PROCEDURES—GENERAL CONSIDERATIONS, METHODS, AND TECHNIQUES

The field procedure for constructing a compacted fill is simple in principle. The fill soil is transported to the area being worked, where it is spread in relatively thin layers, and each layer is then compacted to a predetermined density (unit weight). However, proper accomplishment of these requirements involves consideration of the soil type, its water content, and the type of equipment used for the compacting operation.

On big projects, where the soil fill is obtained from an earth cut or borrow pit on or near the construction site, self-propelled scrapers (Fig. 14-1) can be considered for hauling and spreading the fill. Self-propelled scrapers are widely used because of their excavating

Figure 14-1 Self-propelled scraper hauling from a soil borrow area. (Courtesy of Caterpillar Tractor Company)

and self-loading capabilities and the relatively fast haul speeds (Figs. 14-2 and 14-3). With scraper operations, bulldozers or graders frequently work in the fill area to help keep newly placed uncompacted fill uniformly spread (Figs. 14-4 and 14-5). Compaction equipment follows the spreading equipment.

If the fill material is obtained from an off-site source, trucks are generally used for transportation (Fig. 14-6). At the fill location, the soil is dumped and then spread with dozers and graders. Compaction follows the spreading operation.

Trucks are also commonly used for transporting the soil on cut-and-fill projects, particularly when the fill material is obtained from a cut or borrow pit area where excavation can be efficiently handled by power shovel equipment or when the hauling distance is great (Figs. 14-7, 14-8, and 14-9).

When the site being worked is of limited area and fill is obtained from an on-site cut, bulldozers may be used to perform excavation and to push the soil to the fill area, where it is spread and compacted.

The thickness of layer that can be properly compacted is known to relate to the soil type and the method or equipment of compaction. Typically, granular soils can be ade-

Figure 14-2   Scrapers self-loading in borrow pit. (Courtesy of Terex Division, General Motors Corporation)

Figure 14-3   Scraper being loaded in borrow area with assistance of pushdozer, a procedure to reduce the time required for loading. (Courtesy of Terex Division, General Motors Corporation)

Figure 14-4 Bulldozer utilized to spread soil fill. (Courtesy of Terex Division, General Motors Corporation)

Figure 14-5 Grader of type used on fill projects to spread soil and construct drainage ditches. Photo shows grader working on a roadside ditch. (Courtesy of Galion Division of Dresser Industries, Inc.)

quately compacted in thicker layers than the fine-grained silt and clay soils. Generally, for a given soil type, heavy compaction equipment is capable of compacting thicker layers than light equipment.

The types of equipment commonly used for compaction include pneumatic or rubber tire rollers; drum-type rollers that have projecting feet or lugs, such as the sheep's-foot roller; vibratory compactors, which impart vibrations into the soil; and smooth drum rollers. All types are available in a variety of sizes and weights. The vibratory compactors are most effective on cohesionless soils. The sheep's-foot and similar rollers that have feet or other

Figure 14-6   Front-end loader used in borrow area. Truck type shown is typically used for on-site or highway hauling. (Courtesy of Fiat-Allis Chalmers Construction Machinery, Fiat-Allis Chalmers)

Figure 14-7   Rear dump truck of type used for on-site hauling. (Courtesy of Caterpillar Tractor Company)

projections should be limited to cohesive soils. Pneumatic tire rollers can be used effectively on all soil types. The smooth drum rollers have a very limited effective depth of compaction; hence, their use should be restricted to situations where only thin layers or a surface zone needs to be compacted.

Hand-operated tamping and vibratory compactors are available for working in limited spaces and for compacting soil close to structures, where care is required to prevent damage.

The water content of the fill soil has an effect on its ability to be well compacted with reasonable effort. The fine-grained silt and clay soils, and granular soils containing fine

Figure 14-8 Bottom dump truck, typically utilized for on-site hauling. (Courtesy of Caterpillar Tractor Company)

Figure 14-9 Power shovel loading operation. (Courtesy of Terex Division, General Motors Corporation)

materials, are particularly affected by variations in water content. Soil that is too dry is difficult to compact. Material that is too wet may be difficult to spread properly and to compact. When the soil is too wet at its source, measures should be taken to dry it before compaction is attempted at the fill area. Drying can be accomplished by scarifying and aerating the soil at the borrow pit or cut area, by hauling it to an open area and loosely spreading it to permit aeration, or by letting it aerate after spreading across the fill area. These procedures presume that space and time are available. When the fill soil is too dry, water should be added and mixed throughout. Wetting can be achieved by spraying water at either the cut or fill area with a water truck, followed by a mixing operation performed with the excavating or earth-moving equipment. If the dry soil is spread across the fill area before wetting, dozers and graders can frequently perform some mixing after spraying by "rolling" the earth on the front of their blades as they work to even out the layer thickness. Where the fill source is a large borrow area, ponds can be created and flooded. This procedure permits the water to percolate through the soil before it is excavated and transported. Fine-grained soils will require a longer ponding period than granular soils.

When an area is being filled, the ability of the first fill layers to be properly compacted will depend on the condition of the natural material being covered. If weak or compressible soil exists, it may be very difficult to compact the fill properly. If poor material is left in place and covered over, it may compress over a long period under the weight of the earth fill, causing settlement cracks in the fill or in any structure supported by the fill. Consequently, where the postconstruction settlement and deformation of an earth fill has to be limited, it is common practice to remove poor soil and nonsoil materials (vegetation, garbage, other waste material) before filling commences.

To determine whether the surface zone of the natural soil is adequate for supporting the compacted fill, the area can be "proofrolled." Proofrolling consists of utilizing a piece of heavy construction equipment (typically, heavy compaction equipment or hauling equipment) to roll across the fill site and watching for poor areas to be revealed. Prior areas will be indicated by the development of rutting or ground weaving. Where the height of a structural fill will be limited, even marginally poor soils should be removed. If the height of fill is to be great, marginal materials often can be left in place without future adverse effect.

Placement of a compacted fill may have to begin at an elevation that is below the water table. This requirement frequently develops where the excavation of poor material lowers the beginning working surface. Fill placed under water should consist of coarse, granular material. Fine-grained soil should not be used. The thickness of the initial fill layer may have to be considerably greater than the normal fill layer (several feet thick) in order to create a working pad capable of supporting equipment used for the subsequent placement and compaction of fill. A procedure that has been followed with success in situations where the disturbance to the in-place underwater soil must be minimized is to create a working base of very coarse material, such as cobble sizes. Clean sand and gravel materials are placed over this base. Close to and above the water surface, the granular material can be compacted with conventional equipment. Once above the water table, the placement and compaction of fill layers can proceed in the normal manner; such upper fill can consist of fine-grained soil as well as granular soil.

The working surface of earth fills made with fine-grained soils can deteriorate rapidly if exposed to accumulations of water from rain or other causes. These fills should be constructed so that rapid surface drainage can occur. This need can be satisfied by placing fill so as to achieve a cambered or turtle-backed surface. When the fill area is lower in elevation than the surrounding terrain, as is frequently the situation during early stages of construction, interceptor drainage ditches should be provided around the perimeter of the fill site. Accumulating surface water is normally not a problem in fills constructed of granular soils, because the material's high percolation rate permits rapid dispersing. The major detrimental effect of surface water on granular soil fills is the erosion possibility on slope areas.

Compaction of backfill in narrow trenches and against buildings normally cannot be accomplished by using the compaction equipment utilized for areal fills. Small or hand-operated equipment is required, and progress may be comparatively slow. Soil densification in such confined spaces may be attempted by using flowing water, such as from a hose. This method can be successful where the backfill consists of clean granular material and the material surrounding and underlying the fill is also a coarse soil, since the flowing water percolates rapidly through the soil and does not puddle. This rapid flow causes a re-arrangement of soil particles, and densification results. The method should *not* be attempted where the backfill is a fine-grained soil or where the surrounding soils do not possess a high permeability. With such conditions, the water will not flow rapidly through the fill and will not densify the soil. The water may puddle in the excavation, causing a loose soil structure to remain after drying occurs. For fine-grained backfills, compaction should be achieved through the use of compaction equipment and by working with thin soil layers.

Alternative Methods of Transporting Earth Fill

Most transporting of soil for earthworks projects is achieved through the use of construction equipment and methods discussed in preceding paragraphs. However, where large quantities of material are to be moved, alternate methods of transportation should also be considered because of possible savings in cost or time. Methods that fall into this category are belt-conveyor transportation and hydraulic transportation.

Belt-conveyor systems offer the possibility for economic advantage where large volumes of material are moved for distances of up to several miles. A belt-conveyor system is usually constructed to closely follow the terrain it crosses and typically consists of a series of completely individual belt flights. Normally, each flight has its own drive motor. The transported material passes from one flight to another in order to complete the trip.

The belts of a conveyor system are available in different widths; stock belts run to 5 ft (about 1.5 m) wide. Conveyor speeds can vary, and rates of several hundred feet per minute are practical. When the wider belts are used, it is easily possible to transport several hundred cubic yards (or cubic meters) of material per hour.

On a compacted fill project, earth received at the deposit end of the conveyor is usually moved and spread by using conventional earth fill equipment and procedures.

A major disadvantage of a belt-conveyor system is the expense and time to construct it. Construction costs include the necessity for a maintenance road for servicing during operation as well as the expense of erection and dismantling. Advantages include more independence from the effects of weather, easy round-the-clock operation, and reduced labor operating costs.

Hydraulic transportation refers to the method of mixing the borrow soil with large quantities of water and pumping the resulting slurry through a pipeline to a desired location. The method is highly dependent on an adequate source of water being available. Waterfront and dam-reservoir construction typify those projects that have the advantage of an available and economical source of water. For practical ease of handling at both the supply and the deposit ends, cohesionless soil or soil possessing only slight cohesion is the most desirable, although all soil types can be transported hydraulically.

Where a sand supply is located below water (a popular situation for hydraulic transportation), suction dredging methods are used. The sand is sucked into a flexible pipe extending from a pump into the soil supply and then is passed into the transporting pipe. Underwater cohesive soils are cut into chunks by powered rotating cutters before entering the flexible pipe. The slurry passing through the pipe is usually 80 to 90 percent water and 10 to 20 percent solids.

Above the water table, the borrow soil can be excavated and washed into a ponding area by utilizing high-pressure streams of water, whence it is then piped to the desired location.

Landfills can be created with hydraulically transported soil. Such fill is referred to as *hydraulic fill*. At a free-flowing outlet, the coarse soil particles drop out close to the point of discharge, and the smaller particles are carried a farther distance by the fanning-out water. This is similar to the natural formation of a soil delta where a river enters a lake. If mixed soils are transported, periodic movement of the discharge end of the pipe to planned locations permits fill areas of the desirable soil type (coarse soil areas and fine soil areas) to be obtained.

The natural side slopes created by a flowing coarse soil slurry are usually between five and ten horizontal to one vertical. Hydraulically transported soft clay soils may develop slopes as shallow as fifty to one. If it is necessary to contain the area receiving fill, dikes are constructed or sheetpiling is installed to form a ponding area. Soil particles will settle out of the slurry in this area. The water accumulating in the surface zone of the pond is drawn off. Sizable land areas have been reclaimed or created and relatively large earth dams have been constructed by using hydraulic fill methods.

Sand deposited by hydraulic methods will be in a relatively loose condition. Usually it is necessary to densify the soil if it is intended to provide structural support.Normal compaction procedures can be followed, using rolling equipment (preferably vibratory) to compact the fill in layers. Silt and clay hydraulic fills are difficult to compact after placement. These soil types would not be a choice for a structural fill but are used on land reclamation projects if economical sources of coarse material are not available. Where time permits, such fill is left to consolidate and stabilize naturally. Major structures are supported by piles or other special foundations. If necessary, the more involved and costly methods of densification, such as by pile driving or use of sand or wick drains, could be performed, but results are somewhat unreliable.

14.2   FIELD EQUIPMENT—SURFACE ZONE COMPACTION

The construction of a structural fill usually consists of two distinct operations—the placing and spreading in layers and then the compaction process. The speed of the compaction operation is typically the more critical of the two steps and often controls the rate of a job's progress. The use of adequate and proper compaction equipment becomes a matter of economic necessity for contractors on almost all earth fill projects. Because of this need, various types of special compaction equipment have been developed for the construction industry. Some equipment has been designed to be specially effective for a particular soil type, whereas other equipment is for general or all-purpose use.

Soil compaction or densification can be achieved by different means—by tamping action, by kneading action, by vibrating, or by impact. Compactors operating on the tamping, kneading, and impact principles are effective on cohesive soils. For cohesionless soils, equipment operating on the tamping, kneading, and vibratory principles are effective.

Surface Compaction Equipment

The *sheep's-foot roller* and similar rollers that have projecting studs or feet are examples of equipment that compacts by a combination of tamping and kneading. Typically, these compactors consist of a steel drum manufactured with small projections (Figs. 14-10 and 14-11). With most rollers, the drum can be filled with water or sand to increase the weight. As rolling occurs, most of the roller weight is imposed through the projecting feet. Contact

Figure 14-10 Sheep's-foot rollers in use on a cohesive soil fill, being pushed-pulled by crawler tractors. (Courtesy of Fiat-Allis Chalmers Construction Machinery, Fiat-Allis Chalmers)

Figure 14-11 Close-up of projecting feet on a sheep's-foot roller.

pressures imposed by the projections can be fairly high. Pressures vary from about 100 psi or 700 kN/m² for the lighter equipment to over 600 psi or 4200 kN/m² for the heavier equipment in common use.

When a loose soil layer is initially rolled, the projections sink into the layer and compact the soil near the lowest portion of the layer. In subsequent passes with the roller, the zone being compacted continues to rise until the surface is reached. This continually rising effect experienced by the compactor is referred to as "walking-out." Such equipment is well suited for compacting clay and silt–clay soil.[1] The depth of layer that can be well compacted relates to

---

[1]Projection-type rollers are not recommended for cohesionless soil, because the studs (feet) continuously loosen the fill surface.

the length of the projecting feet and the compactor weight. It is conventionally assumed that the larger, heavy units will properly compact layers on the order of 0.3 m (12 in.) thick in three to five passes. Small, light equipment is limited to working layers less than 150 mm (6 in.) thick where high soil densities are required.

Sheep's-foot-type compactors are available as both self-propelled units and as rollers only. A separate roller requires a tow tractor.

Pneumatic tire rollers compact primarily by kneading. This type of compaction equipment is available in a variety of designs, ranging from the conventional two-wheel-per-axle units to the multiwheel-per-axle unit (Fig. 14-12). Some equipment is provided with a "wobble-wheel" effect, a design in which a slightly weaving path is tracked by the traveling wheel. Another type is provided with an axle construction that permits individual wheels to follow the ground surface so that low spots are not skipped over.

Pneumatic tire compactors are usually outfitted with a weight box or ballast box so that the total compaction load can be easily varied. Ground contact pressure can also be controlled somewhat by varying air pressure in the tires.

Pneumatic tire rollers are available as self-propelled units (Fig. 14-12) and as towed units (Fig. 14-13). These compactors are available in a wide range of load sizes, the heaviest having a capacity of about 200 tons or 1800 kN. However, the heaviest units in common use are in the 50-ton or 450 kN range.

The pneumatic tire rollers are effective for compacting both cohesive and cohesionless soils and are the best type of equipment for general compaction use. Light rollers (about 20 tons, or 20,000 kg) are generally considered to be capable of properly compacting layers on the order of 150 mm (6 in.) thick, with few passes. Equipment in the 40- to 50-ton, or 40,000 to 50,000 kg, category will usually compact layers on the order of 0.3 m (12 in.) thick, with three to five passes. The very heavy equipment is presumed to compact layers up to about 0.5 m or $1\frac{1}{2}$ feet thick, with a limited number of passes.

Figure 14-12   Pneumatic tire roller being used to compact sand base for a highway. (Courtesy of Galion Division of Dresser Industries, Inc.)

Figure 14-13   Fifty-ton pneumatic tire compactor showing soil-filled weight box.

The effectiveness of pneumatic tire compaction is not limited to the specially made compaction equipment. Other heavy-tire equipment, such as trucks, graders, and scrapers, is capable of providing an effective job of compaction (Fig. 14-14). In emergencies, such equipment can be pressed into service.

Vibratory compactors are available as vibrating drum, vibrating pneumatic tire, and vibrating plate equipment. With the vibrating drum equipment, a separate motor drives an arrangement of eccentric weights so that a high-frequency, low-amplitude, up-and-down oscillation of the drum occurs. Smooth drums and sheep's-foot-type drums are available. On the pneumatic tire compactor, the separate vibrating unit is attached to the wheel axle. The ballast box is suspended separately from the axle so that it does not vibrate. The vibrating plate equipment typically consists of a number of small plates, each of which is operated by a separate vibrating unit. The drum and the pneumatic-type equipment are available as either self-propelled or towed models.

Figure 14-14   Scrapers being utilized to haul and spread fill for a dam project. Note compaction effect achieved by rubber tire in photo foreground. (Courtesy of Fiat-Allis Construction Machinery, Fiat-Allis Chalmers)

On some vibratory compactors, the vibrating frequency can be varied by the equipment operator. Frequencies usually range between 1500 and 2500 cycles per minute, values that are within the natural frequency of most soils. The *natural frequency* is that value at which the soil particles tend to oscillate in unison, giving maximum effect to the repeated impact imposed by the compactor, instead of having a random oscillation of particles occur, which would create a damping effect. Particles are thus "shaken" into a more dense arrangement. Usually, for granular soils, the most effective results are achieved when the compactor travels at a slow speed, on the order of 3 to 6 km/hr (2 to 4 mph).

Smooth-drum vibrators (Fig. 14-15) have proven very effective in compacting granular soils having little or no silt- and clay-sized material. Layers on the order of 1 m (3 ft) deep have been compacted to high densities (densities close to the maximum modified Proctor value). As the percentage of fine material increases, the thickness of layer that can be well compacted is reduced.

The *vibratory pneumatic tire equipment* also has been successful in compacting primarily granular soil. It is commonly presumed that at least 0.3-m- (1-ft-) thick layers of predominantly granular soil will be satisfactorily compacted by most vibratory smooth drum and pneumatic tire compactors after a few passes.

The effectiveness of vibratory equipment on cohesive soil (Fig. 14-16) is not nearly so pronounced as the results achieved on granular soils.

The *vibrating plate compactors* generally have limited depth of effectiveness. Their use has been primarily in compacting granular base courses for highway and airfield pavements.

Conventional *smooth-drum rollers* (Fig. 14-17) are not well suited for compacting earth fill. Because of the size of the drum and the large soil contact area, the resulting com-

Figure 14-15   Vibratory smooth-drum compactor, typically used for cohesionless soil fills. (Courtesy of Ingersoll-Rand Corp.)

Figure 14-16 Vibratory compactor with pads on drum, used where soil has cohesion. (Courtesy of Ingersoll-Rand Corp.)

Figure 14-17 Smooth-drum roller compacting a base course for a roadway. (Courtesy of Galion Division of Dresser Industries, Inc.)

paction pressures are relatively low. Smooth-drum rollers can be utilized for compacting limited thicknesses of material, such as granular base course on highway and airfield work. On earth-moving projects, the smooth-drum roller can be used advantageously to "seal" the surface of the fill at the end of each work day. Sealing provides a smooth surface so that rain water will quickly run off from the work area; the water is not given the opportunity to percolate into the upper fill, where it might subsequently create a soft working surface.

Table 14-1 provides a generalized summary that relates soil fill type and compaction characteristics with equipment types considered suitable for achieving compaction.

Table 14-1 Soil Compaction Characteristics and Recommended Compaction Equipment

General Soil Description	Unified Soil Classification	Compaction Characteristics	Recommended Compaction Equipment
Sand and sand–gravel mixtures (no silt or clay)	SW, SP, GW, GP	Good	Vibratory drum roller, vibratory rubber tire or pneumatic tire equipment
Sand or sand–gravel with silt	SM, GM	Good	Vibratory drum roller, vibratory rubber tire or pneumatic tire equipment
Sand or sand–gravel with clay	SC, GC	Good to fair	Pneumatic tire, vibratory rubber tire, vibratory sheep's-foot equipment
Silt	ML	Good to poor	Pneumatic tire, vibratory rubber tire, vibratory sheep's-foot equipment
	MH	Fair to poor	Pneumatic tire, vibratory rubber tire, vibratory sheep's-foot, sheep's-foot-type equipment
Clay	CL	Good to fair	Pneumatic tire, sheep's-foot, vibratory rubber tire, vibratory sheep's-foot equipment
	CH	Fair to poor	
Organic soil	OL, OH, PT	Not recommended for structural earth fill	

Figure 14-18 Hand-maneuvered vibratory compactor for working close to structures. (Courtesy of Dynapac, Stanhope, New Jersey)

Most self-propelled and towed equipment is large and cannot maneuver adequately to properly compact soil in confined areas and against structures. When structures are adjacent to a fill area, there should be a concern that heavy equipment will exert forces that could damage the structure. Consequently, small, portable compaction equipment should be utilized for working in areas of limited space and locations close to structures. Small vibratory drum and vibrating plate equipment is available (Fig. 14-18). Pneumatic tampers and piston-type tampers are also available. It should be realized that with this small equipment, the thickness of the layer that can be compacted to a high density is frequently less than 150 mm (6 in.).

14.3   DEEP-GROUND IMPROVEMENT TECHNIQUES

The surface compaction equipment and procedures discussed in the preceding sections have the capability to improve in-place soils only to shallow depths. Typically, depths below a few feet, or a meter or two, undergo little or no increase in density. However, techniques have been developed that utilize other special equipment and procedures to accomplish in-place improvement of thick zones of soil materials. Certain of these techniques are capable of changing soil properties to considerable depths. These deep soil improvement techniques represent a relatively new specialty area in the geotechnical construction industry. Included within the growing family of deep ground treatment systems that improve soil materials in place are the vibrocompaction procedures, the vibroreplacement procedure, the dynamic deep compaction procedure, an accelerated consolidation procedure through application of wick-drain installations, and stabilization via several methods of in-place grouting (slurry grouting, compaction grouting, chemical grouting, jet grouting). Where deep or thick zones of problem soils or fill materials exist, the deep-ground treatment techniques may offer practical, economical alternatives to the selection of deep foundations or soil-fill replacement for making the site suitable for use. For the purpose of determining or ensuring the extent of a thick or deep zone's improvement, the treated area can be monitored; commonly, a program of in situ testing for the site prior to and subsequent to treatment is undertaken via use of borings or probings (e.g., using the standard penetration test or cone penetration testing).

The vibrocompaction and vibroreplacement procedures, the dynamic compaction procedure, and soil stabilization are described in the following sections; wick drains are discussed in Chapter 10.

Vibrocompaction

The vibrocompaction methods are suited for compacting thick deposits of loose sandy soil. The vibrocompaction equipment operates from a site's ground surface, but it can densify the full depth of granular deposits that are as deep as about 25 m (75 ft). The procedures followed by the specialty contractors vary, but the vibroflotation and Terra-Probe methods described below are indicative of the general methodology.

The vibroflotation method, illustrated in Figures 14-19 and 14-20, first compacts deep zones of soil and then works its way toward the surface. A cylindrical vibrator, weighing about 18 kN (2 tons) and approximately 2 m (6 ft) long and 40 cm (16 in.) in diameter, called the Vibroflot®, is suspended from a crane and jetted to the depth where compaction is to begin. The jetting consists of a pressured stream of water directed from the tip of the Vibroflot into the earth. As the sand is displaced, the Vibroflot simultaneously sinks into the soil. Depths of about 12 m (40 ft) can be reached. At a desired depth, the vibrator is activated. The Vibroflot vibrates laterally, an action that tends to cause the soil to compact in the *horizontal* direction. Typically, the material up to 1.5 m (4 to 5 ft) outward from the Vibroflot is densified. Vibration continues as the Vibroflot is slowly raised to the surface. As the vibrating process occurs, additional sand is continually dropped into the space around the Vibroflot to fill the created void. To improve an entire site, treatment locations at approximately 3-m (10-ft) spacings are usually necessary.

The *Terra-Probe®* (Fig. 14-21) essentially consists of a vibrodriver (similar to the unit that is used for vibratory pile driving) coupled to an open-ended steel tubular probe (such as a 76-cm- or 30-in.-diameter pipe section). Activated, the vibrodriver causes the probe to vibrate in the *vertical* direction. To achieve soil compaction, the probe is vibrated to the planned depth of penetration; then, with the vibrodriver continuing to operate, the probe is slowly drawn upward. The procedure densifies the soil within and closely surrounding the probe. Densification across a site is achieved by driving and extracting the probe at closely spaced intervals (e.g., spacings on the order of 1.5 m or 5 ft are frequently selected). The Terra-Probe operation works best if the groundwater table is within 2 to 3 m (5 to 10 ft) of the ground surface. For sites where the water table is deep, water jets can be incorporated onto the probe to assist the penetration and densification procedure, or the work area surface can be ponded or flooded to provide the desired water.

Some general comparisons of the vibroflotation and Terra-Probe methods are available. The lateral soil zone improved by each vibroflotation penetration is considerably

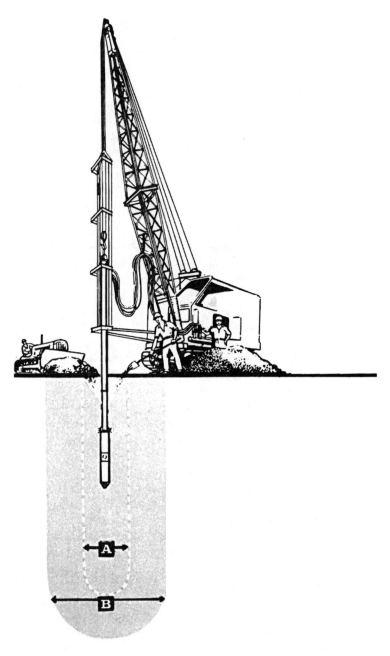

Figure 14-19 Vibroflotation equipment and process. (Courtesy of Vibroflotation Foundation Company, Pittsburgh, Pennsylvania)

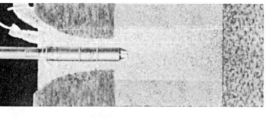

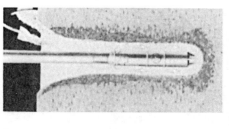

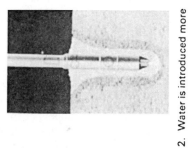

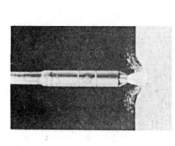

1. At start, lower jet is opened fully.

2. Water is introduced more rapidly than it can drain away. This creates a momentary "quick" condition ahead of the equipment, which permits the vibrating machine to settle to the desired depth.

3. The water from the lower jet is transferred to the top jets and the pressure and volume are reduced just enough to carry the sand to the bottom of the hole.

4. Actual compaction takes place during the intervals between the one-foot lifts which are made in returning the Vibroflot to the surface. The vibrator is first allowed to operate at the bottom of the crater until the desired density around the lower part of the machine is attained. By raising the vibrator step by step and simultaneously backfilling, the entire depth of soil is compacted.

Figure 14-20 Details of vibroflotation: at the location to be compacted the Vibroflot is jetted into the soil. The compaction sequence has four basic steps, as shown. (Courtesy of Vibroflotation Foundation Company, Pittsburgh, Pennsylvania)

(a)   (b)

Figure 14-21 (a) Terra-Probe® equipment at a job site (note water-ponded surface to assist compaction); (b) close-up of vibrodriver and tubular probe. (Photos courtesy of L. B. Foster Company)

larger than that with the Terra-Probe. To improve a given area uniformly, the Terra-Probe requires a more closely spaced penetration pattern than is necessary with vibroflotation. However, the Terra-Probe procedure is considerably faster than the vibroflotation procedure, and the greater number of probes that can be completed in a given time period compensates for the required closer spacing and greater number of probes. Though improvement in density is good with both procedures, the vibroflotation method apparently will achieve somewhat higher relative densities. Overall, in terms of time required to improve most sites approximately equally, the vibroflotation and Terra-Probe methods are comparable. The Terra-Probe method does not require a sand fill and it can be used at offshore locations, important factors for some sites.

Vibroreplacement

The vibroreplacement process uses the equipment and general techniques of the vibrocompaction methods, but crushed stone (or coarse gravel) material rather than sand is added and vibrated into the soil (Fig. 14-22). The method is appropriate for improving sand deposits outside the range considered suitable for vibrocompaction (i.e., for sand deposits

Figure 14-22 Depth vibrator for the vibrocompaction and vibroreplacement processes. Photo is from a vibroreplacement project. Note crushed rock for creating the stone column in the loader bucket. (Photo provided by GKN Hayward Baker Inc.)

with more than 15 percent material finer than the #200 sieve) and for cohesive soils. The installed stone mixes with the in-place soil in the zone affected by the compactor probe; heavy concentration of stone at the immediate probe location results in a "stone column." Stone columns have been found capable of carrying loadings in the 90 to 360 kN (10- to 40-ton) range. The overall load-supporting capability of a treated area will be influenced by the spacing of the columns, typically 1 to 3 m (4 to 10 ft), and the properties of the earth material underlying the improved soil zone.

Dynamic Deep Compaction

Surface and near-surface zones of soil and fill material whose existing condition is considered marginal or inadequate for foundation support have been improved by a field procedure identified by various names, such as *dynamic compaction, pounding,* and *high-energy impact.* The method consists of dropping a heavy weight from a relatively great height; on projects to date, the weights have ranged from 18 to 450 kN (2 to 50 tons), and drops have ranged from about 10 to 45 m (30 to 150 ft). For a given project, the weight and drop utilized relate to the crane equipment available and to the soil depth that requires improvement. Usually, a closely spaced grid pattern is selected for the pounding locations, and multiple poundings are provided at each drop location (typically, 5 to 10 drops). A repetitive series of poundings passes, at intervals of a day or more, may be required at some sites because of the character of the subsurface material. The process can densify loose cohesionless soils, fracture and densify buried building rubble such as that which exists at old building sites, consolidate fine-grained soils, and compact buried garbage fills. The pounding creates a depression at each drop location and also produces an areal settlement (Fig. 14-23).

Realistically, the intent of the pounding process is to economically improve surficial zones of structurally marginal soil materials that exist at a site so as to obtain a foundation-supporting capacity that will be adequate for relatively light structures. When successful, the method precludes the need for the usual alternatives, such as expensive deep foundations or compacted earth fill that replaces the marginal material.

The depths that can be improved relate to the impact equipment and the soil material being treated. Because the method is relatively new, information available for prediction and design is general and approximate. Very compressible, cohesive, and energy-absorbent garbage and fill materials probably realize lesser depths of improvement than do uniform

Figure 14-23 Dynamic deep compaction site, indicating grid pattern for drop locations. (Photo provided by GKN Hayward Baker Inc.)

cohesionless soils. Generally, job experiences have indicated that the improvement depth for soil and rubble materials can be estimated from

$$D = \text{between .3 and .5 times } \sqrt{WH} \qquad (14\text{-}1)$$

where W = falling weight (tons)

$\quad\;\; H$ = height of drop (meters)

$\quad\;\; D$ = depth of zone receiving some improvement (meters)

The lower factor of 0.3 is suggested for rubble and trash landfill sites; the larger factors of 0.4 to 0.5 are suggested for soil sites.

The extent of improvement is greatest near the surface and diminishes with depth. The degree of improvement increases with the number of poundings, up to some limit that seems to be between five and ten drops for the typical conditions where the method is used. An increase in the number of poundings apparently does little to increase the depth of the soil zone experiencing improvement. Where saturated fine-grained soils are involved, a series of poundings, performed at intervals of one or several days to provide time for the dissipation of pore water pressures created by the previous sequence of poundings, may be required to achieve satisfactory results.

If the surface stratum is granular, it may remain or become loosened because of the nature of displacements caused by the poundings. In such a case, surface compaction can be achieved with conventional rolling equipment when the pounding procedure is completed.

Ground vibrations will carry horizontally from the pounding location, and there may be concern about damage when structures are located near an area being treated. On the basis of limited data, it appears that facilities farther than a distance L from the point of pounding will not experience damage, where

$$L \text{ in m} \approx 1.8\sqrt{\text{impact energy in N-m}} \qquad (14\text{-}2\text{a})$$

$$L \text{ in ft} \approx 5\sqrt{\text{impact energy in ft-lb}} \qquad (14\text{-}2\text{b})$$
$$\text{(both very approximate)}$$

Compacting by Explosives

Under proper conditions, explosives can be used to densify loose, sandy deposits. The shock wave and vibrations induced by the explosives' blast produce results similar to those achieved by vibratory compaction equipment. The method is most effective in dry or completely saturated cohesionless soil. Where partial saturation exists, compressive stresses from the presence of air–water menisci act to prevent the soil particle movement necessary for densification. With this method, the depth that can be effectively improved economically is usually limited. Loosest soils experience the most improvement. If variations exist in the original deposit, the blasting would tend to produce a more uniform deposit. Where explosives are used, it should be planned to compact the uppermost 1 m (3 ft) or so by using conventional compaction methods, for little densification of this zone results from the blasting. Costwise, the technique compares well with other procedures for site improvement. However, the involvement of explosives and the concern of nuisance insurance claims limit the practical applications to remote areas.

14.4 SOIL STABILIZATION

Use of Natural and Chemical Additives

Soil stabilization refers to the procedure in which a special soil, a cementing material, or some other chemical material is added to a natural soil material to improve one or more of its properties. One may achieve stabilization by mechanically mixing the natural soil and stabilizing material together so as to achieve a homogeneous mixture or by adding the stabilizing material to an undisturbed soil deposit and obtaining interaction by letting it permeate through the soil voids. Where the soil and the stabilizing agent are blended and worked together, the placement process usually includes compaction.

Many of the stabilization procedures of mixing and then compacting in place are for providing a limited thickness of treated material, such as in bases for road and airfield pavements and beneath floor slabs.

One of the more common methods of stabilization includes the blending of natural coarse aggregate and fine-grained soil to obtain a mixture that possesses some internal friction plus cohesion and provides a material that is workable during placement but will remain stable when in place and subject to a range of temperature and moisture changes. Improvement by the proportioning of coarse to fine soil is commonly referred to as *mechanical stabilization*. The proper combination of coarse- and fine-grained material (the latter is referred to as *binder*) is related to the size and shape of particles in the mixing materials. Generally, the best mixtures will include about one-quarter binder.

Sodium chloride and calcium chloride are additives noted for their water-holding properties. The presence of moisture in a soil creates capillary water in the voids, and the resulting compressive stresses imposed onto the soil particles increase the soil's internal shear strength. Spread on dirt (soil surface roads), such materials help prevent dust. Calcium chloride is an especially preferable material for use on dirt roads, as it is capable of absorbing moisture from the air.

Portland cement and asphalt cement are additives intended to bond soil particles together. The result is a soil cement or asphalt-stabilized soil. In this respect, the cements are utilized as they are in the more familiar portland cement concrete and asphaltic concrete. In the stabilized soils, the range of soil aggregates is not as select as the materials used for the concretes. The strength of a stabilized soil will be less than that of the concretes. Mixing can be in place on site or in a mixing plant. In the construction of a proper soil cement, the soil needs to be at a water content close to the optimum percentage necessary for maximum compaction. After the cement and soil have been mixed, the mixture is compacted in place. Typically, the quantity of cement required ranges from about 7 percent by weight for sandy soils to about 15 percent for clay soils. Requiring only a limited wearing surface, soil cements have been satisfactory as bases for roads carrying low to medium volumes of traffic, provided that wheel loads are not heavy.

Asphalt-stabilized soils are capable of serving as the combination base and wearing surface for low-volume roads not subject to heavy wheel loads or as a quality base material to support pavements of major roads.

Lime and calcium chloride have been used as additives to improve clay soils. A decrease in plasticity results as a base change, or exchanging of cations in the adsorbed water layers, occurs. The resulting material is more friable than the original clay. The lime stabilization process requires mixing, then curing for a period of a few days, followed by remixing and compaction. The stabilized soil will possess strength similar to low-grade concrete. When mixed with expansive-type clays, volume changes are prevented. Lime has been success-fully used to stabilize road materials in areas where expansive clays have been a problem.

Various combinations of commercial and natural chemicals have been successful in acting as sealing or cementing agents for soil. The general categories are the silicate chemi-cals, the polymers, and chrome-lignin. As an illustration, sodium silicate mixed with cal-cium chloride reacts quickly to form calcium silicate, a hard and impervious material. The polymers represent those materials resulting from combining chemically complex organic monomers with a second chemical that acts as a catalyst to cause polymerization (joining together of molecules in the monomer). One of the most widely known soil stabilizers, AM-9, a polymer, has been found to possess hazardous characteristics, and the product AC-400 is being used as a replacement. Chrome-lignin is a slow-acting stabilizing gel formed from the reaction between the lignin in paper manufacture waste and sodium dichromate. The chemicals are typically water soluble and produce the reaction necessary to obtain a bond-ing of particles after being placed in the soil. With most chemical additives, the reaction does not include the soil particle. An advantageous feature of most chemical stabilizers is that their setting or curing time can be varied.

Grouting and injection stabilization are processes wherein a bonding material is forced into or through, or is mixed with, a natural soil deposit in order to improve the strength, water resistance, or other properties. Typically, the method is utilized where it is necessary to improve soil that cannot be disturbed. This procedure is often followed where treatment involves a considerable thickness or depth of earth or where the treated area is close to ex-isting structures or other facilities that cannot be disturbed. Chemical stabilizers are the materials commonly used in the process. Many stabilizers are marketed under special trade names. Injection and grouting are usually handled by specialty contractors that have proper equipment and have developed familiarity with one or more stabilization procedures. The injection or grouting is done under pressure. The stabilizing material moves through the void spaces in the soil. Consequently, the more viscous stabilizers are limited to treating soils that have high coefficients of permeability or are highly fissured. Some chemicals have viscosities comparable to water and can be used for finer soils, such as fine sand and silt deposits. At present, most injection methods are not suitable for clay because of the very low permeability. However, lime slurry injection has been found suitable for improving ex-pansive clays, even though nonhomogeneity results. With all grouting-injection methods, there is some uncertainty in results obtained. The pressurized stabilizer may follow paths through only the more permeable zones in a soil or along cracks and fissures and may not be distributed uniformly through the soil mass as desired. Figure 14-24 depicts the features of the general categories of grouting-injection stabilization.

Freezing techniques (ground freezing or soil freezing) can be appropriate for stabilizing or containing ground for short periods. The technique for depths and volumes considered practical on construction projects requires a refrigeration plant and the temporary installa-tion of a pipe distribution system within the soil mass being treated (for carrying the coolant). When the temperature in the soil mass is sufficiently low, water and possibly other liquid so-lidifies, in effect becoming a bonding material. The frozen ground condition provides

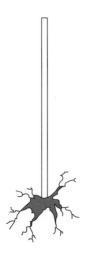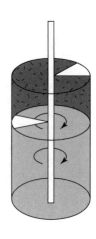

(a) Slurry (intrusion) grouting: Pressurized grout material is forced into seams, joints, cavities of the earth material and through voids of the permeable materials.

(b) Compaction grouting: Low slump grout under high pressure forms a bulb whose volume displaces soil laterally and vertically, causing the soil to densify.

(c) Chemical grouting: Fluid grout under pressure permeates into the earth material; as the grout sets and solidifies, the included soil volume is strengthened.

(d) Jet grouting: High-energy erosive jets of water or air displace soil in the zone requiring treatment, while simultaneously a grout or cementing material is delivered into the created void; a grouted mass or, if mixing of soil and grout or cement is permitted, a cemented soil column results.

Figure 14-24 Features of general categories of grouting-injection stabilization. (After GKN Hayward Baker Inc. [321])

various benefits: Steep excavation walls remain stable without bracing; groundwater flow will be halted or slowed; and migration of hazardous waste liquids can be stopped.

A summary of methods and materials applicable for stabilizing in-place soil of different types is presented in Figure 14-25.

Use of Geosynthetics for Stabilization and Reinforcement

The geosynthetics discussed in Chapter 7 and illustrated in Figure 14-26 are also being used to stabilize and reinforce soil masses. Geotextiles, geogrids, or geowebs integrated within the earth fill behind retaining walls, or in constructed embankments and areal fills, will act as reinforcement (Fig. 14-26a, b, c). In retaining walls and dams, the structure itself has been of earth and geotextile or geogrid construction (Fig. 14-26d, i).

Geosynthetics are widely used for erosion control on soil slopes and in areas exposed to flowing water (Fig. 14-26f, g). Fine-mesh geotextiles, geonets, and geowebs can be used

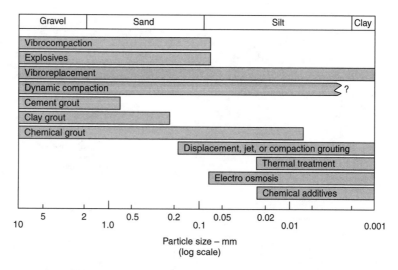

(a) Methods and particle-size range for in-place treatment of soils

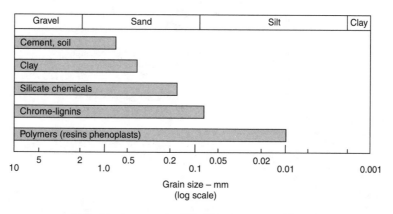

(b) Particle size ranges to which different grout materials apply

Figure 14-25 Summary of various methods for stabilizing in-place soils. (Modified from Mitchell [202])

where the earth slope is newly seeded; the soil surface is adequately protected against erosion from surface flow, while the open mesh permits the vegetation to germinate and grow. The open texture allows water from within the slope to escape and not be trapped beneath the fabric. Erosion control sheets are also made of natural fibers and synthetics that are degradable; the intention is to have the erosion control materials deteriorate after the plant root system has been established. *Fabric mattresses* (Fig. 14-26g) are receiving usage where erosion protection against flowing water is required. Generally, mattresses consist of a double layer of synthetic fabric whose interior cells will subsequently be filled (usually with concrete or grout material). At the project site, the "empty" mattresses are assembled to achieve the necessary cover area and placed on the surfaces to be protected. Typically,

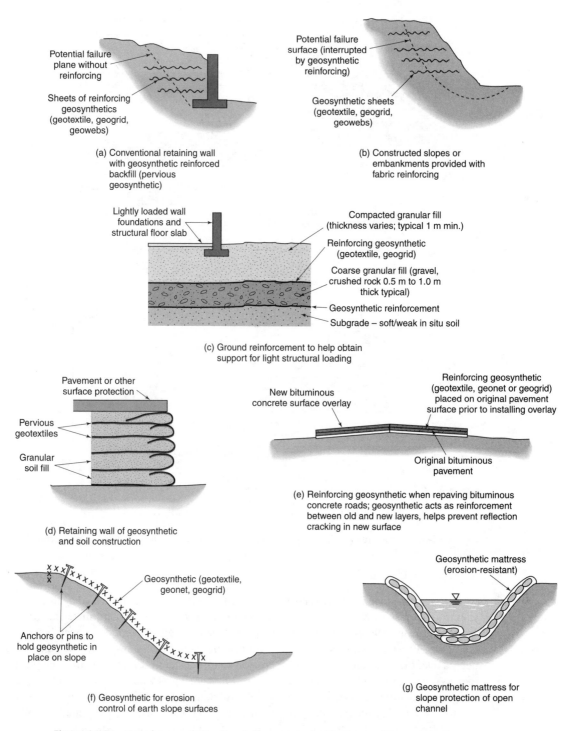

Figure 14-26 Use of geosynthetics for reinforcement, stabilization, and slope protection.

(h) Installation of geotextile for a pavement resurfacing project, to prevent reflection cracking (Courtesy of Hoechst Celanese Corp.)

(i) Retaining wall constructed with use of geosynthetics (Courtesy of Hoechst Celanese Corp.)

Figure 14-26   (Con't)

the filler grout is then pumped into the mattress, to create a protective covering that possesses strength and durability. Advantages of the mattress method include relative ease of installation on slopes and underwater surfaces without the need for forms or dewatering.

The geosynthetics are also being incorporated into road pavement asphalt concrete resurfacing overlays (Fig. 14-26h). Typically, geotextile or geonet sheets are placed on the old pavement surface prior to placement of the new overlay (the geosynthetic is adhered to the original surface with a liquid asphaltic sealant). The procedure helps prevent reflection cracking (the formation of cracks in the new surface above the location of cracks in the original surface) and extends the pavement service life.

14.5   METHODS FOR ESTABLISHING REQUIRED SOIL DENSITY

On structural earth filled projects, job specifications will indicate the soil density (unit weight)[2] or degree of compaction that must be achieved in order for the fill to be considered satisfactory. The job specification requirements are typically based on the results of laboratory compaction tests (more properly described as moisture–density or moisture–unit weight tests) performed on representative samples of soil to be used in the filling operation. The laboratory test determines the maximum density (or unit weight) for the soil and the influence of moisture content on obtaining that density.

The most widely used procedure for moisture–density testing consists of compacting the soil in layers in a cylindrical mold by using a drop hammer (equipment illustrated in Fig. 14-27). For a particular method, the mold will have set dimensions, and the number of

---

[2]The proper term relating weight per unit volume for in-place soil (pcf, kN/m^3) is *unit weight*. In the USA, the term *density* has often been inaccurately applied (recall that density is mass per unit volume, such as Mg/m^3), but usually the terminology does not cause a practical problem so long as test results are identified as pcf or kN/m^3.

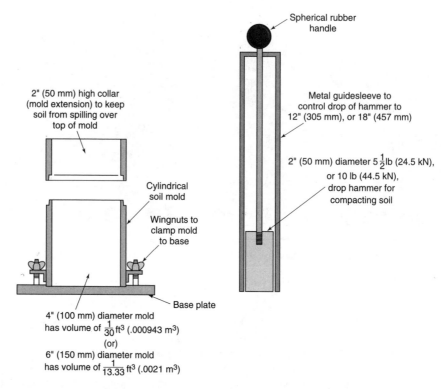

Figure 14-27 Widely used cylindrical mold and drop-hammer type of apparatus for performing moisture–density (compaction) test.

layers used to fill the mold will be specified, as will the weight and drop of the compacting hammer. To establish the moisture–density relationship for a soil, separate samples are each compacted at a different water content. Each sample is compacted in the same manner (same volume, same number of layers, same compaction energy). The compacted dry density and water content for each trial is then determined by weighing and drying the soil.

A comparison of results obtained from all the samples will reveal that the dry densities are different. This is caused by the variation in water content present during the compaction process. If the results obtained from all samples are plotted on dry density versus water content coordinates, a curve as indicated in Figure 14-28 is developed. From such a plot, the maximum dry density is evident. The water content corresponding to the maximum density is termed the *optimum moisture content*. The optimum moisture is the best water content for achieving a high density for the given soil when a compaction energy corresponding to the particular laboratory test method is used.

With sands, the influence of moisture on the compacted density at low water contents is less well defined than for fine-grained soils (Fig. 14-29). A scattering of dry density versus water content points is rather usual at moisture contents below the optimum. However, the information obtained from the overall test is still useful.

It should never be expected that compaction will result in a no-void-space condition for the soil, primarily because of the irregular shapes and various sizes of soil particles.

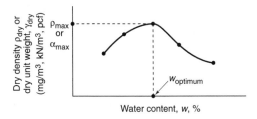

Figure 14-28   Typical moisture–density curve obtained from laboratory compaction test trials.

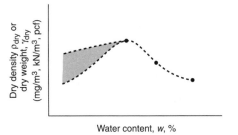

Figure 14-29   Representative moisture–density curve expected for sands, indicating indefinite location of curve at low water contents.

Reference to a moisture–density curve should reaffirm the necessity for the presence of water and void spaces.

In Figure 14-30, the curve representing a typical moisture–density relationship is shown, along with a curve showing the 100 percent saturation (or zero-air-voids) condition. The moisture–density curve approaches but does not overlap the zero-air-voids curve. This indicates that some air will always exist in void spaces during compaction.

The moisture–density curve for one method of compaction, as represented by Figure 14-28, shows the maximum dry density for that test. If the results of a second test, which imparts more compaction energy, are included on the same coordinates, the new curve will be located upward and to the left, indicating that a greater dry density is attained at a lower optimum moisture content (Fig. 14-31). Typically, a greater compaction energy results in a greater dry density, coupled with a reduced dependency on water content in obtaining that maximum density.

The development of the aforementioned general method, which establishes the influence of water content on the ability of a soil to be compacted and provides a value of compacted density to use for field control, is credited to R. R. Proctor. Proctor developed the laboratory test procedure while working with compacted earth dam projects in the early 1930s. The laboratory procedure established by Proctor utilized a 100-mm- (4-in.-) diameter mold having a volume of 0.000943 m³ ($\frac{1}{30}$ ft³) and a 50-mm- (2-in.) diameter, 24.5 N ($5\frac{1}{2}$-lb) hammer having a 0.3 m (1-ft) drop, and indicated that three layers be used to fill the mold, with each layer receiving 25 blows of the drop hammer.[3] This test is commonly referred to as the standard Proctor compaction test. The energy used to compact 1 ft³ of soil is 12,400 ft-lb or 16.74 kN-m. This energy compared favorably with the compaction energy transmitted by construction equipment in use at the time the method was developed.

---

[3]ASTM Test Designation D-698.

Figure 14-30 Zero-aid-voids curve related to moisture–density curve, indicating that soil does not become fully saturated by compaction.

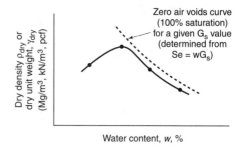

Figure 14-31 Effect of compaction procedure on dry density. Curve 1 represents results from method using heavier compaction hammer and greater compaction energy than used to obtain curve 2.

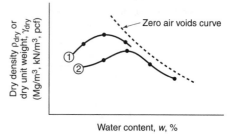

With technological advances have come the development of heavier vehicles and aircraft that need to be supported by compacted earth bases, plus the desire to construct larger earth structures and the intent to support heavy building structures on compacted earth. Larger and heavier earth-moving and soil-compaction equipment has been created. Consequently, laboratory test procedures were developed that would use a greater compactive energy and obtain higher values of dry density than the standard Proctor test. What has become known as the modified Proctor compaction test utilizes a 100-mm- (4-in.-) diameter mold having a volume of 0.000943 m³ ($\frac{1}{30}$ ft³), as before, but it requires a 44.5 N (10-lb) hammer having a drop of 0.46 m (18 in.). Soil to fill the mold is compacted in five layers, with each layer receiving 25 blows.[4] The energy used to compact 1 ft³ of soil in this test is 56,250 ft-lb or 75.94 kN-m, a value comparing favorably with that provided with current construction methods. Laboratory data and the resulting moisture–density (unit weight) curve for a soil compacted in accord with the modified Proctor requirements are shown in Figure 14-32.

If the compaction samples contain significant coarse material, a larger mold should be used, as the presence of coarse aggregates such as gravel in the $\frac{1}{30}$-ft³ mold can give dry densities that are too high to represent field conditions. A 150-mm- (6-in.-) diameter mold having a volume of 0.0021 m³ (0.075 ft³) is commonly used in place of the smaller mold. Layers are compacted with 56 blows of the drop hammer.

Variations of the drop hammer methods described above exist throughout the many organizations involved in developing earthwork specifications. Most commonly, the size of the drop hammer or mold is changed, or the number of layers or blows is varied as necessary to reflect local experiences or preferences. However, other types of laboratory test methods, such as those using kneading forces, have also been developed in attempts to reproduce more closely conditions resembling field compaction procedures. Their use has

[4]ASTM Test Designation D–1557.

Compaction Test Data Sheet

Compaction method

1.   – 1/30 cu. ft. mold, 5-1/2 lb hammer, 12" drops, 3 layers @ 25 blows/layer

2.   – 0.075 cu. ft. mold, 5-1/2 lb hammer, 12" drop, 3 layers @ 56 blows/layer

3.   – 1/30 cu. ft. mold, 10 lb hammer, 18" drop, 5 layers @ 25 blows/layer

4.   – 0.075 cu. ft. mold, 10 lb hammer, 18" drop, 5 layers @ 50 blows/layer

Trial	1	2	3	4	5	6	7
Wet density determinations							
Weight of mold and wet soil	13.84	14.30	14.00	13.89			
Weight of mold	9.32	9.32	9.32	9.32			
Weight of wet soil (W_T)	4.52	4.98	4.68	4.57			
Wet unit weight, pcf	135.6	149.4	140.4	137.1			
Moisture determinations							
Cup identification	B-1	B-2	B-3	B-4			
Weight of cup plus wet soil	39.10	55.30	66.60	75.48			
Weight of cup plus dry soil	38.15	52.81	62.28	68.53			
Weight of cup	15.10	14.21	14.43	14.33			
Weight of dry soil	23.05	38.60	47.85	54.20			
Weight of water	0.95	2.49	4.32	6.95			
Water content – %	4.0	6.5	9.0	12.8			
Dry unit weight – pcf	130.3	140.3	129.0	121.5			
Dry density – Mg/m³	2.09	2.25	2.07	1.95			

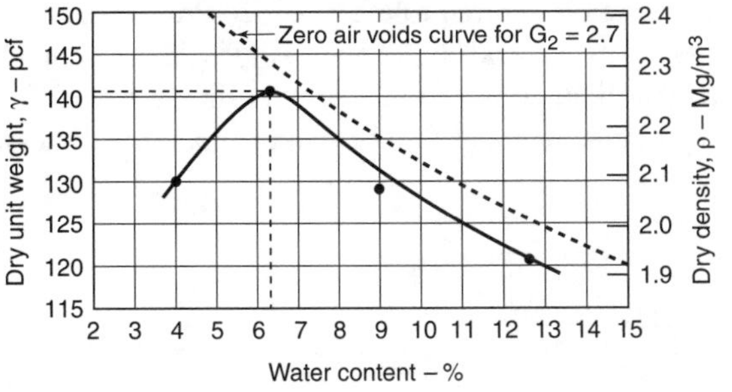

Soil classification:
    Silty sand

Soil sample from:
    Job 152-96
    TP 2 @ 3'

$\gamma_{max} = 140.5$ pcf (or $\rho_{max} = 2.25$ Mg/m³)

$w_{opt} = 6\% - 6\frac{1}{2}\%$

Figure 14-32   Laboratory moisture–density test results.

been limited because of the specialized techniques involved and the establishment of the drop-hammer method described previously. On a comparison basis, results obtained with other methods are not significantly different from the drop-hammer results.

A summary of the details for the laboratory compaction tests in common use is shown in Table 14-2.

Typical earthwork specifications for a structural fill project will require that the soil be compacted to a density equal to at least *X* percent of the maximum density obtainable with the (standard or modified) Proctor laboratory compaction test method. Currently, many

Table 14-2 Summary of Common Compaction Test Equipment and Procedures

ASTM Designation	AASHTO Designation	Mold Size	Hammer Weight and Drop	Number of Layers at Blows per Layer	Upper Particle Size limits or Sieve Size
D-698 (A)[a]	T-99 (A)	4-in. dia, $\frac{1}{30}$ ft³	5.5 lb @ 12 in.	3 at 25	#4 sieve
(B)	(B)	6-in. dia, 0.075 ft³	5.5 lb @ 12 in.	3 at 56	#4
(C)	(C)	4-in. dia, $\frac{1}{30}$ ft³	5.5 lb @ 12 in.	3 at 25	$\frac{3}{4}$ in.
(D)	(D)	6-in. dia, 0.075 ft³	5.5 lb @ 12 in.	3 at 56	$\frac{3}{4}$ in.
D-1557 (A)[b]	T-180 (A)	4-in. dia, $\frac{1}{30}$ ft³	10 lb @ 18 in.	5 at 25	#4 sieve
(B)	(B)	6-in. dia, 0.075 ft³	10 lb @ 18 in.	5 at 56	#4
(C)	(C)	4-in. dia, $\frac{1}{30}$ ft³	10 lb @ 18 in.	5 at 25	$\frac{3}{4}$ in.
(D)	(D)	6-in. dia, 0.075 ft³	10 lb @ 18 in.	5 at 56	$\frac{3}{4}$ in.

[a]Standard Proctor compaction test.
[b]Modified Proctor compaction test.
Note: $\frac{1}{30}$ ft³ = 0.000943 m³
0.075 ft³ = 0.0021 m³
5.5 lb @ 12 in. = 24.5 N @ 0.3 m
10 lb @ 18 in. = 44.5 N @ 0.46 m

compacted fills for buildings, roads, and dams use the modified Proctor as the reference test. For small dams, the standard Proctor may still be the reference. It is common to require that at least 95 percent of the maximum laboratory density be obtained by field compaction (ref. Fig. 14-33). Other typical requirements are 90 percent and 92 percent. The percentage selected is usually determined by the project designer on the basis of the project's requirements (e.g., fill for a parking area or fill beneath a building area) and the designer's experience with various soil types. Occasionally, a requirement of 100 percent or greater is found, particularly for granular soils. Unless the project is unusual, such a high percentage should not be required. Values of 100 percent compaction referred to the modified Proctor test will be close to the maximum relative density for many sands (relative density is described in Chapter 4).

When it is assumed that field compaction energy is similar to laboratory compaction energy, accepting a field compaction density less than the maximum laboratory density re-

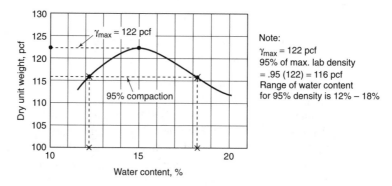

Figure 14-33 Example of method to compute density for a specified percentage of compaction and related range of water content.

Table 14-3 Approximate Range of Optimum Moisture Content Versus Soil Type

Soil Type	Probable Value of Optimum Moisture, %, Modified Proctor Test
Sand	6–10
Sand–silt mixture	8–12
Silt	11–15
Clay	13–21

duces the criticalness of water content in the fill soil. Reference to a moisture–density curve will reveal that it is possible to achieve a field density equal to, say, 95 percent compaction over a comparatively wide range of water content (Fig. 14-33).

Usually, the dry density of the soil being placed is the major item of concern on a compacted fill project. The water content of the compact soil is normally not specified and is left to the choice of the contractor. An exception is on deep fills and earth dam projects involving cohesive soils, where the water content during placement and compaction is limited to the optimum but preferably is slightly lower. If the water content is above optimum, excess pore water pressures may develop in the lower fill zones due to the weight of overlying material. Excess pore pressures interfere with the development of soil shear strength (see Eq. 11-3). Embankment failure due to inadequate shear strength may result.

Results of laboratory compaction tests are frequently included with the job specifications data assembled for a project. Experienced earth-moving contractors will pay close attention to the moisture–density curve(s) and the natural moisture content(s) of the fill soil, for a comparison will indicate if the fill requires wetting or drying before compaction. If laboratory compaction results are not available, the approximate ranges of optimum water contents from modified Proctor tests, presented in Table 14-3, can be used as a guide to evaluate the water content condition of a borrow soil (too wet or too dry).

14.6   FIELD CONTROL AND FIELD DENSITY TESTS

Where soil fill is compacted for a project or deep compaction is performed, it is usual to obtain some type of check or verification of the improvement. Surface zones of soil commonly are checked using an in-place density (or in-place unit weight) test procedure. Deep zones commonly are checked by evaluating soil resistance using the standard penetration test or cone penetrometer test.

On compacted earth fill projects, it is usual practice to have in-place density tests performed on the compacted soil layers to ensure that the desired density is being achieved. Typically, each layer is tested at several random locations after it has been compacted. When tests indicate that satisfactory compaction has been obtained, the contractor can begin placement of the next layer. If the density tests indicate inadequate compaction, the contractor is notified and more compaction rolling is performed. Normally, the personnel doing in-place density testing are directly responsible to the project owner or designer and are independent of the earthwork contractor.

Several methods for determining or checking in-place density are in use. The simpler, more direct methods involve the "destructive testing" principle: A small hole is excavated in the compacted layer and its volume is determined. The wet soil obtained from the hole is weighed and the water content is determined; these values permit the dry soil weight to be calculated. Knowing the dry soil weight or mass and the in-place volume permits the dry density or unit weight to be determined. The *sand-cone method* and the *rubber-balloon method* are widely utilized applications of this type of testing.

A "nondestructive" type of testing is provided with nuclear moisture–density–determining apparatus. With such equipment, a nuclear source emits gamma rays, which are passed into the soil and reflected back to a detector. The gamma photon energy received at the detector is calibrated to indicate the wet density or unit weight of the soil material. The amount of moisture is determined by calibration to a "thermal neutron" count moving through the soil from a fast neutron source. Moisture readouts are expressed as weight or mass of water per unit of volume (i.e., lb/ft^3 or g/cm^3).

Moisture–density determinations using the sand-cone and balloon methods are relatively time consuming. The minimum time required to complete all steps necessary to obtain a dry density and moisture determination at one location could be on the order of half an hour. This is a drawn-out period in which construction progress awaits answers. When a series of in-place density determinations are made for a large area—a usual procedure—the time to obtain results is even greater. The time required for the performance of such testing is frequently cited as a major drawback. In contrast, density and moisture determinations are rapid with a nuclear apparatus. Results for a test location are typically available in one to two minutes. Disadvantages associated with the use of nuclear moisture–density equipment include the relatively high purchase cost and the safety precautions necessary when personnel deal with a radioactive material. Due to the rapid provision of desired information, the nuclear equipment is now in widespread use in North America. The trend for commercial testing firms is to use the sand-cone and the balloon apparatus mainly for calibration of the nuclear apparatus, and for emergencies.

On fill projects involving sufficient area, density testing can be planned to minimize interference with soil placement and compaction. If work operations are scheduled so that one area is receiving a fill layer while a different area is being compacted, the density testing is performed at locations that are believed to have received adequate passes of the compaction equipment. As soon as testing indicates that an area has been properly compacted, the compaction equipment moves over to begin work on a newly placed layer. Placement of a new layer begins in the area just checked.

When the source of a fill soil is a natural deposit, it should be anticipated that the soil and its properties may vary within the deposit. Consequently, when density tests are performed, the tested soil type should be examined for correlation to the soil used to obtain the moisture–density curve. If a difference in soil exists, the laboratory-determined density value may not apply.

Field control of a compacted fill project should involve more than making in-place density tests. The thickness of newly placed soil layers should be watched to the greatest extent possible; proper depth and the possibility of undesirable soil being mixed with the desirable soil should be checked. The behavior of the ground surface as earth hauling or compaction equipment moves over it should also be observed. Rutting, weaving, or other movement of the ground surface is typically an indication that poor soil exists at the surface or is buried below it. Where poor material is observed and it cannot be improved by additional compaction, it should be excavated and removed from the fill area. Restoring the area to grade involves placement and compaction of soil in layers, as is required at all other locations. With experience, field personnel will impose better control over a fill project by observing the performance of equipment traveling over fill areas than by continuously making in-place density tests. Density tests are performed frequently at the beginning of the project to establish a familiarity with the soil types and related moisture. As the project progresses, the amount of density testing can be reduced. The running of in-place density determinations should not be discontinued completely, however, for they represent one of the permanent written records relating to earthwork performed for the project.

Sand-Cone Method

Determinations of in-place density using the sand-cone method involve the use of a sand-cone apparatus for obtaining only the volume of a density test hole. Field density (or field unit weight) will be soil mass, kg (or weight, kN or lb) divided by total volume occupied. The soil mass or weight is determined directly by using the actual soil removed from the test hold. The water content is calculated after obtaining the wet and dried weight of the soil.

The sand-cone apparatus derives its name from the shape of the cone funnel and sand jar, as shown in Figure 14-34. The volume of a test hole is determined by obtaining the mass or weighing the amount of sand originally in the jar that is necessary to fill the test hole. The sand utilized with this equipment is a dry, free-flowing, uniformly graded sand whose "bulk density" (unit weight) is known. Bulk density is the sand weight (voids included) per unit volume. Ottowa sand having a bulk density on the order of 16 kN/m^3 (100 pcf) is commonly used. If the weight of jar sand to fill the test hole is known, the volume of the hole is easily calculated. Test holes up to about 150 mm (6 in.) in diameter and 150 mm (6 in.) deep can be made with conventional equipment. Details of the sand-cone equipment and test procedure are presented in ASTM Test Designation D-1556.

Balloon Method

As with the sand cone, the balloon apparatus is actually limited to determining the volume of a density test hole. The mass or weight of soil necessary to complete a density or unit

Figure 14-34   Sand-cone density test being performed at a construction project.

Figure 14-35 Balloon apparatus to perform in-place density test.

weight calculation is determined by direct measurement of the material dug from the test hole.

The typical balloon apparatus (Fig. 14-35) consists of a water-filled vertical cylinder having a bottom opening over which a rubber membrane or balloon is stretched. The cylinder is constructed so that water levels in it are visible. Graduation marks indicating volumes are etched onto the sides of the cylinder. A small hand pump (bulb or piston type) is attached so that air can be forced into the top of the cylinder. When the apparatus is placed over a density hole, the pumped air forces the balloon and water into the hole. The volume is determined directly by noting the water level in the cylinder before and after the balloon is forced into the hole. The water and balloon are retracted from the test hole by reversing the air pump and evacuating air from the cylinder. Outside atmospheric pressure forces the water and balloon back into the cylinder. The apparatus is then ready for another test location.

Rubber-balloon equipment is available in a range of sizes. The smaller, more typical type is capable of measuring holes approximately 100 mm (4 in.) in diameter and 150 mm (6 in.) deep. Larger equipment can measure larger holes. More information on the balloon-type apparatus and test method is presented in ASTM Test Procedure D-2167.

Nuclear Moisture–Density Method

Surface-type nuclear moisture–density equipment (Fig. 14-36) is currently in widespread use for performing density tests on compacted fill, replacing the sand-cone and rubber-balloon methods, primarily because of the rapid results that can be obtained.

The principal elements in a nuclear density apparatus are the nuclear source, which emits gamma rays, a detector to pick up the gamma rays or photons passing through the tested soil, and a counter or scaler for determining the rate at which the gamma rays reach the detector.

Commonly used nuclear sources include radium–beryllium and cesium–americium–beryllium combinations in equipment where moisture determinations as well as density information are to be provided. The detectors usually consist of gas-filled Geiger-Mueller tubes.

When the equipment is in use, the gamma rays penetrate into the soil, where some are absorbed but some reach the detector by direct transmission or after reflecting off soil mineral electrons. The amount of gamma radiation reaching the detector is inversely propor-

Figure 14-36   Nuclear moisture–density apparatus in use. (Courtesy of Troxler Electronic Laboratories, Inc., North Carolina)

tional to the soil density. Densities or unit weights are determined by obtaining a nuclear count rate received at the detector, then relating such readings to density or unit weight (through comparison to calibration readings made on materials having known properties). Calibrations are provided by the equipment manufacturer. The density (unit weight) determined by this method is a wet, or total, density (unit weight).

Moisture determinations are obtained from a "thermal neutron" count. Alpha particles emitted from the americium or radium source bombard a beryllium target. This bombardment causes the beryllium to emit fast neutrons. These fast neutrons lose velocity if they strike the hydrogen atoms in water molecules. The resulting low-velocity neutrons are thermal neutrons. Moisture results are provided as weight or mass of water per unit of volume. Dry densities (or unit weights) are obtained by subtracting the moisture determination from the wet density (or unit weight) determination. With this method for determining moisture, significant error can result if the soil contains iron, boron, or cadmium.

Several field procedures for making the moisture–density determinations are available. The *direct transmission mode,* as indicated in Figure 14-37a, provides the most accurate results. In this method, the radioactive source is placed into the test material by utilizing a punched or drilled hole. Depths between 50 and 300 mm (2 and 12 in.) can be tested with conventional equipment.

The *backscatter method* operates by locating both the radioactive source and the detector on the surface of the test material (Figure 14-37b). The gamma rays are directed into the soil and some are reflected back to the detector. With this method, accuracy suffers if a gap exists anywhere between the bottom of the device and the soil surface.

A third procedure, the *air-gap method,* requires that the nuclear device sit a distance above the surface of the test material, as indicated in Fig. 14-37c. The accuracy obtained with this method is related to attaining a proper gap height for the equipment.

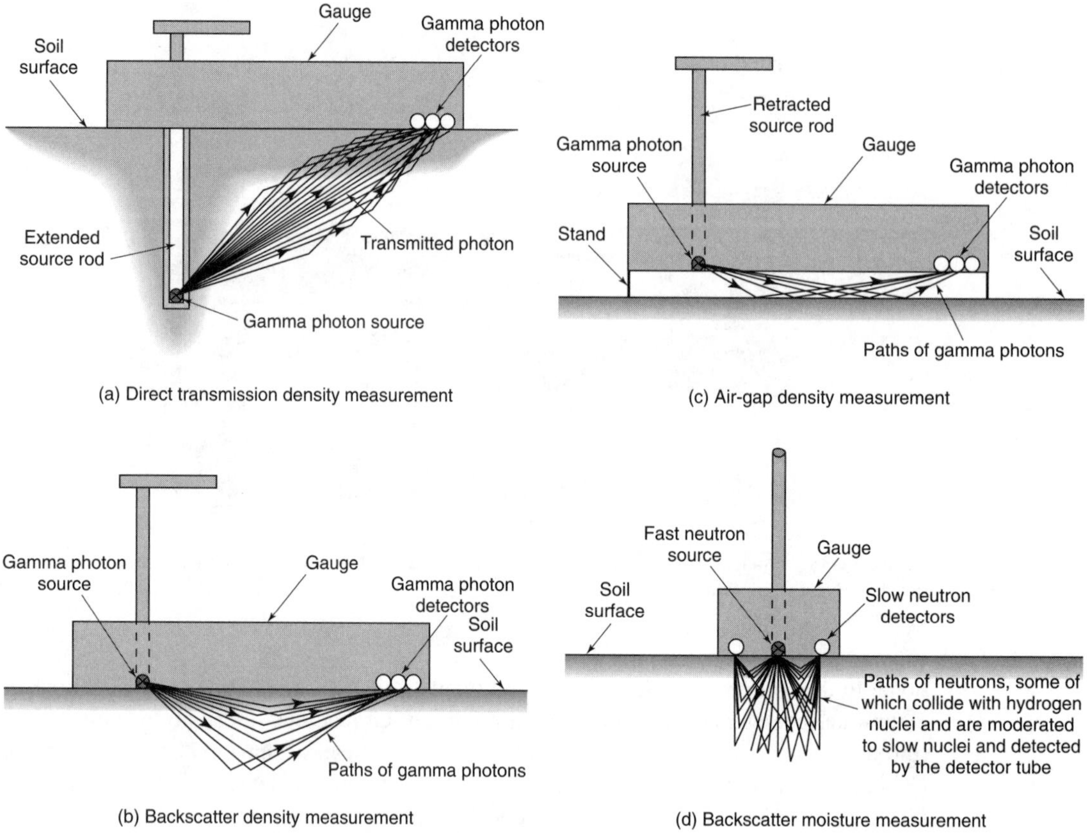

(a) Direct transmission density measurement

(c) Air-gap density measurement

(b) Backscatter density measurement

(d) Backscatter moisture measurement

Figure 14-37 Illustration of different modes for measuring soil density and moisture content by nuclear methods. (Courtesy of Troxler Electronic Laboratories, Inc., North Carolina)

The direct transmission method provides information on the soil volume surrounding the gamma rays' source, whereas the backscatter and air-gap methods provide data on the zone of soil nearest the surface at the test location. More detailed information for making moisture–density determinations by nuclear methods is presented in ASTM Test Procedure D-2922.

PROBLEMS

14-1. What are the practical reasons for requiring the soils that are placed for earth fill projects to be well compacted?

14-2. Relating to the placement of compacted earth fill (structural earth fill), indicate the best type and also other suitable types of surface zone compaction equipment (such as sheep's-foot roller, vibratory roller, etc.) to use for easiest or most effective compaction of the following soil materials:

(a) Well-graded fine to coarse sand
(b) Silty fine to coarse sand, trace clay
(c) Silt of low plasticity, trace clay, and sand
(d) Silty clay of low plasticity
(e) Sand–silt–clay mixture, some gravel

14-3. A contractor has bid on two different earth-moving/compacted fill projects. One project involves cohesionless soil, the other cohesive soil. The volume of soil to be excavated, transported, and compacted is similar for both projects. All working conditions are similar for both projects. The contractor's bid for compacted fill on the cohesive soil project is 50 percent greater than on the cohesionless soil project. Indicate whether the contractor's bidding practice is proper or not, and why.

14-4. For an earth-moving project, the contractor is required to locate a source of off-site borrow soil to use for compacted fill. Two possible sources are found; the cost of purchase and hauling from either will be comparable. At location A, the soil is a silt–clay material and is found to have a natural moisture content in the 10- to 20-percent range. At location B, the soil is also cohesive and is found to have a natural water content in the 30- to 40-percent range. The contractor selects location A. Indicate whether this represents a good choice, and why.

14-5. Indicate the general advantages and disadvantages associated with hydraulic fill placement methods.

14-6. Outline the basic concepts behind the methods in use for accomplishing deep-ground improvement of soil deposits (i.e., what changes or improvements in the soil deposit are being achieved, and physically how do the field procedures accomplish the desired changes?).

14-7. Asphalt, cement, and lime are commonly used to stabilize soil for road bases. Which of these seems most suitable for stabilizing the following:
(a) Sand
(b) Sand–silt mixture
(c) Silt with trace of sand and clay
(d) Clay with silt

14-8. What is the basic objective of the various soil stabilization procedures, and how do the stabilizing additives accomplish this objective?

14-9. It is proposed to stop groundwater from seeping through a building's basement walls by pressure-injecting a bentonite clay slurry into the soil against the exterior of the leaking walls, to seal the walls on the outside. The soil type is silt–clay. Will the method work in the indicated soil conditions?

14-10. The use of geosynthetics has proven to be effective and practical for improving soil conditions for some categories of construction projects. Outline the concept behind the basic purposes (what

is being changed or improved). (*Hint:* Refer to Fig. 14–26 for typical uses and where ground improvement is desired.)

14-11. Outline the main steps in the procedure to place a structural earth fill where in-place density (or unit weight) tests are performed during the soil placement and compaction process.

14-12. Laboratory compaction test results and field-determined in-place density (unit weight) results are reported in terms of dry density or unit weight and *not* wet density or unit weight. Outline the reasons for this procedure of reporting. (*Hint:* Consider the soil properties directly related to density or unit weight.)

14-13. The following moisture–unit weight (density) data are results from laboratory compaction tests:

Water Content, %	Dry Unit Weight, pcf
8	111
11	113
14	115
17	114
20	109

(a) Plot the curve for the moisture–dry unit weight relationship and indicate maximum dry unit weight and optimum moisture content.
(b) What range of water content appears advisable in order to obtain 92 percent (or more) compaction?

14-14. A series of laboratory compaction tests on the same soil as the previous problem provides the moisture and unit weight (density) information shown. Plot the moisture–dry unit weight curve and determine the maximum dry unit weight and optimum moisture content.

Water Content, %	Wet Unit Weight, pcf	Dry Unit Weight, pcf
7.0	112	
9.5	129	
13.5	136	
15.0	132	

14-15. The following information results from a series of laboratory compaction tests to determine the moisture–density relationship for a soil. The test follows ASTM D-1557 (modified Proctor) standards where a 0.000943-m^3 ($\frac{1}{30}$-ft^3) volume compaction mold is used. From the given data, finish the calculations necessary to complete the tabulation, then determine the maximum dry density and the optimum moisture content.

Wet Density Determinations

Trial Number	1	2	3	4
Mass of wet soil plus mold (kg)	6.24	6.45	6.31	6.24
Mass of compaction mold (kg)	4.40	4.40	4.40	4.40
Mass of wet soil (kg)				
Wet density (Mg/m³)				

Small Sample Moisture Determinations

Trial Number	1	2	3	4
Mass of wet sample (g)	106	112	120	120
Mass of dry sample (g)	100	101	105	102
Mass of water (g)				
Water content (%)				
Dry density (Mg/m³)				

14-16. Laboratory compaction test results for a soil are tabulated below. Review the data and determine if there are errors.

Water Content, %	Wet Unit Weight, pcf	Dry Unit Weight, pcf
9	122	112
12	129	117
15	137	119
18	107	109

14-17. Laboratory compaction tests have been performed on several different soils. The maximum dry unit weight and optimum moisture content for each soil are as shown. Determine the probable type (classification) of soil for each test.

Soil	Max. Dry Unit Weight, pcf	Optimum Moisture, %
A	119	9
B	124	18
C	114	14

14-18. What are the standard methods or procedures for evaluating the quality achieved or degree of success

(a) in compacted earth fill projects?

(b) for deep-ground improvement projects?

14-19. Sand-cone equipment is used to determine an in-place unit weight (field density test) on a compacted earth fill. Ottawa sand is used in the cone and is known to have a bulk density of 100 pcf.

(a) From the information given below, determine the in-place dry unit weight of the tested soil, and the water content.

Soil sample dug from test hole, wet weight	4.62 lb
Dried weight of soil sample	4.02 lb
Weight of Ottawa sand (sand cone) to fill test hole	3.60 lb

(b) Determine the percentage of compaction of the tested soil if the laboratory moisture–unit weight curve indicates a dry unit weight of 115 pcf and an optimum moisture content of 13 percent.

14-20. A balloon-type apparatus is used to determine an in-place unit weight for a soil (field density test). The volume of the test hole determined by reading the water level graduations on the apparatus cylinder before and after digging the test hole is 0.025 ft³. The wet weight of soil obtained from the test hole is 3.10 lb. The water content is determined to be 14 percent by drying a small sample on a field stove.

(a) Determine the in-place dry unit weight of the tested soil.

(b) Determine the percentage of compaction for the test result from part a if the soil is part of a compacted earth fill whose maximum unit weight (from laboratory compaction tests) is 118 pcf.

14-21. Outline the basic principles involved in the process that enables the nuclear moisture–density testing equipment to provide data readout on soil density (or unit weight) and moisture content.

CHAPTER 15

Stability of Unsupported Slopes

The stability of sloped land areas and the potential for failure, or landslide, is a concern where movements of existing or planned slopes would have an effect on the safety of people and property or the usability and value of the area. The region affected by slope failure is not necessarily the slope's immediate area; landslides have dammed up rivers and caused upstream flooding, and mass earth movements into large bodies of water also have created huge tidal waves that have devastated areas great distances away.[1]

Existing slopes that have been stable can experience significant movement, termed *slope failures* or *landslides,* when natural or human-induced conditions cause a change in the forces that act at the sloped area. *Changed natural conditions* can result from the occurrence of earthquake, subsidence of underground caverns, erosion, a general slope weakening because of the development of tension cracks or shrinkage cracks followed by water intrusion, a variation in the elevation of groundwater or a change in the slope's subsurface flow, which creates new seepage forces, or the weakening of buried soil or rock seams because of groundwater flow or chemical waste leachate. Typical *human-induced changes* include increased loading on a slope or near its crest, removal of earth below the toe of a slope, and removal of material from a slope so as to make it steeper. Short-term as well as permanent changes can be the cause of slope failures. [2]

Planned slope topography develops where the site grading for a construction project requires a change in elevation from one area to another within the confines of the project and at locations where a construction area is to be joined to the surrounding topography

---

[1]For example, a sudden slide into Italy's Vaiont Reservoir (1963) produced a wave on the order of 100 m (300 ft) high, which then overtopped the reservoir dam and caused great destruction in the region downstream.

[2]Short-term open excavations, common for ongoing construction projects, represent a type of situation which historically has been particularly susceptible to slope sliding problems. Though the soil mass involved may be limited (on a relative scale), the condition is potentially very dangerous for construction workers. Accordingly, construction safety laws are in effect (federal, state, and local government laws). For example, in the USA, Occupational Safety and Health Administration (OSHA) regulations have required since 1990 that excavations greater than 1.5 m (5 feet) deep have some type of protection; excavations including trenches between 1.5 m and 6 m (5 and 20 ft) deep must have sloped, benched, or shored (braced) soil walls (unless an engineering analysis indicates such precautions are not necessary); when deeper than 6 m (20 ft) the means for stabilization must be engineer-designed.

through the creation of new slopes or modification of existing ones. Typical of those jobs that include topography modification are earth-moving, excavation, and embankment projects such as for highways, airfields, railroads, and dams. Figure 15-1 illustrates these various conditions.

15.1 TYPES OF SLOPE MOVEMENTS

Movements of sloped soil masses can be classified into broad categories, depending on the type of motion relative to the adjacent or underlying earth. *Slide* refers to the occurrence where the moving mass is rather well defined and separated from the underlying and adjacent earth by a plane, or a zone comprising a number of adjacent planes, where slippage results. The slippage plane or zone represents the continuous surface where the maximum shear strength of the earth material has been reached, with the result that large displacement occurs. The failure mass remains essentially intact, although it may fracture into subunits.

Slides are classified as rotational or translational in accord with the shape and directional movement of the soil mass. *Rotational slides* are associated with natural slopes and constructed embankments of homogeneous materials possessing cohesion; the failure surface is curved, with the failed mass characteristically slumped in the toe area of the original slope (Fig. 15-2a). Rotation occurs about an imaginary axis in space, one that is aligned parallel to the slope. *Translational slides* are associated with slopes of layered materials where the mechanism of slippage occurs along a weak plane or zone that possesses a downward dip (Fig. 15-2b) and in cohesionless soil slopes where a change in conditions (such as seepage) occurs. Unlike the rotational slide, whose movement tends to cease when the mass reaches the slumped position, the translational slide can continue over a long distance. In the course of its movement, the sliding mass may remain essentially intact or subdivide.

Block or wedge failure refers to the displacement of an intact mass of soil because of the action of an adjacent zone of earth. Distinct blocks and wedges of soil may become separated from the adjoining earth because of the presence of cracks, fissures, or joints, or because materials having different properties are involved. Block failures may be included within translational slides as well as within the lateral spreads described below. The occurrence has also been noted in natural slopes and embankments where the failure block rests on a weak soil plane (Fig. 15-2c).

More complex types of soil mass movements are flows and spreads (Fig. 15-2d, e). A *flow* involves lateral movement of soil having the characteristics of a viscous fluid, although the actual consistency of the moving mass may vary from very wet to dry. A flow tends to retain less definition than does a sliding mass; slip surfaces within the mass are continuously developing and disappearing. *Spread* refers to the occurrence of multidirectional lateral movements by a fractured soil mass. The causes are similar to those for translational slides, involving a buried plane or zone that becomes too weak to resist the tendency for movement due to gravity forces or other factors. Earthquakes are a typical cause of lateral spreads.

Methods of analysis to evaluate the stability of existing and planned slopes using principles of engineering mechanics have concentrated on earth slides (i.e., rotational and translational) and block movements. Means for evaluating an area's susceptibility to a flow or

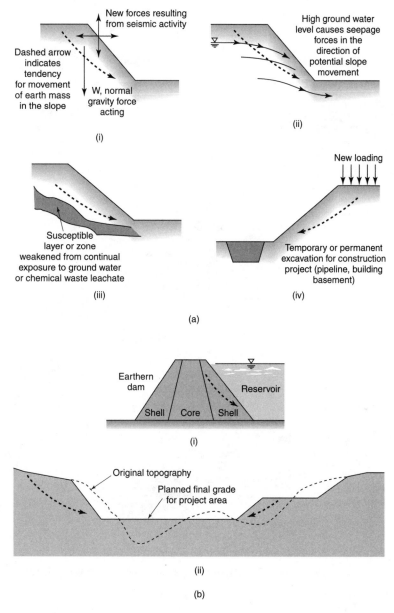

Figure 15-1 (a) Illustrations of changed conditions for existing slopes; (b) illustrations of new slopes resulting from construction.

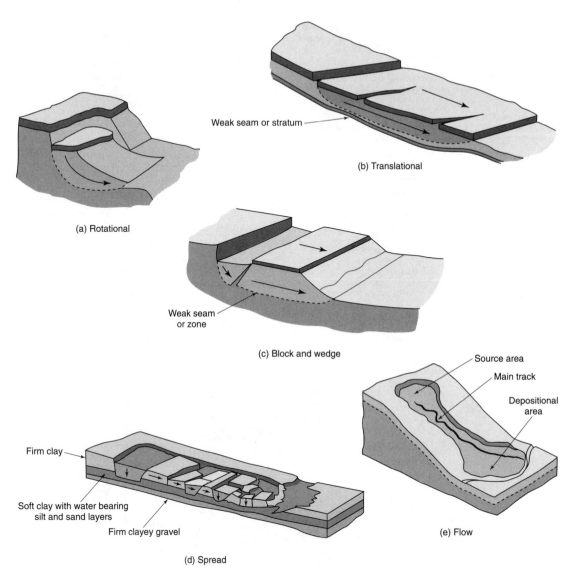

Weak seam or stratum

(b) Translational

(a) Rotational

Weak seam
or zone

(c) Block and wedge

Source area

Main track

Depositional
area

Firm clay

Soft clay with water bearing
silt and sand layers

Firm clayey gravel

(d) Spread

(e) Flow

Figure 15-2 Types of slope movements. (After Varnes [312])

spread occurrence have relied principally on interpretation of geologic conditions or, in the case of earth-moving projects such as hydraulic fills, on the properties and conditions of the fill and foundation materials. Analytical procedures in common use for studying slides are discussed in the following sections. The relatively simple condition of isotropic soils and uniform slopes of infinite extent is discussed initially. The major portion of the chapter then discusses methods for analyzing the more commonly encountered condition of slopes that are of finite extent and are composed of either isotropic, nonisotropic, or layered soils that are cohesionless or that possess cohesion.

15.2   ISOTROPIC SOILS AND UNIFORM SLOPES OF INFINITE EXTENT

For purposes of analysis, slopes extending a great distance in the height and length direction are considered to be of infinite extent. With uniform slopes of infinite extent, the analysis of forces affecting stability is considerably simpler than the analysis for a slope of limited extent. However, the procedures for studying the uniform slope of infinite extent in isotropic soil also serve as an introduction to methods for analyzing the more involved condition where slopes are of limited extent and are composed of nonisotropic (nonuniform) soils.

Cohesionless Soil

For a uniform slope composed of dry cohesionless soil, the forces acting on an incremental slice taken above the plane where sliding is assumed are as shown in Figure 15-3. The shear strength of the soil in the slope is expressed as

$$\tau = \bar{\sigma} \tan \phi \qquad (11\text{-}1b)$$

where τ is the maximum shearing strength or resistance that can be developed on the plane where $\bar{\sigma}$, the effective compressive stress (i.e., intergranular stress), is acting. The angle of internal friction, ϕ, is a property of the soil affected by particle size, shape, and mineralogical composition.

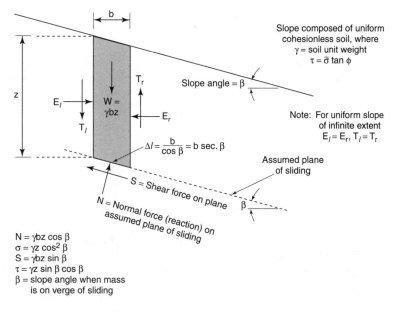

Figure 15-3   Description of forces acting on a representative slice of cohesionless soil in uniform slope of infinite extent.

For a uniform slope of limited extent, the tangential forces (T_l and T_r) and the lateral forces (E_l and E_r) acting on a slice of incremental width are equal. When the slope is on the verge of sliding along the assumed plane, all forces acting on the slice are in equilibrium. Summing forces in the horizontal direction to obtain the relationship between S and N gives

$$\Sigma F_{\text{horizontal}} = 0$$

$$S \cos \beta = N \sin \beta$$

$$S = N \frac{\sin \beta}{\cos \beta} = N \tan \beta$$

The maximum total shearing resistance that can be developed by the soil on the assumed plane of sliding is

$$\tau(\Delta l) = \overline{\sigma}(\Delta l) \tan \phi$$

where $\overline{\sigma}(\Delta l)$ is the maximum effective normal force N that can act on the plane. For the slice to be on the verge of sliding (at the limit of an equilibrium condition), the value of S on the base of the slice is equal to the maximum shearing resistance that can be developed by the soil.

$$S = \tau(\Delta l)$$

or

$$N \tan \beta = N \tan \phi$$

The limit of the equilibrium condition for other slices in the slope exists simultaneously when the relationship among W, N, and S is identical to that just shown. This condition is satisfied only where the failure plane and slope surface are parallel. Therefore, the maximum slope angle before sliding occurs is

$$\tan \beta = \tan \phi$$

or

$$\beta = \phi \text{ (for dry cohesionless soil)} \tag{15-1}$$

Thus, when β equals ϕ, the factor of safety (the ratio of forces resisting sliding to those causing sliding) based on the shear strength of the soil is 1. For slopes that have an inclination less than ϕ, the factor of safety is greater than 1. Where i is the actual slope angle, existing or planned, the factor of safety against sliding is expressed as

$$FS = \frac{\tan \phi}{\tan i} \tag{15-2}$$

Procedures for performing stability analysis based on the concepts just discussed have been termed the *limit equilibrium method* of analysis. In its present scope, the method involves the determination of the total shear strength that is required or mobilized (i.e., utilized) by the soil along an assumed slip surface in order to maintain slope equilibrium; a factor of safety against sliding is then indicated as the ratio of total shear strength available (the maximum strength of the soil) to the total shear strength that is required or mobilized for the equilibrium condition.

If the slope is submerged, the effective unit weight of soil in the slice is reduced to the buoyant or submerged unit weight. The values of N or S on each slice will be reduced proportionately. The shearing resistance that can be developed within the soil is correspondingly reduced because of the decreased value of $\bar{\sigma}$. The result is that the maximum safe slope angle for a completely submerged slope is the same as for an unsubmerged slope of dry soil, provided that no flow of water is occurring through the slope.

For the condition where the submerged slope is subject to flow that travels parallel to the slope's surface, the resulting seepage forces contribute to those forces that act to cause sliding, and the maximum safe slope angle becomes

$$\tan \beta = \frac{\gamma_{sub}}{\gamma_{total}} \tan \phi \tag{15-3a}$$

where γ_{total} is the wet soil unit weight above water and γ_{sub} is the submerged or buoyant soil weight. Since the submerged unit weight for many soils is on the order of half the weight above water, Equation 15-3a becomes

$$\tan \beta = \frac{1}{2} \tan \phi \quad \text{(approximately)} \tag{15-3b}$$

Equations 15-3a and 15-3b indicate the condition for the factor of safety to be 1. The factor of safety for lesser slopes can be determined from

$$FS = \frac{\gamma_{sub} (\tan \phi)}{\gamma_{total} (\tan i)} \quad \begin{array}{l}\text{(for submerged slope} \\ \text{of cohesionless soil} \\ \text{subject to ground} \\ \text{water flow)}\end{array} \tag{15-4}$$

where i is the actual angle of the existing or planned slope.

Relating to safe slopes of cohesionless soils above the water table, certain conditions frequently observed in practice warrant consideration. Where the soil is dense because of past geologic conditions or method of placement, the peak value of ϕ is significantly greater than the ultimate value (refer to Chapter 11). Relatively steep slopes of dense cohesionless soil are susceptible to sliding if external forces, such as from earthquake or blasting operations, disturb the dense condition and ϕ is reduced from the peak to the ultimate value (Fig. 15-4, also ref. Figs. 11-16 and 11-17). A slope stabilized at an inclination angle equal to the *ultimate* value of ϕ is considered to be at the angle of repose, the slope angle resulting when a dry cohesionless soil is loosely dumped. Slopes composed of damp cohesionless soil also require certain cautions. Damp cohesionless soils possess shear strength from apparent cohesion (Fig. 15-5) due to the presence of water menisci and capillary tension in addition to the strength in effect due to ϕ. The ability to obtain steep slopes in new excavations of such cohesionless deposits is a common occurrence. However, this extra strength disappears if

Figure 15-4 Relative failure envelopes (or shear strength values) due to difference in peak and ultimate ϕ values.

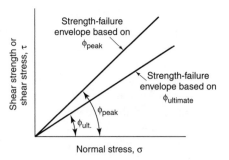

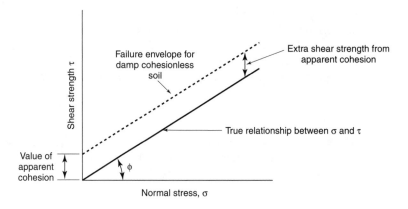

Figure 15-5 Effect of capillary moisture on shear strength of cohesionless soil.

the moisture evaporates or the slope becomes saturated or submerged. The result is that a zone of instability is created. A changed condition of seeping water (e.g., the result of a rising groundwater table or percolating surface water from rain or snow) can be responsible for the failure of steep slopes through the effect of seepage forces and erosion. For the rather common condition of a natural granular deposit containing some silt and clay materials, the movement of water can flush such "fines," which have been contributing a bonding or cohesive strength to the slope mass (Fig. 15-6).

Soils Possessing Cohesion

Soils possessing cohesion have their shear strength expressed as

$$\tau = c + \bar{\sigma} \tan \phi \qquad (11\text{-}3)$$

Referring to Figure 15-3, note that the shear stress acting on the base of the slice is

$$\tau = \gamma Z \sin \beta \cos \beta$$

Figure 15-6 Illustrations of steep slopes and unusual slope configurations that are possible, for a limited time period, in granular deposits. In left photo, note accumulation of fallen material at base of slope.

for the nonsubmerged case. When the soil mass is in equilibrium but on the verge of sliding,

$$\gamma Z \sin \beta \cos \beta = c + \overline{\sigma} \tan \phi$$

By substituting the value of $\gamma Z \cos^2\beta$ for $\overline{\sigma}$, obtain the relationship

$$\gamma Z \sin \beta \cos \beta = c + \gamma Z \cos^2 \beta \tan \phi$$

If the factor of safety against sliding is taken as the ratio of the soil shear *strength available* to the shear *stress acting* on the plane, a value of unity is obtained by substitution of the just-discussed shear strength and shear stress values; that is,

$$FS = 1 = \frac{c + \gamma Z \cos^2 \beta \tan \phi}{\gamma Z \sin \beta \cos \beta} = \frac{c}{\gamma Z \sin \beta \cos \beta} + \frac{\tan \phi}{\tan \beta} \qquad (15\text{-}5a)$$

For a slope of inclination i (where i is less than the angle β), the factor of safety is obtained by

$$FS = \frac{c}{\gamma Z \sin i \cos i} + \frac{\tan \phi}{\tan i} \qquad \text{(nonsubmerged case)} \qquad (15\text{-}5b)$$

For a completely submerged slope, the soil unit weight term in Equation 15-5b must be γ_{sub}. When the full height of the slope is subject to groundwater flow parallel to the surface of the slope, the factor of safety is determined by

$$FS = \frac{c}{\gamma Z \sin i \cos i} + \frac{\gamma_{sub} \tan \phi}{\gamma_{total} \tan i} \qquad \begin{array}{l}\text{(submerged slope} \\ \text{subject to groundwater} \\ \text{flow)}\end{array}$$

In practical encounters, most earth slopes are not isotropic nor of infinite extent. Safe inclinations for finite slopes comprised of isotropic cohesionless soils *can* be determined by utilizing the procedures for infinite slopes. For nonisotropic slopes and slopes composed of soil possessing cohesion (clays and mixed soils) that are of finite extent, the methods presented in the sections that follow are required.

15.3 SLOPES OF FINITE EXTENT—SOILS POSSESSING COHESION

Investigators examining failed slopes that consisted of isotropic deposits possessing cohesion (clays and mixed soils) have noted that, typically,the failure mass moved downward and outward relative to the slope's original position by sliding along a curved plane of slippage. These observed slip surfaces have been approximately but not truly circular in shape, the actual curvature being greatest near the center of the slippage arc and least in the upper section of arc. A result of this information is that many of the methods of analysis that have evolved to study the stability of not-yet-built and not-yet-failed slopes have assumed that a plane of failure would follow along a circular arc because of this shape's probable close approximation to any actual failure plane and, also, for mathematical convenience. Other shape failure planes, such as the logarithmic spiral, have been assumed in attempts to define the failure mass more accurately, but results have not differed significantly from those based upon the circular slip surface. Found to be more crucial to the study of a slope's stability is the proper evaluation of the involved soil's shear strength and the determination of pore water pressures that exist in the slope. Realistically, the practical ability to obtain highly precise subsurface information, such as high definition of the soil profile and related soil strength properties and pore pressure values, does not match the analytical precision possible within available slope stability theories. In recognition of this, approximations and practical assumptions have been incorporated into many analyses to simplify computational procedures. Nevertheless, these methods for slope stability evaluation provide results of good accuracy and practical value.

Types of Rotational Slides in Isotropic Soils

The strength properties of the soil composing the slope have an effect on the general location of the slip surface. Where the soils possess significant internal friction (mixed soils and primarily cohesionless soil), the failure arc is shallower than for the condition where an isotropic mass of soft or normally consolidated clay exists (Fig.15-7). The plane of sliding for slopes of soil having internal friction passes through the face of the slope or involves but little of the zone below the level of the toe. In contrast, failures in isotropic clay deposits involve a slippage plane that extends some distance below the elevation of the toe (classified as a base failure) and is often tangent to a deep hard or firm layer. Generally, and on a relative basis, failure of the flatter slopes involves a deeper failure zone than that of the steeper slopes.

Stability Analysis—General Concepts

Where the stability of a sloped earth mass is to be studied for the possibility of failure by sliding along a circular slip surface, the principles of engineering statics can be applied to determine if a stable or unstable condition exists. When the total sliding mass is assumed to be cylindrical or spoon-shaped (Fig. 15-8) a unit width extending along the face of the slope is taken for analysis, and the slip surface of the slope cross section is the segment of a circle (Fig. 15-9a). Forces that would affect the equilibrium of the assumed failure mass

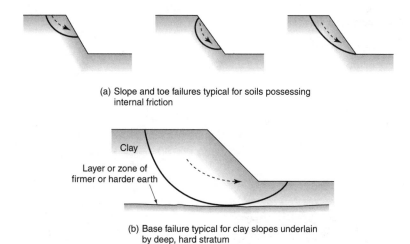

(a) Slope and toe failures typical for soils possessing
internal friction

(b) Base failure typical for clay slopes underlain
by deep, hard stratum

Figure 15-7 Types of rotational slides occurring in slopes comprising homogeneous soils.

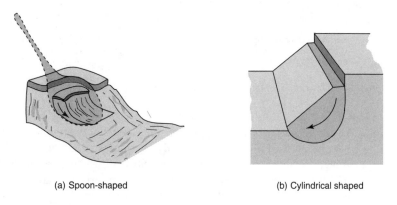

(a) Spoon-shaped

(b) Cylindrical shaped

Figure 15-8 Shapes of rotational slides.

are determined, and *rotational moments* of these forces with respect to a point representing the center of the slip circle arc (the point is actually an axis in space parallel to the face of the slope) are computed. With this procedure, the weight of soil in the sliding mass being considered as well as external loading on the face and top of the slope contribute to the moments acting to cause movement. Resistance to sliding is provided by the shear strength of the soil on the assumed slip surface (Fig. 15-9b).

A computational method used to indicate if failure (sliding) occurs is to compare moments that would resist movement to those that tend to cause movement. The maximum shear strength possessed by the soil is used in the calculation of the resisting moment. Failure is indicated when moments causing motion exceed those resisting motion. The factor of safety against sliding or movement is expressed as

$$\text{FS} = \frac{\text{Moments resisting sliding}}{\text{Moments causing sliding}} \qquad (15\text{-}6a)$$

Calculations for factor of safety

(a) Let $\tau = \tau_{max}$ (Maximum soil shearing strength)

$$FS = \frac{\tau_{max} \times L \times r}{(W_1 \times d_1) + (W_2 \times d_2)}$$

(b) Let τ_{req} = shear strength required for equilibrium

So that $\tau_{req} \times L \times r = (W_1 \times d_1) + (W_2 \times d_2)$

$$\tau_{req} = \frac{(W_1 \times d_1) + (W_2 \times d_2)}{L \times r}$$

$$FS = \frac{\tau_{max}}{\tau_{req}}$$

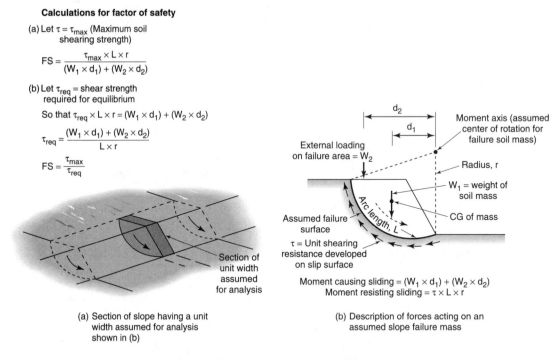

Section of unit width assumed for analysis

External loading on failure area = W_2

d_2

d_1

Moment axis (assumed center of rotation for failure soil mass)

Radius, r

W_1 = weight of soil mass

CG of mass

Assumed failure surface

Arc length, L

τ = Unit shearing resistance developed on slip surface

Moment causing sliding = $(W_1 \times d_1) + (W_2 \times d_2)$
Moment resisting sliding = $\tau \times L \times r$

(a) Section of slope having a unit width assumed for analysis shown in (b)

(b) Description of forces acting on an assumed slope failure mass

Figure 15-9 Description of terms and basic method for slope stability analysis.

A factor of safety of unity implies that the assumed failure mass is on the verge of sliding. A variation to this method for studying slope stability involves determining the shear strength required to have sliding moments and resisting moments balance (equilibrium). The shearing resistance required along the slip surface is compared to the shear strength that can be developed by the soil. If the soil shearing strength that can be developed by the soil is greater than the shearing resistance required for equilibrium, failure does not occur. With this method, the factor of safety against sliding is

$$FS = \frac{\text{Shear strength possessed by the soil}}{\text{Soil shear strength required for equilibrium}} \qquad (15\text{-}6B)$$

The factor of safety indicated by this method is a value based on the soil's shear strength; this method is used in most of the mathematical slope stability theories.

A number of analytical methods have been developed that utilize the basic concepts discussed above but differ in the procedure for handling various forces influencing equilibrium or in assigning practical simplifying assumptions. The methods indicated in Table 15.1 are procedures commonly referred to and are discussed in the following subsections of this chapter.

Computational aids (i.e., charts and curves) that simplify the effort related to determining the stability of slopes composed of isotropic soils are also available, and their use is discussed in later sections of the chapter.

Table 15-1 Partial Listing of Procedures for Analyzing Slope Stability

Procedure	Typical Applications and Limitations	Comments on Usage
Basic method	Circular failure surfaces in isotropic clay slopes	Nonrigorous method; hand-calculator solution practical
Fellenius method of slices	Circular failure surfaces, all soil types	Nonrigorous method; hand-calculator solution practical
Bishop method of slices	Circular failure surfaces, all soil types	Rigorous method; computer program best for solution
Simplified Bishop method	Circular failure surfaces, all soil types	Semirigorous method; hand calculator solution practical but computer solution preferable
Morganstern-Price method of slices	Circular and noncircular failure surfaces, all soil types	Rigorous method; computer required for solution
Spencer method of slices	Circular and noncircular failure surfaces, all soil types	Rigorous method; computer required for solution
Janbu method of slices	Circular and noncircular failure surfaces, all soil types	Widely used for noncircular failure surfaces; computer best for rigorous method but chart aids make hand calculator solutions practical

Strength Properties for Analysis—Clay and Mixed Soils

The shear strength of a *clay* soil is related to the soil's stress history (e.g., normally consolidated or overconsolidated), the effective or consolidation stress presently acting, and the pore water pressure condition existing, along with the opportunity for pore water drainage to occur. Use of the soil strength value determined from the in situ condition—for example, the total stress case, where $\phi = 0$ and $\tau = c$, values as would be obtained from unconsolidated–undrained shear tests (see Figs. 11-22 and 11-23) and unconfined compression tests (see Fig. 11-35)—is considered appropriate for analysis of end-of-construction stability and short-term stability. This total stress analysis, conventionally referred to as the $\phi = 0$ condition, is utilized when dealing with normally consolidated soils but should not be applied to the study of overconsolidated and fissured or jointed deposits. Studies dealing with the long-term stability of slopes and the short-term stability where overconsolidated or jointed–fissured soils exist require use of shear strength properties applicable to the effective stress condition. With the effective stress being the net intergranular compressive stress excluding pore water pressure, the relationship among total stress, effective stress, and pore water pressure in a soil mass is

$$\text{Total stress} = \text{Effective stress} + \text{Pore water pressure}$$

or

$$\sigma = \bar{\sigma} + u \qquad \text{(ref. Eqs. 9-3, 9-4)}$$

A practical reason for using the total stress ($\phi = 0$) analysis to study *newly constructed* slopes relates to the difficulty in selecting a pore pressure because the values are changing as drainage of the newly opened slope takes place. Realistically, it is simpler to work with a concept of total stress where, provided that *excess* pore pressures do not exist, the total stress is defined from

$$\sigma = \gamma_{total}Z \qquad \text{(ref. Eqs. 9-1, 9-3)}$$

where γ_{total} is the total soil weight including pore water and z is the depth to the plane under analysis.

The shear strength of mixed and clay soils (soils possessing cohesion) *based on the effective stress* is

$$\tau = c' + (\sigma - u)\tan\phi_{true} = c' + \bar{\sigma}\tan\phi_{true} \qquad (11\text{-}3)$$

where c' and $\tan\phi_{true}$ are values obtained from either consolidated–drained shear tests or consolidated–undrained tests when pore water pressures u are measured (Fig. 15-10). The magnitude of c' is often small for normally consolidated and lightly overconsolidated clay soils, and a zero value can be conservatively assigned in the slope analysis that uses effective stress parameters.

Basic Methods for Stability Analysis

With an isotropic clay slope, the factor of safety for a selected failure mass and slip surface can be determined by direct use of Equation 15-6a or 15-6b when the condition for the $\phi = 0$ case is under study (see Ill. 15-1).[3] For this case, the total shear strength considered

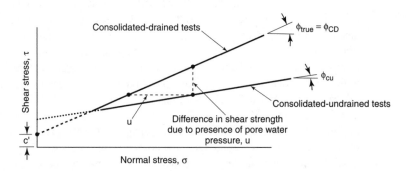

Figure 15-10 Qualitative comparison of shear strength envelope from consolidated–undrained tests.

---

[3]The area (volume) of earth in an assumed failure mass can be obtained by using a plenimeter if a to-scale drawing of the slope is used. The center of gravity of the failure mass can be determined by using principles of engineering mechanics for locating the centroid of an irregular shape.

is the value of cohesion, c, from undrained shear tests (i.e., unconsolidated–undrained tests or unconfined compression tests). Knowledge of the effective stresses existing on the slip surface is not a requirement for analysis, since they do not influence the shear strength value used. However, where a soil possesses internal friction (ϕ is not zero) that contributes to the shearing resistance along a slip surface, computational procedures include determining the effective normal stress at points along the assumed slip surface in order to provide values for shear strength (resistance) available from the $\overline{\sigma} \tan \phi$ factor. A procedure for analyzing slopes that include soils possessing internal friction, the *method of slices,* is discussed in later subsections.

Illustration 15-1

An unsupported slope is planned as indicated by the sketch for an area where a deep homogeneous clay soil deposit exists. What is the factor of safety against sliding for the trial slippage plane indicated?

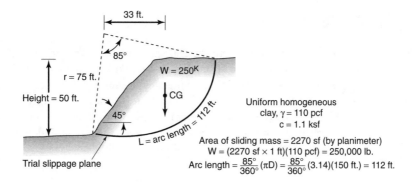

Uniform homogeneous
clay, $\gamma = 110$ pcf
$c = 1.1$ ksf

Area of sliding mass = 2270 sf (by planimeter)
$W = (2270 \text{ sf} \times 1 \text{ ft})(110 \text{ pcf}) = 250,000 \text{ lb.}$
Arc length $= \dfrac{85°}{360°}(\pi D) = \dfrac{85°}{360°}(3.14)(150 \text{ ft.}) = 112 \text{ ft.}$

Solution

(a) FS based on ratio of resisting to causing moments:

$$FS = \frac{cLr}{Wd} = \frac{(1.1 \text{ ksf})(112 \text{ ft})(75 \text{ ft})(1 \text{ ft width})}{(250 \text{ k})(33 \text{ ft})}$$

$$FS = 1.12$$

(b) FS based on soil shearing strength:

$$\text{let } \tau_{req} = \text{shear strength required}$$
$$\text{for slope equilibrium}$$

$$Wd = \tau_{req} Lr$$

$$\tau_{req} = \frac{Wd}{Lr} = \frac{250 \text{ k} \times 33 \text{ ft}}{112 \text{ k} \times 75 \text{ ft}} = .985 \text{ ksf}$$

$$FS = \frac{\tau_{max}}{\tau_{reg}} = \frac{c}{\tau_{req}} = \frac{1.1 \text{ ksf}}{.985 \text{ ksf}} = 1.12$$

The location of the moment axis has a significant effect on the computed factor of safety because it establishes the position and length of the assumed slip surface and the size of the sliding soil mass. It may be necessary to perform the stability computation for a number of assumed slip surfaces to locate the moment axis that identifies the most critical slip circle (the slip surface giving the lowest factor of safety is the one most susceptible to an actual failure; Fig. 15-11a). A grid pattern can be utilized to systematically change the location of moment centers. By labeling each point on the grid with its calculated factor of safety, contours representing equal values for the factor of safety can be established (similar to the procedures that land surveyors use to establish map contours by obtaining ground surface elevations in the field at points on a grid pattern). By plotting these points and sketching contours on an as-you-proceed basis, the locations of the more critical moment centers and slip surfaces become apparent; the procedure helps eliminate unnecessary random attempts to identify the most critical slippage arc (Fig. 15-11b).

Methods of Slices

Fellenius Method Where the slip surface under study for a soil slope passes through soil materials whose shear strength is based upon internal friction and effective stress, the method of slices is recognized as a practical means to account for the expected variation in shearing resistance that develops along the different portions of the assumed slippage arc. The procedure is applicable to slopes in isotropic soils, layered soils, and nonisotropic soils. Credit for the method and its application is given to Wolmar Fellenius as a result of his publications relating to analysis of soil slides [95].

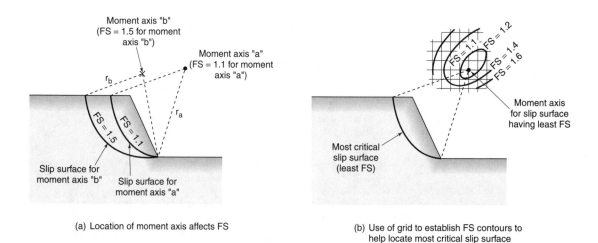

(a) Location of moment axis affects FS

(b) Use of grid to establish FS contours to help locate most critical slip surface

Figure 15-11 Effects of moment axis location on calculated factor of safety and method to establish most critical slip surface.

The procedure involves dividing a mass assumed to be involved in a slope slide into vertical slices, as indicated in Figure 15-12a. The width for each slide is arbitrarily selected, and slices do not have to be of equal width. For practical convenience, the section of base arc for each slice should pass across only one soil type, and the slice width should be limited so that the curved base can be approximated by a straight line.

Figure 15-12b, showing the free-body diagram of a representative slice, serves as the basis for arriving at the expression for determining a slope's factor of safety.

The *total* weight of soil in the slice, including the weight of water even when the slope is submerged, plus any external loading acting on the slope surface, comprises the W_t term shown in Figure 15-12. W_t is resolved into components tangent and normal to the base of the slice. With this method, the effects of the T and E forces on the sides of the slice are neglected. The factor of safety is the ratio of rotational moments, with respect to the center for the sliding arc, resisting sliding to those causing sliding. For the slice of Figure 15-12b, the moment due to shearing resistance on the slip surface is

$$M_{\text{resisting}} = (S)(r)$$

where S is the total shearing force and r is the radius of the circular slip arc. The shearing resistance on the base of each slice is

$$S = \tau l = l(c' + \overline{\sigma} \tan \phi)$$

where $\overline{\sigma} = W \cos \alpha / l$ if the soil is not submerged; therefore,

$$S = c'l + \frac{W \cos \alpha}{l} l \tan \phi = c'l + W \cos \alpha \tan \phi$$

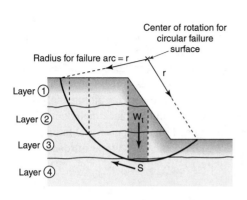

Center of rotation for circular failure surface

Radius for failure arc = r

Layer ①

Layer ②

Layer ③

Layer ④

(a) Selection of slices so base arc passes through only one soil type

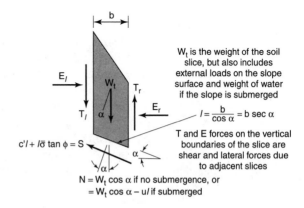

W_t is the weight of the soil slice, but also includes external loads on the slope surface and weight of water if the slope is submerged

$l = \dfrac{b}{\cos \alpha} = b \sec \alpha$

T and E forces on the vertical boundaries of the slice are shear and lateral forces due to adjacent slices

$N = W_t \cos \alpha$ if no submergence, or $= W_t \cos \alpha - ul$ if submerged

$c'l + l\overline{\sigma} \tan \phi = S$

(b) Forces acting on the individual slice

Figure 15-12   Basis of slope stability analysis by method of slices.

and the resisting moment is

$$M_{\text{resisting}} = r(c'l + W \cos \alpha \tan \phi)$$

The normal component of W_t passes through the center of rotation, giving a zero moment. The tangential component of W_t acts to cause sliding, giving

$$M_{\text{causing}} = (W_t \sin \alpha)r$$

The factor of safety for the slope, obtained by summing the resisting and causing moments for all slices, and noting that the radius r appears in both the numerator and denominator, is

$$FS = \frac{r \Sigma [c'l + W_t \cos \alpha \tan \phi]}{r \Sigma W_t \sin \alpha}$$

or

$$FS = \frac{\Sigma[c'b \sec \alpha + W_t \cos \alpha \tan \phi]}{\Sigma W_t \sin \alpha} \tag{15-7}$$

If the slope is submerged or if the groundwater table is at an elevation above the toe, pore water pressures u will act on the base of those slices that extend below water. This neutral pressure u reduces the effective stress and hence the shearing resistance acting along the base of such slices. In applications of the Fellenius method, it has been found most practical to account for the neutral pressure effect by using the submerged unit weight for soil located below water when the value of W_t in the numerator of Equation 15-7 is calculated.[4] The W_t term for the denominator continues to be determined by using the total (wet) weight of soil for each slice.

The analytical procedure for determining the resisting and causing moments is similar with each slice. A tabulated computational procedure, which can include graphical methods to determine the magnitude of a slice's weight components, has been found useful in obtaining values for Equation 15-7 when a slope stability study is performed with "longhand" methods (assisted by a hand calculator). The use of computers for aiding analysis is becoming widespread. The typical software programs in the marketplace include iteration instructions to systematically move the moment centers and establish new trial failure arcs for the checking the effect on the factor of safety. An example of the type of results produced by a computer program is shown later in Figure 15-15.

---

[4]Mathematically, the numerator should be $[c'l + (W \cos \alpha - ul) \tan \phi]$. The ul term could be larger than the $W \cos \alpha$ term if the slope is submerged, indicating a negative effective stress and shear, which is unreasonable. Recall from Chapter 9 that pore pressure $u = \gamma_w z_w$, where z_w is the depth below the water surface.

Illustration 15-2

Using the *Fellenius method,* calculate the factor of safety against sliding for the slope and trial failure surface shown in the sketch.

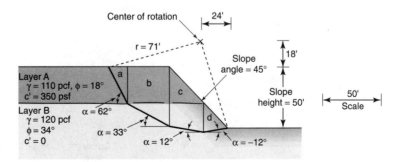

Computing Weight of Slices
(slices are 1 ft wide, perpendicular to page)

$$W_a = \left(\frac{26\ ft \times 14\ ft}{2} \times 110\ pcf\right)(1\ ft\ wide)$$
$$= 20.1^K/ft.\ of\ width$$

$$W_b = (26\ ft \times 27\ ft \times 110\ pcf)(1\ ft\ wide)$$
$$+ \left(\frac{27\ ft \times 19\ ft}{2} \times 120\ pcf\right)$$
$$= 108^K/ft.\ of\ width$$

$$W_c = \left(\frac{26\ ft + 6\ ft}{2} \times 26\ ft \times 110\ pcf\right)(1\ ft)$$
$$+ \left(\frac{19\ ft + 24\ ft}{2} \times 26\ ft \times 120\ pcf\right)$$
$$= 112.8^K/ft.\ of\ width$$

$$W_d = \left(\frac{6\ ft \times 6\ ft}{2} \times 110\ pcf\right)(1\ ft)$$
$$+ \left(\frac{6\ ft + 26\ ft}{2} \times 19\ ft \times 120\ pcf\right)$$
$$+ \left(\frac{26\ ft \times 5\ ft}{2} \times 120\ pcf\right)$$
$$= 40.3^K/ft\ of\ width$$

Slice	Base in Layer	W_t (k/ft)	tan ϕ	c' (psf)	b (ft)	α (deg)	sec α	cos α	sin α	$c'b$ sec α (k/ft)	$W_t\cos \alpha$ tan ϕ (k/ft)	$W_t \sin \alpha$ (k/ft)
a	A	20.1	.325	350	14	62	2.13	.469	.883	10.44	3.06	17.75
b	B	108.0	.675	0	27	33	1.192	.839	.545	0	61.16	58.86
c	B	112.8	.675	0	26	12	1.022	.978	.208	0	74.46	21.54
d	B	40.3	.675	0	26	−12	1.022	.078	−.208	0	26.60	−8.38
									Σ	10.44	165.79	89.77

Solution

$$FS = \frac{\Sigma\,[c'b\,\sec\,\alpha\,+\,W_t\,\cos\,\alpha\,\tan\,\phi]}{\Sigma\,W_t\,\sin\,\alpha} = \frac{10.44\ k/ft\,+\,165.79\ k/ft}{89.77\ k/ft} = 1.96$$

Bishop Method. A method of analysis presented by A. W. Bishop [30] utilizes the method of slices and considers forces to be acting on each slice, as shown in Figure 15-13. The requirements of equilibrium are applied to the slices comprising the slope, with the factor of safety against a slope failure then being defined as the ratio of the maximum shear strength possessed by soil on the trial slippage plane ($\tau_{available}$) to that shearing resistance necessary for equilibrium ($\tau_{mobilized}$), or

$$FS = \frac{\tau_{available}}{\tau_{mobilized}}$$

If the shear strength of the soil is

$$\tau_{available} = c' + (\sigma - u) \tan \phi = c' + \overline{\sigma} \tan \phi$$

the shearing resistance mobilized (i.e., required) for equilibrium is then

$$\tau_{mobilized} = \frac{1}{(FS)}[c' + (\sigma - u) \tan \phi]$$

Equation 15-8 provides the factor of safety for a trial or assumed failure arc and is obtained by summing slice forces in the *vertical* direction; the development of this equation is presented in Appendix C. The final form of the equation is the result of considering that the total influence of tangential forces on the sides of slices is small enough to be neglected without having great effect (i.e., the influence of $\Sigma [T_l - T_r]$ is small). The procedure for

Figure 15-13 Representative slope slice with forces assumed in Bishop method of analysis.

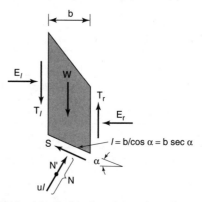

W = Total weight of soil in slice, plus surcharge if present
N = N' + ul
Where N = total normal force
       N' = effective normal force
       ul = force due to pore water pressure
       u = pore water pressure acting on base
           of slice equal to $\gamma_w z_w$, where z_w
           is the depth below the water surface

slope stability analysis that includes incorporating the simplifying assumption is referred to as the *simplified Bishop method*:

$$FS = \frac{\sum \dfrac{1}{m_a}[c'b + (W - ub)\tan\phi]}{\sum W \sin\alpha} \tag{15-8}$$

The m_a factor represents a grouping of terms as defined in the Bishop development presented in Appendix C, "Mathematical Development of the Bishop Equation for Slope Stability." Values of m_a can be obtained from Figure 15-14.

Comparisons between the simplified method and a rigorous solution that includes the effects of T_l and T_r indicate that the simplified method gives results that are slightly conservative (i.e., the simplified Bishop gives a lower factor of safety).

With the Bishop method, the factor of safety appears on both sides of the equation to evaluate the stability of a trial failure mass. The procedure for solution involves *assuming* a value for the factor of safety term on the right side of the equation, using Figure 15-14 to expedite the computations. When the proper factor of safety has been used for the trial, the value for the left side of the equation will equal the value assumed for the right side. Practically, precise agreement is not required to obtain a factor of safety value considered valid for the assumed slip surface. The result is for a particular trial failure mass, however, and, as indicated previously, a series of trials is usually required to identify the slope section and slippage plane most susceptible to actual failure or having the lowest factor of safety.

A limitation to the Bishop method is that is provides unrealistically high factors of safety when the negative angle alpha $(-\alpha)$ for the lower slope area slices approaches a value of about 30°. This condition can develop where an assumed center of rotation is above the vicinity of the slope's crest or where deep failure circles are being investigated.

Comparisons between the simplified Bishop and Fellenius methods indicate that, generally, the Bishop factor of safety will usually be slightly greater than that obtained with the

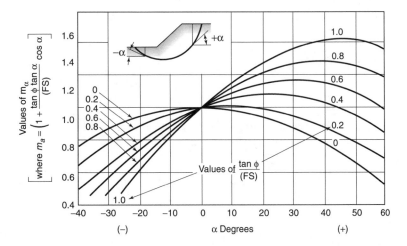

Figure 15-14   Values of m_a for Bishop equation.

Fellenius method. (The Fellenius method gives the smaller, more conservative value.) However, whereas the Bishop method provides overly high and inaccurate factors of safety when the negative angle alpha ($-\alpha$) is large, the Fellenius method tends to provide overly conservative (low) values.

Illustration 15-3

For the soil slope and trial failure surface of Illustration 15-2, calculate the factor of safety against sliding using the *simplified Bishop method.* The factor of safety by the Bishop method is expected to be greater than indicated by the Fellenius method.

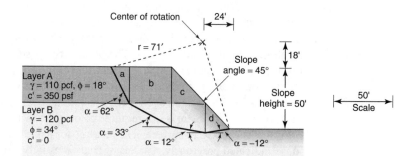

Slice	Base in Layer	W (k/ft)	$\tan\phi$	c' (psf)	b (ft)	α (deg)	$\sin\alpha$	ub (k/ft)	m_a if $FS=2$	$\frac{1}{m_a}[c'b + (W - ub)\tan\phi]$ (k/ft)	$W\sin\alpha$ (k/ft)
a	A	20.1	.325	350	14	62	.883	0	0.82	(1/.82)(4.6 + 6.53) = 13.93	17.75
b	B	108.0	.675	0	27	33	.545	0	1.03	(1/1.03)(72.9) = 70.78	58.86
c	B	112.8	.675	0	26	12	.208	0	1.03	(1/1.03)(76.14) = 73.92	21.54
d	B	40.3	.675	0	26	−12	−.208	0	0.93	(1/.93)(27.20) = 29.20	−8.38
									Σ	187.83	89.77

Figure 1 Tabulation for Illustration 15-3.

Solution

(See tabluation in illustration Fig. 1.) Preliminary calculations indicate a factor of safety of about 2. If a factor of safety equal to 2 is assumed, then

for Layer A,
$$\frac{\tan\phi}{FS} = \frac{.325}{2} = .16$$

for Layer B,
$$\frac{\tan\phi}{FS} = \frac{.675}{2} = .34$$

$$FS = \frac{\Sigma\frac{1}{m_a}[c'b + (W - ub)\tan\phi]}{\Sigma W\sin\alpha} = \frac{187.83 \text{ k/ft}}{89.77 \text{ k/ft}} = 2.09$$

The assumed factor of safety is close to the calculated value; therefore, the calculation is considered adequate. (Were the FS close to unity, additional calculations would be performed to improve the accuracy of the answer.) Actual FS for this trial slip surface is about 2.05.

The Bishop method of slope stability analysis, as just illustrated, is based on having the location for the moment center and a related radial distance as known values (to define the soil mass being checked for sliding). Unfortunately, for many slope stability problems, the moment center and the radius that identifies the soil mass and slip surface most susceptible to sliding (the lowest factor of safety) are not easily determined but can be found by repeating the analysis for an array of trial moment centers and radial distances. Done "longhand," the iteration process can be tedious and time consuming; problems requiring repetitious computations are ideally suited to solution by computer, however. At present, computer software programs for performing slope stability studies are available in the marketplace for a variety of computer languages, including those used by the popular PC equipment. These programs conventionally refer to an X-Y coordinate system for defining the geometry of the slope and the failure surface being considered. Programs typically include iteration instructions to systematically relocate the moment center and establish a new trial failure arc for checking the effect on the factor of safety. An example of results from such a software package run on a PC, outlining the important input and output information, is shown in Illustration 15-4.

Other Methods of Circular Arc Analysis.   A number of investigators have developed analytical procedures for slope analysis that are similar in basic approach to the Bishop method but differ in the handling of interslice forces. Two methods that are widely referred to are from the work of Spencer [280] and Morganstern-Price [210]. Spencer's analysis assumes that the forces on the sides of the slices are parallel (Fig. 15-15a). The Morganstern-Price analysis imposes normal and shear forces on the slice sides and also includes water pressure effects (Fig. 15-15b). Both methods satisfy the requirements of equilibrium in regard to forces and moments that act on a slice. (Recall that the Bishop method required equilibrium of slice forces for only the vertical direction.) Slope stability evaluations utilizing

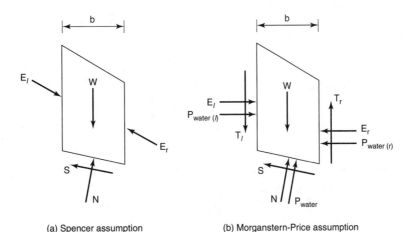

(a) Spencer assumption   (b) Morganstern-Price assumption

Figure 15-15 Description of forces acting on the free-body of a slope slice.

Illustration 15-4

Example of computer software program for a slope stability analysis (Bishop Method). Software: "Interactive Slope Stability Analysis Program," Alpha Tech Engineering Software, Blacksburg, Virginia

---

SLOPE STABILITY ANALYSIS DATA INPUT ECHO

PROJECT TITLE : example

WATER UNIT WEIGHT= 62.40 | **EARTHQUAKE= 0.00**

SUBMERGENCE AT 150.00 FROM 0.00 TO 400.00

POINT	X-COORDINATE	Y-COORDINATE
1	0.00	100.00
2	300.00	100.00
3	400.00	150.00
4	500.00	200.00
5	600.00	250.00
6	1000.00	250.00
7	1000.00	200.00
8	1000.00	150.00

(1) Input: Configuration of slope and soil layers using X-Y coordinates including position of ground water or surface pool

SOIL	UNIT WEIGHT	COHESION	PHI	SATURATED
1	127.0	2000	20	1
2	130.0	1000	33	1
3	130.0	1000	33	0

(2) Input: Soil properties

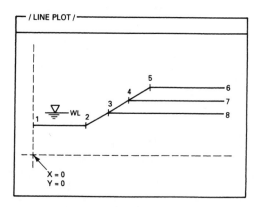

/ LINE PLOT /

(3) Output: Computer plots soil slope and position of soil layers

---

KTYPE SEARCH ENTRY OPTION

KTYPE=1	KTYPE=2	KTYPE=3	KTYPE=4	KTYPE=5
✗	+ + ✗ + +	+++ +✗+ +++	+++++ ++✗++ +++++	+++✗+++

NOTE: ENTER KTYPE=0 FOR SINGLE RADIUS SERACH

ENTER KTYPE SEARCH PARAMETER ---> 4

(4) Input: Select search option (the number of centers for trial circles).

GRID SEARCH OPTION SELECTED

GRID SEARCH OPTION -----> 4
DX= 50.00 DY= 50.00 DR= 40.00

Note: Analyst selects the values for DX, DY, DR.

X= 250.00 Y= 400.00

R=	75.00	CIRCLE DOES NOT INTERSECT SLOPE
R=	115.00	CIRCLE DOES NOT INTERSECT SLOPE
R=	155.00	CIRCLE DOES NOT INTERSECT SLOPE
R=	195.00	CIRCLE DOES NOT INTERSECT SLOPE
R=	235.00	CIRCLE DOES NOT INTERSECT SLOPE
R=	275.00	CIRCLE DOES NOT INTERSECT SLOPE
R=	315.00	--> FACTOR OF SAFETY = 4.35 <--
R=	355.00	--> FACTOR OF SAFETY = 3.35 <--
R=	395.00	CIRCLE EXCEEDS TOP LINE END POINTS

X= 250.00 Y= 450.00

R=	75.00	CIRCLE DOES NOT INTERSECT SLOPE
R=	115.00	CIRCLE DOES NOT INTERSECT SLOPE
R=	155.00	CIRCLE DOES NOT INTERSECT SLOPE
R=	195.00	CIRCLE DOES NOT INTERSECT SLOPE
R=	235.00	CIRCLE DOES NOT INTERSECT SLOPE
R=	275.00	CIRCLE DOES NOT INTERSECT SLOPE
R=	315.00	CIRCLE DOES NOT INTERSECT SLOPE
R=	355.00	--> FACTOR OF SAFETY = 4.06 <--
R=	395.00	--> FACTOR OF SAFETY = 3.03 <--
R=	435.00	CIRCLE EXCEEDS TOP LINE END POINTS

(5) Output: Computer determines FS against sliding for range of trial centers and different length radii. (Output for all combinations of X and Y being analyzed are not included here; outputs shown are representative)

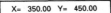

X= 350.00 Y= 450.00

R=	75.00	CIRCLE DOES NOT INTERSECT SLOPE
R=	115.00	CIRCLE DOES NOT INTERSECT SLOPE
R=	155.00	CIRCLE DOES NOT INTERSECT SLOPE
R=	195.00	CIRCLE DOES NOT INTERSECT SLOPE
R=	235.00	CIRCLE DOES NOT INTERSECT SLOPE
R=	275.00	CIRCLE DOES NOT INTERSECT SLOPE
R=	315.00	--> FACTOR OF SAFETY = 2.73 <--
R=	355.00	--> FACTOR OF SAFETY = 1.98 <--
R=	395.00	--> FACTOR OF SAFETY = 2.20 <--
R=	435.00	--> FACTOR OF SAFETY = 2.54 <--
R=	475.00	--> FACTOR OF SAFETY = 2.89 <--
R=	515.00	CIRCLE EXCEEDS TOP LINE END POINTS

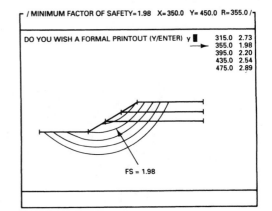

/ MINIMUM FACTOR OF SAFETY=1.98 X=350.0 Y=450.0 R=355.0 /

DO YOU WISH A FORMAL PRINTOUT (Y/ENTER) y

315.0	2.73
355.0	1.98
395.0	2.20
435.0	2.54
475.0	2.89

FS = 1.98

X= 350.00 Y= 500.00

R=	75.00	CIRCLE DOES NOT INTERSECT SLOPE
R=	115.00	CIRCLE DOES NOT INTERSECT SLOPE
R=	155.00	CIRCLE DOES NOT INTERSECT SLOPE
R=	195.00	CIRCLE DOES NOT INTERSECT SLOPE
R=	235.00	CIRCLE DOES NOT INTERSECT SLOPE
R=	275.00	CIRCLE DOES NOT INTERSECT SLOPE
R=	315.00	CIRCLE DOES NOT INTERSECT SLOPE
R=	355.00	--> FACTOR OF SAFETY = 3.07 <--
R=	395.00	--> FACTOR OF SAFETY = 2.04 <--
R=	435.00	--> FACTOR OF SAFETY = 2.20 <--

(5) **Output cont'd**

SLICE	WEIGHT	INCLINATION	COHESION	WIDTH	EFF WEIGHT	PHI	X
1	30168.9	−8.9	1000	9.4	483.2	33	295.3
2	137648.0	−5.4	1000	33.3	26510.3	33	316.7
3	182000.1	0.0	1000	33.3	67600.0	33	350.0
4	212758.9	5.4	1000	33.3	101621.3	33	383.3
5	247075.3	10.8	1000	33.3	145812.5	33	416.7
6	284372.6	16.4	1000	33.3	199873.8	33	450.0
7	306263.1	22.1	1000	33.3	245923.4	33	483.3
8	185718.0	26.8	1000	19.9	164701.2	33	510.0
9	182785.1	30.5	1000	19.9	175248.3	33	529.9
10	261084.7	35.3	1000	30.1	261084.7	33	554.9
11	225463.2	41.5	1000	30.1	225463.2	33	585.0
12	13236.4	45.0	1000	2.0	13236.4	33	601.0
13	102577.8	47.7	2000	20.6	102577.8	20	612.4
14	37585.2	52.9	2000	20.6	37585.2	20	633.0

DRIVING FORCE COUNTER BALANCE OF 73239.5

ITERATION	INITIAL	CALCULATED
1	1000	1.770
2	1.770	1.954
3	1.954	1.980
4	1.980	1.984

(6) **Output:** Computer plots position of various trial slip circles; graphic includes circle radius and related FS. Bishop values for lowest FS are then presented in tubular form.

the mathematically rigorous Spencer and Morganstern-Price methods are complex because of the form of their equations and the application procedure. Practically, computer programming is required for solution. However, slope stability charts that significantly simplify the analytical procedure have been developed from the computer-required methods; such charts are presented in a later section of the chapter.

Under comparison, the methods that handle the slice side forces differently do not differ greatly in the indicated factor of safety. Such results imply that slope stability evaluations are not sensitive to the manner in which the slice forces are considered to be acting.

Though means of analysis such as the Fellenius and simplified Bishop methods are not considered to provide the precision of the rigorous methods, the use of such hand-calculation methods retains a practical importance for performing independent checks on results from complex programs, for use when a computer is not available, and for preliminary evaluations or routine checks to determine if a slope requires greater study.

Noncircular Slip Surfaces

In areas where the topography varies (the slope is not uniform or well defined) or where the subsurface is stratified or otherwise nonisotropic, the soil zone most susceptible to a sliding failure may not be properly represented by a circular arc (Fig. 15-16a). An analytical procedure that permits flexibility in locating the plane of slippage is required. The Spencer and the Morganstern-Price methods referred to for circular arc analysis also have application for the noncircular slip surface but require solution by computer. A generalized method for slice analysis of noncircular slip surfaces advanced by Janbu [17, 18] is for computer solution, but the method is also adaptable to hand-calculation solutions when a simplified procedure (which still provides results of a practical accuracy) is utilized.

The Janbu method assumes that forces acting on each slice will be as indicated in Figure 15-16b. By applying the requirements of moment equilibrium and force equilibrium in the horizontal and vertical directions to each slice, and imposing the necessity for equilibrium in the horizontal direction to the entire assumed failure mass, an expression for the slope factor of safety is written

$$FS = \frac{\Sigma (b\tau \sec^2 \alpha)}{\Sigma[W + (T_l - T_r)] \tan \alpha}$$

where $\tau = \dfrac{\left\{c' + \left[\dfrac{W + (T_l - T_r)}{b} - u\right] \tan \phi\right\}}{\left(1 + \dfrac{\tan \alpha \tan \phi}{FS}\right)}$ (15-9a)

This factor of safety is in reference to the shear strength of the soil along the assumed slippage plane. When the term m_a is included (introduced for the Bishop analysis, Fig. 15-14), the expression for determining the factor of safety for a slope mass under study becomes

$$FS = \frac{\Sigma\{c'b + [W + (T_l - T_r) - ub] \tan \phi\}\dfrac{1}{(\cos \alpha)(m_a)}}{\Sigma[W + (T_l - T_r)] \tan \alpha}$$ (15-9b)

The FS term appears on both sides of Equation 15-9. Trial values are selected, and, as with the Bishop method, the factor of safety has been properly determined when both sides of the equation are in practical agreement. For solution, Equation 15-9 requires values for the tangential forces, T_l and T_r, which also are determined by a trial-and-error process. All other terms of the expression can be obtained from the geometry of each slice and the involved soil properties. However, the simplified method for the Janbu analysis initially neglects the interslice forces and then modifies the calculated factor of safety for their effect by applying an empirically determined correction factor. The final form of the equation with the modification is

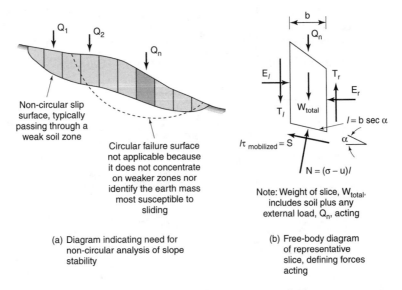

(a) Diagram indicating need for non-circular analysis of slope stability

(b) Free-body diagram of representative slice, defining forces acting

Figure 15-16   Noncircular slippage plane for slope stability analysis, Janbu method.

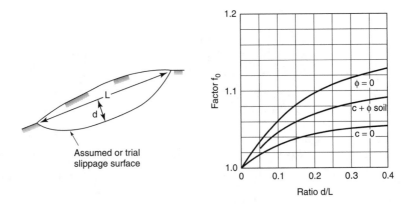

Figure 15-17   Correction factor, f_0, as a function of d/L ratio.

$$FS = \frac{f_0 \, \Sigma \, [c'b + (W - ub) \tan \phi] \, \dfrac{1}{(\cos \alpha)(m_a)}}{\Sigma \, (W \tan \alpha)} \qquad (15\text{-}10)$$

where the correction factor f_0 is obtained from Figure 15-17 (see Illus. 15-4).

Illustration 15-5

For the soil slope and trial failure surface shown in illustration Figure 1, calculate the factor of safety, using the simplified Janbu method.

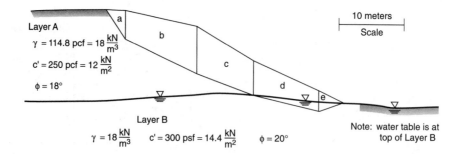

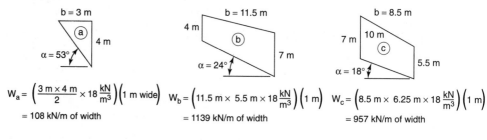

Computing weight of slices
(slices 1 m wide perpendicular to page)

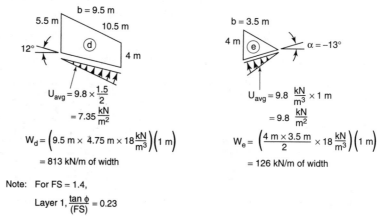

$$W_a = \left(\frac{3\,m \times 4\,m}{2} \times 18\,\frac{kN}{m^3}\right)(1\ m\ wide)$$
$$= 108\ kN/m\ of\ width$$

$$W_b = \left(11.5\,m \times 5.5\,m \times 18\,\frac{kN}{m^3}\right)(1\ m)$$
$$= 1139\ kN/m\ of\ width$$

$$W_c = \left(8.5\,m \times 6.25\,m \times 18\,\frac{kN}{m^3}\right)(1\ m)$$
$$= 957\ kN/m\ of\ width$$

$$U_{avg} = 9.8 \times \frac{1.5}{2}$$
$$= 7.35\,\frac{kN}{m^2}$$

$$W_d = \left(9.5\,m \times 4.75\,m \times 18\,\frac{kN}{m^3}\right)(1\ m)$$
$$= 813\ kN/m\ of\ width$$

$$U_{avg} = 9.8\,\frac{kN}{m^3} \times 1\ m$$
$$= 9.8\,\frac{kN}{m^2}$$

$$W_e = \left(\frac{4\,m \times 3.5\,m}{2} \times 18\,\frac{kN}{m^3}\right)(1\ m)$$
$$= 126\ kN/m\ of\ width$$

Note: For FS = 1.4,

Layer 1, $\dfrac{\tan\phi}{(FS)} = 0.23$

Layer 2, $\dfrac{\tan\phi}{(FS)} = 0.26$

Figure 1 Sketches for Illustration 15-4

Solution

See tabulation in illustration Fig. 2. Preliminary calculations indicate a factor of safety of about 1.5; use FS = 1.4 for first trial. For $d/L \approx \frac{4}{40} \approx 0.1$, get $f_0 = 1.04$ from Figure 15-17. Therefore,

$$FS = f_0\left(\frac{1615\ kN/m}{1178\ kN/m}\right)$$
$$= 1.04(1.37) = 1.42 \pm \qquad \text{OK, compares with assumed FS.}$$

Slice	b (m)	c' (kN/m²)	α (deg)	$\tan \alpha$	$\cos \alpha$	$\tan \phi$	u (kN/m²)	W (kN/m)	m_a	$[c'b + (W - ub) \tan \phi] \dfrac{1}{\cos \alpha \, m_a}$ (kN/m)	$W \tan \alpha$ (kN/m)
a	3.0	12	53	1.327	.602	.325	0	108	0.80	$[(12 \times 3) + (108 \times .325)] \dfrac{1}{(.602 \times .80)} = 152$	143
b	11.5	12	24	.445	.914	.325	0	1139	1.01	$[(11.5 \times 12) + (1139 \times .325)] \dfrac{1}{(.914 \times 1.01)} = 551$	507
c	8.5	12	18	.325	.951	.325	0	957	1.02	$[(8.5 \times 12) + (957 \times .325)] \dfrac{1}{(.951 \times 1.02)} = 414$	366
d	9.5	14.4	12	.213	.978	.364	7.35	813	1.03	$[(9.5 \times 14.4) + (813 - 7.35 \times 9.5)(.364)] \dfrac{1}{(.978 \times 1.03)} = 404$	191
e	3.5	14.4	−13	−.231	.974	.364	9.8	126	0.92	$[(3.5 \times 14.4) + (126 - 9.8 \times 3.5)(.364)] \dfrac{1}{(.974 \times .92)} = 94$	−29
										Σ 1615	1178

Figure 2 Tabulation for Illustration 15.5.

Effect of Tension Cracks

Tension cracks developing along the crest or face of a slope (a condition most often anticipated where cohesive soils exist) can influence the slope's stability. In an analysis, zero soil shearing resistance is assigned to the section of slippage plane affected by tension cracks. If the tension crack(s) could fill with water, a hydrostatic pressure distribution is assumed to exist in the crack, and this pressure contributes to those forces and moments acting to cause slope movement. Frequently, however, tension cracks have minor influence on a slope's computed factor of safety, but these cracks may provide the opportunity for water to reach otherwise buried earth layers whose strength may be weakened by such exposure, an effect requiring consideration in the slope analysis.

Sliding Block Analysis

Natural earth slopes consisting of stratified materials and earth embankment structures constructed on stratified soil foundations can experience failure due to sliding along one or more of the weaker layers (Fig. 15-18). Failures of this type often occur when changed conditions in an area cause susceptible layers to become exposed to, or saturated by, water. Ex-

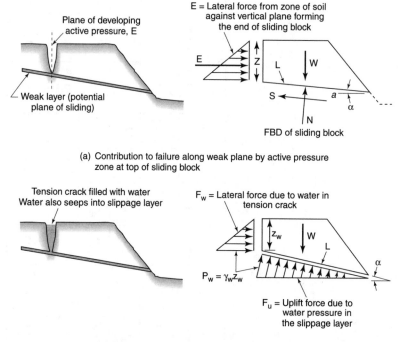

Figure 15-18 Block failure for a soil slope.

posure to moisture can cause physical breakdown and weakening of some earth materials, such as fine-grained sedimentary deposits, and saturation may cause a stratum's shear strength to be reduced because of increased pore water pressures.

Where the potential for the occurrence of a block slide is under study, the factor of safety with regard to the shear strength of the soil on the assumed sliding plane is, for the conditions described in Figure 15-18a, with no pore pressure effects on the block,

$$FS = \frac{cL + W \cos \alpha \tan \phi}{W \sin \alpha + E} \tag{15-11}$$

where E can be approximated as $0.25\gamma_{soil}Z^2$ for cohesionless soil and $0.5\gamma_{soil}Z^2$ for cohesive soil.

If the formation of a tension crack along the top of the slope permits the development of water pressures in the crack and the slippage zone, as described by Figure 15-18b, the factor of safety is expressed as

$$FS = \frac{cL + (W \cos \alpha - F_u - E_w \sin \alpha) \tan \phi}{W \sin \alpha + F_w \cos \alpha} \tag{15-12}$$

where F_w is the force due to water pressures in the tension crack, equal to $0.5\gamma_{water}Z^2_{water}$, and $F_u = 0.5\gamma_{water}Z_{water}L$.

Sections of soil embankments have been known to fail by translation along a weak foundation zone or layer, the force responsible for movement resulting from lateral soil pressures developed within the embankment itself. In the case of earth dams, the zone of slippage may develop only after the dam has impounded water for a period, with seepage through the eventual slippage zone being responsible for weakening to the extent that a failure can occur. For the conditions shown in Figure 15-19, the factor of safety against block translation is evaluated from

$$FS = \frac{S}{E_l - E_r + P_{w_1} - P_{w_2}} \tag{15-13}$$

where $S = cL + (W - U) \tan \phi$ on sliding plan b'-d or b-c
  $E_l \cong$ force of active earth pressure on plane a-b' or a-b
  $E_r \cong$ force of passive earth pressure on plane c-d
  $P_{w_{1,2}} \cong$ force of lateral water pressure on plane b'-b, c-d
  $U =$ uplift force on plane b'-d or b-c

With this method, the uplift and lateral water pressure determinations are evaluated from flow nets or other seepage studies.

Upstream zones as well as downstream zones should be studied for stability. Though the effect of water on the upstream embankment increases the weight W, and the lateral pressures of the impounded water oppose block translation, the uplift force U is considerably greater for upstream zones. As with other categories of sliding failures, determining the size and location of the section most susceptible to movement is typically a trial-and-error procedure, because the most critical zone is not always obvious.

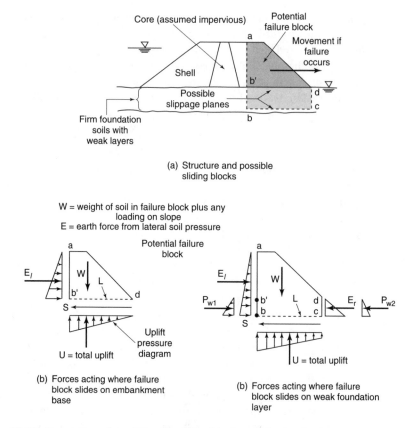

Figure 15-19 Description of conditions for sliding block analysis of embankments.

15.4 COMPUTATIONAL AIDS—SLOPE STABILITY CHARTS

Taylor Charts

Slope stability charts to simplify the computational procedure involved in studying slopes comprised of isotropic cohesive or mixed soils have been developed by a number of investigators. The earliest charts to receive widespread usage are those from the work of Taylor, whose results are based upon application of the friction-circle method of analysis [290, 291].[5] This procedure for analysis follows convention in that it assumes that the resistance to a circular mass's tendency to slide is developed by the cohesion and internal friction of the soil on the slippage plane. In an incremental slices analysis of a slope (Fig. 15-20), the reaction on the slice base results from resolution of the normal (i.e., intergranular) force and the frictional shearing force. At slippage, this resultant reaction acts at an angle ϕ to the normal. The extended line of action for the resultant is a tangent to the friction-circle located at the center of

---

[5]Wolmar Fellenius' 1936 paper, "Calculation of the Stability of Earth Dams" (2nd Congress on Large Dams, Washington, D.C.) also included charted curves relating slope stability to slope geometry and soil properties [96].

rotation for the sliding mass; properties of the friction circle are defined in Figure 15-20. At the limit of the equilibrium condition for the slice, the soil cohesion c, the resultant R (which includes the effect of internal friction ϕ), and the weight of the soil W are in balance (Fig. 15-20c). This requirement for equilibrium is extended to an entire trial failure mass and relates total weight, total cohesion, and total friction-related reactive force. Values for these factors that will provide an equilibrium condition reflect the geometry of the assumed failure mass and the properties of the involved soil. In his work, Taylor introduced a slope stability number N_s, to group factors that affect the safe inclination of a soil's slope.

$$N_s = \frac{c}{\gamma H_{max}} = \frac{c}{\gamma H_{actual} FS} \tag{15-14}$$

where N_s = slope stability number (dimensionless)
c = soil's cohesion (kN/m^2, psf)
γ = soil unit weight (kN/m^3, pcf)
H = slope height (ft, m)

Where a slope is at limit equilibrium (sliding is impending), a unique relationship exists between its height and inclination and the soil properties (γ, c, ϕ). This relationship provides the curves of Figure 15-21a and b. These curves are for slopes where the soils are uniform with depth and isotropic. It is important to recognize that natural deposits of cohesive soils do not tend to possess constant strength and density through large vertical distances; under conditions giving normally consolidated and overconsolidated deposits, the strength and density of such soils typically increase with depth. However, constructed embankments often are homogeneous, whereas natural deposits possessing homogeneity may exist or may be assumed in practical applications if limited heights are considered.

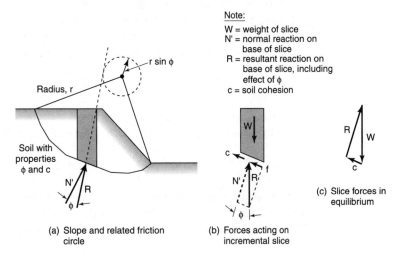

Note:
W = weight of slice
N' = normal reaction on base of slice
R = resultant reaction on base of slice, including effect of ϕ
c = soil cohesion

(a) Slope and related friction circle

(b) Forces acting on incremental slice

(c) Slice forces in equilibrium

Figure 15-20 Basis for the friction-circle analysis.

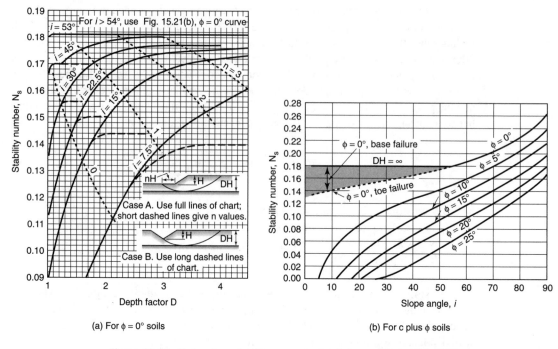

Figure 15-21 Taylor slope stability curves for total stress analysis [290, 291].

The Taylor curves are based upon a *total stress analysis,*[6] and the related soil strength values are obtained from undrained shear tests. These charts, therefore, are applicable for the end-of-construction and short-term stability category of analysis, where slope drainage and resulting changes in pore pressures and intergranular stresses do not have time to occur. In applications, the curves are often utilized to determine the safe inclination for an identified slope height, or the maximum height allowable when a particular inclination is desired. The H term in Equation 15-14 can be used to determine the maximum safe height based upon slope inclination and soil properties, which then can be compared to an actually existing or planned height. When the actual height is less than the maximum, the factor of safety is

$$FS = \frac{H_{max}}{H_{actual}} \tag{15-15a}$$

This factor of safety is actually a value based upon the ratio of maximum cohesion possessed by the soil (c_{max}) to the cohesion required for stability (c_{reqd}):

$$FS = \frac{c_{max}}{c_{reqd}} \tag{15-15b}$$

For slopes composed of ϕ plus c soils, the slope's factor of safety in regard to the soil's internal friction is 1 when a test-determined value of ϕ is used to enter the stability chart. Where it is desired to have an equal factor of safety for both cohesion and internal friction,

---

[6]Total stress equals effective stress plus neutral stress due to pore water pressure.

the actual values of c and ϕ are both divided by the factor of safety, and the charts are then entered with these new values.

Illustration 15-6

A slope 12 m (40 ft) high has a face inclination of 50°. Undrained shear strength tests on the soil indicate that the cohesion is 24 kN/m² (500 psf) and ϕ is 15°. The unit weight is 17.3 kN/m³ (110 pcf). What is the factor of safety (in regard to cohesion only)?

Solution

For $i = 50°$, $\phi = 15°$, get $N_s = 0.095$ (Fig. 15-21b):

$$FS = \frac{c}{\gamma H_{act} N_s} = \frac{24 \text{ kN/m}^2}{(17.3 \text{ kN/m}^3)(12 \text{ m})(.095)} = 1.2$$

The critical height (maximum height) is

$$H_{max} = (FS)H_{act} = (1.2)(12 \text{ m}) \cong 15 \text{ m or } 48 \text{ ft}$$

or, since $N_s = c/\gamma H_{max}$,

$$H_{max} = \frac{c}{\gamma N_s} = \frac{24 \text{ kN/m}^2}{(17.3 \text{ kN/m}^3)(.095)} = 15 \text{ m}$$

Illustration 15-7

What inclination is required where a slope 10 m high is to be constructed and must possess a factor of safety of 1.25 (this factor of safety is to apply to both c and ϕ)? The soil properties are $\gamma = 16.5 \text{ kN/m}^3$, $c = 16.8 \text{ kN/m}^2$, and $\phi = 10°$.

Solution

$$N_s = \frac{c}{\gamma H_{act}(FS)} = \frac{16.8 \text{ kN/m}^2}{(16.5 \text{ kN/m}^3)(10 \text{ m})(1.25)} = .082$$

and

$$\phi_{chart} = \frac{\phi}{FS} = \frac{10°}{1.25} = 8°$$

For $N_s = 0.082$, $\phi = 8°$, obtain slope angle $i \cong 28°$ (Fig. 15-21b).

Illustration 15-8

What is the factor of safety for a 45° slope 40 ft high in a clay soil ($\phi = 0°$, $c = 1000$ psf, $\gamma = 110$ pcf) where a rock stratum exists 40 ft beneath the toe elevation?

Solution

$$D = \frac{DH}{H} = \frac{(40 + 40) \text{ ft}}{40 \text{ ft}} = 2$$

For $i = 45°$, $D = 2$, $\phi = 0°$, obtain $N_s = 0.178$ (Fig. 15-21a).

$$FS = \frac{c}{N_s \gamma H_{act}} = \frac{1000 \text{ psf}}{(.178)(110 \text{ pcf})(40 \text{ ft})} = 1.28$$

Also note, $nH \cong (1.6)(40 \text{ ft}) \cong 64 \text{ ft}$

Figure 15-21a and b can also be used to study submerged slopes and slopes subject to sudden drawdown. For evaluating the stability of a *submerged slope,* the only modification to procedures previously outlined is that the submerged soil unit weight be used in determining the value of N_s. The *sudden drawdown* case occurs where a submerged embankment (e.g., a dam) has a rapid drop in the level of the impounded water. This drawdown condition is often the most critical for an embankment, as seepage forces resulting when water drains out of an embankment are additive to the ever-present gravity forces acting to cause slope sliding.[7] With the sudden drawdown case, a modified value of ϕ, indicated as ϕ_{dd}, is necessary, but other factors remain as defined earlier. For practical purposes, the modified ϕ is obtained from

$$\phi_{dd}^{\circ} = \frac{\gamma_{sub}}{\gamma_{total}} (\phi^{\circ})$$

(15-16)

where γ_{sub} = submerged soil unit weight
γ_{total} = total or wet woil unit weight

Illustration 15-9

(a) Determine the factor of safety (in regard to cohesion only) for a submerged embankment 30.5 m (100 ft high) whose upstream face has an inclination of 30°. The soil properties are γ_{total} = 18.85 kN/m³ (120 pcf), c = 37.6 kN/m² (700 psf), ϕ = 10°.

Solution
For i = 30°, ϕ = 10°, obtain N_s = 0.075 (Fig. 15-21a).

$$FS = \frac{c}{N_s \gamma_{sub} H_{act}} = \frac{700 \text{ psf}}{(.075)\left(\dfrac{120}{2} \text{ pcf}\right)(100 \text{ ft})} = 1.56$$

(b) What is the factor of safety if the embankment experiences the effects of a sudden drawdown?

Solution

$$\phi_{dd} = \frac{\gamma_{sub}}{\gamma_{total}} (\phi) \cong \frac{60 \text{ pcf}}{120 \text{ pcf}} (10°) \cong 5°$$

For i = 30°, ϕ = 5°, obtain N_s = 0.11.

$$FS = \frac{c}{N_s \gamma H_{act}} = \frac{700 \text{ psf}}{(.11)(120 \text{ pcf})(100 \text{ ft})} = .53 \quad \text{(failure)}$$

Taylor's analysis also includes a procedure to locate the slope's most critical failure circle, information that is of practical importance for identifying the horizontal area and vertical zone involved with and affected by a slide. Figure 15-22 provides data necessary to locate the critical circle where relatively steep slopes are under study (where the failure arc passes through the toe of the slope). Data for the flatter slopes, where the critical circle passes through the base and whose location is affected by hard substrata, are not developed. However, for soils having zero internal friction (ϕ = 0), the center of the critical circle is located above the midpoint of the slope.

---

[7]Also see further discussion of drawdown in the next subsection, on Bishop-Morganstern's charts.

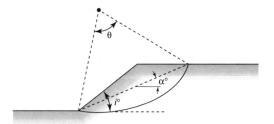

Angle of internal friction	Slope inclination, $i°$									
	$i = 90°$		$i = 75°$		$i = 60°$		$i = 45°$		$i = 30°$	
$\phi°$	$\alpha°$	$\theta°$	$\alpha°$	$\theta°$	$\alpha°$	$\theta°$	$\alpha°$	$\theta°$	$\alpha°$	$\theta°$
0	47.6	30.2	41.8	51.8	35.3	70.8	*	*	*	*
5	50	28	45	50	38.5	69	31.2	84.2	*	*
10	53	27	48	47	41	66	34	79.4	25	88
15	56	26	50	46	44	63	36.1	74.4	27	78
20	58	24	53	44	46.5	60.4	38	69	28	61
25	60	22	56	44	50	60	40	62	29	50

*Critical circle is a base circle, not a toe circle

Figure 15-22 Data for locating center of critical slip circle for toe failures (Taylor analysis, $\phi = 0$ case [290]).

Bishop-Morganstern, Barnes Charts

For the evaluation of long-term stability, slope analyses must be performed in terms of effective stresses and pore water pressures acting on the slippage plane. Soil properties for such analysis (c' and ϕ) are obtained from drained shear tests. A number of investigators have produced stability charts or equivalent tabulations from effective stress studies of slopes exposed to a variety of conditions. Typically, the developments have involved certain assumed conditions relating to slope equilibrium, and may also include simplifying assumptions where effects are small. Commonly, the resulting design charts or tables have application to a wide range of practical conditions and deal in easily calculated parameters.

Charts and tables based on the work of Bishop and Morganstern [31] are in wide usage for evaluating conditions where groundwater or seepage affect the soil shear strength and related slope stability. For the effective stress analysis, pore water pressures and intergranular forces can be handled as separate items. It has been found convenient to handle the influence of pore pressures in terms of a pore pressure ratio, r_u, the ratio of pore water pressure to total soil overburden pressure at a particular depth in the soil mass (typically at the slippage plane), defined as

$$r_u = u/\sigma_v = (\gamma_w h_w)/(\gamma_{soil} Z) \quad \text{(Fig. 15-23)}$$

The Bishop-Morganstern charts and tables have been developed from computer solutions of the simplified Bishop method of analysis described earlier in the chapter and where the calculated factor of safety against the slope failure relates to the shear strength of the soil.

The Bishop-Morganstern procedure provides results in terms of the dimensionless parameter $c'/\gamma H$, the slope inclination i, the pore pressure ratio r_u, and the depth DH, where the arc defining the slip plane is tangent to the surface of a buried firm stratum (ref. Fig. 15-24).

The simplified Bishop method of analysis requires that a series of assumed or trial failure arcs be analyzed in order to determine the least (or actual) factor of safety for the slope. Bishop and Morganstern noted that, for a given soil profile and soil properties, the calculated factor of safety varies in relation to the pore pressure ratio r_u, as illustrated by Figure 15-26. If the slope of the plotted line is $-n$, the factor of safety for any pore pressure value can be determined as the factor of safety at zero pore pressure minus nr_u. If m indicates the factor of safety at zero pore pressure and the slope factor of safety for any pore pressure condition is indicated as FS_{r_u}, then

$$FS_{r_u} = m - nr_u \qquad (15\text{-}17)$$

The m and n coefficients for different combinations of slope geometry, soil properties, and pore water pressures are tabulated in Table 15-2; slope inclinations are expressed

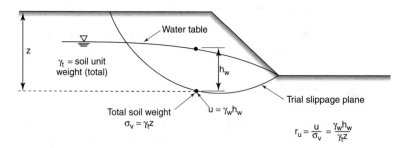

Figure 15-23 Definition of terms for calculating pore pressure ratio, r_u.

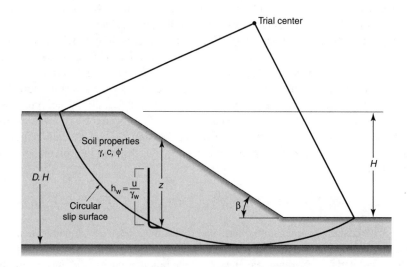

Figure 15-24 Notation and specification of parameters (Bishop-Morganstern).

Figure 15-25 Illustration of linear relationship between factor of safety, FS, and pore pressure ratio, r_u.

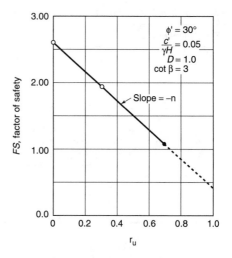

Figure 15-26 Excavated slopes, notation and parameters (Barnes [22]).

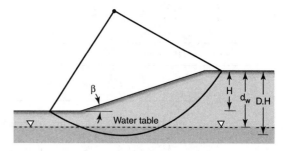

(a) Water level below toe,
critical circle passes below toe

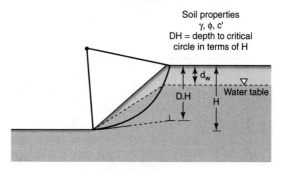

(b) Water level extends into slope,
critical circle located above slope

Table 15-2 Slope Stability *m* and *n* Factors (Bishop-Morganstern Method)

(a) $c'/\gamma H = 0$; all values of D

Slope	0.5:1		1:1		2:1		3:1		4:1		5:1	
	m	*n*	*m*	*n*	*m*	*n*	*m*	*n*	*m*	*n*	*m*	*n*
ϕ												
20°	0.182	0.910	0.364	0.728	0.728	0.910	1.092	1.213	1.456	1.547	1.820	1.892
25°	0.233	1.166	0.466	0.933	0.933	1.166	1.399	1.554	1.865	1.982	2.332	2.424
30°	0.289	1.443	0.577	1.155	1.155	1.443	1.732	1.924	2.309	2.454	2.887	3.001
35°	0.350	1.751	0.700	1.400	1.400	1.751	2.101	2.334	2.801	2.977	3.501	3.639
40°	0.420	2.098	0.839	1.678	1.678	2.098	2.517	2.797	3.356	3.566	4.196	4.362

(b) $c'/\gamma H = 0.025$; $D = 1.00$

Slope	0.5:1		1:1		2:1		3:1		4:1		5:1	
	m	*n*	*m*	*n*	*m*	*n*	*m*	*n*	*m*	*n*	*m*	*n*
ϕ												
20°	0.523	0.733	0.707	0.764	1.124	1.022	1.542	1.347	1.962	1.698	2.380	2.050
25°	0.586	0.740	0.839	0.974	1.356	1.282	1.875	1.696	2.400	2.141	2.921	2.596
30°	0.667	0.875	0.983	1.202	1.606	1.567	2.235	2.078	2.873	2.622	3.508	3.191
35°	0.755	1.010	1.142	1.454	1.880	1.885	2.635	2.505	3.396	3.160	4.156	3.849
40°	0.857	1.183	1.318	1.731	2.190	2.247	3.090	2.993	3.984	3.778	4.885	4.592

(c) $c'/\gamma H = 0.025$; $D = 1.25$

Slope	0.5:1		1:1		2:1		3:1		4:1		5:1	
	m	*n*	*m*	*n*	*m*	*n*	*m*	*n*	*m*	*n*	*m*	*n*
ϕ												
20°	0.996	0.943	1.076	1.036	1.309	1.216	1.618	1.478	1.961	1.775	2.334	2.090
25°	1.224	1.198	1.328	1.321	1.619	1.547	2.007	1.891	2.437	2.269	2.897	2.669
30°	1.472	1.483	1.602	1.631	1.956	1.915	2.431	2.342	2.953	2.806	3.511	3.299
35°	1.737	1.773	1.905	1.974	2.331	2.321	2.901	2.841	3.524	3.400	4.191	3.998
40°	2.045	2.118	2.242	2.362	2.753	2.775	3.431	3.399	4.164	4.064	4.958	4.784

(d) $c'/\gamma H = 0.050$; $D = 1.00$

Slope	0.5:1		1:1		2:1		3:1		4:1		5:1	
	m	*n*	*m*	*n*	*m*	*n*	*m*	*n*	*m*	*n*	*m*	*n*
ϕ												
20°	0.688	0.783	0.912	0.818	1.380	1.071	1.840	1.387	2.333	1.776	2.783	2.091
25°	0.797	1.000	1.069	1.042	1.624	1.338	2.193	1.757	2.778	2.211	3.336	2.651
30°	0.908	1.217	1.222	1.247	1.888	1.630	2.574	2.157	3.261	2.693	3.934	3.259
35°	1.032	1.417	1.379	1.469	2.178	1.958	2.990	2.592	3.803	3.253	4.597	3.927
40°	1.148	1.617	1.559	1.755	2.505	2.332	3.451	3.071	4.425	3.926	5.344	4.668

(e) $c'/\gamma H = 0.050$; $D = 1.25$

Slope	0.5:1		1:1		2:1		3:1		4:1		5:1	
	m	*n*	*m*	*n*	*m*	*n*	*m*	*n*	*m*	*n*	*m*	*n*
ϕ												
20°	1.172	0.988	1.253	1.084	1.509	1.266	1.834	1.493	2.230	1.799	2.643	2.107
25°	1.405	1.242	1.509	1.363	1.822	1.595	2.222	1.897	2.705	2.287	3.211	2.690
30°	1.656	1.518	1.783	1.669	2.161	1.950	2.645	2.342	3.221	2.819	3.829	3.324
35°	1.935	1.830	2.087	2.007	2.535	2.344	3.114	2.839	3.795	3.413	4.511	4.025
40°	2.245	2.174	2.429	2.390	2.953	2.791	3.642	3.400	4.442	4.090	5.273	4.806

(f) $c'/\gamma H = 0.050; D = 1.50$

Slope	0.5:1		1:1		2:1		3:1		4:1		5:1	
	m	n	m	n	m	n	m	n	m	n	m	n
ϕ												
20°	1.491	1.289	1.561	1.343	1.752	1.501	2.011	1.705	2.337	1.993	2.690	2.280
25°	1.826	1.637	1.910	1.709	2.143	1.903	2.467	2.179	2.867	2.534	3.302	2.902
30°	2.187	2.015	2.287	2.104	2.568	2.342	2.964	2.696	3.443	3.120	3.967	3.577
35°	2.587	2.436	2.704	2.541	3.041	2.832	3.515	3.269	4.082	3.771	4.707	4.325
40°	3.040	2.915	3.175	3.036	3.574	3.389	4.136	3.915	4.803	4.507	5.543	5.171

(g) $c'/\gamma H = 0.075; D = 1.00$

Slope	0.5:1		1:1		2:1		3:1		4:1		5:1	
	m	n	m	n	m	n	m	n	m	n	m	n
ϕ												
20°	0.845	0.800	1.088	0.837	1.610	1.100	2.141	1.443	2.664	1.801	3.173	2.130
25°	0.950	1.013	1.245	1.053	1.872	1.386	2.502	1.815	3.126	2.259	3.742	2.715
30°	1.064	1.238	1.416	1.296	2.142	1.686	2.884	2.201	3.623	2.758	4.357	3.331
35°	1.190	1.485	1.605	1.564	2.443	2.030	3.306	2.659	4.177	3.331	5.024	4.001
40°	1.332	1.762	1.798	1.824	2.772	2.386	3.775	3.145	4.785	3.945	5.776	4.759

(h) $c'/\gamma H = 0.075; D = 1.25$

Slope	0.5:1		1:1		2:1		3:1		4:1		5:1	
	m	n	m	n	m	n	m	n	m	n	m	n
ϕ												
20°	1.336	1.023	1.387	1.087	1.688	1.285	2.071	1.543	2.492	1.815	2.954	2.173
25°	1.575	1.284	1.656	1.386	2.004	1.641	2.469	1.957	2.972	2.315	3.523	2.730
30°	1.830	1.560	1.943	1.701	2.352	2.015	2.888	2.385	3.499	2.857	4.149	3.357
35°	2.109	1.865	2.245	2.025	2.728	2.385	3.357	2.870	4.079	3.457	4.831	4.043
40°	2.424	2.210	2.583	2.403	3.154	2.841	3.889	3.248	4.729	4.128	5.603	4.830

(i) $c'/\gamma H = 0.075; D = 1.50$

Slope	0.5:1		1:1		2:1		3:1		4:1		5:1	
	m	n	m	n	m	n	m	n	m	n	m	n
ϕ												
20°	1.637	1.305	1.706	1.349	1.918	1.514	2.199	1.728	2.548	1.985	2.931	2.272
25°	1.977	1.663	2.052	1.708	2.308	1.914	2.660	2.200	3.083	2.530	3.552	2.915
30°	2.340	2.041	2.426	2.100	2.735	2.355	3.158	2.714	3.659	3.128	4.218	3.585
35°	2.741	2.459	2.841	2.537	3.211	2.854	3.708	3.285	4.302	3.786	4.961	4.343
40°	3.193	2.931	3.310	3.031	3.742	3.397	4.332	3.926	5.026	4.527	5.788	5.185

(j) $c'/\gamma H = 0.100; D = 1.00$

Slope	0.5:1		1:1		2:1		3:1		4:1		5:1	
	m	n	m	n	m	n	m	n	m	n	m	n
ϕ												
20°	0.993	0.797	1.263	0.871	1.841	1.143	2.421	1.472	2.982	1.815	3.549	2.157
25°	1.106	1.025	1.422	1.078	2.102	1.430	2.785	1.845	3.358	2.303	4.131	2.743
30°	1.222	1.259	1.592	1.306	2.378	1.714	3.183	2.258	3.973	2.830	4.751	3.372
35°	1.347	1.508	1.781	1.576	2.692	2.086	3.612	2.715	4.516	3.359	5.426	4.059
40°	1.489	1.788	1.995	1.879	3.025	2.445	4.103	3.230	5.144	4.001	6.187	4.831

(k) $c'/\gamma H = 0.100; D = 1.25$

Slope	0.5:1		1:1		2:1		3:1		4:1		5:1	
	m	n	m	n	m	n	m	n	m	n	m	n
ϕ												
20°	1.489	1.036	1.529	1.095	1.874	1.301	2.283	1.558	2.751	1.843	3.253	2.158
25°	1.735	1.313	1.799	1.394	2.197	1.642	2.681	1.972	3.233	2.330	3.833	2.758
30°	1.997	1.602	2.091	1.718	2.540	2.000	3.112	2.415	3.753	2.858	4.451	3.372
35°	2.280	1.908	2.414	2.076	2.922	2.415	3.588	2.914	4.333	3.458	5.141	4.072
40°	2.597	2.253	2.763	2.453	3.345	2.855	4.119	3.457	4.987	4.142	5.921	4.872

(l) $c'/\gamma H = 0.100; D = 1.50$

Slope	0.5:1		1:1		2:1		3:1		4:1		5:1	
	m	n	m	n	m	n	m	n	m	n	m	n
ϕ												
20°	1.778	1.314	1.863	1.371	2.079	1.528	2.387	1.742	2.768	2.014	3.158	2.285
25°	2.119	1.674	2.211	1.732	2.477	1.942	2.852	2.215	3.297	2.542	3.796	2.927
30°	2.489	2.063	2.586	2.122	2.908	2.385	3.349	2.728	3.881	3.143	4.468	3.614
35°	2.892	2.484	3.000	2.553	3.385	2.884	3.900	3.300	4.520	3.800	5.211	4.372
40°	3.347	2.957	3.469	3.046	3.924	3.441	4.524	3.941	5.247	4.542	6.040	5.200

(m) $c'/\gamma H = 0.125; D = 1.00$

Slope	0.5:1		1:1		2:1		3:1		4:1		5:1	
	m	n	m	n	m	n	m	n	m	n	m	n
ϕ												
20°	1.121	0.808	1.425	0.881	2.042	1.148	2.689	1.541	3.263	1.784	3.868	2.124
25°	1.254	1.051	1.596	1.112	2.323	1.447	3.062	1.908	3.737	2.271	4.446	2.721
30°	1.376	1.267	1.769	1.337	2.618	1.777	3.457	2.298	4.253	2.810	5.073	3.368
35°	1.505	1.530	1.956	1.586	2.929	2.115	3.880	2.705	4.823	3.407	5.767	4.048
40°	1.612	1.743	2.171	1.891	3.272	2.483	4.356	3.183	5.457	4.060	6.551	4.893

(n) $c'/\gamma H = 0.125; D = 1.25$

Slope	0.5:1		1:1		2:1		3:1		4:1		5:1	
	m	n	m	n	m	n	m	n	m	n	m	n
ϕ												
20°	1.642	1.057	1.671	1.102	2.054	1.324	2.492	1.579	2.983	1.861	3.496	2.167
25°	1.888	1.326	1.941	1.402	2.377	1.671	2.894	1.993	3.481	2.379	4.078	2.753
30°	2.156	1.626	2.234	1.727	2.727	2.042	3.324	2.431	4.009	2.916	4.712	3.405
35°	2.447	1.948	2.557	2.085	3.110	2.452	3.801	2.928	4.586	3.500	5.414	4.128
40°	2.767	2.295	2.922	2.490	3.542	2.913	4.338	3.494	5.237	4.161	6.207	4.945

(o) $c'/\gamma H = 0.125; D = 1.50$

Slope	0.5:1		1:1		2:1		3:1		4:1		5:1	
	m	n	m	n	m	n	m	n	m	n	m	n
ϕ												
20°	1.920	1.322	2.015	1.385	2.234	1.545	2.565	1.749	2.963	2.004	3.400	2.287
25°	2.261	1.683	2.368	1.754	2.638	1.972	3.028	2.229	3.500	2.550	4.019	2.913
30°	2.631	2.073	2.745	2.145	3.072	2.425	3.529	2.749	4.083	3.149	4.692	3.598
35°	3.039	2.504	3.160	2.577	3.549	2.923	4.084	3.324	4.727	3.813	5.436	4.362
40°	3.497	2.982	3.628	3.065	4.089	3.485	4.712	3.980	5.456	4.566	6.278	5.226

(p) $c'/\gamma H = 0.150$; $D = 1.00$

Slope	0.5:1		1:1		2:1		3:1		4:1		5:1	
	m	n	m	n	m	n	m	n	m	n	m	n
ϕ												
20°	1.248	0.813	1.585	0.886	2.261	1.170	2.895	1.448	3.579	1.806	4.230	2.159
25°	1.386	1.034	1.761	1.126	2.536	1.462	3.259	1.814	4.052	2.280	4.817	2.765
30°	1.525	1.260	1.944	1.370	2.836	1.791	3.657	2.245	4.567	2.811	5.451	3.416
35°	1.660	1.539	2.134	1.619	3.161	2.153	4.098	2.721	5.137	3.408	6.143	4.117
40°	1.805	1.832	2.346	1.901	3.512	2.535	4.597	3.258	5.782	4.083	6.913	4.888

(q) $c'/\gamma H = 0.150$; $D = 1.25$

Slope	0.5:1		1:1		2:1		3:1		4:1		5:1	
	m	n	m	n	m	n	m	n	m	n	m	n
ϕ												
20°	1.796	1.079	1.813	1.107	2.229	1.334	2.701	1.600	3.225	1.873	3.780	2.182
25°	2.042	1.344	2.083	1.409	2.560	1.692	3.107	2.015	3.724	2.384	4.363	2.769
30°	2.309	1.639	2.377	1.734	2.909	2.065	3.542	2.464	4.262	2.941	5.995	3.406
35°	2.605	1.971	2.700	2.094	3.295	2.475	4.018	2.946	4.846	3.534	5.697	4.129
40°	2.934	2.335	3.066	2.449	3.728	2.938	4.556	3.509	5.498	4.195	6.490	4.947

(r) $c'/\gamma H = 0.150$; $D = 1.50$

Slope	0.5:1		1:1		2:1		3:1		4:1		5:1	
	m	n	m	n	m	n	m	n	m	n	m	n
ϕ												
20°	2.061	1.335	2.164	1.391	2.394	1.550	2.748	1.756	3.174	2.020	3.641	2.308
25°	2.402	1.691	2.520	1.768	2.798	1.978	3.212	2.237	3.711	2.561	4.259	2.924
30°	2.772	2.082	2.902	2.168	3.236	2.441	3.718	2.758	4.293	3.156	4.931	3.604
35°	3.181	2.514	3.319	2.600	3.715	2.940	4.269	3.333	4.938	3.819	5.675	4.364
40°	3.643	3.000	3.788	3.088	4.255	3.503	4.896	3.983	5.667	4.569	6.517	5.228

as the ratio of horizontal to vertical distances. This table includes the work of investigators who extended the original Bishop-Morganstern information [54]. The procedure to evaluate the stability of a particular slope should include calculating the factor of safety using the lowest m and n terms for each D value (appropriate to $c'/\gamma H$ and r_u) shown in the table, then selecting the lowest answer. For the situation where the $c'/\gamma H$ value is intermediate to the table values, the factor of safety can be calculated by using the tabulated values that bracket the actual values, then obtaining the answer by linear interpretation.

Normally, the pore pressure r_u is expected to vary within the slope zone being studied; an average value or weighted value should be used for r_u.

The values in Table 15.2 were developed for general usage from studying a variety of depths for the path of the assumed slippage or failure plane. One investigator (Barnes [21]) also developed m and n values for the condition of the slip plane passing through the toe of the slope (Table 15-3) concluding that for many situations the factor of safety is lowest for a toe circle.

Table 15-3 Slope Stability m and n Factors for Toe Circles (Barnes)

ϕ	$c'/\gamma H = 0$		$c'/\gamma H = 0.025$		$c'/\gamma H = 0.050$		$c'/\gamma H = 0.075$		$c'/\gamma H = 0.100$		$c'/\gamma H = 0.125$		$c'/\gamma H = 0.150$	
(°)	m	n	m	n	m	n	m	n	m	n	m	n	m	n
Slope = 0.5:1														
20	0.182	0.910	0.486	0.758	0.660	0.771	0.811	0.785	0.958	0.810	1.094	0.820	1.227	0.830
25	0.233	1.166	0.563	0.986	0.749	0.982	0.909	0.995	1.061	1.015	1.202	1.026	1.340	1.040
30	0.289	1.443	0.649	1.236	0.848	1.212	1.019	1.222	1.174	1.237	1.321	1.249	1.465	1.267
35	0.350	1.751	0.734	1.503	0.943	1.461	1.125	1.479	1.288	1.490	1.442	1.502	1.591	1.520
40	0.420	2.098	0.833	1.807	1.055	1.745	1.250	1.768	1.421	1.773	1.583	1.786	1.737	1.805
45	0.500	2.500	0.952	2.164	1.192	2.076	1.402	2.100	1.580	2.098	1.751	2.112	1.912	2.135
Slope = 1:1														
20	0.364	0.728	0.701	0.766	0.909	0.810	1.101	0.859	1.270	0.878	1.436	0.902	1.600	0.925
25	0.466	0.933	0.829	0.969	1.050	1.013	1.248	1.060	1.427	1.089	1.601	1.117	1.769	1.144
30	0.577	1.155	0.968	1.189	1.203	1.234	1.408	1.277	1.598	1.317	1.779	1.351	1.952	1.381
35	0.700	1.400	1.114	1.431	1.363	1.478	1.578	1.523	1.776	1.563	1.964	1.603	2.143	1.635
40	0.839	1.678	1.280	1.705	1.544	1.753	1.770	1.801	1.978	1.846	2.172	1.887	2.359	1.922
45	1.000	2.000	1.472	2.022	1.754	2.072	1.993	2.123	2.212	2.171	2.414	2.217	2.609	2.255
Slope = 1.5:1														
20	0.546	0.789	0.913	0.880	1.145	0.938	1.352	0.982	1.545	1.018	1.730	1.050	1.912	1.080
25	0.700	1.010	1.093	1.105	1.339	1.169	1.556	1.218	1.758	1.260	1.951	1.298	2.138	1.331
30	0.866	1.251	1.288	1.349	1.550	1.419	1.777	1.474	1.989	1.523	2.190	1.566	2.384	1.604
35	1.050	1.517	1.497	1.621	1.773	1.694	2.011	1.754	2.233	1.808	2.440	1.854	2.642	1.898
40	1.259	1.818	1.734	1.928	2.025	2.005	2.276	2.070	2.508	2.130	2.723	2.180	2.934	2.230
45	1.500	2.167	2.008	2.284	2.317	2.365	2.583	2.436	2.827	2.503	3.051	2.557	3.272	2.615
Slope = 2:1														
20	0.728	0.910	1.121	1.027	1.372	1.094	1.593	1.158	1.804	1.201	2.000	1.225	2.194	1.260
25	0.933	1.166	1.354	1.291	1.620	1.366	1.853	1.432	2.076	1.488	2.278	1.515	2.478	1.554
30	1.155	1.443	1.607	1.578	1.890	1.661	2.133	1.730	2.362	1.786	2.580	1.830	2.787	1.873
35	1.400	1.751	1.879	1.891	2.176	1.981	2.433	2.058	2.673	2.130	2.895	2.165	3.110	2.214
40	1.678	2.098	2.186	2.245	2.499	2.342	2.773	2.430	3.012	2.486	3.252	2.543	3.476	2.599
45	2.000	2.500	2.542	2.655	2.873	2.761	3.166	2.860	3.413	2.921	3.665	2.981	3.900	3.045
Slope = 3:1														
20	1.092	1.213	1.530	1.362	1.808	1.444	2.055	1.516	2.286	1.588	2.501	1.604	2.714	1.649
25	1.399	1.554	1.869	1.716	2.164	1.806	2.426	1.888	2.665	1.958	2.892	1.989	3.112	2.037
30	1.732	1.924	2.236	2.099	2.550	2.199	2.826	2.288	3.076	2.359	3.316	2.407	3.543	2.458
35	2.101	2.334	2.634	2.521	2.965	2.629	3.253	2.730	3.518	2.803	3.765	2.855	4.000	2.908
40	2.517	2.797	3.084	2.997	3.433	3.114	3.737	3.231	4.008	3.303	4.272	3.361	4.516	3.417
45	3.000	3.333	3.605	3.549	3.976	3.677	4.297	3.811	4.582	3.884	4.860	3.947	5.114	4.007

							Slope = 4:1							
20	1.456	1.547	1.932	1.720	2.232	1.811	2.498	1.903	2.748	1.974	2.974	1.998	3.200	2.048
25	1.865	1.982	2.376	2.170	2.695	2.270	2.980	2.361	3.246	2.459	3.477	2.482	3.710	2.535
30	2.309	2.454	2.858	2.658	3.197	2.768	3.496	2.888	3.770	2.961	4.022	3.007	4.264	3.064
35	2.801	2.977	3.381	3.194	3.738	3.312	4.055	3.445	4.339	3.518	4.600	3.571	4.854	3.635
40	3.356	3.566	3.973	3.800	4.350	3.928	4.680	4.061	4.984	4.173	5.254	4.209	5.520	4.281
45	4.000	4.250	4.658	4.502	5.059	4.641	5.409	4.784	5.730	4.915	6.011	4.948	6.292	5.029
							Slope = 5:1							
20	1.820	1.892	2.327	2.082	2.648	2.185	2.934	2.301	3.190	2.361	3.430	2.394	3.665	2.447
25	2.332	2.424	2.876	2.630	3.217	2.744	3.520	2.861	3.796	2.944	4.042	2.979	4.288	3.038
30	2.887	3.001	3.472	3.225	3.834	3.351	4.150	3.461	4.442	3.576	4.707	3.613	4.965	3.680
35	3.501	3.639	4.122	3.880	4.501	4.012	4.846	4.159	5.146	4.249	5.415	4.297	5.685	4.373
40	4.196	4.362	4.856	4.621	5.256	4.760	5.609	4.918	5.923	5.019	6.216	5.070	6.499	5.154
45	5.000	5.200	5.707	5.479	6.130	5.626	6.509	5.818	6.836	5.905	7.144	5.966	7.442	6.060

Illustration 15-10

Determine the slope stability factor of safety for a site where the slope-region conditions are as follows: slope = 4 horizontal to 1 vertical; $c'/\gamma H = 0.05$, $\phi' = 30°$, r_u in slope = 0.5.
(a) Use the Bishop-Morganstern procedure.
(b) Use the Barnes toe failure circle procedure.

Solution
Bishop-Morganstern analysis:

(i) For 4:1 slope, $r_u = 0.5$, $c'/\gamma H = 0.05$, assume trial $D = 1.00$. From Table 15-2d, obtain $m = 3.261$, $n = 2.693$. So,

$$FS = 3.261 - (.5)(2.693) = 1.91$$

(ii) For the same slope conditions, assume trial $D = 1.25$. From Table 15-2e obtain $m = 3.221$, $n = 2.819$. So,

$$FS = 3.221 - (.5)(2.819) = 1.81$$

(iii) For the same slope conditions, assume trial $D = 1.50$. From Table 15-2f, obtain $m = 3.443$, $n = 3.120$. So,

$$FS = 3.443 - (.5)(3.120) = 1.88$$

Therefore, design FS (the lowest calculated value for FS) = 1.81.
(b) Barnes toe failure circle analysis, for the same slope conditions: From Table 15-3, obtain $m = 3.197$, $n = 2.768$. So,

$$FS = 3.197 - (.5)(2.768) = 1.81$$

Barnes Charts/Tables for Excavated Slopes

The Bishop-Morganstern method discussed above and originally developed to evaluate earth dams requires knowledge of the pore water pressure distribution throughout the soil zone associated with the slip plane. Barnes [22] developed an alternative method to evaluate the stability of excavated earth slopes where the depth of groundwater is known (Fig. 15-26), but, as is common with many excavated slopes (highway cuts, etc.), no slope seepage occurs because a steady-state condition for water levels as shown is reached. The method is based on an effective stress analysis using the simplified Bishop method but in final form does not require use of pore water pressure distributions. The Barnes relationship is expressed as

$$FS = a + b \tan \phi \qquad (15\text{-}18)$$

where a and b are stability coefficients presented as tabulated values in Table 15-4, and $\tan \phi$ is the angle of internal friction from drained shear tests (or the drained strength conditions).

Table 15-4 Stability Factors for Excavated Slopes (Barnes)

d_w/H	$c'/\gamma H = 0.005$		$c'/\gamma H = 0.025$		$c'/\gamma H = 0.050$		$c'/\gamma H = 0.100$		$c'/\gamma H = 0.150$	
	a	b	a	b	a	b	a	b	a	b
0	↑	0.16	↑	0.27	↑	0.36	↑	0.45	↑	0.52
0.10		0.51		0.54		0.60		0.67		0.74
0.20		0.71		0.73		0.78		0.85		0.91
0.25	0.06	0.79	0.22	0.82	0.38	0.87	0.68	0.93	0.97	1.00
0.30		0.87		0.90		0.95		1.01		1.08
0.40		1.01		1.06		1.11		1.17		1.24
0.50	↓	1.15(35)		1.21		1.27		1.33		1.40
0.60			↓	1.29(35)		1.36		1.42		1.50
0.70						1.41(40)		1.50		1.60
0.75					↓	1.43(35)		1.54(45)		1.63(45)
0.80							↓	1.56(40)		1.65(40)
0.90									↓	1.70(30)
1.00										
DRY	0.08	1.12	0.27	1.21	0.44	1.32	0.75	1.46	1.04	1.56

Where ϕ is greater than the value shown in brackets treat the slope as 'dry'.
(a) Stability coefficients a and b for slope 1:1.

d_w/H	$c'/\gamma H = 0.005$		$c'/\gamma H = 0.025$		$c'/\gamma H = 0.050$		$c'/\gamma H = 0.100$		$c'/\gamma H = 0.150$	
	a	b	a	b	a	b	a	b	a	b
0	↑	0.88	↑	1.01	↑	1.11	↑	1.27	↑	1.37
0.10		1.24		1.31		1.38		1.50		1.60
0.20		1.46		1.51		1.57		1.69		1.78
0.25		1.56		1.60		1.66		1.78		1.87
0.30		1.64		1.69		1.75		1.86		1.95
0.40		1.81		1.85		1.91		2.02		2.11
0.50	0.06	1.96	0.24	2.00	0.42	2.06	0.75	2.17	1.07	2.26
0.60		2.07		2.13		2.22		2.33		2.41
0.70		2.17(45)		2.22		2.31		2.43		2.52
0.75		2.20(30)		2.27		2.35		2.48		2.57
0.80	↓	2.22(25)		2.30(45)		2.38		2.52		2.62
0.90				2.38(30)		2.44		2.60		2.70
1.00			↓	2.39(25)	↓	2.50(40)	↓	2.67	↓	2.79
DRY	0.10	2.08	0.30	2.24	0.50	2.37	0.86	2.56	1.20	2.69

Where ϕ is greater than the value shown in brackets treat the slope as 'dry'.
(b) Stability coefficients a and b for slope 2:1.

d_w/H	$c'/\gamma H = 0.005$		$c'/\gamma H = 0.025$		$c'/\gamma H = 0.050$		$c'/\gamma H = 0.100$		$c'/\gamma H = 0.150$	
	a	b	a	b	a	b	a	b	a	b
0	↑	1.48	↑	1.65	↑	1.77	↑	1.95	↑	2.09
0.10		1.87		1.95		2.03		2.18		2.30
0.20		2.11		2.16		2.24		2.37		2.48
0.25		2.21		2.26		2.33		2.45		2.56
0.30		2.31		2.35		2.42		2.54		2.65
0.40		2.49		2.53		2.59		2.71		2.81
0.50	0.07	2.66	0.25	2.69	0.44	2.75	0.79	2.87	1.12	2.97
0.60		2.79		2.85		2.92		3.02		3.13
0.70		2.91		2.96		3.03		3.14		3.25
0.75		2.97		3.01		3.09		3.20		3.31
0.80	↓	3.02		3.06		3.14		3.25		3.36
0.90		3.11		3.16		3.23		3.36		3.47
1.00			↓	3.25	↓	3.32	↓	3.45	↓	3.57
DRY	0.11	3.10	0.33	3.27	0.56	3.41	0.95	3.63	1.31	3.80

(c) Stability coefficients a and b for slope 3:1.

d_w/H	$c'/\gamma H = 0.005$		$c'/\gamma H = 0.025$		$c'/\gamma H = 0.050$		$c'/\gamma H = 0.100$		$c'/\gamma H = 0.150$	
	a	b	a	b	a	b	a	b	a	b
0	↑	2.05	↑	2.22	↑	2.35	↑	2.53	↑	2.67
0.10		2.46		2.54		2.63		2.78		2.90
0.20		2.72		2.77		2.84		2.98		3.09
0.25		2.83		2.88		2.94		3.07		3.18
0.30		2.93		2.98		3.04		3.16		3.27
0.40		3.13		3.17		3.22		3.34		3.44
0.50	0.07	3.32	0.26	3.35	0.47	3.40	0.84	3.51	1.18	3.61
0.60		3.49		3.52		3.57		3.67		3.77
0.70		3.62		3.66		3.70		3.81		3.90
0.75		3.69		3.72		3.76		3.87		3.97
0.80	↓	3.74		3.78		3.82		3.93		4.03
0.90		3.86		3.89		3.94		4.05		4.15
1.00		3.97	↓	4.00	↓	4.05	↓	4.16	↓	4.26
DRY	0.12	4.11	0.37	4.29	0.60	4.45	1.01	4.70	1.38	4.90

(d) Stability coefficients a and b for slope 4:1.

The indication *dry* in the table refers to the condition where the water level lies below the critical circle.

Illustration 15-11

Determine the slope stability factor of safety using the Barnes analysis for excavated slopes at a site where the slope is 2 horizontal to 1 vertical, $c'/\gamma H = 0.05$, $\phi' = 30°$, and the ratio d_w/H is 0.3.

Solution
From Table 15-4b for 2:1 slopes, obtain $a = 0.42$, $b = 1.75$.

$$FS = a + b \tan \phi' = .42 + 1.75 \tan 30° = 1.43$$

Spencer Charts

The Spencer method to evaluate slope stability, based on effective stress analysis, assumes a circular slip surface and divides the studied soil slope into a series of vertical slices, similar to the Bishop procedure [280]. The forces acting on each slice are as shown in Figure 15-15. The factor of safety for an analyzed slope relates to the shear strength properties of the soil, c' and ϕ (as determined from drained shear strength tests). The calculation procedure to determine a factor of safety requires achieving equilibrium for forces and moments acting on the slices (for the respective slices, using the center at the base of the slice for the moment axis). The Spencer method to determine the slope zone having the lowest factor of safety is a mathematically rigorous procedure best accommodated by computer because of the complex form of the equations and the required interactions. However, Spencer has developed chart solutions, presented in terms of a stability number N_s [where $N_s = c'/FS (\gamma H)$] (Eq. 15-14) and slope angle i (as used for the Taylor charts) for select pore pressure ratios r_u within the slope (Fig. 15-27). The pore pressures are assumed to be homogeneous throughout the slope. Results are in close agreement with solutions obtained by the simplified Bishop method of analysis.

The curves shown in Figure 15-27 are most easily used for the problem where the height of the slope (embankment height) as well as soil properties and pore pressure conditions are known, but the slope angle is to be determined. The curves provide an inclination angle for the *critical slope* (that is, where the slope stability factor of safety is equal to one and the slope is on the verge of sliding). To apply a desired factor of safety, it is necessary to enter the charts with a modified value for ϕ'_m, where ϕ'_m is ϕ_{actual}/FS (refer Illustration 15-12). If it is necessary *to determine a factor of safety* when the slope inclination and height along with soil properties and pore pressure conditions are known, a trial and error procedure is required, as shown in Illustration 15-13. For either category of problem, slopes having an r_u value between the indicated chart values can be analyzed by assuming that linear interpolation applies.

Illustration 15-12

Use the Spencer slope stability curves to determine the required slope angle i for the embankment 25 m high to have a factor of safety equal to 1.3. Soil properties are $\gamma = 19$ kN/m³, $c' = 28$ kPa, $\phi' = 26°$. The weighted pore pressure r_u for the analyzed condition is 0.50.

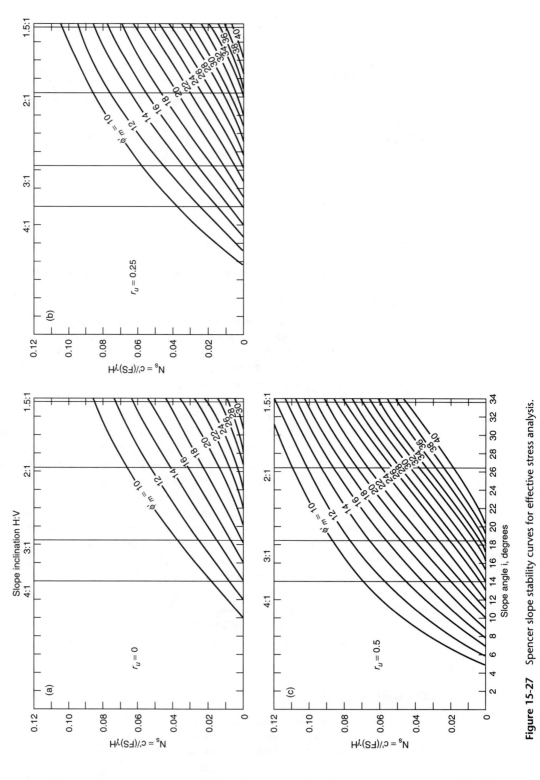

Figure 15-27 Spencer slope stability curves for effective stress analysis.

Solution
For the given values

$$N_s = \frac{c'}{FS(\gamma H)} = \frac{28 \text{ kN/m}^2}{(1.3)(19 \text{ kN/m}^3)(25 \text{ m})} = .045$$

From the curves of Figure 15-27, for $r_u = 0.5$, $\phi_m = 26°/FS = 20°$, obtain $i \approx 20°$.

Illustration 15-13

A 15-m high embankment is required for a construction site where soil properties will be as follows: $\gamma = 18.5 \text{ kN/m}^3$, $c' = 24 \text{ kPa}$, $\phi = 20°$, $r_u = 0.25$. For the condition where a slope inclination of 2.5 horizontal to 1 vertical (or approximately 22°) is planned, what factor of safety is provided? Use Spencer curves and FS $= \frac{c'}{N_s \gamma H}$.

Solution
A trial-and-error procedure is required because the factor of safety affects the value of ϕ'_m selected on the Spencer curves and the value for N_s needed to calculate FS.

> *Trial 1.* Try/assume FS $= 1.5$, then $\phi'_m = 20°/1.5 = 13.3°$; from curve for $\phi'_m = 13.3°$ and 2.5 to 1 slope, obtain $N_s = 0.05$. Next, calculate
>
> $$FS = c'/N_s \gamma H = 24 \text{ kN/m}^2/(.05)(18.5 \text{ kN/m}^3)(15 \text{ m}) = 1.73 \text{ (no good)}$$
>
> *Trial 2.* Try FS $= 1.75$, follow procedure above, calculate FS $= 1.37$ (no good).
>
> *Trial 3.* Try FS $= 1.6$, then $\phi'_m = 11.5°$ and $N_s = 0.063$. Calculate FS $= 1.57$ (close to assumed value; adequate). Therefore, FS ≈ 1.6.

Morganstern Charts for Rapid Drawdown

Stability charts applicable to the case of reservoir embankments subject to drawdown have been developed by Morganstern [208]. The drawdown condition is typically more critical than either the submerged or the nonsubmerged condition for embankments containing cohesive soils. Immediately after the reservoir level is lowered, a combination of potentially dangerous conditions exists: The weight of the embankment soil has changed from a submerged to the much heavier saturated state, while high pore water pressures are retained in the embankment along with the lower effective intergranular stresses as a carryover from the previously submerged condition, and seepage forces from pore water draining out of the slope act in the general direction of other gravity forces that tend to cause sliding. Morganstern's charts are based upon an effective stress analysis utilizing the method of slices. This work assumes that the soil properties are constant throughout the embankment and that the full slope height had been submerged prior to drawdown (Fig. 15-28a). Factors of safety against sliding can be computed for a complete drawdown[8] and for intermediate levels of drawdown. For a complete drawdown, the least factor of safety is for a slip circle passing tangent to the rigid base (Fig. 15-28b). Realistically, complete drawdowns occur

---

[8]Recall that Taylor's charts can also be used to evaluate the stability for a slope subject to a *complete* drawdown.

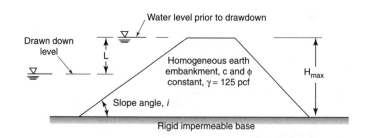

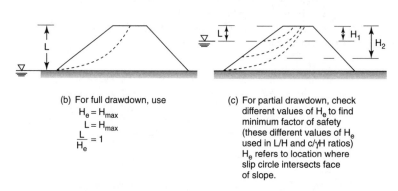

(b) For full drawdown, use
$$H_e = H_{max}$$
$$L = H_{max}$$
$$\frac{L}{H_e} = 1$$

(c) For partial drawdown, check different values of H_e to find minimum factor of safety (these different values of H_e used in L/H and $c/\gamma H$ ratios) H_e refers to location where slip circle intersects face of slope.

Figure 15-28 Terms and assumptions in Morganstern's stability analyses for rapid drawdown.

infrequently. When a partial drawdown is studied, the location of the most critical slip circle is determined by a trial procedure (Fig. 15-28c; see also Ill. 15-13).

The parameter $c'/\gamma H$ used by other investigators has also been applied by Morganstern, and his charts are in terms of this value and slope inclination (Fig. 15-29). Factors of safety are determined directly from the curves. Values of $c'/\gamma H$ differing from those presented can, for practical purposes, be obtained by interpolation.

Illustration 15-14

(a) Find the factor of safety when an 80-ft-high embankment with a three-to-one upstream slope experiences a complete drawdown. Soil properties are $\gamma_{total} = 125$ pcf, $c' = 250$ psf, and $\phi = 30°$

Solution

$$\frac{c'}{\gamma H} = \frac{250 \text{ psf}}{(125 \text{ pcf})(80 \text{ ft})} = .025$$

For a complete drawdown, the most critical value of $L/H_e = 80$ ft/80 ft or 1, obtain FS = 1.2 (see Fig. 15-28b and 15-29).

(b) Find the factor of safety when the drawdown is half of embankment height. (Assume that the slip is tangent to the embankment base.)

Solution

$$H_e = 80 \text{ ft}, \quad \frac{c'}{\gamma H} = .025$$

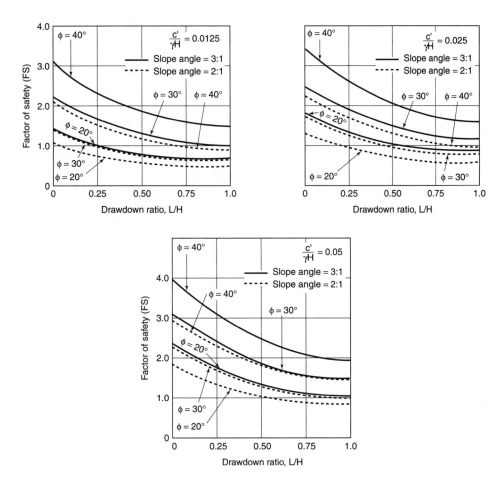

Figure 15-29 Morganstern stability curves for rapid drawdown [208].

For $L/H_e = \dfrac{(.5)(80 \text{ ft})}{(80 \text{ ft})} = .5$, obtain FS = 1.52.

(c) Assume that the slip circle is through the face of the slope and passing tangent to a horizontal plane at the depth L used in part b above (ref. Fig. 15-28c).

Solution

$$H_e = \frac{80 \text{ ft}}{2} = 40 \text{ ft}, \qquad \frac{c}{\gamma H_e} = \frac{250 \text{ psf}}{(125 \text{ pcf})\left(\dfrac{80 \text{ ft}}{2}\right)} = .05$$

For $L/H_e = \dfrac{40 \text{ ft}}{40 \text{ ft}} = 1$, obtain FS = 1.45.

15.5 VARIATION IN SHEAR STRESS AND FACTOR OF SAFETY

The factor of safety computed by the limit equilibrium method of analysis is the average value developed along the length of the slippage arc under study. However, it is recognized that the factor of safety (in regard to the maximum soil shear strength) actually varies at different locations along the slippage plane, because the state of stress within and beneath a slope varies considerably. Figure 15-30 illustrates how the increase in the soil zone stressed so to be at its ultimate shear strength occurs when an slope is heightened by excavating below the toe. Even in natural slopes that have long been stable, zones where the soil is near or at the ultimate shear strength probably exist. Increases in slope loading are then forced into being supported by the soil zones still at low stresses, because at or near the ultimate strength a soil deforms plastically and is unable to develop additional shearing resistance unless large strains (movements) occur. A concern with slopes where low factors of safety exist is that changes in loading conditions may cause the heavily stressed plastic zones to expand through enough of the slope and base region to have slippage occur.

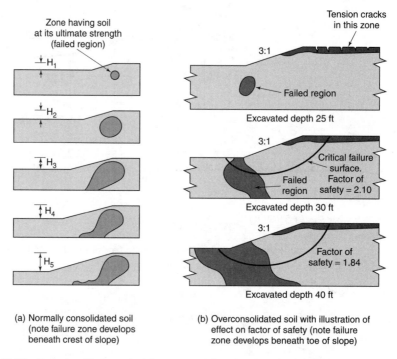

(a) Normally consolidated soil (note failure zone develops beneath crest of slope)

(b) Overconsolidated soil with illustration of effect on factor of safety (note failure zone develops beneath toe of slope)

Figure 15-30 Patterns of increase in failure zone as slope height increases [88].

15.6   IMPROVING STABILITY OF SLOPES

Practical considerations can require that slopes susceptible to sliding have their stability improved to make the area usable and safe. Slope areas that have experienced sliding in the past should be considered likely to undergo further movement if loading conditions change. Ancient slides are viewed with the same or greater concern than presently stable areas where an analysis predicts sliding (e.g., where a very low factor of safety is indicated).

A variety of procedures have been successfully used in the past to stabilize slopes. A history of unsuccessful procedures also exists. Generally, corrective or preventative measures involve reducing the mass or loading that contributes toward sliding, improving the shear strength of the earth in the failure zone, and constructing or installing elements that will provide resistance to movement. The procedure best for a given slope is related to the type of soil in the slope, the thickness and depth of materials involved in sliding, the groundwater conditions, the areal extent requiring stabilizing, the space available to undertake corrective changes, the topographical conditions in the vicinity of the slope, and the tendency for changes such as the advent of seismic and vibratory loadings to occur.

Where area is available, slope flattening, as indicated in Figure 15-31a, can be performed to reduce the weight of the mass tending to slide. When it is anticipated that a base failure could occur, resistance to movement can be increased by the placement of a berm below the toe of the slope (Fig. 15.31b). If the zone below the toe is susceptible to severe erosion, an occurrence that would undermine the toe area and remove earth that provides resistance to sliding, a protective rock fill blanket and riprap can be installed (Fig. 15-31c).

High groundwater and pore water pressures in the slope affect the soil shearing resistance that can be developed. Methods to lower the zone of subsurface water and intercept surface water to prevent infiltration and erosion, and to reduce seepage forces, are depicted in Figure 15-31d.

Where cohesionless soils exist, the shearing strength of the slope material can be improved through densification by use of explosives or vibroflotation or Terraprobe procedures. For cohesive soils, shear strengths have been improved by consolidation, and water-content reduction has been achieved through surcharging (which may include wick drains), electro-osmosis, and thermal drying.

Grouting and injection methods have been utilized to add cementing or bonding agents into specific soil zones where a weak condition was critical to the stability of the slope.

Driven piles, sheetpiling, and retaining walls have been installed to provide lateral support and increase the resistance for slopes tending to slide (Fig. 15-31e and f).

In the interest of economy, improvement and protective methods such as slope flattening and drainage control conventionally are preferred and generally receive first consideration. Soil stabilization methods and the use of piling or retaining walls are expensive procedures, typically utilized where limited areas require stabilizing.

Where building construction is planned for the vicinity of a slope, the procedure depicted by Figure 15.31g should be followed wherever possible to have the effects of building loading act to benefit slope stability.

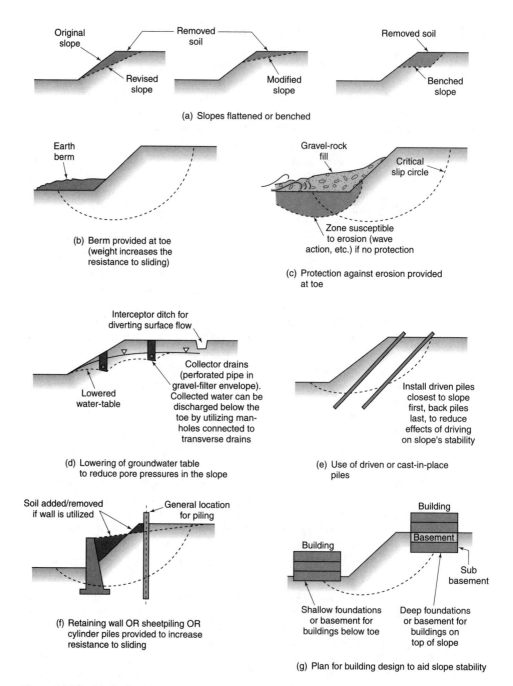

Figure 15-31 Methods to improve and protect slope stability.

PROBLEMS

15-1. List the general categories of slope movements (failures) and provide a brief description of each.

15-2. Indicate the significant differences between (a) the slide and the block-wedge categories of slope movements, and (b) the flow and the spread categories.

15-3. An area of planned construction is underlain by sandy soil where the angle of internal friction is 35°.
   (a) What is the maximum safe slope angle for this soil if there is no groundwater flow?
   (b) If it is assumed that groundwater flow could occur through the slope, approximately what slope angle should be considered as the allowable maximum?

15-4. Indicate probable reasons why slopes in cohesionless soils may exist or be made at inclinations steeper than the soil's angle of repose.

15-5. (a) What would be the factor of safety for an infinitely long slope having an inclination of 25° in an area underlain by firm cohesive soils ($\gamma = 18$ kN/m³) but having a thin weak layer 5 m below and parallel to the slope surface ($\gamma = 16$ kN/m³, $c = 20$ kN/m², $\phi = 15°$ for the weak layer?
   (b) If groundwater rises to the surface of the slope so that flow occurs parallel to the slope, what factor of safety would result?

15-6. Calculate the slope factor of safety for a circular arc trial failure plane that has a 35 m radius with a center of rotation located 20 m vertically above the slope's midpoint. The slope has an inclination of 45° and a vertical height of 20 m. Soil borings indicate that a uniform clay soil with $\gamma = 16.5$ kN/m³ and $c = 55$ kN/m² ($\phi = 0°$) exists in the area. The force (weight) of the failure mass is 9900 kN per meter of length (length perpendicular to the cross section). The horizontal distance between the center of rotation and the center of gravity of failure mass is 11 m. Use the basic method for the stability analysis.

15-7. (a) Utilizing the basic method for a stability analysis, determine a slope's factor of safety against a rotational slide for the following conditions
   • Slope height is 50 ft.
   • Slope inclination is 40°.

• Trial failure arc center of rotation is 40 ft above slope midpoint; radius of failure arc is 80 ft.
• Soil profile determined from top of slope shows clay soil from 0 to 100 ft ($\gamma = 102$ pcf, $c = 940$ psf, $\phi = 0°$).
• A line loading of 3 kips per ft of length of slope acts along the top of the slope 20 ft from the edge.
• The failure mass involves 3725 ft³ of soil per foot of slope, and the horizontal distance from the center of gravity of the failure mass to the center of rotation is 26 ft.
   (b) Find the most critical failure circle and the least factor of safety for the above conditions by systematically changing the point representing the center of rotation by 10 ft horizontally and 10 ft vertically. (A suggested class project in which each student checks one or two different trial failure circles.)

15-8. (a) Using the Fellenius method of slices for slope analysis, calculate the factor of safety against a rotational slide for the following conditions:
   • Slope inclination is 2 horizontal to 1 vertical.
   • Trial failure arc center of rotation is 35 m above slope midpoint; arc passes through toe of slope.
   • Slope height is 25 m.
   • Soils are uniform to great depth (drained shear tests indicate $\phi = 15°$, $c' = 15$ kN/m², $\gamma = 17$ kN/m³; water table is very deep.
   • Use five slices, approximately equal in width.
   (b) Calculate the factor of safety for the above conditions but assuming a fully submerged slope (water level even with the top of the slope).

15-9. The method of slices is a widely used procedure for evaluating the factor of safety for a slope against a sliding failure which involves a circular mass of soil. A number of procedures have been developed, such as the Fellenius method, the Bishop method, the Spencer method, and the Morganstern-Price method discussed in this text. Provide a sketch of a reference slice for each of these four methods, showing related active forces. For purposes of comparison, indicate the major differences between the methods.

15-10. (a) Determine the factor of safety for the slope conditions indicated in Problem 15-7(a) by using the simplified Bishop method, but with soil whose properties are $\gamma = 16$ kN/m³, $c' = 10$ kN/m², $\phi = 20°$.

(b) Using the soil properties in part (a), solve Problem 15-7(b) using the simplified Bishop method. Use grid 3 m by 3 m.

15-11. Using the simplified Janbu method, determine the factor of safety for stability of a slope and potential failure mass identified by the following description:

- Slope inclination is 30°; surface is level at top of slope.
- Slope height is 20 m; soil unit weight γ is 17 kN/m³.
- Buried thin weak layer (the assumed plane of sliding) extends back from toe of slope at an inclination of 20° for a horizontal distance of 40 m, then upward at a 45° inclination until it intersects the ground surface ($c' = 12$ kN/m², $\gamma = 17$ kN/m³, $\phi = 18°$.

15-12. A sliding block analysis is to be made for an earth embankment structure as depicted by Figure 15-19. Determine the factor of safety against sliding along the base of the embankment (Fig. 15-19b) for the following conditions:

- Distance $b'd$ or $L = 130$ m.
- Distance $ab' = 40$ m.
- Slope inclination is 3 horizontal to 1 vertical.
- Unit weight of compacted earth in embankment is 18 kN/m³.
- Active pressure on plane a-b is computed using $K_a = 0.4$.
- Uplift pressure at point $b = 10\gamma_w$ and at point d = 0.
- Shear strength on assumed weak plane b'-d is based on $c = 20$ kN/m², $\phi = 20°$.

15-13. Slope stability charts and tables developed as computational aides apply to either analysis for a total stress condition or for an effective stress condition. Indicate the major differences between the two categories and the conditions where each is appropriate to apply.

15-14. (a) Determine the factor of safety for a 45° slope 30 m high in an area where a uniform clay exists, using $c = 96$ kN/m², $\gamma = 17$ kN/m³, $\phi = 0$. A rock stratum exists 45 m below the toe of the slope. Use Taylor stability charts for the total stress case.

(b) What inclination is required to have a factor of safety of 1.2 for a slope 40 m high in an area where uniform soils exist ($c = 25$ kN/m², $\gamma = 16.5$ kN/m³, $\phi = 12°$)? The factor of safety is to apply to the angle of internal friction as well as cohesion. Use Taylor stability charts.

(c) A reservoir embankment 25 m high consists of compacted earth, where $c = 40$ kN/m², $\gamma = 19$ kN/m³, $\phi = 10°$. The embankment has a slope of 3 horizontal to 1 vertical. What is the factor of safety if a sudden drawdown of the reservoir occurs? Use Taylor stability charts.

15-15. A slope 60 ft high has a face inclination of 30°. Uniform soils comprise the slope and foundation zone and have a unit weight of 115 pcf, an angle of internal friction of 25°, and a cohesion of 400 psf. Use the Taylor stability charts for total stress analysis to:

(a) Determine the slope factor of safety in regard to soil cohesion only.

(b) Determine slope factor of safety in regard to both cohesion and angle of internal friction. (*Hint:* Use a trial-and-error procedure by assuming a value for the FS to apply to ϕ, obtain a value for N_s, then compare the assumed FS with the formula-calculated FS.)

15-16. Use Taylor stability charts for the total stress condition for the following:

(a) Determine the factor of safety for a planned 30-ft-high slope in clay soil (unit weight of 110 pcf, cohesion of 700 psf, zero angle of internal friction). The slope inclination is 1 horizontal to 1 vertical. A stratum of very firm hard clay exists 15 ft below the toe elevation for the slope.

(b) What would the factor of safety become if the soil slope was made at an inclination of 60°?

15-17. Use the Bishop-Morganstern slope stability tables to determine the factor of safety against sliding for a planned earth embankment if the $\frac{c}{\gamma H}$ value is 0.025, ϕ' is 20°, and the sloped face will be constructed at an inclination of 1 horizontal to 1 vertical, when

(a) $r_u = 0$

(b) $r_u = .25$

(c) Conditions are as for part (a) except the slope face is flattened to 1.5 H to 1 V.

15-18. Use the Barnes slope stability tables to evaluate embankment slopes having conditions as described in Problem 15-14, parts (a), (b), and (c).

15-19. (a) Using the Bishop-Morganstern tables for slope stability, determine the factor of safety for a slope where the following conditions exist:
- Slope inclination is 30°, height is 30 m.
- Soil cohesion $c' = 20$ kN/m², $\phi = 20°$, $\gamma = 20$ kN/m³.
- Pore pressure ratio $r_u = 0$.

(b) Determine the factor of safety for the slope described in part (a), except that $r_u = 0.25$.

15-20. Use the Barnes slope stability tables to determine the factor of safety for the slope embankment described in Problem 15-19.

15-21. Use the Bishop-Morganstern slope stability tables to determine the factor of safety for an embankment slope where the following conditions apply:
- Slope height = 55 ft
- Slope inclination = 1 horizontal to 1 vertical
- Soil unit weight = 110 pcf
- Soil cohesion, $c = 600$ psf
- Angle of internal friction = 20°

(a) when $r_u = 0$

(b) when $r_u = 0.30$

(c) Use the Barnes toe circle factors to solve part (b).

15-22. (a) Using the Barnes slope stability tables, determine the steepest slope angle that provides a factor of safety of 1.20 for an earth embankment where the slope geometry and soil properties are as follows:
- Slope height = 16 m
- Soil unit weight = 18.75 kN/m³
- Angle of internal friction = 25°
- Cohesion, $c' = 15$ kPa
- $r_u = 0$

(b) Use the same conditions as for part (a) except the weighted value for $r_u = 0.30$.

15-23. (a) Using the Barnes slope stability tables for excavated slopes, determine if a face inclination of 1 horizontal to 1 vertical is safe for a 16-m-high slope in soils where γ is 18.75 kN/m³, c' is 15 kPa, ϕ is 25°, and the groundwater table is 8 m below the top of the slope (or 8 m above the toe of the slope).

(b) Determine the slope factor of safety if the excavation for the conditions described in part (a) is made with an inclination of 2 horizontal to 1 vertical.

(c) Can this excavated slope be made with an inclination of 1.5 horizontal to 1 vertical and still be safe against a slide failure?

15-24. Use the Barnes stability factors for excavated slopes. An area is to be excavated to establish the alignment desired for a highway project. The necessary depth for the excavation is 10 m. The $\frac{c}{\gamma H}$ value for the slope will be 0.025. The position of the water table is 6 m below the top of the slope. It is desired to have the excavation made at an inclination of 1 horizontal to 1 vertical.

(a) What is the factor of safety if $\phi = 25°$?

(b) What value of ϕ is necessary to have the factor of safety equal to at least 1.15?

(c) For the value ϕ determined in part (b), should the slope analysis be treated as "dry" or "wet"?

(d) What is the actual value for the factor of safety with the ϕ value determined for part (b)?

15-25. Use the Barnes slope stability factors for excavated slopes.

(a) The plans for a new roadway indicate that excavation for one segment will result in an earth slope 45 ft high. The excavation work will occur in a soil formation where the unit weight is 112 pcf, the cohesion c' is 500 psf, and the angle of internal friction is 22°. The groundwater table is 35 ft below the top of the slope. What is the slope factor of safety if the inclination is 1 horizontal to 1 vertical?

(b) What is the slope factor of safety if the natural groundwater table is at the top of the slope $(d_w = 0)$?

(c) Would a slope of 2 horizontal to 1 vertical be safe for the high water table condition of part (b)?

15-26. Use Spencer slope stability curves for effective stress analysis. An earth embankment is to be constructed as a protective levee for a land area bordering the route of a river. The required height is 8 m and the properties of the soil to be used are $\gamma = 18$ kN/m³, $c' = 36$ kPa (kN/m²), $\phi = 20°$. For evaluating the slope stability when flood conditions exist, use $r_u = 0.5$. What slope inclination should be constructed to have a factor of safety equal to 1.25?

15-27. Use the Spencer slope stability curves to check the factor of safety for a planned embankment where the $\frac{c}{\gamma H}$ value is 0.025, ϕ is 20°, and the desired slope inclination is 1.5 horizontal to 1 vertical, if

(a) $r_u = 0$

(b) $r_u = 0.25$

15-28. The conditions listed below apply to an embankment slope. Compare the factor of safety for slope stability determined when using the Barnes toe circle factors and when using the Spencer slope stability curves.

- Slope height = 62 ft
- Slope inclination = 1.5 horizontal to 1 vertical
- Soil unit weight = 112 pcf
- Soil cohesion, $c' = 700$ psf
- Angle of internal friction = 20°
- Pore pressure condition, $r_u = 0.25$

15-29. (a) Using the Morganstern stability curves for rapid drawdown, determine the factor of safety for a reservoir embankment subject to sudden drawdown where the following conditions exist:

- Inclination of upstream slope is $2\frac{1}{2}$ horizontal to 1 vertical.
- Embankment height is 60 m; reservoir level is completely drawn down.
- Embankment soil properties are $\gamma = 20$ kN/m³, $c' = 30$ kN/m², $\phi = 20°$.

(b) Determine the factor of safety if the drawdown was to a level 30 m below the top of the embankment.

CHAPTER 16

Lateral Pressures and Retaining Structures

Structures and components of structures whose major design function is to resist lateral forces are common to private and public works projects. Historically, the magnitude and effect of lateral forces represent one of the earliest structural problems to be studied analytically. Lateral forces most typically develop against structures from the presence of soil or water against vertical sections. Retaining walls and foundation walls are representative of structures that serve principally to support or resist an adjacent mass of earth, whereas dams are identified with facilities used to retain bodies of water. Frequently these structures must be capable of resisting forces resulting from the presence of both soil and water; the design for foundation walls and retaining walls considers water as well as soil pressures where a high groundwater table exists, whereas for earth dams the internal lateral soil pressures are important to the structural stability. Retaining structures for waterfront facilities, such as bulkheads, cofferdams, and quay walls, conventionally must consider the effects of both soil and water pressures in their design. Representative retaining-type structures are depicted in Figure 16-1.

Earth retaining structures are common where a project requires that changes in the ground surface elevation be achieved abruptly because space is not available or cannot be used for the construction of stable earth slopes—for example, retaining walls where highway and railroad routes require an elevation different from the surrounding terrain. For walls that make up the subsurface components of structures, the major portion of the design loading usually results from the lateral pressures of soil or water that exist immediately adjacent to the structure as necessary for its utility or function—for example, walls for basements of buildings and walls for ship-loading facilities. The most familiar water retaining structures are for impounding bodies of water for municipal use and power generation, but they are also used to protect or create land areas.

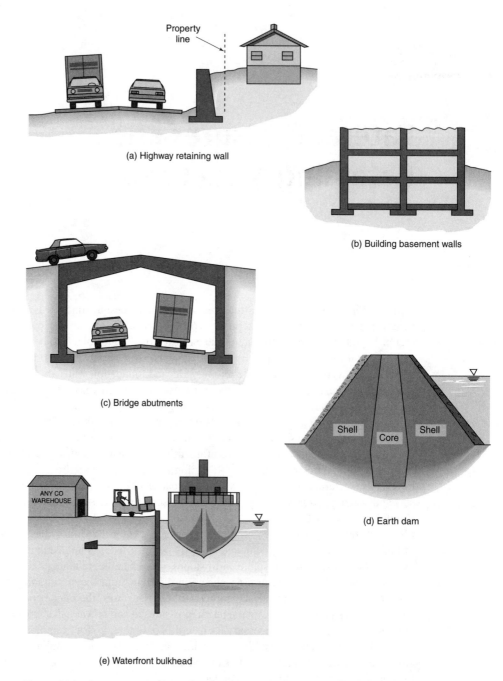

(a) Highway retaining wall

(b) Building basement walls

(c) Bridge abutments

(d) Earth dam

(e) Waterfront bulkhead

Figure 16-1 Representative types of structures subject to lateral soil and water pressures.

16.1 LATERAL EARTH PRESSURE AND RETAINING WALLS

The magnitude of the lateral soil pressure that can exist or develop in a soil mass is related to the strength and stress–strain properties of the material and the deformations that occur within the mass as a result of lateral movement.

At-Rest Pressure

In a homogeneous and isotropic soil mass of infinite extent bounded by a level ground surface, as illustrated in Figure 16-2a, the vertical and horizontal stresses shown on the representative element are principal stresses. For the condition where the soil deposit is normally consolidated (the existing overburden pressure represents the maximum vertical pressure the soil mass has been subjected to in its history), the vertical principal stress is equal to the weight of the overburden, or

$$\sigma_v = \gamma_{soil} Z \quad \text{or} \quad \gamma_t Z \tag{9-1}$$

where the groundwater table is below the depth Z and no seepage or buoyancy forces exist. If the overburden acting as part of the soil mass were to increase, the vertical stress σ_v would

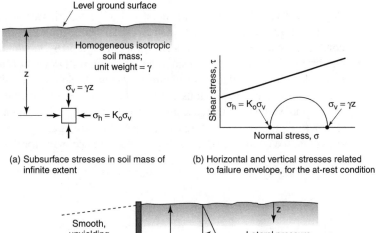

(a) Subsurface stresses in soil mass of infinite extent

(b) Horizontal and vertical stresses related to failure envelope, for the at-rest condition

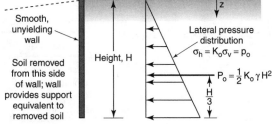

(c) Lateral pressure distribution against smooth, rigid, unyielding wall which replaces part of original soil mass (no lateral movements within soil mass)

Figure 16-2 Subsurface stresses relating to the at-rest condition.

also increase, and the soil would be subject to vertical compression as a result. Deformation in the lateral direction would not occur, however, because, though increasing, the horizontal stress on all volumetric elements would be equal and opposite in magnitude and direction, effecting a lateral restraint. For this condition, the relationship between horizontal stress σ_h and vertical stress σ_v is indicated by

$$\frac{\sigma_h}{\sigma_v} = \frac{v}{1-v} \cong 1 - \sin\phi \qquad (16\text{-}1a)$$

or

$$\sigma_h = \sigma_v\left(\frac{v}{1-v}\right) \cong \sigma_v(1 - \sin\phi) \qquad (16\text{-}1b)$$

where v is Poisson's ratio[1] and ϕ is the angle of internal friction for the soil. The ratio of σ_h to σ_v in a soil mass is termed the coefficient of lateral earth pressure, K. Where no lateral deformations occur in the soil mass, the value expressed by Equation 16-1a is considered the *at-rest* condition for which the earth pressure coefficient is designated K_0. Therefore,

$$\sigma_h = K_0\sigma_v \qquad \text{(at-rest condition)} \qquad (16\text{-}1c)$$

Representative values of K_0 for generalized soil categories are shown in Table 16-1.

The relationship of subsurface stresses to the soil's limiting strength for a typical at-rest condition is shown by the Mohr's circle plot of Figure 16-2b, which indicates that a state of equilibrium is maintained. If a smooth unyielding wall (a wall of infinite rigidity) could be embedded into the soil mass of Figure 16-2a and the soil to one side then removed without causing change in the previously established soil conditions, the lateral pressures acting against the wall would increase uniformly with depth, achieving a pressure distribution as shown in Figure 16-2c. Letting p_0 represent pressure at any depth against the wall that results from the retained soil mass, Equation 16-1c becomes, for the at-rest condition,

$$p_0 = \sigma_h = K_0\sigma_v = K_0\gamma_t Z \qquad (16\text{-}2a)$$

Table 16-1 Representative Values of K_0

Soil Type	K_0
Granular, loose	0.5–0.6
Granular, dense	0.3–0.5
Clay, soft	0.9–1.1 (undrained)
Clay, hard	0.8–0.9 (undrained)

---

[1]Recall from engineering mechanics that Poisson's ratio relates strain in a lateral direction to the strain in a longitudinal direction that occurs as a result of load applied in the longitudinal direction.

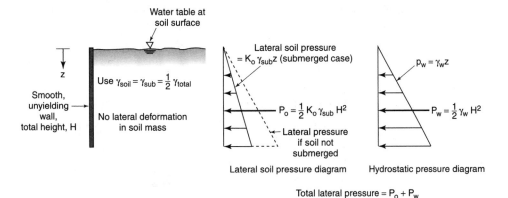

Figure 16-3   Conditions for submerged soil: Lateral pressure is total of soil plus water pressure.

and the total lateral force P_0 per unit of wall length is the area of the pressure diagram, or

$$P_0 = \frac{1}{2} K_0 \gamma_t H^2 \qquad (16\text{-}2b)$$

whose resultant acts at a depth $\frac{2}{3}H$.

Effect of Submergence

Where soil is below the groundwater table, the intergranular (i.e., effective) stress $\overline{\sigma}_v$ between soil particles at any depth is reduced by the magnitude of the water pressure u at the same depth, in accord with Equation 9-4,

$$\overline{\sigma}_v = \gamma_t Z - u = \sigma_v - u \qquad (9\text{-}4)$$

Submergence influences the lateral pressures being exerted on a wall, as in Figure 16-3, in two manners. The submerged soil unit weight or effective vertical stress is used with Equations 16-1c and 16-2, the result being that, compared to the nonsubmerged condition, the lateral soil pressure decreases. However, the presence of water also causes hydrostatic pressures to act against the submerged section of the wall, where the value of such pressure at any depth is a product of the unit weight of water γ_w and the depth below the surface of the water table. The hydrostatic pressure diagram is triangular, and the total lateral force is equal to the area of the diagram. Comparing cases of water existing and no water existing behind a wall, the presence of a water table with the combined effect of hydrostatic pressure and (reduced) lateral soil pressure causes a greater total lateral force to act on the wall.

Illustration 16-1 _____

Calculate the total lateral force acting against a smooth, unyielding wall that retains a cohesionless soil, assuming that no lateral deformation occurs in the soil mass and the at-rest condition applies. Properties of the retained soil and location of the groundwater table are indicated on the sketch.

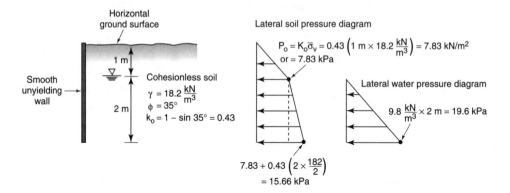

Solution

Total lateral pressure against wall = area of soil pressure and water pressure diagrams

$$= \left(\frac{7.83 \text{ kPa}}{2} \times 1 \text{ m}\right) + \left(\frac{7.83 \text{ kPa} + 15.66 \text{ kPa}}{2} \times 2 \text{ m}\right) + \left(\frac{19.6 \text{ kPa}}{2} \times 2 \text{ m}\right)$$

$$= 3.92 \text{ kN/m} + 23.49 \text{ kN/m} + 19.6 \text{ kN/m} = 47 \text{ kN per meter of wall length}$$

(*Note:* If the water table was below the wall, the total lateral soil pressure would be (18.2 kN/m³ × 3 m × 0.43)(3 m × ½) = 35.2 kN per meter of wall.)

Active and Passive Lateral Pressures—Basic Considerations

If the frictionless, infinitely rigid wall of Figure 16-2c is allowed to move away from the retained soil mass a slight distance (Fig. 16-4a), the soil starts to expand in the lateral direction, following the movement of the wall. Shearing resistance developed within the soil mass because of the soil's shear strength acts opposite to the direction of the expansion, the result being that the lateral soil pressures on the wall *decrease,* becoming less than the at-rest pressures. When lateral movement adequate to develop the soil's maximum shearing resistance occurs (see Fig. 11-16 for a generalized stress–strain relationship), the lateral earth force on the wall will be a minimum for that retained soil mass. The minimum lateral soil pressure condition based on the soil's shear strength is the *active pressure, p_a,* and the active lateral pressure coefficient is K_a. The procedure for determining the theoretical minimum pressure that results with a *dry cohesionless soil* is indicated by the Mohr's circle plot in reference to the soil's failure envelope (Fig. 16-4b), whereby

$$\frac{p_a}{\sigma_v} = \frac{1 - \sin \phi}{1 + \sin \phi} = \tan^2 \left(45 - \frac{\phi}{2}\right) = K_a \tag{16-3a}$$

or

$$p_a = \sigma_v K_a = \sigma_v \left(\frac{1 - \sin \phi}{1 + \sin \phi}\right) = \sigma_v \tan^2 \left(45 - \frac{\phi}{2}\right) \tag{16-3b}$$

where p_a is the active lateral soil pressure at the depth where σ_v acts.

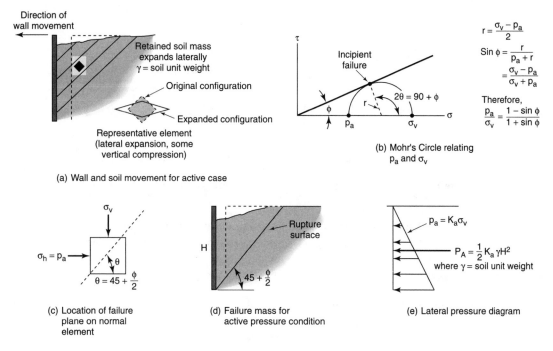

Figure 16-4   Conditions relating to the active lateral earth pressure case for cohesionless soil.

The *total* lateral force P_A (per unit of wall length) resulting from active soil pressures against the wall (Fig. 16-4e) is

$$P_A = \frac{1}{2}\gamma_t H^2 K_a \tag{16-3c}$$

Figure 16-4c also shows the inclination of potential failure planes in the laterally expanding soil mass. When shearing strains reach values corresponding to the soil's maximum shearing strength, incipient planes of failure are at an angle of $45 + \frac{\phi}{2}$ degrees to the horizontal. This angle is used to identify the soil zone that contributes to the lateral force on the wall when the fully active pressure condition is reached (Fig. 16-4d). Within this zone, an infinite number of failure planes (planes of sliding) oriented at the $45 + \frac{\phi}{2}$ degree angle exist. The distribution of lateral pressures with depth is indicated by Figure 16-4e.

Where the soil possesses *internal friction plus cohesion, c,* from the presence of cohesive material or moisture, the Mohr's circle–failure envelope relationship for the active pressure condition is, from Figure 16-5a,

$$p_a = \sigma_v \tan^2\left(45 - \frac{\phi}{2}\right) - 2c \tan\left(45 - \frac{\phi}{2}\right) \tag{16-4a}$$

and the total force against the wall, P_A, per unit of wall length is indicated by the area of the pressure diagram

$$P_A = \frac{1}{2}\gamma_t H^2 \tan^2\left(45 - \frac{\phi}{2}\right) - 2cH \tan\left(45 - \frac{\phi}{2}\right) \tag{16-4b}$$

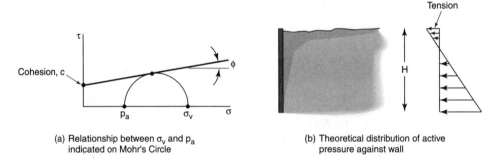

(a) Relationship between σ_v and p_a
indicated on Mohr's Circle

(b) Theoretical distribution of active
pressure against wall

Figure 16-5 Active lateral pressure case for soil possessing cohesion and internal friction.

For saturated clay soils analyzed under conditions where the undrained shear strength applies (refer to U–U tests of Fig. 11-25), ϕ is zero and Equation 16-4a reduces to

$$p_a = \sigma_v - 2c \tag{16-4c}$$

and Equation 16-4b becomes

$$P_A = \frac{1}{2}\gamma_t H^2 - 2cH \tag{16-4d}$$

The resulting pressure distribution is similar to that indicated by Figure 16-5b.

If the wall of Figure 16-2c moves into the retained soil mass, the soil compresses in the lateral direction (Fig. 16-6a), with the soil shearing resistance acting to oppose the lateral compression. When sufficient lateral movement occurs, the maximum shearing strength of the soil is mobilized, and the reaction of the resulting lateral earth pressure on the wall will be a maximum. The maximum lateral pressure condition is the *passive pressure,* and the lateral pressure coefficient is K_p. The relationship between vertical stress and passive lateral stress where a level, dry, cohesionless soil mass exists is shown on the Mohr's circle presentation of Figure 16-6b. For a *dry cohesionless soil,* the ratio of lateral to vertical stress for the passive pressure condition is

$$\frac{p_p}{\sigma_v} = \frac{1 + \sin\phi}{1 - \sin\phi} = \tan^2\left(45 + \frac{\phi}{2}\right) = K_p \tag{16-5a}$$

or

$$p_p = \sigma_v K_p = \sigma_v\left(\frac{1 + \sin\phi}{1 - \sin\phi}\right) = \sigma_v \tan^2\left(45 + \frac{\phi}{2}\right) \tag{16-5b}$$

where p_p is the passive lateral soil pressure at the depth where σ_v acts.

The force of the passive pressure P_p against the wall, per unit of wall length (Fig. 16-6b), becomes, for *cohesionless* soil

$$P_p = \frac{1}{2}\gamma H^2 K_p \tag{16-5c}$$

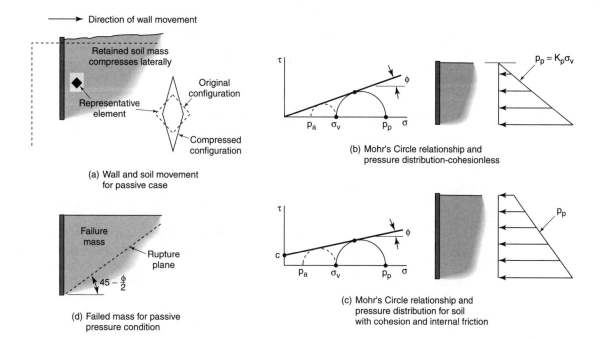

Figure 16-6   Conditions for passive pressure case.

For soils possessing *internal friction and cohesion, c,* the passive pressure (Fig. 16-6c) is

$$p_p = \sigma_v \tan^2\left(45 + \frac{\phi}{2}\right) + 2c \tan\left(45 + \frac{\phi}{2}\right) \qquad (16\text{-}6a)$$

and the force of the passive soil pressure against the wall (Fig. 16-6c) becomes

$$P_p = \frac{1}{2}\gamma_t H^2 \tan^2\left(45 + \frac{\phi}{2}\right) + 2cH \tan\left(45 + \frac{\phi}{2}\right) \qquad (16\text{-}6b)$$

For saturated clay soils being analyzed for the condition of undrained shear ($\phi = 0$), the passive pressure is

$$p_p = \sigma_v + 2c \qquad (16\text{-}6c)$$

and the total lateral soil force P_p becomes

$$P_p = \frac{1}{2}\gamma_t H^2 + 2cH \qquad (16\text{-}6d)$$

For the passive pressure condition, the maximum shearing stresses develop on planes that are at an angle of $45 - \frac{\phi}{2}$ degrees to the horizontal. This angle defines the passive soil zone or failure wedge associated with passive resistance as well as the orientation of the infinite number of incipient failure planes within the passive zone.

If the groundwater table is within the soil mass retained by a wall, the submerged soil unit weight is used in the calculations to determine the active or passive soil pressure, and the lateral water pressure is additive, as discussed in the preceding subsection and as shown in Illustration 16-1.

16.2 EARTH PRESSURES AGAINST RETAINING WALLS

Classical Theories

Two classical theories concerning earth pressure against retaining walls, the Rankine[2] and the Coulomb[3] analyses, still serve as the basis for determining lateral soil pressures as they affect retaining wall design.

Rankine's theory, for the general case of a sloping cohesionless backfill, obtains a relationship between lateral and vertical stress such as that which exists in a soil mass of infinite extent having a uniform slope (Fig. 16-7a), and assumes that a similar stress condition would exist against the back of a retaining wall used to support the slope. This assumption requires that friction between a wall and the retained soil be nonexistent or have insignificant effect on the shearing stresses that develop in the soil behind the wall. This creates conditions identical to those used as the basis for Equations 16-3 and 16-4, and for a level soil mass, the Rankine expression corresponds. For a wall supporting a uniformly sloping backfill of *dry cohesionless soil,* the *Rankine active pressure p_a* against the wall is

$$p_a = \sigma_v \cos i \, \frac{\cos i - \sqrt{\cos^2 i - \cos^2 \phi}}{\cos i + \sqrt{\cos^2 i - \cos^2 \phi}} \tag{16-7a}$$

where i represents the slope inclination angle, and p_a acts parallel to the slope surface.

The force P_A against a retaining wall of height H that results from the *Rankine active* condition (Fig. 16-7b) becomes, per unit of wall length,

$$P_A = \frac{1}{2} \gamma H^2 K_a = \frac{1}{2} \gamma_t H^2 \cos i \, \frac{\cos i - \sqrt{\cos^2 i - \cos^2 \phi}}{\cos i + \sqrt{\cos^2 i - \cos^2 \phi}} \tag{16-7b}$$

The *passive* pressure by the Rankine theory is

$$p_p = \sigma_v \cos i \, \frac{\cos i + \sqrt{\cos^2 i - \cos^2 \phi}}{\cos i - \sqrt{\cos^2 i - \cos^2 \phi}} \tag{16-8a}$$

---

[2]"On the Stability of Loose Earth," V. J. M. Rankine, Philosophical Transactions of the Royal Society, Vol. 147, London, 1857.
[3]"Essai sur une Application des Regles de Maximis et Minimus à Quelques Problemes de Statique Relatifs à Architecture," C. A. Coulomb, presented 1773, published Memoires de Mathematiques et de Physique presentes a l'Academie des Sciences par divers savents, et lus dans ses Assemblées, Vol. 7, 1776.

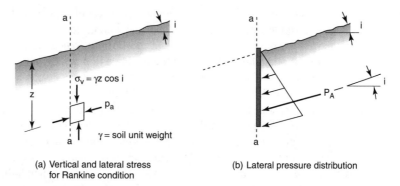

(a) Vertical and lateral stress
for Rankine condition

(b) Lateral pressure distribution

Figure 16-7   Lateral pressures defined for Rankine active pressure condition.

and the force P_p imposed against a wall of height H that results from the *Rankine passive* condition is, per unit of wall length,

$$P_p = \frac{1}{2}\gamma_t H^2 K_p = \frac{1}{2}\gamma_t H^2 \cos i \frac{\cos i + \sqrt{\cos^2 i - \cos^2 \phi}}{\cos i - \sqrt{\cos^2 i - \cos^2 \phi}} \qquad (16\text{-}8b)$$

The *Coulomb theory,* for the active lateral pressure resulting from a retained mass of *cohesionless* soil, considers that a failure wedge forms behind the wall by sliding on a plane generally similar to that indicated by Figure 16-4d. As the retaining structure moves away from the soil mass, permitting its lateral expansion, a resulting vertical compression of the soil failure wedge and the relative movement between the wall and soil causes friction to develop on the back of the wall. The resultant of this friction and lateral soil pressure force P_A acts at an angle ψ measured normal to the face of the wall, as indicated by Figure 16-8. This resultant also establishes the magnitude and direction of the wall *reaction* against the soil failure wedge. When the failure wedge of cohesionless soil is satisfactorily retained by

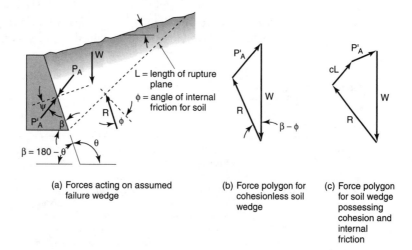

(a) Forces acting on assumed
failure wedge

(b) Force polygon for
cohesionless soil
wedge

(c) Force polygon
for soil wedge
possessing
cohesion and
internal
friction

Figure 16-8   Description of terms relating to Coulomb active pressure condition.

a wall, three forces acting on the wedge—the effect of gravity on the wedge's volume (W, the weight of the wedge), the reaction of the wall on the wedge (P_A), and the soil reaction on the wedge's failure plane (R)—are in equilibrium. When the unit weight and angle ϕ for the retained soil and the angle ψ for the wall reaction are known (Table 16-2), the force P_A imposed on the wall as a result of the active pressure wedge can be determined for any assumed failure wedge by a simple vector addition, as shown in Figure 16-8b. The magnitude of the force P_A acting on the wall will be affected by the size of the assumed failure wedge but also by the slope of the wall face and the angles ϕ and ψ. The actual active pressure condition for a given wall results from that unique failure wedge that provides the greatest numerical value for the force P_A. Determining the proper value for P_A can be accomplished by the trial-and-error procedure of analyzing a series of assumed failure wedges having different sizes. However, where the slope of the soil backfill is uniform and the back of the retaining wall is a plane surface such as indicated in Figure 16-8, the critical failure wedge is triangular, and the following expression can be used to directly calculate the *Coulomb active pressure condition* force resulting from active soil pressures against the wall. For each unit length of wall,

$$
P_A = \frac{1}{2}\gamma_t H^2 K_A \left[\frac{\sin(\theta - \phi)\left(\dfrac{1}{\sin\theta}\right)}{\sqrt{\sin(\theta + \psi)} + \sqrt{\dfrac{\sin(\phi + \psi)\sin(\phi - i)}{\sin(\theta - i)}}} \right]^2 \tag{16-9}
$$

When the retained soil mass possesses cohesion as well as internal friction, the vector addition representing equilibrium for forces acting on the failure wedge must include the total cohesive resistance, cL, acting on the rupture plane (Fig. 16-8c). The procedure for determining the actual force imposed on the wall reverts to the trial-and-error process of assuming different failure wedges and then selecting that wedge that produces the largest P_A value.

Whereas the Coulomb theory assumes a plane failure surface for the active soil zone, actual rupture surfaces tend to be slightly curved (Fig. 16-9). The effect is that the lateral force calculated by Equation 16-9 is slightly low. Typically the discrepancy is minor, and considering the variations that tend to exist in a retained soil mass (which affect soil prop-

Table 16-2 Representative Values for Wall Friction Angles

Wall Material	Soil Type	Wall-Friction Angle ψ (degrees)
Concrete or masonry	Fine to coarse sand, sand and gravel (clean)	30±
	Fine sand, silty sand, clayey sand	25±
	Silt, sandy silt, silt–clay, clay	20±
Poured concrete (formed)	Fine to coarse sand, sand and gravel (clean)	25±
	Fine to medium sand, sand–silt–gravel mixture	20±
	Sand–silt, sand–clay mixture	15–18
Steel sheeting	Coarse sand, sand–gravel mixture (clean)	22
	Fine to medium sand, sand–silt–gravel mixture	17
	Silt, sand–silt, sand–clay mixture	12–14
Wood	Cohesionless (sands and silts)	15±

Figure 16-9   Assumed and actual failure surfaces for active pressure condition
in Coulomb analysis.

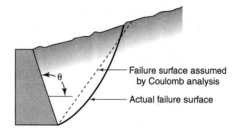

erties and wall friction), the Coulomb procedure for arriving at the effects of active pressure provides a practical accuracy.

The value of P_A computed from the Coulomb theory is slightly greater than the value
obtained from the Rankine theory for similar conditions. However, the lateral pressure resultant for the Rankine condition acts parallel to the wall's retained slope, whereas the resultant from the Coulomb analysis acts at a greater inclination because of the wall friction
factor. The consequence is that the Rankine analysis typically creates the more severe condition when the stability of a retaining wall is evaluated, an effect that results in a slightly
more conservative wall design.

For the *passive* pressure case, the movement of the retaining wall is toward the soil
mass, and the resultant of the lateral pressure and wall friction P_p is inclined downward.
The reaction on the plane of rupture within the soil mass is at an angle opposing movement
(Fig. 16-10). When the back of the retaining wall is planar and the retained soil has a uniform inclination so that the failure mass is triangular in shape, the *Coulomb* equation for
the force resulting from the *passive pressure* P_p of a *cohesionless* soil becomes, for each
unit of wall length,

$$P_p = \frac{1}{2}\gamma H^2 K_p = \frac{1}{2}\gamma_t H^2 \left[\frac{\sin(\theta + \phi)\left(\dfrac{1}{\sin\theta}\right)}{\sqrt{\sin(\theta - \psi)} - \sqrt{\dfrac{\sin(\phi + \psi)\sin(\phi + i)}{\sin(\theta - i)}}} \right]^2 \qquad (16\text{-}10)$$

The shape of the actual rupture surface within the retained soil mass is influenced by
the soil friction developing on the back of the retaining wall. Where the magnitude of

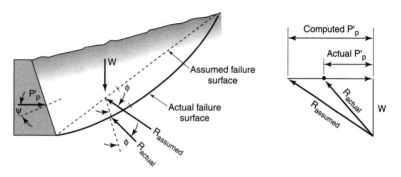

Figure 16-10   Assumed and actual failure surfaces for passive pressure condition and effect on
wall—Coulomb analysis.

friction is low so that the angle ψ is small, the rupture surface is approximately planar. As the angle ψ increases, however, the lower zone of the failure wedge becomes curved. For values of ψ up to about one-third of ϕ, Equation 16-10 closely approximates the actual lateral pressure force, but as ψ becomes larger, the error in the computed lateral force becomes increasingly greater, whereby the actual passive resistance is less than the computed value (Fig. 16-10). For the larger wall friction angles, analysis of the force resulting from passive pressure should be based on a curved surface of rupture [291].

16.3   LATERAL PRESSURES RELATED TO WALL MOVEMENT

Practical Considerations

Walls for buildings and structures that support an adjacent zone of soil as well as free-standing retaining walls normally are designed to resist an at-rest or active lateral soil pressure. In designs, passive soil pressures typically are considered only as reactionary loading, resisting movement where a structural element is moved toward or into a soil mass as an effect of the primary design load; for example, soil along the front of a retaining wall *foundation* tends to develop passive resistance if the supported soil mass causes the wall to move laterally outward (Fig. 16-11). Since lateral movement in a retained soil mass is necessary for the active and passive pressure conditions to develop, structural elements, such as basement walls and bridge abutments whose design requires or assumes that the wall remains rigid without deflection or translation and where connecting framing components act to keep the structure stationary, need to consider lateral soil pressures corresponding to the *at-rest* condition. At-rest pressures for the case where backfill against the wall is placed and compacted in layers tend to be greater than with uncompacted fill and natural deposits.

Figure 16-12 shows the general relationship between retaining wall movements and lateral pressures. This presentation illustrates the important fact that lateral pressures change gradually in accord with wall movement and reach the fully active or passive condition only when adequate movement has occurred. Until such movement is achieved, the lateral pressure acting is intermediate between the at-rest and active or at-rest and passive value. The ex-

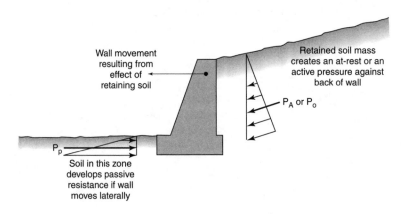

Figure 16-11   Zones tending to develop active and passive lateral pressures.

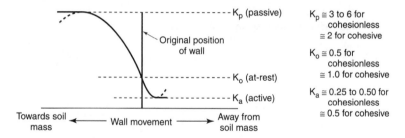

Figure 16-12   Lateral pressure condition related to wall movement.

tent of movement required for a soil mass to reach the passive condition is considerably greater than for the active condition. The pressure variations indicated by Figure 16-12 are for the short-term condition; changes back toward the at-rest pressure can occur over an extended duration, particularly with cohesive soils but also with some cohesionless soils.

For free-standing retaining walls, movement adequate to develop a lateral force corresponding to the active pressure or passive pressure condition can occur by rotation or translation (Fig. 16-13).

It is common to find that retaining wall design has been on the basis of the theoretical active soil pressure condition indicated by Figure 16-13a. However, because of the typical modern procedure of placing and compacting retaining wall backfill, and factors that influence the retained soil mass's shear stress–strain behavior and its ability to help support itself over an extended duration, this conventional or established design procedure does not apply to all wall construction. Where wall movement must be limited or where continuous wall movement cannot be tolerated, design on the basis of a lateral pressure condition in the at-rest range is more appropriate than assuming an active pressure condition. Further, if a limited zone behind the wall represents the excavation that is subsequently backfilled, the full Rankine–Coulomb failure wedge may not have space to develop. Additionally, the natural soil slope adjacent to the work area behind the wall often has earlier experienced some lateral movement as a result of having been unsupported temporarily, and that zone may not be capable of providing the further shearing resistance necessary for the full active pressure condition to develop against the wall. For this situation and with the backfill compacted in layers (Fig. 16-14), the lateral pressures assumed for design should correspond to at least the at-rest pressure condition.

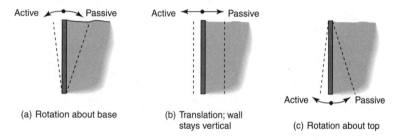

Figure 16-13   Wall movements that produce the active and passive pressure conditions.

Figure 16-14 Limited backfill zone affects lateral pressures against wall.

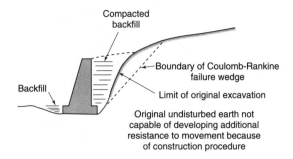

Cohesionless Soil Backfill

Active case lateral pressure distributions that result with a *cohesionless backfill* and wall rotation or translation are shown in Figure 16-15. Since retaining wall behavior conventionally expected corresponds to the condition represented by Figure 16-15a or b, these cases serve as the basis for most designs. The condition for Figure 16-15c could result where unusual areal settlement occurs or where a rotational slide that is deep seated and below the base of the wall occurs (a condition discussed in Chapter 15).

Methods of backfill placement and the category of granular soil have an important effect on the lateral pressures that eventually act on the wall. Over the long term, lateral pressures will correspond to the theoretical active condition when coarse granular backfill is used but not compacted. When the backfill directly behind the retaining wall is compacted, a common procedure intended to increase the angle of internal friction ϕ and theoretically to reduce the K_a value, final lateral pressures may actually reach or exceed at-rest values.

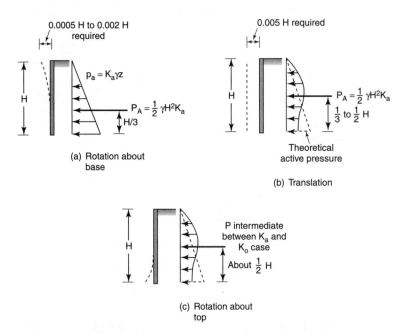

Figure 16-15 Pressure distributions for cohesionless soil related to type of wall movement.

The procedure of placing and compacting the soil in layers causes lateral squeezing and progressive wall deflection as the height of backfill is being increased, one result being that the postconstruction movement necessary to achieve the active condition does not occur. There is also evidence to indicate that even when lateral pressures resulting from a retained granular soil decrease to theoretical active values as wall movement occurs, the pressures subsequently increase with time and approach at-rest values [51]. Further wall movement may occur as a result, and the repetitive occurrence of pressure increase and wall movement can lead to wall failure. The extent of change is greatest with the finer cohesionless soils.

Goh Analysis

Analytical studies performed by A. Goh [109] to determine lateral earth pressures acting against cantilever-type retaining walls where backfill is a lightly compacted cohesionless soil provide insight into the distribution and magnitude of lateral pressures where the wall and retained backfill undergo limited lateral movement (translation) and limited vertical movement (due to settlement of the base). Cantilever retaining walls will tend to deflect more at the top than near the base because of the rotation effect resulting from the combination of horizontal and vertical movements (and possibly structural bending of the stem). An important effect is that, with limited wall movement, the lateral earth pressures decrease from the initial at-rest condition to the active condition for only the upper section of the wall height, while pressures against the lower wall section decrease but to values intermediate between the at-rest and active case. There are also indications that lateral pressures tend toward higher values when a wall surface is smooth (compared to a rough surface). Construction materials commonly used for retaining walls are rough surfaced, but if the wall is backed with a geosynthetic (such as for drainage), the behavior of the soil backfill approaches the condition for a smooth wall surface. Lateral pressure distributions proposed for the analysis and design of cantilever walls retaining cohesionless soil backfill are shown in Figure 16-16; Figure 16-16b applies to the structural design of the wall stem, and Figure 16-16c is appropriate for studying the overturning and sliding stability of the wall (refer Illus. 16-3b). The indicated lateral pressure coefficients K_0 and K_A are calculated from the Rankine–Coulomb equations.

Cohesive Soil Backfill

The expression from theory for the active lateral pressure resulting from a retained *cohesive soil* is given by Equation 16-4a, whereby

$$p_a = \gamma_t Z \tan^2\left(45 - \frac{\phi}{2}\right) - 2c \tan\left(45 - \frac{\phi}{2}\right) \tag{16-4a}$$

Where conditions for the undrained shear strength apply, ϕ is zero and the above equation reduces to

$$p_a = \gamma_t Z - 2c \tag{16-4c}$$

These expressions imply that the lateral pressure imposed on a wall by the uppermost zone of a retained clay is zero. The internal cohesion enables the soil mass to resist shearing

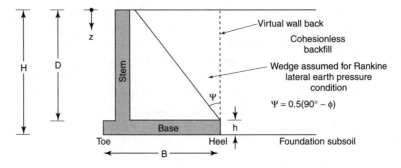

(a) Retaining-wall geometry.

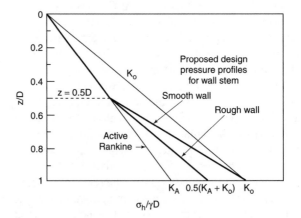

(b) Proposed lateral pressure diagrams for structural analysis of wall stem.

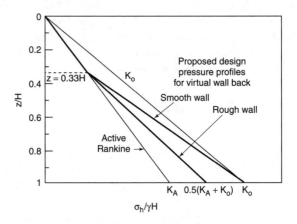

(c) Proposed lateral pressure diagrams for stability analysis of wall.

Figure 16-16 Goh analysis: Lateral pressure diagrams proposed for use in design of cantilever retaining walls [109].

stresses developing as lateral support is removed, in effect providing some self-support. The theoretical active pressure is negative (the soil is considered to be in tension) to a depth z where p_a in Equation 16-4a is zero, or

$$z = \frac{2c}{\gamma_t \tan\left(45 - \dfrac{\phi}{2}\right)} \qquad (16\text{-}11a)$$

and for the condition where undrained shear strength applies ($\phi = 0°$),

$$z = \frac{2c}{\gamma} \qquad (16\text{-}11b)$$

Only below the depth z does a soil pressure develop against the wall.

The active pressure distribution against a retaining wall resulting from application of Equation 16-4a is shown in Figure 16-17b. This diagram helps explain why excavations in clay soils can be maintained with vertical banks without lateral supports being provided. The pressure distribution of Figure 16-17b can be assumed for supporting a cohesive soil

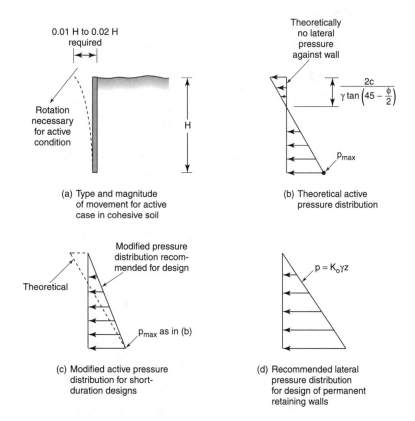

(a) Type and magnitude of movement for active case in cohesive soil

(b) Theoretical active pressure distribution

(c) Modified active pressure distribution for short-duration designs

(d) Recommended lateral pressure distribution for design of permanent retaining walls

Figure 16-17 Lateral pressure conditions for cohesive soil.

over a *short time period*. However, two important factors limit the apparent benefit indicated by this figure and Equation 16-4a for realizing a reduced lateral force acting on a retaining structure over an *extended* duration. Tension cracks can develop in the upper soil zone, and the soil may pull away from the wall, to the depth z, affecting the theoretical pressure distribution. Should the open cracks against and near to the wall fill with water, a hydrostatic pressure can result against the wall. To compensate for the possible effect of such water pressures, a modified active pressure diagram can be assumed as indicated in Figure 16-17c for short-duration designs. For a long-term design and permanent retaining structures, the stress–strain behavior of clay in shear becomes significant. During sustained shear near a clay's maximum strength, most will creep or flow. Internal shearing resistance is retained only if progressive deformation can occur. Applying this principle of soil behavior to the force resulting against a retaining wall, lateral pressures from a clay will increase above the initially active value when movement is prevented and the soil's internal shearing resistance diminishes. For pressures that correspond to the active case to remain in effect against a retaining wall, continuous wall movement must occur. Walls designed for only the active pressure condition are subject to movement when lateral pressures increase above that magnitude. After the additional wall movement does occur, the lateral pressures decrease to the active value but subsequently will again increase. The sequence of increasing pressure followed by wall movement is ongoing, and if enough wall movement occurs, failure will result. In recognition of this phenomenon, when cohesive soil must be used for backfill behind a retaining wall (the use of such soil should be avoided wherever possible), the wall should be designed to resist at least an at-rest lateral pressure, as indicated by Figure 16-17d.

Wall behavior can be further affected if a clay backfill is subject to large volume changes as its water content varies (e.g., expansive clays, see Chap. 3, 4). Lateral pressures against the wall increase in periods when moisture is available and swelling occurs. Pressure changes are greatest if shrinkage cracks from dry periods have become filled with soil or debris. Where highly expansive clays have been used as backfill against a fixed wall, such as a building basement wall, and the clay subsequently swells, limited experience indicates development of lateral pressures that are close to values calculated for the passive pressure condition [56].

Recommended Design Criteria for Gravity Walls

As a conclusion to studying effects of soil factors and wall translations–rotations which influence the magnitude and distribution of earth pressures acting against gravity retaining walls, Duncan, et al. [87], have proposed that earth load conditions as summarized in Figure 16-18 be applied for analysis.

Effects of Groundwater and Freezing

Lateral soil pressures against a wall are computed by using the *effective* unit weight of the retained soil. If the groundwater level rises into the backfill behind a retaining wall because

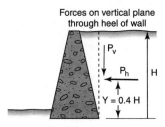

Forces on wall

Forces on vertical plane
through heel of wall

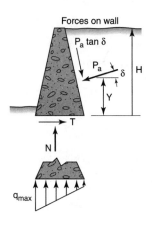

Stability criteria:

(1) N within middle third of base

(2) $q_{allowable} \geq q_{max}$

(3) Safe against sliding

(4) Settlement within tolerable limits

Earth loads:

P_a or P_h calculated using Coulomb
active earth pressure theory

δ or P_v estimated using judgment,
with allowance for movement of
backfill relative to wall.
(e.g., consider $\delta = \psi$, Table 16-2,
or $0.5\phi < \delta < \phi$)

(a) Granular backfills and foundations of sand or gravel.

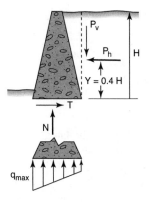

Stability criteria:

(1) N within middle third of base

(2) $q_{allowable} \geq q_{max}$

(3) Safe against sliding

(4) Settlement within tolerable limits

(5) Safe against deep-seated foundation failure.

Earth loads:

P_h and P_v based on experience,
with allowance for creep
(or refer Fig. 16-29 values
for type 4 soil)

(b) Clay soils in the backfill or foundation.

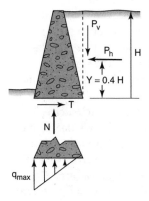

Stability criteria:

(1) N within middle half of base

(2) $q_{allowable} \geq q_{max}$ for foundation rock, and allowable
compressive stress in concrete $\geq q_{max}$

(3) Safe against sliding

Earth loads:

P_h based on at-rest pressure.
$K_o = 0.45$ for compacted backfill,
and $K_o = 0.55$ for uncompacted backfill.
P_v estimated using judgment

(c) Granular backfills, foundations on rock.

Figure 16-18   Earth loads and stability criteria for retaining walls [87]

Figure 16-19 Lateral pressures affected by water behind wall.

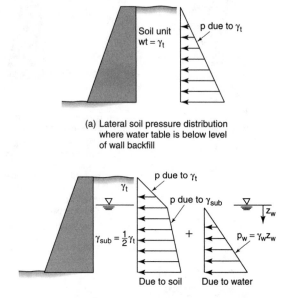

(a) Lateral soil pressure distribution where water table is below level of wall backfill

(b) Lateral pressures against wall when water level behind wall is above base of wall

of either a changing groundwater condition or percolating surface water, the lateral pressures against the wall are changed (Fig. 16-19). For soil below the water table, a submerged unit weight is used, and the related lateral soil pressure becomes approximately half of the nonsubmerged value. Additionally, however, a hydrostatic pressure develops against the back of the wall. The combined effect of soil and water pressures causes overturning moments and sliding forces to be greater than for the condition of no water (ref. Ill.16-1 on p. 617). To eliminate the opportunity for a great height of water to build up behind a wall, through-the-wall weep holes or a collector–drainage system, or both, are commonly provided and are considered to be an essential part of the design and construction (Fig. 16-20). Weep holes should be at least 75 mm (3 in.) in diameter and provided with a granular soil filter or filter fabric at the back of the wall to prevent backfill erosion. The horizontal spacings normally selected range from about 1.5 to 3 m (4 to 10 ft). For tall walls, two or more rows of weep holes may be used, with a typical vertical spacing of about 1.5 m (5 ft).

Freezing temperatures can cause ice lens formation behind the wall if water is available in the backfill. The backfill zones around weep holes are particularly vulnerable because of the hole opening and temperature conduction of the weep hole pipe if metal is used. The conduction properties of solid concrete and masonry are also high, the effect being that the back sides of such retaining walls have temperatures close to air temperatures. Coarse, non-frost-susceptible material behind the wall (Fig. 16-20b) serves to prevent the development of ice lenses in the backfill zone. A coarse drainage blanket located as indicated by Figure 16-20c acts to prevent deep groundwater from rising by capillary action into the freezing zone behind the wall. Drainage geosynthetics also are in use for removing water which otherwise tends to accumulate behind retaining walls (Fig. 16-20d) (refer to discussions on geosynthetics in Chapters 7 and 14).

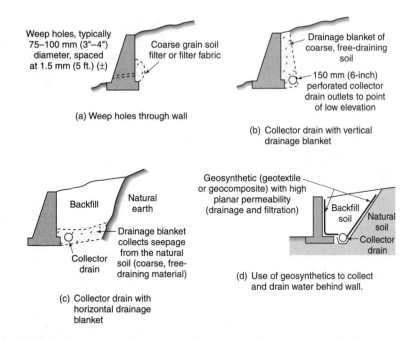

Figure 16-20   Drainage methods to prevent development of water pressures behind retaining walls.

16.4   RETAINING WALL DESIGN REQUIREMENTS

Common Types of Walls, Construction Considerations

Retaining walls have been constructed of plain and reinforced concrete, masonry, laid-up stone (dry construction), timber, and metal materials. Simple walls are the gravity type, with the weight of the wall providing the required stability against the effects of a retained soil. Representative types of gravity walls are shown in Figure 16-21. The practical use of gravity retaining walls is controlled by height limitations, whereby the required wall cross section starts to increase significantly with tall heights because of effects of the approximately triangular soil pressure distribution behind the wall. Cantilever retaining walls (Fig. 16-22) incorporate a design feature that enables some of the retained soil to contribute to the stability of the structure, thereby permitting a design that requires less wall material than the equivalent-height gravity type. Such walls require reinforcing steel because of the relative slenderness of the stem (i.e., vertical section) and base.

Retaining walls have also been constructed of timber and steel sheet piling, closely spaced, large-diameter concrete piles, and spaced *H*-piling with horizontal lagging of concrete or timber planks (Fig. 16-23). Cube-shaped wire mesh baskets filled with crushed rock (gabions) have been used to form gravity walls in a manner similar to that achieved by using laid-up stone (dry, no mortar) construction, although their prevalent use has been for slope protection. Figure 16-32 and the related text further discuss gabions.

Retaining walls are long structures. The usual design procedure is to analyze a section one unit in length, with the unit length selected being an increment where the wall

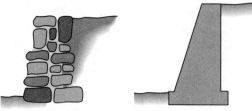

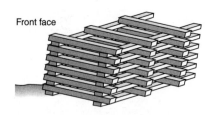

(a) Masonry or laid-up stone. May also be gabions (stone-filled wire baskets).

(b) Concrete or masonry

(c) Crib (members are wood, concrete, or steel). Coarse soil is used to fill the crib interior and contribute to the weight of the wall. Structures are frequently tilted backward to improve stability. Except for the exposed front face, crib walls often are completely covered with soil so the cribbing is not visible.

Figure 16-21 Types of gravity retaining walls.

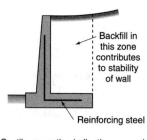

(a) Cantilever section indicating general principle whereby backfill assists stability

(b) Semi-gravity type

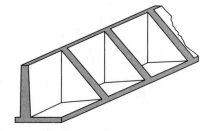

(c) Cantilever

(d) Counterfort or buttress wall (counterfort wall has buttress located on backfill side, while buttress wall has buttresses along front side)

Figure 16-22 Cantilever-type retaining walls.

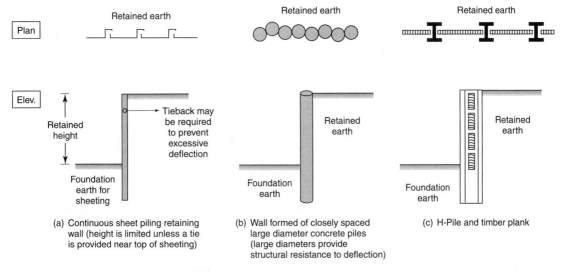

Figure 16-23   Types of retaining walls using piling.

section begins to repeat; walls whose general cross section is constant for a long length would suggest a segment one meter or one foot in length, whereas a counterfort or buttressed wall would suggest a section that extends the center-to-center distance between buttresses. The height of soil to be retained will usually vary along the length of the wall, and with homogeneous backfill and foundation conditions the most severe design loading occurs where the height is greatest. The wall cross section required for this crucial location is also assumed for adjacent locations, although the actual height constructed may be lower. For long walls of varying height requirements, a separate design can be provided for the different segments in order to economize on materials. Conversely, for concrete walls of limited length that retain an approximately constant height of soil, the critical cross section may be used for the entire wall, because savings related to labor and forming procedures associated with constructing the wall offset the extra material used. A benefit resulting from this latter approach is that the average theoretical structural factor of safety across the full length of retaining wall is greater than the design factor of safety.

Basic Design Considerations

A proper retaining wall design satisfies the following requirements:

1. The structural components of the wall (the base and the stem) are capable of resisting the internal shears and bending moments developing as a result of soil and other loading.

2. The wall structure is safe against overturning.

3. The wall structure is safe against sliding.

4. The bearing capacity of the foundation material supporting the wall is not exceeded.

5. Settlement and distortion of the wall due to compression of the foundation soil are limited to a tolerable value.

All factors relating to the stability of a retaining wall are affected by the magnitude of the lateral earth pressure. For design, then, it is important to know the properties of the backfill and to be assured that wall movement, if necessary for the design lateral pressures to develop, can actually occur. For instance, if a gravity or cantilever wall is founded on rock or a very firm or compact soil (Fig. 16-25a), rotational movement (Fig. 16-15a) will probably not occur; such walls should be designed with the assumption of an at-rest pressure (see Fig. 16-18c). In a design analysis, the position of the lateral pressure resultant at the back of a wall is as indicated in Figure 16-24 (also ref. Fig. 16-18).

The analysis for shear forces and bending moments acting on the vertical section of a wall is similar to the procedure for analyzing a structural beam carrying a distributed loading (Fig. 16-25b). The stability against overturning is evaluated by comparing moments about an assumed axis of rotation (typically the toe of the wall) that result from forces acting to cause overturning and forces acting to resist overturning (Fig. 16-25c). The ratio of moments resisting to those causing overturning is the factor of safety against overturning; a value of 2 or greater is normally desired. The stability against sliding is evaluated by comparing forces causing sliding to those resisting sliding (Fig. 16-25d). The ratio of the forces resisting to the forces causing is the factor of safety against sliding; a desired minimum value of 1.5 is typical.

Foundation bearing pressures are computed by following the procedures for determining stress distributions where an axial loading and bending moment act in combination on a structural section. The position of the foundation reaction is first located from

$$d = \frac{\Sigma M_{toe}}{\Sigma V} \quad \text{or} \quad \frac{\Sigma M_{heel}}{\Sigma V}$$

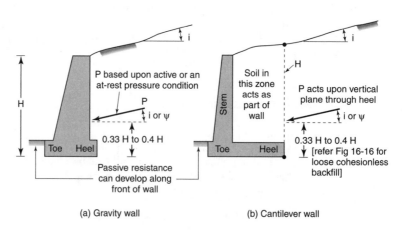

(a) Gravity wall   (b) Cantilever wall

Figure 16-24 Position of lateral pressure resultant for stability analysis.

where ΣM = algebraic sum of moments about the wall's toe or heel
ΣV = vertical component of the foundation reaction including the weight of the wall and backfill, and vertical component of the lateral soil pressure
d = distance from toe or heel to point where foundation reaction is applied

Desirably, the d distance is located within the middle third of the base so that the wall loading is spread across the entire base (theoretically, no tensile stresses will develop). An

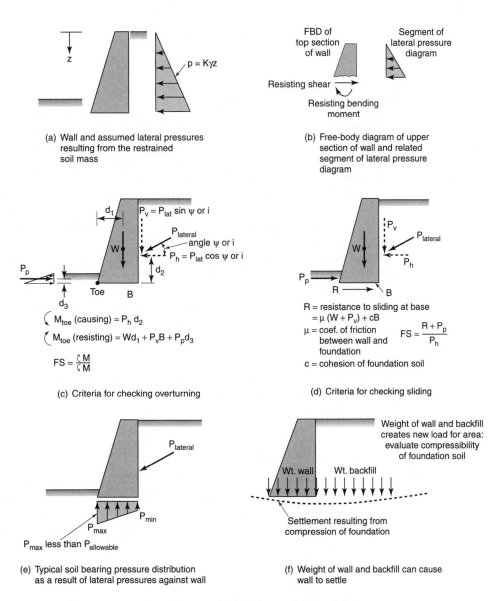

(a) Wall and assumed lateral pressures resulting from the restrained soil mass

(b) Free-body diagram of upper section of wall and related segment of lateral pressure diagram

M_{toe} (causing) $= P_h\, d_2$

M_{toe} (resisting) $= Wd_1 + P_vB + P_pd_3$

$$FS = \frac{\Sigma M}{\Sigma M}$$

(c) Criteria for checking overturning

R = resistance to sliding at base
$= \mu\,(W + P_v) + cB$
μ = coef. of friction between wall and foundation
c = cohesion of foundation soil

$$FS = \frac{R + P_p}{P_h}$$

(d) Criteria for checking sliding

P_{max} less than $P_{allowable}$

(e) Typical soil bearing pressure distribution as a result of lateral pressures against wall

(f) Weight of wall and backfill can cause wall to settle

Figure 16-25 Summary of conditions to check during retaining wall design.

eccentricity distance, e, which is the distance from the center of the wall base to the position where the foundation reaction acts, is then determined from

$$e = \frac{d}{2} - d$$

where b = width of wall base.

The foundation pressure p_f at the toe and at the heel can be calculated by using

$$p_f = \frac{V}{(b)(1)} \pm \frac{Ve\left(\frac{b}{2}\right)}{\left(\frac{1}{12}\right)(1)(b^3)} = \frac{V}{b} \pm \frac{6Ve}{b^2} \qquad (16\text{-}12)$$

If the foundation reaction V is located to the toe side of the base's center, the plus sign is used in Equation 16-12 to compute pressure at the toe while the negative sign is used to calculate the pressure at the heel, and vice versa. The computed values are then compared to allowable bearing pressures determined from bearing capacity equations or other information. Wall settlement or movement expected as a result of the weight of the wall and backfill loading being imposed on an area can be evaluated from settlement analysis based on the compressibility of the foundation soils (Fig. 16-25f and Chapter 10).

Relating to consideration of foundation pressure and settlement, the capability of the foundation soils underlying the location of a retaining wall *and* its backfill may actually be the aspect most crucial to achieving a satisfactory performance in areas where clay exists. Studies of retaining wall performances have indicated that clay foundations were involved in the majority of failures [227]. The undrained shearing strength and bearing capacity of the foundation deposit should be evaluated to ensure that the in situ conditions will be capable of supporting the loading to result from the wall and backfill. Strength gain can be expected when foundation clays are subject to new loading, but changes associated with consolidation normally occur very slowly.

Illustration 16-2

A simple retaining wall of masonry supports a sand backfill as shown by the sketch. Assuming an active lateral pressure condition, determine the stability of the wall.

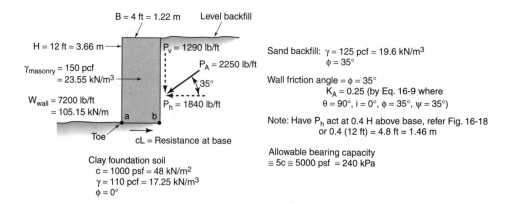

B = 4 ft = 1.22 m   Level backfill

H = 12 ft = 3.66 m P_v = 1290 lb/ft

$\gamma_{masonry}$ = 150 pcf = 23.55 kN/m³ P_A = 2250 lb/ft   35°

W_{wall} = 7200 lb/ft = 105.15 kN/m P_h = 1840 lb/ft

a   b

Toe   cL = Resistance at base

Clay foundation soil
c = 1000 psf = 48 kN/m²
γ = 110 pcf = 17.25 kN/m³
ϕ = 0°

Sand backfill: γ = 125 pcf = 19.6 kN/m³
ϕ = 35°

Wall friction angle = ϕ = 35°
K_A = 0.25 (by Eq. 16-9 where
θ = 90°, i = 0°, ϕ = 35°, ψ = 35°)

Note: Have P_h act at 0.4 H above base, refer Fig. 16-18
or 0.4 (12 ft) = 4.8 ft = 1.46 m

Allowable bearing capacity
≅ 5c ≅ 5000 psf = 240 kPa

Solution

$$P_A = \frac{1}{2}\gamma H^2 K_a = \frac{1}{2}(125 \text{ pcf})(12 \text{ ft} \times 12 \text{ ft})(.25) = 2250 \text{ lb per ft of length}$$

$$P_v = P_A \sin 35° = 1290 \text{ lb/ft} = 18.86 \text{ kN/m}$$

$$P_h = P_A \cos 35° = 1840 \text{ lb/ft} = 26.91 \text{ kN/m}$$

1. ΣM_{toe} to determine stability against overturning:

$\curvearrowleft$ (1840 lb/ft)(4.8 ft) = 8832 ft·lb per ft of wall length,

or, (26.91 kN/m)(1.46 m) = 39.4 kN·m per m of wall length

$$\curvearrowright(1290 \text{ lb/ft} \times 4 \text{ ft}) + \left(150 \text{ pcf} \times 4 \text{ ft} \times 1 \text{ ft} \times 12 \text{ ft} \times \frac{4}{2} \text{ ft}\right) =$$

$$= 5160 + 14{,}400 = 19{,}560 \text{ ft·lb per ft of wall length,}$$

$$\text{or, } \left(18.86 \frac{\text{kN}}{\text{m}}\right)(1.22 \text{ m}) + \left(23.55 \frac{\text{kN}}{\text{m}^3}\right)(1 \text{ m})(1.22 \text{ m})(3.66 \text{ m})\left(\frac{1.22}{3} \text{ m}\right) =$$

$$= 87.14 \text{ kN·m per m of wall length}$$

$$\text{FS against overturning} = \frac{19{,}560 \text{ ft-lb}}{8832 \text{ ft-lb}} = \frac{87.14 \text{ kN·m}}{39.4 \text{ kN·m}} = 2.2 \pm$$

2. ΣF in horizontal direction to determine stability against sliding:

$$\leftarrow F_{\text{causing}} = 1840 \text{ lb/ft} \quad \text{or} \quad 26.91 \text{ kN/m}$$

$$\rightarrow F_{\text{resisting}} = cB = (1000 \text{ psf})(4 \text{ ft}) = 4000 \text{ lb per ft of wall length,}$$

$$\text{or, } \left(48 \frac{\text{kN}}{\text{m}^2} \times 1.22 \text{ m}\right) = 58.56 \text{ kN per m of wall length.}$$

$$\text{FS against sliding} = \frac{F_{\text{resisting}}}{F_{\text{causing}}} = \frac{4000 \text{ lb}}{1840 \text{ lb}} = \frac{58.56 \text{ kN}}{26.91 \text{ kN}} = 2.2 \pm$$

3. Determining foundation pressures:

$$\Sigma M_b = (1840 \text{ lb/ft} \times 4.8 \text{ ft}) + (7200 \text{ lb/ft} \times 2 \text{ ft})$$

$$= 23{,}230 \text{ ft-lb per ft of length,}$$

$$\text{or, } (26.91 \text{ kN/m})(1.46 \text{ m}) + (23.55 \text{ kN/m}^3)(3.66 \text{ m})(1.22 \text{ m})(1.22 \text{ m}/2)$$

$$= 103.43 \text{ kN·m per m of length.}$$

$$\Sigma V = 7200 \text{ lb/ft} + 1290 \text{ lb/ft} = 8490 \text{ lb/ft,}$$

$$\text{or, } (105.1 + 18.9)\text{kN/m} = 124 \text{ kN per m of length.}$$

$$d = \frac{\Sigma M}{\Sigma V} = \frac{23{,}230 \text{ ft-lb/ft}}{8490 \text{ lb/ft}} = \text{or, } 2.74 \text{ ft, or } (103.43 \text{ kN·m/m})/(124 \text{ kN/m}) = .84 \text{ m.}$$

The computed negative pressure p_b (sketch) indicates that the heel section is not bearing on soil; accordingly, the bearing area is less than 4 ft² per foot of wall length, and p_a is greater than calculated.

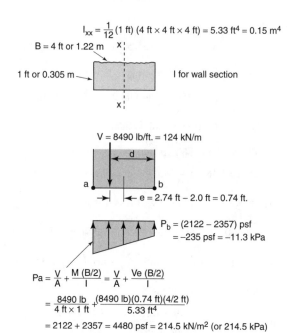

$$I_{xx} = \frac{1}{12}(1 \text{ ft})(4 \text{ ft} \times 4 \text{ ft} \times 4 \text{ ft}) = 5.33 \text{ ft}^4 = 0.15 \text{ m}^4$$

B = 4 ft or 1.22 m

1 ft or 0.305 m

I for wall section

V = 8490 lb/ft. = 124 kN/m

d

a   b

e = 2.74 ft – 2.0 ft = 0.74 ft.

$P_b = (2122 - 2357)$ psf
$= -235$ psf $= -11.3$ kPa

$$P_a = \frac{V}{A} + \frac{M \, (B/2)}{I} = \frac{V}{A} + \frac{Ve \, (B/2)}{I}$$

$$= \frac{8490 \text{ lb}}{4 \text{ ft} \times 1 \text{ ft}} + \frac{(8490 \text{ lb})(0.74 \text{ ft})(4/2 \text{ ft})}{5.33 \text{ ft}^4}$$

$$= 2122 + 2357 = 4480 \text{ psf} = 214.5 \text{ kN/m}^2 \text{ (or } 214.5 \text{ kPa)}$$

Recalculate, using a smaller bearing area to determine the actual p_a and p_b. Try B dimension as 3.78 ft (based on a trial iteration procedure).

$$\text{new } I \text{ value} = \frac{1}{12}(1 \text{ ft})(3.78 \text{ ft})^3 = 4.50 \text{ ft}^4$$

$$\text{new } e \text{ value} = \frac{3.78 \text{ ft}}{2} - (4.0 \text{ ft} - 2.74 \text{ ft}) = .63 \text{ ft}$$

$$\text{new } p_a = \frac{8490 \text{ lb}}{(3.78 \times 1) \text{ ft}^2} + \frac{(8490 \text{ lb})(.63 \text{ ft})(3.78/2 \text{ ft})}{4.50 \text{ ft}^4}$$

$$= 2246 \text{ psf} + 2246 \text{ psf} = 4492 \text{ psf} = 21.5 \text{ kPa}$$

$$\text{new } p_b = (2246 - 2246) \text{ psf} = 0 \text{ psf} \quad \text{(also for a distance 0.22 ft inward from heel)}$$

Assumption $B = 3.78$ ft is correct, and actual $p_a = 4492$ psf $= 215$ kPa. (Bearing pressure diagram will be triangular shaped, ranging from 0 psf at a distance 0.22 ft from the heel to the maximum 4492 psf at the toe.)

Illustration 16-3

(a) Check the stability against sliding and overturning and determine the bearing pressures beneath the base for the cantilever retaining wall shown, using standard-classical method of analysis, U.S. customary units.

(b) Use Goh analysis, SI units, to check stability against sliding and overturning.

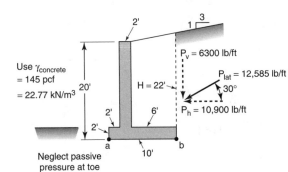

Sand backfill, $\phi = 35°$, $\gamma = 130$ pcf $= 20.4$ kN/m³
Use K = 0.4 since backfill is placed and
compacted in layers and limited
work area behind wall interferes
with development of active wedge

$P_v = 6300$ lb/ft

$P_{lat} = 12,585$ lb/ft

$H = 22'$ 30°

$P_h = 10,900$ lb/ft

Use $\gamma_{concrete}$ = 145 pcf = 22.77 kN/m³

20'

Silty sand foundation
$\gamma = 115$ pcf $= 18$ kN/m³, $\phi = 30°$
Assume $\mu = 0.5$ for sliding

Neglect passive pressure at toe

10'

Solution

(a) Standard analysis (U.S. customary units)

$$P_{lateral} = \frac{1}{2}\gamma H^2 K = \frac{1}{2}(130 \text{ pcf})(22 \text{ ft} \times 22 \text{ ft})(.4)$$

$$= 12,585 \text{ lb per ft of wall length}$$

(i) ΣF in horizontal direction to check stability against sliding:

$$\leftarrow F_{causing} = 10,900 \text{ lb per ft of wall length}$$

$$\rightarrow F_{resisting} = .5[(2 \text{ ft} \times 10 \text{ ft} \times 1 \text{ ft} \times 145 \text{ pcf})$$

$$+ (2 \text{ ft} \times 18 \text{ ft} \times 1 \text{ ft} \times 145 \text{ pcf})$$

$$+ (6 \text{ ft} \times 19 \text{ ft} \times 1 \text{ ft} \times 130 \text{ pcf}) + 6300 \text{ lb}]$$

$$= 15,320 \text{ lb per ft of wall length}$$

$$\text{FS against sliding} = \frac{15,320 \text{ lb}}{10,900 \text{ lb}} = 1.4 \pm$$

(ii) ΣM_{toe} to check stability against overturning:

$$\left(10,900 \text{ lb} \times \frac{22 \text{ ft}}{3}\right) = 79,935 \text{ ft·lb per ft of wall length}$$

$$(6300 \text{ lb})(10 \text{ ft}) + (145 \text{ pcf})(1 \text{ ft})(10 \text{ ft})(2 \text{ ft})\left(\frac{10 \text{ ft}}{2}\right)$$

$$+ (145 \text{ pcf})(1 \text{ ft})(18 \text{ ft})(2 \text{ ft})(3 \text{ ft}) + (6 \text{ ft})(1 \text{ ft})\left(\frac{18 + 20}{2} \text{ ft}\right)(7.1 \text{ ft})(130 \text{ pcf})$$

$$= 198,385 \text{ ft·lb per ft of wall length}$$

$$\text{FS against overturning} = \frac{198,385 \text{ ft·lb}}{79,935 \text{ ft·lb}} = 2.5 \pm$$

(iii) Determining foundation pressures:

$$\Sigma M_a = 198{,}385 \text{ ft·lb} - 79{,}935 \text{ ft·lb} = 118{,}450 \text{ ft·lb}$$

$$\Sigma V = (145 \text{ pcf})[(2 \text{ ft})(1 \text{ ft})(18 \text{ ft}) + (2 \text{ ft})(1 \text{ ft})(10\text{ft})]$$

$$+ \left(\frac{20 + 18}{2}\right)\text{ft } (6 \text{ ft})(1 \text{ ft})(130 \text{ pcf}) + 6300 \text{ lb} = 29{,}240 \text{ lb}$$

$$d = \frac{\Sigma M_a}{\Sigma V} = \frac{118{,}450 \text{ ft·lb}}{29{,}240 \text{ lb}} = 4.05 \text{ ft}$$

$$e = \frac{b}{2} - d = \frac{10 \text{ ft}}{2} - 4.05 \text{ ft} = .95 \text{ ft}$$

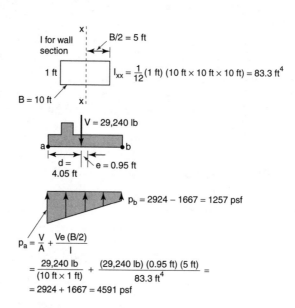

$$p_a = \frac{V}{A} + \frac{Ve \, (B/2)}{I}$$

$$= \frac{29{,}240 \text{ lb}}{(10 \text{ ft} \times 1 \text{ ft})} + \frac{(29{,}240 \text{ lb}) \, (0.95 \text{ ft}) \, (5 \text{ ft})}{83.3 \text{ ft}^4} =$$

$$= 2924 + 1667 = 4591 \text{ psf}$$

(b) *Goh analysis (SI units)*

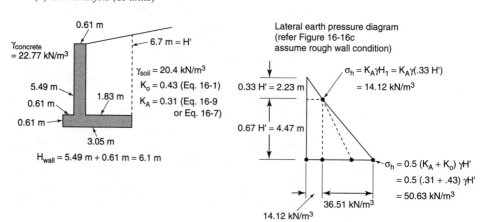

(i) ΣF in horizontal direction to check stability against sliding:

$$-F_{causing} = \frac{1}{2}(14.12 \text{ kN/m}^2 \times 2.23 \text{ m}) + (14.12 \text{ kN/m}^2 \times 4.47 \text{ m})$$

$$+ \frac{1}{2}(36.51 \text{ kN/m}^2 \times 4.47 \text{ m}) = 160.45 \text{ kN/m}$$

$$-F_{resisting}[\text{ref. part (a)(i)}] = [15,320 - (.5 \times 6300)]\text{lb/ft} \times 3.28 \text{ ft/m} \times 4.45 \text{ N/lb}$$

$$= 177.6 \text{ kN/m}$$

$$\text{FS against sliding} = \frac{177.6 \text{ kN/m}}{160.45 \text{ kN/m}} = 1.11$$

[*Note:* This computation did not assign a vertical component of the lateral earth force onto the virtual wall line; compare to part (a).]

(ii) ΣM_{toe} to check stability against overturning:

$$\circlearrowleft\left(\frac{1}{2} \times 14.12 \frac{\text{kN}}{\text{m}^2} \times 2.23 \text{ m}\right)\left(\frac{2.23}{2} \text{ m} + 4.47 \text{ m}\right) + \left(14.14 \frac{\text{kN}}{\text{m}^2} \times 4.47 \text{ m} \times \frac{4.47}{2} \text{ m}\right)$$

$$+ \left(\frac{1}{2} \times 36.51 \frac{\text{kN}}{\text{m}^2} \times 4.47 \text{ m}\right)\left(\frac{4.47}{3} \text{ m}\right) = 314.95 \text{ kN·m per m of wall length}$$

$$\curvearrowleft(3.05 \text{ m} \times .61 \text{ m} \times 22.77 \text{ kN/m}^3)\left(\frac{3.05}{2} \text{ m}\right) + \left(5.49 \text{ m} \times .61 \text{ m} \times 22.77 \frac{\text{kN}}{\text{m}^3}\right)(.91 \text{ m})$$

$$+ (1.83 \text{ m} \times 5.8 \text{ m} \times 20.4 \text{ kn/m}^3)(2.16 \text{ m}) = 601.69 \text{ kN·m per m of wall length}$$

$$\text{FS against overturning} = \frac{601.69 \text{ kN·m} \curvearrowright}{314.95 \text{ kN·m} \circlearrowleft} = 1.91$$

Surface Loads near Wall

Loadings placed on the ground surface behind a retaining wall cause lateral pressures against the wall that are additive to those pressures created by the backfill and, if applicable, a high water table. If an areal load of uniform intensity q occurs, a uniform *increase* in the lateral pressures acting against the wall, equal to qK, develops (Fig. 16-26). The lateral pressure coefficient K used for the wall backfill also applies to the areal load.

A concentrated load Q will cause a lateral pressure distribution against the wall which is generally as indicated by Figure 16-27. The lateral pressure resulting at any point along the height of the wall directly in front of the load can be closely approximated by

$$p_Q = \left(\frac{Q}{H^2}\right)\frac{(.28n^2)}{(.16 + n^2)^3} \quad \text{for } m \leq .4 \tag{16-13a}$$

$$p_Q = \left(\frac{Q}{H^2}\right)\frac{(1.77m^2n^2)}{(m^2 + n^2)^3} \quad \text{for } m > .4 \tag{16-13b}$$

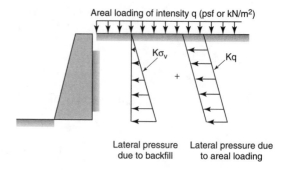

Figure 16-26 Influence of areal load-ing on wall pressures.

If a line load of intensity q acts behind a wall, a pressure distribution against the back of the wall is created, as generally shown by Figure 16-28. The lateral pressure at points along the height of the wall can be approximated from

$$p_q = \left(\frac{q}{H}\right)\frac{(.20n)}{(.16 + n^2)^2} \text{ for } m \leq .4 \qquad (16\text{-}14a)$$

$$p_q = \left(\frac{q}{H}\right)\frac{(1.28\ m^2 n^2)}{(m^2 + n^2)^2} \text{ for } m > .4 \qquad (16\text{-}14b)$$

In these equations, the units for p_Q and p_q are in accord with the units used for Q or q, and H.

Terzaghi-Peck Charts for Backfill Pressures

Where small, routine projects are involved, the designer may have little or no information about a wall's backfill, or the source of backfill may be subject to change. Job economics may limit or preclude investigation of backfill properties and control of its placement. For low retaining walls such as those bordering highways and railroads, where stability is the principal concern but wall movements and backfill settlements are not critical, Terzaghi and Peck have presented charts for estimating backfill pressures as related to general soil cate-gories. Their information, evolving from a coupling of theoretical concepts and evaluation of the performance of existing walls, is shown in Figure 16-29. These curves are consid-

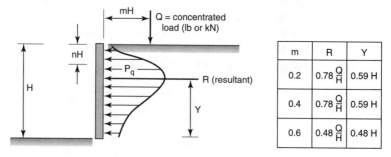

Figure 16-27 Pressure distribution against vertical wall resulting from point load, Q.

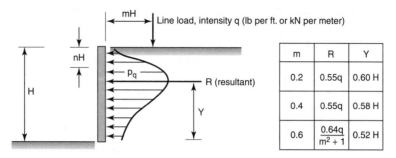

m	R	Y
0.2	0.55q	0.60 H
0.4	0.55q	0.58 H
0.6	$\dfrac{0.64q}{m^2 + 1}$	0.52 H

Figure 16-28   Pressure distribution against vertical wall resulting from line load of intensity *q*.

ered suitable for the design of walls up to about 6 m (20 ft) in height. Such charts are limited to providing realistic but safe values of lateral pressures due to soil backfill. The designer must still include provisions for drainage and consider the possibility of soil freezing and swelling behind the wall. Importantly, the designer must independently obtain information about the character and strength of the area's foundation soils to evaluate their adequacy for supporting the wall and backfill.

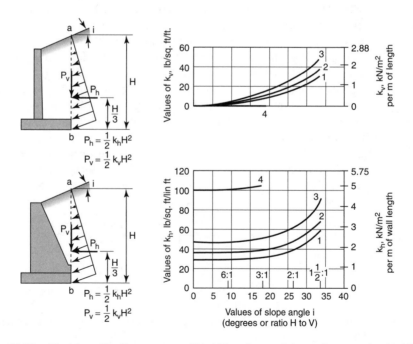

Figure 16-29   Chart for estimating pressure of backfill against retaining walls supporting backfills with plane surface. Use of chart limited to walls not over about 6 m (20 ft) high. Curves refer to different soil types: (1) Backfill of coarse-grained soil without admixture of fine particles, very permeable, as clean sand or gravel. (2) Backfill of coarse-grained soil of low permeability due to admixture of particles of silt size. (3) Backfill of fine silty sand, granular materials with conspicuous clay content, and residual soil with stones. (4) Backfill of very soft or soft clay, organic silt, or silty clay. (After Terzaghi and Peck. From *Foundation Engineering* by Peck, Hanson, and Thornburn, by permission of John Wiley and Sons, Inc., publisher [225].)

Earthquake Forces and Gravity Retaining Walls

During earthquakes, retaining walls become subject to seismic-related horizontal and vertical inertial forces that act on both the wall and the soil retained by the wall, as indicated by Figure 16-30.

The seismic active lateral earth force resulting against the wall, P_{AE}, conventionally is taken from the studies of Mononobe and Matsuo[4] and Okabe[5] providing

$$P_{AE} = \frac{1}{2}\gamma H^2(1 - k_v)K_{AE}$$

where $K_{AE} = \left(\dfrac{\cos^2(\phi - \theta - \beta)}{\cos\theta \cos^2\beta \cos(\delta + \beta + \theta)}\right)\dfrac{1}{\left[1 + \sqrt{\dfrac{\sin(\phi + \delta)\sin(\phi - \theta - i)}{\cos(\delta + \beta + \theta)\cos(i - \beta)}}\right]^2}$

and values of k_v, ϕ, θ, β, δ, and i are as identified on Figures 16-30 and 16-31.

For studies of seismic stability concerned with sliding and overturning, the positions of earth forces acting against a wall are assumed as shown in Figure 16-31.

Studies of retaining wall stability by Richards and Elms [240] have indicated that walls subject to earthquake conditions first tend to translate or slide (often in incremental movements as the array of seismic waves pass through the area), and overturning is unusual unless the foundation soils experience bearing capacity failure. The angle defining the soil failure surface which forms behind the wall for a static loading condition (with reference to a horizontal plane) decreases during earthquake activity, and an active soil zone of greater lateral extent develops; the angle of the seismic failure surface can be half the static condition angle, an important factor for locating structural elements behind the wall. Where cohesionless backfill and foundation soils exist, Richards-Elms have proposed that Equation 16-15 be applied to calculate the wall weight or mass required to maintain stability against sliding during earthquakes for the condition where a limited translation can be tolerated. A factor of safety for the seismic condition is then the ratio of the actual wall weight (or mass) to the weight (or mass) calculated by Equation 16-15.

$$W_{\text{wall}} = \frac{1}{2}\gamma H^2 K_{AE}\left[\frac{\cos(\delta + \beta) - \sin(\delta + \beta)\tan\phi_b}{\tan\phi_b - \tan\theta}\right] \qquad (16\text{-}15a)$$

$$M_{\text{wall}} = \frac{1}{2}\rho H^2 K_{AE}\left[\frac{\cos(\delta + \beta) - \sin(\delta + \beta)\tan\phi_b}{\tan\phi_b - \tan\theta}\right] \qquad (16\text{-}15b)$$

where $\tan\theta = k_h/(1 - k_v)$, and for the condition causing wall movement,

---

[4]"On the Determination of Earth Pressures during Earthquakes," N. Mononobe and H. Matsuo, Proceedings of World Engineering Congress, 9, 1929.

[5]"General Theory of Earth Pressures," S. Okabe, Journal of Japan Society of Civil Engineering, 12(1), 1926.

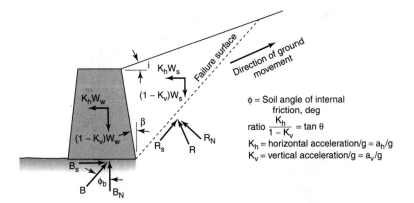

Figure 16-30 Combined force diagram: Seismic forces for active case.

$$k_h = A\left(\frac{.087\ V^2}{\Delta Ag}\right)^{.25} = A_a\left(\frac{.2A_v^2}{A_a\Delta}\right)^{.25}$$

(16-16)

where γ = unit weight of soil retained
   ρ = density of soil retained behind the wall
A, A_a = acceleration coefficient for site location (see Figs. 1-26b, 1-17c)
   A_v = velocity-related acceleration coefficient (see Fig. 1-27b)
   Δ = maximum tolerable or design allowable wall displacement (in.)
   g = acceleration of gravity, 32.2 ft/sec²
   V = maximum ground velocity (in./sec)(see relationship to A_v below, ref. Fig. 1-27b)

Effective Peak Velocity, V (in./sec)	Velocity-related Acceleration Coefficient, A_v
12	0.4
6	0.2
3	0.1
1.5	0.05

Figure 16-31 Position of earth forces acting on wall, for determination of seismic stability.

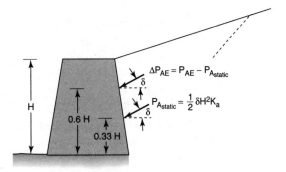

Illustration 16-4

The cross section for a retaining wall and related site conditions are shown in the sketch below. Determine the required wall weight to establish a factor of safety of 1.1 for the given seismic conditions.

$i = 20°$

$H = 5m$

$P_{A\ eq}$

$\delta = \dfrac{\phi}{2} = 17°$

$\beta = 15°$

Backfill soil:
$\gamma = 17.3$ kN/m³
$\phi = 34°$, c = 0

Foundation soil: $\gamma = 17.3$ kN/m³
$\phi_b = 34°$, c = 0

$i = 20°$
$\beta = 15°$
$\delta = 17°$
$\gamma = 17.3$ kN/m³ (110 pcf)

$\left.\begin{array}{l} A_v = 0.3 \\ A_a = 0.3 \end{array}\right\}$ Fig. 1-27b,c

Allowable wall displacement
$\Delta = 100$ mm = 4 inches assume $k_v = 0$

Solution

$$k_h = A_a\left(\frac{.2\,A_v^2}{\Delta A_a}\right)^{.25} = .3\left[\frac{(.2)(.3 \times .3)}{(4\ \text{in.})(.3)}\right]^{.25} = .105$$

$$\tan\theta = k_h/(1 - k_v) = (.105)/1 = .105$$

$$\theta = 6°$$

$$K_{AE} = \frac{\cos^2(\phi - \theta - \beta)}{\cos\theta\cos^2\beta\cos(\delta + \beta + \theta)\left[1 + \sqrt{\dfrac{\sin(\phi + \delta)\sin(\phi - \theta - i)}{\cos(\delta + \beta + \theta)\cos(i - \beta)}}\right]^2}$$

$$= \frac{\cos^2(34° - 6° - 15°)}{\cos 6°\cos^2 15°\cos 38°\left[1 + \sqrt{\dfrac{(\sin 51°)(\sin 8°)}{(\cos 38°)(\cos 5°)}}\right]^2}$$

$$= \frac{.949}{1.375} = .69$$

(*Note:* The magnitude of K_{AE} is significantly influenced by backfill slope and face angle of the retaining wall.)

The calculated weight for the wall becomes

$$W = \frac{1}{2}\gamma H^2 K_{AE}\left[\frac{\cos(\delta + \beta) - \sin(\delta + \beta)\tan\phi_b}{\tan\phi_b - \tan\theta}\right]$$

$$= \frac{1}{2}(17.3\ \text{kN/m}^3)(5\ \text{m})^2(.69)\left[\frac{\cos(17° + 15°) - \sin(17° + 15°)\tan 34°}{\tan 34° - \tan 6°}\right]$$

$$= \frac{1}{2}(17.3\ \text{kN/m}^3)(5\ \text{m})^2(.69)(.86) = 513\ \text{kN per m of wall length}$$

To have a factor of safety against sliding equal to 1.1, the required or actual wall weight should be

$$W_{\text{req'd}} = 1.1(513\ \text{kN/m}) = 565\ \text{kN per m of wall length}$$

For comparison, the active earth pressure coefficient K_A for the static condition (no earthquake effects) can be calculated from the K_A term included in Equation 16-9, or the following

$$K_{A_{static}} = \frac{\cos^2(\phi - \beta)}{\cos^2\beta \cos(\delta + \beta)\left[1 + \sqrt{\dfrac{\sin(\delta + \phi)\sin(\phi - i)}{\cos(\delta + \beta)\cos(\beta - i)}}\,\right]^2}$$

$$= .52, \quad \text{compare with } K_{AE} = .69$$

16.5   ALTERNATE TYPES OF RETAINING STRUCTURES

Gabion Retaining Walls

Gabions are large steel wire mesh baskets filled with stone (Fig. 16-32a). The units[6] can be assembled, like building blocks, to construct retaining wall, slope protection, and water-front protection structures. Gabions are assembled and filled with the desired size and gradation of rock material at the job site. Conventionally, a wall structure is built by placing an empty basket in position, filling it with rock, then placing and filling the next basket, and so on (Fig. 16-32b). Each newly placed basket is attached by wire to the adjacent and underlying gabions.

A gabion structure obtains most of its stability and resistance to erosive forces from the gabion's rockfill. Virtually all of the resistance to earth or water forces acting against a

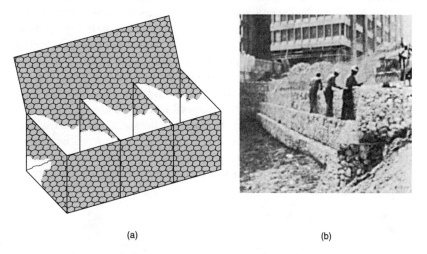

(a)   (b)

Figure 16-32   Illustration of gabion unit and installation procedure: (a) gabion basket; (b) wall under construction. (Courtesy of Maccaferri Gabions, Inc.)

---

[6]Basket sizes range from approximately 2 to 4 m (6 to 12 ft) in length and are 1 m (3 ft) in width and 1 m (3 ft) in depth. Internal diaphragms or dividers are included so that the interior cells do not become too large; small cells limit internal movement of the rockfill.

completed structure is provided by the friction and interlocking which develops within a basket's rockfill and between the rockfill surfaces of the individual gabion units. Structurally, the main purpose of the gabion basket is to keep the rock materials assembled in place.

A gabion retaining wall offers the advantage that it is a flexible structure that can tolerate foundation settlement and lateral movements while still retaining stability. The open structure of the rockfill permits water to move through the wall, a feature that prevents the occurrence of high water pressures behind the wall. However, tall walls may still be constructed with a drain system behind the wall to prevent seepage effects that could reduce stability. Compared to concrete construction, the gabion structure may show economy in construction time and overall cost.

A gabion retaining wall is analyzed as a gravity wall. The conventional design procedures discussed previously in the chapter are used to determine the lateral forces that will act against the structure. Resistance to sliding and overturning is checked for each section of wall (along each row of gabions) and for the full wall height as with the analysis of a masonry retaining wall. Walls usually follow either the straight-front-face, stepped-rear-face design or the stepped-front-face, straight-rear-face design (Fig. 16-33). Walls usually will be tilted slightly backward, say to a slope of 1 horizontal to 10 vertical, to improve the stability insofar as being affected by the angle of the lateral soil pressure resultant.

Reinforced Earth

Retaining walls and embankments have been built of earth reinforced with strips, fabrics, or fibers of various materials so as to increase the shear strength and supporting properties of the constructed mass. The use of reinforced earth for engineered structures as pioneered by the Reinforced Earth Company® involves reinforcing metal or fabric strips integrated with a granular earth fill that is placed and compacted in layers. A facing material, typically of thin metal or concrete panels, is used to form the exposed boundary or face of the reinforced earth structure and prevent the loss of fill material. The facing is mechanically fastened to the reinforcing strips. A reinforced earth structure is constructed systematically, layer by layer. Reinforcing strips are placed on the surface of each fill layer and connected to the face paneling, then covered with the next soil layer, a procedure that ensures the design number and location of reinforcing strips. Facing panels are placed as the height of fill progresses. The schematic diagram of Figure 16-34a illustrates the basic components of re-

Figure 16-33 Typical designs for gabion retaining walls.

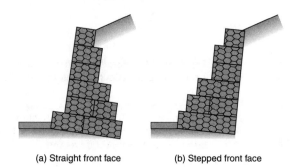

(a) Straight front face   (b) Stepped front face

Figure 16-34   (a) Basic components of reinforced earth construction. (b) Reinforced earth embankment under construction. (Courtesy of the Reinforced Earth Company)

inforced earth construction, while Figure 16-34b shows a project in progress. Variations of the *tied-back face panel retaining wall* concept exist and are available from other commercial organizations; usually the differences relate to the type of embedded strip or to the type and connection of the face paneling.

The friction or bond developed between the soil fill and strip surfaces provides the resistance required to hold the face panels in place. In terms of design, the lateral pressures exerted on the facing by the granular fill are opposed by the frictional resistance developed along the surface of the reinforcing strips. In effect, then, the reinforced earth mass retains itself, even though a sharp discontinuity (the vertical facing) exists. The lateral soil pressure imposed on the facing is approximately equal to the active lateral pressure indicated by conventional earth pressure theory. The required total surface area for each layer of reinforcing can be determined by computing the increment of lateral pressure developed by each soil layer. Typically, the width of fill and length of strips perpendicular to the wall facing is on the order of 0.8 times the fill height. However, the fill width and strip length must extend back beyond any potential failure plane that could result from outward movement of the facing.

Reinforced earth structures that function as retaining walls and for stabilizing earth slopes (including earth dam structures) can offer a cost advantage compared to conventional retaining structures. Part of this cost advantage relates to the condition that, generally, reinforced earth structures do not require special foundations, often being built directly on natural ground. The structures normally retain stability where foundation settlement or wall deflections occur, provided that movements are not excessive.

The design of a reinforced earth structure requires consideration of the effective life of material used for the reinforcing. Metals that experience only limited corrosion or nondegradable geosynthetics (fabrics) are preferred. (Geosynthetic retaining walls are designed using the basic concepts outlined here for reinforced earth walls; ref. Fig. 14-26d and i for illustrations of fabric walls.) Design must also ensure that the reinforced earth structure will not be part of or cause an areal failure, such as a slope stability failure that passes outside the limits of the reinforced earth structure.

16.6 EXCAVATION BRACING

Types and Installation of Excavation Bracing

Excavations for subsurface construction must consider procedures for preventing cave-in of the perimeter earth and protection of adjacent outlying areas against ground subsidence and lateral movement during the construction period. Where shallow excavations and excavations in open or undeveloped areas are made, the periphery of the excavation can simply be sloped at a safe inclination to ensure stability of the resulting embankment. With deep excavations, however, the safe sloping of soil banks may involve considerable distance beyond the limits of the work area and require that large volumes of earth be removed and subsequently replaced; such procedures can be expensive and time consuming, and may create a problem of storage. In developed or built-up regions, there simply may not be adequate space for sloping outside the excavation perimeter; in urban locations, excavations frequently extend to the property line or are required to be adjacent to an existing structure. Protection of the excavation perimeter and surrounding region for these situations then necessitates some system for vertically bracing the earth walls of the excavated area. Sheeting and bracing is frequently used.

A braced sheeting system when completed consists of a relatively flexible sheeting material, installed vertically against the excavation walls, and bracing that provides lateral support to hold the sheeting in place. Either partially or wholly, the sheeting support is installed prior to excavating. Bracing is provided as the excavation proceeds and the face of the sheeting components becomes exposed. A variety of materials, methods, and procedures have been used to achieve the required support bracing, the choices being influenced by factors such as subsurface conditions, excavation depth, working space, conditions surrounding the construction site, availability and cost of materials, equipment and labor force available, climate and season, construction timetable, and ingenuity of the construction contractor or designer. There are three basic categories of braced sheeting in common use—vertical wood sheeting, interlocking steel sheeting, and soldier beam with horizontal timber sheeting. Each grouping offers both advantages and disadvantages.

Wood and *steel* sheeting (Fig. 16-35) is driven to a depth that will be below the base of the excavation prior to excavating in the construction area. Lateral support to resist the pressure of earth behind the sheeting is provided by horizontal bracing termed *struts* or diagonal bracing termed *rakers* that are positioned transverse to the sheeting line and within the excavated area. Lines of continuous horizontal members extending along the face of the sheeting, termed *wales,* are provided at the time the struts or rakers are installed (and are also supported by them) to reinforce the sheeting between the strut or raker locations. The installation of strut and wale bracing takes place as excavation proceeds; that is, the upper section of sheeting is the first to be braced, with subsequent bracing for the lower sheeting provided as the excavation progresses (Figs. 16-36 and 16-38). Diagonal bracing (rakers) is used where horizontal struts cannot extend across the width of a large excavation because of the great span or because a too-cumbersome interior framework would be required to support the struts. Diagonal bracing obtains its interior support at the base elevation of the excavation (Figs. 16-37 and 16-38). This requires at least some deep excavation before bracing is provided, a procedure that can be safely accomplished by initially excavating at a slope only the central area of the construction site. A berm or wedge of earth is left against the sheeting. The rakers can then be installed as removal of the berm progresses, or, alternatively, in trenches dug between the sheeting and interior excavation. A conventional soldier beam and timber lagging system consists of vertical steel H-piles (the soldier beam) which are driven to a penetration below the excavation's planned base elevation before digging begins. Piles may be positioned with the flanges parallel or perpendicular to the excavation wall. As excavation proceeds, timber plank lagging is inserted between or behind the piles if the flanges are parallel to the excavation (similar to the representation shown in Figure 16-23c) or held by clips against the front of the pile when the flanges are perpendicular to the excavation. The strut or raker and wale bracing is placed as for a sheeting installation.

Vertical wood sheeting, consisting of planks 75 to 100 mm (3 in. to 4 in.) thick, is, practically speaking, used for excavations extending no deeper than about 6 m (20 ft). Where the soil has properties that permit an excavation wall of limited depth to support itself temporarily, the wood sheeting can be installed in the open or perhaps partially completed excavation to provide support as required by law and to protect adjacent ground. With deep excavations, the sheeting typically is driven in place with sheet piling hammers, and the size of this equipment must be limited to prevent overstressing and damaging the wood. Soil displacement tends to be rather great during the driving of wood sheeting, one effect being that very firm and dense soils as well as gravel or boulder materials cannot be penetrated. In deep excavations, the sheeting bracing must be closely spaced because of the relatively low strength of the wood, an arrangement that subsequently may interfere with construction work within the excavation.

Interlocking steel sheeting offers the advantages of being able to be driven to great depths through many soil types without suffering damage. As an excavation wall, it acts to prevent the inflow of groundwater from behind it (for such cases the sheeting and bracing must be designed to resist water pressures as well as soil pressures). Steel sheeting cannot

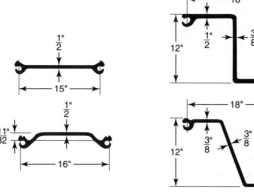
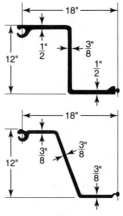
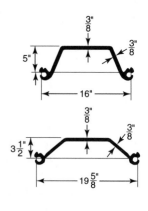

Straight sections

Used where tension value of interlock is important, particularly in cellular construction. Piling should be ordered with required interlock strength specified and type of structure indicated.

Z sections

High ratio of beam strength to weight means maximum economy in steel-sheet piling. Used primarily in filled bulkheads or other walls requiring high beam strength.

Archweb sections

Designed essentially for applications where a combination of beam strength and interlock tightness is required. Used generally in light bulkheads, internally braced single wall cofferdams, cut-off walls.

Figure 16-35   Representative steel sheeting sections.

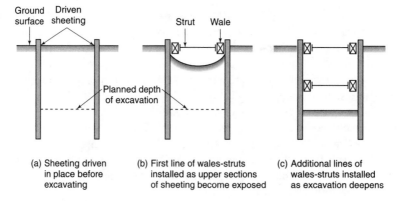

(a) Sheeting driven in place before excavating

(b) First line of wales-struts installed as upper sections of sheeting become exposed

(c) Additional lines of wales-struts installed as excavation deepens

Figure 16-36   Procedure for bracing sheeting with struts and wales.

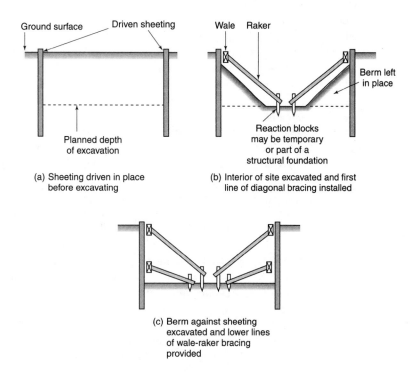

Figure 16-37 Procedure for bracing sheeting with rakers and wales.

(a) Sheeting driven in place before excavating

(b) Interior of site excavated and first line of diagonal bracing installed

(c) Berm against sheeting excavated and lower lines of wale-raker bracing provided

Figure 16-38 Steel sheeting with crosslot or strut bracing (upper section of photo) and raker bracing (lower section of photo). (Courtesy Spencer, White, and Prentis)

be used to penetrate earth that includes boulders, and there can be difficulty with gravelly material. Similarly, fill containing heavy debris or large objects such as broken concrete and timbers cannot be penetrated. A major disadvantage is cost on those projects where the sheeting cannot be withdrawn for use elsewhere; for example, with projects in urban areas, pulling of the sheeting when the substructure is completed could result in damage to the new construction or could disturb the ground and structures surrounding the construction site.

Steel soldier beam and horizontal timber sheeting is considered to be a flexible system with respect to installation. The H-pile soldier beams can penetrate problem soils more easily than sheeting; some freedom in spacing is permitted when local obstructions are encountered. Struts and wales normally can be the same-size member as the soldier beam, simplifying on-site storage and handling of these members. The lagging [timber planks 75 to 100 mm (3 to 4 in.), thick] is wedged between the pile flanges or behind the back flange as the excavation deepens. Commonly, the soldier beams are left in place to become an integral part within a poured concrete foundation wall for the permanent structure. When lagging is installed behind the back flange, it acts as the back form for construction of the concrete wall. During its installation, the lagging may be placed so as to create open horizontal spacings 25 to 50 mm (1 to 2 in.) wide, to permit drainage of the surrounding area and prevent hydrostatic pressures from developing against the back of the sheeting. Hay is often used in the space between lagging when seeping water tends to erode or wash in soil from behind the sheeting. An advantage to having drained earth behind the sheeting is the higher soil shear strength and lesser lateral pressure. Disadvantages include the effects of seepage forces, which contribute to lateral earth movement and consolidation–settlement of the retained earth mass.

Generally, the use of continuous steel sheeting is preferred for excavations deeper than about 8 m (25 ft), especially where soft clays exist at and below the lower zones of the planned excavation, because the embedded sheeting improves the stability of the excavation foundation.

Bracing within an excavation interferes with the construction activities that must take place inside the excavated area. A method that eliminates the need for interior bracing involves the use of tiebacks (Fig. 16-39), which extend into the earth behind the sheeting. Essentially, the tieback is a tensioned rod or cable connected to the sheeting or lagged wall plus an anchor embedded into soil or rock *outside* of the excavation area for establishing resistance to lateral forces imposed against the sheeting. Figure 16-40 illustrates a soldier beam–lagging system secured with tiebacks.

Alternatives to the use of driven sheeting for protecting the perimeter of an excavated area are available. Concrete walls constructed to support the earth excavation can also serve as all or part of the permanent foundation wall for the permanent structure. Unformed concrete walls can be created by utilizing drilled, cast-in-place pile installation procedures. Alternatively, where firm cohesive soils exist, unsupported narrow trenches can be excavated and filled with concrete. When trenches are deep, or will extend through granular or soft clay soils as well as the water table, *slurry-trench* construction can be performed (Fig. 16-41). This procedure involves excavating the trench while simultaneously keeping a heavy, viscous slurry (typically a bentonite clay–water mixture) present in the excavation to exert a lateral pressure that retains the trench's soil wall. Concrete is then tremied into the trench from the bottom up in a manner that displaces the slurry and prevents mixing. Reinforcing steel often is included, being placed into the slurry-filled trench before the concrete is

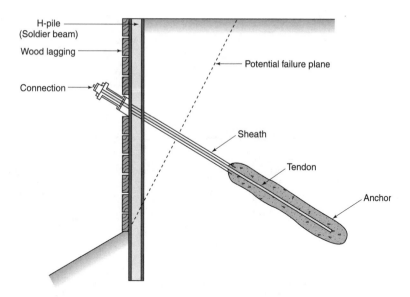

Figure 16-39 Typical earth tieback. Anchor is located outside potential failure zone.

Figure 16-40 Tiebacks used to support steel soldier beam and timber lagging installation. (Photo courtesy of Spencer, White, and Prentis)

poured. A tieback system or interior bracing is used to support the new concrete walls temporarily when the interior excavation is made.

Lateral Pressures for Designing Bracing

The top struts for sheeting bracing are normally installed before much earth is removed from the excavation area. The struts are wedged or jacked into position, a procedure that places each brace in compression and forces the sheeting outward from the excavation,

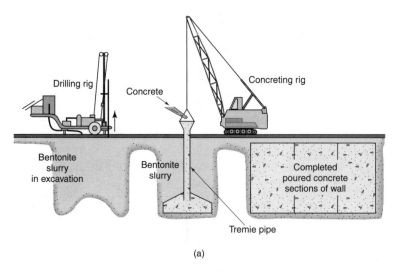

Drilling rig

Concrete

Concreting rig

Bentonite
slurry
in excavation

Bentonite
slurry

Completed
poured concrete
sections of wall

Tremie pipe

(a)

(b)

Figure 16-41. (a) Sequence of slurry-trench (slurry wall) construction. (Diagram courtesy of Spencer, White, and Prentis) (b) Tied-back concrete walls constructed by slurry-wall procedure. (Photo courtesy of Bencor Corporation of America foundation Specialist)

acting to prevent a subsequent inward deflection. The compression in the strut before it receives sheeting load is termed *prestress.* As excavation proceeds, some inward deflection (toward the excavation) of the lower sheeting occurs before the lower struts can be placed. This described sequence of excavation and bracing results in a lateral soil pressure distribution against the sheeting (Fig. 16-42), that generally is similar to the parabolic distribution that occurs when a retaining wall rotates about its top.

Forces developed in the struts of braced excavations provided for sand and clay soils have been measured. Pressure distributions against the sheeting have been approximated by assuming that each strut supports a sheeting area extending vertically and horizontally halfway to adjacent struts. The results of such studies have established patterns of pressure distribution but also show marked variation in magnitudes, even when measured in the same soil deposit. Factors found to influence the pressure distribution include the speed at which excavation progresses and the care taken when bracing is installed, as well as soil type and properties. The effects of all variables have not been clearly established, however. Nevertheless, the results of field studies serve as the basis for developing earth pressure diagrams that can be used to design bracing for new work. The pressure distribution diagrams recommended for design (Fig. 16-43) actually represent an envelope that encompasses the field-determined pressure diagrams. Such design pressure diagrams are conventionally referred to as *apparent pressure diagrams* to differentiate them from test-determined diagrams. Use of the diagrams provides conservative but not necessarily uneconomical design of struts.

The procedure for the design of struts used to brace excavation sheeting reverses the process used to develop the apparent pressure diagram, as shown by Figure 16-44. This figure indicates that the effect of the lower embedded segment of sheeting does not require consideration when the bracing is designed by reference to the apparent pressure diagram.

The apparent pressure diagram represents conditions for a completed bracing system. Struts, and rakers as well, are conventionally designed on the assumption that the maximum load could exceed the planned load at some time during installation of the bracing or after its completion if a construction mishap results in damage to some bracing or if unforeseen loading conditions occur in the area surrounding the excavation. A 50 percent overload is a common assumption.

Relating to the surface elevation at the bottom of an excavation, a surcharge pressure effect exists at this elevation behind the sheeting from the height of the retained soil. There is a resulting tendency for soil beneath the excavation base to be pushed upward, as occurs with a bearing capacity failure (Fig. 16-45). The occurrence is most likely where clay soils exist. A heaving of the base is accompanied by lateral movement of the lower sheeting. With large movement, significant settlement of the ground surrounding the excavation also occurs.

With excavation in clay, the strength for the case of undrained shear applies to the conditions that exist during the construction period. Since the bearing capacity of clay in undrained shear is controlled by its cohesion, *c,* the safe excavation depth in regard to bottom heave is similarly related.[7] Heave is expected to be small when $\gamma H/c$ values (the ratio of surcharge pressure to soil shear strength) are less than 6. Where the ratio exceeds 8, significant and possibly uncontrolled heave is anticipated, and a standard bracing system

---

[7]Recall that the ultimate bearing capacity of clay, q_{ult}, when a condition of undrained shear prevails is close to $6c$.

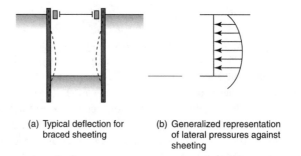

(a) Typical deflection for braced sheeting

(b) Generalized representation of lateral pressures against sheeting

Figure 16-42 Mode of deflection and lateral pressure distribution for sheeting supporting a vertical cut.

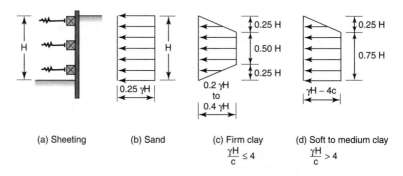

(a) Sheeting

(b) Sand

(c) Firm clay
$\dfrac{\gamma H}{c} \le 4$

(d) Soft to medium clay
$\dfrac{\gamma H}{c} > 4$

Figure 16-43 Apparent pressure diagrams for designing strut loads for braced vertical sheeting. (After Peck) 296

Figure 16-44 Pressure zones assumed for calculating strut loads.

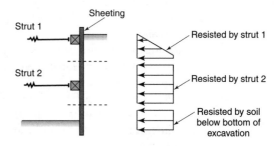

Figure 16-45 Loss of stability in bottom of excavation when surcharge effect is excessive.

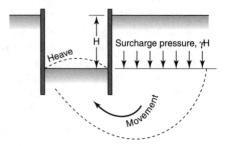

should not be used. The ratio of excavation depth to clay shear strength also affects the lateral pressures developing against the sheeting, as indicated by the design recommendations of Figure 16-43c and d.

Where a firm stratum underlies the excavation area and is reasonably close, the danger of bottom heave can be eliminated by extending sheeting close to or into the firm material. If soft clays extend a great distance below the excavation, increasing the sheeting penetration offers little benefit unless the excavation is narrow. Excavations in sand typically do not experience a problem with bottom heave. However, instability of the base can occur in a pumped excavation where a groundwater table rises high behind the sheeting and the resulting seepage into the excavation area causes high uplift pressures.

Compared to strut bracing, raker bracing and tiebacks involve differences in method that affect lateral pressures against the sheeting. Where the uppermost section of sheeting is not directly braced, some freedom for movement exists, and the lateral pressures may initially be less than indicated by the pressure diagrams obtained from measurements of strut bracing. However, lateral pressures may subsequently increase because of soil creep and the effects of the surface activities adjacent to the excavation area. In the absence of data relevant to the design of diagonal or tieback support for a particular project, the apparent pressure diagrams of Figure 16-43 can be referred to for estimating the horizontal component of the axial force in rakers and tiebacks. Because tiebacks are typically provided at spacings closer than interior braces, an overload to one tie easily results in redistribution to or sharing of load with adjacent ties. It has been found that tiebacks do not require the conservatism that interior bracing does. Consequently, tiebacks commonly are designed to resist 80 percent of the loading indicated by the earth pressure envelopes developed for interior bracing (i.e., 80 percent of the loading indicated by diagrams such as in Figure 16-43). A further comparative advantage of tieback construction is that the close spacing results in smaller shear forces and bending moments being developed in the sheeting, an occurrence that permits sheeting with lighter cross section to be used.

The horizontal and vertical spacing of braces has an effect on the deflection that sheeting experiences. Large deflections are possible where the bracing is widely spaced. It is important to recognize that lateral movement of a retained soil also results in ground settlement. For excavation work in developed areas, where settlement or shifting of ground will endanger nearby facilities, the bracing system should be designed to keep the sheeting rigid and movement minimized.

16.7 ANCHORED BULKHEADS

The category of structure termed the *anchored bulkhead* is commonly used for marine and waterfront structures such as wharfs, piers, waterfront retaining walls, floodwalls, and breakwaters but may also be used inland. The anchored bulkhead, when structurally used for supporting a soil mass that is adjacent to a body of water, consists of a vertical wall having its lower section embedded in earth and a tie or anchor near its top for support and to limit deflection. Steel sheet piling is the material most typically used for the bulkhead wall, but timber and concrete are also common. For marine work, interlocking sheeting has the advantage that it can be driven into place through an existing depth of water, eliminating the requirement to first isolate and dewater an area to permit construction of the bulkhead.

When an area is to be used as a docking facility for shipping, the location for the bulkhead is often dredged first to provide the desired water depth (draft). If soft soils exist below dredging depth, they should be removed and replaced with granular fill to a depth that will be below the sheeting's tip elevation, so that better support is provided. Soil placed behind the sheeting to create a land surface should be cohesionless material. Weak natural soil behind the bulkhead should be removed prior to placement of the granular fill if future settlement would create a problem in the use of that area. (For example, buildings, utilities, and rail or vehicle roads to be located on the fill area could be damaged by settlement due to compression of weak natural soil left in place beneath the fill.)

Various methods of providing the anchor system for supporting the upper section of sheeting have been developed. As most typically designed, the upper support consists of horizontal wales that extend the plan length of the sheeting plus steel tie rods that connect the wales to anchors capable of providing the necessary resistance to pull imposed by the sheeting (Fig. 16-46). Blocks and walls that serve as anchors develop their resistance primarily from passive soil pressures.

A number of procedures are in use for representing the lateral pressures acting against wood and steel sheetpiling, a requirement for determining the required sheeting embedment and strength and the anchor resistance. The *free-earth-support method,* considered the most conservative of the methods shown in Figure 16-47, assumes that the sheeting acts as a ver-

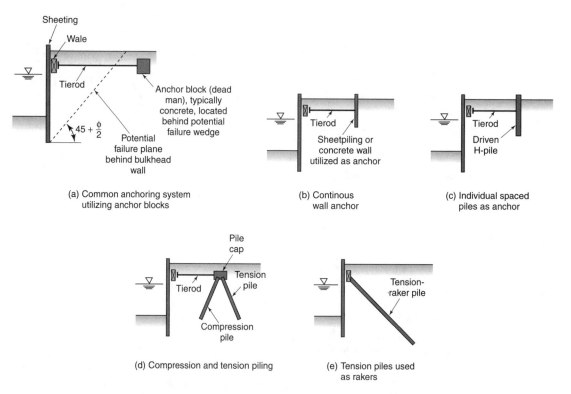

Figure 16-46 Methods and procedures for providing tiebacks.

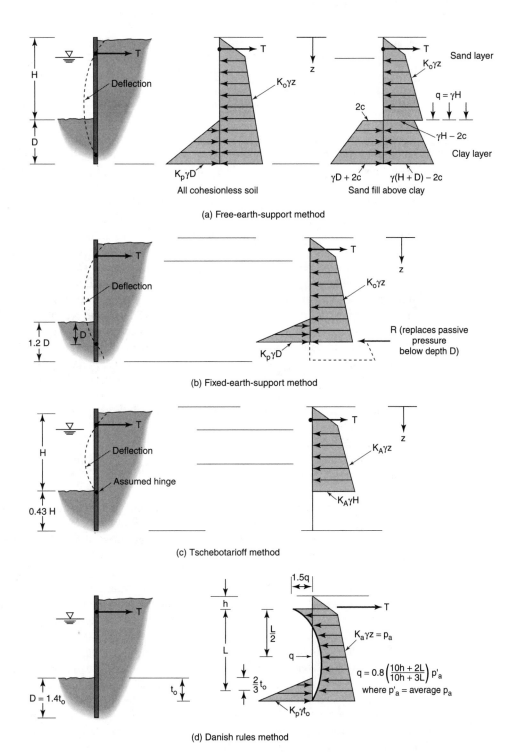

Figure 16-47 Methods to represent lateral pressures against sheeted bulkheads.

tical beam that receives its top support from the anchor and its bottom support from soil in front of the bulkhead (Fig. 16-47a). Forces act on the sheeting only from lateral soil pressures and the anchor pull; it is assumed that water pressures in front of and behind the sheeting will balance. The lateral pressures due to a retained cohesionless soil fill conventionally have been assumed to be active pressures, but recent studies suggest that the at-rest pressure distribution may be more appropriate. In a design analysis, the anchor pull on the sheeting is taken as a moment center, and moments due to the lateral pressure of the retained earth are compared to moments of passive soil pressure in front of the sheeting. An adequate depth of penetration D is achieved when these moments balance, and the theoretical factor of safety is 1. The required anchor tension is subsequently determined as the difference between the lateral soil forces acting behind and in front of the sheeting. For construction, the actual penetration is commonly increased to be 1.5 to 2 times the calculated value of D. The sheeting section is designed to resist the shear forces and bending moments that are indicated as a result of the anchor pull and lateral pressures acting along its length.

Usual designs call for drainage holes to be included in the sheeting so that water levels behind and in front of the sheeting are always similar even though tidal changes occur. If means for rapid drainage are not provided, the design should consider the additional lateral force that would develop because of unbalanced hydrostatic pressures.

The *fixed-earth-support method* considers the lowest section of sheeting to be fixed in the earth, similar to the manner in which a cantilever beam is fixed at its wall support. For analysis or design, an extra lateral resistance based on passive pressures developing against the land side of the sheeting's lower section is assumed to exist and to be additional to the conventional at-rest and passive soil pressures, as shown in Figure 16-47b. To design a bulkhead, an estimated sheeting length is analyzed for deflections, and a proper embedment has been reached when the deflection at the depth D is zero. The procedure typically requires a number of trials and is a laborious and time-consuming method unless some computational aid or computer program is available. Compared to the free-earth-support method, greater sheeting penetrations are required, but the use of lighter and more economical sheeting sections usually results.

The *Tschebotarioff method* (Fig. 16-47c) recommends a penetration equal to 0.43 times the sheeting support height. The sheeting section and anchor system is then designed on the basis that active soil pressures exist above the dredge line and a hinge is present at the dredge line. The *Danish rules method*, based on an "active-arching" lateral pressure distribution acting against the sheeting (Fig. 16-47d), was developed from studies on flexible bulkheads that indicated that bending was less than the values predicted by the free-earth-support method. However, the pressure distribution is appropriate only for the case where there is no anchor yield. This method is considered the least conservative of the four procedures discussed.

The occurrence of a bulkhead failure because the steel sheeting was overstressed in bending is rare. Where stability problems have occurred, the cause usually has been failure of the anchor system or has been related to inadequate soil support along the embedded lower section. Anchor failures often originate with the tie rod. If the bulkhead fill settles, as is typical, the tie rod deflects. Although steel tie rods normally can deflect without problem, the connection at the anchor, or more probably at the sheeting, may be overstressed. Hinged connections represent one method used to prevent the overstress problem. Bulkhead failure due to inadequate penetration can result during construction as the procedure of placing fill behind the sheeting causes the various effects of the retained material to increase. Instability can occur if the passive pressure resistance below the dredge line is inadequate, or if a rotational slide precipitated by the weight of the fill (discussed in Chapter 15) develops behind the bulkhead and passes beneath the sheeting. Failure due to similar causes can also occur after construction is completed if the site's foundation includes weak cohesive soil. Projects that require dredging and filling after the bulkhead sheeting is in place should have filling on both sides progress evenly. For existing facilities, protection against scour and overdredging is necessary to prevent loss of material that provides embedment for the sheeting.

16.8 WATER-RETAINING STRUCTURES: DAMS

Dams, structures that serve to retain water for storage and utilization purposes or to protect land areas, represent one of civilization's oldest categories of major constructed facilities. In current times, dams are an essential component of water supply systems, hydroelectric power facilities, and flood control projects; they also serve to create reservoirs for recreational and navigational use and for sediment retention. Frequently, a dam provides multipurpose service.

At present, there are more than 60,000 known dams in service in the United States. Over 95 percent of this number have been constructed during the twentieth century. Early dams typically were for water supply systems or hydroelectric usage, whereas many of the recently built dams have been for flood control and recreational use. For the future, new dams will be needed for water supply, flood control, and recreational use; additionally, with the concern over the availability and cost of energy, dams for providing new sources of hydroelectric power are being studied and proposed for many areas.

Dams for permanent facilities (as compared to temporary dams involved with construction projects) typically have been of masonry, concrete, earth fill, or rockfill construction, although timber and steel materials have also been used. Of the United States dams, some 7,000 are masonry or concrete, and more than 53,000 are earth fill or rock fill.[8] Certain advantages and disadvantages are associated with the construction, function, and operation of dams built of each of these materials. The selection for a given project is influenced by the dam's purpose, the location and site conditions including area available, construction conditions including materials and labor available as well as climate, and cost. Concrete, masonry, and rockfill dams require strong foundations such as rock or firm incompressible soil, whereas earth fill dams can be designed for locations where the foundation material is less firm.

A dam facility must include a spillway and outlet works (Fig. 16-48). The spillway enables excess reservoir water—water above the reservoir's design elevation but lower than the crest of the dam's impounding section—to move safely past the dam. Outlet works permit water to be drawn from the normal reservoir pool as required for a water supply, for generating hydropower, or for navigation purposes, and also enables the reservoir to be drained for dam repair or other purposes. Concrete and masonry dams often are designed so that part of the length will also serve as the spillway; the construction material can withstand the effects of flow over the section. In contrast, earth fill and rockfill dams should not be topped because of the probability that erosion and loss of section would result. For these dams, special spillway structures are required so that the reservoir's excess water will flow around or through the dam

---

[8]Personal communication, J. A. Gomez. Stetson-Harza Engineering Company (Harza Northeast), Utica, N.Y.; derived from U.S. Army Corps of Engineers data.

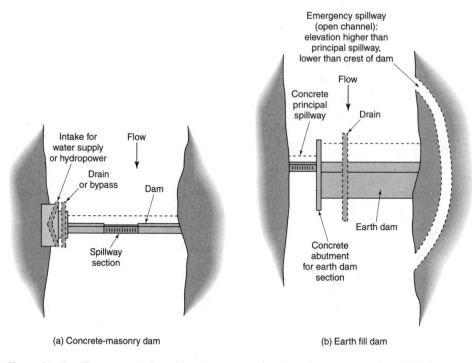

Figure 16-48 Plan representation of components and locations for concrete and earth fill dams.

before cresting (i.e., topping) can occur. Because of such requirements, dams for sites where flow is great, as along rivers, commonly are concrete or masonry, since the structure can be easily designed to spill large volumes of water without danger to the structure itself. Earth fill and rockfill dams typically are used for locations where the spill volume under normal operation is limited, such as with storage reservoirs for water supply and recreational lakes. These projects often include a principal spillway possessing limited capacity, while a relatively inexpensive emergency spillway, such as a grassed open channel, is provided to bypass large excess flows, such as those that occur in time of flood. Bypass channel spillways may involve large volumes of excavation that desirably could be used for constructing the dam itself.

Concrete and masonry dams can be gravity type, buttress type, or arch type (Fig. 16-49). *Gravity* dams, owing their stability against overturning and sliding forces that act on the structure to the mass of the dam section, are usually most practical where limited heights of water are to be impounded.[9] *Buttress* and *arch* dams are considered appropriate for medium-height and high dams; compared to the gravity type, buttress and arch dams will economize on materials but require advanced design and construction techniques.

Earth fill and rockfill dams obtain their overall structural stability from the large mass of soil or rock comprising the structure. These structures are expected to permit at least

---

[9]Designations according to height are somewhat arbitrary, but a height of 25 m (80 ft) can be assumed as an order-of-magnitude division between the low and moderate height categories.

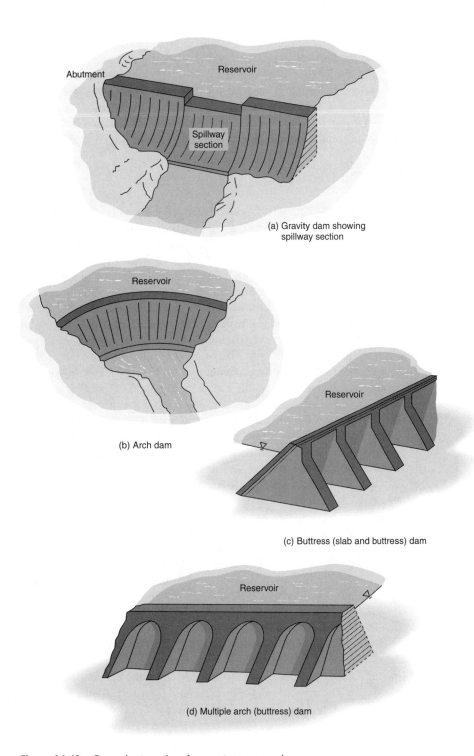

(a) Gravity dam showing spillway section

(b) Arch dam

(c) Buttress (slab and buttress) dam

(d) Multiple arch (buttress) dam

Figure 16-49   General categories of concrete-masonry dams.

some leakage to occur. Means are necessary to keep such flow within the limits required for stability and protection of downstream areas, or in arid regions to control the loss of valuable water. To restrict through-the-dam flow, impervious barriers or zones are incorporated into the dam's cross section. Drains and filters are also designed into the dam so that seepage that does occur can be collected and carried off without affecting stability. A barrier or cutoff is also extended into the zone below the dam if the site includes a soil or rock foundation that would otherwise permit an undesirable volume of seepage.

Earth Fill Dams

Earth fill dams are classified as homogeneous or zoned (Fig. 16-50). A homogeneous section is most appropriate where the soil used for its construction has a relatively low permeability after placement (e.g., soils with adequate silt or clay content). Zoned earth fill dams generally consist of coarse-soil outer zones or shells (e.g., predominantly sand, gravel, and possibly cobble materials) and a relatively impermeable internal zone or core of fine-grained soil.[10] Where a very coarse material can be provided separate from the smaller coarse-grained material (e.g., cobbles or gravels can be separated from gravels or sands), a zoning of the shell should be planned, with the coarsest material being used for the outermost section. Practically speaking, even the materials provided for large homogeneous dams show a variation in grading, and for that occurrence the finer material is placed near the center of the dam while the coarser materials are used for the shell zones. When a dam is designed as a homogeneous earth fill section, a blanket of cobbles or large gravel, or a slab of Portland cement concrete, bituminous concrete, or soil cement, is provided for the *upstream slope* as protection against the erosive effects of wave action, fluctuations in reservoir level including intentional drawdowns, and ice and rainfall. Downstream slopes of homogeneous dams can be provided with a filter blanket of coarse material so rain and snow melt will not cause surface erosion. Grasses and other low-height vegetation are also used on the downstream slope as erosion protection. Trees should not be permitted on slopes, because an uprooting during storm periods could begin a serious erosion. Deep tree roots could encourage through-dam seepage by shortening the seepage path.

For zoned designs, the moderately sloping fine-grained core (Fig. 16-50e and f) is considered to provide a more stable section than the severe sloping or vertical core. Failures that do occur in earth fill sections often are slip-circle slides (see Chapter 15) which pass through one of the slopes. A core zone is commonly the weakest material in the dam, and the moderately sloping core represents the optimum shape for obtaining benefit from the weight of water against the dam's upstream slope while limiting adverse effects from seepage moving through the core.

Earth dams as total structures are subject to settlement as the lower sections and probably also the foundation material compress under the weight of the overlying embankment. Upon reservoir filling, additional settlement and some lateral movement occur because of the effect of the impounded water above and against the dam. The soil mass making up the dam is deformed as a result of these movements. If the dam is properly designed and

---

[10]The design of zoned earth dams in the early twentieth century frequently showed a "puddled" core consisting of a blend of clay and sand–gravel materials to be placed and rolled in layers.

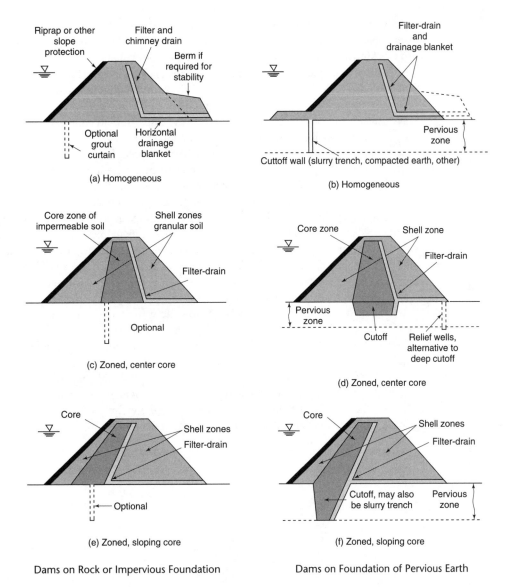

Figure 16-50   Typical earth dam sections.

constructed, movements between adjacent sections are limited, and no adverse effect on the structural stability results. Cracks may form,[11] particularly near the interface of zones, because of the different reaction to loading by the different materials. However, properly designed soil transition zones will permit self-healing to occur in areas where seepage begins to move through cracks.

---

[11]The condition where cracks develop within the dam section because of reservoir water pressures acting against the upstream slope, or because seepage under pressure enters a small crack and enlarges it, is referred to as *hydraulic fracturing*.

Concrete core walls to serve as the internal barrier to seepage are no longer being recommended for use, except possibly for very low dams. Concrete core walls are susceptible to cracking as settlement of the dam structure occurs.

For homogeneous and zoned dams, drains such as those represented in Figure 16-50 are included to collect and carry away any through-the-dam seepage that develops. Alternatively, for the zoned design represented in Figure 16-51, the downstream zoning of properly graded pervious materials can serve as the filter drain to collect seepage for harmless disposal downstream. Internal drains also should be zoned to act as filters[12] so that the seeping water does not carry soil particles from the core into the drain and shell zone (e.g., the filter zone closest to a silt–clay core may consist of a cohesionless silt or fine sand, while the outer zones of the filter drain consist of clean, coarse sand).

Cutoff walls into the foundation strata to prevent seepage beneath the dam can be an extension of the impermeable core zone if the cutoff excavation to the desired depth can be made dry, a requirement to permit proper placement including compaction of the soil cutoff and its filter drain. Steel sheetpiling has been widely used as a cutoff in the past, although sheeting installed into coarse soils has proven to be a risky choice susceptible to developing open joints and tears that permit locations for high-gradient seepage to occur. Concrete cutoff walls similarly are not recommended.

Where a foundation zone extends below the water table and a cutoff excavation cannot be adequately dewatered to permit dry construction, slurry trench methods can be considered. A bentonite clay slurry is used during the excavation process to temporarily support the trench walls against cave-in; then a fine-grained–coarse-grained soil mixture including some slurry material is placed to obtain a wall of impermeable backfill. Such slurry trench backfill is not highly compacted and is subject to settlement under loading. Consequently, the best location for the slurry trench cutoff is in the vicinity of the upstream toe to minimize the weight of the dam and water overburden. Steel sheeting can be installed within the backfilled slurry trench as added protection.

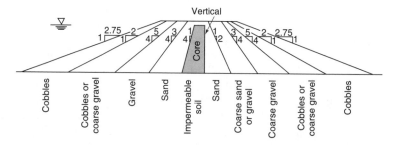

Figure 16-51   Illustration of zoned design for a granular earth fill dam. The downstream sand and gravel zones can also serve as drains for seepage.

---

[12]Filters will need to satisfy the gradation requirements for filter material discussed in Chapter 7.

An alternative to the foundation zone cutoff for controlling *deep* under-dam seepage is to use a system consisting of drains and relief wells located beneath the downstream slope, near the toe, to intercept and collect seepage and prevent the development of piping through the foundation and erosion at the toe (Fig. 16-50d).

Earth fill dams should be constructed by placing and compacting the soil in relatively thin layers.[13] Hydraulically placed earth fills are not recommended because of the relatively loose, weak, and unstable condition of the fill. Almost any soil type can be used for the compacted earth fill with the exception of material with organic content. Highly plastic clays are not recommended for use, however, because of a practical difficulty with handling and compacting. Similarly, dispersive clays (Chapter 3) should not be used because of the great susceptibility to erosion and piping, which can lead to loss of embankment once leakage commences.

Experiences have shown that satisfactory strength and performance are achieved if the compacted density of the earth fill is in the range of 95 to 100 percent of the maximum density obtainable from the standard Proctor moisture–density relationship (i.e., compaction test). Design slopes range from about 2.5 horizontal to 1 vertical where granular soil shell zones are used to about 4 horizontal to 1 vertical for homogeneous silt–clay embankments. Fine-grained soil should be placed at a water content 1 to 2 percent drier than the optimum moisture indicated by the compaction test results, particularly for the lower zones of high dams, to prevent the development of excess pore water pressures when the weight of the overlying dam and impounded water causes an expected soil compression. In the early history of earthwork construction, with embankments of limited height, it was common to compact the earth fill at water contents slightly above the optimum, anticipating that a moist fill would be more flexible and less brittle than a dry mass and therefore less subject to cracking when settlement occurred. Compared to crack formation, however, there is greater concern over effect on stability if high pore water pressures develop in the dam section (a condition that limits the soil shear strength that can develop). It is felt that the proper design of soil filter transition zones prevents the danger of serious crack–leakage problems.

Earth fill dams may be exposed to ice loading and earthquake forces. Ice formed at the reservoir surface normally does not have great direct structural effect on a dam's stability, because it tends to slide upward on the embankment's upstream slope. The danger associated with ice lies in its potential for eroding or otherwise damaging the riprap or other face material.

Analytical methods that predict the behavior of earth dams when subject to earthquake are not precise.[14] Dam failures have resulted when sliding occurred on a weak foundation or when direct structural damage resulted because fault movement affected the foundation. Failures have also been caused by embankment sloughing or cracking followed by a dam break, or by overtopping where large settlement occurred or when slides into the reservoir area created a water wave that topped the dam. However, evaluation of the performance of dams that have experienced earthquake indicates that, generally, rolled earth embankments

---

[13]The layer thickness that can be properly compacted relates to the soil type and compaction equipment. Fine-grained soils probably will require that layers be no greater than 0.3 m (1ft) in thickness, whereas clean granular soils may permit layers approaching 0.6 m (2 ft) in thickness where vibratory compactors are used.

[14]The paper "Considerations in the Earthquake-Resistant Design of Earth and Rockfill Dams," H. Bolton Seed, *Geotechnique,* September 1979, provides a good evaluation of factors and procedures that apply to earthquake d sign for dams.

that include plastic cohesive soils and are located on firm foundations retain stability when subject to moderate earthquake forces. The effects of repeated seismic loads are not well established, however.

Rockfill Dams

Rockfill dams include features that make them generally similar to earth fill dams. A rockfill dam is one where placed rock is responsible for providing all or most of the necessary strength and stability required for proper performance of the structure. Rockfill dams may include soil zones.

Rockfills use large-size quarried or natural rock but also include smaller soil-size materials such as gravels and sands. Rockfill dams have been constructed in thick lifts by loosely dumping the material in place but also by spreading and compacting the material in layers of limited thickness. Loosely dumped thick lifts—various projects have used lifts anywhere from about 3 to 18 m (10 to 60 ft) thick—that included large materials in the 2 m (6- to 8-ft) size range have proved to create stable structures, provided that the rockfill is sluiced with large volumes of high-pressure jetted water. (The volume of water used has typically been three to four times the volume of the placed rockfill.) Apparently the great volume of water flushes fines from within the dumped rock mass so better edge-to-edge contact for the large fragments develops, while, simultaneously, the prolonged inundation softens rock edges and corners so that fracture occurs, which causes settlement but also produces a greater interlocking of materials with greater strength for the rockfill mass. Compared to dry placement, the sluicing procedure reduces postconstruction settlement. Typically, the postconstruction settlement for sluiced rockfills has been on the order of 1 percent of the embankment height.

The alternative procedure of constructing a rockfill by spreading and compacting the fill in relatively thin layers is preferred for dams where stability or leakage problems would constitute a hazard to life or valuable property. Better control over the fill materials and the resulting in-place condition is achieved with compacted layer techniques and, where zoning is required, provides a practical approach for constructing what is designed. Layers on the order of 1 m (3 ft) thick, comprising materials whose maximum size does not exceed layer thickness, are considered practical for achieving desired compaction densities. Vibratory compactors on the order of 10 tons in size have proved to be effective for compacting such rockfill layers. Moisture can be added during placement, but moisture content variations are less crucial for compaction of rockfill than for earth fill. In fact, one general advantage of rockfill construction is that it is less affected by adverse weather than earth fill or concrete construction. With either sluiced or compacted layer construction, rockfills can be constructed to relatively steep slopes (on the order of 1 or 1.5 horizontal to 1 vertical), although flatter slopes are often provided.

As with earth fill embankments, some settlement and lateral movement of a compacted rockfill dam are expected when the reservoir is filled. Settlements for properly constructed compacted rockfills are expected to be less than 1 percent of the embankment height. Where the length is limited, rockfill dams have been constructed with a slight upstream curvature to the longitudinal axis, so lateral movements occurring as a result of reservoir water load would cause compression within the structure, an effect that would tend to close developed vertical cracks.

Rockfill dams require an impervious upstream face or an impermeable core to control leakage. When provided with an impervious upstream face, the embankment is referred to as a *diaphragm section*. Surfaces of reinforced portland cement concrete and bituminous concrete are typical for present-day projects, but wood timbers and steel plates have also been used successfully. If possible, the concrete diaphragm is placed toward the end of the construction period to minimize the effect of embankment movements, which could cause cracks to occur in the facing.

Where an impermeable core is to serve as the barrier to seepage, soil of low permeability is typically selected. Bituminous concrete has also been used. Internal cores can be sloping or vertical zones which are compacted in thin layers as for earth fill dams. Soil core zones require adjacent soil transition filter zones, to prevent particle migration when seepage flows occur. Similarly, a bituminous concrete core zone should be bounded by transition filter zones in the event cracking occurs. Portland cement concrete cores are not recommended for high dams because of their rigidity and the probability of extensive cracking when embankment settlement occurs. The portland cement concrete core, if used, should be bounded by transition filter soil zones. Because of embankment settlement, it is expected that cracking will occur within the soil or concrete cores and the adjacent soil transition zones, but such cracks will fill with soil (self-heal) where the transition filters have been properly designed.

Using an upstream face for the impervious barrier offers the advantage that the vertical component of weight from the overlying reservoir water acts to increase the stability of the dam structure. A disadvantage is that postconstruction settlement will cause cracking in the facing. If repair becomes necessary, however, relatively easy access is available. The facing also effectively provides necessary upstream slope protection, but its exposed location tends to hasten deterioration and the need for repair.

Rockfill dams sited on previous soil foundations can have cutoffs similar to those used for earth fill dams to control underdam seepage. Diaphragm sections constructed on rock foundations use a footwall anchored into the rock or a grout curtain of closely spaced drilled-and-pressure-grouted holes at the upstream toe. Earth-core rockfill dams on rock foundations can use a grout curtain beneath the earth core to control foundation seepage. Where the surficial zone of foundation rock is fractured, a grout blanket can be used as a seal in place of or to complement the grout curtain.

Concrete and Masonry Dams

Concrete and masonry dams are appropriate where part of the dam must also function as a spillway, as with a run-of-river structure, or where the area taken up by the cross section must be limited, or where earth fill or rockfill material is scarce, or where site topography makes a concrete or masonry structure a more economical choice than earth fill or rockfill (such as where a narrow rock canyon exists and the side channel spillway necessary with an earth fill dam cannot be economically built).

Categories of concrete or masonry dams include the gravity dam, the arch dam, and the buttress dam. In modern construction, reinforced concrete is most conventionally used. The mass of the gravity dam section provides the resistance to the overturning and sliding effects that develop from the loadings that act on the structure, similar to gravity retaining walls. Gravity dams are suitable where the required height is not exceptionally great. Arch dams, relatively thin in section, retain stability against reservoir loading because of the curved alignment which effectively puts the longitudinal axis of the dam in compression. Arch dams are considered appropriate for narrow valley locations, where the ratio of length to height is small, say 5 or less. The buttress dam essentially consists of a sloping upstream slab supported by buttresses that transmit the effects of reservoir loading to the foundation soil or rock. The sloping upstream face, at a relatively steep angle, such as 45°, enables part of the reservoir loading to contribute to the dam's stability. Buttress dams are of two basic designs—the flat slab section with buttress and the multiple arch. With the slab–buttress structure, each flat slab section and each buttress can be considered to be a separate component. In contrast, the multiple-arch design results from buttresses that thicken as they get closer to the dam's upstream face and crest, in effect blending into the face slab and crest. The downstream elevation of this dam shows adjacent buttresses integrated to create the multiple-arch structure. The multiple-arch design produces a rigid structure. The basic flat slab–buttress dam is considered to be a flexible structure where adjacent buttress spans can tolerate some differential foundation settlement without harmful effect, provided that slab joints falling along each buttress centerline remain watertight if movement does occur. Buttress dams are considered most appropriate for long dams of moderate to high height. Compared to gravity dams, the buttress structure economizes on materials but involves greater labor and forming costs.

In addition to loading resulting from impounded reservoir water, dams are also designed to resist the effects of other forces that may act, including ice load, reservoir silting, uplift against the base, and earthquake.

Ice loading represents a winter condition that can occur when a frozen reservoir surface experiences temperature change. As an ice sheet expands, a lateral force is imposed against the upstream side of the dam. The magnitude of the ice loading relates to the thickness of ice that undergoes temperature change (which can be less than the total ice thickness) as well as to the degrees of temperature change. For winter climates similar to those in the northernmost United States, the maximum ice load is commonly assumed as 135 kN per meter (10 kips per foot) of dam length, corresponding to a temperature change through approximately a 0.7 m (2 ft) thickness of ice. When a dam's upstream side is at a significant slope, it is conventional to assume that an expanding ice sheet will slide up the face, with little of the lateral force acting to affect structural stability.

Silting behind a dam results as sediment carried into the reservoir area by the feeding water flow accumulates behind the dam. Silting is analyzed as a lateral earth pressure problem when the effect on dam stability is evaluated. A submerged unit weight appropriate for loose soil (e.g., 6 kN/m^3 or 40 pcf) and a lateral pressure coefficient equal to 0.5 can be used when computations are performed.

An uplift force will develop against the base of a dam if underdam seepage occurs. The uplift pressure acting at different points along the base relates to the pattern of seepage through the foundation strata and the water levels behind and in front of the dam, as can be studied from a flow net analysis (see Chapter 7). For the design of a dam situated on a foundation of soil or permeable rock, it is also common to assume an uplift force based on a full headwater hydrostatic pressure acting at the structure's upstream corner and a hydrostatic pressure based on the tailwater elevation acting at the downstream corner; pressures along the base vary uniformly between these limits, resulting in a trapezoidal uplift pressure pattern. Uplift computed by this simplified procedure will be somewhat greater than actual, but the result has a conservative effect on design.

Where a dam's foundation is to be embedded in sound rock, the magnitude of uplift that will develop is uncertain. If the rock is very sound and impervious, seepage would be very low and uplift pressures would require a very long period of time to develop. Alternatively, if the rock is layered or jointed, uplift could act on sections or slabs of the foundation rock that are larger than the base width of the dam and cause a total uplift force greater than would develop directly against the dam's base. Nevertheless, it is common *during design* to assume that uplift pressures for dams situated on rock will relate directly to upstream and downstream water levels, as previously discussed.

With buttress dams, uplift pressures between buttress locations can be reduced or eliminated by use of relief wells. Since the plan area of a buttress foundation is usually small compared to the total plan area of the dam, such a procedure can remove most of the effect of uplift.

Earthquakes create ground motions that produce excitations in a structure. A result is that the equivalent of vibratory-type horizontal and vertical loads, additional to those resulting from static loading, act on the structure. For *small* dams, the effect of seismic or shock forces on the overall structural stability can be evaluated by assuming that a static-type force P_{eq} having an effect equivalent to the earthquake acts at the dam's center of gravity, with

$$P_{eq} = Ma = \frac{W}{g}\alpha g = \alpha W \qquad (16\text{-}17)$$

where M = mass of the structure
  g = acceleration of gravity
  W = weight of the structure
  a = acceleration or tendency for motion resulting from the earthquake or other seismic shock
  α = ratio of a to g, an empirically determined factor that, for horizontal earthquake accelerations, ranges from about 0.03 for areas where seismic activity is light to 0.40 where activity could be severe. The vertical factor is assumed to be one-half to three-fourths the horizontal factor.

In a stability analysis, the seismic forces are taken for that instant in time when they act in the direction that creates the least stable condition for the dam; that is, the horizontal component acts in the downstream direction while the vertical component acts upward.

Earthquake effects also cause the pressure of the impounded water acting against the dam to increase. The magnitude of the increase is related to the dam's upstream slope. The total increase in the horizontal water force against the back of the dam H_{ew} for an impounded height of water h is

$$H_{ew} = (.73)C\alpha h^2\,\gamma_w \qquad (16\text{-}18)$$

where γ_w = unit weight of water
  C = coefficient relating to the dam's upstream slope, measured with respect to the vertical (Table 16-3).

The total increase in a dam's overturning moment M_{ew} with reference to an axis on a plane corresponding to the ground level behind the dam is

$$M_{ew} = (.3)C\alpha h^3\,\gamma_w \qquad (16\text{-}19)$$

Table 16-3 Values of Coefficient C for Determining Earthquake Effect of Reservoir Water

Slope of Dam's Upstream Face (deg)	Coefficient, C
0	0.73
15	0.64
30	0.53
45	0.42

Stability Analysis for Gravity Dams. A dam must retain stability when subject to the various forces that tend to cause sliding or lateral movement and overturning. Similarly, the dam's construction materials and foundation must not be overstressed. The stability of gravity dams can be studied through application of the basic laws of engineering mechanics (see Ill. 16-5). The stability of buttress dams can be evaluated approximately by using the methods applied to gravity dams. Stability studies for arch dams require advanced analytic procedures that are considered outside the scope of this text.

The stability analysis for gravity dams is similar to the analysis for earth retaining walls. A unit length of dam is isolated from its adjoining sections, and the design loads are applied. Where an instability against overturning exists, the overturning tends to occur by rotation about a longitudinal axis extending along the downstream toe. A factor of safety against overturning is computed by relating moments of forces (with reference to the moment axis at the toe of the dam section) that resist overturning to those that cause overturning, or

$$\text{FS against overturning} = \frac{\Sigma \text{ moments resisting overturning}}{\Sigma \text{ moments causing overturning}}$$

In general, the weight of the dam and reservoir water that can exert a *vertical* load on the dam's sloping upstream face will produce moments that resist overturning. Lateral water and soil pressures against the dam's upstream side, and ice, uplift, and earthquake effects, create loadings that act to cause overturning. Factors of safety computed by this method should be 2 or greater.

The magnitude and direction of stresses developed at the base of the dam require checking to guard against overstressing of the dam material or foundation.[15] The foundation pressure distribution diagram typically will be triangular or trapezoidal, with the maximum pressure at the downstream toe. To determine the pressure distribution, the position of the foundation reaction is first located from

$$d = \frac{\Sigma M_{\text{toe}}}{\Sigma V}$$

where ΣM_{toe} = moments resisting overturning minus moments causing overturning, for one unit of dam length

V = vertical component of the foundation reaction, including the weight of the dam and water on a sloping upstream face but less any uplift force, for one unit of dam length

d = distance from toe to point where the foundation reaction is applied

---

[15]Structurally, of course, the dam section must be capable of resisting the internal shear forces and bending moments that develop as a result of applied loads.

Desirably, the location for V falls within the middle third of the base; such a condition theoretically ensures that compressive stresses act across the entire base. An eccentricity distance, e, which is the distance from the center of the dam's base to the position where the foundation reaction acts, is then determined from

$$e = \frac{b}{2} - d$$

where b = width of dam base. The foundation pressure at the toe and at the heel of the dam is calculated from

$$p_{\text{toe}} = \frac{V}{b} + \frac{6Ve}{b^2} \qquad \text{(from Eq. 16-12)}$$

$$p_{\text{heel}} = \frac{V}{b} - \frac{6Ve}{b^2} \qquad \text{(form Eq. 16-12)}$$

The foundation pressures should not exceed the allowable compressive strength of the dam material or foundation material. The pressure at the heel should not be negative, a condition that would indicate tensile stress in the dam or a tendency for uplifting to occur. Some small tensile resistance can be developed within concrete or where a dam's concrete is bonded to a rock foundation, but in design such tensile resistance is normally not relied on. An indication of tension at the heel typically would warrant a design change, such as enlarging the dam section, or a modification, say to include a supplementary anchoring system such as steel dowels extending from the dam into a rock foundation.

A dam's ability to resist sliding along its base is evaluated by comparing horizontal forces that act to cause movement and those that resist movement. A factor of safety against sliding is the ratio of forces resisting movement to those causing movement, or

$$\text{FS against sliding} = \frac{\Sigma \text{ forces resisting lateral movement}}{\Sigma \text{ forces causing lateral movement}}$$

Typically, water and soil pressures behind the dam, and ice and earthquake loading, act to cause sliding. Resistance is provided by friction, shear, and bond developed on the potential plane of sliding at the base of the dam. Where a gravity dam is supported on a soil foundation, the resistance to sliding is computed from

$$\text{Sliding resistance} = \mu W + cbl \qquad (16\text{-}22)$$

where W = weight imposed by a unit length of the dam structure in contact with the foundation, including reservoir water lying above the dam's sloping faces. If uplift acts, W is the weight of the dam less the uplift force.

μ = coefficient of friction between the dam material and foundation soil (μ equals approximately $\tan \phi$).

c = cohesion of a cohesive foundation soil

b = width of dam base in contact with the foundation

l = length of dam section under analysis (typically 1 m or 1 ft)

The factor of safety computed by this method should be 1.5 or greater.

Where the dam is supported on rock, the friction–shear method for determining resistance to sliding can be applied, where

$$\text{Sliding resistance} = \mu W + lb\tau \qquad (16\text{-}23)$$

where μ = coefficient of friction between dam material and the foundation rock. Suggested values of μ range between 0.65 and 0.80 for concrete and masonry.

τ = bond or shear strength developing between the dam material and foundation rock. A value of 1380 kN/m² (200 psi) can be assumed where modern portland cement concrete is bonded to sound clean rock. If the rock is stratified or laminated with weak seams, the shear strength of the seams should be compared to the concrete–rock bond value, and the least value assumed.

Factors of safety computed by this method should be 4 or greater.

Illustration 16-5

A concrete river dam also functions as a spillway. The gravity spillway section is indicated by the sketch. Determine the stability against overturning and sliding for

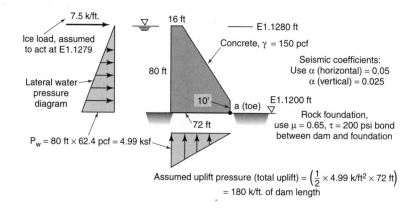

(a) The normal winter operations condition, where the reservoir level is at the spillway elevation and ice loading acts.

(b) The normal operations condition, excluding ice but with seismic effects included.

Solution

$$\text{Weight of dam} = (.150 \text{ kcf})[(16 \text{ ft})(70 \text{ ft})(1 \text{ ft}) + \left(\frac{1}{2}\right)(70 \text{ ft})(56 \text{ ft})(1 \text{ ft})$$
$$+ (10 \text{ ft})(72 \text{ ft})(1 \text{ ft})] = 570 \text{ kips per foot of length}$$

$$M_{toe} \text{ due to weight of dam} = (.150 \text{ kcf})\left[(10 \text{ ft})(72 \text{ ft})(1 \text{ ft})\left(\frac{72 \text{ ft}}{2}\right) \right.$$
$$+ \left(\frac{1}{2}\right)(56 \text{ ft})(70 \text{ ft})(1 \text{ ft})\left(\frac{2 \times 56 \text{ ft}}{3}\right)$$
$$\left. + (16 \text{ ft})(70 \text{ ft})(1 \text{ ft})\left(\frac{16 \text{ ft}}{2} + 56 \text{ ft}\right) \right]$$

$$= 25{,}616 \text{ ft-kip per foot of length}$$

M_{toe} causing overturning due to upstream water, ice, and uplift

$$= (4.99 \text{ ksf})(1 \text{ ft})\left(\frac{80 \text{ ft}}{2}\right)\left(\frac{80 \text{ ft}}{3}\right) + (7.5 \text{ k/ft})(1 \text{ ft})(79 \text{ ft})$$

$$+ (4.99 \text{ ksf})(1 \text{ ft})\left(\frac{72 \text{ ft}}{2}\right)\left(\frac{2}{3}\right)(72 \text{ ft}) = 5325 + 539 + 5751$$

$$= 11,669 \text{ ft-kip per foot of length}$$

(a) Normal Winter Operations Case

1. FS against overturning $= \dfrac{\Sigma M_a \text{ (resisting)}}{\Sigma M_a \text{ (causing)}} = \dfrac{25,616 \text{ ft-kip/ft}}{11,669 \text{ ft-kip/ft}} = 2.2$ (adequate)

Position of resultant measured from toe:

$$d = \frac{\Sigma M_a}{\Sigma V} = \frac{(25,616 \text{ ft-kip} - 11,669 \text{ ft-kip})}{\text{dam weight} - \text{uplift}} = \frac{13,947 \text{ ft-kip/ft}}{(570 \text{ k/ft} - 180 \text{ k/ft})}$$

$$= 35.8 \text{ ft from (a) axis at toe} \qquad \text{(satisfactory, within middle third of base)}$$

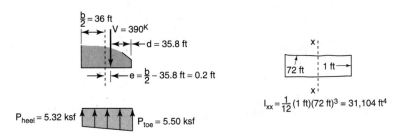

$$p_{toe} = \frac{\Sigma V}{A} + \frac{Ve\left(\dfrac{b}{2}\right)}{I}$$

$$= \frac{390 \text{ k}}{(72 \text{ ft})(1 \text{ ft})} + \frac{(390 \text{ k})(.2 \text{ ft})(36 \text{ ft})}{31,104 \text{ ft}^4}$$

$$= 5.41 + .09 = 5.50 \text{ ksf}$$

$$p_{heel} = 5.41 - .09 = 5.32 \text{ ksf}$$

2. FS against sliding (shear–friction method) $= \dfrac{\mu(\Sigma V) + \tau bl}{\text{lateral water force and ice force}}$

$$= \frac{(.65)(570 \text{ k} - 180 \text{ k}) + (.200 \text{ ksi})(72 \text{ ft})(1 \text{ ft})(144 \text{ in.}^2/\text{ft}^2)}{\left(\dfrac{1}{2}\right)(4.99 \text{ ksf})(80 \text{ ft}) + (7.5 \text{ k/ft})}$$

$$= \frac{253 \text{ k} + 2072 \text{ k}}{208 \text{ k}} = 11 \pm \quad \text{(adequate, greater than 4)}$$

(b) Seismic Effects Added to Normal Nonwinter Operations Case

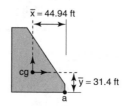

Additional overturning M_a due to seismic effects on dam mass

$$= .05\, W\bar{y} + .025\, W\bar{x} = (.05)(570\text{ k})(31.4\text{ ft}) + (.25)(570\text{ k})(44.94\text{ ft})$$
$$= 1535 \text{ ft·kip per foot of length}\curvearrowleft$$

Additional M_a due to reaction between dam and reservoir water (ref. Eq. 16-19)

$$= (.30)(.73)(.0624\text{ kcf})(.05)(80\text{ ft})(80\text{ ft})(80\text{ ft})\,(1\text{ ft wide})$$
$$= 350 \text{ ft·kip}\curvearrowleft$$

1. FS against overturning $= \dfrac{25{,}616\text{ ft·kip/ft}}{(5325 + 5751 + 1535 + 350)\text{ ft·kip/ft}} = 1.98$ (low)

2. FS against sliding $= \dfrac{\mu(\Sigma V) + \tau bl}{(\text{lat. water force}) + (\alpha W_{\text{dam}}) + \left(\begin{smallmatrix}\text{seismic reaction between dam}\\ \text{and water, from Eq. 16-18}\end{smallmatrix}\right)}$

where lateral water force $\simeq 200$ k, $\alpha W_{\text{dam}} = .05\,(570$ k$) \simeq 29$ k, and seismic reaction from Eq. 16-18 $= (.73)(.05)(.0624\text{ kcf})(.73)(80$ ft$)(80$ ft$)(1$ ft$) = 11$ k. The substitution of values gives

$$\text{FS} = \frac{.65[570\text{ k} - (.025)(570\text{ k}) - 180\text{ k}] + 2072\text{ k}}{200\text{ k} + 29\text{ k} + 11\text{ k}} = 9.7 \text{ (adequate)}$$

PROBLEMS

16-1. (a) Define *lateral pressure,* and indicate the requirements for the at-rest lateral soil pressure condition.

(b) For a retaining wall, indicate the requirements for wall behavior so that the active lateral soil pressure condition will apply.

(c) For a retaining wall, describe how a passive lateral soil pressure condition could develop against the wall.

16-2. Relating to vertical walls having contact with soil on one side of the wall (to restrain or hold back the soil mass):

There are three possible design conditions for determining the lateral soil pressure against the wall

as affected by wall movement that is permitted (or occurs). Using words and sketches, describe the three conditions.

16-3. Assume that a rigid, unyielding wall can be inserted into a soil mass where the at-rest lateral pressure condition exists. Explain the effect that soil submergence will have on the *total* lateral force (the summation of lateral pressures) that acts against one side of the wall.

16-4. Referring to the basic concepts for studying lateral earth pressures, calculate the magnitude of lateral force resulting from an active soil pressure against the back of a frictionless wall 6 m high. The surface behind the wall is level, with the retained soil having an angle of internal friction equal to 37° and a unit weight of 18 kN/m³.

16-5. Calculate the lateral force resulting where an active soil pressure condition develops against the backface of a frictionless wall 20 ft high if the backfill surface is level and the retained soil is cohesionless (unit weight of 115 pcf and angle of internal friction of 35°).

16-6. Calculate the lateral force created by soil pressures acting against a retaining wall having a vertical frictionless backface for the following conditions (ref. Fig. 16-4):
- Wall height is 6 m, backface is vertical.
- Cohesionless soil backfill is level, soil unit weight is 17 kN/m³, angle of internal friction is 35°.
  - (a) Assume the at-rest soil pressure condition.
  - (b) Assume the active soil pressure condition.
  - (c) Express the difference between (a) and (b) as a percentage related to the active pressure condition.

16-7. Assume that a frictionless retaining wall 15 ft high having a vertical backface supports a sandy soil with a unit weight of 118 pcf and an angle of internal friction of 34°. The soil backfill is level. Calculate the difference in the lateral force imposed onto the wall (per foot of wall length) when the at-rest lateral pressure condition applies compared to when the active lateral pressure condition applies (express the difference as a percentage increase related to the active pressure condition).

16-8. A 10-m-high concrete retaining wall (gravity type) having a vertical backface supports a cohesionless soil backfill sloping at an inclination of 30° with the horizontal. The retained soil has the following properties: $\gamma = 17$ kN/m³, $\phi = 35°$.
  - (a) Calculate P_A using the Rankine theory
  - (b) Calculate P_A using the Coulomb theory, assuming a wall-friction angle equal to 30°.

16-9. A 10-m-high gravity-type retaining wall is required to support a clay backfill having a level surface (where γ equals 17 kN/m³ and the cohesion equals 45 kN/m². Referring to Figure 16-17, compare the total lateral force resulting from the retained soil, for
  - (a) the theoretical active pressure condition
  - (b) the modified active pressure distribution recommended for short-term design
  - (c) the pressure distribution recommended for permanent retaining wall design

16-10. Explain why at-rest lateral soil pressure conditions and not active lateral pressures should be assumed for the design of basement walls and permanent retaining walls where lateral movement must be limited.

16-11. Use the Goh analysis to check a cantilever retaining wall (reinforced concrete construction) for stability against overturning and sliding, given the following conditions:
- Total height of wall is 8 m, stem is 0.7 m thick.
- Wall base is 4 m long, thickness is 1 m.
- Backface of stem is 1.5 m from toe of base.
- Unit weight for the concrete (wall construction) is 22.75 kN/m³.
- Cohesionless soil backfill is level, soil unit weight is 17 kN/m³, angle of internal friction is 34°, coefficient of friction between base and foundation soil is 0.60.
- Assume conditions for the wall stem having a rough surface.

16-12. Assume a cantilever retaining wall having the cross-section dimensions indicated for the wall shown in Illustration 16-3. The soil retained behind the wall is level. The soil backfill surface along the front (toe) of the wall is 4 ft above the base, and level. The wall backfill consists of a granular soil which is compacted in layers so a unit weight of 128 pcf and an angle of internal friction equal to 36° results. Perforated drain pipes are installed along the top of the wall's base section, so that a high water table condition cannot develop. Relating to the backfilling procedure, assume that a lateral pressure coefficient equal to 0.45 applies. Refer to Figure 16-24b; assume that the resultant force acts 30° from horizontal. For the backfill against the front toe section of the wall, assume a well-compacted granular soil having a unit weight of 133 pcf and an angle of internal friction of 37°. An erosion control surface will be installed to prevent washout of soil along the front of the wall, and a passive pressure can be assumed for this soil. For this wall, determine the factor of safety against overturning and against sliding (use $\mu = 0.55$), assuming
  - (a) The classical-standard method of analysis applies
  - (b) The Goh analysis applies (assume $K_o = 0.45$).

16-13. A 5-m-high concrete retaining wall has a cross section as depicted in Figure 16-25. The top of the wall is 1 m wide, and the base of the wall is 2 m wide. The backfill is a granular soil, with a unit weight equal to 18 kN/m³ and an angle of internal friction equal to 35 degrees. The backfill surface is level. Using criteria shown in Figure 16-18a, calculate the factor of safety against overturning and sliding for this wall. Assume that the lateral force due to the backfill acts at an angle of 25° from the horizontal (for computing the horizontal and vertical components of the earth pressures force). To check sliding among the base, assume that the coefficient of friction is equal to the value for the tangent of the angle of internal friction. The unit weight for concrete in the wall is 22.75 kN/m³.

16-14. A concrete retaining wall having a cross section as shown by Figure 16-25 is 20 ft high. The top of the wall is 2 ft wide, and the base of the wall is 5 ft wide. The backfill is a cohesionless soil and is level with the top of the wall. The backfill soil has a unit weight of 116 pcf and an angle of internal friction of 36°. Use the criteria indicated by Figure 16-18a to determine the wall stability against overturning and sliding. Assume that the force from the retained backfill acts at an angle of 25° from the horizontal. To check sliding along the base, assume a coefficient of friction equal to the tangent of the angle of internal friction. Use the unit weight for concrete as 145 pcf.

16-15. A 6-m-high retaining wall having a cross section as indicated by Figure 16-25 is to be designed to support a backfill consisting of predominantly coarse-grained soil, containing some silt but no clay. The surface of the retained soil has a slope of 3 horizontal to 1 vertical. Using the Terzaghi-Peck charts, determine the magnitude of P_h and P_v that can be assumed for designing the wall. Provide a sketch of the wall, indicating the magnitude and location of these two forces.

16-16. A masonry gravity retaining wall 5 m high has a cross section as shown by Figure 16-25 and supports a clay soil backfill that is level with the top of the wall. The top of the wall is 0.75 m wide and the base of the wall is 1.75 m wide. Check the stability of this wall against overturning and sliding using criteria indicated by Figure 16-18b. Use a unit weight for the masonry equal to 22 kN/m³. Use Terzaghi charts, Figure 16-29, to obtain values for P_h and P_v. Assume that the unit resistance

to sliding along the base is equal to the value for the clay cohesion, determined from shear strength testing as 35 kPa.

16-17. A 20-ft retaining wall is to be used to support a Terzaghi-Peck type 1 backfill (Fig. 16-29). The wall cross section is similar to that shown in Illustration 16-3, except that the base is 8 ft instead of 10 ft. The backfill surface is level. Determine the factor of safety against overturning and against sliding (using a coefficient of friction between wall base and supporting soil, $\mu = 0.5$) and the foundation pressure distribution (i.e., reaction against the base).

16-18. A retaining wall is to be checked for stability under seismic conditions. The wall has a rectangular cross section 6 m high and 1.5 m wide. The wall retains a cohesionless soil (unit weight of 17.5 kN/m³, angle of internal friction of 33°). The foundation soil under the wall is similar. The backfill is sloped, extending upslope from the wall at an inclination of 30°. The wall is located in an area where the values of A_v and A_a equal 0.20 (ref. Fig. 1-27b). For this study, assume that the angle of the lateral earth force acts 16° from the horizontal (approximately half of the value of the angle of internal friction) and that k_v is half of k_h.
(a) Determine the value of the earthquake active lateral pressure coefficient, K_{AE}.
(b) The wall is located in an area where the backfill region is undeveloped so that the allowable wall displacement is 150 mm. What weight of wall should be provided to have a factor of safety against sliding equal to 1.25?
(c) Determine the factor of safety against overturning for this earthquake condition (ref. Fig. 16-31 for position of lateral forces). Assume that the foundation soils are adequately compact so that liquefaction during earthquake will not occur.

16-19. A furniture moving truck parks in the driveway provided for a residential building, to deliver new appliances and furniture. The driveway is located close to the building. Estimate the resultant lateral force R imposed on the building's masonry basement wall by a truck wheel if the wall extends 2 m deep below the driveway level, the wheel is positioned 1 m from the building, and the wheel load is 20 kN.

16-20. A commercial truck parks in a house driveway to complete a delivery. The driveway borders one side of the house. Estimate the resultant lateral force R imposed onto the masonry basement wall by the truck's rear wheel closest to the house if the wall is 5 ft deep, the axle wheel is positioned 2 ft from the house, and the wheel load is 3500 lb.

16-21. Braced vertical sheeting is to be used for an open-trench excavation undertaken for the construction of a section of subway tunnel that extends through medium-stiff clay ($c = 42\,\text{kN/m}^2$, $\gamma = 17\,\text{kN/m}^3$). The excavation will be 12 m deep and supported by three rows of struts. The top strut will be level with the ground surface, with the second strut 3 m down and the third strut 7 m down from the ground surface. The struts will be positioned at 3.5 m center-to-center spacings (plan view spacing). Calculate the force in each strut (ref. Figs. 16-42 and 16-43).

16-22. Braced vertical sheeting is used to brace a deep open trench excavation made for the purpose of installing a large-diameter water main (pipe) for a municipal supply system. The excavation will be 20 ft deep, extending through sandy soils (unit weight of 120 pcf). The sheeting will be supported by three rows of struts. The top strut will be level with the ground surface, the second strut 7 ft down, and the third strut 14 ft down from the ground surface. In plan view, the struts are positioned 12 ft apart. Calculate the force in each strut (ref. Figs. 16-42 and 16-43).

16-23. A sheetpile bulkhead for a shipping canal loading dock is used to support a cohesionless soil fill that extends 10 m higher than the canal bottom. The soil fill and the canal foundation soil are similar; soil unit weight is 18 kN/m³ and angle of internal

friction is 34°. The wall anchor (tieback) will be 1.5 m below the top of the fill. The design depth for the canal water is 7 m. Determine the required depth for the sheetpile embedment, using the free-earth-support method (ref. Fig. 16-47). Assume an at-rest lateral pressure coefficient k_o equal to 0.5 and a passive pressure coefficient K_p equal to 3.5.

16-24. A sheetpile bulkhead for a shipping channel is used to support a fill that extends 25 ft higher than the channel bottom. The fill and channel bottom consist of granular soil where γ is 120 pcf and ϕ is 35°. Determine the required depth of sheetpile embedment, using the free-earth-support method (ref. Fig. 16-47). The wall anchor (tieback) will be 5 ft below the top of the fill, and the design depth for the channel water is 18 ft.

16-25. A concrete gravity dam 25 ft high is used to create a lake in an area where a small river flows. The dam will also function as a spillway, and the design height for the water behind the dam can be assumed level with the crest. The dam has a cross-sectional shape similar to that shown in Illustration 16-5, with the top width equal to 4 ft and the bottom width equal to 22 ft. The downstream face slopes uniformly to intersect the indicated base dimension. The dam is situated on a rock foundation, with the downstream water level considered to be at the rock surface. Applying a 5-kip-per-lineal-foot ice loading at the crest of the dam and using μ equal to 0.75, determine
(a) the factor of safety against overturning
(b) the factor of safety against sliding
(c) the location of the resultant passing through the base.

Application of LaPlace Equation to Flow Nets

FLOW LINES AND EQUIPOTENTIAL LINES

The LaPlace equation provides an application from potential theory for energy systems to relate flow of water that occurs through a porous media (such as a soil deposit) to energy difference.

Where a soil deposit is saturated and submerged, the pressure gradient responsible for flow is constant, the soil deposit is isotropic (coefficient of permeability is equal in all directions), and the flow velocity is in the laminar range, then the conditions for a steady-state analysis exist and Darcy's law applies (Eq. 6-7). Within a soil deposit, subsurface flow can occur in any of three general directions—represented as the x, y, and z directions on a rectangular coordinate system—but for many naturally occurring systems (say the subsurface system beneath a dam), the flow pattern is similar in parallel vertical planes and analysis need involve only a two-dimensional flow.

For a submerged soil deposit, the water pressure head at any point is equal in all directions. Darcy's law, $q = k(\Delta h/l)A = kiA$, applies for flow in any direction. Hydraulic gradient, i, equals change in energy head Δh that occurs over a related flow distance.

The conditions for analysis of flow that occurs across a differential element are defined by the sketch. The $(+)x$ and $(+)y$ directions relate to the direction in which flow occurs.

$$q_{x_{in}} = k_x i_x (dydz) = k_x \frac{\partial h}{\partial x}(dydz)$$

$$q_{x_{out}} = k_x i_{(x+dx)}(dydz) = k_x(dydz)\left[\frac{\partial h}{\partial x} + \frac{\partial}{\partial x}\left(\frac{\partial h}{\partial x}\right)dx\right]$$

$$= k_x(dydz)\left(\frac{\partial h}{\partial x} + \frac{\partial^2 h}{\partial x^2}dx\right)$$

$$q_{y_{in}} = k_y i_y (dxdz) = k_y \frac{\partial h}{\partial y}(dxdz)$$

$$q_{y_{out}} = k_y i_{(y+dy)}(dxdz) = k_y(dxdz)\left[\frac{\partial h}{\partial y} + \frac{\partial}{\partial y}\left(\frac{\partial h}{\partial y}\right)dy\right]$$

$$= k_y(dxdz)\left(\frac{\partial h}{\partial y} + \frac{\partial^2 h}{\partial y^2}dy\right)$$

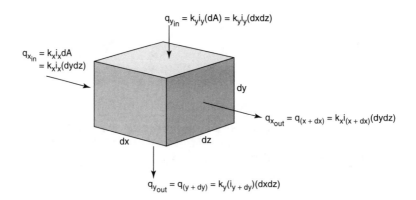

For the steady state, flow into the element equals flow leaving the element, so

$$(q_{x_{out}} - q_{x_{in}}) + (q_{y_{out}} - q_{y_{in}}) = 0$$

Substituting items,

$$k_x(dydz)\left(\frac{\partial h}{\partial x} + \frac{\partial^2 h}{\partial x^2} dx\right) - k_x(dydz)\left(\frac{\partial h}{\partial x}\right)$$

$$+ k_y(dxdz)\left(\frac{\partial h}{\partial y} + \frac{\partial^2 h}{\partial y^2} dy\right) - k_y(dxdz)\left(\frac{\partial h}{\partial y}\right) = 0$$

or

$$k_x(dydz)\left(\frac{\partial^2 h}{\partial x^2} dx\right) + k_y(dxdz)\left(\frac{\partial^2 h}{\partial y^2} dy\right) = 0$$

Since the volume of the element, $dx \cdot dy \cdot dz$, is not zero, the equation becomes

$$k_x\left(\frac{\partial^2 h}{\partial x^2}\right) + k_y\left(\frac{\partial^2 h}{\partial y^2}\right) = 0 \qquad \text{[LaPlace equation for 2-dimensional flow]}$$

Referring to the following sketch, flow lines refer to the travel path for the water flow, and equipotential lines refer to locations of similar total energy head.

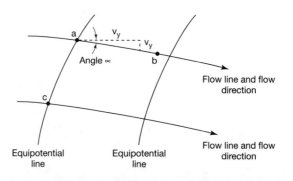

For Darcy's law, $q = kiA$, the product of ki represents flow velocity v. Directional velocities become

$$v_x = k_x \frac{\partial h}{\partial x}$$

$$v_y = k_y \frac{\partial h}{\partial y}$$

and the direction of flow represented by change from position a to position b along the flow line is

$$\tan \alpha = \frac{v_y}{v_x} = \frac{k_y \partial h / \partial y}{k_x \partial h / \partial x} = \frac{dy}{dx}$$

and, when $k_y = k_x$, becomes

$$\frac{v_y}{v_x} = \frac{\partial h / \partial y}{\partial h / \partial x} = \frac{dy}{dx}$$

The flow gradient (or energy head difference) between points such as a and c on the same equipotential line is zero, or $dh = 0$. Therefore

$$dh = \frac{\partial h}{\partial x} dx + \frac{\partial h}{\partial y} dy = 0$$

or

$$\frac{\partial h}{\partial y} dy = -\frac{\partial h}{\partial x} dx$$

and

$$\frac{dy}{dx} = -\frac{\partial h / \partial x}{\partial h / \partial y} = -\frac{v_x}{v_y}$$

This last relationship is the negative reciprocal of $\tan \alpha$ identified as the slope for a flow line at a point; therefore, equipotential lines and flow lines intersect at 90 degrees, or right angles.

Laboratory Procedure
to Determine
Coefficient of Consolidation

The typical laboratory compression test is performed by placing a carefully trimmed soil sample into a ring-shape apparatus that functions to provide lateral confinement during the testing procedure. (Fig. 10-8) Compressive loadings perpendicular to the surface plane of the ring are applied in increments (the loads are controlled and of known magnitude), and the resulting changes in soil sample thickness are recorded. It is common to have loadings increased (or decreased) in a sequence where the new loading is double (or half) the previous loading (e.g., if the previous load was 50 kN/m^2 or kPa, the applied new load is 100 kN/m^2 or kPa). The sample thickness at the end of each loading period (just prior to changing the loading) is determined and the related soil void ratio calculated; values of void ratio corresponding to compressive stress can then be plotted as a summary of test results.

A value for the coefficient of consolidation c_v is determined for a desired loading range (e.g., for the increment where loading has been increased from 50 kPa to 100 kPa, as used for the illustration). When applied to foundation settlement studies, the consolidation test load range to be used for the computation of c_v should match the loading increment expected to occur in the soil deposit at the construction site.

The determination of the coefficient of consolidation for a test load range requires measurement of the compression that results at particular times referenced to when the new load increment was first applied. That is, measurements indicating the amount of soil sample compression, and related time for that compression to occur, are recorded. Two procedures are in common use to determine values for the coefficient of consolidation, c_v: (1) a *logarithm of time* fitting method is used to determine c_v for 50 percent consolidation, and (2) a *square root of time* fitting method is used to determine c_v for 90 percent consolidation.

LOGARITHM OF TIME METHOD

For the studied loading range, values of compression are plotted on arithmetic coordinates (compression dial gage readings will suffice) versus the logarithm of the time for the related compression to result (Fig. 1). Draw tangents to the two straight-line portions of the resulting curve. The intersection of the tangents occurs at what is taken as the compression

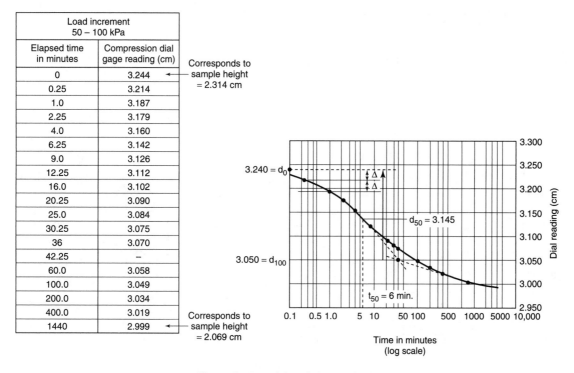

Load increment 50 – 100 kPa	
Elapsed time in minutes	Compression dial gage reading (cm)
0	3.244
0.25	3.214
1.0	3.187
2.25	3.179
4.0	3.160
6.25	3.142
9.0	3.126
12.25	3.112
16.0	3.102
20.25	3.090
25.0	3.084
30.25	3.075
36	3.070
42.25	–
60.0	3.058
100.0	3.049
200.0	3.034
400.0	3.019
1440	2.999

Corresponds to sample height = 2.314 cm

$3.240 = d_0$

$d_{50} = 3.145$

$3.050 = d_{100}$

$t_{50} = 6$ min.

Corresponds to sample height = 2.069 cm

Figure 1   Log of time fitting method.

for 100 percent primary consolidation. The compression dial reading is indicated on the plot as d_{100}. Determine what is assumed as the point of zero percent consolidation (indicated on the plot as d_0) by taking the amount of compression between 0.25 minute and 1.0 minute, and adding this value *above* the reading for 0.25 minute. The compression corresponding to 50 percent consolidation is midway between d_0 and d_{100}. The coefficient of consolidation, c_v, for the applied range of loading is then

$$c_v = \frac{(T_v)(H_{avg}/2)^2}{t_{50}} = \frac{(.20)(H_{avg}/2)^2}{t_{50}}$$

where H_{avg} = average sample thickness for the range of loading considered (cm, mm, or in.)
t_{50} = time for 50 percent consolidation to occur (time corresponding to d_{50} on the logarithim of time plot) (min)
c_v = coefficient of consolidation (cm²/min, mm²/min, or in.²/min)
T_v = time factor for 50 percent consolidation (refer Figure 10-23)

SQUARE ROOT OF TIME METHOD

For the desired loading range, values of compression (dial gage readings) are plotted on arithmetic coordinates versus the square root of time for the related compression to occur (e.g., the dial reading for 4 minutes is plotted against the $\sqrt{4}$, or 2; the dial reading for 9

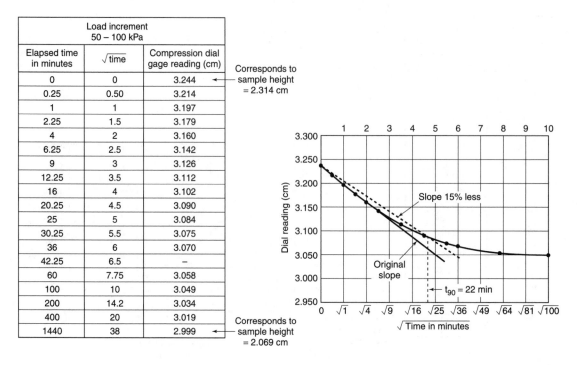

Figure 2 Square root of time fitting method.

minutes is plotted against $\sqrt{9}$, or 3, etc.) (Fig. 2). Draw a tangent to the straight-line portion of the plotted curve and extend it back to zero time. Through the point where the tangent crosses zero time, draw another line having a slope 15 percent less than the original tangent line. The place where this new line crosses the plotted curve is taken as the time for 90 percent consolidation to occur. The time t_{90} is obtained by projecting the point down to the square root of time scale, then squaring the time value to convert from square root of time back to true time. The coefficient of consolidation, c_v, for the applied range of loading is then

$$c_v = \frac{(T_v)(H_{avg}/2)^2}{t_{90}} = \frac{(.85)(H_{avg}/2)^2}{t_{90}}$$

where H_{avg} = average sample thickness for the range of loading considered (cm, mm, or in.)

t_{90} = time for 90 percent consolidation to occur (from the square root of time plot) (min)

T_v = time factor for 90 percent consolidation (Figure 10-23)

c_v = coefficient of consolidation (cm²/min, mm²/min, or in.²/min)

Typically, the values of c_v will be different for each loading range. Values of T_v in the following calculations are obtained from Chapter 10, Figure 10-23.

Calculating c_v

| c_v *for t_{50} (or 50% consolidation)* | c_v *for t_{90} (or 90% consolidation)* |

$H_{start} = 2.314$ cm
$H_{finish} = 2.069$ cm

Sample heights same as
for t_{50} calculation

$$c_v = \frac{T_v(H_{avg}/2)^2}{t_{50}}$$

$$c_v = \frac{T_v(H_{avg}/2)^2}{t_{90}}$$

$$= \frac{(.20)[(2.314 + 2.069)/2]^2}{6 \text{ min}} \text{ cm}^2$$

$$= \frac{(.85)[(2.314 + 2.069)/2]^2}{22 \text{ min}} \text{ cm}^2$$

$$= .04 \text{ cm}^2/\text{min} = 4 \times 10^{-6} \text{ m}^2/\text{min}$$

$$= .046 \text{ cm}^2/\text{min} = 4.6 \times 10^{-6} \text{ m}^2/\text{min}$$

Mathematical Development of the Bishop Equation for Slope Stability

Refer to the sketch of the earth slope and representative slice. For equilibrium, moments about the center of rotation which act to cause sliding and those which act to resist sliding must balance:

$$\Sigma\ Wx = \Sigma\ Sr = \Sigma\tau lr$$

Letting

$$FS = \frac{\tau_{available}}{\tau_{mobilized}}$$

get

$$\Sigma\ Wx = \Sigma\ \frac{1}{FS}\left[c' + \left(\frac{N}{l} - u\right)\tan\phi\right](rl)$$

$$FS = \frac{\Sigma\ r}{\Sigma\ Wx}(c'l + (N - ul)\tan\phi)$$

since

$$x = r\sin\alpha$$

$$FS = \frac{\Sigma(c'l + N'\tan\phi)}{\Sigma\ W\sin\alpha} \qquad Ⓐ$$

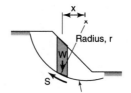

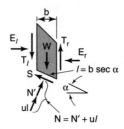

Summing vertical forces acting on the slice, obtain

$$W = N' \cos \alpha + ul \cos \alpha + S \sin \alpha + (T_r - T_l)$$

Since $S = \Sigma \tau l$,

$$W = N' \cos \alpha + ul \cos \alpha + (T_r - T_l) + \frac{1}{FS}\left[c' + \left(\frac{N}{l} - u\right) \tan \phi\right]/\sin \alpha$$

$$W = N' \cos \alpha + ul \cos \alpha + (T_r - T_l) + \frac{C'l \sin \alpha}{FS} + (N - ul)\frac{\tan \phi \sin \alpha}{FS}$$

$$W = N' \cos \alpha - ul \cos \alpha + (T_r - T_l) + \frac{c'l \sin \alpha}{FS} + N'\frac{\tan \phi \sin \alpha}{FS}$$

$$W - ul \cos \alpha + \frac{c'l \sin \alpha}{FS} - (T_r - T_l) = N'\left[\cos \alpha + \frac{\tan \phi \sin \alpha}{FS}\right]$$

Since $b = l \cos \alpha$,

$$N' = \frac{W - ub - \dfrac{c'l \sin \alpha}{FS} - (T_r - T_l)}{\cos \alpha + \left(\dfrac{\tan \phi \sin \alpha}{FS}\right)}$$

Substituting N' into Equation Ⓐabove, obtain

$$FS = \frac{\Sigma\left\{c'l + \tan \phi\left[\dfrac{w - ub - \dfrac{c'l}{FS} \sin \alpha - (T_r - T_l)}{\cos \alpha + \dfrac{\tan \phi \sin \alpha}{FS}}\right]\right\}}{\Sigma W \sin \alpha}$$

If the effect of $\Sigma (T_r - T_l)$ is assumed to be small and is neglected, and letting

$$m_a = \cos \alpha + \frac{\tan \phi \sin \alpha}{FS} = \left(1 + \frac{\tan \phi \tan \alpha}{FS}\right)\cos \alpha, \text{ obtain}$$

$$FS = \frac{\Sigma + \left\{c'b \sec \alpha + \left[\dfrac{(W - ub) \tan \phi - \dfrac{c'b}{FS} \tan \phi \tan \alpha}{m_a}\right]\right\}}{\Sigma W \sin \alpha}$$

$$FS = \frac{\Sigma\dfrac{1}{m_a}\left[(c'b \sec \alpha)\left(\cos \alpha + \dfrac{\tan \phi \sin \alpha}{FS}\right) - \dfrac{c'b}{FS} \tan \phi \tan \alpha + (W - ub) \tan \phi\right]}{\Sigma W \sin \alpha}$$

$$FS = \frac{\Sigma\dfrac{1}{m_a}\left[c'b \sec \alpha \cos \alpha + \dfrac{c'b \sec \alpha \tan \phi \sin \alpha}{FS} - \dfrac{c'b \tan \phi \tan \alpha}{FS} + (W - ub) \tan \phi\right]}{\Sigma W \sin \alpha}$$

Since $\sec \alpha = \dfrac{1}{\cos \alpha}$,

$$FS = \frac{\Sigma \frac{1}{m_a} \left\{ c'b \left[\frac{\cos \alpha}{\cos \alpha} + \left(\frac{\tan \phi}{FS} \right) \left(\frac{\sin \alpha}{\cos \alpha} \right) - \frac{\tan \phi \tan \alpha}{FS} \right] + (W - ub) \tan \phi \right\}}{\Sigma W \sin \alpha}$$

and noting that $\dfrac{\sin \alpha}{\cos \alpha} = \tan \alpha$, obtain

$$FS = \frac{\Sigma \frac{1}{m_a} [c'b + (W - ub) \tan \phi]}{\Sigma W \sin \alpha} \tag{15-8}$$

Glossary

Adsorbed water   Water bound to soil particles because of the attraction between electrical charges existing on soil particle surfaces and (dipole) water molecules.

Aquifer   Soil or rock layer (stratum) in which groundwater flows easily (i.e., a stratum which is a carrier of, or subsurface travel route for, groundwater). Typically, aquifers consist of coarse-grained soils or fractured, seamy rock.

Aquitard   Type of soil or rock layer (stratum) that restricts or prevents the movement of subsurface water. Typically, equitards consist of fine-grained soils such as silts and clays or sound rock (solid, not fractured or seamy).

Arching   The transfer of stress from a yielding part of a soil mass to an adjacent, less yielding or restrained part of the mass.

Atterberg limits   The liquid limit, plastic limit, and shrinkage limit for soil. The *water content* where the soil behavior changes from the liquid to the plastic state is the liquid limit; from the plastic to the semisolid state is the plastic limit; and from the semisolid to the solid state is the shrinkage limit.

Backfill   Soil material placed back into an area that has been excavated, such as against structures and in pipe trenches.

Bearing capacity   The pressure that can be imposed by a foundation onto the soil or rock supporting the foundation.

Boring   The method of investigating subsurface conditions by drilling into the earth. Frequently, soil or rock samples are also extracted from the boring for classification and testing.

Borrow   Soil or rock material obtained from an off-site source for use as fill on construction and projects.

Caisson   A large structural chamber used to keep soil and water from entering into a deep excavation or construction area. Caissons may be installed by being sunk in place or by systematically excavating below the bottom unit to the desired depth.

Capillarity   The movement of water, due to effects other than gravity, through very small void spaces that exist in a soil mass. Water movement occurs in very small channels such as capillary-sized openings because of the affinity between soil and water, which acts to increase the boundary of contact between the two materials, and the surface tension property developed by water in contact with air. Capillary flow can occur in a direction opposite to that of the pull of gravity.

Chemical weathering.   The process of weathering whereby chemical reactions such as hydration, solution, oxidation, and ion exchange break down and possibly change rock and soil materials.

Clays (clay minerals)   Very small soil particles having a crystalline (layered) structure, created as the result of the chemical alteration of primary rock minerals. Most clay particles, because of their mineralogical composition, are flat or platelike in shape, with a large surface area to mass ratio. Clay particle dimensions are often smaller than 2μ.

Coarse-grained soil   Those soil types having particles large enough to be seen without visual assistance. The coarse-grained materials include the sand and gravel (or larger) soil particles.

Cohesion   The bonding or attraction between particles of fine-grained soil that creates shear strength.

Compaction   The process of increasing the density or unit weight of a soil (frequently fill soil) by rolling, tamping, vibrating, or other mechanical means.

Compressibility   The change, or tendency for change, that occurs in the thickness of a soil mass when it is subjected to compressive loading.

Conduit   Pipe that is buried in a soil mass or passes through a soil embankment and carries water or other fluid materials, electrical cables, and the like.

Consolidation   The process by which compression of a newly stressed clay soil occurs simultaneously with the expulsion of water present in the soil void spaces. Initially, the newly imposed stress acting on the clay is imparted onto the water in the soil voids (pore water), and not onto the soil particles. Because of the increased pressure, the water is gradually forced out of the soil. As the pore water pressure is reduced, the magnitude of stress being imposed onto the soil particles is correspondingly increased. Compression of the clay layer occurs only as rapidly as pore water can drain from the soil, and this is related to the permeability of the soil layer.

Contamination (groundwater, soil)   A degradation or lowering of the quality (of groundwater or soil) which affects suitability for conventional usage. The condition occurs because of exposure to, inclusion of, or mixing with undesirable biological, chemical, or radioactive substances.

Density   The mass per unit of volume. In reference to soil, the term often also indicates weight per unit volume and is synonymous with unit weight.

Dewatering   The procedure used to remove water from a construction area, such as pumping from an excavation or location where water covers the planned working surface; the procedure used to lower the groundwater table in order to obtain a "dry" area in the vicinity of an excavation that would otherwise extend below water.

Dispersive clays   Clay soils that deflocculate in still water and erode when exposed to a low-velocity flow of water. A clay-pore water system that has a high concentration of sodium ions tends to have high dispersivity.

Ditch conduits   Conduits installed in narrow ditches or trenches that are subsequently backfilled.

Drawdown   The lowering of the level of the groundwater table that occurs in the vicinity of a water well (on dewatering equipment) when it is pumped.

Dynamic compaction   The procedure whereby surface and near-surface zones of soil or fill are compacted by dropping a heavy weight (commonly 5 to 15 tons) from a relatively great height (drops of 10 to 30m (30 to 100 ft) are typical). Multiple poundings are provided at each drop location, and closely spaced drop locations are utilized to improve a construction site.

Earth pressure   Normally used in reference to the lateral pressure or force imposed by a soil mass against an earth-supporting structure such as a retaining wall or basement wall, or on a fictitious vertical plane located within a soil mass. The *coefficient of earth pressure* refers to the ratio of lateral pressure to vertical pressure existing at a point in a soil mass.

Earthquake   The shaking and movement of the earth which results when a release of energy occurs because of deep rock fracturing or shifting, volcanic eruption, or a large explosion.

Earthquake intensity   A numerical scale of reference to indicate the damage and other effects resulting at a given geographical location because of an earthquake.

Earthquake magnitude   Value used to indicate the relative severity of earthquake events; the numeric value is obtained by taking the logarithm (base 10) of the maximum seismic wave amplitude (in 0.001 mm) recorded by a seismograph positioned 100 km from an earthquake epicenter.

Effective stress   The actual particle-to-particle contact stress (or pressure) existing between soil grains. This stress compensates for the possible buoyancy influence of water pressure. Effective stress relates directly to the shear strength possessed by a soil.

Expansive clays   Clay soils that experience significant volume expansion in the presence of water and shrink upon drying. Clays including the montmorillonite mineral are especially noted for their volume-change characteristics.

Fill   Earth placed in an excavation or other area to raise the surface elevation. Also referred to as *earth fill* or *soil fill*. *Structural earth fill* refers to the material that is placed and compacted in layers in order to achieve a uniform and dense soil mass which is capable of supporting structural loading.

Fines or **Fine-grained**   Refers to silt- and clay-sized particles that exist in a soil mixture.

Flow line   The path of travel traced by moving water as it flows through a soil mass.

Flow net   A pictorial method used to study the flow of water through a soil. Used to indicate the paths of travel followed by moving water and the subsurface pressures resulting from the presence of the water.

Footing   Type of foundation typically installed at a shallow depth and constructed to provide a relatively large area of bearing onto the supporting soil.

Fricton, internal   The particle (solid-to-solid) friction developed by cohesionless soils, and the property responsible for most of the shear strength that this type of soil can develop. The angle of internal friction, ϕ, refers to the arc tan value of the sloped line that results from graphing the relationship between shear strength and the imposed normal stress for a soil. This strength property also applies to fine-grained soils when the effect of consolidation or drainage on the shearing resistance of those soils is being studied.

Gabions   Stone-filled steel wire baskets that can be assembled or stacked like building blocks to act as retaining walls or provide slope and erosion protection.

Geosynthetics   Sheetlike materials along with other shapes manufactured of durable, nondegradable synthetics, such as polyester, polypropylene. polyethylene, polyvinylchlorides, butyl (synthetic) rubber, etc. to be used in construction assemblies that are in contact with the ground or included in earth fill, to provide functions which enhance the performance of the construction, such as reinforcement or drainage.

Groundwater table   The surface of the underground supply of water. Also referred to as the *phreatic surface.*

Head   Shortened form of the phrase *pressure head,* referring to the pressure resulting from a column of water or elevated supply of water. Pressure would be computed from $\gamma_w h$, where γ_w is the unit weight of water and h is the height or elevation of the water supply. The h term is the pressure *head.*

Heave   Upward movement of soil and foundations supported on soil, caused by expansion occurring in the soil as a result of such factors as freezing or swelling due to increased water content. *Frost heave* refers to the vertical soil movement that occurs in freezing temperatures as ice layers or lenses form within the freezing soil and cause the soil mass to expand.

Hydraulic gradient   Mathematical term indicating the difference in pressure head existing between two locations divided by the distance between these same locations. Given the designation i.

In situ   Refers to soil when it is at its natural location in the earth and in its natural condition.

Isotropic   Pertaining to a soil whose properties are the same in all directions.

Landslide   The relatively rapid lateral and downhill movement of a generally well-defined earth mass or land form due to gravitational forces.

Laterite   The category of residual soil formed from the weathering of igneous rock in tropical regions that, through the process of its formation, will include high concentrations of iron and aluminum sesquioxides with low concentrations of silica.

Leachate   Liquid flowing from a solid waste landfill (usually, mostly water) which includes contaminants or becomes polluted because of contact with the solid waste.

Limit equilibrium A method of analysis used to evaluate the stability of soil mass (such as in a slope or foundation support) that could be involved in movement associated with failure. The method involves determining the soil shear strength on an assumed failure surface as required to maintain equilibrium or stability, and compares this value with the actual shear strength of the soil; this comparison indicates if equilibrium will exist or if the limits of equilibrium will be exceeded.

Liquefaction Loss of strength occurring in saturated cohesionless soil exposed to shock or vibrations when the soil particles momentarily lose contact. The material then behaves as a fluid.

Love waves Earthquake-related shear-type seismic waves which travel along the earth's surface zone.

Mechanical weathering The process of weathering whereby physical forces, such as frost action and temperature changes, break down or reduce rock to smaller fragments without involving chemical changes.

Mineral A naturally formed chemical element or compound having a definite chemical composition and usually a characteristic crystal form.

Penetration test Term generally applied to subsurface investigative methods for determining a strength-related property of a soil by measuring the resistance to advancement of penetration or boring equipment.

Permafrost The permanently frozen ground located in the northern regions of the earth.

Permeability The ability of water (or other fluid) to flow through a soil by traveling through the void spaces. A high permeability indicates that flow occurs rapidly, and vice versa.

Pier Category applied to columnlike concrete foundations, similar to piles. The pier is generally considered the type of deep foundation that is constructed by placing concrete in a deep excavation large enough to permit manual inspection. *Pier* is also used frequently to indicate heavy masonry column units that are used for basement-level and substructural support.

Pile The relatively long, slender, columnlike type of foundation that obtains supporting capacity from the soil or rock some distance below the ground surface.

Pipe bedding Preparation of the surface that is to support a buried conduit. Established bedding classes consider effects of contouring the supporting surface to the shape of the pipe, extent of compaction, and placement of a supporting material such as concrete or gravel. Generally, preparation of bedding reduces the stresses that will develop in the pipe (conduit) from an overlying fill.

Piping Erosion by subsurface water moving through a soil zone, which results in the formation of continuous tunnels or "pipes" through which water then travels rapidly. Progressive erosion or cave-in of the ground results. The condition is associated with the movement of water through permeable dam foundations.

Plane strain A state of strain in which all displacements that arise from deformation are parallel to one particular plane.

Plasticity Term applied to fine-grained soils (particularly clays) to indicate the soils' (plus included water's) ability to flow or be remolded without raveling or breaking apart.

Plate tectonics The concept that the earth's outer zone consists of a small number (10 to 25) of large thick plates that "float" on a viscous underlayer and can move more or less independently. The continents are carried on the plates and move with them; oceans are similarly carried on the plates and expand or shrink as the distances between continents change.

Poisson's ratio The ratio of lateral unit strain to the longitudinal unit strain in a body that has been stressed longitudinally within its elastic limit.

Pore pressure   Water pressure developed in the voids of a soil mass. *Excess pore pressure* refers to pressure greater than the normal hydrostatic pressure expected as a result of position below the water table.

Porosity   The relative volume of open space (pores or voids) existing within the total volume occupied by a soil or rock mass. Mathematically, porosity n is equal to the volume of void spaces divided by total volume of the soil or rock material (the total volume consists of the volume of void spaces plus the volume of solids).

Pressuremeter   An instrument used to determine the in situ strength of a soil zone through measurement of the pressure-related lateral expansion of a flexible cylinder that is at a known depth in a borehole.

Primary (P) wave   Earthquake-related seismic or shock wave that travels through the earth with a compression–dilation (push–pull) type of oscillation.

Projecting conduits   Conduits in areas where earth fill or earth embankment will be placed above the installation.

Radon   An invisible, odorless, tasteless gas formed from the decay of uranium, radium, and polonium found in some rock, soil, and waste materials.

Rayleigh waves   Earthquake-related seismic waves that travel along the earth's surface zone with a rolling or rippling motion.

Reinforced earth   Earth structures such as embankments, retaining walls, and dams that are constructed in layers reinforced with fabrics, strips, or fibers to increase the strength of the soil mass.

Relative density   Term applied to sand deposits to indicate a relative state of compaction compared to the loosest and most dense conditions possible.

Retaining wall   A vertical structure designed to resist the lateral pressure of soil and water behind it.

Revetment   A protective wall, assembly or facing of stone, concrete, or other durable material built to separate an embankment or shore structure from wave erosion.

Riprap   The layer of boulders or crushed rock materials, typically ranging from 150 to 600 mm (6-in to 24-in) in size, placed as a covering to protect the surface of earth dams and earth slopes, against erosion.

Rollers, compaction   The category of construction equipment utilized to compact (or densify) soil by rolling it. The compaction force typically results from the heavy weight of the equipment and/or vibrations transmitted from the equipment into the soil.

Sand   The category of coarse-grained soil whose particle sizes range between about 0.07 mm and 5 mm in diameter.

Secondary (S) waves   Earthquake-related shear type seismic waves that travel through the earth, causing up-and-down and side-to-side oscillations.

Seepage   Generally refers to the quantity of water flowing through a soil deposit or soil structure such as an earth dam. Also may refer to the quantity of subsurface water leaking into a building's underground (basement) area.

Seismic   Pertaining to an earthquake or earth vibration, including one that is artificially induced.

Seismic exploration   The method of determining subsurface soil and rock conditions (without excavation) by inducing a shock wave into the earth and measuring the velocity of the wave's travel through the earth material. This *seismic velocity* indicates the type of earth material.

Seismogram   The oscillating-line record of ground movements measured by the seismograph during an earthquake event.

Seismograph   Instrument to measure the horizontal and vertical movements or vibrations that occur within the earth or at the surface because of earthquake or other seismic waves.

Settlement   The downward vertical movement experienced by structures or a soil surface as the underlying supporting earth compresses.

Shear strength   The ability of a soil to resist shearing stresses developed within a soil mass as a result of loading imposed onto the soil.

Sheetpiling   A pile with a generally flat cross section, made to interlock with adjoining sections to form a thin diaphragm wall or bulkhead; used to resist the lateral force of retained earth or water when part of temporary and permanent structures.

Sieve   Pan or traylike equipment having a screen or mesh bottom; used in laboratory or field work to separate particles of a soil sample into their various sizes.

Silt   The category of fine-grained soil particles (individual soil grains whose particle size is smaller than 0.07 mm or too small to be seen without visual aid) whose mineralogical composition remains similar to the rock they were derived from.

Soil sampler   The equipment used to extract soil samples from borings or test pits made in a subsurface investigation.

Soil stabilization   Treatment of soil to improve its properties; includes the mixing of additives and other means of alterations such as compaction or drainage.

Solid waste   Solid throwaway or disposable materials such as garbage and trash (nonliquids) generated by the residential and commercial sector of civilization, and the nonusable residue or remains of mining and manufacturing operations.

Solid waste (or sanitary) landfill   Area or project where solid waste is buried as a means of disposal. Typically, small volumes of solid waste such as a daily supply are discharged into an open excavation, then covered with a layer of soil; a landfill site consists of many buried zones or cells of waste, each enveloped with soil. Modern solid waste landfills will include a synthetic bottom liner to function as a barrier to prevent the solid waste from contaminating the surrounding earth, and a synthetic capping cover when the site is filled to function as a top seal.

Sump   Small excavation or pit provided in the floor of a structure, or in the earth, to serve as a collection basin for surface water and near-surface underground water.

Terra-probe   A method of compacting thick surficial and sand zones through the use of a vertically tubular probe; the probe is vibrated to the desired depth, then slowly withdrawn while continuing to vibrate. Closely spaced probe locations are utilized to compact a construction site.

Till   Description given to glacially transported formations consisting of a heterogeneous mixture of fine-grained and coarse-grained material.

Unit weight   The weight per unit volume of a material such as soil, water, concrete, etc. Typically expressed as pounds per cubic foot (lb/ft^3), grams per cubic centimeter (g/cm^3), or kilonewtons per cubic meter (kN/m^3).

Vadose zone   The subsurface soil zone that lies directly above the position of the groundwater table. Usually, this zone is partially saturated with water that has migrated upward via capillary movement from the water table.

Vibroflotation   A method of compacting thick surficial sand zones through the use of a horizontally vibrating cylinder termed a Vibroflot®. The Vibroflot is jetted to a desired depth and slowly brought to the surface while it continues to vibrate. The process typically compacts the soil in a 5-ft zone surrounding the Vibroflot, and a series of penetrations are provided to compact a construction site. Sand is added to the zone surrounding the Vibroflot to make up the volume lost by compaction.

Void ratio   The total volume occupied by a soil mass includes the soil particles plus void spaces (which in nature always exist between the particles because of their irregular shape). The void ratio is the ratio of the void space volume to the volume of soil solids.

Water content   The ratio of the quantity of water in a soil (by weight) to the weight of the soil solids (dry soil), typically expressed as a percentage.

Well point   The perforated end section of a wellpipe that permits the groundwater to be drawn into the pipe for pumping.

Bibliography

1. Abdun-Nur, E. A. *A Standard Classification of Soils as Prepared by the Bureau of Reclamation.* ASTM Special Technical Publication No. 113, 1950.
2. Abramson, L., T. Lee, S. Sharma, and G. Boyce. *Slope Stability and Stabilization Methods.* New York: Wiley, 1996.
3. Acker, W. L., III. *Basic Procedures for Soil Sampling and Core Drilling.* Scranton, Penna.: Acker Drill Company, 1974.
4. Agerschou, H. A. "Analysis of the Engineering Pile Formula." *Journal of Soil Mechanics and Foundations Division* American Society of Civil Engineers (ASCE) 88, No. SM5, October 1962.
5. Aldrich, H. P. "Precompression for Support of Shallow Foundations." *Journal of Soil Mechanics and Foundations Division* (ASCE) 91, No. SM2, March 1965.
6. Aldrich, H. P. "Selection of Foundation Systems." *Soil Mechanics Lecture Series.* Foundation Engineering, Soil Mechanics and Foundations Division, Illinois Section, ASCE; and Department of Civil Engineering, Northwestern University, Evanston, Ill., 1969.
7. American Cyanamid Company. *AM-9 Chemical Grout, Technical Data.* Wayne, N.J.: Author.
8. American Society of Civil Engineers, "Uplift in Masonry Dams," *Transactions* (ASCE) 117, 1952.
9. American Society of Civil Engineers. "Ice Pressure against Dams, A Symposium." *Transactions* (ASCE) 119, 1954.
10. American Society of Civil Engineers (ASCE). *Design of Pile Foundations.* ASCE, 1993.
11. Andersland, O. B., and B. Ladanyi. *An Introduction to Frozen Ground Engineering.* New York: Chapman and Hall, 1994.
12. Anderson, R. D. "Foster Vibrator." *Proceedings, Design and Installation of Pile Foundations and Cellular Structures.* Lehigh Valley, Penna.: Envo Publishing Company, Lehigh University, 1970.
13. Andrews, D. E., and D. J. Cosler. "Preventing and Coping with Water Pollution." *Journal of Testing and Evaluation* (ASTM), March 1989.
14. Antevs, E. *Maps of the Pleistocene Glaciations.* Geological Society of America, Bulletin 40, 1929.
15. Applied Technology Council. *Tentative Provisions for the Development of Seismic Regulations for Buildings.* ATC Publication ATC 3-06; National Bureau of Standards Publication 510, 1978.
16. Applied Technology Council. *Seismic Design Guidelines for Highway Bridges.* ATC Publication ATC-6, Redwood City, Calif., 1981.
17. Armento, W. J. "Criteria for Lateral Pressures for Braced Cuts." *Proceedings, Performance of Earth and Earth-Supported Structures.* Soil Mechanics and Foundations Division, ASCE; and Purdue University, 1, Part 2, June 1972.
18. Baguelin, F., J. F. Jezequel, and D. H. Shields. *The Pressuremeter and Foundation Engineering.* Clausthal, Germany: Trans Tech Publications, 1978.
19. Baker, C. N., and F. Kahn. "Caisson Construction Problems in Chicago." *Journal of Soil Mechanics and Foundations Division* (ASCE) 97, No. SM2, February 1971.
20. Barcilona, M., J. F. Keely, A. Wehrmann, and W. A. Pettyjohn. *Contamination of Ground Water—Prevention, Assessment, Restoration.* Park Ridge, N.J.: Noyes Data Corporation, 1990.

21. Barnes G. E. "A Simplified Version of the Bishop-Morganstern Slope Stability Charts," *Canadian Geotechnical Journal* 28, No. 4, 1991.

22. Barnes, G. E. "Stability Coefficients for Highway Cutting Slope Design." *Ground Engineering,* May 1992.

23. Barron, R. "Consolidation of Fine-Grained Soils by Drain Wells." *Transactions* (ASCE) 113, 1948.

24. Basham, P. W., D. J. Weichart, F. M. Anglin, and M. J. Berry. "New Probabilistic Strong Ground Motion Maps of Canada." *Journal of Seismological Society of America* 75, No. 2, April 1985.

25. Basore, C. E., and J. D. Boitano. "Sand Densification by Piles and Vibroflotation." *Journal of Soil Mechanics and Foundations Division* (ASCE) 95, No. SM6, November 1969.

26. Bell, J. M. "Dimensionless Parameters for Homogeneous Earth Slopes," *Journal of Soil Mechanics and Foundations Division* (ASCE) 92, No. SM5, September 1966.

27. Berezantzev, V. G., V S. Kristoforov, and V. N. Golubkov. "Load Bearing Capacity and Deformation of Piled Foundations." *Proceedings, Fifth International Conference on Soil Mechanics and Foundation Engineering.* Paris, 1961, Vol. 2.

28. Berner, E. K., and R. A. Berner. *The Global Water Cycle—Geochemistry and Environment,* Englewood Cliffs, N.J.: Prentice Hall, 1987.

29. Bertram, G. E. "Design Requirements and Site Selection." *Design and Construction of Earth Structures, Soil Mechanics Lecture Series.* Soil Mechanics and Foundation Division, Illinois Section, ASCE; and Civil Engineering Department, Illinois Institute of Technology, Chicago, 1966.

30. Bishop, Alan W. "The Use of the Slip Circle in the Stability Analysis of Slopes" *Geotechnique* 5, No. 1, 1955.

31. Bishop, A. W., and N. R. Morganstern. "Stability Coefficients for Earth Slopes." *Geotechnique* 10, No. 4, 1960.

32. Bolt, B. A. *Nuclear Explosions and Earthquakes: The Parted Veil.* San Francisco: W. H. Freeman, 1976.

33. Bolt, B. A. *Earthquakes.* New York: W. H. Freeman, 1988.

34. Brennan, T., and B. Turner. "Radon's Thrust Can Be Subdued." *Solar Age,* May 1985.

35. Brennan, T., and B. Turner, "Defeating Radon." *Solar Age,* March 1986.

36. Broms, Bengt. "Lateral Earth Pressures due to Compaction of Cohesionless Soils." In A. Kedzi, ed., *Proceedings, Fourth Budapest Conference on Soil Mechanics and Foundation Engineering.* Budapest: Akademiai Kiado, 1971.

37. Brown, D. R., and J. Warner. "Compaction Grouting." *Journal of Soil Mechanics and Foundations Division* (ASCE) 99, No. SM8, August 1973.

38. Brown, R. E. "Vibroflotation Compaction of Cohesionless Soils." *Journal of Geotechnical Engineering Division* (ASCE) 103, No. GT12, December 1977.

39. Brown, R. E., and A. J. Glenn. "Vibroflotation and Terra-Probe Comparison," *Journal of Geotechnical Engineering Division* (ASCE) 102, No. GT10, October 1976.

40. Building Officials and Code Administrators International. *BOCA National Building Code.* Homewood, Ill.: BOCA, 1994.

41. Bureau of Reclamation. *Design of Small Dams,* 3rd ed. Washington, D.C.: U.S. Department of the Interior, 1987.

42. Burland, J. B., B. B. Broms, and V. F. B. DeMello. "Behavior of Foundations as Structures, State of the Art Report." *Proceedings, Ninth International Conference on Soil Mechanics and Foundation Engineering,* Tokyo, 1977.

43. Burland, J. B., and M. C. Burbridge. "Settlement of Foundations on Sand and Gravel." *Proceedings, Institution of Civil Engineers* 78, 1985.

44. Bussey, W. H. "Foundation Evaluation and Treatment." *Design and Construction of Earth Structures, Soil Mechanics Lecture Series.* Soil Mechanics and Foundation Division, Illinois Section, ASCE; and Civil Engineering Department, Illinois Institute of Technology, Chicago, 1966.

45. Canter, L. W., R. C. Knox, and D. M. Fairchild. *Ground Water Quality Protection.* Chelsea, Mich.: Lewis, 1987.

46. Carpenter, J. C., and E. S. Barker. "Vertical Sand, Drains for Stabilizing Muck-Peat Soils." *Transactions* (ASCE) 124, 1959.

47. Carson, A. B. *General Excavating Methods.* New York: McGraw-Hill, 1961.

48. Casagrande, A. "Research on the Atterberg Limits of Soils." *Public Road* 13, October 1932.

49. Casagrande, A. "Classification and Identification of Soils." *Transactions* (ASCE) 113, 1948.

50. Casagrande, L. "Electro-Osmotic Stabilization of Soils." *Journal of the Boston Society of Civil Engineers,* January 1952.

51. Casagrande, L. "Comments on Conventional Design of Retaining Structures." *Journal of Soil Mechanics and Foundations Division* (ASCE) 99, No. SM2, February 1973.

52. Caslowica M. "Spin Control," *Earth.* Kalmbach Publishing, Co., Wankesha, Wisc., December 1966.

53. Cedergren, H. R. *Seepage, Drainage and Flow Nets.* New York: Wiley, 1967.

54. Chandler, R. J., and Peiris. "Further Extensions to the Bishop-Morganstern Slope Stability Charts." *Ground Engineering,* May 1989.

55. Chellis, R. D. "The Relationship between Pile Formulas and Load Tests." *Transactions* (ASCE) 114, 1949.

56. Chen, F. J. *Foundations on Expansive Soils.* Amsterdam: Elsevier Science. 1988.

57. Chen, W. F., ed., *The Civil Engineering Handbook*, Section III. Geotechnical Engineering, Boca Raton, Fla.: CRC Press, 1995.

58. Choudbury, R. N. *Slope Analysis.* New York: Elsevier Scientific Publishing Company, 1978.

59. Christian, J. T., and W. F. Swiger. "Statistics of Liquefaction and SPT Results." *Journal of Geotechnical Engineering Division* (ASCE) 101, No. GT11, November 1975.

60. Clough, C. W. "Deep Excavations and Retaining Structures." *Analysis and Design of Building Foundations.* Lehigh Valley, Penna.: Envo Publishing Company, Lehigh University, 1976.

61. Cornell University, Geotechnical Group (F. H. Kulhawy, Principal Investigator). *Transmission Line Structure Foundations for Uplift–Compression Loading.* Palo Alto, Calif.: Electric Power Research Institute, February 1983.

62. Cousins, B F. "Stability Charts for Simple Earth Slopes." *Journal of Geotechnical Engineering Division* (ASCE) 104, No. GT2, February 1978.

63. Coyle, H. M., and L. C. Reese. "Load Transfer for Axially Loaded Piles in Clay." *Journal of Soil Mechanics and Foundations Division* (ASCE) 92, SM2, March 1966.

64. Coyle, H. M., and R. R. Castello. "New Design Correlations for Piles in Sand." *Journal of Geotechnical Engineering Division,* (ASCE) 107, No. GT7, July 1981.

65. Coyle, H. M., and I. H. Sulaiman. "Bearing Capacity of Foundation Piles: State of the Art." *Highway Research Record No. 333.* Washington, D.C.: Highway Research Board, National Academy of Sciences—National Academy of Engineering, 1970.

66. Crouch, M. S. "Check Soil Contamination Easily." *Chemical Engineering Progress* (American Institute of Chemical Engineers) 86, September 1990.

67. Cummings, A. E., G. O. Kerkhoff, and R. B. Peck. "Effect of Driving Piles into Soft Clay." *Transaction* (ASCE) 115, 1950.

68. D'Appolonia, D. J. "Soil-Bentonite Slurry Trench Cutoffs." *Journal of Geotechnical Engineering Division* (ASCE) 106, No. GT4, April 1980.

69. D'Appolonia, D. J., E. D'Appolonia, and R. F. Brissette. "Settlement of Spread Footings on Sand." *Journal of Soil Mechanics and Foundations Division* (ASCE) 94, No. SM3, May 1968.

70. D'Appolonia, D. J., R. V. Whitman, and E. D'Appolonia. "Sand Compaction with Vibratory Rollers." *Journal of Soil Mechanics and Foundations Division* (ASCE) 95, No. SM1, January 1969.

71. D'Appolonia, E. "Load Transfer–Pile Clusters." *Soil Mechanics Lecture Series, Foundation Engineering.* Soil Mechanics and Foundations Division, Illinois Section, ASCE; and Department of Civil Engineering, Northwestern University, Evanston, Ill., 1969.

72. Davisson, M. T. "Design Pile Capacity." *Proceedings, Design and Installation of Pile Foundations and Cellular Structures.* Lehigh Valley, Penna.: Envo Publishing Company, Lehigh University, 1970.

73. DeBeer, E. E. "Bearing Capacity and Settlement of Shallow Foundations on Sand.: *Proceedings, Bearing Capacity and Settlement of Foundations Symposium.* Durham, N.C.: Duke University, 1967.

74. DeBeer, E. E. "Experimental Determination of the Shape Factors and the Bearing Capacity Factors of Sand." *Geotechnique* 20, No. 4, London, 1970.

75. Deere, D. V., and F. D. Patton. "Slope Stability in Residual Soils." *Proceedings, Fourth Pan American Conference on Soil Mechanics and Foundation Engineering.* San Juan, P.R., June 1971, Vol. 1 (pub. ASCE).

76. deMello, V. F. B. "The Standard Penetration Test." *Proceedings, Fourth Pan American Conference on Soil Mechanics and Foundation Engineering.* San Juan, P.R., June 1971, Vol. 1 (pub. ASCE).

77. de Rutter, J. "Electric Penetrometer for Site Investigations." *Journal of Soil Mechanics and Foundation Division* (ASCE) 97, No. SM2, February 1971.

78. *Design of Small Dams.* Denver, Colo.: U.S. Bureau of Reclamation, 1960.

79. Devinny, J. S., L. G. Everett, J. C. S. Lu, and R. L. Stoller. *Subsurface Migration of Hazardous Wastes.* New York: Van Nostrand-Reinhold, 1990.

80. Dewey, J. F. "Plate Tectonics." *Scientific American,* May 1972.

81. Dietz, R. L., and J. C. Holden. "The Breakup of Pangaea." *Scientific American,* October 1970.

82. *Dispersive Clays, Related Piping, and Erosion in Geotechnical Projects.* ASTM Special Technical Publication 623. J. L. Sherard and R. S. Decker, eds. Philadelphia: American Society for Testing and Materials, 1976.

83. Dobson, T. "Vibro Techniques." *Proceedings, Conference on Geotechnical Engineering Practice.* Central Pennsylvania ASCE and Pennsylvania Department of Transportation, Harrisburg, April 1986.

84. Drumright, E. E., C. W. Pfingsten, and R. G. Lukas, "Influence of Hammer Type on SPT Results." *Journal of Geotechnical Engineering Division* (ASCE) 122, No. GT7, July 1996.

85. Duell, R. W., I. A. Leone, and F. B. Flower. "Effect of Landfill Gases on Soil and Vegetation." *Pollution Engineering* 18, June 1986.

86. Duncan, J. M. "State of the Art: Limit Equilibrium and Finite-Element Analysis of Slopes." *Journal of Geotechnical Engineering Division* (ASCE) 122, No. GT7, July 1996.

87. Duncan, J. M., C. W. Clough, and R. M. Ebeling. "Behavior and Design of Gravity Earth Retaining Structures." *Design and Performance of Earth Retaining Structures.* Geotechnical Special Publication No. 25, ASCE, 1990, pp. 251–77.

88. Dunlop, P., and J. M. Duncan. "Development of Failure around Excavated Slopes." *Journal of Soil Mechanics and Foundations Division* (ASCE) 96, No. SM2, March 1970.

89. *Earth Manual.* Denver, Colo.: U.S. Bureau of Reclamation, 1960.

90. *Earth Resistivity Manual.* Chicago: Soiltest, Inc., 1968.

91. Elms, D. G., and R. Richards. "Seismic Design of Retaining Walls." *Design and Performance of Earth Retaining Structures.* Geotechnical Special Publication No. 25, ASCE, 1990, pp. 854–71.

92. "Engineering Properties of Lateritic Soils." *Proceedings, Specialty Session, Seventh International Conference on Soil Mechanics and Foundations,* Mexico City, Mexico. Sponsored by Asian Institute of Technology, Bangkok, Thailand, August 1969.

93. *Engineering Seismograph Instruction Manual.* Chicago: Soiltest, Inc., 1975.

94. Fang, H-Y., ed. *Foundation Engineering Handbook,* 2nd ed. New York: Van Nostrand-Reinhold, 1991.

95. Fellenius, B. H. "Test Loading of Piles and New Proof Testing Procedure." *Journal of Geotechnical Engineering Division* (ASCE) 101, No. GT9, September 1975.

96. Fellenius, W. "Calculation of the Stability of Earth Dams." *Proceedings, Second Congress on Large Dams,* Washington, D.C., 1936.

97. Ferrians, O. J., Jr., R. Kachadoorian, and G. Greene. *Permafrost and Related Engineering Problems in Alaska.* Geological Survey Professional Paper 678. Washington, D.C.: U.S. Department of the Interior, 1969.

98. Fletcher, G. "Standard Penetration Test, Its Uses and Abuses." *Journal of Soil Mechanics and Foundations Division* (ASCE) 91, SM4, July 1965.

99. *Foundation Piling.* Report No. 4 for the Federal Construction Council Building Research Advisory Board, National Academy of Sciences—National Research Council Pub. 987, Washington, D.C., 1962.

100. Fowler, J. W. "Pile Installation Case Histories." *Proceedings, Design and Installation of Pile Foundations and Cellular Structures.* Lehigh Valley, Penna.: Envo Publishing Company, Lehigh University, 1970.

101. Fredlund, D. G., and J. Raherdjo, *Soil Mechanics for Unsaturated Soils.* New York: Wiley, 1993.

102. Friels, D. R. "Pile Capacity in Cohesive Soils Computed from SPT N-Values." *Piletips.* Clifton, N.J.: Associated Pile and Fitting Corp., September–October 1980.

103. *Frost Action in Roads and Airfields.* Highway Research Board, Special Report No. 1, National Academy of Sciences—National Research Council Pub. 211, Washington, D.C., 1952.

104. Frye, S. C. "The Protection of Piling." *Proceedings, Design and Installation of Pile Foundations and Cellular Structures.* Lehigh Valley, Penna.: Envo Publishing Company, Lehigh University, 1970.

105. Galloway, J. D. "The Design of Rockfill Dams." *Transactions* (ASCE) 104, 1939.

106. Gendron, G. J. *Pile Driving: Hammers and Driving Methods.* Highway Research Record No. 333, Highway Research Board, National Academy of Sciences—National Academy of Engineering, Washington, D.C., 1970.

107. Gibbs, H. J., "Standard Penetration Test for Sand Denseness." *Proceedings, Fourth Pan American Conference on Soil Mechanics and Foundation Engineering.* San Juan, P.R., June 1971, Vol. 2 (pub. ASCE).

108. Gibbs, H. J., and W. G. Holtz. "Research on Determining the Density of Sands by Spoon Penetration Testing." *Proceedings, Fourth International Conference on Soil Mechanics and Foundation Engineering,* London, 1957.

109. Goh, A. T. C. "Behavior of Cantilever Retaining Walls." *Journal of Geotechnical Engineering Division* (ASCE) 199, No. GT11, November 1993.

110. Golder, H. Q. "State-of-Art of Floating Foundations," *Journal of Soil Mechanics and Foundations Division* (ASCE) 91, No. SM2, March 1965.

111. Golder, H. Q. "The Allowable Settlement of Structures." *Proceedings, Fourth Pan American Conference on Soil Mechanics and Foundation Engineering.* San Juan, P.R., June 1971, Vol. 1 (pub. ASCE).

112. Golze, A. R., ed. *Handbook of Dam Engineering.* New York: Van Nostrand-Reinhold, 1977.

113. Gould, J. P. "Lateral Pressures on Rigid Permanent Structures." *Proceedings, 1970 Specialty Conference on Lateral Stresses in the Ground and Design of Earth Retaining Structures.* ASCE, Cornell University, 1970.

114. Grand, B. A. *Types of Piles: Their Characteristics and General Use.* Highway Research Record No. 333, Highway Research Board, National Academy of Sciences—National Academy of Engineering, Washington, D.C., 1970.

115. Gray, H. "Field Vane Shear Tests of Sensitive Cohesive Soil." *Transactions* (ASCE) 122, 1957.
116. Grim. R. E. "Physico-Chemical Properties of Soils: Clay Minerals," *Journal of Soil Mechanics and Foundation Division* (ASCE) 85, No. SM2, April 1959.
117. Grim, R. E. *Clay Mineralogy,* 2nd ed. New York: McGraw-Hill, 1968.
118. Hanna, A. M., and G. G. Meyerhof, "Experimental Evaluation of Bearing Capacity of Footings Subjected to Inclined Loads." *Canadian Geotechnical Journal* 18, No. 4, 1981.
119. Hansen, J. "A Revised and Extended Formula for Bearing Capacity." *Danish Geotechnical Institute Bulletin* No. 28, Copenhagen, 1970.
120. Harr, M. E. *Groundwater and Seepage.* New York: McGraw-Hill, 1962.
121. Hausmann, M. R. *Engineering Principles of Ground Modification.* New York: McGraw-Hill, 1990.
122. Hays, W. W. *Procedures for Estimating Earthquake Ground Motions.* Geological Survey Professional Paper 1114. Washington, D.C.: U.S. Govt. Printing Office, 1980.
123. Hays, W. W., ed. *Facing Geologic and Hydrologic Hazards, Earth Science Considerations.* Geological Survey Professional Paper 1240-B. Washington D.C.: U.S. Govt. Printing Office, 1981.
124. Healy, K. A. "Evaluation and Repair of Stonewall-Earth Dams." *Proceedings, Safety of Small Dams, ASCE Engineering Foundation Conference,* New England College, Henniker, N.H., August 1974.
125. Hedges, C. S. "Standard Test Boring with Drilling Mud." *Proceedings, Fourth Pan American Conference on Soil Mechanics and Foundation Engineering.* San Juan, P.R., June 1971, Vol. 2 (pub. ASCE).
126. Hirsch, T. J., L. L. Lowery, H. M. Coyle, and C. H. Samson. *Pile Driving by One-Dimensional Wave Theory: State of the Art.* Highway Research Record No. 333, Highway Research Board, National Academy of Sciences—National Academy of Engineering, Washington, D.C., 1970.
127. Hoechst Fibers Industries Manual. *Engineering Fabrics for Geotextile Applications.* Spartansburg, N.C.: American Hoechst Corporation.
128. Holtz, W. G., and H. J. Gibbs. "Engineering Properties of Expansive Clays." *Transactions* (ASCE) 121, 1956.
129. Holtz, W. G., and H. J. Gibbs. "Discussion, Settlement of Spread Footings on Sand." *Journal of Soil Mechanics and Foundations Division* (ASCE) 95, SM3, May 1969.
130. Horne, R. A. *Marine Chemistry: The Structure of Water and the Chemistry of the Hydrosphere.* New York: Wiley, 1969.
131. Hough, B. K. "Compressibility as the Basis for Soil Bearing Value." *Journal of Soil Mechanics and Foundations Division* (ASCE) 85, No. SM4, August 1959.
132. Housel, W. S. *Checking Up on Vertical Sand Drains.* Highway Research Board Bulletin 90, National Academy of Sciences—National Research Council, Washington, D.C., 1954.
133. Housel, W. S. "Michigan Study of Pile Driving Hammers." *Journal of Soil Mechanics and Foundations Division* (ASCE) 91, No. SM5, September 1965.
134. Housel. W. S. "Pile Load Capacity: Estimates and Test Results." *Journal of Soil Mechanics and Foundations Division* (ASCE) 92, No. SM4, July 1966.
135. Hunt, C. B. *Geology of Soils, Their Evolution, Classification and Uses.* San Francisco: W. H Freeman, 1972.
136. Hunt, H. W. *Piletips, Design and Installation of Pile Foundations.* New Jersey: Associated Pile and Fitting Corporation, 1974.
137. Hvorslev, M. J. *Subsurface Exploration and Sampling of Soils for Civil Engineering Purposes.* Vicksburg, Miss.: U.S. Waterways Experiment Station, 1949.
138. Illinois Institute of Technology. *Design of Structures to Resist Earth Pressures.* Soil Mechanics and Foundation Division, Illinois Section, ASCE; and Civil Engineering Department, Illinois Institute of Technology, 1964.

139. International Conference of Building Officials. *Uniform Building Code.* Whittier, Calif.

140. Janbu, N. *Stability Analysis of Slopes with Dimensionless Parameters.* Harvard Soil Mechanics Series, No. 46, 1954.

141. Janbu, N. "Dimensionless Parameters for Homogeneous Earth Slopes: Discussion." *Journal of Soil Mechanics and Foundations Division* (ASCE) 93, No. SM6, November 1967.

142. Janbu, N. "Slope Stability Computations." *Embankment Dam Engineering.* Casagrande Volume. New York: Wiley, 1973.

143. Johnson, B. "Methane Gas Seeps into Seattle Community Houses." *The Management of World Wastes* 29, December 1986.

144. Johnson, S. J. "Foundation Precompression with Vertical Sand Drains." *Journal of Soil Mechanics and Foundations Division* (ASCE) 96, No. SM1, January 1970.

145. Johnson, S. J. "Precompression for Improving Foundation Soils." *Journal of Soil Mechanics and Foundations Division* (ASCE) 96, No. SM1, January, 1970.

146. Jumikis, A. R. *Soil Mechanics.* New York: Van Nostrand-Reinhold, 1962.

147. Jumikis, A. R. *Thermal Soil Mechanics.* New Brunswick, N.J.: Rutgers University Press, 1966.

148. Jurgenson, L. "The Application of Theories of Elasticity and Plasticity to Foundation Problems." *Journal, Boston Society of Civil Engineers,* July 1954.

149. Justin, J. D. "The Design of Earth Dams." *Transactions* (ASCE) 87, 1924.

150. Kansas, University of. *Proceedings, Eighteenth Annual Soil Mechanics and Foundations Engineering Conference,* Lawrence, Kans., 1969.

151. Karol, R. H. "Chemical Grouting Technology." *Journal of Soil Mechanics and Foundations Division* (ASCE) 94, SM1, January 1968.

152. Kehew, A. E. *General Geology for Engineers.* New York: Prentice-Hall, 1988.

153. Keller, E. A. *Environmental Geology,* 5th ed. Columbus: Merrill, 1988.

154. Kerfoot, H. B. "Soil Gas Measurement for Detection of Groundwater Contamination by Volatile Organic Compounds." *Environmental Science and Technology* (American Chemical Society) 21, No. 10, October 1989.

155. Kerisel, J. L. "Vertical and Horizontal Bearing Capacity of Deep Foundations in Clay." *Proceedings, Bearing Capacity and Settlement of Foundations Symposium.* Durham, N.C.: Duke University, 1967.

156. Kerisel, J. L. The Bicentennial of C. Coulomb's Theory of Loose Media." *Proceedings, Eighth International Conference on Soil Mechanics and Foundations,* Moscow, 1973.

157. Kezdi, A. *Handbook of Soil Mechanics.* Amsterdam: Elsevier, 1974.

158. Kjellman, W. "Mechanics of Large Swedish Landslides." *Geotechnique* 5, No 1, 1955.

159. Koerner, R. M. *Designing with Geosynthetics,* 3rd ed. Englewood Cliffs, N.J.: Prentice Hall, 1994.

160. Koerner, R. M., and J. P. Welsh. *Construction and Geotechnical Engineering Using Synthetic Fabrics.* New York: Wiley, 1980.

161. Kovacs, W. D. "Effects of SPT Equipment and Procedures on the Design of Shallow Foundations on Sand." *Vertical and Horizontal Deformations of Foundations and Embankments.* Geotechnical Special Publication No. 40, ASCE, 1994, pp. 121–31.

162. Kraft, L. M., J. A. Focht, and S. F. Amerasinghe. "Friction Capacity of Piles Driven into Clay." *Journal of Geotechnical Engineering Division* (ASCE) 107, No. 11, November 1981.

163. Kreb, R. D., and R. D. Walker. *Highway Materials.* New York: McGraw-Hill, 1971.

164. Krinitzsky, E. L., J. P. Gould, and P. H. Edinger. *Fundamentals of Earthquake Resistant Construction.* New York: Wiley, 1993.

165. Krynine, D. P., and W. R. Judd. *Principles of Engineering Geology and Geotechnics.* New York: McGraw-Hill, 1957.

166. Kulhawy, F. H. "Analysis of a High Gravity Retaining Wall." *Proceedings, Analysis and Design in Geotechnical Engineering.* Geotechnical Engineering Division, ASCE, University of Texas, 1, June 1974.

167. Kulhawy, F. H., and T. M. Gurtowski. "Load Transfer and Hydraulic Fracturing in Zoned Dams." *Journal of Geotechnical Engineering Division* (ASCE) 102, No. GT9, September 1976.

168. Lambe, T. W. "The Structure of Inorganic Clay," *Journal of Soil Mechanics and Foundations Division* (ASCE) 79, October 1953.

169. Lambe, T. W. "The Engineering Behavior of Compacted Clay." *Journal of Soil Mechanics and Foundations Division* (ASCE) 84, No. SM2, May 1958.

170. Lambe, T. W., et al. "Compacted Clay—A Symposium." *Transactions* (ASCE) 125, 1960.

171. Lee, I. K., and P. J. Moore. "Stability Analyses Application to Slopes, Rigid and Flexible Retaining Structures." In I. K. Lee, ed. *Soil Mechanics, Selected Topics.* New York: American Elsevier, 1968.

172. Lee, K. L., B. D. Adams, and J-M J. Vagneron. "Reinforced Earth Retaining Walls." *Journal of Soil Mechanics and Foundations Division* (ASCE) 99, No. SM10, October 1973.

173. Leggett, R. F. *Geology and Engineering,* 2nd ed. New York: McGraw-Hill, 1962.

174. Leggett, R. F., and A. W. Hathaway. *Geology and Engineering.* New York: McGraw-Hill, 1988.

175. Leonards, G. A., W. A. Cutter, and R. D. Holtz. "Dynamic Compaction of Granular Soils." *Journal of Geotechnical Engineering Division* (ASCE) 106, No. GT1, January 1980.

176. Liao, S. S. C., and R. V. Whitman, "Overburden Correction Factors for SPT in Sand." *Journal of Geotechnical Engineering Division* (ASCE) 112, No. 3, March 1986.

177. Lindahl, H. A. "Steel Pile Retaining Walls." *Proceedings, Eighteenth Annual Soil Mechanics and Foundation Engineering Conference,* University of Kansas, March 1969.

178. Lukas, R. G. "Densification of Loose Deposits by Pounding." *Journal of Geotechnical Engineering Division* (ASCE) 106, No. GT4, April 1980.

179. Lukas, R. G., and B. L. deBussey. "Pressuremeter and Laboratory Test Correlations for Clays." *Journal of Geotechnical Engineering Division* (ASCE) 102, No. GT9, September 1976.

180. Maccaferri Gabions Inc. *Gabion Retaining Walls.* Williamsport, Md.: Author.

181. Malasheskie, G. J. "Prefabricated Drainage Systems." *Proceedings, Conference on Innovations in Geotechnical Engineering.* Central Pennsylvania ASCE and Pennsylvania Department of Transportation, Harrisburg, April 1985.

182. Marangos, C. N. "Vertical and Horizontal Deformations for Foundations on Sand: An Experimental Study." *Vertical and Horizontal Deformations of Foundations and Embankments.* Geotechnical Special Publication No. 40, ASCE, 1994, pp. 179–89.

183. Marchetti, S. "In-Situ Tests by Flat Dilatometer." *Journal of Geotechnical Engineering* (ASCE) 106, No. GT3, March 1980.

184. Marrin, D. L., and G. M. Thompson. "Gaseous Behavior of TCE Overlying a Contaminated Aquifer." *Groundwater* 25, No. 1, January–February 1987.

185. Marrin, D. L., and H. B. Kerfoot. "Soil Gas Surveying Techniques." *Environmental Science and Technology* (American Chemical Society) 22, No. 7, July 1988.

186. Matthes, G., and H. C. Harvey. *The Properties of Groundwater.* New York: Wiley, 1982.

187. McClelland, B. "Design and Performance of Deep Foundations." *Proceedings, Specialty Conference on Performance of Earth and Earth-Supported Structures.* Purdue University, ASCE, June 1972.

188. McClelland, B. "Design of Deep Penetration Piles for Ocean Structures." *Journal of Geotechnical Engineering Division* (ASCE) 100, No. GT7, July 1974.

189. McClelland, B., J. A. Focht, and W. J. Emrich. "Problems in Design and Installation of Off-Shore Piles." *Journal of Soil Mechanics and Foundations Division* (ASCE) 95, No. SM6, November 1969.

190. Meyerhof, G. G. "The Ultimate Bearing Capacity of Foundations." *Geotechnique* 2, No. 4, London, 1951.

191. Meyerhof, G. G. "The Bearing Capacity of Footings under Eccentric and Inclined Loads." *Proceedings, Third International Conference on Soil Mechanics and Foundation Engineering,* Zurich, 1953.

192. Meyerhof, G. G. "The Influence of Roughness of Base and Ground Water on the Ultimate Bearing Capacity of Foundations." *Geotechnique* 5, No. 3, London, September 1955.

193. Meyerhof, G. G. "Penetration Tests and Bearing Capacity of Cohesionless Soils." *Journal of Soil Mechanics and Foundations Division* (ASCE) 82, No. SM1, January 1956.

194. Meyerhof, G. G. "The Ultimate Bearing Capacity of Foundations on Slopes." *Proceedings, Fourth International Conference on Soil Mechanics and Foundation Engineering,* London, 1957.

195. Meyerhof, G. G. "Compaction of Sands and Bearing Capacity of Piles." *Journal of Soil Mechanics and Foundations Division* (ASCE), 85, No. SM6, December 1959.

196. Meyerhof, G. G. "Some Recent Research on the Bearing Capacity of Foundations." *Canadian Geotechnical Journal* 1, No. 1, September 1963.

197. Meyerhof, G. G. "Shallow Foundations." *Journal of Soil Mechanics and Foundations Division* (ASCE) 91, No. SM2, March 1965.

198. Meyerhof, G. G. "Bearing Capacity and Settlement of Pile Foundations." *Journal of Geotechnical Engineering Division* (ASCE) 102, No. GT3, March 1976.

199. Millar, C. E., L. M. Turk, and H. D. Foth. *Fundamentals of Soil Science.* New York: Wiley, 1958.

200. Miller, D. W. "Sources of Ground-Water Pollution." *EPA Journal* 10, July–August 1984.

201. Mitchell, J. K. "Fundamental Aspects of Thixotropy in Soils." *Journal of Soil Mechanics and Foundations Division* (ASCE) 86, No. SM3, June 1960.

202. Mitchell, J. K. "In-Place Treatment of Foundation Soils." *Journal of Soil Mechanics and Foundations Division* (ASCE) 96, No. SM1, January 1970.

203. Mitchell, J. K. *Fundamentals of Soil Behavior,* 2nd ed. New York: Wiley, 1993.

204. Moore, W. S. "Experiences with Predetermining Pile Lengths." *Transactions* (ASCE) 114, 1947.

205. Moorhouse, D. C. "Shallow Foundations," *Proceedings, Specialty Conference on Performance of Earth and Earth-Supported Structures.* Purdue University, ASCE, June 1972.

206. Moorhouse, D. C., and G. L. Baker. "Sand Densification by Heavy Vibratory Compactor." *Journal of Soil Mechanics and Foundations Division* (ASCE) 95, No. SM4, July 1969.

207. Moretrench Corporation. *Moretrench Wellpoint System.* Rockaway, N.J.: Author, 1967.

208. Morganstern, N. R. "Stability Charts for Earth Slopes during Rapid Drawdown." *Geotechnique* 13, No. 2, 1963.

209. Morganstern, N. R. "Slopes and Excavations." *Proceedings, Ninth International Conference on Soil Mechanics and Foundation Engineering,* Tokyo, 1977.

210. Morganstern, N. R., and V. E. Price. "The Analysis of the Stability of General Slip Surfaces." *Geotechnique* 15, No. 1, March 1965.

211. Mosley, E. T., and T. Raamot. *Pile Driving Formulas.* Highway Research Board, National Academy of Sciences—National Academy of Engineering, Washington, D.C., 1970.

212. National Research Council of Canada. *Supplement to the National Building Code of Canada.* Ottowa, 1994.

213. Navy, Department of the. *Design Manual, Soil Mechanics, Foundations and Earth Structures.* NAVFAC DM-7, Rev. 1974.

214. Nordlund, R. L. "Bearing Capacity of Piles in Cohesionless Soils." *Journal of Soil Mechanics and Foundations Division* (ASCE) 89, No. SM3, May 1963.

215. Nordlund, R. L. "Pressure Injected Footings." *Proceedings, Design and Installation of Pile Foundations and Cellular Structures.* Lehigh Valley, Penna.: Envo Publishing Company, Lehigh University, 1970.

216. O'Connor, M. J., and R. J. Mitchell. "An Extension of the Bishop-Morganstern Slope Stability Charts." *Canadian Geotechnical Journal* 14, 1977.

217. Office of Technical Assessment. *Protecting the Nation's Groundwater from Contamination.* Washington, D.C.: Congress of the United States, 1984.

218. Osterberg, J. O. "Drilled Caissons—Design, Installation, Application." *Soil Mechanics Lecture Series, Foundation Engineering.* Soil Mechanics and Foundations Division, Illinois Section, ASCE; and Department of Civil Engineering, Northwestern University, Evanston, Ill., 1969.

219. Page, G. W. *Planning for Groundwater Protection.* Orlando, Fla.: Academic Press, Harcourt Bruce Jovanovich, 1987.

220. Parola, J. F. "Mechanics of Impact Pile Driving," Ph.D. dissertation, University of Illinois, Urbana, 1970.

221. Parsons, J. D. "Piling Difficulties in the New York Area." *Journal of Soil Mechanics and Foundations Division* (ASCE) 92, No. SM1, January 1966.

222. Paz, M. ed. *International Handbook of Earthquake Engineering.* New York: Chapman and Hall, 1994.

223. Peck, R. B. "Pile and Pier Foundations." *Journal of Soil Mechanics and Foundations Division* (ASCE) 91, No. SM2, March 1965.

224. Peck, R. B., and A. R. S. Bazaraa. "Discussion Settlement of Spread-Footings on Sand." *Journal of Soil Mechanics and Foundations Division* (ASCE) 95, No. SM3, May 1969.

225. Peck, R. B., W. E. Hanson, and T. H. Thorburn. *Foundation Engineering,* 2nd ed. New York: Wiley, 1973.

226. Peck, R. B., and H. O. Ireland. "Full Scale Lateral Load Test on a Retaining Wall Foundation." *Proceedings, Fifth International Conference on Soil Mechanics and Foundation Engineering,* 1961.

227. Peck, R. B., H. O. Ireland, and C. Y. Teng. "A Study of Retaining Wall Failures." *Proceedings, Second International Conference on Soil Mechanics and Foundation Engineering,* 1948.

228. Peuchen, J., S. J. Plasman, and R. Van Stereninck. "Core Penetration Testing in Tropical Residual Soils." *Ground Engineering,* January–February, 1966.

229. Peurifoy, R. L. *Construction Planning, Equipment and Methods,* 2nd ed. New York: McGraw-Hill, 1970.

230. Portland Cement Association. *Small Concrete Dams.* Portland, Ore.: Author, 1971.

231. Prandtl, L. "Uber die Harte, Plastischer Korper" ("On the Hardness of Plastic Bodies"). *Nachr. Kgl. Ges. Wiss.* Gottingen, Math, Phys. Kl., 1920.

232. *Procedures for Testing Soils.* Philadelphia: American Society for Testing and Materials.

233. Proctor, R. R. "Fundamental Principles of Soil Compaction." *Engineering News Record,* Aug. 31, Sept. 7, Sept. 21, Sept. 28, 1933.

234. Rad, N. S., and T. Lunne. "Gas in Soil, Part 1: Detection and η Profiling," *Journal of Geotechnical Engineering Division* (ASCE) 120, No. GT4, April 1994.

235. Ramaswamy, S. D., et al. "Treatment of Peaty Clay by High Energy Impact." *Journal of Geotechnical Engineering Division* (ASCE) 105, No. GT8, August 1979.

236. Reese, L. C., and M. W. O'Neill. "Criteria for the Design of Axially Loaded Drill Shafts." Research Report 189-11F, Center for Highway Research, University of Texas, Austin, 1971.

237. Reese, L. C., F. T. Touma, and M. W. O'Neill. "Behavior of Drilled Piers under Axial Loading." *Journal of Geotechnical Engineering Division* (ASCE) 102, No. GT5, May 1976.

238. Reginatto, A. R. "Standard Penetration Tests in Collapsible Soils." *Proceedings, Fourth Pan American Conference on Soil Mechanics and Foundation Engineering,* San Juan, P.R., June 1971, Vol. 2 (pub. ASCE).

239. Reissner, H. "Zum Erddruck-problem" (The Earth Pressure Problem). *Proceedings, First International Congress on Applied Mechanics,* 1924.

240. Richards, R., and D. G. Elms. "Seismic Behavior of Gravity Retaining Walls." *Jourrnal of Geotechnical Engineering Division* (ASCE) 105, No. GT4, April 1979.

241. Richards, R., D. G. Elms, and M. Budhu. "Seismic Bearing Capacity and Settlements on Sands." *Journal of Geotechnical Engineering Division* (ASCE) 199, No. GT4, April 1993.

242. Richards, R., K. L. Fishman, and R. L. Vivitok. "Threshold Accelerations for Rotation or Sliding of Bridge Abutments." *Journal of Geotechnical Engineering Division* (ASCE) 122, No. GT9, September 1996.

243. Richart, F. E. "A Review of the Theories for Sand Drains." *Journal of Soil Mechanics and Foundations Division* (ASCE) 83, No. SM3, July 1957.

244. Roberts, D. V. "Notes on Predetermination of Pile Capacities." *Dames and Moore Engineering Bulletin* 5, Dames and Moore, Los Angeles, Calif., 1959.

245. Robertson, P. K., R. G. Campanella, and A. Wightman. "SPT–CPT Correlations." *Journal of Geotechnical Engineering Division* (ASCE) 109, No. GT11, 1983.

246. Robinsky, E. I., and K. E. Bespflug. "Design of Insulated Foundations." *Journal of Soil Mechanics and Foundations Division* (ASCE) 99, No. SM9, September 1973.

247. Rodin, S. "Experiences with Penetrometer, with Particular Reference to the Standard Penetration Test." *Proceedings, Fifth International Conference on Soil Mechanics and Foundation Engineering,* Paris, 1961.

248. Rollins, K. M., R. L. Rollins, T. D. Smith, and G. H. Beckwith. "Identification and Characterization of Collapsible Gravels." *Journal of Geotechnical Engineering Division* (ASCE) 120, No. GT3, March 1944.

249. Roscoe Moss Company. *Handbook of Ground Water Development.* New York: Wiley, 1990.

250. Rose, E. "Thrust Exerted by Expanding Ice Sheet." *Transactions* (ASCE) 112, 1947.

251. Rosenquist, I. T. "Physico-Chemical Properties of Soils: Soil–Water Systems." *Journal of Soil Mechanics and Foundations Division* (ASCE) 85, No. SM2, April 1959.

252. Rutledge, P. C. "Construction Methods and Inspection." *Soil Mechanics Lecture Series, Foundation Engineering.* Soil Mechanics and Foundations Division, Illinois Section, ASCE; and Department of Civil Engineering, Northwestern University, Evanston, Ill., 1969.

253. Rutledge, P. C. "Utilization of Marginal Lands for Urban Development." *Journal of Soil Mechanics and Foundations Division* (ASCE) 96, No. SM1, January 1970.

254. Sanglerat, G. *The Penetrometer and Soil Exploration* (English). London–New York: Elsevier, 1972.

255. Schmertmann, J. H. "Static Cone Penetrometers for Soil Exploration." *Civil Engineering* (ASCE) June 1967.

256. Schmertmann, J. H. "Static Cone to Compute Static Settlement over Sand." *Journal of Soil Mechanics and Foundations Division* (ASCE) 96, No. SM3, 1970.

257. Schmertmann, J. H. "Introduction to the Marchetti Dilatometer Test." *Proceedings, Conference on Innovations in Geotechnical Engineering.* Central Pennsylvania ASCE and Pennsylvania Department of Transportation, Harrisburg, April 1985.

258. Schmertmann, J. H., J. P. Hartman, and P. R. Brown. "Improved Strain Influence Factors Diagrams." *Journal of Geotechnical Engineering Division* (ASCE) 104, No. GT8, 1978.

259. Schmid, W. E. "Low Frequency Pile Vibrators." *Proceedings, Design and Installation of Pile Foundations and Cellular Structures.* Lehigh Valley, Penna.: Envo Publishing Company, Lehigh University, 1970.

260. Schnable, H., Jr., and J. W. Siquorney. *Design and Installation of Earth Tiebacks.* Washington, D.C.: Schnabel Foundation Company.

261. Seed, H. B. "Design Problems in Soil Liquefaction." *Journal of Geotechnical Engineering Division* (ASCE) 113, No. GT8, August 1987.

262. Seed, H. B., K. Tokimatsi, L. F. Harder, and R. M. Chung. "Influence of SPT Procedures in Soil Liquefaction Resistance Evaluations." *Journal of Geotechnical Engineering Division* (ASCE) 111, No. 12, December 1985.

263. Sherard, J. L. "Embankment Dam Cracking." In R. C. Hirschfeld and S. J. Poulos, eds. *Embankment Dam Engineering.* Casagrande Volume. New York: Wiley Interscience, 1973.

264. Sherard, J. L., R. S. Decker, and N. L. Ryker. "Piping in Earth Dams of Dispersive Clays." *Proceedings, Specialty Conference on Performance of Earth and Earth Support Structures.* Purdue University and Soil Mechanics and Foundations Division, ASCE, June 1972.

265. Sherard, J. L, L. P, Dunnigan, and R. S. Decker. "Identification and Nature of Dispersive Soils." *Journal of Geotechnical Engineering Division* (ASCE) 102, No. GT4, April 1976.

266. Sherard, J. L., L. P. Dunnigan, and J. R. Talbot. "Basic Properties of Sand and Gravel Filters." *Journal of Geotechnical Engineering Division* (ASCE) 110, No. 6, June 1984.

267. Sherard, J. L. "Filters for Silts and Clays." *Journal of Geotechnical Engineering Division* (ASCE) 110, No. 6, June 1984.

268. Sherard, J. L., R. J. Woodward, S. F. Gizienski, and W. A. Clevenger. *Earth and Earth-Rock Dams.* New York: Wiley, 1963.

269. Skempton, A. W. "The Colloidal Activity of Clays." *Proceedings, Third International Conference on Soil Mechanics and Foundation Engineering,* Zurich, 1953.

270. Skempton, A. W. "Soil Mechanics in Relation to Geology." *Proceedings, Yorkshire Geological Society.* 29, Part 1, No. 3, April 1953.

271. Skempton, A. W., and J. Hutchinson. "Stability of Natural Slopes and Embankment Foundations." *Procedures, Seventh International Conference on Soil Mechanics and Foundation Engineering,* Mexico, 1969.

272. Smith, E. A. L. "Pile-Driving Analysis by the Wave Equation." *Transactions* (ASCE) 127, 1962.

273. Smith, R. E. "Guide for Depth of Foundation Exploration." *Journal of Soil Mechanics and Foundations Division* (ASCE) 96, No. SM2, March 1970.

274. *Soil Cement Construction Handbook.* Chicago, Ill.: Portland Cement Association.

275. *Soil Cement Laboratory Handbook.* Chicago, Ill.: Portland Cement Association.

276. *Soil Density Control Methods.* Highway Research Board, Bulletin 149, National Academy of Sciences—National Research Council, Pub. 498, Washington, D.C., 1957.

277. Sorenson, T., and B. Hansen. "Pile Driving Formulae—An Investigation Based on Dimensional Considerations and a Statistical Analysis." *Proceedings, Fourth International Conference on Soil Mechanics and Foundation Engineering,* London, 1957.

278. Sowers, G. F. "Fill Settlement despite Vertical Sand Drains." *Journal of Soil Mechanics and Foundations Division* (ASCE) 90, No. SM5, September 1964.

279. Sowers, G. F. "Soil Stress–Strain, Strength and Earth Pressure." *Design of Structures to Resist Earth Pressures, Soil Mechanics Lecture Series.* Soil Mechanics and Foundations Division, Illinois Section, ASCE; and Civil Engineering Department, Illinois Institute of Technology, Chicago, Ill., 1964.

280. Spencer, E. "A Method of Analysis of the Stability of Embankments Assuming Parallel Interslice Forces." *Geotechnique* 17, No. 1, March 1967.

281. Spencer, E. "Effect of Tension on Stability of Embankments." *Journal of Soil Mechanics and Foundations Division* (ASCE) 94, No. SM5, September 1968.

282. Spencer, E. "Circular and Logarithmic Spiral Slip Surfaces." *Journal of Soil Mechanics and Foundations Division* (ASCE) 95, No. SM1, January 1969.

283. Spencer, E. "Thrust Line Criterion in Embankment Stability Analysis." *Geotechnique* 23, No. 1, March 1973.

284. Stearns, R. *Permafrost (Perennially Frozen Ground).* Cold Regions Science and Engineering, Part 1, Section A2, U.S. Army Material Command, Cold Regions Research and Engineering Laboratory, Hanover, N.H., August 1966.

285. *Stresses and Deflections in Foundations and Pavements.* Soil Mechanics and Bituminous Materials Research Laboratory, Department of Civil Engineering, University of California, Berkeley, 1965.

286. Strom, J. A. "Development of the Earth Deformation Recorder." *Dames and Moore Engineering Bulletin* 21, Dames and Moore, Los Angeles, 1962.

287. Suchomel, K. H., D. K. Kreamer, and A. Long. "Production and Transport of Carbon Dioxide in a Contaminated Vadose Zone: A Stable and Radioactive Carbon Isotope Study." *Environmental Science and Technology* (American Chemical Society) 24, No. 12, December 1990.

288. Swatek, E. P., S. P. Asrow, and A. M. Seitz. "Performance of Bracing for Deep Chicago Excavation." *Proceedings, Performance of Earth and Earth-Supported Structures.* Soil Mechanics and Foundations Division, ASCE, and Purdue University, I, Part 2, June 1972.

289. Taylor, A. W. "Physico-Chemical Properties of Soils: Ion Exchange Phenomena." *Journal of Soil Mechanics and Foundations Division* (ASCE) 85, No. SM2, April 1959.

290. Taylor, D. W.. "Stability of Earth Slopes." *Journal of the Boston Society of Civil Engineers,* July 1937.

291. Taylor, D. W. *Fundamentals of Soil Mechanics.* New York: Wiley, 1949.

292. Teng, W. C. *Foundation Design.* Englewood Cliffs, N.J.: Prentice-Hall, 1962.

293. Terzaghi, K. *Theoretical Soil Mechanics.* New York: Wiley, 1943.

294. Terzaghi, K. "Effect of Minor Geologic Details on the Safety of Dams" (1929), reproduced in *From Theory to Practice in Soil Mechanics.* New York: Wiley, 1960.

295. Terzaghi, K. *From Theory to Practice in Soil Mechanics.* New York: Wiley, 1960.

296. Terzaghi, K., and R. B. Peck. *Soil Mechanics in Engineering Practice,* 2nd ed. New York: Wiley, 1968.

297. Tillman, N., K. Ranlet, and T. J. Meyer. "Soil Gas Surveys: Part 1." *Pollution Engineering* 21, July 1989.

298. Tillman, N., K. Ranlet, and T. J. Meyer. "Soil Gas Surveys: Part II, Procedures," *Pollution Engineering* 21, August 1989.

299. Tomlinson, M. J. "The Adhesion of Piles Driven in Clay." *Proceedings, Fourth International Conference on Soil Mechanics and Foundations in Engineering,* London, 1957.

300. Tomlinson, M. J. *Foundation Design and Construction.* New York: Wiley–Interscience, 1969.

301. Touma, F. T., and L. C. Reese. "Behavior of Bored Piles in Sand." *Journal of Geotechnical Engineering Division* (ASCE) 100, No. GT7, July 1974.

302. Tourtelot, H. A. "Geologic Origin and Distribution of Swelling Clays." *Bulletin of Association of Engineering Geologists* XI, No. 4, 1974.

303. Townsend, F. C. "Geotechnical Characteristics of Residual Soils." *Journal of Geotechnical Engineering* (ASCE) 111, No. 1, January 1985.

304. Tschebotarioff, G. P. *Foundations, Retaining and Earth Structures,* 2nd ed. New York: McGraw-Hill, 1973.

305. Tschebotarioff, G. P., Chairman. "Lateral Pressures of Clayey Soils on Structures." Specialty Session 5, *Proceedings, Eighth International Conference on Soil Mechanics and Foundation Engineering,* Moscow, 1973.

306. Turnbull, W. J., J. R. Compton, and R. G. Ahlvin. "Quality Control of Compacted Earthwork." *Journal of Soil Mechanics and Foundations Division* (ASCE) 92, No. SM1, January 1966.

307. Turnbull, W. J., and C. I. Mansur. "Compaction of Hydraulically Placed Fills." *Journal of Soil Mechanics and Foundations Division* (ASCE) 99, No. SM11, November 1973.

308. Tuttle, J., "Experiences with Piles in Coarse Granular Soils." *Dames and Moore Engineering Bulletin* 6, Dames and Moore, Los Angeles, Calif., 1959.

309. U.S. Dept. of the Interior, Water and Power Resources Service. *Groundwater Manual.* New York: Wiley, 1981.

310. U.S. Environmental Protection Agency. *Procedures Manual for Ground Water Monitoring at Solid Waste Disposal Facilities.* Publication SW-616. Cincinnati, Oh., 1977.

311. U.S. Navy. *Soil Mechanics.* Naval Facilities Engineering Command (NAVFAC) Design Manual 7.01, rev. 1986.

312. Varnes, D. J. "Slope Movement and Type and Processes." *Landslide Analysis and Control.* Special Report 176. Washington, D.C.: Transportation Research Board, National Research Council, 1978.

313. Vesic, A. S. "Ultimate Loads and Settlement of Deep Foundations in Sand." *Proceedings, Bearing Capacity and Settlement of Foundations Symposium,* Duke University, Durham, N.C., 1967.

314. Vesic, A. S. "Load Transfer in Pile–Soil Systems." *Proceedings, Design and Installation of Pile Foundations and Cellular Structures.* Lehigh Valley, Penna.: Envo Publishing Company, Lehigh University, 1970.

315. Vesic, A. S. "Test on Instrumental Piles, Ogeechee River Site." *Journal of Soil Mechanics and Foundations Division* (ASCE) 96, No. SM2, March 1970.

316. Vesic, A. S. "Analysis of Ultimate Loads of Shallow Foundations." *Journal of Soil Mechanics and Foundations Division* (ASCE) 96, No. SM1, January 1973.

317. Vesic, A. S. "Bearing Capacity of Shallow Foundations," In H. F. Winterkorn and H-Y. Fang, eds. *Foundation Engineering Handbook.* New York: Van Nostrand-Reinhold, 1975.

318. Vijayvergiya, V. N., and J. A. Focht. "A New Way to Predict Capacity of Piles in Clay." *Proceedings, Fourth Annual Offshore Technology Conference,* Houston, Tex., May 1972.

319. Webb, D. L., and R. I. Hall. "Effects of Vibroflotation on Clayey Sands." *Journal of Soil Mechanics and Foundations Division* (ASCE) 95, No. SM6, November 1969.

320. Wegman, E. *The Design and Construction of Dams.* New York: Wiley, 1918.

321. Welsh, J. P. "In-Situ Testing for Ground Modification Techniques." *Proceedings, In-Situ '86, ASCE Specialty Conference,* Virginia Technical Institute, Blacksburg, June 1986.

322. Whitman, R. V. "Hydraulic Fills to Support Structural Loads." *Journal of Soil Mechanics and Foundations Division* (ASCE) 96, No. SM1, January 1970.

323. Whitman, R. V. "Seismic Design and Behavior of Gravity Retaining Walls." *Design and Performance of Earth Retaining Structures,* Geotechnical Special Publication No. 25, ASCE, 1990, pp. 854–71.

324. Wilson, S. D., and R. Squier. "Earth and Rockfill Dams." *Proceedings, Seventh International Conference on Soil Mechanics and Foundation Engineering,* Mexico, 1969.

325. Winterkorn, H. F., and H. Y. Fang, eds. *Foundation Engineering Handbook.* New York: VanNostrand-Reinhold, 1975.

326. Woodward, R. J., W. S. Gardner, and D. M. Greer. *Drilled Pier Foundations.* New York: McGraw-Hill, 1972.

327. Wu, T. H. "Relative Density and Shear Strength of Sands." *Journal of Soil Mechanics and Foundations Division* (ASCE) 83, No. SM1, January 1957.

328. Yancheski, T. B. "Suburban Superfund," *Civil Engineering* (ASCE) 59, No. 4, April 1989.

329. Yegian, M. K., V. G. Ghahraman, and G. Gazetas. "1988 Armenia Earthquake, Part I: "Seismological, Geotechnical and Structural Overview" and Part II: "Damage Statics vs. Geologic and Soil Profiles." *Journal of Geotechnical Engineering Division* (ASCE) 120, No. GT1, January 1994.

Answers to Selected Problems

Chapter 1

1-36. M_L 5.5; $E \approx 8.28 \times 10^{12}$ ft lb
M_L 7.5; $E \approx 8.28 \times 10^{15}$ ft lb
(1000 × more energy)

1-37. (a) mag 4.2 (±)   (b) 375 km (approx)

1-38. $V = 957$ kips

1-39. $V = 737$ kips

1-40. $V = 3930$ kN

1-41. $V = 1310$ kips

1-42. UBC: $V = 795$ kips
BOCA: $V = 500$ kips
NEHRP: $V = 1041$ kips

Chapter 2

2-1. $w = 23.6\%$

2-2. $w = 13.5\%$

2-3. $w = 15.4\%$, $\rho_{wet} = 2038.5$ kg/m³, $\rho_{dry} = 1766.5$ kg/m³

2-4. $w = 20\%$, $\gamma_{wet} = 123.5$ pcf, $\gamma_{dry} = 103$ pcf

2-5. $G_s = 2.63$

2-7. $e = 0.57$

2-8. $e = 0.57$, $n = 36\%$

2-9. $\rho_{dry} = 1442$ kg/m³, $\gamma_{dry} = 90$ pcf $= 14.13$ kN/m³, $e = 0.87$

2-10. $e = 0.58$, $\gamma_{dry} = 108$ pcf $= 16.96$ kN/m³

2-11. $e = 0.62$, $G_s = 2.68$, $\gamma_{dry} = 103.3$ pcf $= 16.22$ kN/m³
$\gamma_{wet} = 127.3$ pcf $= 19.99$ kN/m³

2-12. $\gamma_{dry} = 77$ pcf $= 12.09$ kN/m³

2-13. $e = 0.51$, $w = 12.7\%$, Sat $= 69.4\%$

2-14. increase $= 18.5$ pcf $= 2.90$ kN/m³

2-16. $w = 16.6\%$, Sat $= 97.3\%$, $e = 0.467$

2-17. $w = 23.5\%$, $e = 0.62$, Sat $= 100\%$

2-18. $\rho_{wet} = 2.0$ Mg/m³ $= 124.8$ pcf
$\rho_{dry} = 1.75$ Mg/m³ $= 109.2$ pcf
$w = 14.3\%$, $e = 0.556$, Sat $= 70\%$

2-19. $w = 85\%$, $e = 2.31$, Sat $= 100\%$

2-20. $\gamma_{wet} = 112.2$ pcf $= 17.62$ kN/m³
$\gamma_{dry} = 102$ pcf $= 16.01$ kN/m³
$\rho_{wet} = 1792$ kg/m³
$\rho_{dry} = 1630$ kg/m³
Sat $= 42\%$

2-21. $G_s = 2.57$

2-22. $e = 0.54$ (±)

2-24. (a) $\gamma_{sub} \approx 60$ pcf   (b) $\gamma_{sub} = 65$ pcf

2-25. (a) $\rho_{sub} = 1.11$ Mg/m³   (b) $\rho_{sub} = 1.27$ Mg/m³

Chapter 4

4-5. (a) $\rho = 1812.5$ kg/m³
(b) $\gamma = 17.76$ kN/m³ $= 113$ pcf

4-6. (a) $\gamma_{wet} = 130.8$ pcf
(b) $\gamma_{dry} = 113.7$ pcf (dense)

4-7. $D_R\% = 76.4\%$

4-8. approx. 13,000 m³, or 8% more volume

4-9. $D_R\% = 69\%$

4-10. $D_R\% = 74\%$

4-11. (b) PI $= 32\%$ (c) MH   (d) A-7-5

4-12. (a) LI $= 0.33$ (b) CL-CH, A-7-6

4-13. (a) LI $= -0.22$ (b) CH, A-7-5

4-14. $V_T = 83.9$ cm³

4-15. $G_s = 2.66$

4-16. Activity $= 0.88$, probably kaolinite

4-17. PI = 44, Activity ≈ 1, VCP very high

4-18. A-4, silty loam

4-19. SM-ML, A-4

4-20. 60% or more (low as 40% some mixtures)

Chapter 6

6-13. $R_H = 0.5$ r

6-14. $R_H = 1.2$

6-21. (a) $k = 0.225$ mm/sec   (b) $k = 0.13$ mm/sec

6-24. $k = 2.56$ mm/min = 0.043 mm/sec

6-25. $k = 1.5$ mm/sec

6-26. $k = 0.021$ ft/min ≈ 0.11 mm/sec

6-27. 80 ml = 0.08 liter

6-28. $k = 3.75 \times 10^{-4}$ mm/sec

6-29. $k = 1.93 \times 10^{-3}$ mm/sec ≈ 4×10^{-4} ft/min

6-30. $k = 9.6 \times 10^{-4}$ ft/min

6-33. $h_c = 3.1$ m

6-34. $p_w = -375$ psf

6-35. $p_w = -304$ kPa

6-36. $h_c = 15$ m

Chapter 7

7-1. $Q = 56.25$ m³/day per km length

7-2. $Q = 32$ ft³/hr = 5700 gal/day

7-6. 2.7 m³/day per m wide

7-7. (a) 8.5 gal/day per ft wide
(b) 5.6 gal/day per ft wide

7-8. $i = 1.03$

7-9. $i\gamma_w = 0.95\ \gamma_w$ (no quicksand, close)

7-13. distance approx. 350 ft

Chapter 8

8-2. $\tau_{max} = 15$ kPa, $\sigma_n = 35$ kPa (C)

8-3. $\tau_{max} = 13$ kPa, $\sigma_n = 31$ kPa (T)

8-4. $\tau_{max} = 1.8$ ksi, $\sigma_n = 5.4$ ksi (C)

8-5. $\tau_{max} = 25$ kPa, $\sigma_n = 11$ kPa (C)

8-6. $\tau_{max} = 5.4$ ksi, $\sigma_n = 1.8$ ksi (C)

8-7. (b) $\tau_{max} = 40$ kPa, $\sigma_n = 80$ kPa
(c) $\tau = 34.64$ kPa, $\sigma_n = 100$ kPa

8-8. (b) $\tau_{max} = 1500$ psf, $\sigma_n = 3500$ psf
(c) $\sigma_n = 2750$ psf, $\tau = 1300$ psf

8-10. (b) $\tau_{max} = 30$ psi, $\sigma_n = 50$ psi (C)
(c) $\tau = 26$ psi, $\sigma_n = 65$ psi (C)

8-12. (b) $\tau_{max} = 50$ psi, $\sigma_n = 30$ psi
(c) $\tau = 43$ psi, $\sigma_n = 55$ psi

8-15. (b) $\sigma_1 = 8600$ psf (C), $\sigma_3 = 1400$ psf (C)
(c) $\theta = 16.9°$
(d) $\tau_{max} = 3600$ psf

8-16. (b) $\sigma_1 = 127$ psi (C), $\sigma_3 = 33$ psi (C)
(c) $\theta = 16°$
(d) $\tau_{max} = 47$ psi

8-17. $\tau_{max} = 30$ kPa, $\sigma_n = 70$ kPa

8-18. $\sigma_1 = 130$ kPa

8-20. $\tau_{max} = 500$ psf

8-22. (a) $\sigma_1 = 1200$ psf, $\sigma_3 = 600$ psf
(b) $\tau_{max} = 300$ psf

8-23. $\sigma_1 = 80$ kPa, $\sigma_3 = 40$ kPa

8-25. (a) $\sigma_1 = 940$ psf, $\sigma_3 = 424$ psf
(b) $\tau_{max} = 258$ psf

Chapter 9

9-2. (a) $\overline{\sigma}_v = 78$ kPa
(b) $\overline{\sigma}_v = 39$ kPa
(c) $\sigma_v = 58.5$ kPa

9-3. (a) $\overline{\sigma}_v = 1476$ psf
(b) $\overline{\sigma}_v = 1110$ psf
(c) $\sigma_v = 738$ psf

9-4. $\overline{\sigma}_v = 107.9$ kPa, σ_v (total) = 225.6 kPa

9-5. $\overline{\sigma}_v = 1390$ psf

9-7. (a) $K_o = 0.5$, $\sigma_h = 1150$ psf (deep wta),
$\sigma_h = 1828$ psf (high wta)
(b) $K_o = 0.4$, $\sigma_h = 920$ psf,
$\Delta\sigma_h = 230$ psf (deep wta), 120 psf (high wta)

9-8. granular $K_o = 0.5$, $\sigma_h = 27.8$ kPa
clay $K_o ≈ 1.0$, $\sigma_h = 55.5$ kPa

9-9. $\Delta\sigma_h = 364$ psf = 17.43 kPa

9-13. (a) $\Delta\sigma_v = .576$ kPa at CL, $z = 5$ m
$\Delta\sigma_v = .144$ kPa at $r = 5$ m
$\Delta\sigma_v = .108$ kPa at CL, $z = 10$ m
$\Delta\sigma_v = .087$ kPa at r = 5 m

9-15. (a) $\Delta\sigma_v = 1.08$ ksf (center), 0.90 ksf (edge)
(b) $\Delta\sigma_v = 0.72$ ksf (center), 0.54 ksf (edge)

9-16. $\sigma_{v_0} = 36$ kPa, $\Delta\sigma_v = 24$ kPa, total $\sigma_v = 60$ kPa

9-18. (a) $\Delta\sigma_v = 2.7$ ksf at -4 ft
$\Delta\sigma_v = 1.5$ ksf at -8 ft
$\Delta\sigma_v = 1.05$ ksf at -12 ft

9-20. σ_v (total) $= 73$ kPa

9-22. $\sigma_v = 3125$ psf (empty)
$\sigma_v = 4730$ psf (center, full), 4025 (edge, full)

9-24. $\Delta\sigma_v = 1.54$ ksf and 2.20 ksf

9-26. (a) $\Delta\sigma_v = 2400$ psf (center), 1520 psf (edge)
(b) $\Delta\sigma_v = 1580$ psf

9-28. (a) $\Delta\sigma_v = 540$ psf
(b) $\Delta\sigma_v = 840$ psf

9-30. $\Delta\sigma_v = +760$ psf

9-32. $\Delta\sigma_v = 3650$ psf

Chapter 10

10-7. $S \approx 25$ mm

10-10. (a) estimated $S = 20$ to 30 mm
(b) estimated S = 14 to 20 mm
(c) estimated $S = 18$ to 25 mm

10-11. $K_v = 32.14$ kN/m^3

10-12. $K_v = 240$ kcf

10-14. $S = 0.75$ in.

10-15. $S_{vd} \approx 6$ mm

10-18. $S_{vd} \approx 7$ to 8 mm

10-25. $a_v = .0014$ m^2/kN, $C_c = 0.465$

10-31. (a) $S_{pc} = 42$ mm   (b) $S_{pc} = 75$ mm

10-33. (a) $S_{pc} = 1.9$ in.
(b) $S_{pc} = 3.2$ in.
(c) $S_{pc} \approx 1.0$ in.

10-36. $S_{pc} = 17$ in.

10-41. (a) $t = 860$ days
(b) $S = 27$ mm
(c) $t = 200$ days

10-45. $t \approx 200$ days

Chapter 11

11-1. (a) $\phi = 34°$ (b) $\tau = 97.9$ kPa

11-2. (a) $\sigma = 102$ kPa, $\tau = 68.8$ kPa
(b) $\phi = 34°$
(c) med. dense

11-3. $\phi \approx 31.5°$

11-5. $\phi \approx 31°$, apparent $c = 14$ kPa

11-6. $\phi \approx 27°$, apparent $c = 1000$ psf

11-7. $\phi = 33°$

11-8. $\sigma_1 = 154$ kPa

11-9. $\phi \approx 37.5°$

11-10. $\sigma_1 = 74$ psi

11-12. $\sigma_1 \approx 3400$ psf

11-16. $\phi = 33.7°$, $\tau = 3340$ psf

11-17. $c = 33$ kPa, $q_u = 66$ kPa

11-18. $c = 1000$ psf, $q_u = 2000$ psf

11-21. $\tau = 1366$ psf

11-24. $\phi_{CD} = 29°$

11-25. $\phi_{CD} = 30°$

11-26. $\phi_{CD} = 30.6°$, $\phi_{CU} = 21.5°$

11-28. $\alpha = 22.1°$

11-29. start $\overline{\sigma}_1 + \Delta\sigma_1 = 43.3$ kPa

Chapter 13

13-3. (a) $q_{des} = 432$ kPa   (b)$q_{des} = 318$ kPa

13-4. (a) $q_{des} = 5.6$ ksf
(b) $q_{des} = 7.3$ ksf (30% change)
(c) $q_{des} = 5.0$ ksf

13-5. (a) $I_r = 152$ (b) $I_r = 4.3$

13-6. (a) $I_r = 44.2$
(b) $I_r = 5.2$
(c) $I_r = 90$

13-7. $q_{des} = 72$ kPa, $Q_{allow} = 90$ kN per m length

13-9. $q_{des} = 197.2$ kPa, $B = 2.25$ m

13-11. $q_{des} = 3.77$ ksf, $B = 11.25$ ft ($\pm$)

13-12. $B = 1.94$ m

13-13. $B \approx 10.75$ ft

13-15. $Q_{allow} = 385$ kip (slightly low)

13-16. $B \approx 1.5$ m

13-18. $Q_{allow} = 106$ kips

13-20. $B = 11.3$ ft ($\pm$), $L = 22.6$ ft ($\pm$)

13-22. $B = 6.2$ ft, $L = 26$ ft

13-23. $B_1 = 1.69$ m, $B_2 = 2.09$ m, $L = 5$ m

13-24. $Q_{allow} = 162$ kN per m length

13-26. $B = 4.6$ ft

13-28. $Q_{allow} = 8.7$ kip per ft length

13-30. $Q_{allow} = 11.6$ kip per ft length

13-31. (a) $B = 6$ ft   (b) $B \approx 7.5$ ft

13-32. $B = 5.5$ to 6 ft

13-33. $c \approx 3$ kg/cm², $q_{des} = 5.2$ kg/cm³ ≈ 5.2 tsf ≈ 500 kPa

13-34. q (static) $= 525$ kPa, q (seismic) $= 230$ kPa

13-35. $S_E = 3$ mm

13-39. $Q_{des} = 85$ kip (for K = 3)

13-40. Capacity $= 937$ k or 340 k (for $K = 3$)

13-42. $Q_{des} = 77$ kip

13-44. $Q_{des} = 43$ kip

13-46. $Q_{des} = 117$ kips

13-48. $Q_{down} = 106$ kips, $Q_{up} = 94$ kip

13-50. $Q_{des} = 316$ kips

13-52. Q_{des} (down) $= 295$ kip

13-54. (a) $Q_{des} = 280$ kips
(b) $Q_{des} = 246$ kips

13-55. $Q_{des} = 100$ kips

13-56. $s = 0.33$

13-57. Q ("failure") ≈ 170 tons

Chapter 14

14-13. (a) $\gamma_{max} \approx 115$ pcf, $w_{optimum} = 15\%$ ($\pm$)
(b) 7% to 20%

14-14. $\gamma_{max} = 120$ pcf, $w_{opt} = 13$ to 14%

14-15. $\rho_{max} = 1960$ kg/m³, $w_{opt} \approx 11\%$

14-19. (a) $\gamma_{dry} = 112$ pcf, $w = 15\%$
(b) 97.5 pct. compaction

14-20. (a) $\gamma_{dry} = 109$ pcf
(b) 93 pct. compaction

Chapter 15

15-3. (a) $\beta = \phi = 35°$
(b) $\beta \approx 19°$

15-5. (a) $FS = 1.15$ (b) $FS = 0.86$

15-6. $FS = 1.11$

15-7. (a) $FS = 1.17$ (b) min. $FS = 1.01$

15-8. (a) $FS = 1.15$ (b) $FS = 0.70$

15-10. (a) $FS \approx 1.50$ (b) min. $FS = 0.93$

15-11. $FS \approx 1.25$

15-12. $FS \approx 3.2$ (no sub.), $FS \approx 2.7$ (lower zone sub)

15-14. (a) $FS = 1.05$
(b) $i = 18°$
(c) $FS = 1.05$

15-16. (a) $FS = 1.22$ (b) $FS = 1.11$

15-19. (a) $FS = 1.09$ (b) $FS = 0.85$

15-20. (a) $FS = 1.10$ (b) $FS = 0.85$

15-23. (a) fails
(b) $FS = 1.38$
(c) $FS = 1.17$

15-24. (a) $FS = 0.82$
(b) $\phi = 35.8°$
(c) dry
(d) $FS = 1.14$

15-25. (a) $FS = 1.30$
(b) $FS = 0.86$
(c) $FS = 1.26$

15-26. $i \approx 14°$ or 4:1 slope

15-28. Barnes $FS = 1.29$, Spencer $FS = 1.28$

15-29. (a) $FS \approx 0.8$ (fails)
(b) $FS \approx 0.9$ (fails)

Chapter 16

16-3. $P_A = 81$ kN per m length

16-5. $P_A = 6230$ lb per ft

16-7. $P_o = 6640$ lb per ft

$P_A \approx 3760$ lb per ft., 77 pct. diff.

16-8. (a) $P_A = 374$ kN per m length
(b) $P_A = 378$ kN per m

16-9. (a) $P_A = 188$ kN per m length
(b) $P_A = 400$ kN per m
(c) $P_A = 765$ kN per m

16-12. (a) for $K = 0.45$: $FS = 2.77$ (ovt), $FS = 1.96$ (sliding)
(b) Goh: $FS = 3.6$ (ovt), $FS = 2.37$ (sliding)

16-15. $P_h = 7762$ lb per ft length, $P_v = 1940$ lb per ft

16-17. $FS = 1.97$ (ovt), $FS = 1.38$ (sliding)
$p_{toe} \approx 4680$ psf, $p_{heel} \approx 0$

16-20. $R = 546$ lb per ft wall length; $y = 0.59$
H = 2.95 ft

16-21. $F = 47$ kN (top strut), $F = 394$ kN (center), $F = 567$ kN (lower)

16-24. calculated $D \approx 8$ ft, recommend $1.5\,D = 12$ ft

16-25. (a) $FS = 1.31$
(b) $FS = 27$
(c) R at 5.25 ft from toe

Index

Q

R

S

Other terms derived from the fundamental units to identify various physical properties, conditions, and occurrences are in common use in the geotechnical and other scientific fields (e.g., area, volume, stress, pressure, velocity, acceleration, energy, density, unit weight). The following compilation of terms with usually applied units, and the relationship between the SI and U.S. Customary systems, should prove useful:

Property or Term	SI Unit	U.S. Customary Unit	Equivalencies
Length (L)	meter [also micron (μ) and Angstrom (A°)]	foot (also inch, yard, mile)	1 m = 3.281 ft 1 mm = .0394 in. = .00328 ft 1 ft = .3048 m 1 in. = .0254 m = 25.4 mm 1 yd = .91 m 1 km = .621 mile 1 mile = 1610 m = 1.61 km 1 μ = 1 × 10^{-6} m 1 A° = 1 × 10^{-10} m = 3.28 × 10^{-10} ft
Area (L^2)	m^2 hectare	ft^2 (also in.^{2}, yd^2, sq. mile, acre)	1 m^2 = 10.764 ft^2 = 1.196 yd^2 1 mm^2 = 10.764 × 10^{-6} ft^2 = .00155 in.^{2} 1 ft^2 = .093 m^2 1 in.^{2} = 645.16 mm^2 1 yd^2 = .836 m^2 1 km^2 = .386 sq. mile 1 sq. mile = 2.59 km^2 1 acre = 43,450 ft^2 = 4047 m^2 = .4047 hectare 1 hectare = 2.471 acres 1 hectare = 10 000 m^2
Volume (L^3)	m^3 [also liter, (l) where 1 l = 1000 cm^3, or 1 ml = 1 cm^3]	ft^3 (also yd^3, gallon)	1 m^3 = 35.32 ft^3 = 1.31 yd^3 = 61,023.4 in^3 1 cm^3 = .061023 in^3 1 ft^3 = .0283 m^3 = 28.3 l 1 in.^{3} = 16.39 cm^3 1 yd^3 = .765 m^3 1 l = 1000 cm^3 = .263 gal = .0354 ft^3 1 gal = 3.8 l = .0038 m^3 = .134 ft^3 1 m^3 = 263 gal
Mass (M)	kilogram (gram in metric)	slug (unit is lb-sec^2/ft) [also lb-mass (lbm)]	1 slug = 14.59 kg 1 kg = .0685 slug